# THE CHEMICAL BASIS
OF DEVELOPMENT

Contribution No. 234 of the McCollum-Pratt Institute

*A Symposium on*
# THE CHEMICAL BASIS OF DEVELOPMENT

*Sponsored by*

THE
McCOLLUM-PRATT INSTITUTE
OF
THE JOHNS HOPKINS UNIVERSITY
with support from
THE NATIONAL SCIENCE FOUNDATION

*Edited by*

WILLIAM D. McELROY AND BENTLEY GLASS

BALTIMORE
THE JOHNS HOPKINS PRESS
1958

PREVIOUS SYMPOSIA OF THE McCOLLUM-PRATT INSTITUTE

*Copper Metabolism—A Symposium on Animal, Plant and Soil Relationships.*

*Phosphorus Metabolism—A Symposium on the Role of Phosphorus in the Metabolism of Plants and Animals, Vol. I.*

*Phosphorus Metabolism—A Symposium on the Role of Phosphorus in the Metabolism of Plants and Animals, Vol. II.*

*The Mechanism of Enzyme Action.*

*Amino Acid Metabolism.*

*Inorganic Nitrogen Metabolism.*

*The Chemical Basis of Heredity.*

© 1958, The Johns Hopkins Press. Distributed in Great Britain by: Oxford University Press, London. Printed in the United States of America by The Colonial Press Inc., Clinton, Mass.

*The Library of Congress catalog entry for this book appears at the end of the text.*

## PREFACE

A Symposium on the Chemical Basis of Development was held at The Johns Hopkins University under the sponsorship of the McCollum-Pratt Institute on March 24–27, 1958. This volume consists of the papers and informal discussions presented during the Symposium.

The present volume is in many ways a logical sequel to the preceding volume on the Chemical Basis of Heredity which was the published account of the Symposium held in 1956. Developmental biology encompasses a wide range of problems dealing with both the initiation and the progression of sequences of events that lead to the emergence of adult characteristics. The genes, at the focus of studies in heredity, partake directly or indirectly in all these events involved in development. Thus the chemical basis of development can be viewed as an elaboration or expression of the chemical basis of heredity.

In the planning of the present Symposium we attempted to bring together those biologists and biochemists who had made observations which might help in explaining some of the chemical aspects of growth and differentiation. By facilitating an exchange of information among the various disciplines represented at the Symposium we can surely expect to increase the rate of progress in this important area of biology.

I should like to acknowledge the active participation of the members of the McCollum-Pratt Institute and of the Biology Department in the planning of the Symposium in honor of Professor B. H. Willier for his distinguished contributions as teacher and investigator in developmental biology. It is also a pleasure to acknowledge the valuable contributions of the following moderators: Dr. Jean Brachet, Dr. Johannes Holtfreter, Dr. Elmer Butler, Dr. Louis Flexner and Dr. Viktor Hamburger. We also wish to thank Dr. Paul Weiss for his summation and evaluation of the contributions as well as of the role of the Symposium in furthering progress in the science of developmental biology.

Through the generous aid of the National Science Foundation it was possible to support a number of foreign investigators as participants in the Symposium. Unfortunately, there were many that we were unable to invite because of inadequate funds. We hope, however, that the published book will be of value to all who are interested in developmental biology and in particular to those who were unable to attend the Symposium.

W. D. McElroy
Director of McCollum-Pratt Institute

## CONTENTS

|  | page |
|---|---|
| Preface | v |
| Additional Participants | xiii |

### I. Developmental Cytology

| | |
|---|---|
| Chemical Concepts of Cellular Differentiation | 3 |
| —Clement L. Markert | |
| The Initiation of Development | 17 |
| —R. D. Allen | |
| Discussion | 67 |
| Phases of Dependent and Autonomous Morphogenesis in the So-Called Mosaic-Egg of Tubifex | 73 |
| —Fritz E. Lehmann | |
| Discussion | 85 |
| The Role of the Cell Nucleus in Development | 94 |
| —A. E. Mirsky and Vincent Allfrey | |
| Discussion | 99 |
| Chromosomal Differentiation | 103 |
| —Joseph G. Gall | |
| Chromosomal Replication and the Dynamics of Cellular Proliferation—Some Autoradiographic Observations with Tritiated Thymidine | 136 |
| —W. L. Hughes | |
| Discussion | 152 |
| Origin and Structure of Intercellular Matrix | 157 |
| —Mac V. Edds, Jr. | |

|  |  | page |
|---|---|---|
| Discussion | | 169 |
| Cytoplasmic Particulates and Basophilia | | 174 |
| —Hewson Swift | | |
| Discussion | | 210 |

II. CELLULAR AND TISSUE INTERACTIONS IN DEVELOPMENT

| Embryonic Induction | 217 |
|---|---|
| —Tuneo Yamada | |
| Inductive Specificity in the Origin of Integumentary Derivatives in the Fowl | 239 |
| —John W. Saunders, Jr. | |
| Discussion | 253 |
| A Developmental Analysis of Cellular Slime Mold Aggregation | 264 |
| —Maurice Sussman | |
| Enzyme Patterns during Differentiation in the Slime Mold | 296 |
| Barbara E. Wright and Minnie L. Anderson | |
| Discussion | 313 |
| Exchanges Between Cells | 318 |
| M. Abercrombie | |
| Some Problems of Protein Formation in the Sclera and Cornea of the Chick Embryo | 329 |
| —Heinz Herrmann | |
| Cell Migration and Morphogenetic Movements | 339 |
| —Robert L. DeHaan | |
| Discussion | 374 |

III. PROBLEMS OF SPECIFICITY IN GROWTH AND DEVELOPMENT

| Comparative Cytochemistry of the Fertilized Egg | 381 |
|---|---|
| —J. J. Pasteels | |

CONTENTS

*page*

A Cytochemical Study of the Growth of the Slug Oocyte . . 404
—Ronald R. Cowden

Discussion . . . . . . . . . . . . . 414

Changes in Enzymatic Patterns During Development . . . 416
—S. C. Shen

Biochemical Patterns and Autolytic Artifact . . . . . . 433
—J. Lee Kavanau

Enzyme Induction in Dissociated Embryonic Cells . . . . 448
—Richard N. Stearns and Adele B. Kostellow

Discussion . . . . . . . . . . . . . 454

On the Differentiation of a Population of *Escherichia Coli* with Respect to $\beta$-galactosidase Formation . . . . . . . 458
—Melvin Cohn

Feed-Back Control of the Formation of Biosynthetic Enzymes . 469
—Luigi Gorini and Werner K. Maas

Comment on the Possible Roles of Repressers and Inducers of Enzyme Formation in Development . . . . . . . 479
—Henry J. Vogel

The Metabolic Regulation of Purine Interconversions and of Histidine Biosynthesis . . . . . . . . . . 485
—Boris Magasanik

Discussion . . . . . . . . . . . . . 490

Biosynthesis of Urea in Metamorphosing Tadpoles . . . . 495
—G. W. Brown, Jr. and P. P. Cohen

Patterns of Nitrogen Excretion in Developing Chick Embryos . 514
—Robert E. Eakin and James R. Fisher

Discussion . . . . . . . . . . . . . 523

*CONTENTS*

*page*

Immunochemical Analysis of Development . . . . . 526
—*James D. Ebert*

Immuno-Histological Localization of One or More Transitory Antigens in the Optic Cup and Lens of *Rana Pipiens* . . 546
—*George W. Nace and William M. Clarke*

Influence of Adult Amphibian Spleen on the Development of Embryos and Larvae: An Immune Response? . . . . 562
—*Louis E. DeLanney*

Discussion . . . . . . . . . . . . . . . 568

Actively Acquired Tolerance and Its Role in Development . . 575
—*R. E. Billingham*

Discussion . . . . . . . . . . . . . . . 592

Selective Uptake and Release of Substances by Cells . . . . 596
—*Morgan Harris*

Discussion . . . . . . . . . . . . . . . 623

IV. CONTROL MECHANISM IN DEVELOPMENT

Chemical Control of Development . . . . . . . . 629
—*Nelson T. Spratt, Jr.*

Discussion . . . . . . . . . . . . . . . 642

Chemical Stimulation of Nerve Growth . . . . . . 646
—*Rita Levi-Montalcini*

A Nerve Growth-Promoting Protein . . . . . . . 665
—*Stanley Cohen*

Discussion . . . . . . . . . . . . . . . 676

Neoplastic Development . . . . . . . . . . . 680
—*Leslie Foulds*

## CONTENTS

|  | page |
|---|---|
| Discussion | 700 |
| Organ Differentiation in Culture | 704 |
| —E. Borghese | |
| A Study of Sex Differentiation in the Fetal Rat | 774 |
| —Dorothy Price and Richard Pannabecker | |
| Discussion | 777 |
| Role of Genes in Developmental Processes | 779 |
| —Ernst Hadorn | |
| Discussion | 791 |
| Hormonal Regulation of Insect Metamorphosis | 794 |
| —Carroll M. Williams | |
| The Metabolic Control of Insect Development | 807 |
| —G. R. Wyatt | |
| Discussion | 811 |
| The Mechanism of Liver Growth and Regeneration | 813 |
| —André D. Glinos | |
| Discussion | 839 |
| Summation and Evaluation | 843 |
| —Paul Weiss | |
| V. SUMMARY | 855 |
| —Bentley Glass | |
| AUTHOR INDEX OF PARTICIPANTS | 923 |
| SUBJECT INDEX | 925 |

## ADDITIONAL PARTICIPANTS

B. Afzelius
R. Airth
M. Avron
R. Babcock
S. Bacher
R. Ballentine
L. A. Bard
R. Beard
S. Bessman
S. Bieber
D. Bishop
D. Bodenstein
D. Bodian
E. J. Boell
B. Boving
J. Brachet
A. Brown
E. Burpi
E. Butler
F. Carlson
M. W. Childe
A. C. Clement
N. S. Cohn
J. Coleman
S. P. Colowick
A. Colwin
A. J. Coulombre
G. Crowley
R. Davenport
G. de la Haba
H. Eagle
D. R. Evans, Jr.
L. Flexner
G. Fankhauser

W. Fredricks
J. Gaffinet
J. Gallant
D. Gambal
J. Giovanelli
H. B. Glass
A. Goldin
E. Goldberg
H. Gordon
C. T. Grabowski
P. Grant
V. Hamburger
H. Hamilton
J. R. Harrison
P. E. Hartman
Y. Hayashi
J. Holtfreter
H. Holtzer
T. E. Hunt
A. T. Jagendorf
A. Kaji
R. E. Kane
F. T. Kenney
J. Klingman
I. Koenigsberg
N. Kretchmer
C. A. Lang
H. Laufer
H. Lenhoff
D. Lindsay
F. Lipmann
J. C. Loper
M. Margulies
W. D. McElroy

E. V. McCollum
A. Mehler
D. Mellon, Jr.
T. Merz
L. Mettler
A. Millerd
W. C. Mohler
K. J. Monty
F. Moog
W. B. Muchmore
V. Najjar
A. Nason
J. Nicholas
H. Nitowsky
M. C. Niu
G. Novelli
A. B. Novikoff
J. Oppenheimer
J. Papaconstantinou
A. M. Pappenheimer
P. Pearson
D. Peck
R. Popp
Van Potter
D. Prescott
T. C. Puck
E. Racker
S. Reichard
C. Reimer
H. Reiter
R. Roberts

G. Roussos
D. Rudnick
B. Ruth
A. San Pietro
D. Shapiro
A. Siger
M. Singer
H. Schneiderman
S. Spiegelman
M. R. Spurbeck
K. Stein
M. Steinberg
B. Strehler
S. Suskind
C. P. Swanson
C. A. Thomas, Jr.
K. Trayser
H. E. Umbarger
W. Van der Klott
F. Vasington
W. S. Vincent
S. G. Waelsch
R. Watterson
R. Weber
C. E. Wilde
F. Wilt
M. Wolbarsht
S. Wright
C. Wyttenbach
P. Zamecnik
E. Zwilling

# Part I

*DEVELOPMENTAL CYTOLOGY*

# CHEMICAL CONCEPTS OF CELLULAR DIFFERENTIATION

CLEMENT L. MARKERT

*Department of Biology*
*Johns Hopkins University*

IN A SYMPOSIUM devoted to the chemical basis of development, it is perhaps appropriate that the first report should make a general assessment of the significance of our topic. Biological phenomena, though closely interrelated, are for purposes of study and investigation assignable to one or more of the major disciplines of biology, each concerned with a significant attribute of living systems: their structure, function, ecological relationships, evolutionary history, heredity, and development. Of these, the field of development has been relatively slighted despite the tremendous importance of developmental processes in characterizing living organisms. No doubt part of the lack of attention stems from the manifest difficulty inherent in any analysis of development. The nature of this problem was stated very well by Ross G. Harrison (16) more than 30 years ago in his presidential address to the American Society of Zoologists. He said:

"The living embryo changes continually; its form, its mechanisms and its functions change; its parts function while changing. These transformations are themselves functions. We have, then, superposed on the ordinary functions of nutrition, respiration, protoplasmic and nervous transmission, action of internal secretions circulating in the internal medium, etc., a whole system of developmental functions, which, as far as we have been able to find out, are totally different from the former. The embryologist has, therefore, a problem of a higher order of complexity—a superproblem —to contend with than has he who directs his attention to the study of the structure or function of the finished organism.

"This is usually overlooked. Embryology, from its close relation to comparative anatomy and from the employment of schemata to represent its processes, came to bear the reproach of physiologists that it was a morphological science and on that account dealt with statics and not kinetics. A moment's consideration shows this view to be altogether erroneous. The organism never reaches a state of rest until it has run its course or is securely preserved in a bottle. The physiologist accepts the finished organism as given and endeavors to find out how it works. The embryologist, on the other hand, attempts to show the origin of the mechanisms which the physiologist is content to accept ready made for study. May not the embryologist, then, return the reproach and say that the physiologist is merely looking for something easy to do?"

These words of Harrison properly emphasize the difficulties inherent in embryological investigations, but they do not make clear the indispensable role of physiologists and biochemists in providing the essential tools and information with which to analyze development in chemical terms.

Since the development of the organism is dependent upon the organization and development of its parts, it is possible to study development and differentiation on several levels of physical and chemical organization. These levels of organization are reflected in the titles of the reports to be delivered here. We may not forget, however, that no description of structure or function, however sophisticated it may be, is truly developmental biology, for development is primarily concerned with the transformation of one condition into another. Although the structural and functional characteristics of each successive embryonic state are inseparable from the mechanisms that transform them, it is the vectorial qualities of embryonic states on which developmental biologists must focus attention. It is doubtful that transformations at these various levels of organization can be comprehended within a single mechanism, and the relationships that must exist between changes on the simpler levels of organization—on a molecular level, for example—and developmental processes at the more complex levels of organization of cells, tissues, and organs, are at present completely obscure.

Nevertheless, there is a strong desire to interpret complex developmental events in terms of simpler antecedent events and eventually in terms of molecular changes (41, 42). The closely allied field of genetics affords us hope in aspiring to make such interpretations because identified genetic characters (like those concerned in development) range through all levels of physical and chemical organization, from the entire organism to cytoplasmic inclusions and chromosome morphology. Furthermore, in developmental genetics we have been able on occasion to trace gross effects back to simpler morphological and/or chemical events. Some genetic characters may be traced to the synthesis of, or to the failure to synthesize, specific protein molecules, and thence to the gene. Genetically controlled complex characters are attributed to antecedent interactions at lower levels of organization, and therefore analysis involves the unravelling of these patterns of interaction which depend upon such variables as effective concentrations of substances, relative rates of reactions, and the timing or programming of events. Although the problems of interpretation may prove extraordinarily complicated in physiological genetics, no serious conceptual difficulties appear to exist. Granted initial qualitative differences at the gene and molecular levels, subsequent variations follow logically.

We may note that the analysis of such patterns of interaction involve two distinct aspects which may be as applicable to development as to genetics: (1) the characteristics of individual genes and the initial elementary events for which they are responsible, and (2) the interactions of products of genic activity which yield the wide array of metabolic patterns characterizing the specialized cells and tissues of the emerging adult organism. In physiological genetics the initial gene changes are the raw materials that make the science possible, but in the study of normal development no such simple, causal, fundamental event can yet be discerned. However, an effort has been made to overcome this difficulty, conceptually at least, by making the activity of genes during development dependent upon the physical and chemical environment in the cell—an environment for which the genes themselves are largely responsible. A sort of cyclical feedback is visualized between the genome and its ever-changing chemical environment so that the changing interactions between gene and environment provide the motive power for driving embryonic cells along their diverse paths of differentiation (48). Although this way of viewing embryonic development seems logically sound, it is not very informative and reduces studies of embryonic change to problems in metabolic complexity. Even so, there are numerous points of observational and experimental attack within such a conceptual framework.

## Ontogeny of Metabolic Patterns

The characteristic metabolic pattern of cells and tissues at any stage of differentiation may be analyzed in terms of the substances present at any given time (8) or in terms of the substances produced during a defined period of development (24). A wide variety of substances—simple and complex—have been identified and measured in various parts of embryos in an effort to provide a biochemical description of development. Enzymes have been prime objects of study (10, 15, 27, 34) because of their decisive position in the metabolic machinery of the cell, but the products of enzymatic activity have also been studied both for their own significance and as indicators of enzyme activity. To a substantial degree the state of differentiation of a cell is a reflection of the pattern of its enzymatic composition (35), and the progress of differentiation depends upon mechanisms for changing effective enzyme concentrations in cells. Two problems appear here: (1) the origin of enzymatic activity; and (2) the regulation of the amount of activity. These two problems may relate to different aspects of a single mechanism, but at present this seems unlikely.

*Adaptive Enzyme Synthesis in Embryos*

One notion commonly advanced to account for the effective origin of new enzymatic activity is that of adaptive enzyme synthesis (7, 26). This notion is attractive because it makes simple variations in the chemical environment of cells adequate to account for their changing enzymatic content and therefore for their differentiation. Unfortunately, there is very little direct experimental evidence from embryos to support this notion (11, 12, 19, 20). Its general significance in embryos is severely limited by the fact that embryonic cells of different types do not develop the same additional enzyme even when placed in identical chemical or tissue environments. Thus, though mechanisms for enzyme induction may exist in embryos, such mechanisms do not account for the diverse capacities of cells to respond. The origin of these divergent capacities for enzyme synthesis is the more fundamental problem in development, with the elicitation of response of secondary significance.

*Embryonic Induction*

The origin of metabolic capacities is in part due to that interaction between tissues in development commonly known as embryonic induction. This type of interaction probably involves the passage of specific substances from inducing to reacting tissues (14, 29), but more than simple chemical exchange may be required, at least in some types of induction (13, 44). The inductive stimuli are of limited duration and are exerted over short distances by specific tissues at specific periods in their development. Only particular tissues at receptive stages in their development are competent to react to a given stimulus. Other tissues remain indifferent. The transmitted substances involved in some types of induction appear to be proteins, as shown by the work of Toivonen (39), Kuusi (22, 23), Niu (28, 29), Yamada (49), and others. Although embryonic induction may result in the synthesis of new enzymes, the mechanism of embryonic induction does not seem equivalent to that of enzyme induction.

## Role of the Chemical Environment in Differentiation and Maintenance

Leaving aside the specific problem of adaptive enzyme synthesis, there is little doubt that the chemical environment of differentiating and already differentiated cells is of great importance in the development and maintenance of specific characters. Embryonic cells explanted to cultures in vitro do not readily continue their differentiation, and mature cells cultivated in vitro (45) commonly lose much of their specialized metabolic machinery. Furthermore, the nutritional requirements and synthetic activities of cells in vitro are to some extent dependent upon the structure of the population of which they are a part (33)—thus indicating the importance of mutual interaction in regulating the chemical environment for developing cells in vivo. The interaction of cells is to a substantial degree mediated through surface contact and is dependent upon the specific properties of the cell surface (41, 42, 43, 44). These surface specificities are reflections of the metabolic pattern of the cell and are responsible for defining the structure of the cell population of any tissue and for regulating the relationships between tissues. The particular association into which any cell enters of course limits the type of cellular (and chemical) environment to which it will be exposed; the changing chemical properties of the environment then further modify cellular metabolism to alter surface specificities which compel the regrouping of cells into new associations.

This sensitivity of the metabolic pattern in cells to the chemical environment has stimulated extensive research on the developmental effects of a wide variety of chemical substances—nutrients, antimetabolites, enzyme inhibitors, antibodies, tissue extracts, etc. These substances are generally used to produce a kind of metabolic deletion or addition experiment. If development is significantly altered by such chemical intervention, then one may hope to identify chemical events that are decisive for the type of development studied. The rationale for this type of investigation assumes that the metabolic pattern of the cell is a steady state that may be shifted to alternate steady states by enhancing or repressing strategically placed reactions. Although mechanisms for transforming steady states have been presented in plausible form by several investigators, none of these model mechanisms has been demonstrated to operate in developing embryos. However, several biochemical mechanisms have been described that help to maintain a metabolic steady state (48). Specifically, metabolic stability may be achieved by feedback controls involving products and enzymes within a single chain of reactions (40) and by similar, though more complex, interactions between different chains of reactions. Such mechanisms offer exciting prospects for analyzing chemical processes in developing embryos, although the four-dimensional web of events that constitutes embryonic development is far more complex than the metabolic patterns studied so successfully in microorganisms.

GENETIC CONTROL OF DEVELOPMENT

Let us turn now to the primary material of physiological genetics mentioned previously—that is, the genome of the cell. Just as adult characters are gene-controlled, so also of necessity are the developmental steps which produced them (37). A goal in developmental genetics is the identification of the primary effect of a gene change that leads to an alteration in development. Enough progress has been made to demonstrate the multifarious consequences in development of changing the initial activity of a gene (36), but the primary or even derived chemical consequences of gene function still escape detection in almost every case. All stages of development at various levels of organization and all major components of an interacting system, as well as the identified mechanisms of their interaction, have been shown to depend upon genic activity. The capacity of cells to respond or to elaborate inductive stimuli, or to synthesize a given enzyme and carry out a biochemical reaction, or even to acquire particular asymmetries of organization, have all been related to single gene changes. Does not the truly impressive data of developmental genetics enable us to assume that integrated developmental events are no more than the expression of a corresponding integrated pattern of primary genic activity (48)? If so, developmental phenomena will ultimately be analyzed in terms of mechanisms controlling genic activity. All else would then be mere derivative superstructure. An alternative and still tenable view is that developmental phenomena are superimposed upon a foundation of uniform primary genic activity in all cells. The effectiveness of a given gene would then depend upon the characteristics of its chemical environment rather than upon the gene itself. This view receives some support from the precision of mitotic phenomena that seem to guarantee each metazoan cell an identical genetic endowment from the fertilized egg. That is, all the cells of an organism, through this mechanism, should have the same genes.

*Gene Function*

The concept that all potential genes in every cell function continuously or even intermittently seems very doubtful, since numerous cellular functions known to be genetically controlled are apparently restricted to only certain specialized cells and cannot be detected in most cells of an organism. The gene controlling the ability to synthesize the enzyme tyrosinase will serve as an example. Animals with mutant, inactive forms of this gene cannot synthesize tyrosinase, therefore make no melanin pigment, and consequently become albinos. However, in a normal animal only the melanin-synthesizing cells give any evidence of possessing a gene for tyrosinase synthesis. The remaining cells are indistinguishable from the corresponding cells of an albino animal. If geneticists had only mammalian eggs to work with, it is doubtful that they ever would have recognized the gene for tyrosinase synthesis. This points up the fact that apparent gene function is dependent upon the state of differentiation of the cell and in its turn the state of differentiation at any time is dependent upon the integrated pattern of previous gene activity.

An illustration of this reciprocal relationship is provided by the genes for pink eye in the house mouse (25). When genes for pink eye are present in the pigment-forming cells of the mouse (that is, in the migratory melanoblasts that produce melanin pigment when the cell resides in an appropriate cellular environment), the melanin granules formed have an abnormal structure. Thus these genes control the characteristic expression, pigment granule synthesis, of this type of cellular differentiation. More important, however, is the fact that these same genes prevent the harderian gland of the mouse from providing its usual favorable environment for melanoblast differentiation. In all other respects the harderian gland appears normal, but no pigment cells are able to reach maturity within this gland of pink-eye mice. Other tissues of these mice, despite the presence of genes for pink eye, do retain the capacity to elicit melanoblast differentiation—the cells of the hair papilla, for example. These observations indicate that the effect of a gene on a particular step in differentiation is dependent upon the metabolic pattern of the cell (its state of differentiation)—a pattern which in its turn is dependent upon all the previous effective influences exerted upon the cell both from within and without. The role of genes in melanoblast differentiation and in the formation of pigment patterns has previously been clearly analyzed and demonstrated also in the domestic fowl (46, 47).

In mice, we do not know the initial function of the genes for pink eye, but this function, whatever it may be, ultimately affects quite different steps in the differentiation of a variety of cell types. On the other hand, the gene for tyrosinase synthesis (and this may be its primary function) affects only a terminal step in differentiation of a single cell type. No other effect of this gene has been observed, and the simplest interpretation is that this gene does not function in other cell types. In this example we have been relating a functional gene to the production of a specific molecule—specifically, a protein enzyme. A vast amount of genetic data supports this gene–enzyme relationship (17). And a growing body of data now relates the enzymatic content of embryonic cells to their state of differentiation (27, 34). We will hear more of this later in the symposium, but I should like briefly to present the results of one such investigation on the esterases of developing mouse tissues.

*Ontogeny of Esterases*

The esterases investigated are concerned with the hydrolysis of carboxylic acid esters of phenols. There is a large number of such esterases, perhaps twenty, that together constitute a family of enzymes of overlapping specificities. These esterases may readily be separated from one another by electrophoresis in starch gels and individually identified by their electrophoretic mobility, and their substrate and inhibition specificities (18). Each adult tissue exhibits its own distinctive repertory of these esterases arranged in characteristic relative concentrations. During embryonic and early post-natal development the esterases appear one after another as the tissue reaches new stages of differentiation, until finally the repertory is completed at the time the tissue reaches maturity. Furthermore, electrophoretically homologous esterases from different tissues are alike in their substrate and inhibition specificities. If each of these esterases represents the primary activity of a different gene, then the characteristic esterase pattern of different tissues reflects the particular array of esterase genes activated in that tissue during the course of differentiation. Since the relative and absolute concentration of any particular esterase varies in different tissues, the specific effects of the esterase genes must be subject to quantitative modification by metabolic activities of the tissue.

## BIOPHYSICAL ASPECTS OF DEVELOPMENT

Before concluding this discussion of the genic control of the chemical basis of development we should note another important aspect of developing organisms—an aspect more properly classified under biophysics than under biochemistry. This aspect relates to the movement and organization of organs, tissues, cells, subcellular entities, and intercellular materials. This changing physical organization of developing embryos and their parts cannot at present be explained in chemical terms, nor have we even elucidated the primary physical factors involved. One need only recall our inability to explain the movement of chromosomes, or of migratory cells, despite extensive investigation, in order to emphasize both the importance and the difficulty of analyzing the physical changes in differentiating cells and tissues. But even in this biophysical area of development the genes have indicated their masterful role. The organization of the eggs of the snail, *Lymnaea,* is such that cleavage is asymmetrical, the spindle being tilted to one side. As the egg divides, the planes of cleavage are alternately tilted in opposite directions to produce a spiral organization of the resulting blastomeres. This cleavage pattern is reflected in the direction of spiral in the shell of the resulting adult snail and may be either left-handed or right-handed. The potential direction of spiral is fixed in the egg organization by the maternal organism in accord with her own genotype. Only a single gene appears to be involved (5, 9, 38). This example, though it tells us nothing about primary gene activity, indicates gene control of basic cell structure, perhaps of course through controlling elementary chemical reactions.

## Differential Gene Activation in Differentiation

To return now to the general relationship between genes and developmental events, the preferred viewpoint in this presentation has pictured a cell, when subjected to differentiating stimuli, as responding through the activation of a gene which then initiates a new reaction. Perhaps, on occasion, genes are also deactivated. At least differentiated cells, when explanted to tissue culture, after repeated multiplication in this disorganized environment lose some of their enzymes (31). This loss may be interpreted as a failure to maintain the responsible genes in an activated state during repeated cell divisions. An adequate description of differentiation must account not only for the origin of differences among cells but also for the persistence of these differences through many cell generations. Although relatively persistent, the characteristics of differentiated cells are not uniformly stable. When put to the test by growth in abnormal tissue culture environments some properties are readily lost while others persist, perhaps in conformity with the new pattern of stimuli from the culture environment. However, it should be noted that even after prolonged culture for many months in identical media, originally distinct cell types remain distinct from one another. The cell is not just a reflection of the current attributes of the environment, whatever they may be, but rather the cell reacts to changed conditions in a distinctive fashion determined by its own characteristics. In the absence of cell division the differentiated properties of cells are more stable and this is not surprising if genic activation is fundamental to cellular differentiation, for the replication of genes would pose additional problems in the retention and replication of the activated state of the gene as contrasted with the inactive state.

A more serious problem in making genic activation fundamental is the mutual exclusiveness of diverging paths of differentiation. The acquisition of one character commonly excludes an alternative. On the level of genes it is difficult to visualize a simple mechanism by which the activation of one gene would preclude the activation of another. Perhaps there are also mechanisms for specifically inhibiting genes so as to render them inaccessible to activating mechanisms.

If differential genic activation does occur during cellular differentiation, then this should be reflected in a gradual differentiation of the chromosomes. Such a mechanism of differentiation should be most welcome, since it would provide developmental biologists with a simple elementary event —genic activation—with which to work in analyzing development. Part of the great success of genetics and biochemistry must surely stem from the fact that these disciplines are able to focus on elementary events—gene changes or biochemical reactions—and are not bemused by the complexities flowing from these elementary events.

## Chromosomal Differentiation

But do we have any evidence for believing that genic activation or chromosomal differentiation does in fact occur? Later reports will deal specifically with differentiation at subcellular levels, and no extended comments of mine would be in order now. However, at least three lines of evidence can be brought to bear on this issue, and I should like to describe them briefly. First, the recent work of King and Briggs (21) on nuclear transplantation lends strong support to the view that the nucleus (and inferentially the chromosomes) differentiate during development. These investigators transplanted nuclei from frog embryonic cells at various stages of differentiation into enucleated frog eggs. The subsequent development of these eggs measures the capacities of the transplanted nuclei. Nuclei from cells in early stages of development proved equivalent in developmental capacity to an egg nucleus—that is, normal animals developed from eggs containing such transplanted nuclei. However, nuclei from progressively later stages of development gave rise to an increasing proportion of abnormal embryos. Most significantly, the nature of these abnormalities was correlated with the state of differentiation of the cell furnishing the donor nucleus. Upon serial retransplantation of nuclei from these abnormal embryos the same pattern of abnormal development was observed in the succeeding generations. Thus differentiated nuclear characteristics had become fixed and transmissible through repeated nuclear division. It seems likely that non-replicating parts of the nucleus would have been diluted out during these many divisions, leaving the chromosomes as most probably the bearers of the fixed, differentiated characteristics of the nucleus.

The second line of evidence stems from the observations of Pavan (6, 30) and of Beermann (4) on the chromosomes in the larval tissues of certain insects. These chromosomes display an intricate and characteristic banding pattern. During development this banding pattern is modified by the temporary appearance along the chromosome of greatly enlarged areas known as Balbiani rings. The occurrence and distribution of these rings along the chromosomes is specific for both the tissue and its stage of development. Probably these rings indicate areas of more active chromosomal function, but in any event, the chromosome morphology varies, and in a fashion that is characteristic of the stage of cellular differentiation. We may thus conclude that these chromosomes are themselves differentiated.

The third line of evidence is based upon direct chemical analysis of nuclei and chromosomes from various differentiated cells or tissues (1, 2, 3). These analyses reveal four principal constituents in chromosomes: (1) deoxyribonucleic acid, (2) ribonucleic acid, (3) histones, and (4) residual protein. The DNA remains constant, as would be expected, but the RNA varies considerably in amount not only between different tissues but also within the same tissue at different levels of physiological activity. Analysis of histones reveals no striking differences (except in gametes); but residual protein does vary in different tissues. Taken together, these analyses strongly suggest that chromosomes undergo chemical differentiation in accord with the functional requirements of the differentiating cell (32).

## Hypothesis of Development

This sketchy survey of various aspects of the chemical basis of development, with particular emphasis on the role of the genes, may be concluded with the presentation of a working hypothesis that compounds old and new speculation to explain the changing sequence of metabolic patterns that characterizes cellular differentiation. The hypothesis may be described in four parts:

(1) The newly fertilized egg contains chromosomes that are undifferentiated with reference to those functions that will later distinguish and

characterize differentiated cells. The egg is endowed at the outset with an unstable pattern of metabolic activities that begins to change in distinctive ways in different regions of the egg.

(2) As cleavage proceeds, the chromosomes in various regions of the embryo are subjected to different chemical environments. By some still obscure mechanisms specific metabolic patterns in the cell impose correspondingly specific proteins on the DNA of the chromosomes. The coupling of these proteins to specific regions of the DNA produces a functional nucleoprotein or activated gene. Previously this portion of the DNA was non-functional because of the absence of an essential protein complement.

(3) The creation of a functional gene during development would be reflected in a specific synthesis—an enzyme, for example. This new synthetic activity would change the metabolic pattern of the cell and would result in the activation of additional genes in the same cell or in adjacent cells. This cyclic interaction would stabilize when the metabolic pattern of the cell ceased to change or when the chromosomes were completely differentiated.

(4) Functional genes would display varying degrees of stability. Under conditions of artificially induced rapid proliferation, as in tissue culture, the maintenance of the differentiated structure of the chromosome through numerous replications would not always occur, and the cell would appear to dedifferentiate, to become simpler in the variety and specializations of its syntheses. In non-dividing adult cells a high order of stability would easily be maintained even in the face of chemical provocation to the cell, because the activated genes would remain unchanged by nonspecific variations in their chemical environment.

The observations and conclusions that I have discussed do not of course critically test this hypothesis. Perhaps pertinent evidence will be forthcoming in later papers, and in any event, if we are stimulated to take a closer look at chromosomes during development the hypothesis will have served a useful purpose. Speculation has frequently proved valuable, and we may confidently agree with Darwin when he said in a letter to Wallace, "I am a firm believer that without speculation there is no good or original observation."

## REFERENCES

1. Allfrey, V. G., Mirsky, A. E., and Stern, H., *Advances in Enzymol.*, **16**, 411-500 (1955).
2. Allfrey, V., Mirsky, A. E., and Osawa, S., in *The Chemical Basis of Heredity* (McElroy, W. D., and B. Glass, eds.), p. 200-231, The Johns Hopkins Press, Baltimore (1956).

3. Allfrey, V., in *The Chemical Basis of Heredity* (McElroy, W. D., and B. Glass, eds.) p. 186-194, The Johns Hopkins Press, Baltimore (1956).
4. Beermann, W., *Cold Spring Harbor Symposia Quant. Biol.*, **21**, 217-232 (1956).
5. Boycott, A. E., and Diver, C., *Proc. Roy. Soc. (London)*, B, **95**, 207-213 (1923).
6. Breuer, M. E., and Pavan, C., *Chromosoma*, **7**, 371-386 (1955).
7. Cohn, M., *Bacteriol. Revs.*, **21**, 140-168 (1957).
8. Deuchar, E. M., *J. Embryol. Exptl. Morphol.* **4**, 327-346 (1956).
9. Diver, C., Boycott, A. E., and Garstang, S., *J. Genet.*, **15**, 113-200 (1925).
10. Ephrussi, B., in *Enzymes: Units of Biological Structure and Function* (Gaebler, O. H., ed.), p. 29-40, Academic Press, New York (1956).
11. Gordon, M. W., in *Neurochemistry* (Korey, S. R., and Nurnberger, J. I., eds.), p. 83-100, Hoeber-Harper, New York (1956).
12. Gordon, M., and Roder, M., *J. Biol. Chem.*, **200**, 859-866 (1953).
13. Grobstein, C., *Advances in Cancer Research*, **4**, 187-236 (1956).
14. ———, and Dalton, A. J., *J. Exptl. Zool.*, **135**, 57-74 (1957).
15. Gustafson, T., *Intern. Rev. Cytol.*, **3**, 277-327 (1954).
16. Harrison, R. G., *Science*, **85**, 369-374 (1937).
17. Horowitz, N. H., *Federation Proc.*, **15**, 818-822 (1956).
18. Hunter, R. L., and Markert, C. L., *Science*, **125**, 1294-1295 (1957).
19. Jones, M., Featherstone, R. M., and Bonting, S. L., *J. Pharmacol. Exptl. Therap.*, **116**, 114-118 (1956).
20. Kato, Y., and Moog, F., *Science*, **127**, 812-813 (1958).
21. King, T. J., and Briggs, R., *Cold Spring Harbor Symposia Quant. Biol.*, **21**, 271-290 (1956).
22. Kuusi, T., *Ann. Zool. Soc. Zool. Botan. Fennicae Vanamo*, **14**, 1-98 (1951).
23. ———, *Arch. Soc. Zool. Botan. Fennicae Vanamo*, **12**, 73-93 (1957).
24. Markert, C. L., *Cold Spring Harbor Symposia Quant. Biol.*, **21**, 339-348 (1956).
25. ———, and Silvers, W. K., *Genetics*, **41**, 429-450 (1956).
26. Monod, J., in *Enzymes: Units of Biological Structure and Function* (Gaebler, O. H., ed.), p. 7-28, Academic Press, New York (1956).
27. Moog, F., *Ann. N. Y. Acad. Sci.*, **55**, 57-66 (1952).
28. Niu, M. C., *Anat. Record*, **122**, 420 (1955).
29. ———, in *Cellular Mechanisms of Differentiation and Growth* (Rudnick, D., ed.), p. 155-172, Princeton U. Press (1956).
30. Pavan, C., and Breuer, M. E., *Symposium on Cell Secretion* (G. Schreiber, ed.), p. 90-99, Univ. Minas Geraes, Belo Horizonte (1955).
31. Perske, W. F., Parks, Jr., R. E., and Walker, D. L., *Science*, **125**, 1290-1291 (1957).
32. Ris, H., in *The Chemical Basis of Heredity* (McElroy, W. D., and B. Glass, eds.), p. 23-62, Johns Hopkins Press, Baltimore (1956).
33. Sato, G., Fisher, H. W., and Puck, T. T., *Science*, **126**, 961-964 (1957).
34. Shen, S. C., in *Biological Specificity and Growth* (Butler, E. G., ed.), p. 73-92, Princeton Univ. Press (1955).
35. Spiegelman, S., *Symposia Soc. Exptl. Biol.*, **2**, 286-325 (1948).
36. Stern, C., *Am. Scientist*, **42**, 213-247 (1954).
37. ———, in *Analysis of Development* (Willier, B. H., Weiss, P., and Hamburger, V., eds.), p. 151-169, W. B. Saunders Co., Philadelphia (1955).
38. Sturtevant, A. H., *Science*, **58**, 269-270 (1923).
39. Toivonen, S., *J. Embryol. Exptl. Morphol.*, **2**, 239-244 (1954).
40. Vogel, H. J., in *The Chemical Basis of Heredity* (McElroy, W. D., and B. Glass, eds.), p. 276-289, The Johns Hopkins Press, Baltimore (1957).
41. Weiss, P., *Yale J. Biol. and Med.*, **19**, 235-278 (1947).
42. ———, in *The Chemistry and Physiology of Growth* (Parpart, A. K., ed.), p. 135-186, Princeton Univ. Press (1949).

43. ———, *Quart. Rev. Biol.*, **25,** 177-198 (1950).
44. ———, *J. Embryol. Exptl. Morphol.* **1,** 181-211 (1953).
45. White, P. R. (ed.), Decennial Review Conference on Tissue Culture, *J. Natl. Cancer Inst.,* **19,** 467-843 (1957).
46. Willier, B. H., and Rawles, M. E., *Genetics,* **29,** 309-330 (1944).
47. ———, and ———, *Yale J. Biol. and Med.,* **17,** 319-340 (1944).
48. Wright, S., *Am. Naturalist,* **79,** 289-303 (1945).
49. Yamada, T., and Takato, K., *Embryologia,* **3,** 69-79 (1956).

# THE INITIATION OF DEVELOPMENT

R. D. ALLEN

*Department of Biology*
*Princeton University*

ALTHOUGH THE GENERAL title of this paper would indicate a comprehensive coverage of fertilization, and of artificial and natural parthenogenesis, only certain limited aspects of these fields will be considered. In the last decade a number of reviews have appeared on the role of interacting substances in fertilization (116, 163, 202), the acrosome reaction (31, 46), cortical changes and the physiology of activation in the sea urchin (12, 174, 175, 177, 178, 181, 182) and in fish (222), and the block against polyspermy (63b, 162). The recent book, *Fertilization,* by Lord Rothschild (163), contains a wealth of information on many aspects of the initiation of development. The present account will cut across many of the subjects listed above in an effort to bring into focus some relationships between two of the central groups of processes leading to development: those events and processes leading toward syngamy, and those leading toward activation.

The present state of knowledge regarding fertilization and activation is such that the vast majority of papers deal with echinoderms. Only the barest beginnings have been made on investigation of other forms, except for fish (cf. 222 for references). For this reason, the present account will deal principally with the work done on sea urchins. However, an effort has been made to include considerable comparative material where promising leads may provide a stimulus to further research. In addition to sections on syngamy and activation, material has been included on the structure and chemistry of the surface layers of the egg, since this is basic to a discussion of surface changes preceding activation.

I. Processes Leading to Syngamy

## 1. The Acrosome Reaction

Although there have been scattered accounts of filamentous connections between the egg and spermatozoon prior to contact of the gametes at fertilization (22-24, 53), it was J. C. Dan who in a series of studies on a variety of echinoderms, mollusks, and annelids showed unequivocally that the acrosome region of the mature sperm could give rise to an acrosome filament (42-48, 215). Excellent reviews have appeared recently by Dan (46) and by Colwin and Colwin (31) in which the morphological details of acrosomal changes are discussed, along with information on some of the physiological conditions which elicit an acrosome reaction. No attempt will be made here to duplicate either the historical treatment or the descriptive accounts presented in these reviews. Rather, the present aim will be to summarize the principal features of the acrosome reaction which may have a direct bearing on the processes leading to syngamy and activation of the egg. Particular attention will be given to work which has appeared since the last reviews.

The main features of the acrosome reaction (46) are: (1) formation of a fibril from the acrosome region of the spermatozoon; (2) the release of some substance which immediately disperses; and (3) the lateral displacement of the middle-piece. Ultrathin sections viewed in the electron microscope have shown that the acrosome consists of two parts, an apical globule of electron-dense material, and a basal stalk which extends into an invagination of the nucleus (1; also 43, p. 58). During the acrosome reaction, two separate events seem to occur: (1) discharge of the substance of the apical globule; and (2) formation of the filament from the body of the acrosomal stalk (Afzelius et al., 3, 4). A similar conclusion was reached regarding the dual nature of the acrosome region in *Hydroides* by Colwin, Colwin, and Philpott (32, 33), who found a peripheral component which collapsed during the acrosome reaction, and a central component involved in the formation of the filament. In sections of heavily inseminated eggs, Afzelius and Murray (4) observed that all of the attached spermatozoa had undergone at least a partial reaction, with the acrosomal globule expelled; more often a complete reaction had occurred, including filament formation. The mechanism of the acrosome reaction is still unknown, but it is now certain from measurements of the nuclear invagination that the latter is not everted during the acrosome reaction (4).

The physiological conditions which call forth or are necessary for the acrosome reaction have been listed in the review of Dan (46); they are summarized here with newer information added.

1. *Physiological maturity of the spermatozoa* seems to be an essential prerequisite for the acrosome reaction, as ascertained by studies of *Mytilus*, in which the time of natural spawning could be determined (48).

2. *The presence of calcium* in the medium is required for echinoderm sperm to exhibit an acrosome reaction; however, the sperm agglutinates if jelly is added. Agglutinated sperm did not show reduced fertilizing capacity, an observation suggesting that the normal action of the jelly was due not to "muzzling" of sperm by fertilizin, but to premature response of the acrosome (4, 45). Yanagimachi (223) has observed that sea urchin spermatozoa do not attach radially at the surface of eggs in the absence of calcium. This is probably a consequence of the failure of the acrosomes to react under these conditions. The importance of calcium is further shown by the fact that the acrosome reaction can be elicited by excess calcium alone in *Mytilus* (215).

3. *The presence of species egg water* can evoke an acrosome reaction, especially in the echinoderms studied by Dan (43, 44, 46). This effect has also been shown quantitatively with living *Thyone* sperm (37). Similar results have been reported for *Asterias* and *Nereis* (117). However, Rothschild and Tyler did not find convincing evidence for a stimulating action of egg water on sperm of *Strongylocentrotus purpuratus* (168), and Afzelius and Murray are of the opinion that in several species of European sea urchins the effect of egg water was more injurious than specifically stimulating (4). In contrast to the situation in echinoderms, spermatozoa of gastropods and of the oyster (which has no jelly layer) are comparatively insensitive to egg water treatment (46).

4. *Alkaline sea water* ($pH$ 9 or over) causes breakdown of the acrosome in sea urchins (43, 46), in the holothurian *Thyone briareus* (37), and in various mollusks (46, 48, 202). According to Dan, the action of alkaline sea water has physiological significance, since it can act favorably on fertilization as well as on the acrosome reaction (46).

5. *Contact of a non-specific nature,* such as to glass or collodion, sometimes elicits an acrosome reaction in *Hemicentrotus* (43), but not in *Echinocardium cordatum* (169). Oyster sperm respond readily to contact with glass, as when stirred with a glass rod (46, 48). From the fact that sperm sometimes attach to, penetrate, and even activate heterologous eggs, it could be inferred that non-specific contacts of this sort might elicit an acrosome response.

6. *Contact with one of the surface layers of the egg* can result in an

acrosome response. A distinction should be made between those species in which the sperm normally swim only to the outer border of the jelly and those in which it can penetrate the jelly to the vitelline membrane. In the first group, which includes the starfish and holothurian eggs, the acrosome reaction can be seen to have occurred at the outer border of the jelly [cf. especially Chambers' observations (22-24) when reinterpreted (29, 31, 35-37)]. However, in the second group, for which the sea urchin, the worm *Hydroides,* and various mollusks will serve as examples, the attached sperm nearly always exhibit at least a partial reaction at the egg surface proper (3, 4, 33, 48, 214).

It is perhaps too early to draw conclusions regarding the relative importance in various species of dissolved jelly, intact jelly, or the egg surface proper in initiating the acrosome reaction. However, the fact that sea urchin eggs show an increased rate of fertilization following *careful* removal of the jelly by mechanical means or at a $pH$ no lower than 5.5 (59, 63a, 67), would seem to indicate the absence of an essential role for the jelly, either intact or in solution, for the acrosome reaction. The fact that the jelly can in high concentration induce an acrosome reaction might be attributed to certain chemical similarities between the jelly and mucopolysaccharide substances believed to be present in the cortex (84, 125, 174, 179, 183).

## 2. *The Mechanism of Sperm Entry*

There is at present no quantitative information regarding the relative contributions of the egg surface and of the sperm itself in sperm entry. Differences of opinion in the older literature are exemplified by the term "perforatorium" applied to the acrosomes of certain species, on the one hand, and on the other by the analogy sometimes drawn between sperm "engulfment" on the part of the egg and phagocytosis. In Dan's recent study of sperm entry in various echinoderms (41), it was shown that while sperm movements occurred before attachment, and even during and after penetration in some species, in others (especially starfish) movements of the spermatozoon either stop during entry or clearly play no role (cf. 22-24). In sea urchins, the spermatozoa often continue moving while attached. The sperm head appears to twist, but it is not known whether this is rotation (182) or yawing (57); probably it is the latter, since attachment by the acrosome would probably impede rotatory movements. Yanagimachi (223) has studied these so-called "boring movements" of sea urchin sperm at the egg surface, and has shown that under conditions of calcium deficiency the sperm do not "bore" or orient themselves radially, although they maintain motility. This is probably explained by the more recent discovery of the necessity of calcium for the acrosome reaction (45). In spite of the interesting relationship which Yanagimachi found between the degree of successful cross-fertilization and the ability of the fertilizing spermatozoa to exhibit "boring movements," there is no evidence that the movements themselves play any role in sperm entry. In certain mollusks, the sperm head is held fast by its attachment point so that neither rotatory nor yawing movements could occur (214).

The initial phase of sperm entry is passage through the vitelline membrane of the egg. In some mollusks, this membrane is exceedingly thick and tough. It was shown several years ago by Tyler (199) that an egg-membrane lysin could be extracted from spermatozoa of *Megathura* and *Haliotis* which dissolved membranes of the same species. Since then, Wada, Collier, and Dan have demonstrated the presence of a similar egg-membrane lysin in *Mytilus edulis* sperm, and have shown its probable localization in the acrosome (215). It is released during the acrosome reaction and causes localized dissolution of the egg membrane at the point of sperm entry. Although egg-surface lysins have been described for sea urchins (191), there is so far no evidence for dissolution of the vitelline membrane in the neighborhood of sperm attachment in thin sections viewed with the electron microscope (4).

In those species in which a prominent fertilization cone forms, it is apparent that this reaction on the part of the egg is a response to attachment of the acrosome filament (29, 30, 31, 35-37). In oocytes, especially those of the sea urchin and sand dollar, it is well known that several exaggerated fertilization cones form, often in response to almost every spermatozoon which attaches to the egg surface (81, 167, 219). During the time that a block against polyspermy is gradually established, the number of reception cones which appear on insemination decreases (187). At the same time, the size of the reception cone becomes reduced.

In the most complete and detailed observations on sperm entrance to date, Colwin and Colwin have shown in *Holothuria, Asterias,* and *Thyone* that the fertilization cone, or a projection thereof, creeps up the attached acrosome filament at the same time that the sperm sinks into the egg. Once within the egg, the sperm (complete with tail) moves deeper into the egg cytoplasm through the conspicuous hyaline fertilization cone and funnel (29, 30, 31, 35-37) (Fig. 1). The hyaline funnel was recognized by cytologists because of its staining properties (cf. 219, 221). Three questions are raised by the above observations on sperm entry: (1) what elicits fertilization cone formation; (2) why the membrane of the fertilization

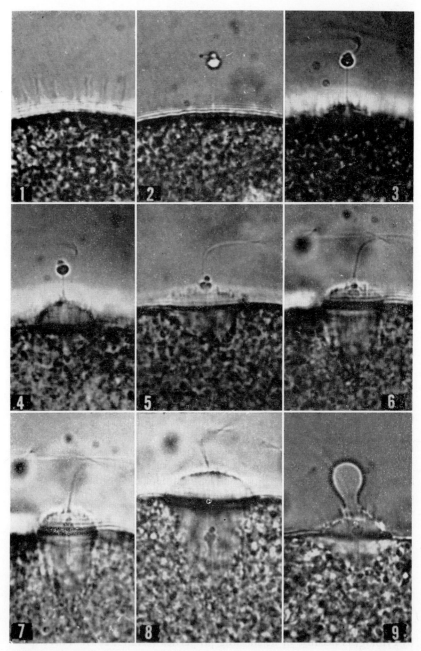

Fig. 1. Unretouched photographs of sperm entry in living *Holothuria atra* (From Colwin and Colwin, 29, 31). Frame 1 shows filamentous radiations from the egg surface extending into the jelly layer. Frame 2 shows a spermatozoon with its acrosome filament extending to the egg surface; no cone formed and the sperm did not enter. Frames 3-8 show successive stages of sperm entry in different eggs. Frame 9 shows a cone with an exudate from a ruptured region. The spermatozoon appeared unaffected.

cone "creeps" up the acrosomal filament to engulf the spermatozoon; and (3) what force pulls the sperm through the hyaline "funnel" region of the cone into the endoplasm. Since the fertilization cone forms but once except in material destined to become polyspermic, it is reasonable to suppose that it represents a specific egg response to some chemical or mechanical action of the acrosome of the fertilizing spermatozoon. Monroy (128) has speculated from his observations on the action of various enzymes on the surface of *Pomotoceros* eggs that a protease from the sperm might elicit cone formation. The fact that the cone forms as an external projection which grows in volume during sperm entry suggests that it could be formed either by an influx of water or by expulsion of fluid from the egg interior. Either might result from the weakening (by the action of local enzymes or surface-active agents) of the membrane in the region of the cone. It is perhaps most noteworthy that the granular material nearest to the attached acrosome recedes, leaving the hyaline funnel through which the sperm enters. If the cytoplasm in the funnel and surrounding the bordering granules were fluid, the funnel would rapidly fill up again with granules by Brownian movement. Since this does not occur, it is likely that the cytoplasm either of the funnel or surrounding the funnel is gelated. If the latter were true, the acrosome might prove to be anchored at the region of the cytoplasm which later became the apex of the funnel, and be pulled down as the cytoplasm in that region contracted away from the surface (Fig. 1). An anchorage at the apex of the funnel is suggested by the following observation: ". . . in a few cases it was found that if enough pressure were applied to the cover-slip to cause the cone to rupture and form an exudate, the spermatozoon within the cone was not expelled" (31, p. 163). One of the other possible mechanisms of sperm penetration that has been suggested (169), namely, that the acrosome might contract and pull the sperm in, appears to be incorrect, for the length of the acrosome remains constant during entry (31). In *Mytilus edulis,* Wada has described what appears to be a dynamic system within the egg for pulling in the fertilizing spermatozoon while rejecting supernumerary ones; he believes this is a morphological representation of the block against polyspermy (214). The sperm possess a 12-micron acrosome filament which apparently pierces the egg on attachment; at least this occurs when sperm attach to isolated "vitelline membrane and vitelline layer preparations" (214). Available evidence so far suggests the acrosome filament may function as a "handle" by which the sperm is drawn into the egg.

## 3. Early History of the Spermatozoon Inside the Egg

Until recently, the events taking place with respect to the spermatozoon immediately following its penetration into the egg were largely unknown except for observations on fixed material (e.g. 219-221). Much information has been provided by the observations of J. C. Dan with the phase contrast microscope on the Japanese echinoderm eggs, some of which are noted for their extraordinary transparency. Among the species studied, variation was found in the activity of the sperm tail during entrance; in *Asterina pectinifera,* for example, the tail was motionless, while in *Pseudocentrotus* vigorous writhing movements were observed. In all species she studied, Dan found that the entire spermatozoon entered, including the tail (41). Older observations (cited by Wilson, 221) were not in agreement as to whether the sperm tail entered, in sea urchins as well as in other animals. It is interesting to note that Dan's observations have been confirmed for *Thyone* and *Holothuria* (29, 31), and for *Paracentrotus* (165). In *Clypeaster* and *Mespilia,* the sperm tail does not cease its movements, for Dan has seen corresponding movements of the head and tail, along with a centripetally directed flow of cytoplasm, combining to bring about early translatory movements of the sperm head from the point of sperm entry (41). These movements came to a halt as the sperm tail was completely drawn into the cytoplasm; formation of the sperm aster soon followed (41).

Classical accounts of the fate of the sperm head within the egg stress the rather puzzling rotation of the sperm head which either precedes or coincides with formation of the aster. The observations of Dan with regard to autonomous movements of the sperm head soon after entry suggest a possible mechanism of rotation. If the sperm tail, on being drawn into the egg, were freer to move in the cytoplasm than the sperm head, the former would be expected to bend into a broad loop (see Wilson, 221, p. 445, Fig. 207), which would rotate the sperm head by as much as 180 degrees. This was suggested by the otherwise perplexing observation that, in spite of the length of the sperm tails (40-55 microns), in no case did a sperm head move more than 20 microns from its entrance point before aster formation (41, p. 409). E. Chambers (21) and R. Chambers (24) have also noted that the site of sperm aster formation was frequently laterally displaced from the position of the fertilization cone remnant.

Rothschild (164) has recently published a remarkable electron micrograph showing a sperm three minutes after penetration into an egg, and lying only a micron or two beneath the cortex. This picture revealed: (1)

the dissolution of the nuclear membrane except in the region of the centriole; (2) the double nature of the centriole; (3) that the sperm tail is clearly visible in the cytoplasm of the egg; and (4) that changes have already occurred in the middle-piece structure. Within four or five minutes after entry, the sperm becomes invisible in living sea urchin eggs, but reappears again as a hyaline "pronucleus" which is led by the growing sperm aster to the center of the cell. The disappearance of the sperm nucleus is caused by its swelling, during which it changes from a phase-retarding to a phase-advancing object when viewed in the cytoplasm.

## 4. The Migration of the Sperm Nucleus

In 1925, Wilson (221) summarized the evidence then available on the mechanism of migration of the sperm head within the egg cytoplasm and pointed to two components, the penetration path, which is nearly radial, and the copulation path, in which the sperm nucleus veers toward the egg nucleus. E. Chambers reexamined the problem, with the added precaution of avoiding compression of the eggs under observation (21). Under these conditions, only one component was found in *Lytechinus* eggs, the radially directed penetration path which is independent of the presence of the egg nucleus, since it occurs in enucleate egg fragments. When a sperm penetrates a flattened egg surface, it penetrates perpendicularly to that surface (21). In eggs made cylindrical by confining them in glass capillaries, the depth of penetration of the sperm is roughly equal to the radius of the capillary (6). These facts have suggested that the growth of the sperm aster might propel the sperm head toward the egg center, particularly since the rays of the aster extend to the cortex (21). Such an astral mechanism for migration is also suggested by observations made by A. Brachet (20) and Allen (6) on sperm movements in dispermic eggs, in which the sperm nuclei push one another apart by growth of their asters.

It has been suggested that the formation of the sperm aster and sperm migration both depend on some kind of interaction between the sperm and the mature cortex (6, 185, 186). In oocytes which are unable to exhibit a cortical response other than exaggerated fertilization cones, the spermatozoa which cause these cones enter, but neither rotate nor migrate to the center (219). In the oocyte this could also be due to cytoplasmic unripeness (51); however, in mature sea urchin gametes, sperm which penetrate a region of egg surface which is injured so that it lacks cortical granules will not form an aster (6, p. 409), whether or not a cortical response is initiated. Sperm passing through the cortex of eggs treated with redox dyes sometimes also fail to form an aster (180). As will be shown in

Section III 7, the occurrence of a cortical reaction is a prerequisite for normal nuclear movements preparatory to syngamy.

In spite of the inferences which can be drawn from the sea urchin work regarding the role of the sperm aster in the migration of the sperm nucleus, it is quite possible that other mechanisms play some part. For example, Dan's observation of centripetally directed cytoplasmic streaming suggests that similar processes might occur later during the astral growth period and should be looked for. Probably time-lapse cinematography of sperm nucleus migration in sufficiently transparent eggs could permit a decision between possible astral growth and cytoplasmic streaming mechanisms. None of the studies carried out so far have had this end in view. It is interesting to note that in the medusan, *Spirocodon,* the sperm characteristically enters so near the egg nucleus that no astral mechanism is employed to bring about syngamy (42). The same has been reported to occur in the eggs of *Mespilia* when the sperm enters near an eccentrically situated egg nucleus (41).

## 5. *Migration of the Egg Nucleus*

Movement of the egg nucleus to its point of fusion or association with the sperm nucleus seems to be more complicated than the migration of the sperm nucleus. Even in artificially activated eggs a "centering" of the egg nucleus occurs (134), showing that such movements can occur independently of the presence of the sperm nucleus. However, in artificially activated eggs, centering may possibly be mediated by growth of a "monaster." In fertilized eggs, the egg nucleus typically migrates without the formation of any conspicuous astral rays. Chambers (21) concluded from his analysis of 36 camera lucida records in which both the nuclei were in the same focal plane that the egg nucleus is carried to the "astral lake" (centrosome) by a current of centripetally directed cytoplasm. Although Chambers found that the adjacent granules accompanied the nucleus, it would be desirable to have cinematographic records of the behavior of a considerable portion of surrounding cytoplasm before deciding on cytoplasmic streaming as the sole mechanism for movement of the egg nucleus. The need for such a study is emphasized by the fact that A. R. Moore decided on the basis of observations quite similar to those of Chambers that the egg nucleus of *Temnopleurus* and of *Dendraster* moved without disturbing the surrounding granules and therefore must be pulled by fibrillar connections (134).

The present skepticism regarding cytoplasmic streaming as a sole basis for movement of the egg nucleus stems from observations of *Psammechi-*

*nus* eggs fertilized in glass capillaries (6). When an egg is taken in so that the capillary approaches the egg symmetrically with respect to the position of the egg nucleus, the latter remains stationary in the cylindrically shaped cell. However, if the egg is rotated so that the capillary approaches it asymmetrically, its nucleus later "creeps" to another position, as if it were satisfying unequal tensions between hypothetical stretched fibers holding the nucleus from all sides. This should not be taken to mean that the nucleus is held by fibers, for other mechanisms could undoubtedly account for this behavior. Following fertilization, the egg nucleus, irrespective of its distance from the point of sperm entry, seeks the axis of the capillary and elongates in this axis before it begins to migrate. Its migration is accompanied by a bulging out in the direction of the sperm nucleus. This was also noted by Chambers, but only just before syngamy (21). In spherical eggs, Chambers noted that the rate of nuclear migration was slow at its beginning and speeded up to a maximum in the middle of its journey. Progress was again slow during entrance into the "astral lake." In cylindrical eggs, the rate of nuclear migration is more or less constant, irrespective of the distance between the nuclei (6) (Fig. 2). Besides rendering eggs cylindrical in capillaries, other devices have been used in order to alter experimentally the distance over which the nuclei must migrate. Chambers (21) obtained his fastest migration when the egg nucleus passed through the rather wide neck of a flask-shaped cell, one lobe of which was an exovate in which the egg nucleus was originally situated. On the other hand, E. B. Harvey found that centrifugation caused a separation of the nuclei by throwing the sperm nucleus to the heavy pole and the egg nucleus to the light pole (72); under these conditions nuclear fusion was delayed by as much as an hour, but it is likely that the effects of centrifugal force were due not only to separation of the nuclei, but to disruptive effects on cytoplasmic structure as well.

Besides nuclear migration, other changes occur at the same time which lead to syngamy and cleavage. One of these is the swelling of the nuclei, which is especially exaggerated if syngamy is delayed (72); the sperm nucleus may even come to exceed the egg nucleus in size before they eventually fuse. If fusion fails to occur, the nuclei may "break down" (i.e., lose their limiting membranes). In this case, the sperm nucleus often forms the center of a mitotic division, while the egg nucleus develops a monster (72). Sometimes in cases of delayed fusion the sperm aster divides to form the amphiaster before the nuclei have come together; this occurs particularly in etherized eggs (220) and following centrifugation (72). The ability of the sperm centriole to divide and to give rise to an

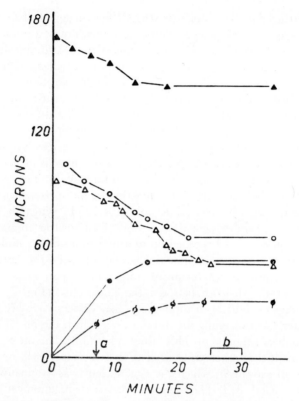

Fig. 2. Migration of the egg nucleus toward, and the sperm nucleus away from the point of sperm entrance (O on the ordinate). *a* is the approximate time of elongation of the egg nucleus, and *b* is the approximate time when the sperm aster disappears. Black circles represent the migration of a sperm nucleus in an egg contained in a wide capillary; black circles barred represent a parallel experiment in a much narrower capillary. Each curve represents a different experiment (Allen, 6).

amphiaster probably explains the fact that when eggs are divided mechanically by a cotton fiber following fertilization, in such a way as to separate the sperm and egg nuclei into different cell fragments, the piece with the sperm nucleus divides, but the other fragment containing the egg nucleus does not (224). The same is true of eggs divided by centrifugal force, except for an interesting case noted by E. B. Harvey (72): in *Paracentrotus lividus,* when eggs were divided by centrifugal force 15 minutes after insemination, the amphiaster associated, as usual, with the sperm nucleus. However, at 17 minutes after insemination the amphiaster remained with the egg nucleus in some, but not all, eggs. By 19 minutes, the nuclei could not be separated. It would appear that the division center, which has apparently been attached physically to the sperm nucleus, acquires, just prior to syngamy, a firm attachment with the egg nucleus. This mechanical association may be important in the last stages of nuclear fusion.

## II. Structure and Chemistry of the Egg Surface

### 1. The Jelly Layer

*a. The sea urchin egg.* The jelly layer of the sea urchin egg is initially rather compact, but swells on hydration to form a loose coat separated from the egg surface by 3-5 microns (67). Several observers have noted that protozoa can swim in the space between the jelly and the egg surface proper (67, 212). This is true not only in the European species *Psammechinus miliaris*, but also in *Strongylocentrotus purpuratus* from California (Mrs. J. Laties, pers. commun.). Due to its low refractive index, the jelly cannot be seen even with the interference microscope, a fact indicating that its concentration is less than about 0.005% organic material (124). The jelly can be visualized, however, (1) after staining with Janus green, (2) when outlined by squid or India ink, or (3) as an invisible barrier separating crowded eggs (73). The jelly does not ordinarily reveal any molecular structure in the polarizing microscope, but when vitally stained with Janus green B, it becomes intensely birefringent due to the lining up of the dye molecules tangentially to the egg surface (123) (Fig. 3). Similar birefringence has been seen in the jelly after treatment with sperm extracts (188). At present, insufficient information is available concerning the site of attachment of the dye or extract molecules to make any interpretation in terms of molecular structure of the jelly. In the starfish egg, Chambers (24) and Fol (53) saw evidence of radial structure in the jelly, to judge by the alignment of ink particles which entered.

The jelly layer undoubtedly provides the egg cell with considerable protection against mechanical injury, for eggs without jelly are sticky and fragile (185, 188). The jelly slowly dissolves in sea water to form "egg water" or "fertilizin," which in sufficient concentration has the power to agglutinate homologous (and sometimes heterologous) sperm (63, 116, 175, 200, 202). As was first shown by Kupelwieser (97), the jelly layer can be removed with acid sea water. At first it appeared that the jelly was essential for fertilization, since jelly-less eggs were difficult to fertilize unless large numbers of sperm were present (200). However, more recently Hagström et al. have shown that jelly removal by more careful means

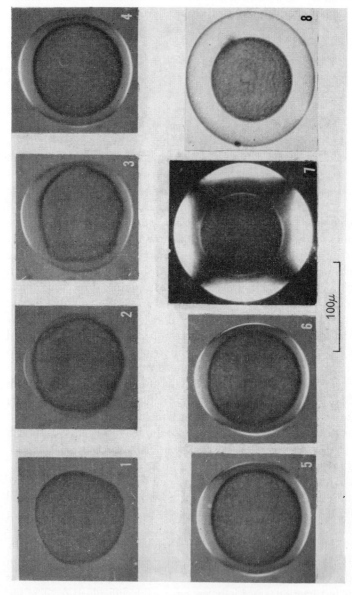

Fig. 3. Frames 1-6 show successive stages in the elevation of the fertilization membrane in *Paracentrotus lividus* photographed at 1 min. intervals in polarized light (compensated). Frame 7 shows the polarization cross in the jelly of an egg vitally stained with Janus green (uncompensated polarized light). Frame 8 shows the same in ordinary light (Mithison and Swann, 123).

actually enhances fertilization (see Fig. 4) and decreases the number of sperm necessary to obtain either fertilization or polyspermy (59, 62, 63a, 67). Previous difficulties were due to injury to the egg surface by sea water of a too acid pH. This was shown by the fact that "cytofertilizin" (60) and one of the proteolytic enzymes (EII) (108-110) of the egg leak out on acid treatment. Although it might seem that the presence of a small amount

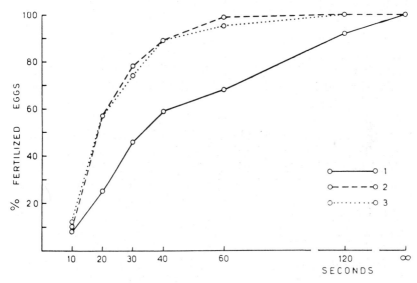

Fig. 4. Acceleration of the fertilization rate by jelly removal in eggs of *Psammechinus microtuberalatus*. 1, Control eggs with intact jelly. 2, Eggs deprived of their jelly coats by treatment with acid sea water at pH 5.8. 3, Eggs deprived mechanically of their jelly coats. Points represent counts of about 250 eggs (Hagström and Markman, 67).

of jelly on the surface could be essential, this is rendered improbable by the fact that substances which destroy or precipitate the jelly (periodate, and jelly-precipitating-factor or "antifertilizin") do not inhibit fertilization under these conditions (63a).

Due largely to the studies of Vasseur (209-213) and Tyler (cf. 205 for references), there is considerable information regarding the chemical nature of the dissolved sea urchin egg jelly. In solution, this material has a molecular weight of 280,000-300,000 with an axial ratio of 28:1 in *Arbacia* (205), and it behaves as an electrophoretically homogeneous acid substance migrating to the anode even at pH 2 (192). However, Messina and Monroy have separated *A. lixula* jelly into two components, one precipitable with protamine sulfate and containing all the fucose and some galactose, whereas the remaining fraction contains only galactose and is devoid of agglutinating ability (115).

On hydrolysis the jelly yields various different monosaccharides, depending on the species (210, cf. 212 for references), sulfate, and amino acids. It appears that the jelly is a highly sulfonated glycoprotein free of hexosamine and glucuronic acid. Except for the lack of the latter two substances, the jelly, or at least its carbohydrate moiety, bears considerable resemblance to heparin, both in its high content of sulfate esters of monosaccharides and in its anticoagulant activity (82). Like heparin, dissolved jelly is metachromatic; this property has proved a useful tool for the study of the interaction between jelly and sperm, in which the jelly material decreases in viscosity and is apparently depolymerized, although possibly not enzymatically (114).

*b. Other invertebrate eggs.* Compared to the sea urchin, very little is known about the structure and composition of the jelly of other invertebrates. To cite a few examples, the egg of the clam *Spisula solidissima* has a thin jelly which stains metachromatically with toluidine blue (9, 198). *Chaetopterus* eggs likewise have a jelly layer, but it is not metachromatic in toluidine blue (90). An interesting case is the egg of *Nereis limbata,* which is free of external jelly prior to fertilization; soon after successful sperm attachment, the jelly is extruded from precursor granules in the thick cortical layer at all points except where the sperm has attached. In *Nereis,* both the formed jelly layer and its precursor granules are birefringent, suggesting crystalline orientation of their molecules (85). What information is available suggests chemical properties not too different from those of the sea urchin egg (39). *Sabellaria vulgaris* eggs are also free of external jelly when shed, but extrusion of the jelly accompanies spontaneous breakdown of the germinal vesicle in this species, and fertilization may occur at any time with respect to the events of jelly extrusion (150).

*c. Amphibian egg jelly.* Frog egg jelly seems to have chemical properties somewhat similar to that of the sea urchin, at least in its content of mucopolysaccharide and amino acid residues. However, according to the study of Folkes, Grant, and Jones, 90 per cent of the sulfate of frog jelly is accounted for by cystine and methionine, a proportion suggestion that most of the carbohydrate is not esterified with sulfate (54). Minganti (120) has also shown the presence of various monosaccharides which show the same kind of species differences noted in echinoderm jellies. The presence of hexosamines and the absence of acid-hydrolyzable sulfate were confirmed.

It is interesting to note that sperm-agglutinating activity has been found in frog egg jelly (18).

For interesting comparisons of the layers the spermatozoa must penetrate before reaching the eggs of other chordates, see the paper of Leghissa (101) for ascidians, and the review of Austin and Bishop (16) for mammals.

## 2. The Cortex of the Mature Egg

Many eggs, in common with a number of other cells, possess an outer differentiated layer or "cortex." Although this layer is difficult to observe microscopically under ordinary conditions, especially in pigmented or yolky eggs, it can be made plainly visible by centrifugation (74, 136). In the eggs in which fertilization has been most commonly studied, three main types of cortex have been described: (1) the "labile" cortex which undergoes profound reorganization or breakdown at fertilization or maturation, and is characteristic of most echinoderms, several chordates, and some members of other groups; (2) the "stable" cortex which contains the same morphological features as occur in the first type, yet remains visibly unaltered at fertilization; and (3) those showing no clearly differentiated cortical layer of appreciable thickness or granular structure such as is to be seen in sea urchins. Ascidian eggs may belong to this type (101).

*a. The sea urchin egg.* The cortical layer of the sea urchin egg was first observed by Harvey following centrifugation, which exposed the single layer of distinctive cortical granules distributed somewhat unevenly over the surface (74) (see Fig. 5). In tangential view, there appears to be a clear layer outside the cortical granules. This clear layer is bounded on the exterior by a fine line which has been assumed to mark the position of the vitelline membrane. Although the cortex is difficult to see in tangential optical section, it can be better visualized by raising the refractive index of the observation medium, as was shown photographically by Motomura (140). The clear layer so well demonstrated by this method corresponds in apparent thickness and location to the light-scattering layer described originally by Runnström (170, cf. also 12, 58). For this reason, the clear layer has been referred to as the "luminous hyaline layer," not to be confused with the hyaline (ectoplasmic) layer which appears on the surface some time after fertilization.

Recently, Mitchison and Swann (123) and Mitchison (121, 122) have raised the question of the reality of some of these surface layers "observed" with the light microscope. Mitchison, for example, has failed to

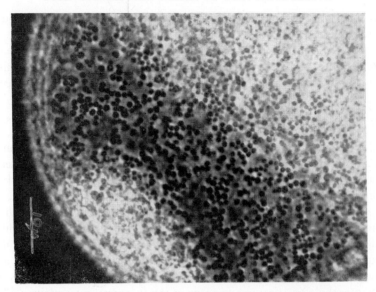

Fig. 5. An unfertilized egg of *Strongylocentrotus purpuratus* photographed in phase contrast after centrifugation (Allen, 8).

demonstrate the vitelline membrane, luminous hyaline layer, or cortical granules in fixed material, but was apparently unaware that the fixation of some of these structures had been successfully achieved by earlier workers. The cortical granules were first described in fixed material by Hendee (76) and later photographed elegantly by Motomura (140) after fixation in Champy's fluid, treatment with 0.5 $N$ NaOH, and staining in Janus green. Cortical structure has also been preserved by Runnström, Monné, and Wicklund (190), and by Monné and Hårde (125) for cytochemical work. Although it is not certain, it seems probable from his photographs that Motomura did, as he believed, demonstrate the vitelline membrane or its precursor in the unfertilized egg of *Strongylocentrotus* (*Hemicentrotus*) *pulcherrimus*. Mitchison's failure to demonstrate some of these surface layers in fixed material was due probably not only to his choice of fixatives, but also to choice of material, as *Paracentrotus* and *Arbacia* are distinctly inferior objects for cortical studies compared to *Psammechinus*, which is also available in Naples. The other aspect of the skepticism regarding the surface layers is, however, quite legitimate; there certainly are optical difficulties involved in observing the surfaces of eggs, especially at low working apertures, because of the diffraction rings produced by the complex light-scattering surface of the egg. It is therefore fortunate that other lines of evidence are now available with which to test the reality and fine structure of the egg surface.

Nearly all eggs have been regarded as having some kind of *primary membrane* (i.e., one formed by the egg), which, when closely applied to the egg surface, has been called a *vitelline membrane* (221). Although there was much dispute a generation ago as to whether sea urchin eggs possess a vitelline membrane, more recent evidence has been taken to mean that it does (for references to the early controversies, see 26, 136, 140; also 175, 178, 182). Present interest centers around the question of whether the fine line situated at the outer border of the luminous hyaline layer is actually a membrane which elevates at fertilization, or whether the fertilization membrane is formed de novo at fertilization (see also Section III 3). Mitchison is quite correct in stating that the evidence from microdissection experiments for the presence of a membrane proves only that such a membrane can be formed from the surface, and not that one existed there before treatment (121). One approach to the problem has been the use of enzymes. Runnström et al. (172, 181, 187, 188), Minganti (118), and Monroy and Runnström (132) have shown that pretreatment with trypsin results in subsequent failure of the fertilization membrane to elevate. Once formed, the fertilization membrane is resistant to trypsin, but not to the hatching enzyme (94, 132). These results have been interpreted as showing not only the presence of a vitelline membrane on the unfertilized egg surface, but also its protein nature. Recently, Tyler and Spiegel have shown that the proteolytic enzyme papain does not destroy the vitelline membrane, since the latter deflates after elevation in the enzyme solution and merges again with the egg surface. They conclude on the basis of the available evidence that the question of the existence of a vitelline membrane on the unfertilized egg is still open (208). The matter seems now to have been solved by the electron micrographs of Afzelius, in which the vitelline membrane has been distinguished as a continuous bordering structure 100 Å thick (2). The same structure has been observed as the outer component of the fertilization membrane, thus confirming the role of the vitelline membrane in the "dual origin of the fertilization membrane" (cf. 140, 171-173, 175, 187-190). The question of whether the vitelline membrane dissolves when treated with proteolytic enzymes appears also to have been solved by the observation of Afzelius and Minganti (reported in 181) that the vitelline membrane is still present in electron micrographs after trypsin pretreatment before fixation. Apparently trypsin, like low temperature (12), weakens the structure of the membrane so that the cortical granules can be expelled through it (181), causing "fenestrated membranes" (119).

The chemistry and physiology of the vitelline membrane remain somewhat obscure. Because of its low thickness and irregular contour, it is inaccessible to the polarizing microscope until after membrane elevation, when it becomes altered (see Section III 3). Lipid content would be suggested by its osmophilia in electron micrographs, and protein content by its sensitivity to trypsin and chymotrypsin. Although this membrane has generally been assumed to play no significant part in the permeability of the cell, it is worthy of note that trypsin treatment increases the permeability of *Psammechinus* eggs to water (119).

The cortical granules of the sea urchin egg are of particular interest because of the role that they play in the cortical reaction and membrane elevation (see also Section III, 3 and 4). Their structure has been described in *Arbacia* by McCulloch (113) and Lansing, Hillier, and Rosenthal (100), and in several species of European sea urchins by Afzelius (2). Typically, these granules appear to be lamellar, with what appear to be peripheral inclusions, and are surrounded by a double membrane 75-90 Å thick. The number and thickness of the lamellae are characteristically different for each species studied, although the size range is fairly constant (0.5-1.0 microns) (see Fig. 6). The chief limitation of electron microscopy seems to be fixation; although Afzelius has tried an impressive array of fixatives, and has found essentially the same internal structure for the cortical granules in all, it must nevertheless be kept in mind that the cortical granules are extremely labile, and it is possible that the peripheral inclusions might represent the beginnings of abortive granule breakdown due to injurious effects of the fixatives.

The chemistry of the cortical granules is a subject to which an increasing number of workers will turn for an explanation of fertilization in chemical terms. The investigation of the chemical properties of small cell inclusions is limited to two main types of approach, cytochemical staining techniques and chemical examination of isolated material. Both approaches have been applied with limited yet promising results. With cytochemical techniques, Monné and Hårde (125), Immers (84), and Runnström and Immers (183) have found evidence for mucopolysaccharide in the cortex and probably the cortical granules and interesting changes in staining properties at fertilization. Several attempts have been made to isolate cortical material from living or fixed eggs. Some of Moser's observations on the stability of oxalate-treated eggs have suggested the possibility of obtaining cortical material (138). Motomura (146) has reported isolated fragments obtained by rupturing acid-fixed eggs; the method is open to obvious disadvantages, and apparently no chemical work on this material has been published at the time of this writing. Nakano and Ohashi have collected and analyzed the carbohydrate present in the "cortical granular material" after the latter

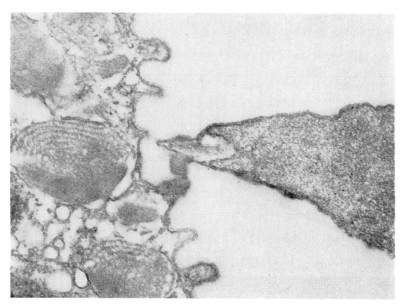

Fig. 6. An electron micrograph of a thin section of an attached sperm (probably a supernumerary one) at the surface of an egg of *Strongylocentrotus droebachiensis*. The acrosome globule has been expelled; part of the filament can be seen in contact with the egg. Note the lamellar structure of the cortical granules. (Courtesy of Dr. B. Afzelius.)

had been expelled from the cortex by urea treatment (149). Their results are of interest, as they find only fucose in the carbohydrate component released, but as a means for studying the chemical composition of the cortical granules the method is not ideal, for it may cause release of other substances as well (as, for example, the vitelline membrane material, 140). A more promising approach was suggested by the fact that strips of cortex can be torn from mature, jelly-free eggs of *Psammechinus* by virtue of their adhesiveness to glass surfaces (7) (see Fig. 7). Later it proved possible to isolate strips or even hulls of cortical material from living eggs by high centrifugal forces (8). This method worked satisfactorily on eggs of *Strongylocentrotus purpuratus* but was not applicable to *Arbacia* because the pigment, which had a similar high density, could not be separated from

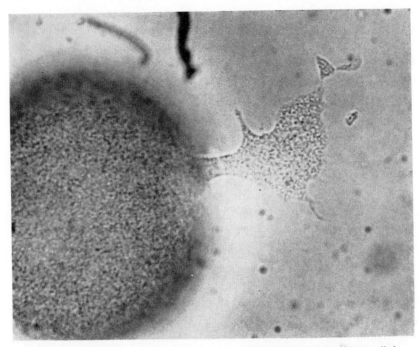

Fig. 7. An unfertilized jelly-free egg of *Psammechinus miliaris* which had been rolled after having adhered to a cover-slip. Note the strip of cortex partly pulled off the egg.

the cortical material. Although the centrifugation method for the removal of cortical material will require further refinement before it can be most profitably applied to chemical studies of the cortex, certain information was gained during the development of the method. The fact that the cortical material could be removed without dispersing the endoplasm added weight to the hypothesis of a subcortical membrane (7, 8, 156). At forces close to 170,000 $\times$ gravity, the cortical material itself separated into "membranes" (presumed to be natural or altered vitelline membranes) and "carpets" of cortical gel (Fig. 8).

The physical nature of the cortical gel in which the cortical granules are embedded has attracted much interest because of the changes undergone in this region during fertilization or artificial activation. Several attempts have been made to approach its molecular structure with the polarizing microscope. There is general agreement that the cortex of the sea urchin exhibits positive radial birefringence before fertilization, and that this disappears at fertilization (49, 122, 123, 126, 127, 130, 131, 175,

188-190). Interpretations, however, vary. Runnström et al. (reviewed in 175), and Monroy et al. (122-131) have conceived of the cortex as consisting of a primarily lipoid or lipoprotein structure similar to a liquid crystal. On the other hand, Mitchison has proposed a looped protein structure (cf. 122 for discussion and references), and Dan and Okazaki have clearly shown that at least part of the cortical birefringence behaves like strain birefringence because it is increased by stretching and is diminished by compression. The arguments in favor of a lipid structure are not supported by the electron microscope, which should be capable of detecting any significant amount of oriented lipid material by its osmophilia. Electron micrographs have, moreover, added another complication: the egg surface has been found to be covered by submicroscopic "microvilli" (or a brush border) which would render interpretations of data from the polarizing microscope very uncertain at the present time (cf. 2, 27, 100). In the future it will probably prove most profitable to combine studies of the sea urchin cortex in the polarizing microscope with parallel observations with the electron microscope under various experimental conditions. At present too little is known about the specificity of agents which have been thought to remove either proteins or lipids selectively.

The "luminous hyaline layer" is a part of the egg surface about which little is known, but it merits some attention because of the striking changes

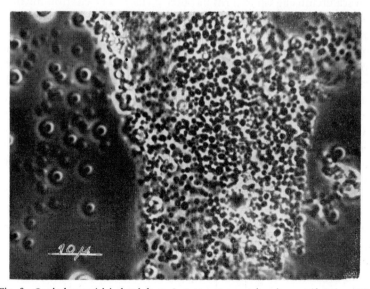

Fig. 8. Cortical material isolated from *S. purpuratus* eggs by ultracentrifugation (8).

in light-scattering it undergoes at fertilization (12, 58, 170). This layer may well be an undifferentiated part of the cortical gel lying outside the borders of the cortical granules; such a zone certainly can be seen in electron micrographs (2) (see Fig. 6). Under normal conditions, the color of this layer changes only when the cortical granules are discharged (see Section III 3), a fact suggesting that the light might be scattered from the outer surfaces of the cortical granules (136). This appears not to be the case, however, since Hagström (58) has found treatments which influence the light-scattering layer without visibly affecting the cortical granules.

Until recently, it was assumed that the permeability barrier of the mature egg lay beneath the vitelline membrane and outside the cortical granules. Two independent lines of evidence suggest a subcortical permeability barrier, however: (1) Parpart and Laris (156) have shown that the cortex is apparently permeable to certain non-electrolytes which do not enter deeper into the cell: and (2) cortical material can be stripped from living eggs without dispersing the endoplasm (7, 8). The above evidence is somewhat weakened by the fact that some non-electrolytes are activating agents for *Arbacia* eggs (138), and it is not yet known whether this is because of their penetration (156) or because they initiate a propagated wave of activation (cf. 195, 196). There is also little information on how rapidly a torn sea urchin egg can reform its outer surface. The electron microscope has not proven as decisive in settling the location of the plasma membrane as might have been anticipated. One of the published electron micrographs of Afzelius (2, Fig. 2) and several unpublished micrographs of Parpart (pers. commun.) show what might be interpreted as a subcortical membrane of wavy outline, but the matter cannot be decided on the basis of present evidence. It would not be surprising if a delicate membrane of the order of thickness of 50-60 Å might be very difficult to detect in sea urchin eggs, where the distance for fixatives to penetrate is great, and where shrinkage would be likely to disrupt delicate structures.

*b. Other eggs with a "labile" cortex.* A similar cortical layer has been seen in the eggs of other echinoderms besides the sea urchin. Just, who thoroughly investigated the cortical reaction in a number of marine eggs, failed to see the cortical granules in any of the species. However, in *Echinarachnius* (the sand dollar) he observed their products, the cortical globules in the perivitelline space during membrane elevation (87, cf. also 26). The presence of cortical granules in this species has been confirmed (9). Eggs of the starfish *Asterina pectinifera* also contain labile cortical granules (140), as do those of the American starfish *Asterias forbesii,* and of the

Japanese crinoid *Comanthus japonicus* (47). Little is known regarding the chemistry, physiology, and ultrastructure of the cortical layers in these forms.

Cortical changes remarkably similar to, but on a larger scale than, those in sea urchin eggs have been reported in a number of fish. Okkelberg (155) was apparently the first to describe such changes in the egg of the brook lamprey, *Entosphenus wilderi,* where the disappearance of cortical alveoli precedes the formation of the perivitelline space. More recently, the physiology of cortical conduction has been studied intensively, especially by Yamamoto, whose most recent paper gives many useful references (222). The organization of the cortical region of the fish egg is somewhat different from that in the smaller invertebrate eggs. In the first place, the cortical alveoli are much larger than cortical granules (4–25 microns compared to 0.5–1.0), and alveoli can be dislodged by moderate centrifugation (99). However, like the sea urchin granules, the fish alveoli discharge their contents (mucopolysaccharides) into the perivitelline space, and the alveolar membrane apparently becomes part of the new cell surface (93, 99). According to Kemp (93), no characteristic ultrastructure appears in electron micrographs of fish cortical alveoli; evidently they form from the primitive "yolk vesicles" during oogenesis and migrate outward.

Frog eggs show an entirely different cortical structure, including a layer two or three deep of cortical granules which disappear at fertilization (91, 92, 143) (Fig. 9). These are also lacking in distinctive ultrastructure; they apparently form either in the cortical layer or just below it during oogenesis (93). A labile cortex containing cortical granules has also been described for the hamster egg (15).

A labile cortex is also found in *Nereis limbata* eggs; however, it is of a different kind. Instead of being thin, with a single layer of granules, as in sea urchins, it is thick and tightly packed with jelly-precursor granules. Costello (40) believes that jelly extrusion at fertilization is basically similar to the cortical reaction in echinoderms.

*c. Eggs with an apparently "stable" cortex.* It has been noted in studies with various annelid or mollusk eggs that although a differentiated cortex was present, no striking changes of the type noted in sea urchin or fish eggs occurred on fertilization. This was first noticed in *Spisula* eggs (5). According to Rebhun (159), the cortical granules of *Spisula* are formed in situ early in oogenesis and differ only slightly in appearance in electron micrographs from yolk (Fig. 10). The rigidity of this cortex is sufficiently great that the granules remain in place even after 10 minutes at 200,000 ×

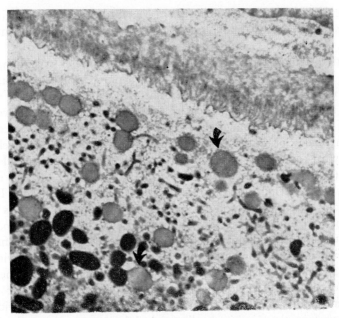

Fig. 9. Electron micrograph of a thin section of the cortex of the maturing egg of the frog *Rana pipiens*, showing the cortical granules (arrow). Exterior to the cortex is the zona radiata, showing microvilli. Courtesy of Dr. Norman E. Kemp (cf. 92).

gravity (9); strips of cortex can be peeled away in this species in the same manner as in the sea urchin (7).

*Chaetopterus* eggs also exhibit cortical granules which can be seen upon centrifugation (Fig. 11). These are stable on fertilization. It would probably be of great value to examine in detail with combined electron and polarization microscopy the alterations in ultrastructure which occur during fertilization in eggs with a stable cortex; in this way some change common both to the eggs with labile and with stable cortices might be found. Preoccupation with the sea urchin egg has led to a failure so far to separate the structural events involved in membrane elevation from those attending activation. The latter may well be masked in labile cortices.

### III. Processes Leading to Activation

#### 1. *The Activating Function of the Acrosome*

The starting point for a discussion of the processes leading to activation should perhaps be a consideration of the possible activating function of the

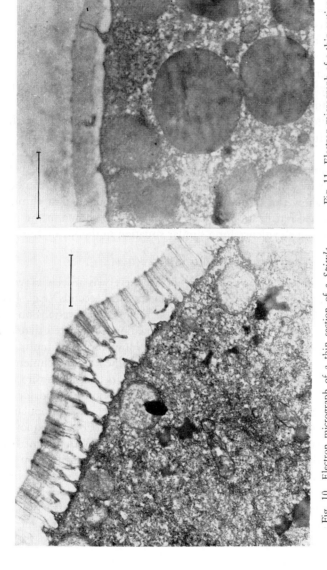

Fig. 10. Electron micrograph of a thin section of a *Spisula* egg 29 minutes after fertilization, showing the presence of cortical granules which differ somewhat from the yolk granules seen in the endoplasm. Note the thick vitelline membrane or layer into which run microvilli. Courtesy of Dr. L. I. Rebhun.

Fig. 11. Electron micrograph of a thin section of a *Chaetopterus* egg showing the continued presence of cortical granules 51 minutes after fertilization. Note here, as in the case of *Spisula*, the thick vitelline membrane or layer and the microvilli. Courtesy of Dr. L. I. Rebhun.

acrosome (see Section I 1). Although there have been no experimental studies pointed in this direction since the discovery of the acrosome reaction, there is ample evidence in the descriptive literature on fertilization to suggest such a function. In *Nereis,* it is well known that jelly extrusion and initiation of maturation, manifestations of activation in this species, occur while the sperm is attached only by its acrosomal region. Although Lillie (103) was the first to remove the attached sperm and show that egg development proceeded as well as in artificial parthenogenesis, it was Goodrich (56) who showed by careful micromanipulation experiments that *Nereis* sperm could be pulled away from their attachment at the egg surface as early as two minutes after insemination, and yet activation (but no cleavage) followed.

If some of the observations of R. Chambers (23, 24, 25) on the starfish *Asterias* are reinterpreted in the light of the findings of J. C. Dan and of Colwin and Colwin (28-37, 43-48), it becomes almost certain that the cortical reaction and membrane elevation are initiated while the sperm is attached only by its long acrosome filament. In fact, some of Chambers' line drawings show the fertilizing spermatozoon about to pass through the already-formed fertilization membrane. As will be shown in the next section, there is some evidence for a similar situation in the sea urchin.

## 2. The Latent Period Before the Cortical Reaction

A number of observers have recognized that there was a delay between insemination and membrane elevation of about 30-35 seconds (e.g., Just, 88). Moser (136), who was the first to make detailed observations on the disappearance of the cortical granules at fertilization, recognized that this process took from 10 to 20 seconds to begin. However, the time of insemination does not permit a reliable estimate of when the attachment of the fertilizing spermatozoon took place. Several years ago, Hagström and Hagström (64) introduced the lauryl-sulfate interruption method for determining the time of successful sperm attachment in eggs of a population by instantaneously inactivating free swimming spermatozoa in aliquots of fertilizing gametes at different times after insemination. With this method, it is possible to determine with considerable accuracy from fertilization rate curves (cf. Figs. 4, 12) the time at which 50 per cent of the eggs had been fertilized. By combining this method with a second interruption of the same batch of fertilizing gametes in 4 per cent formalin, which kills the eggs immediately and arrests cortical propagation but not membrane elevation over areas already covered by the cortical reaction, it was possible to establish the time at which 50 per cent of the same eggs had begun to show cortical granule breakdown (Fig. 12). The period between sperm attachment and the beginning of the visible cortical changes has been designated the *latent period* (10).

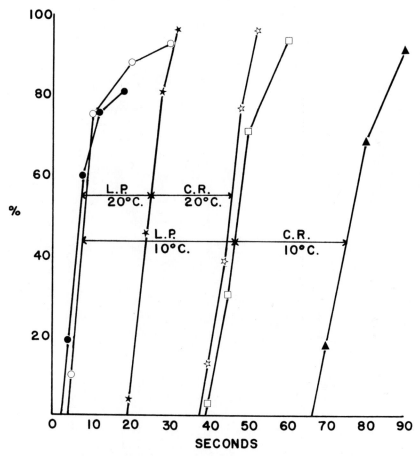

Fig. 12. Graphed data from which the duration of the latent period and of the cortical reaction were determined (explanation in the text). The two left-hand curves show the time of attachment of the successful spermatozoa at 10 (open circles) and 20 (black circles) degrees. The middle curves (marked with open squares or black stars) show the time of initiation of cortical granule breakdown at these respective temperatures. The right-hand curves (black triangles and open stars) represent the completion of the cortical reaction (Allen and Griffin, 10).

46    THE CHEMICAL BASIS OF DEVELOPMENT

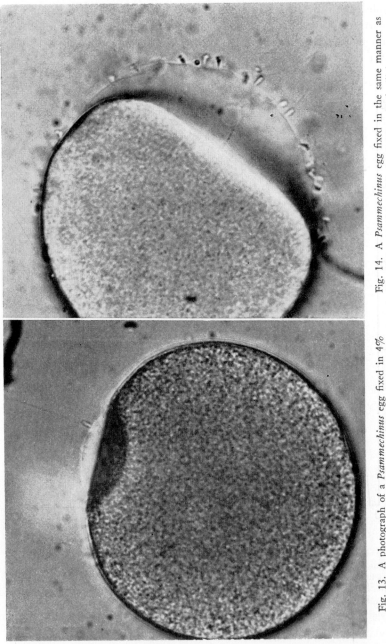

Fig. 13. A photograph of a *Psammechinus* egg fixed in 4% formalin within a few seconds after the beginning of the cortical reaction. Note that the fertilizing spermatozoon has been pulled from the cortex by post-fixation membrane elevation (Allen and Griffin, 10).

Fig. 14. A *Psammechinus* egg fixed in the same manner as in Fig. 13 but several seconds later. Note the fertilizing spermatozoon now embedded in the cortex (Allen and Griffin, 10).

Contrary to earlier reports (87, 88), the fertilizing spermatozoon begins to penetrate the egg surface not a few seconds after attachment, but after the cortical reaction has proceeded part way around the egg, or at least 25 seconds after attachment. This was shown clearly by the fact that eggs fixed soon after the beginning of the cortical reaction nearly always showed the fertilizing spermatozoon pulled out of the cortex by post-mortem elevation of the fertilization membrane (Fig. 13). Eggs fixed in the same way after the cortical changes had covered between a third and a half of the egg surface, however, showed the sperm embedded in the cortex and unaffected by elevation of the membrane after fixation (Fig. 14). Runnström (180) and Runnström and Kriszat (186) have studied the solidifying action of the redox dyes porphyrindin and porphyrexid, which so harden the egg surface that the sperm cannot enter, although development to the monaster stage is initiated by its mere attachment. These observations suggest that in the sea urchin, as well as in other animals mentioned above, the cortical reaction is initiated while the sperm is attached only by its acrosome. The latent period, then, covers a substantial part of this period. The time sequence of some of the known early events in fertilization of *Psammechinus* eggs is shown in Fig. 15.

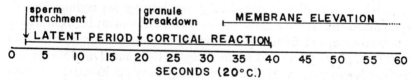

Fig. 15. A schematic representation of the time-sequence of early events in the fertilization of *Psammechinus* eggs (Allen and Griffin, 10).

Although it is known that pricking can initiate a cortical response in some eggs, it is doubtful that the function of the acrosome filament is purely mechanical. From Moser's photographs of the response of *Arbacia* eggs to pricking (137), one might conclude that the cortical reaction induced was of a different character from that normally initiated by the sperm. Response to injury (6) and to alternating current (13) are non-propagating in *Psammechinus* eggs (Fig. 16). There is, therefore, reason to suppose that the spermatozoon introduces, presumably by its acrosome, some specific stimulus, possibly mechanical but more likely chemical, which initiates the cortical reaction. The probability that chemical reactions precede the cortical reaction is increased by the finding that the temperature coefficient (10-20°C.) for the duration of the latent period

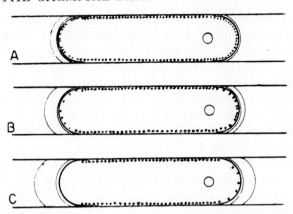

Fig. 16. A diagram showing the gradual nature of an electrically induced cortical response in an egg taken into a capillary in such a way that its nucleus was much closer to one end. A, B, and C are progressive steps in the cortical reaction produced by increase in current density (Allen, Lundberg, and Runnström, 13).

is about 2.3, or within the range expected for chemical reactions (10) (see Fig. 12). Concerning the specificity of the activating action of the acrosome, it is worth mentioning the finding of Hagström (61) that the height of membrane elevation in cross-fertilization between sea urchins depends upon the sperm. When a cross-reaction occurs occasionally between widely separated groups of animals [for example, mollusk sperm and sea urchin eggs (97), *Ciona* sperm and *Phallusia* eggs (160), and *Triton* sperm and *Bufo* or *Pelodytes* eggs (17)], the sperm may carry out its activating function without otherwise participating in development. On the other hand, the sperm may not carry out any activating function at all in some species. Kusa (98) has shown that sperm can enter salmon eggs in Ringer solution without activating them or participating further in development unless the eggs are subsequently activated by fresh water. Fresh water, even without the sperm, can initiate some parthenogenetic developmental changes.

The discovery of a latent period of substantial duration has complicated the approach to the polyspermy problem. One might at first suppose that the first sperm could initiate a propagating wave of granule breakdown while inhibiting such changes in the neighborhood of its own attachment, as appears to be the case in *Nereis*. However, since in the sea urchin many presumably fertile spermatozoa arrive at the egg surface almost simultaneously, why do these all not initiate a cortical reaction at their point of attachment? Simultaneous initiation of two or more cortical reactions by different sperm is an exceedingly rare occurrence, as has been determined by observation of several thousand eggs fixed in formalin at various times after insemination (9). This must be due to some resistance on the part of the egg, for after pretreatment with nicotine, many of these attached sperm do enter (63b, 161, 162, 168). It seems necessary then to agree tentatively with Rothschild and Swann that there exists some kind of "fast block" against polyspermy, if not to exclude supernumerary sperm directly, then to prevent more than one sperm from initiating a cortical response. There is evidence that as the sea urchin passes through maturation from the oocyte stage, it changes from a multiple-response pattern to the production of one fertilization cone of considerably reduced size (187). This acquisition of a "selecting mechanism" on the part of the egg can be reversed by the sulfhydryl inhibitor N-ethylmaleimide. This agent returns the egg to the condition in which it responds to many attached sperm by the production of multiple fertilization cones, so that it becomes polyspermic (68).

## 3. The Role of the Cortical Reaction in Membrane Elevation

It is now generally recognized that activation in the sea urchin involves dramatic changes in the surface layers of the egg, changes which have been referred to as the *cortical reaction*. Loeb (106) was the first to grasp the potential importance of these changes when he proposed his "superficial cytolysis" theory of activation. The nature of this process was first revealed in an observation of E. N. Harvey, who commented in a footnote on the presence in *Arbacia* eggs of ". . . numerous minute stained granules, quite unmoved by the centrifuge. At the time of fertilization, these disappear, apparently going to form the substance which passes out of the egg and hardens to a fertilization membrane" (74). The "minute . . . granules," later called the cortical granules, were rediscovered independently three times (76, 105, 136). Moser (136) was the first to demonstrate the wave-like nature of cortical granule breakdown in the sea urchin egg (although similar changes had been described in fish (155; for references cf. 99, 221)), and to show that it coincided with the wave of surface "roughening" observed by earlier authors; he also compared this wave with the wave-like dark-field color change discovered by Runnström (170) and concluded that the latter was an expression of the disappearance of the cortical granules, which he believed reflected the light in dark-field. Hagström (58) later showed, however, that the two phenomena could be experimentally separated. Moser also showed that granule breakdown, although not always propagating, resulted from a wide variety of chemical and physical stimuli (saponin, toluol, pricking, ultraviolet light, direct current, urea, glycerol, and thiourea) (137, 138). Because none of these agents were effective in breaking down cortical granules in oxalated eggs, he interpreted his results as indicating the primary necessity of calcium in the light of the "calcium-release theory" (cf. 75 for discussion). It has not been shown, however, that oxalate (or citrate) selectively binds calcium without side effects on cells. Moser suggested that membrane elevation might come about by the formation of vesicles or vacuoles at the surface which might fuse at the outer border to form the fertilization membrane and at the inner border to form the new cell surface.

Prior to 1940, there had been two opposing views with respect to the origin of the fertilization membrane (cf. 136 for references): (1) that the membrane was "formed" de novo from a precursor at the time of fertilization; or (2) that a membrane preexisted on the egg surface and became elevated. These opposing views were beautifully reconciled by the work of Motomura (140), who showed that the cortical granules, which could be followed by their staining with Janus Green, merged with the vitelline membrane of the egg, thus establishing the dual origin of the fertilization membrane. The view of Motomura concerning the fate of the cortical granules, as shown in Fig. 17, is essentially that held today, except for a few modifications. It is probable that some of the material from the cortical granules contributes colloids to the perivitelline space and hyaline layer (157, 158). The dual origin of the fertilization membrane provides an explanation for the fact that centrifuged *Asterias* (38) and *Arbacia* (71) eggs give rise to fertilization membranes which are thicker and more widely separated from the egg surface at the centrifugal pole. The "membrane precursor" substance which had been displaced centrifugally was the cortical granular material, which was at that time unknown (cf. 8).

Through the work of Runnström et al. (171-173, 188-190) and of Endo (52), we have a rather complete and detailed account of the morphological events in the development of the fertilization membrane. The wave of cortical granule breakdown which precedes membrane elevation occurs randomly in a zone which spreads over the surface (6, 52). Endo observed swelling of the cortical granules before expulsion, but this was not observed in parallel observations with *Psammechinus* eggs, possibly because the process proceeds too rapidly (6). Runnström observed globular and rod-shaped particles in trypsin-treated and underripe eggs, and has recorded the details of the delayed merging of some of this material with the vitelline membrane. Endo has confirmed these observations in general and has illus-

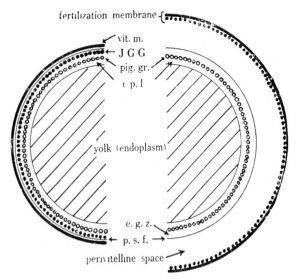

Fig. 17. A diagram showing the disposition of cortical granular material before (left half) and after (right half) fertilization. e.g.z., extragranular zone; i.p.l., inner protoplasmic layer; J.G.G., Janus green granules (cortical granules); pig. gr., pigment granules; p.s.f., protoplasmic surface film; vit. m., vitelline membrane (Motomura, 140).

trated his paper with remarkable photographs showing the steps involved in the merging of the granular material with the vitelline membrane. During this process, the membrane begins to thicken, harden, and become resistant to enzymes except for the hatching enzyme (see Section II 2). The hardening has been shown to be due to the release from the cortex of a "third [hardening] factor" which can be isolated by extraction of living eggs with acid sea water (142). (The hardening factor is not the jelly.) Extracted eggs exhibit fragile fertilization membranes which can be hardened by adding back the neutralized acid extract. Recently Motomura has succeeded in isolating and probably localizing (just beneath the cortex) the substance involved. He has called this substance "colleterin," but it is probably identical with "antifertilizin."

Endo (52) studied the birefringent rods which Runnström discovered in the perivitelline space and showed that they probably consist of the substance that hardens the fertilization membrane, for both substances are attacked by the hatching enzyme but not by trypsin, and are precipitated by calcium. Furthermore, the sign of birefringence is the same with respect to the axis of crystalline or membrane orientation. As the membrane elevates, its birefringence increases, presumably due to the addition of oriented material from the inside (123, 187) (cf. Fig. 3). However, the suggestion of Dan and Okazaki (49) that this may be partly strain birefringence is plausible, since the fertilization membrane is undoubtedly stretched somewhat by the colloid osmotic pressure of material in the perivitelline space (78, 79, 80, 107).

The origin of the colloids of the perivitelline space has not been established; it is probable that this material comes from or with the cortical granules (157). Its concentration is about 0.07% dry organic matter (124), and it very likely contains a mucoprotein or mucopolysaccharide precipitable by calcium (158, 183). The fact that the volume of the perivitelline space can be experimentally altered by the $pH$ of the medium (81) or can be doubled or tripled by non-electrolytes which simultaneously interfere with the hardening of the membrane (135, 138) suggests that the materials which come out of the cortex can serve either to harden the membrane or to provide colloid osmotic pressure. The results of Dean and Moore (50) suggest that the molecular weight of the colloidal particles must exceed 10,000, for smaller molecules pass through the fertilization membranes of *Dendraster* eggs.

In addition to the hyaline (ectoplasmic) layer, Hiramoto has recently shown the existence of a new gel layer in the perivitelline space of *Hemicentrotus* and *Clypeaster* eggs. Since the same methods were used by Chambers (25) to demonstrate the absence of gel in the perivitelline space of other echinoderm ova, we must conclude that species differences exist. Hiramoto's gel layer (78, 79) apparently does not exist in the eggs of *Psammechinus,* for these slide back and forth inside their fertilization membrane when held in a capillary (6).

### 4. The Function of the New Surface Layers of the Fertilized Egg

Although it would seem at first sight that the fertilization membrane would most likely serve as a mechanical barrier to the entrance of supernumerary sperm, it has been shown repeatedly that eggs with their membranes mechanically removed could not be refertilized (cf. 194 for references). Sugiyama (194) has confirmed these results and has demonstrated that the sperm barrier beneath the fertilization membrane is broken down to permit refertilization in sea water lacking divalent ions. Hagström and Hagström (65) confirmed and extended these results and demonstrated a high degree of coincidence between refertilizability and damage to the hyaline layer. Since the hyaline layer is a product of the cortical reaction and the only apparent physical barrier to the sperm beneath the fertilization membrane, it seems justified to accept Hagström's view that the hyaline layer can function as a barrier to the entrance of supernumerary sperm in the absence of a fertilization membrane. Nakano (148) has held a similar view on the function of the hyaline layer from experiments on the fertilization of activated eggs following treatment with sea water lacking divalent ions. Recently, however, Tyler, Monroy, and Metz (207) have reported the possibility of refertilizing mechanically demembranated eggs of two species of *Lytechinus*. Although, as the authors point out, these results are probably partly explained by the thinness of the hyaline layer in this species and by the slowness with which this layer arises, it is quite likely that the hyaline layer was not completely formed at the time the membranes were removed. There are indications that the hyaline layer is formed, at least in part, from materials precipitated from the perivitelline space (52, 133, 157, 158). In the case of membrane removal before hyaline layer formation, these materials may be permanently lost, resulting in underdeveloped hyaline layers. Hagström and Hagström (65) have shown that basic substances such as lysozyme and clupein cause polyspermy, probably by precipitating the acid mucopolysaccharides in the cortical granules, and preventing this material from taking part in the formation of the hyaline layer. Since the formation of the hyaline layer proceeds at different rates in different species, it would have been of interest to test the refertilizability of *Lytechinus* eggs which had been demembranated after full growth of the hyaline layer. It is interesting to note that mechanical as well as chemical means can be employed to remove the hyaline layer; E. B. Harvey showed that the hyaline material could be peeled off the egg by centrifugation (72). So far, there have been no published attempts at refertilizing eggs with mechanically removed hyaline layers. Aside from the sperm-barrier role of the hyaline layer, this structure is well known to hold the blastomeres together during cleavage, although it is not necessary for cleavage itself (77).

## 5. The Nature of the Propagated "Fertilization-Impulse"

The literature contains a great many reports on treatments which will cause membrane elevation (cf. 73, 201 for references); however, relatively few of these cause further developmental changes. Sugiyama (195-197) has made a very significant advance by showing that various agents which he tried could be divided into two groups: (1) those which acted directly on the cortical granules to bring about their localized breakdown (including saponin, "monogen," wasp-venom, and "lipon"); and (2) those which initiate a propagating wave of cortical granule breakdown in what is presumably the same manner as the sperm (distilled water, urea, and butyric acid). The different action of these groups of agents was revealed by ingenious experiments in which the surfaces of eggs were exposed only partially to the reagent. It would have been very valuable, particularly in experiments using capillary tubes, if the rates of propagation of the "fertilization-wave" or "impulse" had been measured under the stimulating action of the different substances to determine whether their effect was actually similar to that of the sperm.

There have been a few attempts to approach the nature of the propagating wave by study of its kinetics. Rothschild and Swann (167) analyzed cinematographic records of a light-scattering change propagated around the largest optical section of *Psammechinus miliaris* eggs. Because only equatorially fertilized eggs exhibit the true rate of propagation, only a few of many records taken could be used. Two hypotheses were dealt with, namely, that the sperm might release a substance which diffused either through the interior or through the cortex to cause the observed surface changes. Kacser (89) later reanalyzed and replotted the same record and decided that the curve did not indicate either diffusion mechanism, but rather was compatible with an autocatalytic mechanism. This suggestion is more in accord with the results of Sugiyama (197) and of Allen (6) indicating that the egg has the ability to exhibit a self-propagating impulse in response to artificial activating agents which presumably do not diffuse through the cell to cause their effect. A later attempt to extend the cinematographic data on cortical propagation (9) revealed one new fact: if the rates of propagation were plotted for the two hemispheres of the egg cut by the penetration path of the sperm, they were found not to be equal, but sometimes half as rapid on the surface nearest to the egg nucleus as on the opposite side. This may have some relation to the fact noted by Allen, Lundberg, and Runnström (13) that the part of the egg surface nearest to the egg nucleus had the highest threshold to electrical stimulation (see Fig. 16). It is doubtful whether the nucleus itself exerted any effect, as the opposite results were obtained with respect to the nucleus in centrifuged eggs. It is not known whether the threshold effect and the differential rates of propagation were related to egg polarity or to the distribution of some endoplasmic constituent. Moser noted an influence of the distribution of endoplasmic materials on the cortical response to artificial agents (138). It is apparent that much more work is needed to clarify some of the extraneous factors affecting the kinetics of cortical conduction; such work should be carried out on an egg such as *Echinocardium*, which has a clearly marked polarity and an easily followed dark-field change. It is worth mentioning that differential rates of cortical conduction at different parts of the egg surface have been noted in fish (222).

An attempt to study the kinetics of propagation in cylindrically shaped eggs in capillaries yielded unexpected and puzzling results (6). The rate of propagation was initially similar to that in spherical eggs (Fig. 18), but it soon decreased and often died out entirely, particularly in the more

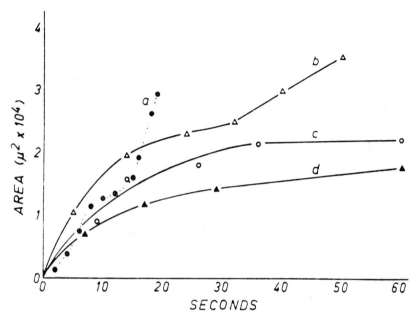

Fig. 18. The rate of propagation of the fertilization impulse as studied *a* by the dark field change in spherical eggs (data replotted from Rothschild and Swann, 167) and *b*, *c*, and *d* by the breakdown of cortical granules in cylindrical eggs in capillaries. Egg *b* was treated with ATP, length 188 $\mu$; egg *c* was a dispermic egg (length 423 $\mu$); and egg *d* was monospermic (Allen, 6; see also 12).

stretched eggs. Temperature coefficients for propagation rates in cylindrical eggs were between 2 and 3, except for impulses initiated by parthenogenetic concentrations of periodate, for which the temperature coefficient dropped to nearly 1 (12). This fact suggested that a chemical component of the wave had been short-cut by periodate treatment, which is believed to free the egg from the influence of a natural mucopolysaccharide inhibitor believed to be present in the cortex (174, 175, 181-183). More recently, approximate temperature coefficients have been measured for the durations of the latent period and of the cortical reaction as 2.3 and 1.4 respectively. The latter value is characteristic for a diffusion process in spherical eggs, yet conduction in cylindrical eggs showed a $Q_{10}$ characteristic of chemical processes. It is possible that the capillary had the effect of slowing whatever chemical component of the cortical reaction would normally occur during the latent period, with the result that this would then act as a rate-limiting process in the later visible phase of cortical granule breakdown. This suggestion is put forth tentatively and may require modification as more data become available. The kinetic approach to the fertilization impulse has so far been disappointing; this is largely due to the technical difficulties in obtaining sufficient data.

Aside from somewhat unsatisfactory information on its kinetics, the other information available about the propagated "fertilization impulse" concerns what will initiate it, favor its conduction, or block it. Among chemically defined initiating substances are those previously listed which Sugiyama has studied; to these should be added periodate (6, 12). One would be tempted to believe from this list that the egg is poised to respond by some autocatalytic mechanism to any sort of non-specific "trigger." However, this appears not to be the case, for not only do many activating agents fail to pull the trigger (cf. 195), but response to insemination seems to depend on the origin and physiological state of the spermatozoa (61, 175, 204). We must conclude, therefore, that the spermatozoon introduces a certain "charge" of activating substance which, depending on its amount or composition, can initiate qualitatively different impulses. It has been shown that some eggs in a population show marked differences in their ability to conduct a fertilization impulse (6, p. 416, Tables I and II).

Among the experimental conditions which favor propagation of the fertilization wave are reduced temperature, glycine, adenosine triphosphate (ATP), and periodate pretreatment; at least this is the case for eggs fertilized in capillaries. Among the agents which tend to block the fertilization impulse are elevated temperature, either alone or in combination with other conditions, the non-specific poison nicotine, and either deformation, stretching, or extensive contact with glass surfaces involved in the capillary experiments. However, in spite of reports to the contrary (185), there is no evidence that contact or even attachment to glass can inhibit the passage of the cortical reaction (cf. 7). Okazaki has recently reported blocking the cortical reaction with the uncoupling agents dinitrophenol and azide, but these acted only at elevated temperatures and were not antagonized by ATP. It seems reasonable to suppose that the effect Okazaki found (154) was due primarily to elevated temperature (cf. 11) in synergism with some effect of the chemical agents which need not necessarily have been that of uncoupling phosphorylation from oxidative processes. The sensitivity of the fertilization impulse to elevated temperature is not unlike heat narcosis in nerve, for it is effective only during conduction and causes no injury when applied before or after conduction (11, 12). Inhibition by deformation, as would be suggested from the capillary experiments, is also reminiscent of nerve. For this reason it is of great interest that Tyler, Monroy, Kao, and Grundfest (206) have succeeded in demonstrating a change in membrane potential within a half minute after insemination. Sugiyama (197) has recently demonstrated two properties of the activation process in sea urchin eggs that are characteristic of irritable processes in general: (1) the ability of an egg to conduct a fertilization wave is blocked on narcosis with urethane; and (2) summation of two subthreshold stimuli will produce a fertilization wave if given within a few minutes of one another. The latter suggests the build-up of some substance within the egg which would trigger the fertilization wave or impulse.

## 6. Chemical Theories of Activation

The early chemical theories of activation are not very useful in interpreting the mechanism of the fertilization impulse and its role in activating the egg. The earliest, F. R. Lillie's "fertilizin" theory, cannot be reconciled with most of the information available at present (for a discussion of the fertilizin theory in the light of recent findings, see 63a).

Prof. J. Runnström, his students and collaborators have approached the physiology of activation in a systematic manner by studying the action of various substances which enhance or inhibit various aspects of the fertilization process. Among the inhibitors found have been various mucopolysaccharides, including heparin, blood group substances, an extract of *Fucus*, and even the dissolved jelly of the egg (63a, 69, 70, 95, 176, 179, 181, 182, 216-218). Periodate ions were found capable of removing the inhibiting action of these substances, and in addition caused certain changes in the eggs indicative of incipient activation, or "preactivation," which was accompanied by a lowered threshold to artificial activation (96, 184). In sufficient concentration, periodate was effective in causing artificial activation. For this reason, it was of particular interest when Lundblad detected the successive activation of three proteolytic enzymes at the time of fertilization (108-110). The first of these to be activated, E II, was found on isolation to be capable of being activated by periodate, a fact suggesting that it occurred naturally in combination with a mucopolysaccharide inhibitor, perhaps the same one which seemed to act in an inhibitory manner on fertilization, especially in under-ripe eggs (see especially 175, 179).

On the basis of these and other results, Runnström has proposed a "kinase theory" of activation based on available facts. It is proposed that the kinase is originally bound to a natural inhibitor, and perhaps also to ATP and $Mg^{++}$ ions (174, 175, 179). The introduction of a substance from the spermatozoon (originally suggested to be antifertilizin) would bind the inhibitor and release the kinase, which would in turn activate various proenzymes in the cell. In the past, R. S. Lillie also proposed the involvement of an enzyme (protease) in the early stages of activation (104). Lundblad's proteolytic enzyme E II has been proposed to act as the kinase by virtue of its original binding with a carbohydrate inhibitor. It might turn out that the inhibitor was identical with "cytofertilizin," since there is evidence that both this material and E II may be localized in the cortex, since they can apparently "leak out": E II during jelly removal in acid sea water below $pH$ 5.5 (109, 110), and cytofertilizin upon aging (58). It is possible that we have only begun to find enzymes at the surface which might participate in the process of activation. For example, Numanoi (151-153) has suggested a role for sulfatases in fertilization and activation. It will be interesting to follow the future of work on such enzymes, for they might provide a direct enzymatic means for destruction of the proposed natural carbohydrate inhibitor within the egg. In order to interpret the available evidence on the conduction of the fertilization impulse in terms of the kinase theory, it is necessary to assume a two-step process, in which first an enzymatically controlled change spreads over or through the egg on stimulation from the sperm; this would lower the threshold of the surface for the second phase, presumably a diffusion process involving a small molecule over a short distance. The value of such speculations as these is only in planning experiments, for there are many questions left unanswered, such as the nature of the stimulating substance from the sperm, the chain of enzymes, inhibitors, etc., which are involved, and the time sequence of their action in relation to visible events.

A somewhat simpler theoretical approach to the chemical mechanism of activation has been undertaken by Monroy, who has assumed from observations with the polarizing microscope (see Section II 2 b) that the sea urchin egg cortex is a lipoprotein system with the properties of a liquid crystal (126, 127, 129), which undergoes structural changes due mostly to changes in the lipids themselves. Accordingly, he has designed interesting model experiments on the interaction between living sea urchin sperm and a lipoprotein extracted from hens' eggs. His results show that sperm can cause the release of free phospholipids which later break down into substances with hemolytic activity. Monroy has supposed that something like this sequence of events takes place during activation of the egg by the sperm, and possibly in the propagation of the surface changes. These interesting experiments may lead to a better understanding of some of the metabolic pathways of phospholipid utilization, since these materials constitute a major endogenous source of substrate for sperm cells (166). For a number of reasons, however, the theoretical basis of these experiments is at present not very attractive. Although there is no question that the unfertilized egg is birefringent, and that this birefringence disappears dramatically at fertilization, there has been much disagreement as to the significance of this observation. There seems to be, at the present time, no clear evidence for the localization in the cortex, below the vitelline membrane, of oriented lipid layers; otherwise these would show in the electron micrographs. Therefore, interpretation of the cortical changes in terms of changes in lipid molecules may be premature. Furthermore, in view of the fact that wasp-venom, saponin, and various detergents, which attack lipoid and lipoprotein structures, exert only a localized effect on the egg surface (195, 197), it seems unlikely that a lipoprotein-splitting process could be the basis of propagation of the fertilization impulse. However, lipoprotein-splitting processes could certainly be important in the passage of the sperm through the vitelline membrane.

## 7. *The Significance of the Cortical Reaction in Relation to the Processes Leading to Syngamy*

The discovery of methods to block the passage of the fertilization impulse has provided the possibility of weighing the relative significance of the entrance of the sperm and the passage of the fertilization impulse for later development (6, 11, 63b). Eggs with a blocked fertilization impulse have been designated "partially fertilized" because a portion of their surface remains unaltered and susceptible to refertilization. Not only are the surfaces of the fertilized and unfertilized parts different in these eggs, but the endoplasm is different as well. In partially fertilized eggs obtained by the capillary method (Fig. 19), the light-scattering properties of the endoplasm underlying the fertilized surface showed the same decrease as upon fertilization in spherical eggs, and the division between unfertilized and fertilized endoplasm could be seen to persist (6). However, when spherical eggs were made partially fertilized by blocking the fertilization impulse with heat, the initial differences between unfertilized and fertilized endoplasm did not always remain, suggesting that a mixing had occurred. The following nuclear changes occurring in partially fertilized eggs represent deviations from the normal: (1) growth of the sperm aster and migration of the sperm nucleus were restricted to fertilized cytoplasm; (2) the egg

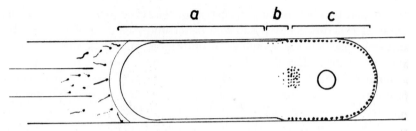

Fig. 19. Diagram of a partially fertilized egg in a capillary. *a*, fertilized portion; *b*, intermediate zone; and *c*, unfertilized portion (Allen, 6).

nucleus migrated only if originally located in fertilized cytoplasm; however, it was able to undergo swelling or even break down in unfertilized cytoplasm; and (3) sometimes nuclear fusion failed to occur; when it did take place, cleavage was delayed, and often the cleavage furrow did not pass entirely through the cell (Fig. 20).

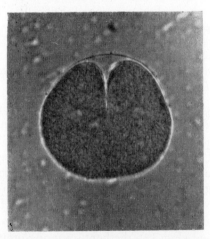

Fig. 20. A partially fertilized *Psammechinus* egg showing an incomplete cleavage furrow. The fertilization impulse was blocked by elevated temperature (Allen and Hagström, 11).

The observations on partially fertilized eggs have raised a new and important problem: what is the action of fertilized cortex on its underlying endoplasm? Some observations which bear on this point, but do not answer the question, have been made in the *Arbacia punctulata* egg, where fertilization is followed in about 10 minutes by a mass migration of nearly all of the echinochrome pigment granules (chromatophores) to the surface (112). In partially fertilized eggs the pigment granules migrate only to fertilized parts of the cortical layer, to produce a red patch at the point of sperm entry (14). In experiments with eggs in capillaries, it was shown that fertilized cortex could "induce" pigment granule migration from a distance of about 26 microns within unfertilized cytoplasm (Fig. 21). The

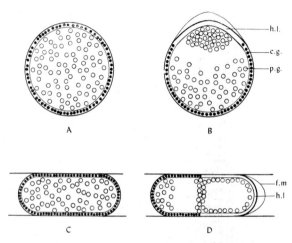

Fig. 21. A highly schematic representation of the behavior of pigment granules in *Arbacia* eggs when the fertilization impulse is blocked, leaving the eggs partially fertilized. A and B are spherical eggs, unfertilized and 10 minutes after partial fertilization; C and D are cylindrical eggs, unfertilized and after partial fertilization. c.g., cortical granules; f.m., fertilization membrane; h.l., hyaline layer; p.g., pigment granules (cf. Allen and Rowe, 14).

mechanism of the interaction between fertilized cortex and the pigment granules is unknown, but it provides a visible manifestation of endoplasmic activation as a result of influences emanating from the cortical layer after fertilization.

The results presented so far in this section leave us with at least two possible interpretations: either the presence of cortical granules in the cortex imposes some kind of inhibition on developmental processes from which the egg must be "released" in order to develop, or else the fertilization impulse or wave is itself an activating influence and the disappearance of cortical granules is incidental to activation. Fortunately, there is evidence on this point in observations made long ago by Motomura (139, 140). Optimal exposure to butyric acid, followed by return to sea water, results in membrane elevation and endoplasmic activation. However, prolonged exposure to butyric acid stabilizes the cortical granules so that on subsequent return to sea water and fertilization the granules remain intact through cleavage, indicating that they are not in themselves deterrents to cleavage. Motomura has also shown more recently that it is possible to activate eggs by other means without disturbing the cortical granules (143, 145). We can conclude tentatively from these observations that the event of primary importance in the initiation of development, at least in echinoderm eggs, is the fertilization wave, which evidently transmits a state of excitation not only around the cortex, but presumably into the endoplasm as well. The cortical granules might seem at first to lose some of their former importance, but it should be stressed that their most essential role seems to be in establishing the new surface layers after fertilization. They also serve quite importantly as the only markers we have for the progress of the fertilization impulse.

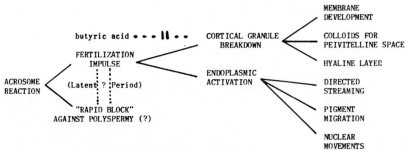

Fig. 22. Explanation in the text.

In Fig. 22, which is based on the work discussed above, of many authors, an attempt has been made to point out what now appear to be some of the causal relationships in the early events in the initiation of development in the sea urchin. On examination of this scheme, it will be seen that although the processes involved in syngamy, on the one hand, and activation, on the other, can be separated experimentally by artificial activation, nonetheless, under normal conditions of fertilization, the two sets of processes undoubtedly interact, as shown by the failure of the nuclei to migrate and fuse if the fertilization impulse is blocked. There are many problems in the physiology and biochemistry of activation which still await solution, although promising leads have been uncovered in many directions. It is hoped that in the future more attention will be directed toward fertilization in eggs with a stable cortex, for in these it may be easier to distinguish the processes which are involved in activation without the interference of the membrane elevation mechanism.

I wish to express my appreciation to the following individuals for kindly contributing illustrations for this article: Drs. B. Afzelius, A. and L. Colwin, N. E. Kemp, J. M. Mitchison, and L. I. Rebhun. I am also grateful to Prof. John Runnström for allowing me to read his manuscript, *Fertilization*, which will be published in *The Cell* (182).

## REFERENCES

1. Afzelius, B., *Z. Zellforsch. u. mikroskop. Anat.*, **42**, 134-148 (1955).
2. ———, *Exptl. Cell Research*, **10**, 257-285 (1956).
3. ———, in *Electron Microscopy, Proceedings of the Stockholm Conference, September 1956* (F. S. Sjöstrand and J. Rhodin, eds.), p. 167-169, Academic Press, New York (1957).
4. Afzelius, B. A., and Murray, A., *Exptl. Cell Research*, **12**, 325-337 (1957).
5. Allen, R. D., *Biol. Bull.*, **105**, 213-239 (1953).
6. ———, *Exptl. Cell Research*, **6**, 403-422 (1954).
7. ———, *Exptl. Cell Research*, **8**, 397-399 (1955).
8. ———, *J. Cell. Comp. Physiol.*, **49**, 379-394 (1957).
9. ———, unpub.
10. Allen, R. D., and Griffin, J. L., *Exptl. Cell Research* (in press).
11. Allen, R. D., and Hagström, B., *Exptl. Cell Research*, **9**, 156-167 (1955).
12. ———, and ———, *Exptl. Cell Research*, **Suppl. 3**, 1-15 (1955).
13. Allen, R. D., Lundberg, A., and Runnström, J., *Exptl. Cell Research*, **9**, 174-180 (1955).
14. Allen, R. D., and Rowe, E. C., *Biol. Bull.* (in press).
15. Austin, C. R., *Exptl. Cell Research*, **10**, 533-540 (1956).
16. Austin, C. R., and Bishop, M. W. H., in *Beginnings of Embryonic Development* (A. Tyler, R. C. von Borstel, and C. B. Metz, eds.), p. 71-107, A.A.A.S., Washington, D. C. (1957).
17. Bataillon, E., *Arch. Zool. exptl. et gén.*, **46**, 101-135 (1910).
18. Bernstein, G. S., *Biol. Bull.*, **103**, 285 (1952).
19. Bowen, R. H., *Anat. Rec.*, **28**, 1-13 (1924).
20. Brachet, A., *Wilhelm Roux' Arch. Entwicklungsmech. Organ.*, **30**, 261-303 (1910).
21. Chambers, E. L., *J. Exptl. Biol.*, **16**, 409-424 (1939).
22. Chambers, R., *J. Gen. Physiol.*, **5**, 821-829 (1923).
23. ———, *Biol. Bull.*, **58**, 344-369 (1930).
24. ———, *J. Exptl. Biol.* **10**, 130-141 (1933).
25. ———, *J. Cell. Comp. Physiol.*, **19**, 145-150 (1942).
26. Chase, H. Y., *Biol. Bull.*, **69**, 159-184 (1935).
27. Cheney, R. H., and Lansing, A. I., *Exptl. Cell Research*, **8**, 173-180 (1955).
28. Colwin, A. L., and Colwin, L. H., *Biol. Bull.*, **109**, 357 (1955).
29. ———, and ———, *J. Morphol.*, **97**, 543-568 (1955).
30. ———, and ———, *Biol. Bull.*, **109**, 357-358 (1955).
31. ———, and ———, in *Beginnings of Embryonic Development* (A. Tyler, R. C. von Borstel, and C. B. Metz, eds.), p. 135-168. A.A.A.S., Washington, D. C. (1957).
32. Colwin, A. L., Colwin, L. H., and Philpott, D. E., *Biol. Bull.*, **111**, 289 (1956).
33. ———, ———, and ———, *J. Biophys. Biochem. Cytol.*, **3**, 489-510 (1957).
34. Colwin, L. H., and Colwin, A. L., *J. Morphol.*, **95**, 351-372 (1954).
35. ———, and ———, *Biol. Bull.*, **109**, 357-358 (1955).
36. ———, and ———, *Biol. Bull.*, **109**, 357 (1955).
37. ———, and ———, *Biol. Bull.*, **110**, 243-257 (1956).
38. Costello, D. P., *Physiol. Zool.*, **8**, 65-72 (1935).
39. ———, *J. Gen. Physiol.*, **32**, 351-366 (1949).

40. ———, *Biol. Bull.*, **113**, 341-342 (1957).
41. Dan, J. C., *Biol. Bull.*, **99**, 399-411 (1950).
42. ———, *Biol. Bull.*, **99**, 412-415 (1950).
43. ———, *Biol. Bull.*, **103**, 54-56 (1952).
44. ———, *Biol. Bull.*, **107**, 203-218 (1954).
45. ———, *Biol. Bull.*, **107**, 335-349 (1954).
46. ———, *Intern. Rev. Cytol.*, **5**, 365-393 (1956).
47. Dan, J. C., and Dan, K., *Jap. J. Zool.*, **9**, 565-573 (1941).
48. Dan, J. C., and Wada, S. K., *Biol. Bull.*, **109**, 40-55 (1955).
49. Dan, K., and Okasaki, K., *J. Cellular Comp. Physiol.*, **38**, 427-435 (1951).
50. Dean, R. B., and Moore, A. R., *Biodynamica*, **6**, 159-163 (1947).
51. Delage, Y., *Arch. zool. exptl. et gén.*, **29**, 285-326 (1901).
52. Endo, Y., *Exptl. Cell Research*, **3**, 406-418 (1952).
53. Fol, H., *Arch. zool. exptl. et gén.*, **6**, 145-169 (1877).
54. Folkes, B. F., Grant, R. A., and Jones, J. K. N., *J. Chem. Soc.*, **440**, 2136-2140 (1950).
55. Godlewski, E., *Wilhelm Roux' Arch. Entwicklungsmech. Organ.*, **33**, 196-254 (1912).
56. Goodrich, H. B., *Biol. Bull.*, **38**, 196-201 (1920).
57. Gray, J., *J. Exptl. Biol.*, **32**, 775-802 (1955).
58. Hagström, B. E., *Exptl. Cell Research*, **9**, 407-413 (1955).
59. ———, *Exptl. Cell Research*, **10**, 24-28 (1956).
60. ———, *Exptl. Cell Research*, **11**, 160-168 (1956).
61. ———, *Exptl. Cell Research*, **11**, 507-510 (1956).
62. ———, *Arkiv Zool.*, **10**, 307-315 (1956).
63a. ———, The role of the jelly coat and the block to polyspermy in the fertilization of sea urchins (thesis). Uppsala (1956).
63b. Hagström, B. E., and Allen, R. D., *Exptl. Cell Research*, **10**, 14-23 (1956).
64. Hagström, B., and Hagström, Britt, *Exptl. Cell Research*, **6**, 479-484 (1954).
65. ———, and ———, *Exptl. Cell Research*, **6**, 491-496 (1954).
66. ———, and ———, *Exptl. Cell Research*, **6**, 532-534 (1954).
67. Hagström, B., and Markman, B., *Acta Zool.*, **38**, 219-222 (1958).
68. Hagström, B., and Runnström, J., *Protoplasma*, **44**, 154-164 (1954).
69. Harding, C. V., *Exptl. Cell Research*, **2**, 403-415 (1951).
70. Harding, C. V., and Harding, D., *Arkiv Zool.*, **3**, 357-361 (1952).
71. Harvey, E. B., *Biol. Bull.*, **65**, 389-396 (1933).
72. ———, *Biol. Bull.*, **66**, 228-245 (1934).
73. ———, *The American Arbacia and Other Sea Urchins*. Princeton Univ. Press, Princeton (1956).
74. Harvey, E. N., *J. Exptl. Zool.*, **10**, 507-556 (1911).
75. Heilbrunn, L. V., *An Outline of General Physiology*. W. B. Saunders Co., Philadelphia (1952).
76. Hendee, E. C., *Papers Tortugas Lab. Carnegie Inst. Wash. Publ.*, **27**, 101 (1931).
77. Herbst, K., *Wilhelm Roux' Arch. Entwicklungsmech. Organ.*, **9**, 424-463 (1900).
78. Hiramoto, Y., *Jap. J. Zool.*, **11**, 227-243 (1954).
79. ———, *Jap. J. Zool.*, **11**, 333-344 (1954).
80. ———, *Annot. Zool. Jap.*, **28**, 183-192 (1955).
81. Hobson, A. D., *Proc. Roy. Soc. Edinburgh*, **47**, 94-117 (1927).
82. Immers, J., *Arkiv Zool.*, **42A**, 1-9 (1949).
83. ———, *Arkiv Zool.*, **3**, 367-371 (1952).
84. ———, *Arkiv Zool.*, **9**, 367-375 (1956).
85. Inoué, S., *Biol. Bull.*, **97**, 258 (1949).
86. James, S. B., *Biochem. J. (London)*, **49**, liv (1951) (abstract).
87. Just, E. E., *Biol. Bull.*, **36**, 1-10 (1919).
88. ———, *The Biology of the Cell Surface*. Blakiston, Philadelphia (1939).

89. Kacser, H., *J. Exptl. Biol.*, **32**, 451-467 (1955).
90. Kelly, J. W., *Protoplasma*, **39**, 386-388 (1950).
91. Kemp, N. E., *J. Biophys. Biochem. Cytol.*, **2**, 281-292 (1956).
92. ———, *J. Biophys. Biochem. Cytol.*, **2** (Suppl.), 187-191 (1956).
93. ———, pers. commun.
94. Kopac, M. J., *J. Cellular Comp. Physiol.*, **18**, 215-220 (1941).
95. Kriszat, G., *Exptl. Cell Research*, **7**, 103-110 (1954).
96. Kriszat, G., and Runnström, J., *Exptl. Cell Research*, **3**, 597-604 (1952).
97. Kupelwieser, H., *Wilhelm Roux' Arch. Entwicklungsmech. Organ.*, **27**, 434-462 (1909).
98. Kusa, M., *Annot. Zool. Jap.*, **24**, 22-28 (1950).
99. ———, *Embryologia*, **3**, 105-129 (1956).
100. Lansing, A., Hillier, J., and Rosenthal, T. B., *Biol. Bull.*, **103**, 294 (1952).
101. Leghissa, S., *Riv. biol. (Perugia)*, **43**, 187-197 (1951).
102. Lillie, F. R., *Problems of Fertilization*, Univ. Chicago Press, Chicago (1919).
103. ———, *J. Morphol.*, **22**, 362-390 (1911).
104. Lillie, R. S., *Biol. Bull.*, **32**, 131-158 (1917).
105. Lindahl, P. E., *Protoplasma*, **16**, 378-386 (1932).
106. Loeb, J., *Proc. Soc. Exptl. Biol. and Med.*, **7**, 120-121 (1910).
107. ———, *Wilhelm Roux' Arch. Entwicklungsmech. Organ.*, **26**, 82-88 (1908).
108. Lundblad, G., *Arkiv Zool.*, **6**, 387-415 (1953).
109. ———, *Arkiv Kemi*, **7**, 127-157 (1954).
110. ———, Proteolytic activity in sea urchin gametes (Thesis). Uppsala (1954).
111. Mazia, D., *J. Cellular Comp. Physiol.*, **10**, 291-303 (1937).
112. McClendon, J. F., *Science*, **30**, 454-455 (1909).
113. McCulloch, D., *Exptl. Cell Research*, **3**, 605-607 (1952).
114. Messina, L., *Pubbl. staz. zool. Napoli*, **25**, 454-458 (1954).
115. Messina, L., and Monroy, A., *Pubbl. staz. zool. Napoli*, **28**, 266-268 (1956).
116. Metz, C. B., Specific egg and sperm substances and activation of the egg, in *Beginnings of Embryonic Development* (A. Tyler, R. C. von Borstel, and C. B. Metz, eds.), p. 23-69, A.A.A.S., Washington, D. C. (1957).
117. Metz, C. B., and Morrill, J. B., Jr., *Biol. Bull.*, **109**, 349 (1955).
118. Minganti, A., *Exptl. Cell Research*, **5**, 492-499 (1953).
119. ———, *Exptl. Cell Research*, **7**, 1-14 (1954).
120. ———, *Exptl. Cell Research*, **Suppl. 3**, 248-251 (1955).
121. Mitchison, J. M., *Quart. J. Micr. Sci.*, **97**, 109-121 (1956).
122. ———, *Exptl. Cell Research*, **10**, 309-315 (1956).
123. Mitchison, J. M., and Swann. M. M., *J. Exptl. Biol.*, **29**, 357-362 (1952).
124. ———, and ———, *Quart J. Microscop. Sci.*, **95**, 381-389 (1953).
125. Monné, L., and Hårde, S., *Arkiv Zool.*, **1**, 487-497 (1951).
126. Monroy, A., *Experientia*, **1**, 335-336 (1945).
127. ———, *J. Cellular Comp. Physiol.*, **30**, 105-110 (1947).
128. ———, *Arkiv Zool.*, **40A**, 1-7 (1948).
129. ———, *Exptl. Cell Research*, **10**, 320-323 (1956).
130. Monroy, A., and Monroy-Oddo, A., *Pubbl. staz. zool. Napoli*, **20**, 46-60 (1946).
131. Monroy, A., and Montalenti, G., *Biol. Bull.*, **92**, 151-161 (1947).
132. Monroy, A., and Runnström, J., *Arkiv Zool.*, **40A**, 1-6 (1948).
133. Moore, A. R., *Protoplasma*, **3**, 524-530 (1928).
134. ———, *Protoplasma*, **27**, 544-551 (1937).
135. ———, *Scientia*, **86**, 195-199 (1951).
136. Moser, F., *J. Exptl. Zool.*, **80**, 423-440 (1939).
137. ———, *J. Exptl. Zool.*, **80**, 447-464 (1939).
138. ———, *Biol. Bull.*, **78**, 68-79 (1940).
139. Motomura, I., *Science Repts. Tôhoku Univ., Fourth Ser.*, **9**, 33-45 (1934).

140. ——, *Science Repts. Tôhoku Univ. Fourth Ser.*, **16**, 345-364 (1941).
141. ——, *Science Repts. Tôhoku Univ., Fourth Ser.*, **18**, 554-560 (1950).
142. ——, *Science Repts. Tôhoku Univ., Fourth Ser.*, **18**, 561-570 (1950).
143. ——, *Science Repts. Tôhoku Univ., Fourth Ser.*, **19**, 203-206 (1952).
144. ——, *Annot. Zool. Jap.*, **25**, 238 (1952).
145. ——, *Science Repts. Tôhoku Univ., Fourth Ser.*, **20**, 213-217 (1954).
146. ——, *Science Repts. Tôhoku Univ., Fourth Ser.*, **20**, 318-321 (1954).
147. ——, *Science Repts. Tôhoku Univ., Fourth Ser.*, **23**, 167-181 (1957).
148. Nakano, E., *Jap. J. Zool.*, **11**, 245-251 (1954).
149. Nakano, E., and Ohashi, S., *Embryologia*, **2**, 81-85 (1954).
150. Novikoff, A. B., *J. Exptl. Zool.*, **82**, 217-237 (1939).
151. Numanoi, H., *Sci. Papers Coll. Gen. Educ., Univ. Tokyo*, **3**, 55-66 (1953).
152. ——, *Sci. Papers Coll. Gen. Educ., Univ. Tokyo*, **3**, 67-70 (1953).
153. ——, *Sci. Papers Coll. Gen. Educ., Univ. Tokyo*, **3**, 71-80 (1953).
154. Okazaki, R., *Exptl. Cell Research*, **10**, 476-504 (1956).
155. Okkelberg, P., *Biol. Bull.*, **26**, 92-99 (1914).
156. Parpart, A. K., and Laris, P. C., *Biol. Bull.*, **107**, 301 (1954).
157. ——, and ——, *Biol. Bull.*, **109**, 350 (1955).
158. Parpart, A. K. and Cagle, J., *Biol. Bull.*, **113**, 331 (1957).
159. Rebhun, L. I., pers. commun.
160. Reverberi, G., *Pubbl. staz. zool. Napoli*, **15**, 175-193 (1935).
161. Rothschild, Lord, *J. Exptl. Biol.*, **30**, 57-67 (1953).
162. ——, *Quart. Rev. Biol.*, **29**, 332-342 (1954).
163. ——, *Fertilization*. Methuen & Co., London (1956).
164. ——, *Discovery*, **18**, (2), (1956).
165. ——, *Endeavour*, **25**, 79-86 (1956).
166. Rothschild, Lord, and Cleland, K. W., *J. Exptl. Biol.*, **29**, 66-71 (1952).
167. Rothschild, Lord, and Swann, M. M., *J. Exptl. Biol.*, **26**, 164-176 (1949).
168. ——, and ——, *J. Exptl. Biol.*, **27**, 400-406 (1950).
169. Rothschild, Lord, and Tyler, A., *Exptl. Cell Research*, **Suppl. 3**, 304-311 (1955).
170. Runnström, J., *Acta Zool.*, **4**, 285-311 (1923).
171. ——, *Arkiv Zool.*, **40A**, 1 (1947).
172. ——, *Arkiv Zool.*, **40A**, (17), 1-16 (1948).
173. ——, *Arkiv Zool.*, **40A**, (19), 1-6 (1948).
174. ——, *Pubbl. staz. zool. Napoli*, **21** (Suppl.), 9-21 (1949).
175. ——, *Advances in Enzymol.*, **9**, 241-328 (1949).
176. ——, *Exptl. Cell Research*, **1**, 304-308 (1950).
177. ——, *Symposia Soc. Exptl. Biol.*, **6**, 39-88 (1952).
178. ——, *Harvey Lectures*, **46**, 116-152 (1952).
179. ——, *Zool. Anz.* **156**, 91-101 (1956).
180. ——, *Exptl. Cell Research*, **12**, 374-394 (1957).
181. ——, in *Festschrift Arthur Stoll*, 850-868, Sandoz A. G., Basel (1957).
182. ——, Fertilization, in *The Cell* (in press).
183. Runnström, J., and Immers, J., *Exptl. Cell Research*, **10**, 254-263 (1956).
184. Runnström, J., and Kriszat, G., *Exptl. Cell Research*, **1**, 335-340 (1950).
185. ——, and ——, *Exptl. Cell Research*, **3**, 419-426 (1952).
186. ——, and ——, *Exptl. Cell Research*, **12**, 526-536 (1957).
187. Runnström, J., and Monné, L., *Arkiv Zool.*, **36A** (18), 1-25 (1945).
188. Runnström, J., Monné, L., and Broman, L., *Arkiv Zool.*, **35A** (3), 1-32 (1944).
189. Runnström, J., Monné, L., and Wicklund, E., *Nature*, **153**, 313-314 (1944).
190. ——, ——, and ——, *J. Colloid Sci.*, **1**, 421-452 (1946).
191. Runnström, J., Tiselius, A., and Lindvall, S., *Arkiv Zool.*, **36A** (22), 1-25 (1945).

192. Runnström, J., Tiselius, A., and Vasseur, E., *Arkiv Kemi Mineral. Geol.*, **15A** (16) (1942).
193. Seifritz, W., *Protoplasma*, **1**, 1-14 (1926).
194. Sugiyama, M., *Biol. Bull.*, **101**, 335-344 (1951).
195. ———, *Biol. Bull.*, **104**, 210-215 (1953).
196. ———, *Biol. Bull.*, **104**, 216-223 (1953).
197. ———, *Exptl. Cell Research*, **10**, 364-376 (1956).
198. Thomas, L. J., *Biol. Bull.*, **106**, 124-138 (1954).
199. Tyler, A., *Proc. Natl. Acad. Sci. U. S.*, **25**, 317-323 (1939).
200. ———, *Biol. Bull.*, **81**, 190-204 (1941).
201. ———, *Biol. Revs.*, **16**, 291-336 (1941).
202. ———, *Physiol. Revs.*, **28**, 180-219 (1948).
203. ———, *Am. Naturalist*, **85**, 195-219 (1949).
204. ———, *Biol. Bull.*, **99**, 324 (1950).
205. ———, *Exptl. Cell Research*, **10**, 377-386 (1956).
206. Tyler, A., Monroy, A., Kao, C. Y., and Grundfest, H., *Biol. Bull.*, **111**, 153-177 (1956).
207. Tyler, A., Monroy, A., and Metz, C. B., *Biol. Bull.*, **110**, 184-195 (1956).
208. Tyler, A., and Spiegel, M., *Biol. Bull.*, **110**, 196-200 (1956).
209. Vasseur, E., *Arkiv Kemi Mineral. Geol.*, **25** (6) (1947).
210. ———, *Acta Chem. Scand.*, **2**, 900-913 (1948).
211. ———, *Acta Chem. Scand.*, **6**, 376-384 (1952).
212. ———, Chemistry and physiology of the jelly coat of the sea urchin egg (Thesis), Stockholm (1952).
213. Vasseur, E., and Immers, J., *Arkiv Kemi*, **1**, 39-41 (1949).
214. Wada, S. K., *Mem. Fac. Fisheries, Kagoshima Univ.*, **4**, 105-112 (1955).
215. Wada, S. K., Collier, J. R., and Dan, J. C., *Exptl. Cell Research*, **10**, 168-180 (1956).
216. Wicklund, E., *Arkiv Zool.*, **6**, 485-501 (1954).
217. ———, Improving and inhibiting agents action on the gametes and on their interaction at fertilization in sea urchins (Thesis), Uppsala (1954).
218. Wicklund, E., Kriszat, G., and Runnström, J., *J. Embryol. Exptl. Morphol.*, **1**, 319-325 (1953).
219. Wilson, E. B., *An Atlas of the Fertilization and Karyokinesis of the Ovum*, Macmillan, New York (1895).
220. ———, *Wilhelm Roux' Arch. Entwicklungsmech. Organ.*, **12**, 529-596 (1901).
221. ———, *The Cell in Development and Heredity*, Macmillan, New York (1925).
222. Yamamoto, T., *Exptl. Cell Research*, **10** (2), 387-393 (1956).
223. Yanagimachi, R., *Dobusugaki Zasshi* (*Zool. Mag.*), **62**, 220-224 (1953).
224. Ziegler, H. E., *Wilhelm Roux' Arch. Entwicklungsmech. Organ.*, **6**, 249-293 (1898).

## DISCUSSION

Dr. Novikoff. I would like to put a conceptual question to Dr. Markert, and a specific one to Dr. Allen.

I wonder, Dr. Markert, if two levels of organization, cell and tissue, were distinguished sufficiently in your presentation of the esterase story. When you follow different esterases in homogenates of the whole embryo, to what extent are you describing shifting cell populations and to what extent enzymatic differentiation within the individual cells?

To the degree that specific staining methods reveal the true levels of enzyme activities, it may be said that different cell types of a tissue almost invariably show marked quantitative differences. And even among the parenchymatous liver cells, a more homogeneous tissue than most, such quantitative enzyme differences are always found. Non-specific esterase and lipase are more active in the centrolobular cells than in the cells near the lobule periphery. The same is true for DPNH-diaphorase activity; with succinic dehydrogenase, however, the situation is reversed: its activity is high in the peripheral cells and low in the centrolobular cells.

Have you used such staining methods on your material?

My question to Dr. Allen concerns the possibility that in some eggs the egg filaments described by earlier embryologists may also play an important role in sperm entry. Does the egg play a more significant role where it possesses conspicuous filaments, as in *Sabellaria* and *Nereis*?

DR. ALLEN: As far as I know, there are no reports of filaments from the egg which could be confused with the type of sperm filament which I have discussed this morning. There have been reports, however, especially the work by Colwin and Colwin in *Holothuria,* of egg filaments as well as sperm filaments. However, the sperm filaments can always be distinguished from the others. Monroy showed many years ago in *Pomatocerus* that there were many filaments emanating out of the egg. However, this was before the acrosome reaction had been clearly described. He did mention, I believe, that the filament from the sperm did look different from those of the egg. Possibly Dr. Colwin would like to comment on this problem.

DR. COLWIN: I think I agree with you. There is really a difference between the filaments from the egg and the sperm. In some species the acrosome filament of the spermatozoon can be produced even when no eggs are present so that obviously it cannot come from the egg. The egg filaments are probably akin to microvilli. With regard to *Sabellaria,* which I know you have studied a good deal, Dr. Novikoff, I believe that the filaments you saw were certainly egg filaments. But in our studies of *Sabellaria* we found that there were acrosome filaments as well—the situation was *fundamentally* the same as in the other species we have studied.

DR. MARKERT: I should like to ask Dr. Allen a question but first perhaps I should comment on Dr. Novikoff's question. In studying the development of enzymes it is necessary, as Dr. Novikoff suggests, to correlate information from homogenates with data from intact or sectioned tissues. Particularly in complex tissues where more than one type of cell contains esterase and where homogenate analyses reveal several distinct esterases, it is quite possible that the various esterases are distributed non-uniformly among the active cells. The duodenum, in fact, provides such an example. The cells of the villi are strongly positive when alpha naphthyl butyrate is used as substrate in tissue sections, and only faintly positive when naphthyl AS acetate is used as substrate. Just the reverse is true for the Brunner's glands in the wall of the duodenum. By applying these two substrates to electrophoretically resolved esterases in starch gels it is apparent that three esterases, designated A, G, and H, show the greatest specificity for naphthyl AS acetate while the remaining esterases are relatively more specific for alpha naphthyl butyrate. Thus the Brunner's glands probably contribute relatively more of the A, G, and H esterases and the cells of the villi more of the remaining esterases to homogenates of the duodenum. From this type of data we can identify the successive addition of new esterases by particular cell types during differentiation. I don't believe that the evidence can be interpreted merely on the basis of shifts in cell populations.

And now I should like to address a question to Dr. Allen. Dr. Allen, you raised the question of whether the cortical reaction was an autocatalytic one or not, or whether it might be possible also that the acrosome reaction was a consequence of something provided by the sperm cell itself. It wasn't clear to me how variations in the physiology of the sperm could be reconciled with the idea of an autocatalytic reaction.

DR. ALLEN: The answer to this question is not very clear to me either, Dr. Markert. One possibility is that the sperm has to inject a particular quantity of some "initiating substance" into the egg. This amount injected might be greater in some cases, and the fertilization wave would then proceed at a more rapid rate and stand less chance of dying out. We do know that the fertilization wave can proceed at different rates under different physiological conditions. This suggests that different sperm might introduce a different quantity or quality of substance into the egg cortex. There is really not much basic information pertinent to this point.

DR. WILLIER: I would like to ask you a question about Just's old observation on the sand dollar, *Echinarachnius*. He felt that in this form there was a wave of negativity spreading over the egg surface.

DR. ALLEN: Just made some interesting observations on *Echinarachnius*. His observations, in fact, first suggested that there might be a latent period. He remarked that after sperm attachment there seemed to be a period when nothing happened. Then, he observed a wave of "turbidity" which he commented on in other papers as a "wave of negativity." The turbidity was probably the same as the "roughening" that Libby Hyman saw and the change in surface color that Runnström observed coincident with the breakdown of cortical granules. The term "negativity" is not very meaningful today. I think that Just himself probably wasn't too clear on what he observed.

DR. WEISS: I would like to make some remarks on the very important question raised by Dr. Novikoff. In general, one must check experiments with homogenates against more intact preparations. Any change in an organ such as the liver which appears later must have come about by changes either in the admixture of immigrant cells or from a change in the native cell population itself. Immigrant cells in this particular case, except for the case of blood cells, are ruled out, and so I think the evidence is clear that we are dealing with a progressive change in the enzyme pattern of the local liver cell line. Perhaps the problem will be a little easier if we raise our sights from the cell individual and the enzymes of the individual cell and begin to look at the continuum of cell generations or the cell line. I think when we look at it from this point of view the evidence seems quite conclusive.

I would like, however, to ask a question of Dr. Markert. Dr. Markert ruled out the very attractive hypothesis of the differentiation of extrachromosomal material. He argued that at each step during the differentiation of the cell when it divided, if the extrachromosomal material were differentiated, it would be diluted out as the nuclear content merges with the cytoplasm. However, this can be looked at in exactly the opposite way. When the nucleus is reconstituted after division, it reacquires a part of the cytoplasm which lies around it. So, the nucleus could, so to speak, bite off a part of the cytoplasm to which it had previously contributed some of its own products. I would be a little cautious about concentrating solely on the chromosomes.

DR. MARKERT: Your point is well taken. It really is quite difficult to distinguish between the two possibilities. I probably just prefer the chromosomes due to my own genetic background.

DR. BRACHET: I would like to ask Dr. Allen if there is any known correlation between increases in enzymatic activity and the breakdown of cortical granules. We might have a situation comparable to that of the lysosomes in the liver, in which there is a great concentration of hydrolytic enzymes. If that would be true, and if there would be a release of these enzymes when fertilization occurred, it would make a very nice and complete picture.

DR. ALLEN: Yes, I think we are coming back to some of Loeb's old ideas about "superficial cytolysis" in newer terms. However, it is extremely difficult to assign to one enzyme or to many enzymes the responsibility for the events in fertilization which we see. This is because the time scale is so very short. Of course, it would be very nice to show that some enzyme is activated at this time, but I am not sure how best to approach this problem. In the few seconds in which fertilization occurs the application of histological or biochemical techniques to detect these changes may well in themselves lead to activation, so the problem has technical difficulties of considerable magnitude. The information which we have on this is the work of Lundblad showing that a proteolytic enzyme, E II, becomes active "at fertilization." However, he did not define precisely what he meant on the time axis by "at fertilization." As you know, a difference between the exact time of sperm attachment and a minute or so later might make a considerable difference. I think we are just getting some idea now as to what the temporal sequence of events is. The important changes for the cortical reaction and for a number of other events taking place at this time may lie within the latent period. These important events may well be all over by the time we actually see the cortical reaction.

DR. HARRIS: Dr. Markert made a very interesting observation in discussing the esterases, namely that cells cultured in vitro tend to lose these enzymes to some extent. I am wondering whether this may be in part an adaptive phenomenon. With respect to cholinesterase, Burkhatter, Jones, and Featherstone have shown a direct relationship between the amount of acetylcholine in culture and the level of this particular enzyme. I would like to know if Dr. Markert has tried to reverse the loss of esterases in culture by the addition of appropriate substrates. If it were possible to restore the level of certain of these enzymes, we might then have a means of contrasting adaptive and constitutive enzymes in these cells. It would be an example of modulation in the sense that Weiss and others have used this term. If, on the other hand, the loss is irreversible, it might have to do then with the loss of certain genes or a change in their action during the evolution of cell strains.

DR. MARKERT: No, I haven't tried to restore esterase activity by the addition of substrates for these enzymes, partly because the substrates used in testing for these esterases are not normal constituents of the cell, and I doubt that these substrates would act as inducers. Nevertheless, the experiment you suggest should be performed.

DR. PASTEELS: I was quite interested in Dr. Allen's observations on the fertilization reaction in mollusks. I should like to ask him a question regarding mollusk eggs. In the case of a mollusk such as *Barnea,* or *Gryphaea,* there is no visible elevation of the fertilization membrane, yet there are large cortical granules containing mucopolysaccharides. These granules remain intact after fertilization just as in the marine anellid *Chaetopterus.* It would be interesting to know what happens in the absence of the elevation of the fertilization membrane.

DR. ALLEN: Yes, I think observations with the electron microscope would be interesting in this regard. Last summer Dr. Rebhun started to look at this in the *Spisula* egg. We have high hopes that he might find something of interest on this.

DR. EBERT: I note that Dr. Allen emphasized fertilization, not simply egg activation, and that he advanced arguments for the possible role of the sperm. Yet we know of many cases of artificial activation; does one need to argue that the sperm contributes anything at all in the activation process? Do the cortical granules disappear in parthenogenesis?

DR. ALLEN: It was shown in the early days of the century that agents which cause parthenogenesis by and large cause membrane elevation. At that time cortical granules were not generally known about. The reverse situation, however, seems not to be true; it is apparently possible to get membrane elevation without complete activation of the egg. Part of this might be due to failure of the egg centrosome to function in cell division. There is evidence that Sugiyama's "fertilization impulse" must be initiated for complete activation. As he showed, many "membrane-elevating" substances by-pass this process. Also, some activating agents undoubtedly injure the egg. Much of the work in parthenogenesis should be done over now that we know a little more about the cortical reaction.

DR. BODIAN: It seems to me that sometimes negative evidence can be as intriguing as the beautiful positive data that you have presented. When Dr. Brachet brought up the question of the existence of lytic enzymes, possibly in the granules, I was reminded of the fact that you passed very quickly over your difficulties in handling separated cortical granules. Would it be too embarrassing for you to tell us what the difficulties were?

DR. ALLEN: Not at all. I had three weeks in which to go to Dr. Daniel Mazia's lab where he had both a preparative ultracentrifuge and unpigmented sea urchin eggs. In ten days I succeeded in getting a method to isolate cortical granules, and then spent the rest of the time in bed with the flu. The method has not been published and I hope perhaps Dr. Immers in Sweden may take this up.

DR. P. P. COHEN: Are there any extracts of sperm which simulate the effect of the fertilization process?

DR. ALLEN: Many such extracts have been described. The first attempt was probably that of Just who boiled sperm in oxalate and showed that he could get activation. Since then, a number of reports have appeared, and I think probably most of these are due to basic proteins such as protamines which probably precipitate some of the acid mucopolysaccharides in the cortex. This is somewhat suggested by the fact that eggs agglutinate very strongly at the same time that they become activated. Dr. Wicklund in Sweden has studied the effect of protamines on the activation of sea urchin eggs; I have some doubt as to whether this is initiation of the fertilization wave as is caused by sperm and butyric acid. More likely, it is an interference in some other reaction that attends fertilization, for the breakdown of cortical granules in protamine is extremely gradual and takes place in a manner which is quite different from that in the cortical reaction attending fertilization.

DR. GRANT: The fact that you have species of eggs which have cortical granules that do not participate in the formation of the fertilization membrane suggests that these cortical granules must have some other function during development. Now, is there any breakdown of cortical granules in these eggs in later stages of development that is correlated with any morphogenetic event?

DR. ALLEN: My studies of embryology usually stop after the membrane is elevated so I am afraid I cannot answer your question. I don't know anything about the role of these things. I have, however, just learned from Dr. Afzelius that he once in a while finds cortical granules in quite late stages, blastulae or later.

# PHASES OF DEPENDENT AND AUTONOMOUS MORPHOGENESIS IN THE SO-CALLED MOSAIC-EGG OF TUBIFEX

Fritz E. Lehmann

*Zoology Department, University of Berne, Switzerland*

### The Transformation of Oocyte Organization into the Architecture of the Embryo as a Principal of Embryology

In the days of Roux, Driesch, Morgan and Spemann (see Seidel, 38) it seemed fully justified to distinguish two categories of development: regulative and mosaic-like morphogenesis. With the method of isolation of fragments of embryos or eggs only one main problem was accessible: the question, namely, whether fragments are capable of developing into whole embryos, as in the case of the sea-urchin, or into mosaic-like half embryos, as in the case of the ascidian egg. This original question lost much of its alternative character during the subsequent periods of embryological research in which Spemann and Hörstadius (see Seidel, 38) started to work with refined transplantation methods. It has become evident that different regions of one and the same embryonic system show different degrees of regulative or mosaic-like development. Independent and self-organizing regions coexist with more or less dependent areas which are governed by the dominant regions. In addition to this it has been recognized that the various egg types differ in their characteristic patterns of organization. However, the morphodynamic organization of any embryonic system cannot be unraveled by direct light microscopical examination, but is as a rule only indirectly deduced from the various results of experimental data. The organization of an egg appears as a carrier of different action and reaction systems which are released in the moment of activation of the egg. From this time on local processes of independent self-organization, chemical induction of dependent regions by distant or contact actions, and topogenetic activities are interlinked in a very complex manner. The later formative processes of development depend very much upon the egg organization. Therefore it is possible to deduce from anormogenetic developmental types typical features of the egg organization.

At present the morphodynamic organization of the echinoderm egg, some spirally cleaving eggs of molluscs and annelids, several ascidian eggs and the amphibian egg is rather well known. All these eggs possess two structurally distinct regions: the egg cortex, the structure of which appears to be very stable in centrifuged eggs; and a complex endoplasm with several particulate inclusions. The experimental results have disclosed the fact that in the morphogenetic organization of the various eggs cortex and endoplasm are involved in a very different manner.

The organization of the spirally cleaving egg of *Tubifex* is intermediate between that of the echinoderm and that of the amphibian type. There is a cortex with polar fields the action of which determines the localization of the polar plasms, and there is an endoplasm consisting of two sorts of cytoplasm, the polar plasm and the yolk-containing cytoplasm. From the polar plasms there originates the cytoplasm later forming the germ bands. The yolk-containing endoplasm is the precursor of the intestine.

Let us consider here two extremely differing egg types. The echinoderm egg possesses a cortex which carries a double system of antagonistic gradient fields (Runnström and Hörstadius). The endoplasm contains only traces of a morphogenetic pattern. Before the formation of the gray crescent, the amphibian egg is devoid of a clearly demonstrable cortical pattern, but shows different endoplasmic regions with characteristic morphogenetic potentialities: an animal-marginal zone which later forms ectoderm and mesoderm, and a vegetative zone which gives rise to the intestine.

From this brief sketch it can be seen that animal eggs are characterized by cytoplasmic patterns the parts of which carry different morphogenetic functions. In the following section we have to discuss how the intricate cooperation of dependent and independent germinal regions contributes to the formation of the embryonic architecture, especially in the development of *Tubifex*.

## From Self-Organizing and Inducing Cellular Areas to Mosaic-like Patterns in the Tubifex Embryo[1]

### The Epigenetic Formation of the Cytoplasmic Pattern During Maturation and Cleavage

Shortly before the formation of the first cleavage spindle the cytoplasmic pattern of the *Tubifex* egg is completed. Both poles of the egg carry a well-

[1] For illustrations the reader is referred to Lehmann (17) and Lehmann and Mancuso (19).

delimited condensation of polar plasm rich in mitochondria and giving a positive Nadi-reaction. The egg center contains a large cytoplasmic area with the mitotic apparatus, which is surrounded by the large mass of endoplasm carrying yolk. Thus the final cytoplasmic pattern of the egg is formed relatively late. This is to be observed not only in *Tubifex* but also in eggs of several molluscs, annelids, ascidians, and even amphibians. In these different egg types analogous processes are seen in the segregation of several sorts of cytoplasm. These phenomena have rightly been called by Costello "ooplasmic segregation" (5).

As far as *Tubifex* is concerned these delimited cytoplasmic areas are not at all uniform gels containing but a single sort of macromolecule or even "formative stuffs" (E. B. Wilson, 40, p. 1065). The main regions of the *Tubifex* zygote (plasmalemma, polar plasms, yolk-containing endoplasm, nuclear apparatus) consist of very *characteristic populations of particulates*. This is revealed by electron microscopy of ultrathin sections (Weber, 42; Lehmann and Mancuso, 19). Some of the particulates mentioned are to be considered as the carriers of vital phenomena, namely, mitochondria, and the various components of the hyaloplasm (microsomes of vesicular or granular type and cytoplasmic fibrils). We have designated these "vital particulates" as "biosomes" (14) because presumably they possess a vital property, namely, *structural and biochemical continuity*. In the oocyte of *Tubifex* there occurs another category of particulates, "nutrient particulates," such as lipid droplets and yolk spheres. As all these "nutrient particulates" carry cytoplasmic membranes, they may play also an important role in metabolism and eventually in morphogenesis. Vital and nutrient particulates constitute the *important particulate level of living embryonic cells*. This level is distinguished by a typical structural differentiation closely connected with specific cellular activities (7, 19). The polar plasms possess numerous mitochondria and basophilic bodies of the form and size of chromidia or larger microsomes. The endoplasm is poor in endoplasmic reticulum and mitochondria, rich in lipid and yolk spheres, all of which are coated by a delicate cytoplasmic membrane. The plasmalemma or cortex is not yet too well known as regards its fine structure.

Our centrifuging experiments have allowed us to concentrate the particulate components of the two polar plasms in one single layer. If the living eggs are centrifuged in a well-stabilized position, accumulation of one large polar plasm at one pole can be brought about. We have observed the development of eggs with one polar plasm at the animal pole. These eggs cleaved normally and developed later on into embryos with normal germ bands. Eggs with a single polar plasm at the vegetative pole produced a smaller percentage of embryos because in this experiment cleavage proved to be disturbed more frequently. It is remarkable that the mass of polar plasm concentrated at one pole does not show any tendency to move to the other pole devoid of polar plasm. Polar plasm concentrated in the equatorial region of the egg behaves very differently. Immediately after centrifugation, the condensed polar plasm spreads along the cortex in two opposite directions and begins to reach the original animal and vegetative poles. The cortex of the animal and vegetative poles probably contains fields which possess a specific attraction for polar plasm.

Selective adhesion between different cell types plays an important role in the morphogenesis of amphibians, as Townes and Holtfreter (39) have shown. In our case, however, we deal with a case of *intracellular* and not intercellular *adhesiveness between different cytoplasmic regions*. In fact, the segregation and concentration of polar plasms which takes place during the maturation divisions shows some analogies to the movement and the changing selective adhesiveness of whole embryonic blastemas. Before the meiotic divisions begin, the polar plasm material is spread out in nearly the whole of the subcortical area of the egg. At the period of visible polar plasms it is totally concentrated in two regions and has considerably augmented its mass.

This can be demonstrated by centrifuging and stratifying the various meiotic stages. Especially during the phase of the second meiotic lobulations, the concentration of polar plasms seems to proceed rapidly. The more the polar plasms are concentrated underneath the polar cortex, the more they are delimited from the neighboring yolk-containing cytoplasm. The quinones of naphthalene and phenanthrene interfere strongly with the interaction of polar plasms and polar cortex. Treatment of young maturation stages inhibits the concentration of polar plasms at the poles, with the result that there remain irregular subcortical sheets of polar plasm in wide areas of the egg surface. At the same time the formation of regular and conspicuous meiotic lobulations is inhibited and distorted (37).

Our experiments demonstrate that the development of the cytoplasmic pattern of the *Tubifex* zygote represents a kind of "ooplasmic segregation" which is brought about by complex interactions between egg cytoplasmic particulates and egg cortex. The egg endoplasm segregates during meiosis into two separate cytoplasmic territories, the polar plasm with its population of many mitochondria and chromidia, and an endoplasm rich in nutrient particulates. *The particles* of these territories must be endowed with a *characteristic adhesiveness* which keeps them together even after treatment with the centrifuge. But this ooplasmic segregation leads to the typical final pattern only if the condensation of the polar plasm is governed by the action of the polar cortical fields.

The morphodynamic processes during meiosis of the *Tubifex* egg obviously show features of interdependent development and not an independent behavior of the egg parts. This conclusion is corroborated by the fact that the two-cell stage of *Tubifex* can still give rise to double embryos. Another feature is also in favor of our interpretation. As we have shown, it is possible to fuse the whole mass of the polar plasms either at the animal or the vegetative poles. In both cases the zygote with a fused polar plasm can give rise to a normal embryo. All these facts mean that the organization of the one-cell stage does not represent a mosaic-like pattern, the parts of which are fully determined.

## *The Role of Early Cleavage in the Differential Distribution of the Polar Plasms*

As mentioned above, the segregation of the polar plasms in the zygote does not determine definitely the pattern of embryo formation. This is only the case after the large somatoblasts $2d$ and $4d$, containing two main portions of the polar plasms, have been formed. These seem to be the principal blastomeres necessary for embryo formation. Immediately after the first cleavage the two separate polar plasms fuse in the blastomere *CD* into one large, well-delimited territory. This is transferred partly to the ectoblast $2d$, and partly to the mesoblast $4d$. As soon as the somatoblasts are segregated, they show, according to Penners' experiments, separate and mosaic-like potencies (25-29). $2d$ is only capable of producing the ectodermal germ bands, and $4d$ only gives rise to the mesodermal germ bands. The fused mass of polar plasm present in the cell *CD* must have undergone in the following cleavage steps a segregation into two different types of cytoplasm, an "ectodermal" and a "mesodermal" plasm.

In the past there was much discussion about "organ-forming stuffs" (40), which might be responsible for the different potencies of the somatoblasts. But the possibility was not considered that the two somatoblasts might develop differences not on the molecular but on the particulate or bisomatic level. It has been possible to prove for the embryo of *Tubifex* (see Table 1) that in fact striking cytoplasmic differences of ultrastructure are present in the two somatoblasts (19). In the ectoblast $2d$ there occurs a dense cytoplasmic reticulum with many fibrous and vesicular elements, and with rather few mitochondria, fat droplets, and yolk spheres. The mesoblast shows a looser cytoplasmic reticulum but a very rich population of mitochondria and a considerable territory with numerous lipid droplets and yolk spheres. In the somatoblasts of *Ilyanassa* a similar differential distribution of particulates has been observed (Clement and Lehmann, 3). The cell 2d is rich in mitochondria, but poor in lipid droplets and devoid of yolk spheres. In contrast to this, 4d not only possesses mitochondria, but has many lipid droplets and some yolk globules.

TABLE 1

DIFFERENCES IN THE PARTICULATE POPULATION OF THREE MAIN BLASTOMERES OF THE 21-CELL EMBRYO OF *Tubifex*

(Lehmann and Mancuso, 1957)

|  | Somatoblast 2d Telectoblast | Somatoblast 4d Mesoblast | Entoblast 4D |
|---|---|---|---|
| Hyaloplasmic Reticulum | +++ | ++ | + |
| Mitochondria | Nadi reaction positive ++ | Nadi reaction positive +++ | Nadi reaction negative; only perinuclear + |
| Lipid droplets | + | +++ | ++ |
| Yolk spheres | + | ++ Only in vegetative part of 4d | ++++ |

In both these cases of spiralian eggs we find rather striking quantitative differences in the distribution of particulates. In the better investigated case of *Tubifex* the differential distribution of the basophilic reticulum and the mitochondria is especially remarkable. This certainly indicates differences in metabolic pattern, because metabolic activities of the reticular (microsomal) and the mitochondrial particulates are probably very different. Some other morphodynamic processes which finally differ qualitatively are first determined by systems of factors which only show quantitative differences (genetic determination of sex, determination of properties of the lateral mesoderm in amphibians by quantitative factors, gradients of sea urchin). From these cases the assumption may be derived for the somatoblasts of *Tubifex* that the differences of the two somatoblasts may primarily reside in different proportions of the respective populations of particulates. This may induce secondarily a divergent protein metabolism (Weber, 43) and bring about development of qualitative morphogenetic differences.

## Inhibition of Somatoblastic Differentiation by Abnormal Distribution of the Polar Plasms

If the idea is correct that definite proportions of particulates within a given cell determine the final metabolism and by this means the morphogenetic character of the somatoblasts, it is to be expected that disturbances in the distribution of polar plasm to single cleavage cells may bring about anormogenetic development of the embryo. In the case of *Tubifex* this expectation was easy to realize, because cleavage can be distorted by several factors. Centrifuging, as well as different chemical treatments, disturbs the cleavage pattern seriously. In many cases abnormally cleaving cells form several medium-sized cells instead of the large somatoblasts $2d$ and $4d$. These abnormal cells contain polar plasm, as can be shown by a positive Nadi reaction. But they are unable to form germ bands. They live nearly as many days as the controls require to reach the stage of hatching. But the abnormal germs possess only entoderm cells and several small groups of undifferentiated ectoderm cells. Differentiated and segmented organs are completely lacking. In these cases obviously the embryo did not lose embryonic material. But some processes must have been suppressed which are necessary for the determination of the ectodermal and mesodermal germ bands. This break in the sequence of the normal developmental processes is presumably produced by the abnormal cleavage. The normally occurring condensation of the polar plasms in the large somatoblasts is not possible because of the formation of several smaller cells. Concomitantly, the differential distribution of the various cytoplasmic constituents is made impossible and the mosaic pattern of somatoblasts cannot appear.

It seems a very remarkable feature in the development of a spiralian egg that a simple disturbance in the distribution of polar plasm can bring about such a far-reaching effect upon embryonic differentiation. Here the early cleavage of *Tubifex* displays characteristic epigenetic features which are not in accord with the general concept of mosaic development. This statement might sound surprising if we consider the many results in embryological literature which seem to be mostly in favor of a pronounced mosaic-like character of spiralian development, especially in marine forms. But the marine spiralians are distinguished from *Tubifex* by two principal features: they possess a remarkable mechanical resistance, and very often they develop planktonic larvae.

The difference in mechanical resistance is very striking. Marine eggs like those of *Ilyanassa* can be stratified by strong centrifugation, but the stratification disappears rapidly afterwards. On the contrary, the eggs of *Tubifex* are easily stratified and keep the induced order of layers for longer periods. Therefore the dislocation experiments in *Tubifex*, leading to a changed cytoplasmic pattern, are impossible in other spiralian eggs because of the instability of the experimentally induced layers.

Planktonic larvae of marine spiralians possess larval organs which show very strong mosaic characters already during early cleavage. As larval organs are lacking in *Tubifex*, we are mainly dealing with cells responsible for the later embryonic pattern. Therefore in the development of the somatoblasts epigenetic factors seem to remain active during early cleavage. Apparently they have not yet been investigated in detail in other spiralians with well-developed larval forms.

The development of specific somatoblastic cytoplasms appears as a central problem of spiralian development. Especially the isolation of the whole mass of ectoderm and mesoderm plasms in an early period is a peculiarity which does not occur in echinoderms or amphibians. The fact of a transitory separation of the ectoderm-forming and mesoderm-forming cytoplasms from all other neighboring cells might be plausibly explained by our assumption (above). This postulates that the proportions of the various particulates in the two somatoblasts determine finally the morphogenetic character of the somatoblasts. During the period of isolation of these organ-forming cytoplasms the necessary stabilization of the divergent morphogenetic characters of the somatoblasts takes place. And it is only during this short period of the isolated somatoblasts that the development of *Tubifex* shows characteristic mosaic-like features, as proven by *Penners'* radiation experiments.

Hence we distinguish two principal periods in the development of *Tubifex*. A first period lasts from the first maturation division till the formation of the two somatoblasts. This period is characterized by numerous epigenetic activities which are closely interlinked. The second period begins with the embryo possessing the two somatoblasts and lasts till the formation of germ bands. In this period development depends in a mosaic-like mode on the somatoblasts.

*The Interaction of Several Factors in Determining the Decisive Cleavage Pattern of* Tubifex

As we have seen, the cleavage of the *Tubifex* egg is to a high degree responsible for the sorting out and the determination of the different cytoplasms of the somatoblasts. Some centrifugation experiments on *Tubifex* eggs indicate the kind of interaction of cortex, endoplasm, and cleavage spindle which are responsible for the cleavage pattern. During meiosis, size and position of the minute polar bodies are determined by the position of the meiotic spindle in the cortex. Location of the spindle on either pole produces a radially symmetrical pattern of lobulations. Attachment of the translocated spindle in the equatorial cortex induces a bilaterally symmetrical pattern of lobulation. Similar patterns can be produced by translocating the first cleavage spindle to different regions of the cortex of the zygote.

The cleavage rhythm itself, i.e., the rhythmical sequence of interphasic state of nucleus and cytoplasm and cytoplasmic motility connected with mitotic nuclear activity, seems to be mainly located in the cytoplasm. Analogous rhythmical changes of interphase state and motility occur even in enucleated parts of cells (isolated polar lobes of *Ilyanassa*) and in cells of *Tubifex* chemically deprived of the nuclear apparatus (Woker, 41). This change between interphase and mitotic state of the embryonic system is best illustrated by Pasteels' observations on activated but not cleaving eggs of *Chaetopterus* (24). These eggs differentiate after activation without cytoplasmic divisions. The nuclear apparatus produces a characteristic monastral cycle. The cortex of the monastral embryos continues with the same formative activities as the normally cleaving control embryos. During the phases of micromere formation there appear small and numerous lobulations in the activated eggs. In the period of the formation of the somatoblasts the experimental eggs developed large lobe-like protrusions. These changes of form indicate for the egg of *Chaetopterus* a sequence of tendencies, first to form smaller micromere-like lobulations, and then to produce larger somatoblast-like lobes.

The spiralian egg seems to contain a metabolic system developing in a characteristic rhythm of interphase and mitotic states. A progressive change of the fine structure seems to condition a sequence of small and large lobulations. The large mitotic apparatus of the spiralian embryo determines by its final position and the production of anaphasic action substances the positions of the furrows.

This complex interaction of cleavage factors is responsible for the formation of the somatoblasts. Whereas abnormalities of cleavage do not seriously interfere with development, either in amphibian or in sea-urchin eggs, disturbance of cleavage in *Tubifex* may suppress somatoblast formation. This means that in *Tubifex* cleavage plays an important role as an epigenetic factor of embryo formation.

### Autonomous versus Dependent Development or Intracellular versus Blastematic Morphodynamics

[See Table 2]

The original aim of this review was to contrast the autonomous differentiation of spiralian eggs with the dependent differentiation occurring in sea urchin and vertebrate development. But a closer analysis of *Tubifex* development has shown conclusively that the organization of the egg-pattern as well as the formation of somatoblasts is governed by groups of interacting epigenetic factors. Only the final stage of the somatoblast embryo represents a period of mosaic-like behavior of the somatoblasts. The power of regulation and the interaction between blastomeres seem to be extremely low. Autonomy of single blastomeres is in this phase very conspicuous and has been known for a considerable time.

### TABLE 2
#### Morphogenetic Reaction Systems During Embryogenesis

1. *Intracellular reaction patterns*

    *Spiralians:* a) Ooplasmic segregation
    b) Intrablastomeric segregation of cytoplasm during spiral cleavage, and separation of autonomous organ-forming cell types
    *Vertebrates:* Ooplasmic segregation

2. *Intrablastematic reaction patterns*

    *Spiralians:* Segmentation of mesodermal germ bands in *Tubifex*
    *Vertebrates:* Self-organization of chordamesoderm and entoderm, and separation of autonomous organ-forming blastematic areas

3. *Interblastematic reaction systems*

    *Spiralians:* Pattern formation in ectodermal germ bands induced by mesodermal germ bands
    *Vertebrates:* Induction of neural plate by the roof of archenteron

But the whole of the processes leading to the mosaic-like architecture has remained obscure for a long time. This may be explained by the peculiar fact that in mosaic types processes of segregation and determination are of a purely intracellular character and therefore rather difficult to attack experimentally. In contrast to this, morphodynamic events in sea urchins and vertebrates are generally blastematic and hence accessible to the classical transplantation methods.

In this period of developing histochemistry and electron microscopy the intracellular type of morphodynamics becomes especially attractive to the biologist. The stepwise sorting out of specific cytoplasms within cells is often accompanied by histochemical events and changes in the particulate populations. At the moment these well-delimited organ-forming cytoplasms are easier to investigate than the gradient fields of amphibians and sea urchins. So it might be expected that the analysis of the intracellular mode of differentiation may furnish new facts and ideas for the research of the intercellular or blastematic type of morphogenesis.

## REFERENCES

1. Boell, E. J., and Weber, R., Cytochrome oxidase activity in mitochondria during amphibian development. *Exptl. Cell Research,* **9,** 559-567 (1955).
2. Bretschneider, L. H., and Raven, C. P., Structural and topochemical changes in the egg cells of *Limnaea stagnalis* during oogenesis. *Arch. néerl. zool.,* **10,** 1-31 (1951).
3. Clement, A. C., and Lehmann, F. E., Ueber das Verteilungsmuster von Mitochondrien und Lipoidtropfen während der Furchung von *Ilyanassa obsoleta* (Mollusca, Prosobranchia). *Naturwiss.,* **43,** 578-79 (1956).
4. Costello, D. P., Experimental studies of germinal localization in *Nereis*. I. The development of isolated blastomeres. *J. Exptl. Zool.,* **100,** 19-66 (1945).
5. ———, Segregation of ooplasmic constituents, *J. Elisha Mitchell Sci. Soc.,* **61,** 277-289 (1945).
6. Dalcq, A., Etude micrographique et quantitative de la mérogonie double chez *Ascidiella scabra*. *Arch. biol.* (*Liége*), **49,** 397-568 (1938).
7. Eakin, R. M., and Lehmann, F. E., An electron microscopic study of developing amphibian ectoderm. *Wilhelm Roux' Arch. Entwicklungsmech. Organ.,* **150,** 177-198 (1957).
8. Huber, W., Ueber die antimitotische Wirkung von Naphthochinon und Phenanthrenchinon auf die Furchung von Tubifex. *Rev. suisse zool.,* **54,** 61-154 (1947).
9. Krause, G., Induktionssysteme in der Embryonalentwicklung von Insekten. *Ergeb. Biol.,* **20,** 159-198 (1957).
10. Lehmann, F. E., Polarität und Reifungsteilungen bei zentrifugierten Tubifex-Eiern. *Rev. suisse zool.,* **47,** 177-182 (1940).
11. ———, Die Indophenolreaktion der Polplasmen von Tubifex. *Naturwiss.,* **29,** 101 (1941).
12. ———, Die Zucht von Tubifex für Laboratoriumszwecke. *Rev. suisse zool.,* **48,** 559-561 (1941).
13. ———, Mitoseablauf und Bewegungsvorgänge der Zellrinde bei zentrifugierten Keimen von Tubifex. *Rev. suisse zool.,* **53,** 475-480 (1946).
14. ———, Ueber die plasmatische Organisation tierischer Eizellen und die Rolle vitaler Strukturelemente der Biosomen. *Rev. suisse zool.,* **54,** 246-251 (1947).
15. ———, Chemische Beeinflussung der Zellteilung. *Experientia,* **3,** 223-232 (1947).
16. ———, Zur Entwicklungsphysiologie der Polplasmen des Eies von Tubifex. *Rev. suisse zool.,* **55,** 1-43 (1948).
17. ———, Plasmatische Eiorganisation und Entwicklungsleistung beim Keim von Tubifex. *Naturwiss.,* **13,** 289-296 (1956).
18. ———, and Hadorn, H., Vergleichende Wirkungsanalyse von zwei antimitotischen Stoffen, Colchicin und Benzochinon, am Tubifex-Ei. *Helv. Physiol. et Pharmacol. Acta,* **4,** 11-42 (1946).
19. ———, and Mancuso, V., Verschiedenheiten in der submikroskopischen Struktur der Somatoblasten des Embryos von Tubifex. *Arch. Julius Klaus-Stift. Vererbungsforsch. Sozialanthropol. u. Rassenhyg.,* **32**.
20. ———, and Wahli, H. R., Histochemische und elektronenmikroskopische Unterschiede im Cytoplasma der beiden Somatoblasten des Tubifexkeimes. *Z. Zellforsch. u. mikroskop. Anat.,* **39,** 618-629 (1954).

21. Meyer, A., Die Entwicklung der Nephridien und Gonoblasten bei *Tubifex rivulorum* Lam., nebst einigen Bemerkungen zum natürlichen System der Oligochaeten. *Z. wiss. Zool.*, **133**, 517-562 (1929).
22. Novikoff, A., Morphogenetic substances or organizers in annelid development. *J. Exptl. Zool.*, **85**, 127-155 (1940).
23. Novikoff, A. B., Embryonic determination in the Annelid, *Sabellaria vulgaris*. I. The differentiation of ectoderm and endoderm when separated through induced exogastrulation. II. Transplantation of polar lobes. *Biol. Bull.*, **74**, 198-234 (1938).
24. Pasteels, J., Recherches sur la morphogénèse et le déterminisme des segmentations inégales chez les Spiralia. *Arch. anat. microscop.*, **30**, 161-197 (1934).
25. Penners, A., Experimentelle Untersuchungen zum Determinationsproblem am Keim von *Tubifex rivulorum* Lam. *Arch. mikroskop. Anat. u. Entwicklungsmech.*, **102**, 51-100 (1924).
26. ———, Experimentelle Untersuchungen zum Determinationsproblem am Keim von *Tubifex rivulorum* Lam. II. Die Entwicklung teilweise abgetöteter Keime. *Z. wiss. Zool.*, **127**, 1-137 (1926).
27. ———, Experimentelle Untersuchungen zum Determinationsproblem am Keim von *Tubifex rivulorum* Lam. III. Abtötung der Teloblasten auf verschiedenen Entwicklungsstadien des Keimstreifs. *Z. wiss. Zool.*, **145**, 220-260 (1934).
28. ———, Regulation am Keim von *Tubifex rivulorum* Lam. nach Ausschaltung des ektodermalen Keimstreifs. *Z. wiss. Zool.*, **149**, 86-130 (1936).
29. ———, Abhängigkeit der Formbildung vom Mesoderm im Tubifex-Embryo. *Z. wiss. Zool.*, **150**, 305-357 (1938).
30. Raven, Chr. P., The development of the egg of *Limnaea stagnalis* L. from the first cleavage till the trochophore stage, with special reference to its "Chemical Embryology." *Arch. néerl. zool.*, **7**, 496-506 (1946).
31. ———, Morphogenesis in *Limnaea stagnalis* and its disturbance by lithium. *J. Exptl. Zool.*, **121**, 1-78 (1952).
32. ———, and Bretschneider, L. H., The effect of centrifugal force upon the eggs of *Limnaea stagnalis* L. *Arch. néerl. zool.*, **6**, 255-278 (1942).
33. Reverberi, G., The mitochondrial pattern in the development of the ascidian egg. *Experientia*, **12**, 55-56 (1956).
34. ———, and Pitotti, M., Ricerche sulla distribuzione delle ossidase e perossidasi, il "cellineage" di uova a mosaico. *Pubbl. staz. zool. Napoli*, **18**, 250-263 (1940).
35. Ries, E., Histochemische Untersuchungen über frühembryonale Sonderungsprozesse in zentrifugierten Eiern von Aplysia. *Biodynamica*, **40**, 1-8 (1938).
36. ———, Versuche über die Bedeutung des Substanzmosaiks für die embryonale Gewebedifferenzierung bei Ascidien. *Arch. exptl. Zellforsch. Gewebezücht.*, **23**, 95-121 (1939).
37. Roetheli, A., Chemische Beeinflussung plasmatischer Vorgänge bei der Meiose des Tubifex-Eies. *Z. Zellforsch. u. mikroskop. Anat.*, **35**, 62-109 (1950).
38. Seidel, F., Geschichtliche Linien und Problematik der Entwicklungsphysiologie. *Naturwiss.*, **42**, 275-286 (1955).
39. Townes, P., and Holtfreter, J., Directed movements and selective adhesion of embryonic amphibian cells. *J. Exptl. Zool.*, **128**, 53-119 (1955).
40. Wilson, E. B., *The Cell in Development and Heredity*. Macmillan, New York (1925).
41. Woker, H., Die Wirkungen des Colchicins auf Furchungsmitosen und Entwicklungsleistungen des Tubifex-Eies. *Rev. suisse zool.*, **51**, 109-172 (1944).
42. Weber, R., Zur Verteilung der Mitochondrien in frühen Entwicklungsstadien von Tubifex. *Rev. suisse zool.*, **63**, 277-288 (1956).
43. ———, Ueber die submikroskopische Organisation und die biochemische Kennzeichnung embryonaler Entwicklungsstadien von Tubifex. *Wilhelm Roux' Arch. Entwicklungsmech. Organ.* (in press).

## DISCUSSION

Dr. Pasteels: I am not convinced by the arguments of Dr. Lehmann about the differences in developmental mechanism of the *Tubifex* egg and other forms. It is true that centrifugation affects severely the development of the unsegmented egg of *Tubifex* and demonstrates an interaction of "plasms"; but I have myself shown that the same holds true for the egg of amphibians.

Dr. Lehmann: Yes, I admit that in some of your experiments there occurred in amphibian embryos inhibition of differentiation. For the case of *Tubifex* I was surprised that the block of differentiation occurred without loss of important materials of the egg cell. The cleavage just runs the wrong way and there results consequently a wrong distribution of cytoplasm to the cells. Finally the differentiation cannot proceed further. I should expect this also in the case of some insects. The eggs of *Diptera*, for instance, show a very high degree of cytoplasmic pattern in the egg cell. If it would be possible to disturb the genesis of this pattern, distorted differentiation might be predicted.

Dr. Clement: I should like to ask Dr. Lehmann to what extent he thinks visible segregation patterns may be symptomatic of differentiation rather than themselves being the primary factor. Also would you attach more significance to one species of particulate things than to another. I am thinking of the centrifuge experiments of E. B. Wilson on *Chaetopterus*, for example, where the eggs were fragmented, and from small hyaline or nearly hyaline fragments normal or fairly normal differentiation could be obtained. I have seen similar things myself in the eggs of a mollusk, *Physa*, where most of the yolk and lipid can be thrown out and still normal differentiation occurs.

Dr. Lehmann: This question cannot be answered entirely. Let's talk about the reticulum and the mitochondria. It can be said today that these two elements are carriers of different types of metabolism and we could not exchange mitochondria with reticulum or vice versa. We believe that they are biochemically very different things. If one were really able to change the proportions of these materials, it might have great consequences. In regard to your other point, it has been known for a long time that if you take out yolk or lipid it doesn't seem to disturb the development on the average; however, we cannot totally disregard these granules. Since we have seen that these yolk granules have membranes and that digestion is going on, these granules cannot be metabolically inert. The same is, of course, true for the lipid droplets. I personally believe that these elements are in their behavior especially important. Another point is this. Marine eggs are especially well constructed and their resistance to mechanical damage is great. The *Tubifex* egg is buried in the sand and is by no means resistant to mechanical damage. If one centrifuges an *Ilyanassa* egg, for example, after a very short time the original cytoplasmic pattern is restored. In contrast, after centrifugation the cytoplasm of the *Tubifex* egg cannot restore itself to its original pattern, and this makes it easy to obtain an abnormal distribution in the cells. My question would be to the other people interested in mosaic eggs, whether it would be possible to alleviate the abnormal cleavage by chemical treatment.

Dr. WILLIER: Would you care to comment on Penners' observations that twinning can be produced in *Tubifex* eggs by increased temperatures or anaerobiosis?

Dr. LEHMANN: Yes, I have observed this readily when the pole-plasm is equally divided at the first cleavage. When this happens, the two cells can go on to form a whole sequence of somatoblasts and finally two little worms.

Dr. MARKERT: I believe Dr. Lehmann pointed out that one of the important mechanisms of differentiation involves the movement of particles within the cell. I wonder if you would care to comment on the possible mechanisms by which the particles can be moved?

Dr. LEHMANN: I asked Dr. Allen the same question just a few days ago and he told me he didn't know anything. Too bad!

Dr. NOVIKOFF: I would like to return to the question raised by Dr. Clement. As Dr. Lehmann indicated, things like gradients are more difficult to study than mitochondria and endoplasmic reticulum. Might it be that morphogenetic determinants exist in the pole-plasm which cannot yet be visualized in the electron microscope? Obviously the mitochondria are of great metabolic importance, and it would be of much interest if an egg fragment could be freed of *all* cortical mitochondria and still develop normally. The endoplasmic reticulum and ribonucleoprotein granules, too, must be important in cell metabolism. But is there any evidence yet that any of these organelles are the primary morphogenetic determinants?

Dr. LEHMANN: May I just answer to this point. We have also taken into account the possibility that mass proportions of the various particulates may be one factor, of course the cortex is another one and the third one may be that from the interreaction also results a different protein metabolism which is not so obvious. We must never forget that the proteins are specific and this we cannot visualize in the electron microscope.

Dr. WILLIER: Are you implying, Dr. Novikoff, that the mitochondria might be a causal factor in differentiation?

Dr. NOVIKOFF: They might be. I was trying to emphasize another possibility, that a morphogenetic material might exist in the pole-plasm whose distribution was not disturbed by centrifugation and whose structure has not yet been described. Perhaps there is an aspect of "hyaloplasm" still not resolved in the electron micrographs of today.

Dr. COHEN: Is the centrifugal effect a reversible one? If you were to spin these cells in the direction which has already given you abnormal effects and were then to rotate them 180° and spin again, would these effects be reversed?

Dr. LEHMANN: No, we didn't do such experiments but that would be possible, I think.

Dr. WEBER: Observations of *Tubifex* eggs by electron microscopy show that it is very difficult to get rid of cortical mitochondria by centrifugation. These stick very firmly to the cortex and perhaps are of physiological importance for the cleavage pattern. In this connection I would like to ask Professor Lehmann what his idea is as to what fixes the cleavage pattern, for this seems to be very important for the arrangement of cytoplasmic structures in different cleavage cells.

Dr. Lehmann: Well, that goes beyond my knowledge, but I mentioned earlier in my paper that the cortex can be varied somewhat and it is certain that local variations exist within the cortex. Then there is another point: namely, from the centrifugation of the mitotic apparatus, which is very large, we can conclude, as some English workers have done, that the mitotic apparatus is giving off at these poles substances which have influence here on the surface. And then we have a surface "bubbling" which Pasteels has worked out in the other paper I quoted. During the whole cleavage there is a sort of change in the physiological state of these structural materials which brings about either large lobulations or small lobulations; that is a causative factor, and we have no idea where it is localized. But in general I came to the conclusion that many factors are localized in the cytoplasm and rather few in the nucleus concerning these conditions.

Dr. Willier: Is it true, Dr. Lehmann, that in the first division cleavage is unequal?

Dr. Lehmann: Yes.

Dr. Willier: Since it is true that the first two blastomeres of the *Tubifex* egg are unequal in size and are anterior and posterior, features that are said to be due to the organization of the cytoplasm, the organization of the cytoplasm determines the position of the spindle and this in turn determines the position of the first cleavage plane which results in an unequal first division. We still talk about the organization of the egg, Dr. Lehmann, but precisely what do we mean by that term?

Dr. Lehmann: We mean by organization of the egg the characteristic pattern of some cytoplasmic regions: the cortex with two invisible fields on the vegetative and animal poles, the polar plasms with many mitochondrial and microsomal particulates suspended in a hyaloplasmic reticulum, the endoplasm containing many yolk spheres and little cytoplasmic reticulum which has to be considered as the main part of the cytoplasmic ground-substance, besides the interstitial fluid and the nuclear region with nucleus, perinuclear cytoplasm and centrospheres.

The cleavage pattern of the *Tubifex* egg might be determined by an unalterable fine structural pattern of the cellular cortex, for which we have some factual evidence. But this structural factor of the cortex requires the synergistic activities of the oriented mitotic apparatus and a rhythmically changing activity of the hyaloplasmic reticulum which is at the same time an important carrier of contractile properties of the embryonic cell.

Dr. Ebert: I am surprised to hear Novikoff's statement indicating that he retains a vestige of the "mysticism" of the embryologist, despite the fact that he has been working in biochemistry for several years. I can't see any need for invoking hyaloplasm or other "formative stuffs" until we have examined carefully the role of the microsome fraction, the endoplasmic reticulum. Perhaps, we will have to search beyond current concepts of cytoplasmic organization, but I see no real need to look for mysterious "stuffs" in eggs at this time.

Dr. Novikoff: I am sorry if I left the impression that the determinants of which I spoke were mystical. Even the gradient concept or others of more classical embryology need not be viewed as mystical. I was simply cautioning against permitting the concreteness and brilliance of biochemical analyses and of electron micrographs to hide from us the important gaps in our knowledge. I expect that neither the lists of biochemical properties found in mitochondria, microsomes and other subcellular particles nor the details of their fine structure revealed by electron microscopy exhaust all important phenomena in the cell. These are neither mystical forces nor entelechies; rather, they are aspects of concrete reality still to be described.

Dr. Cowden: Have you attempted to study the distribution of ribose nucleic acid in the developing mosaic embryo of *Tubifex* and *Ilyanassa*?

Dr. Lehmann: No, we haven't tried that. I can only tell you that we can visualize this cytoplasmic reticulum, and this is strongly basophilic. I might remark in addition about these things which Dr. Novikoff felt might be in the cell but which we had not yet visualized. I have recently talked to Dr. Allen about the amoeba. He told me the cytoplasmic reticulum of these particular cells is a biphasic system. There is the true reticular material and an interstitial material. This fluid interstitial material may be centrifuged off and presumably is composed of low molecular weight compounds.

Dr. Spratt: I think it might not be remiss to point out a possible danger. This has been pointed out by others before with regard to interpreting results of centrifugation experiments on eggs. I think it is necessary to have a whole spectrum of centrifugal forces in order to make a safe interpretation of the data. For, under certain centrifugal forces certain materials will be moved and, of course, others will not be moved. This will have an important bearing on the interpretation. I also have another question which refers to a matter already suggested by one of you. I refer to the return to the normal condition following centrifugation. If the egg is left alone, you get a normal redistribution of the egg components under normal gravity forces, and thus the interval of time between the end of the centrifugation and the onset of cleavage plays an important role in this respect.

Dr. Lehmann: The experiments Dr. Spratt is proposing were already published in 1948 (*Revue Suisse de Zool.* **55**, 1-43). We have applied graded series of centrifugal forces and different stages of ovocytes have been subjected to centrifuging. Here I only referred to centrifugation of meiotic stages with strong centrifugal forces. Under these conditions the whole mass of presumptive pole-plasm can be concentrated irreversibly on one pole, and on the other pole there is no polar plasm formed later on. Certainly a few mitochondria may stick to the cortex devoid of polar plasm but that does not seem to have any effect on reformation of polar plasm. These results are most easily explained to be a question of the movement of particulate and partly coherent cytoplasmic materials and not of molecular materials.

Dr. Yamada: Dr. Lehmann, in your recently published electron micrographs of amphibian embryos and also those we have obtained at Nagoya we could assume that the segregation of amphibian germ layers is accompanied by the appearance of different populations of mitochondria. We find a high concentration of active mitochondria in and about the organizer and apparently there occurs an increase of concentration of mitochondria here than in the ectoderm. Do you think there is an activation of mitochondria in different regions of the ectoderm during their differentiation? This might be the case in so-called regulative eggs. Now in the case of the so-called mosaic eggs according to your results there are quite different mechanisms by which these differences in mitochondrial populations are brought about; that is, segregation is established before cleavage. But I don't think that the mitochondria are the only particles alone responsible for segregation.

Dr. Weiss: I don't think I would like to let this occasion pass by and let the discussion between Dr. Novikoff and Dr. Ebert go unresolved. We have developed in our history and approach as ideate animals in which seeing is believing. First, we believed only in the things which we saw in the microscope and now we believe only in things which we see in the electron microscope and which we denied in fact at times when we couldn't see them. Now, as a matter of fact, we believe in things under the electron microscope that may not even have been in there in the living animal so long as we can create and make some something visible. I think Dr. Novikoff made a very good point. I don't think that we are at the stage where we should monopolize a particular thing just because it is visible and then say that this is the seat of the ghosts to which we then allocate all of the various properties. I think the system is a multiphasic system and I think all its components may be equally important; our task is to identify just what role and function each one of them plays. I think we are sterilizing our future if we deny simply *per primam* that they don't exist. Now, I make this comment because we may not come back to this spiralian egg, and it has some properties which are unique to it, and that is the property Dr. Lehmann alluded to. Namely, the cleavage is very well regulated. It has been shown that this regulation after the four-cell stage alternates so that the spindles turn one way in one division and then turn around the other way in the next division, and this alternation of the slant of the spindles is characteristic of these eggs. In some experiments and observations by Schachs in the old days he indicated that this can be somehow referred to some asymmetric spiral organization in the cytoplasm. This is what Dr. Willier referred to in his question. I don't think anyone has seen any substratum for that spiraling orientation in the electron microscope or in any other way. Now, if it is present, the next question is, is it significant? And I think the significance of this asymmetric organization in the cytoplasm is evidenced by the fact that Crampton and others, by reversing that orientation by pressure and other means, were able to turn the whole symmetry of the developing snail around. Thus, a left-turning snail became a right-turning snail just by the simple trick of altering the first cleavage plane. Now, we have at present I would say no conceptual approach to this type of crystalline arrangement in the cytoplasm. We don't see any morphological expression of it. We have to take these signals such as we get from these observations. I think this question ought to be kept in mind because it links up with the question that Dr. Markert asked: what directs currents or streams of substance? They are oriented, they are organized, they are not random diffusions. We have no conceptual foundation for assuming what orients them, or what directs them, and so I think the future is wide open along this line. I think we are just clouding our views if we think that only the things which we can see in the electron microscope, such as the granules, the microsomes, even only the ones that contain RNA, should be taken into consideration.

Dr. Nace: Miss Laurel Glass and I have some information which may help to remove some of the mysticism about materials below the resolution of the electron microscope and which may have some bearing on the organization of the oocyte (Laurel Glass, Ph.D. Thesis, Duke University, 1958).

Antisera against adult female frog sera (*Rana pipiens*) were produced in rabbits, conjugated with fluorescein and used as immunohistochemical reagents on eggs collected throughout the growth phase of frog oogenesis. The procedures were similar to those described elsewhere in this volume (Nace and Clarke).

During the stages of pre-vitellogenesis oocytes smaller than 15 $\mu$ did not take the stain. But, as shown in the accompanying diagram, the cytoplasm of later stages stained with this reagent, whereas the nuclei and follicle-theca layers did not. The cytoplasmic fluorescence became greater as growth continued and was particularly intense in oocytes of about 100 $\mu$ diameter. In the last phases of pre-vitellogenesis the fluorescence gradually localized to a cytoplasmic rim corresponding to the cortex of the 200 $\mu$ oocytes.

During the period of vitellogenesis the distribution of staining changed. In the $Y_1$ oocyte, defined by Kemp as the stage in which yolk granules begin to appear below the cortex of the egg, low intensity fluorescence was visible in the follicle-theca complex. The cortical layer became negative and the cytoplasm stained little if at all. The yolk platelets did not stain. However, between the nucleus and cell membrane marked staining was seen in structures which may correspond to the so-called yolk nuclei.

In the $Y_2$ and $Y_3$ stages of vitellogenesis, during which the band of yolk granules gradually extends from the cortex toward the nucleus, the follicle-theca complex fluoresced at about the same intensity as in the $Y_1$. Irregular patches

# DISCUSSION

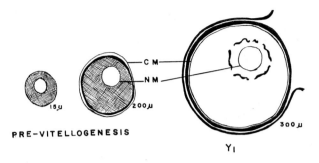

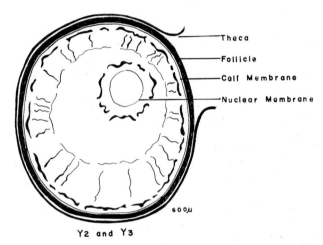

Diagrammatic representation of staining observed with fluorescent anti-adult frog sera at various stages of frog oogenesis. Except for the lines representing the nuclear membrane and the cell membrane, all lines and shadings represent fluorescence. Stage designations according to Kemp.

of fluorescence were seen in the egg cortex. The cytoplasm showed positive staining between the yolk platelets and appeared as "streamers" of fluorescence extending from the cortically-located fluorescent patches toward the inner periphery of the band of yolk platelets. At this stage, a thin band of fluorescence appeared in the cytoplasm just inside the band of yolk platelets. It retained this position as the band of yolk platelets widened to fill the area between the cell membrane and the nucleus and was distinct from the heavily-fluorescing structures closer to the nucleus which seemed to correspond to the "yolk nuclei." Nuclear staining was not seen.

In $Y_4$ eggs the pattern of $Y_3$ staining continued. The cortical patches fluoresced intensely and the cytoplasmic "streamers" of fluorescence between yolk platelets extended to the advancing edge of the yolk platelet band. Some of the smaller yolk platelets began to fluoresce independently of the cytoplasmic matrix in which they were imbedded. The follicle-theca complex stained more intensely than in the previous stage. Nucleolar staining was first seen at this stage.

In the mature ovarian oocyte, $Y_5$, the follicle-theca complex did not stain as heavily as in the $Y_4$. The cortex of the egg showed a peripheral region which did not stain and an inner granule-filled region which did stain. The surface of large yolk platelets stained intensely, perhaps as a result of the inability of the antibody to penetrate the platelets. The smaller yolk platelets also fluoresced. At this stage nearly all nucleoli stained with these fluorescent antisera. Very prominent fluorescent sites were seen as large, or in some cases, dispersed, patches lying against the nuclear membrane.

Thus, it would seem that macromolecules possessing the serological specificity of adult frog serum assume typical patterns of localization in the growing oocyte and that these patterns are characteristic of specific stages. If the situation in the amphibian corresponds to that found by Schechtman *et al.* in the bird and Telfer in the insect, it might be supposed that these antigenic reactive groups are of maternal origin.

There was some indication of an asymmetric distribution of the fluorescence in the oocyte, although this is uncertain as yet. In connection with this discussion it would be of interest to show that this material was not associated with EM-visible structures but rather constituted part of the so-called hyaloplasm.

Dr. CLEMENT: Professor Lehmann has pointed out that the egg of *Tubifex* does not fit the classical conception of a mosaic system. In this connection I would like to describe some recent experiments on the embryonic determination of the foot in the marine gastropod *Ilyanassa obsoleta*. They show that the egg of this form passes through a regulative phase before assuming the final determined state, and that intercellular dependencies are involved in the determination of the foot.

The foot in *Ilyanassa* appears to arise mainly from two third quartet micromeres, 3c and 3d, which are situated at first some distance apart on the posterior side of the egg. During gastrulation, and later, the descendants of these cells evidently converge to a position just behind the blastopore where the single, median foot, with its pair of statocysts, makes its appearance. If, at the 24-cell stage, the 3d micromere is surgically removed, the left half of the foot either fails to develop at all or is represented in the larva by only a trace; the right half of the foot and other larval structures develop well. Removal of the 3c micromere at the same stage results in absence or severe reduction of the right half of the foot. If both 3c and 3d are removed, the egg produces a footless larva. There are thus two determined primordia of the foot at the 24-cell stage; the egg cannot compensate for the loss of either of them, and is in this respect a mosaic.

The conditions are quite different at the 4-cell stage. At this time the D macromere is the prospective source of the cell (3d) which produces the left half of the foot, and the C macromere is the prospective source of the cell (3c) which produces the right half of the foot. However, if the C macromere is removed at the 4-cell stage, the remaining ABD combination will give rise to a larva with a complete foot; in the presence of the D macromere, the egg is able to undergo regulation and produce the right half of the foot from a source other than the missing 3c micromere. On the other hand, if the D macromere is removed at the 4-cell stage, the ABC combination never, so far as I have observed, produces foot structure at all. The ability of the C macromere to differentiate foot tissue appears to depend upon the presence of the D macromere.

Dr. Willier: The one concluding remark I would like to make is that we are still faced with the problem of how much of development is preformistic and how much of it is epigenetic.

# THE ROLE OF THE CELL NUCLEUS IN DEVELOPMENT

A. E. MIRSKY AND VINCENT ALLFREY

*The Rockefeller Institute,*
*New York, N. Y.*

WHEN THE ROLE of the cell nucleus in development is considered, a question which soon arises is whether all the genetic factors present in the chromosomes of the germ cells are present in all cell types of the organism. In discussing this question we should remember that the hereditary factors in the chromosomes are potentialities, and that the realization of these potentialities depends upon conditions both within the nucleus and in the surrounding cytoplasm. The problem of the role of the nucleus in development was stated in such terms as early as 1894 by Hans Driesch (14).[1] He took the position that although nuclear activity varies in different cells, nuclear potentialities remain unchanged.

Since Driesch's time there have been some observations that support the view that all the genetic factors are present in all cells. It is well known that the genetic loci of the germ cells of *Drosophila* are represented in the bands of the salivary chromosomes. An important component of the germinal material, deoxyribonucleic acid (DNA), is present in a constant amount for each set of chromosomes in germ cells and in the various somatic cells of many organisms (25, 10). Several of the histone proteins, which are attached to DNA in the chromosome, are remarkably similar, if not identical, in different somatic cells of an organism (11, 12). Interesting as these observations are, they are inconclusive in demonstrating that all chromosomal genetic factors are retained in the course of development.

Let us turn to the evidence that nuclear activity varies. Before doing so, however, an important condition for nuclear activity should be mentioned. Within the nucleus, the synthesis of both ribonucleic acids (RNA) and proteins requires a source of energy, and thus depends on the presence of phosphorylated nucleosides, such as adenosine triphosphate (ATP). These high-energy phosphates have been shown to be present in a wide variety

---

[1] We are indebted to Professor Viktor Hamburger for this reference.

of isolated nuclear types (27). Moreover, a study of isolated nuclei prepared from the thymus has shown that the nucleus has its own system for phosphorylation, and that the presence of DNA is essential for the aerobic synthesis of ATP. Nuclei deprived of their DNA by treatment with deoxyribonuclease fail to synthesize ATP, but if DNA is restored to them, ATP synthesis resumes (6). Thus one may consider the DNA of the nucleus to be a cofactor in this aerobic phosphorylating system, and one may conclude that an essential non-specific function of DNA is to mediate ATP synthesis. Much of the activity of the nucleus, including its capacity to synthesize RNA and protein, will depend upon the quantity of mononucleotides present in the nucleus and the extent to which these compounds are present as di- or as tri-phosphates. An active nucleus, such as that of a thymus lymphocyte, contains a much greater quantity of mononucleotides than does an inactive nucleus, such as that of an erythrocyte.

Variations in nuclear activity may be expected to depend upon changing conditions of the surrounding cytoplasm. An example of decisive cytoplasmic influence on the state of the nucleus is the observation by Karl Sax on the pollen grains of *Tradescantia* (28). Here, in the first pollen-grain mitosis, the orientation of the mitotic spindle places the daughter nuclei in different regions of the cytoplasm, and it is this difference that causes one daughter nucleus to be vegetative and the other generative. Similar examples of nuclear differentiation have been more recently observed in protozoa.

Biochemical studies of nuclear activity show how rapidly nuclei may respond to cytoplasmic changes (4). This has been shown in a study of protein synthesis in pancreatic acinar cell nuclei during the cycle of enzyme synthesis and secretion. After a prolonged fast, feeding produces intense secretory activity of the acinar cells. Due to discharge of stored digestive enzymes, the cells of the pancreas immediately begin to synthesize protein to replace the secreted material. This synthesis of digestive enzymes occurs in the cytoplasm, and within 30 minutes after feeding there is a marked increase in the uptake of isotopically labeled amino acids by the cytoplasmic proteins (3). At the same time, the cytoplasm influences the nucleus so that there occurs an increase in the amino acid uptake by histones and other proteins of the chromosomes. The response of the nucleus to changing conditions in the cytoplasm can also be demonstrated over longer intervals by first labeling the proteins of the nucleus and then comparing the rates of loss of nuclear isotope in fed and fasted animals. In such isotope retention experiments one finds that the rate of "turnover" of chromosomal proteins is much higher when the animals are fed and protein synthesis in the cytoplasm is thereby accelerated (4).

The activity of chromosomes may be expected to depend upon the internal environment of the nucleus. The intranuclear environment has special characteristics which set it apart from the rest of the cell. An example of this is the role of sodium ions in nuclear processes. It was found recently that isolated thymus nuclei synthesize proteins rapidly only if they are present in a medium containing sodium ions; an equivalent concentration of potassium ions is ineffective (5). This observation suggests that although potassium ions preponderate in the cytoplasm, there may be a relatively high concentration of sodium ions in the nucleus. Direct support for this view is provided by autoradiographs made in 1949 by Abelson and Duryee, in which $Na^{24}$ ions are seen to accumulate in the nuclei of living frog oocytes which are placed in a medium containing $Na^{24}Cl$ (1).

In addition to such characteristics as a high sodium ion concentration, which may be found in all nuclei, are there nuclear components that are restricted to nuclei of special cell types? Differentiation of a cell is recognized by presence in it of proteins not found in other cells. Hemoglobin, myoglobin, and myosin are examples of proteins which identify certain types of cells. Other proteins, enzymes especially, are notably more highly concentrated in certain cells than in others. Are there signs that in a similar way nuclei may be said to be differentiated? The most reliable data on the nuclear content of enzymes and other special proteins come from observations on nuclei prepared in non-aqueous media (2). The isolation procedures employed are modifications of the method introduced by Martin Behrens in 1932 (9). This is a scheme which prevents water-soluble materials from moving between nucleus and cytoplasm during the course of the isolation. Its success depends on a rapid freezing and removal of water from the tissue, which is then ground and fractionated in non-aqueous solvents (e.g. mixtures of cyclohexane and carbon tetrachloride). When the specific gravity of the nucleus differs sufficiently from that of other subcellular components it can be separated from them by centrifugation. This separation is facilitated by selecting a medium of density lower than the nuclear specific gravity and higher than that of possible cytoplasmic contaminants. The nuclei of many tissues were prepared in this way and were shown by chemical, immunological, and enzymatic tests to be essentially free of cytoplasmic contamination. Such nuclear preparations show that some enzymes which "differentiate" a cell type are absent from the cell nucleus, while others are present in the nucleus. In general it may be said that the nucleus itself is differentiated. An example is the presence of hemoglobin in the avian erythrocyte nucleus (30, 2), indeed in the nucleus of an erythroblast (13); myoglobin, on the other hand, is absent from the nucleus of the heart muscle cell.

Nuclear differentiation has also been demonstrated in the superb nuclear transplantation experiments of Briggs and King (24). Nuclear differentiation as it occurs in the course of cell differentiation during development is recognizable in the work of these investigators because under the conditions of their experiments nuclear differentiation is not reversible. It is possible that reversal does not occur because in these experiments the necessary conditions are not realized. This should be said because the basic changes that bring about differentiation in nucleus and cytoplasm are not yet understood. If differentiation under all conditions studied is always found to be an irreversible process, such a summary of experience would lead to implications about the nature of the basic molecular changes involved in differentiation—for example, to implications about changes in the genetic material of the chromosomes. We should, therefore, consider in a broad way what is known of differentiation to see whether there are observations suggesting that under some conditions differentiation is reversible.

The best attested examples of reversal of differentiation in animal cells are those that come from experiments on the eyes of salamanders. There are two such examples: in one (the so-called Wolffian regeneration) removal of the lens causes iris cells to lose some of their differentiated characteristics and then develop into lens cells; in the other, removal of the retina causes retinal pigment cells to lose some of their special characteristics and then to become retinal nerve cells. These examples of what appear to be "dedifferentiation" and "redifferentation" are best described in the experiments of L. S. Stone (31, 32).

Such well-studied cases of what appears to be a reversal of differentiation are rare in animals. There are many examples in plants (16). Investigations of "dedifferentiation" cannot at present be said to have led to very definite conclusions. Enough, however, is known to show that the essential irreversibility of differentiation should be questioned. And this means that it cannot at present be said that differentiation implies changes in the genetic constitution of chromosomes, although there can be little doubt that differentiation is associated with changes in nuclear activity.

Nuclear activity is known to produce substances which have decisive influence on the cytoplasm. One such substance, diphosphopyridine nucleotide (DPN), is a coenzyme. An enzyme required for its synthesis is found only in the nucleus, at least in those cells so far studied (21, 26).

Since the quantity of this enzyme varies considerably in nuclei of different cells, there are probably variations in the quantity of DPN supplied to the cytoplasm. Such variations would cause changes in rates of metabolism and so modify the course of development.

A release of DPN to the cytoplasm represents a relatively non-specific way in which the nucleus can modify the course of cell development. Yet the nucleus exerts characteristically specific effects which determine the nature and activity of the cell, and the structure and composition of its proteins. This controlling role of the nucleus in the formation of cell proteins, now so well attested in the case of the hemoglobins (22, 23), is believed to be mediated by ribonucleic acids. There are several good reasons for this belief. The first is the clear requirement for ribonucleic acid in cytoplasmic protein synthesis (3, 20); the second is the demonstration that the ribonucleic acid of the tobacco mosaic virus determines what type of protein is made in virus infection (17, 18); the third is based on the repeated observation in tracer experiments that the rate of incorporation of precursors into ribonucleic acid of the nucleus far exceeds the corresponding rate for RNA in the cytoplasm. This naturally led to the speculation that all the RNA of the cell is made in the nucleus. What has been shown is that DNA participates in the synthesis of RNA in the nucleus, and that RNA synthesis proceeds most actively in the nucleolus (5, 7). Although there is no conclusive evidence that all the RNA of the cell is made in the nucleus—indeed, kinetic studies (8), and differences in composition (29) argue against too simple an hypothesis—there is good evidence suggesting that nuclear RNA can pass to the cytoplasm. The experiments of Goldstein and Plaut (19), in which nuclei bearing labeled RNA were transferred to unlabeled *Amoeba* and the isotope distribution subsequently determined by the radioautographic technique, show that nuclear RNA can, in some form, pass to the cytoplasm. Another very suggestive observation has recently been made by Edström and Eichner (15), who found a direct proportionality between nucleolar volume and the cell body content of ribonucleic acid in supraoptic neurons. Observations of this sort lead strongly to the conclusion that the specific aspects of development, such as the synthesis of special proteins characteristic of differentiation, are due in large measure to the kinds of RNA (or ribonucleoprotein) that come into the cytoplasm from the nucleus.

Biochemical investigation of how the nucleus acts upon the cytoplasm is now well started. Nuclear activity, we have seen, is much influenced by the state of the cytoplasm. Biochemical investigation of this aspect of nuclear-cytoplasmic interaction, as yet hardly begun, will clarify much that is now obscure in differentiation and development.

## REFERENCES

1. Abelson, P. H. and Duryee, W. R., *Biol. Bull.*, **96**, 205 (1949).
2. Allfrey, V. G., Stern, H., Mirsky, A. E. and Saetren, H., *J. Gen Physiol.*, **35**, 529 (1952).
3. ———, Daly, M. M. and Mirsky, A. E., *J. Gen. Physiol.*, **37**, 157 (1953).
4. ———, ——— and ———, *J. Gen. Physiol.*, **38**, 415 (1955).
5. ———, Mirsky, A. E. and Osawa, S., *Nature*, **176**, 1042 (1955).
6. ———, and ———, *Proc. Natl. Acad. Sci. U. S.*, **43**, 589 (1957).
7. ———, and ———, *Proc. Natl. Acad. Sci. U. S.*, **43**, 821 (1957).
8. Barnum, C. P., Huseby, R. A. and Vermund, H., *Cancer Research*, **13**, 880 (1953).
9. Behrens, M., *Hoppe-Seyler's Z. physiol. Chem.*, **209**, 59 (1932).
10. Boivin, A. R., Vendrely, R. and Vendrely, C., *Compt. rend. acad. sci.*, **226**, 1061 (1948).
11. Crampton, C. E., Stein, W. H. and Moore, S., *J. Biol. Chem.*, **225**, 363 (1957).
12. Davison, P. F., *Biochem. J. (London)*, **66**, 703 (1957).
13. de Carvalho, S. and Wilkins, M. F. H., *Proc. 4th Intern. Congress Intern. Soc. Hematol.*, p. 119-120, Grune & Stratton, N. Y. (1954).
14. Driesch, H., *Analytische Theorie der organischen Entwicklung*, Leipzig (1894).
15. Edström, J. E. and Eichner, D., *Nature*, **181**, 619 (1958).
16. Esau, K., *Plant Anatomy*, John Wiley and Sons, N. Y. (1953).
17. Fraenkel-Conrat, H., *J. Am. Chem. Soc.*, **78**, 882 (1956).
18. Gierer, A. and Schramm, G., *Z. Naturforsch., Pt. b*, **11**, 138 (1956).
19. Goldstein, L. and Plaut, W., *Proc. Natl. Acad. Sci. U. S.*, **41**, 874 (1955).
20. Hoagland, M. B., Stephenson, M. L., Scott, J. F., Hecht, L. I. and Zamecnik, P. C., *J. Biol. Chem.*, **231**, 241 (1957).
21. Hogeboom, G. H. and Schneider, W. C., *J. Biol. Chem.*, **197**, 611 (1952).
22. Hunt, J. A. and Ingram, V. M., *Nature*, **181**, 1062 (1958).
23. Ingram, V. M., *Nature*, **180**, 326 (1957).
24. King, T. J. and Briggs, R., *Cold Spring Harbor Symposia Quant. Biol.*, **21**, 271 (1956).
25. Mirsky, A. E. and Ris, H., *Nature*, **163**, 666 (1949).
26. Morton, K. C., *Nature*, **181**, 540 (1958).
27. Osawa, S., Allfrey, V. G. and Mirsky, A. E., *J. Gen. Physiol.*, **40**, 491 (1957).
28. Sax, K., *J. Arnold Arbor.*, **16**, 301 (1935).
29. Smellie, R. M. S., McIndoe, W. M. and Davidson, J. N., *Biochim. et Biophys. Acta*, **11**, 559 (1953).
30. Stern, H., Allfrey, V. G., Mirsky, A. E. and Saetren, H., *J. Gen. Physiol.*, **35**, 559 (1952).
31. Stone, L. S., *Anat. Rec.*, **106**, 89 (1950).
32. ———, and Steinitz, H., *J. Exptl. Zool.*, **124**, 435 (1953).

## DISCUSSION

DR. SPIEGELMAN: Dr. Mirsky, how specific is the restoration of nuclear RNA synthesis by DNA? Will any type of DNA do it?

DR. MIRSKY: Yes, DNA from many sources will do it. DNA from many sources and also synthetic polynucleotides are effective; however, dinucleotides are ineffective. The question of specificity here is rather hard to answer. We do not know whether the same products are formed when a polynucleotide is substituted for the original DNA.

DR. MARKERT: Dr. Mirsky, you spoke of the constancy of the DNA and histones in the nucleus. I wonder if you would care to extend your remarks to include the residual proteins and RNA?

DR. MIRSKY: The amount of RNA varies from one tissue to another, and it also varies in a single tissue under different physiological conditions. This variability is an interesting phenomenon which we, and of course many others, are studying. The subject is too broad and big to treat exhaustively in this discussion. An interesting aspect of this question is the relative significance of nucleolar and other nuclear RNA. As for the "residual protein," and by that I mean non-histone protein of the chromosomes, it is difficult to distinguish between "nuclear" and "chromosomal" protein. In the large nuclei of oocytes the distinction may be clear, but in nuclei of lymphocytes it is neither semantically nor physically clear what we mean by this distinction. In our system, once sucrose is removed from the medium, some proteins are immediately leached out of the nucleus. This protein is being actively synthesized and its synthesis depends upon the presence of DNA. It might therefore be associated with DNA and we are tempted at the moment to consider it to be part of the chromosome.

DR. HADORN: Dr. Mirsky, do you feel that all of these functional activities are directed by and under the control of so-called Mendelian genes?

DR. MIRSKY: Well, Dr. Hadorn, that is an extremely interesting question and I am not sure that I can answer it. I think that much of what happens during the interphase period of a cell's life is of nucleolar origin. For example the synthesis of DPN is a nuclear function, and yet it was found in Dr. Brachet's laboratory that most of the DPN-synthesizing enzyme is in the nucleolus. It is very difficult for us to answer your question because much of the biochemical activity of the interphase nucleus is located in the nucleolus; and so the answer to your question depends upon how the Mendelian genes function in formation of the nucleolus.

DR. LEHMANN: There are some indications that RNA-proteins inside and outside the nucleus are associated with saturated and especially with unsaturated lipids. Considerable changes of the level of different lipids may occur, when the physiological state of a tissue is changing for example from a well fed to a starving state. Do you know any facts about association of lipids with RNA-proteins and their change in different physiological states?

DR. MIRSKY: Yes, lipids are surely important. The lipids in all of these nuclear fractions should be investigated. You know the biochemist keeps something and throws away something else. We usually discard the lipid fraction and I agree that this in unfortunate.

DR. P. COHEN: You refer to the fact that the arginase of the liver nucleus is different from the arginase in the kidney nucleus. Do you have any experimental evidence that might support the idea that the arginase of the nucleus differs from that of the cytoplasm? I think there is some reason to believe that they might be different.

DR. MIRSKY: We have no information on that. I might say, however, that the hemoglobins found in the nucleus and cytoplasm seem to be the same. It would be interesting to know whether this is true of nuclear and cytoplasmic arginase.

DR. MCELROY: When you say that DNA is a cofactor for ATP synthesis by the nucleus, what do you mean?

DR. ALLFREY: Well, Dr. McElroy, we have used the term "cofactor" in a very broad sense. Experimentally, one finds that under the proper conditions a thymus nucleus will make ATP, and that if you remove the DNA from the nucleus it can no longer phosphorylate its AMP to make ATP. Now, if to a DNAase-treated nucleus, you restore thymus DNA or other DNAs, RNAs, or even polyadenylic acid, ATP synthesis is restored. All of the polynucleotides we have tested have this capacity to somehow serve as "cofactors" for ATP synthesis. Mononucleotides, however, and adenylic acid dinucleotides, do not have this capacity, so we can say that there is a size restriction on polynucleotide structure which somehow determines their capacity to serve in ATP synthesis by the nucleus.

DR. MCELROY: Is this a catalytic function or could it be serving as a substrate for ATP synthesis?

DR. ALLFREY: Well, we cannot eliminate the possibility that the DNA might be a substrate but the reason we feel that it has a catalytic function is because ATP synthesis in the nucleus is aerobic. Now if DNA or other polynucleotides were breaking down in some way, supplying through their hydrolysis or phosphorolysis the energy needed for ATP synthesis, one would expect this to also go on in a nitrogen atmosphere. But one actually finds that there is an absolute requirement for oxygen.

DR. MIRSKY: DNA is acting as a cofactor for ATP synthesis in the nucleus. From a biological point of view it should be emphasized that this is an important function of DNA.

DR. SWIFT: Since you raised the question of the constancy of DNA, I wonder if you would care to comment on the localized increase in DNA found in salivary gland chromosome puffs by Pavan and Ficq, and also by Rudkin and Stich. Also, I wonder how you feel about the decrease in DNA after cold treatment found in specific chromosome regions of *Trillium* and some other plants by LaCour and others.

DR. MIRSKY: I think the work of LaCour is in one category and that Pavan's is in an entirely different one. At this time there are so many examples of the constancy of DNA that apparent exceptions should be examined critically. In the experiments of LaCour and his colleagues the DNA per cell, as measured by the Feulgen procedure, diminishes in plants during cold treatment and then returns to its original value when the temperature is restored. From the present experiments it cannot be said whether the DNA actually decreases or whether some of it is depolymerized sufficiently to be lost in the Feulgen procedure. If DNA is lost, then new DNA is synthesized when the temperature is restored. This could be recognized by uptake of labeled thymidine.

As for the work of Pavan and his coworkers, I think it is interesting in many respects. Professor Pavan showed me his preparations several years ago. It was clear that Feulgen staining showed marked increases in DNA in some bands of the chromosomes and not in others. In a paper published soon thereafter I remarked that this was an exception to the rule of DNA constancy. Pavan thought at first that polytenization had stopped in these chromosomes. In fact the recent experiments by Ficq and Pavan show that polytenization is going on very rapidly. We have here a cell that is very highly polytenized, which is continuing to increase in polyteny and which will soon die. Such a cell may well afford to get out of "genic balance." The point I would like to make is that this may be an exceptional situation in many ways, so that it would not be surprising if an empirical rule like DNA constancy does not apply.

Dr. Brachet: I do not feel that the results obtained by Pavan could be entirely explained on the basis of polytenization. There is a differential activity among the bands and obviously the ones which become very active are those which behave somewhat like a nucleolar organizer.

Dr. Mirsky: I quite agree. Even so, it is likely that continuing polytenization is a factor in these cells.

# CHROMOSOMAL DIFFERENTIATION *

JOSEPH G. GALL

*Department of Zoology, University of Minnesota*

## INTRODUCTION

The rediscovery of Mendel's laws and the proof of the chromosome theory of inheritance in the early decades of this century brought a resurgence of interest in the fine structure of chromosomes. As early as 1905, C. E. McClung (42) expressed the hope "that it will be possible to establish the relations that exist between body characters and individual chromosomes." He was thinking not only of what we call today genetic mapping, but also of a more direct relation whereby morphological features of particular chromosomes might be related to the expression of certain genes. The remarkable studies of this period quickly laid the foundation for modern views of chromosome structure. Still, however, doubt grew that the ultimate gene differences would ever be directly visualized. The outlook vastly improved when attention shifted from typical mitotic chromosomes to the giant structures found in the larval tissues of certain Diptera, particularly to the salivary gland chromosomes of *Drosophila*. Rapid progress resulted in the localization of known genes within very short segments of the chromosomes and in the demonstration of various structural changes which occur naturally or result from radiation. The culmination of the cytogenetic studies on *Drosophila* came with the detailed studies of the late 1930's which correlated the well-known genetic maps of *Drosophila* with the newly prepared cytological maps.

Throughout this period, however, emphasis remained on the genetic aspects of chromosomes, and practically no attention was given to alterations of genetically identical chromosomes which might accompany diverse functional states. Perhaps it was felt that any functioning of the genes would be expressed only as the release of invisible molecules. But in retrospect we can also see a more subtle restraint which limited the

---

* Some of the studies reported here were supported by funds from the National Science Foundation, the National Institutes of Health of the U.S.P.H.S., and the Graduate School of the University of Minnesota.

chances of discovering physiological activity in chromosomes. This restraint was simply the types of material studied. For obvious practical reasons chromosomes were observed in their most "condensed" state, usually during the meiotic or mitotic divisions. And the giant chromosomes, furnishing a powerful tool for the geneticist, were popularly studied in the salivary glands of mature *Drosophila* larvae. It now seems that chromosomes in division are in a metabolically inert state, that we must seek them in their most "diffuse" states to see them actively functioning. In the case of the giant banded chromosomes of the Diptera, some do, in fact, show remarkable changes when one looks at them in different tissues or in different periods during larval development. A few papers appearing in the 1930's did lay stress on morphological changes in chromosomes associated with their physiology. For instance, Koltzoff (40) in Russia and Duryee (24) here in the United States emphasized the metabolic aspects of the peculiar lampbrush chromosomes of the vertebrate oocyte. And Poulson and Metz (57), studying the "puffs" and "Balbiani rings" in the salivary gland chromosomes of *Sciara* and *Chironomus,* concluded that these structures might reflect the physiological state of the cell.

Recent studies on these two materials, the vertebrate oocyte chromosomes and the giant larval chromosomes of the Diptera, permit us, I believe, to make tentative suggestions about the differences between "active" and "inactive" chromosomes or chromosome regions. Both materials, at first sight so strikingly different, point to rather similar conclusions, and so it may be that these conclusions can be generalized to other cases.

The general argument which I shall present here concerns both structural and functional aspects of these chromosomes, although in a certain sense these two must be considered together. From the structural standpoint I shall give reasons for believing that the chromatid of classical cytology is at most only a few hundred Ångstroms in width, but millimeters or even centimeters in length in some organisms. The chromatid probably contains deoxyribose nucleic acid (DNA) and basic protein throughout its length. It is capable of complex coiling or folding so that in its contracted state it is visible in the light microscope, either as a whole chromosome or as part of a chromosome. The expanded or unwound state is correlated with synthetic activity, both in terms of reduplication and the elaboration of ribonucleoprotein products. These latter products may accumulate to such an extent around the delicate DNA-protein strand that the whole structure is visible in the light microscope. The unraveling process may involve much or all of the chromosome at once, or it may be limited to very specific regions. It is essentially reversible and correlated with non-genetic changes in the physiological activity of specific chromosome loci. The latter feature permits us to speculate about the role of the chromosomes in embryonic differentiation, and it is at this point that the discussion of chromosomal differentiation ties in most closely with the topic of the present Symposium.

It should be clear that the scheme outlined here contains much that is common knowledge, while other aspects are quite debatable and may prove to be ultimately untenable. I feel, however, that there is a certain advantage in developing a particular point of view rather than in attempting a general summary of chromosome cytology. Since information on the finest differentiations of chromosomes has come largely from studies on dipteran larval tissues and the oocytes of amphibians, I shall confine most of my discussion to these two tissues.

### Definitions

Discussions of chromosome structure are frequently confused by the variety of terms used and the different shades of meaning attached to them. In the recent literature we find, in addition to many others, the words chromonema, genonema, microfibril, chromofibril, and chromatid, as well as half-chromatid, quarter-chromatid, etc. At times these words carry some implication about chemical constitution; at others they are simply morphological. Since any definition must, of necessity, change with increasing knowledge, the important point perhaps is to be consistent and to delimit the implications carefully.

Throughout this discussion, I shall use the terms *fiber* or *strand* in a simple descriptive sense to refer to something seen either in the conventional or electron microscope. The statement, "a fiber connects two granules" means only that at the resolution employed one sees a single line between the granules. The term *chromonema* is sometimes used in this noncommittal fashion, but often it carries genetic or chemical overtones.

On the other hand, I shall use the term *chromatid* with distinct theoretical implications. In the definition I shall assume that the theory of DNA "constancy" is essentially correct as applied to individual chromosomes (12, 46, 64); that is, that any particular chromosome contains a certain amount of DNA or some integer multiple of this amount (the multiple usually some power of 2). A *chromatid* will then be a chromosome or longitudinal fraction thereof which contains the same amount of DNA as an entire anaphase chromosome from the second meiotic division. Thus each prophase chromosome in a typical diploid cell will consist of two chromatids, the anaphase chromosome will contain one chromatid, and the bivalent chromosomes of meiotic prophase will be made up of four chromatids in two sets of two. The rationale of this definition is simply that the anaphase chromosome of the second meiotic division seems to contain the least common multiple of DNA.

In practice the above definition corresponds to generally accepted usage. It does exclude, however, the simple morphological definition of a chromatid as a "half chromosome." Some authors (e.g., 48) have in the past referred to half an anaphase chromosome as a chromatid. We now know, however, that chromosome duplication, or at least DNA doubling, occurs during interphase (55, 64). Since there is thus no likely structural difference between a chromatid of prophase and an anaphase daughter chromosome, it seems advisable to refer to both by the same term. A definition based on DNA content has the further advantage, at least in theory, of providing an objective measure of the number of chromatids present at any time. Finally, of course, the chromatid is the unit of gene segregation and crossing over.

### Polytene Chromosomes

The giant chromosomes found in dipteran larval tissues need no introduction here, as they are one of the best-known cytological objects. Their structure has been subject to much debate since their "rediscovery" in 1933 by Heitz and Bauer (35) and by Painter (49). Actually they were quite accurately described as early as 1881 by Balbiani (5); and a series of papers from that time to the 1930's attests to a continued, albeit minor interest in their structure. Several recent reviews (2, 7, 8) make it unnecessary for me to discuss in detail the numerous papers dealing with these chromosomes. Here I shall take up briefly some of the evidence for their multistranded or polytene condition and shall discuss in more detail the recent interesting studies on the "puffs" and "Balbiani rings" which characterize these chromosomes.

*Structure*

The polytene chromosomes are of course tremendously longer and thicker than the somatic chromosomes of the same organism. In *Drosophila,* for instance, the total length of the chromosome set is about 7.5 $\mu$ in the spermatogonial metaphase, but 1180 $\mu$ in the salivary gland (15). One of the first suggestions made to account for their great size was that they consist of a bundle of ordinary chromosomes more or less completely stretched out and lined up accurately side by side (6, 39, 50). A modern and well-documented statement of this polytene or multistrand theory is found in the recent papers of Bauer and Beermann (7, 8). Briefly, the giant chromosome is supposed to consist of a bundle of hundreds or thousands of fine fibrils accurately aligned with one another along the major axis of the chromosome. A polytene chromosome can be pictured as a typical mitotic chromosome which has both uncoiled and undergone repeated duplication. The bands which cut transversely across the whole chromosome result from the lateral apposition of bead-like enlargements (chromomeres) on the constituent fibrils. Evidence favoring the polytene theory comes from a number of sources, only a few of which will be considered here. In many instances direct visual observation of fixed and stained chromosomes shows evidence of longitudinal fibrillation; fig. 17 in the paper of Beermann (8) shows this to particular advantage. Long ago, however, it was pointed out, particularly by Metz (45), that such fibrils visible in the conventional microscope are too large to correspond to single stretched-out chromatids of a normal chromosome. Fortunately, however, we do not need to rely on this kind of evidence alone, which is at best open to several interpretations. In his careful study of *Chironomus,* Beermann (8) has shown that certain restricted regions of a polytene chromosome may break up into smaller and smaller bundles of fibers, thus affording the most direct possible evidence of their multistranded condition (Figs. 1-3). More indirect evidence comes from a consideration of the nuclear volume and DNA changes which accompany the development of the polytene condition. Hertwig (36) pointed out that the various-sized nuclei in the larval organs of *Drosophila* fall into a rough geometrical series, the largest nuclei having about 512 times the volume of the smallest. More recently, Swift and Rasch (67) found an even more precise geometrical series by estimating the DNA content spectrophotometrically from the intensity of the Feulgen reaction (Fig. 4). These data can be interpreted as reflecting a series of nine to ten "rounds" of chromosome duplication and suggest that the largest salivary gland chromosomes of *Drosophila* contain between 512 and 1024 chromatids. Finally, an examination of many sorts of insect tissues suggests that polyteny may be simply a special case of a more general phenomenon of chromosome duplication without accompanying mitosis (endopolyploidy). Many insect nuclei show very high amounts of DNA (64) and in some cases it is possible to count an exceedingly large number of individual chromosomes, as shown by Geitler (33) in his study of endopolyploidy. In other instances, the reduplicated chromosomes may remain together in loose bundles (51). In view of the wide occurrence of such phenomena in insects generally, it seems likely that a polytene chromosome is an extreme situation where the strands

108 THE CHEMICAL BASIS OF DEVELOPMENT

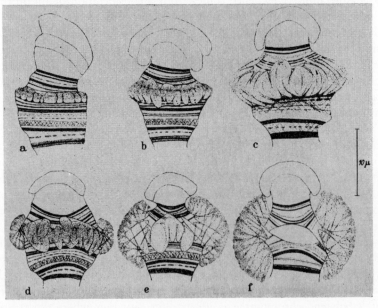

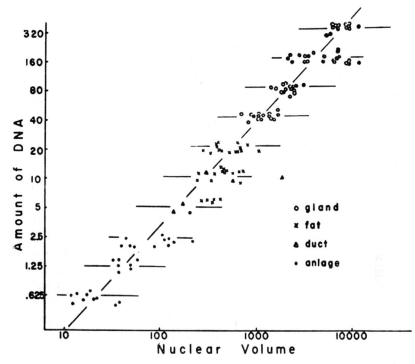

Fig. 4. DNA content as a function of nuclear volume in cells of larval *Drosophila melanogaster*. Estimated spectrophotometrically from the intensity of the Feulgen reaction. The values fall into a regular geometric series, with the largest gland nuclei containing 512 times as much DNA as the smallest anlage nuclei. Data from Swift and Rasch (67).

formed by repeated chromosome duplication remain intimately associated. White (71) has even described a case where a polyploid number of polytene chromosomes are found in a nucleus. Presumably the strands from earlier duplications separated, but those formed later remained together in bundles. These examples by no means exhaust the evidence for polyteny, but indicate some of the strongest arguments in its favor.

Several recent studies of the polytene chromosomes have used the increased resolution of the electron microscope. In some cases surface repli-

Fig. 1. Chromosome 4 from the salivary gland of *Chironomus tentans*.

Fig. 2. The same chromosome showing two typical Balbiani rings and a prominent puff. Note particularly that the banding pattern can be followed into the smaller subdivisions of the chromosome; the origin of the rings and puff can be traced to single bands in the unexpanded chromosome (*BR1* and *BR2* in Fig. 1).

Fig. 3. Different degrees of expansion of the same chromosome locus, varying from a small puff in *a* to a fully formed Balbiani ring in *f*. Note the constancy of the banding pattern in the subdivisions of the chromosome. Figs. 1, 2, 3 from W. Beermann (8).

cas have been made (73, 53), while in others, conventional osmium-fixed, thin-sectioned material (11, 32, 37) has been used. In most of these studies some indication of longitudinal fibrillation is evident, although it is very difficult to follow individual fibers for any great length, especially in the sectioned material. The organization of fibers in these chromosomes is not nearly so striking or regular as in such fibrous materials as muscle, collagen, and sperm tails. Indeed, the continuity of individual fibers over any but the shortest lengths (fractions of a micron) is largely a supposition in so far as the electron microscope pictures are concerned. Most authors describe the finest fibers as a few hundred Ångstroms in diameter, the range of reported values being from about 100 Å to 500 Å.

There are, of course, a very large number of such minute strands in the giant chromosomes. On the basis of their electron microscopic observations on *Chironomus* chromosomes, Beermann and Bahr (11) concluded that they "may contain more than 10,000 longitudinal elements." This estimate was made by counting the number of strands in a thin section and extrapolating to include the whole chromosome. A similar degree of polyteny is suggested by the fact that the largest salivary gland nuclei have about 16,000 times the volume of embryonic nuclei of the same organism (diameters in the ratio of about 100 $\mu$ to 4 $\mu$). Finally, direct measurements showed that the volume of a giant salivary gland chromosome is about 16,000 times that of its corresponding metaphase pairs (8).

Gay (32) has also estimated the number of constituent strands in polytene chromosomes, in this case from *Drosophila melanogaster*. Taking 500 Å as the diameter of the unit strand, she calculates that the salivary gland chromosome may contain between 1000 and 2000 strands. Her estimate compares nicely with the volume and DNA measurements of Swift and Rasch (Fig. 4) on nuclei of the same fly. As already mentioned, these workers found that the largest gland nuclei contain 512 times the amount of DNA in the smallest anlage nuclei (presumably 1024 times the haploid amount).

One point should be stressed about polyteny as related to the general question of multistrandedness of chromosomes. Both Beermann and Gay find that the total number of strands seen in the electron microscope approximates closely the number of times the volume and/or DNA content of the diploid nucleus has increased. If this numerical relationship were strictly true, then of course we could say that each smallest fibril corresponds to an extended chromatid of a mitotic chromosome. I do not wish to "force" these data, since the estimates of strand number could easily be in error by a factor of two or more in either direction. Nevertheless, taken at face value, the evidence does not necessarily point to a high degree of multiplicity in the chromatid itself. I shall return to this problem when discussing the lampbrush chromosomes.

*Functional Aspects*

The polytene chromosomes of many species show local variations in the clarity of the banding, some bands being discrete and sharply bounded, others appearing as diffuse puffs. These modifications have recently been the object of a thorough investigation by Bauer, Beermann, and Mechelke (7-11, 44) and have also been studied independently by Breuer and Pavan (14, 54). Much evidence points to the conclusion that specific modification of the bands occurs in different tissues of an organism and at different times in the life history, and that furthermore these modifications are reversible and are probably related to the functional activity of the cell.

Beermann was the first to show that puffing is not an invariant characteristic of particular bands, but rather that many bands may be either sharp or puffed depending upon the tissue examined. As an example, we may cite a case from *Chironomus tentans* (Fig. 5), where a short segment of chromosome 3 was studied in detail in four different tissues. In these tissues—salivary gland, Malphighian tubules, rectum, and midgut—attention was paid to six specific bands. In each tissue, there was a characteristic pattern of puffing such that any particular locus might be puffed in one,

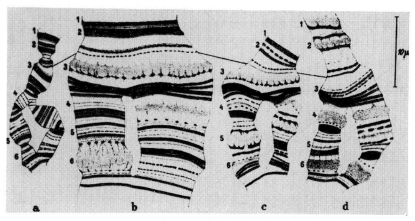

Fig. 5. Drawings of a short region of chromosome 3 in four different tissues of the same individual (*Chironomus tentans*): a. midgut, b. salivary gland, c. Malpighian tubule, d. rectum. Note the bands numbered 1-6, which are distinct in some of the tissues, puffed in others. From W. Beermann (8).

two, or three of the four tissues (puffing at these loci did not occur in the midgut). Beermann's conclusions are as follows: "Diffuse swelling, or 'puffing' of the chromosome is a phenomenon arising from single bands or interband spaces. This, along with the finding that adjacent loci may puff independently, constitutes the first direct morphological demonstration of the fact that giant chromosomes consist of linearly arranged, independently reactive genetic units of the order of magnitude of the bands, in other words, that bands or less probably interbands are gene loci."

Puffs on a vastly more spectacular scale, but still possessing certain basic similarities to those just described, have been known for many years as "Balbiani rings," in reference to the fact that they were seen and described by the first investigator of polytene chromosomes. The typical Balbiani ring appears at first glance to be an annular mass of material surrounding an otherwise normal chromosome. Closer analysis shows, however, that in the region of the Balbiani ring the chromosome actually breaks up into a large number of fine strands which extend out laterally into a more diffuse area (see Figs. 1, 2, 3). Certain features suggest that these rings may differ in their functional significance from the more typical puffs. So far, they have been found only in the salivary gland cells, where there may be from one to three per chromosome set. Beermann has assumed that they are related to the very high synthetic activity of the larval salivary gland, perhaps supplying some sort of precursor needed in the production of the salivary secretion. This view is very much strengthened by observations on *Trichocladius* and *Acricotopus,* each of which has two distinct parts to the salivary gland. In *Acricotopus,* which was studied by Mechelke (44), one lobe of the salivary gland produces a brown secretion at the beginning of metamorphosis, while the remainder of the gland, which differs also in other respects, does not contain the secretion. In the brown lobe are found two typical Balbiani rings and a very large puff; in the other cells, only two Balbiani rings. Most significantly, however, the loci showing these structural modifications are quite different in the two cell types. The situation in *Trichocladius* (9) is quite comparable, the salivary gland in this case possessing six very large specialized cells with their own pattern of Balbiani ring formation. Thus, in both forms specific and prominent chromosomal differentiations are correlated with cellular modifications.

The periphery of the Balbiani ring contains Feulgen-negative materials, sometimes in the form of droplets, which seem to be the actual product or products of this locus. Beermann (8) showed that a short exposure of *Chironomus* larvae to a temperature of 5°C, followed by a return to room temperature, caused a marked accumulation of droplets around these regions of the chromosome. He interpreted this to mean that cold affects the rate of dispersal of the product more than its rate of formation. In any event, the situation shows clearly that something is produced by the locus at a fairly high rate. In their study of *Rhynchosciara*, Breuer and Pavan (14) noted that the region of large puffs is surrounded by material which stains with the pyronin component of Unna's pyronin methyl green mixture. That much of this material contains ribose nucleic acid (RNA) can be shown quite simply by enzymatic digestion used in conjunction with basic-dye staining (65). Figs. 6 and 7 show the Balbiani ring on the small nucleolar chromosome of *Chironomus* sp. In Fig. 7 we see the distribution of basic dye, azure B used at $pH$ 4 (27), after treatment of the sections with RNAase. The staining is limited to parts containing DNA. In Fig. 6, the same section observed with the phase contrast microscope, we see the unstained nucleolus (large mass) and the Balbiani ring. Both these latter structures are intensely basophilic in control sections which were not treated with the enzyme.

Using the electron microscope Beermann and Bahr (11) have demonstrated the presence of numerous fine granules, about 300 Å diameter, in a Balbiani ring region of *Chironomus*. Similar granules, although usually somewhat smaller, are, of course, well known as a component of the microsome fraction of homogenized cells (52) and are presumably identical with the small granules associated with the lamellae of the ergastoplasm or endoplasmic reticulum. Because of their RNA content, these granules are responsible for much of the usual cytoplasmic basophilia. Similar granules are a frequent component of nucleoli, which likewise are known to contain RNA. It now seems possible that the RNA of the chromosomes, like that of the nucleoli and the cytoplasm, is organized into small granules.

That chromosomes contain RNA has been known for some time on the basis of staining characteristics and chemical analysis of isolated material (38, 47). More recently, several studies have shown that salivary gland chromosomes incorporate adenine-C-14 and P-32 into their RNA. Pelc and Howard (56) showed differential uptake of adenine-C-14 by certain bands in the salivary gland chromosomes of *Drosophila*. Control slides involving enzymatic digestion with RNAase prior to making the autoradiographs proved that the incorporation was acutally into RNA and was not simply a reflection of DNA synthesis associated with an increasing degree of polyteny. These same chromosomes showed differential uptake of methionine-S-35 into certain bands, although no attempt was made to see if the same bands simultaneously incorporated both adenine and methionine.

Taylor et al. (68, 69), using autoradiographic techniques, have carried out an extensive and careful set of studies on the RNA metabolism of

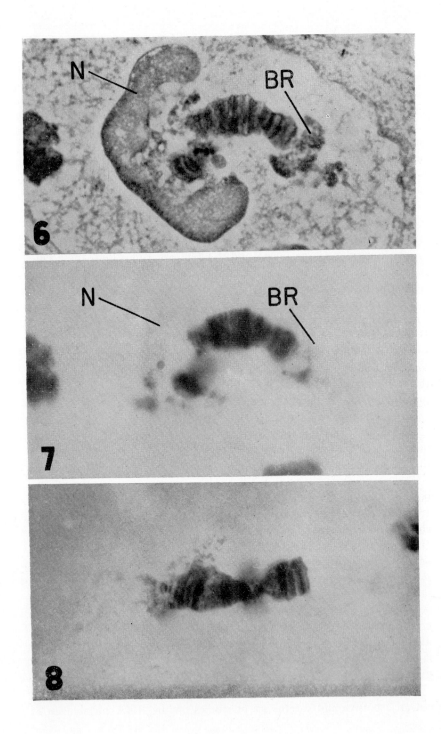

*Drosophila* salivary gland cells. By counting silver grains in the exposed emulsion these authors have been able to get quantitative information of the rates of incorporation of P-32 into several RNA fractions. In general, they find that the nucleolus initially incorporates P-32 at a higher rate than any other part of the cell. If larvae which have fed for a short time on P-32 are transferred to non-radioactive food, their nuclei gradually lose radioactivity, while at the same time P-32 appears in the cytoplasmic RNA. The rates of incorporation are consistent with the idea that nucleolar RNA is a precursor of cytoplasmic RNA, although these two fractions could have a common precursor.

The most recent study of McMaster-Kaye and Taylor (43) gives further quantitative data on rates of P-32 incorporation into chromosomes, nucleoli, and cytoplasm and suggests that each of these fractions has its own pattern of incorporation.

Taken together, the evidence indicates that the chromosomes contain RNA in various bands and particularly in the Balbiani ring regions. At least some of this RNA is metabolically active, and it seems safe to assume that it is being synthesized in situ. It may well be that the individual gene products are ribonucleoprotein in nature, although we are a long way from demonstrating locus specificity of these products. It hardly seems necessary to point out here the many lines of evidence which implicate ribonucleoproteins as intermediates between the DNA of the genes and the active sites of protein synthesis in the cytoplasm (13, 21).

## LAMPBRUSH OR LATERAL LOOP CHROMOSOMES

The giant meiotic bivalents found in the oocytes of many animals have been named "lampbrush" or "lateral loop" chromosomes in reference to the peculiar looped projections which extend laterally from their main axis. Their great size poses certain problems of observation, as recognized long ago by Rückert (63), who studied them in whole isolated nuclei. Duryee (24, 25, 26) was the first to urge observation of the unfixed chromosomes, and he showed the ease with which these giant structures may be isolated manually from the oocyte nucleus. Under phase contrast the

---

Fig. 6. Salivary gland chromosome of *Chironomus*, phase contrast photograph of sectioned material to show the nucleolus (N) and a prominent Balbiani ring (BR). 1500 ×.

Fig. 7. The same preparation by conventional illumination. Stained with azure B at *p*H 4 after treatment with RNAase. Only the DNA of the bands stains. In control slides not treated with the enzyme the nucleolus and Balbiani ring are strongly basophilic due to their RNA content. 1500 ×.

Fig. 8. Aceto-orcein squash of the same chromosome from another individual. Note how the chromosome breaks up into smaller fibers in the Balbiani ring and nucleolus regions. Cf. Figs. 1, 2, 3. 950 ×.

freshly isolated chromosomes are a beautiful sight indeed, and clearly show points of fine structure that can hardly be guessed from fixed or sectioned material. As an aid to observation, an inverted optical system is almost a necessity, since with this one may observe the chromosomes as they rest freely on the surface of a coverslip (17, 29, 72).

The main structural features, as observed in the light microscope, are now quite clear and may be summarized briefly (16-20, 23-26, 28-31, 34, 72; cf. Figs. 9 and 10). These chromosomes are the longest known, ranging

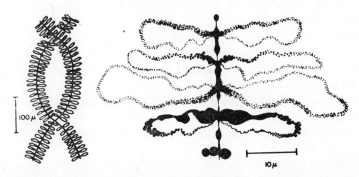

Fig. 9. *Left,* a very diagrammatic representation of a lampbrush chromosome showing the homologues held together at two chiasmata. Hundreds of loop pairs extend laterally from the axis of each homologue. *Right,* a somewhat schematic view of a lampbrush chromosome to show the central axis of chromomeres connected together by a very delicate strand (usually invisible in the light microscope). Loop pairs of differing morphology project laterally from the axis. The loops are more closely spaced on the actual chromosome.

in some species up to a millimeter or more in length. They are actually found in a variety of oocytes, including both vertebrates and invertebrates (18), although they reach their most spectacular dimensions in the salamander. The lampbrush condition is found during the meiotic prophase, so we are in fact discussing meiotic bivalents. When I speak of "chromosome" I shall usually be referring to one homologue of the pair only. The central axis of each homologue consists of a single row of granules (chromomeres in the descriptive sense) connected together by a strand of such fine dimensions that it is generally invisible in the light microscope. The chromomeres themselves are a few micra in length, but generally only 0.5 to 2.0 $\mu$ in diameter. Attached laterally to the chromomeres are *pairs* of loops (34) which display an amazing variety of form from locus to locus. The most common loop is an indistinctly fuzzy or granular strand, some 10 to 50 or more micra in total length, frequently showing a thicker and a thinner end at the points of attachment to the chromomere. Some-

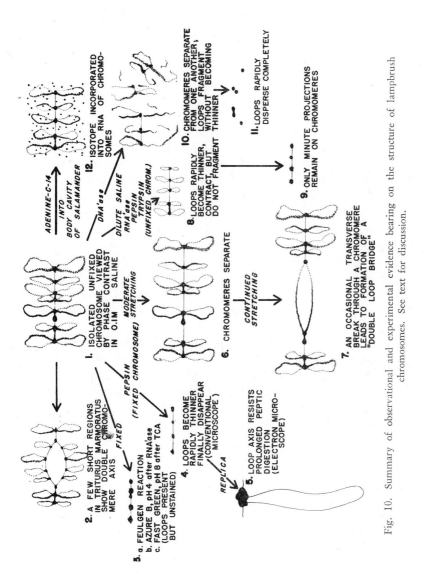

Fig. 10. Summary of observational and experimental evidence bearing on the structure of lampbrush chromosomes. See text for discussion.

times a single pair of loops arises from one chromomere, although particularly in the later stages of oocyte development, when the chromomeres are fewer in number and larger, more than one pair may arise from the same chromomere. The chromomeres were long claimed to be "optical illusions," but they actually differ from the loops in a number of important respects. Thus, the chromomeres are Feulgen-positive, stain by the fast green technique of Alfert and Geschwind (3), and are more or less resistant in the unfixed state to dilute saline, pepsin, trypsin, and RNAase; whereas the loops are negative to the Feulgen reaction and the fast green stain, contain demonstrable amounts of RNA, and are readily dissolved or digested by dilute saline and the enzymes listed.

Of the greatest importance in understanding the structure of the lampbrush chromosomes is the very delicate strand which connects successive chromomeres. I should like to distinguish clearly the evidence that such a strand must exist and must have dimensions generally speaking below the resolving limit of the conventional microscope, and the evidence concerning its actual dimensions, multiplicity, etc., as viewed in the electron microscope. These are of course intimately related questions, but lack of agreement about the fine structure should not obscure the fact that all recent workers, with the exception of Lafontaine and Ris (41), concur that the connections between successive chromomeres are essentially submicroscopic.

When one observes freshly isolated lampbrush chromosomes in a saline solution, the numerous loops tend to obscure the details of the axial structure. However, under three conditions the chromomeres and the regions between them are particularly clear: in stretched regions of the chromosomes, in chromosomes from older oocytes where the loops are much reduced, and in chromosomes isolated in very dilute saline where the loops dissolve. Under these conditions one has no trouble noting that successive chromomeres are separated by a gap which is not transversed by any visible structure (Fig. 11). Such chromosomes are free in the isolating medium

---

Fig. 11. A short segment of a slightly stretched lampbrush chromosome. Note the spaces between chromomeres where the continuity of the axis seems interrupted. Phase contrast view of material fixed in formaldehyde vapor. 1900 ×.

Fig. 12. Low power electron micrograph to show the fine fiber which connects successive chromomeres. The bulk of the loop material has been digested with pepsin to reveal the chromomere axis to advantage. Chromium-shadowed. 5000 ×.

Fig. 13. Higher power view of the interchromomeric fiber from a pepsin-treated chromosome. The fiber diameter is not uniform but ranges from 200-400 Å, with an occasional region of somewhat larger dimensions. Chromium-shadowed. 33,000 ×.

Fig. 14. Similar to Fig. 13, but material dried according to Anderson's (4) critical point method. The fiber diameter here seems more uniform and ranges from 300-400 Å. Chromium-shadowed. 35,000 ×.

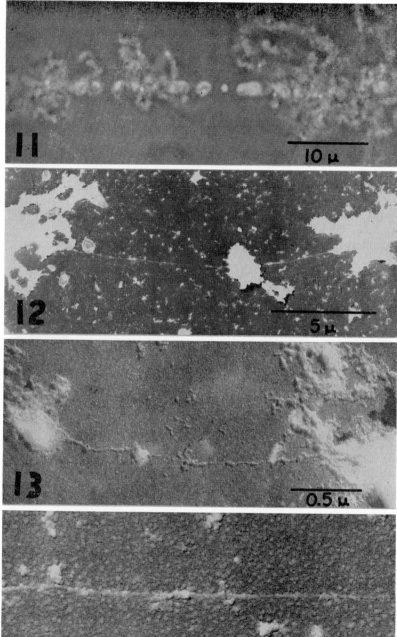

and of course would come apart immediately were there not some sort of submicroscopic strand connecting successive chromomeres. It should be added that neither pepsin, trypsin, nor RNAase destroys this fine connection, although as we shall see, it is rapidly attacked by DNAase.

The strand has been photographed in the electron microscope in three different laboratories. Tomlin and Callan (70) found a single strand of about 200 Å diameter in what are probably somewhat stretched preparations. Guyénot and Danon (34) report two connections each 100-150 Å diameter, but often coated with accessory materials. My own original observations (28) indicated a strand of "less than 1000 Å diameter," but later pictures of material dehydrated according to Anderson's (4) critical point method, suggest that 200-400 Å is more nearly correct for the finest connections (Figs. 12-14). However, it is equally clear in all photographs presented to date that the strand is somewhat variable in diameter. My own guess is that the strand (or strands) tend to contract and thicken rather easily and are perhaps seen at true diameter only in slightly stretched regions.

Under certain exceptional conditions the axial continuity of the chromosome is maintained not by such a fine strand as described above, but rather by what is obviously a stretched out pair of loops. Such "double loop bridges" were first described by Callan (17), and they have furnished strong support for a theory of lampbrush structure recently proposed by Callan (17) and by myself (31). Double loop bridges arise as follows. When a lampbrush chromosome is stretched, as may occur accidentally during isolation of the chromosomes from the nucleus, or which can be carefully controlled with microneedles, the chromomeres at first separate from one another and the interchromomeric gaps become evident (17, 26). Further stretching generally causes breaks in the axis. But amazingly enough, these breaks do not involve the interchromomeric connections, but rather occur transversely through the middle of a chromomere. In this fashion a pair of loops is stretched out in the axis of the chromosome to form a "double loop bridge" (see Fig. 10). In these cases the two ends of a loop are separated by a very long gap in the chromosome axis, the length of the gap being simply the total length of the involved loops.

These various observations have led to the theory of lampbrush chromosome structure illustrated in Fig. 15. According to this theory a lampbrush chromosome consists basically of two chromatids of a few hundred Ångstroms diameter. In the region of the loops these chromatids are completely separated from each other and are individually visible only because of their coating of ribonucleoprotein. In the region of the chromomeres the chro-

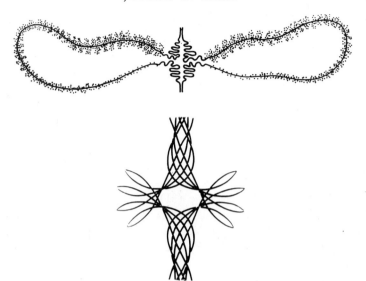

Fig. 15. Diagram of the postulated structure of a lampbrush chromosome.

Fig. 16. Beermann's (8) interpretation of the course of the chromosomal fibers in the region of the Balbiani rings. Compare the lateral looping of the fibers in this case with the loop axes of a lampbrush chromosome.

matids are complexly coiled or folded, but for short distances between chromomeres run more or less straight.

That the loops form part of very long and continuous chromosome strands is an hypothesis first put forward by Rückert (63) in his classic account of lampbrush chromosomes. More recently Ris (58-61) has been the main proponent of this view. The studies of Duryee (26) and my original observations (28, 29) led me at first to reject Ris' formulation, but the stretching experiments of Callan (17) and electron microscopical observations to be discussed shortly have convinced me that the essence of the hypothesis must be retained.

The presence of two chromatids in regions other than the loops is a postulate only and does not rest on experimental evidence. However, we know that in general DNA duplication is completed before prophase of meiosis (55, 64), and so we may suppose that two chromatids are present throughout the length of the lampbrush chromosomes. Furthermore, in *Triturus marmoratus* (17) a few short regions of the lampbrush chromosomes show complete lateral separation not only of the two loops of a pair, but also of the chromomere axes (Fig. 10). For simplicity therefore,

I have assumed that the usual optically single chromomeres are, in fact, double.

Crucial to the theory is the demonstration of an axis in the loops with the same properties as the fine strand (or strands as the case may be) between chromomeres. Fortunately, we have both observational and experimental evidence on this point. As mentioned previously, the loops may be "removed" by several agents which leave the chromomeres and the interchromomeric connections intact. If the theory presented here is correct, then the loop axis should be unaffected by these agents. In point of fact, peptic digestion, which removes all traces of the loops as seen in the light microscope, leaves a very delicate loop axis which has been photographed in the electron microscope (31). The diameter of this axis is somewhat variable in the available pictures but is generally less than 500 Å. We therefore have direct evidence not only for the existence of this strand, but also for the fact that it, like the interchromomeric strand, is resistant to pepsin.

Recently, Callan and Macgregor (20) have studied the action of DNAase on unfixed lampbrush chromosomes, with results which bear on the existence of the loop axis and its chemical composition. DNAase causes a unique and dramatic disintegration of the chromosome structure (Fig. 10). A few minutes after application of the enzyme, the chromosome falls apart into large segments due to breakage of many interchromomeric strands. The process continues until one has only freely scattered chromomeres or short lengths consisting of a few chromomeres each. At the same time the loops fragment into shorter and shorter pieces with no immediate decrease in diameter. Eventually all trace of the loops is lost. It should be emphasized that pepsin, trypsin, and RNAase do not cause breakage of the interchromomeric strands, nor do the loops ever fragment into large pieces in these enzymes. Rather, the loops become thinner and thinner, simultaneously contracting until they are represented by very small projections on the chromomeres (Fig. 10). These observations mean that the continuity not only of the interchromomeric strands but also of the loops is dependent upon the integrity of their contained DNA.

On the basis of the electron microscope evidence and the DNAase experiment, we may conclude that the strands which make up the interchromomeric connections and the axes of the loops are essentially similar in dimensions as well as in composition. The chromomeres themselves resemble the interchromomeric strands and the loop axis in the following features: they are resistant to the same reagents which leave these strands unaffected (dilute saline, RNAase, pepsin, trypsin), and, of course, they contain DNA as shown by a positive Feulgen reaction. That they consist of a more tightly wound portion of the same strands is an inference based largely on the developmental history of the loops. As this developmental history is also the basis of speculation about the functional role of the loops, I shall discuss them both together in a later paragraph.

The picture of lampbrush chromosome structure which I have outlined here differs somewhat from that presented recently by Ris (59-61) and by Lafontaine and Ris (41). Ris' views on lampbrush chromosome structure have been guided by two general postulates about chromosomes. The first of these is that the chromatid is essentially multiple-stranded (the degree of multiplicity being a species characteristic), and the second that the strands making up a chromosome are uniform throughout their length. In a very general sense, these postulates could be applied to Fig. 15, although a high degree of multiplicity seems unlikely because of the fine dimensions of the chromatid. But as a matter of fact, Ris has used the postulates in quite a different fashion.

Originally (58, 62) he held that the lampbrush chromosome contains a bundle of uniform, independent strands (8 or more) each of which could form separate loops and whose central overlap produced the chromomeres as optical illusions. The demonstration of the paired nature of the loops, implying that a loop corresponds to a chromatid and not some smaller unit, and the numerous lines of evidence proving the existence of the chromomeres, has brought revision of these views. Ris still feels that the chromatid is multiple-stranded and that the strands are structurally uniform. Now, however, the multiple strands are found as very much finer (500 Å) threads making up the entire substance of the loops and presumably continuous with similar, tightly coiled threads in the chromomere. Ris supposes that the 500 Å threads contain RNA in the loop regions but DNA in the chromomeres, and that the loops are formed by intercalary growth within certain regions of the fibers. The evidence for these fibers consists of electron micrographs of whole isolated chromosomes and sections of chromosomes.

The possible existence of fibrous material in the loops will not be denied, although I personally feel that many of the pictures presented could as well be interpreted in terms of granular material. But the question of structural continuity of such fibers throughout the length of the chromosome seems to me clearly refuted by evidence at hand. In particular, the results of digestion experiments, already described, suggest that the structural integrity of the loops is dependent upon DNA. The RNA material may be removed either with dilute saline, proteolytic enzymes, or RNAase without causing fragmentation of the loops or disruption of the interchromomeric strands. The material which holds the chromosome together is clearly quite different from that making up the bulk of the loops.

The dimensions of the interchromomeric strand provide an almost insuperable difficulty to Ris' view: if the loops are made up of 8 or more fibers each 500-Å in diameter, then, of course, these fibers cannot be continuous through a portion of the chromosome itself less than 500 Å in diameter. It is not surprising that Lafontaine and Ris (41) discount the evidence for the existence of such a strand. Rather, they claim that successive chromomeres are connected either by two short strands having dimensions similar to the loop bases (i.e., readily visible in the light microscope), or by successive loops. In the latter case, ends of a loop are supposed to be inserted in two successive chromomeres separated by a gap of a few micra (their Fig. 1). In citing the double loop bridges described by Callan as evidence for their view, these authors confuse the issue, since successive chromomeres in an unstretched chromosome or for that matter in most stretched chromosomes are *not* connected by loops. The gaps which they describe are nothing more than the usual interchromomeric gaps, and careful observation of unfixed material would show that they are not traversed by loops. The same fact can be demonstrated with fixed material (Fig. 11), although here the loops are generally much more tangled and difficult to interpret.

## Functional Aspects of the Lampbrush Chromosomes

Evidence has already been presented to show that the loop consists of a very delicate axis about which ribonucleoprotein material is accumulated. It seems very likely that this loop axis is a permanent structural part of the chromosome, unwound during periods of synthesis, folded back into the chromosome axis when inactive. The various loops may correspond rather directly to gene loci. The evidence for these statements will be considered under three headings: structure, synthesis, and genetics.

*Structure.* The origin and disappearance of the loops should provide us with some evidence bearing on the problem of chromosome activity. The early stages in loop formation are difficult to study because of the small size of the nuclei, but loop regression is easily followed in the later stages of oogenesis. The chromosomes at the first metaphase are essentially typical meiotic chromosomes; that is, they are more or less homogeneous, Feulgen-positive rods with few signs of internal differentiation. They are derived from the lampbrush condition by a gradual reduction in the length of the chromosome accompanied by a decrease in the number and size of the loops, and a concomitant decrease in the total number of chromomeres. Most significant for the present interpretation is the fact that the loops never detach as such from the chromosome. Rather, it seems that loop regression is accomplished by a discharge of the ribonucleoprotein from the loop and a gradual folding of the loop axis into the chromomere. Essentially the same process occurs during the dissolution of loops in saline of low concentration or in dilute solutions of pepsin or RNAase. Parenthetically it should be remarked that the electron microscopical demonstration of the loop axis was possible only by first fixing the chromosomes in such a way that the loops adhered firmly to a glass slide. In this way many of the loops can be digested by pepsin without the concomitant contraction; they simply become thinner and thinner. The axis alone remains after prolonged digestion and may be visualized in the electron microscope by replica techniques.

Circumstantial evidence for the idea of raveling and unraveling of the loop axis comes from a comparison of chromosomes from young oocytes with those from later stages. The young oocytes contain chromosomes with very large loops but generally small chromomeres. The opposite situation holds later: the chromomeres are large, but the loops are reduced. Such a reciprocal size relationship suggests that the chromomere may form at the expense of the loop axis and vice versa.

One final theoretical argument should be mentioned. It now seems clear that DNA is present in the loops, probably in the loop axis. If the concept of DNA constancy is in fact correct, then it follows that the DNA of the loops must somehow get back into the main body of the chromosome before metaphase. The coiling up of the loop axis into the chromomere would provide such a mechanism.

*Synthesis.* In all amphibian oocyte nuclei during the later stages of oogenesis the nuclear sap contains an abundance of small granules of various sizes which correspond to quite similar granules found on various lampbrush loops (17, 26, 29). Like the RNA-protein material of the loops, these granules contain cytochemically demonstrable RNA. In some species of salamanders, notably in *T. cristatus carnifex* studied by Callan (17), the loops show an amazing heterogeneity of structure from one locus to the next. Callan has shown that all of the larger granules of characteristic morphology found free in the nuclear sap correspond closely to similar material present on one or more pairs of loops on the chromosomes. The conclusion is almost forced upon us that these free materials are ultimately derived from the loops by some sort of budding process. In *Amblystoma tigrinum,* the nucleolus offers a particularly striking case of the production of RNA-protein materials from a specific chromosome locus, although here the relationship to a loop is not so clear-cut. The oocyte nucleus of *A. tigrinum,* like the nucleus of all salamanders, contains many hundreds of free nucleoli located some distance from the chromosomes. There has always been considerable question as to whether these nucleoli are really homologous to the usual somatic nucleoli, which are of course attached to a chromosome at a specific locus. In *A. tigrinum* the locus of the somatic nucleoli is known from a careful study made some years ago by Dearing (22). When one looks at this same locus in the lampbrush chromosome, a nucleolus or group of nucleoli, identical to the free extra-chromosomal nucleoli, is found attached laterally to the chromosome axis. The exact mode of attachment is not clear because the highly refractile nature of the nucleolar material obscures the fine detail. But it is quite certain that this locus, responsible for the production of the somatic nucleolus, also produces bodies identical to the free oocyte nucleoli. What we see here is probably an extreme example of RNA-protein production by a particular chromosome region. In point of fact, many of the loops are composed of very large lumps or granules of materials, so that from a morphological point of view, there seems to be a gradation from the typical nucleoli through various sorts of heavily granular materials, to the very faintly granular and "amorphous" loops (cf. 17, 29, 30 for details). Whether one wishes to call all of the larger granules attached to the chromosomes and free in the nuclear sap "nucleoli," as has been customary in the past, seems to be a relatively unimportant question of terminology. They all represent materials derived from the chromosome.

More direct evidence for synthetic activity in the loops is indicated by studies using adenine-C-14 as a precursor for RNA (Gall, unpub.). A few hours after injection of the labeled compound into the body cavity of a female salamander, the RNA of the oocyte nuclei shows strong radioactivity (Fig. 17). Although the concentration of RNA in the cytoplasm is much greater than in the nucleus, as indicated by the intensity of azure B staining, the nucleus is the most radioactive part of the cell. The specific

Fig. 17. Autoradiograph of a section of an immature oocyte of the newt, *Triturus,* eight hours after injection of adenine-C-14 into the coelom. The activity in the oocyte proper is due to incorporation into RNA, as shown by control sections pretreated with RNAase. RNA stained by azure B at $p$H 4, photographed in blue light to reduce the extreme contrast in staining between cytoplasm and nucleus. Note the intense nuclear incorporation despite the relatively low RNA concentration. 400 ×.

Fig. 18. Isolated lampbrush chromosome from same animal used in Fig. 17. Conventional illumination to show grains in the autoradiographic film. Extraction experiments show that the incorporation is into RNA. 550 ×.

Fig. 19. Same as Fig. 18, phase contrast showing the outline of the chromosomes. 550 ×.

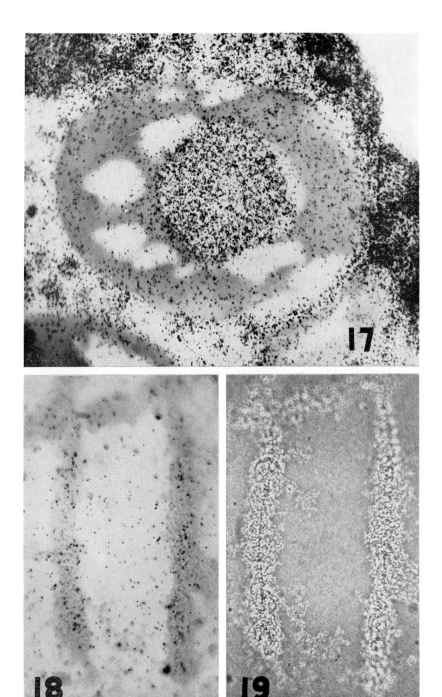

activity of the nuclear RNA is therefore very much higher than that of the cytoplasm. Chromosomes and nucleoli have also been isolated from such radioactive nuclei and fixed onto glass slides before being covered with stripping film. It is clear from the autoradiographs that activity is present not only in the nucleoli, but also in the loops of the chromosomes (Figs. 18, 19). We may reasonably infer, therefore, that synthesis of RNA is taking place directly in the loops.

*Genetic.* Most of the loops on a typical lampbrush chromosome are rather similar in general morphology. As already indicated, however, characteristic loops differing in total length, and especially in granularity of the ribonucleoprotein materials, occur at specific loci along the length of the chromosome. These loops furnish landmarks by which one may recognize particular chromosomes of the set, and indeed the whole chromosome can be mapped at least partially in terms of loop morphology. These variations suggest that perhaps each loop, or loop pair, represents a different genetic locus responsible for the production of some particular material.

A rather more direct demonstration of the genetic nature of differences in loop morphology has recently been made by Callan and Lloyd (19), who studied particularly characteristic loop pairs on the chromosomes of *T. cristatus carnifex*. Three different conditions have been found with respect to certain loci: loops may be present on both homologues, present on only one, or missing entirely. Significantly, all the oocytes of any one female salamander show the same condition for any one locus. Furthermore, the proportions of individuals with the indicated conditions (homozygous present, homozygous absent, heterozygous) follow expectation on the basis of random distribution of a gene in the population. Thus, we have very strong circumstantial evidence that the directly visualized chromosomal states express gene differences. These same workers have now made crosses between individuals of known constitution to see if the presence or absence of loops segregates in the appropriate Mendelian fashion.

## Discussion

### Nature of the Chromatid

The evidence from electron microscopy indicates that the interchromomeric fiber of the lampbrush chromosome is only a few hundred Ångstroms in diameter; furthermore, it constitutes the *entire* chromosome in those short regions where it occurs. Now by commonly accepted definition each homologue of a meiotic bivalent consists of two chromatids. Assuming that the interchromomeric strand is double, as indeed some of the pictures of Guyénot and Danon (34) suggest, we are forced to conclude that the chromatid in these regions is no more than a few hundred Ångstroms in diameter. Callan and Tomlin's data suggest 100 Å, Guyénot and Danon 100-150 Å, and my own pictures, 150-200 Å. It may well be that smaller units exist in these chromatids; the fact that the interchromomeric fiber usually appears single but should correspond to two chromatids is evidence in itself that fibers may be too closely joined for resolution even in the electron microscope. But the strands we are discussing are themselves already near molecular dimensions and it seems most unlikely that any high degree of multiplicity will be demonstrated.

In the case of the salivary gland chromosomes, multiple-strandedness seems well established on a number of grounds. However, the only multiplicity which is certain is that of chromatids, i.e., the DNA content is higher than the diploid amount by some power of 2. We find that the number of strands estimated by Beermann and Gay corresponds closely to the degree of polyteny suggested by volume and DNA measurements. Hence, the component fibril seen in the electron microscope may be a chromatid and not a smaller unit. The evidence, of course, is not so clear in this case as for the lampbrush chromosome. It should be stressed, however, that the presence of multiple strands in the salivary gland chromosomes cannot be used as a general argument for multiple-strandedness of the *chromatid,* unless it is clearly shown that the number of strands exceeds the degree of polyteny. Alfert (2) has stressed essentially this same point.

Fine fibers having a diameter of a few hundred Ångstroms more or less have been described from ordinary mitotic chromosomes and from sperm heads. Figs. 20-21 show such fibers in the spermatid of the grasshopper *Melanoplus,* where they appear on closer inspection to be fine tubules. The mere presence of a number of transversely cut fibers seen in a single electron microscope section, however, tells us nothing about the number of strands in a chromatid. If the chromatids are as long and thin as the evidence just cited would lead us to think, then they must be very complexly woven back and forth in both mitotic chromosomes and sperm heads. We must know the exact geometry of the fiber arrangement before interpreting such pictures in terms of the number of strands. Kaufmann and McDonald (37) claim that mitotic chromosomes consist of a hierarchy of pairs of strands, starting with pairs of the finest (125 Å) threads and proceeding up to quarter chromatids, half chromatids, and finally to the whole chromosome, which presumably contains 32 or more strands.

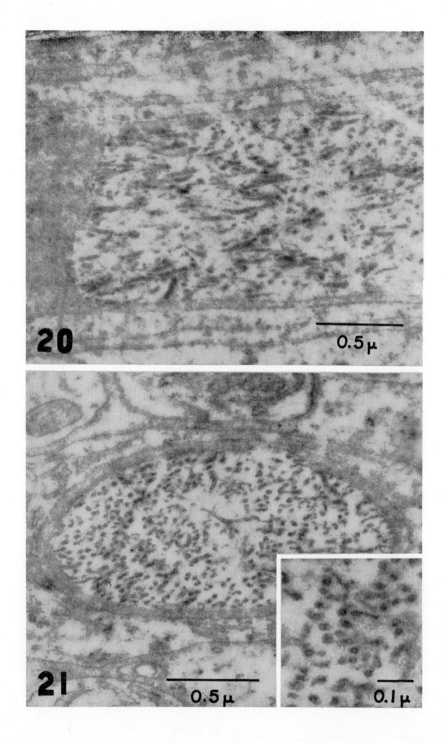

I agree with these authors that their sections of chromosomes show minute fibrils, but I cannot agree so readily that the geometry of their arrangement is at all evident from the published pictures.

Not only are chromatids of extremely minute diameter, but they are also enormously long. Calculations based on the number of lampbrush loops per chromosome and their average length lead us to believe that the total length of a salamander (*Triturus*) chromosome is of the order of 5 centimeters (31). This calculation completely ignores any length of the chromatid which is wound up in the chromomeres and so is presented as a minimal estimate of length. Beermann and Bahr (11) have estimated that the total length of the strands spun out in a Balbiani ring is of the order of 4 micra, and Beermann (8) has also presented the evidence that this material forms at the expense of a single cross-band. If we count the total number of bands on a salivary gland chromosome, say 1000, and assume that any locus is potentially able to unravel strands several micra in length, we arrive at a figure of several millimeters for the total length of the chromosome. In the salivary gland chromosome we must assume this potentiality for more than a few loci; in the lampbrush chromosomes we can see the expanded regions directly.

## Nature of the Functional Unit

The recent elegant studies on polytene chromosomes provide strong reason for believing that the functional unit of the chromosomes is of the general order of magnitude of the bands seen in the light microscope. In the lampbrush chromosome, the functional unit is probably the loop. The similarity between loops and Balbiani rings or puffs is quite striking and has been pointed out already on several occasions (2, 17, 31).

In both, a Feulgen-positive region spins out fibers several micra in length, an estimated four micra for the salivary gland chromosome, up to 200 micra for the longest loops. Ribonucleoprotein accumulates around these fibers but later is lost when the expanded locus regresses to its normal dimensions. Since the fibers which are spun out either demonstrably con-

Fig. 20. Longitudinal section of a spermatid from the grasshopper *Melanoplus*. The centriole is the diffusely granular structure to the left; fibers oriented parallel to the axis of the sperm head extend from the centriole anteriorly for a short distance. However, most of the nucleus at this stage is filled with lamellae (not shown in this picture) which are apparently formed by the lateral association of these fibers. 44,000 ×.

Fig. 21. Transverse section of a comparable nucleus showing dotlike cross sections of the fibers. 46,000 ×. Inset: higher magnification of a small region from the transverse section. Here one sees that the fibers are in fact fine tubules whose dimensions are approximately 170 Å total diameter and 70 Å core diameter. 90,000 ×.

tain DNA (lampbrush) or develop at the "expense" of a DNA-containing region (salivary), the process is very likely a change in the degree of folding of the basic chromosome strands and does not involve intercalary growth.

The differences between the Balbiani rings and the lampbrush loops are perhaps related to the type of chromosome in which the changes occur and to the general significance of the cells in which those chromosomes are found. Salivary gland chromosomes contain some thousand or more chromatids; the lampbrush chromosomes, only two. Therefore, when the salivary gland locus unravels, a large number of component loops spread out laterally; when the same state is assumed by the lampbrush locus, only two loops are formed (cf. Figs. 15 and 16).

Beermann argues that the Balbiani rings and puffs are closely related to the activity of the cells in which they occur. He imagines that the pattern of puffing reflects in some fashion the pattern of physiological functions exhibited by each cell type. Many cell types would have certain functions in common and would therefore show greater or lesser activity in some of the same chromosomal loci. Other functions peculiar to individual cell types would be represented by specific chromosomal modifications. The Balbiani rings, for instance, have so far been seen only in salivary gland cells and might be related to the specific and severe demands on these glands involved in the production of the salivary secretion. Somewhat analogous arguments can be brought forward for the egg cell and the lampbrush chromosome. Here, however, we must suppose that a very large number of loci, certainly hundreds, are functional at once. This extreme activity of the chromosomes may well be related to the establishment of many synthetic functions which will be required of the growing and differentiating egg.

## Chromosomal Differentiation and Cellular Differentiation

In the preceding discussion I have tried to demonstrate that the lampbrush chromosomes and the polytene chromosomes of the Diptera, at first sight so strikingly different, do in fact exhibit both structural and functional similarities. Both the loops and the puffs provide us with simple observational evidence that chromosomes may undergo differentiation not only as a whole but at rather specific loci. The same question can be asked about the differentiation of these chromosomes that one asks in general about the differentiation of whole cells. Are the changes essentially reversible and therefore in a broad sense of primarily physiological significance, or are they permanent genetic changes?

The polytene chromosomes cannot answer this question directly for the same reason that somatic cells cannot generally answer the question: they are not suitable for breeding experiments. Beermann has ably defended the thesis, however, that the changes in the polytene chromosomes are reversible and are to be interpreted in terms of differential gene activity. His argument rests essentially on those cases in which striking Balbiani rings or puffs may appear at one time in the larval life history and subsequently regress so that the normal banding pattern of the chromosome is restored.

On the other hand, the lampbrush chromosomes offer direct evidence on the question of genetic versus physiological change. Here we see the regression of many hundreds of loop pairs during the latter part of oogenesis, culminating in the appearance of typical meiotic chromosomes at the first metaphase. But most significant is the simple fact that these chromosomes, after the meiotic divisions, furnish half of the genetic material of the next generation. No one would argue that chromosomes undergo permanent genetic changes during the course of meiosis; and so we may conclude that the formation of the loops and their subsequent regression are simply reversible physiological changes.

The two cases thus complement one another. The polytene chromosomes show obvious differentiations related to the cell type in which they are found, as well as to the functional state of that cell. The lampbrush chromosomes, so far studied in only one cell type, show that these changes are reversible and non-genetic.

If we picture the formation of puffs and loops as visible expressions of gene function, then we should expect to find variations in this expression in different strains. Heterozygosity for puffs and for particular loops has been demonstrated by Beermann (8) and by Callan and Lloyd (19). These findings do not contradict the idea that puffing or loop formation *per se* are reversible physiological changes; but rather they add weight to the belief that the functional units seen in the chromosome are indeed also genetic units. We should not expect, of course, all heterozygosity to express itself visibly, since many changes in the genes might affect only the rate of their action or produce some chemical alteration, with no morphological concomitants. The "phenotype" of the chromosome, like the phenotype of individual cells or the organism as a whole, is only a rough guide to its genotype.

Long ago the notion of gene segregation was advanced by Weismann to explain embryonic differentiation. The rise of modern genetics, and especially the evidence from experimental embryology, failed to support his concept. In its place came the idea that all cells of a differentiated organism contain the same complement of genes, with such obvious exceptions as chromosome elimination. But the latter view left at least two broad possibilities. All the genes might act concurrently in all cells but against different cytoplasmic backgrounds. On this view, cellular differentiation would be largely the result of prior cytoplasmic changes. On the other hand, genes might act at different times or to varying degrees in different types of tissues. The behavior of genes might itself be dependent in part upon cytoplasmic conditions, but in any event the genes would have a more "active" role in the final production of cytoplasmic differentiation. The evidence from the cytology of giant chromosomes seems generally to favor this latter view.

# REFERENCES

1. Alfert, M., *J. Cellular Comp. Physiol.*, **36,** 381 (1950).
2. ———, *Intern. Rev. Cytol.* **3,** 131 (1954).
3. ———, and Geschwind, I. I., *Proc. Natl. Acad. Sci. U. S.*, **39,** 991 (1953).
4. Anderson, T. F., *Trans. N. Y. Acad. Sci.*, **13,** 130 (1951).
5. Balbiani, E. G., *Zool. Anz.*, **4,** 637 (1881).
6. Bauer, H., *Zool. Jahrb.*, **56,** 239 (1936).
7. ———, and Beermann, W., *Chromosoma*, **4,** 630 (1952).
8. Beermann, W., *Chromosoma*, **5,** 139 (1952).
9. ———, *Z. Naturforsch.*, **7b,** 237 (1952).
10. ———, *Cold Spring Harbor Symposia Quant. Biol.*, **21,** 217 (1956).
11. ———, and Bahr, G. F., *Exptl. Cell Research*, **6,** 195 (1954).
12. Boivin, A., Vendrely, C., and Vendrely, R., *Compt. rend acad. sci.*, Paris, **226,** 1061 (1948).
13. Brachet, J., *Biochemical Cytology*, Academic Press, New York (1957).
14. Breuer, M. E., and Pavan, C., *Chromosoma*, **7,** 371 (1955).
15. Bridges, C. B., *J. Hered.*, **26,** 60 (1935).
16. Callan, H. G., *Symposia Soc. Exptl. Biol.*, **6,** 243 (1952).
17. ———, *Symposium on the Fine Structure of Cells* (Leiden), Intern. Union Biol. Sci. Publ., **B 21,** 89 (1956).
18. ———, *Pubbl. staz. zool. Napoli*, **29,** 329 (1957).
19. ———, and Lloyd, L., *Nature*, **178,** 355 (1957).
20. ———, and Macgregor, H. C., *Nature* **181,** 1479 (1958).
21. Caspersson, T., *Cell Growth and Cell Function*, W. W. Norton, New York (1950).
22. Dearing, W. H., *J. Morphol.*, **56,** 157 (1934).
23. Dodson, E. O. *Univ. Calif. Publ. Zool.*, **53,** 281 (1948).
24. Duryee, W. R., *Arch. exptl. Zellforsch.*, **19,** 171 (1937).
25. ———, in *Cytology, Genetics, and Evolution:* Univ. Penn. Bicentennial Conference, p. 129, Univ. Pennsylvania Press, Philadelphia (1941).
26. ———, *Ann. N. Y. Acad. Sci.*, **50,** 920 (1950).
27. Flax, M. H., and Himes, M. H., *Physiol. Zool.*, **25,** 297 (1952).
28. Gall, J. G., *Exptl. Cell Research*, **Suppl. 2,** 95 (1952).
29. ———, *J. Morphol.*, **94,** 283 (1954).
30. ———, *Symposia Soc. Exptl. Biol.*, **9,** 358 (1955).
31. ———, *Brookhaven Symposia Biol.*, **8,** 17 (1956).

32. Gay, H., *J. Biophys. Biochem. Cytol.*, **2** (Suppl.), 407 (1956).
33. Geitler, L. Endomitose und endomitotische Polyploidisierung. *Protoplasmatologia*, **6** (1953).
34. Guyénot, E., and Danon, M., *Rev. suisse zool.*, **60**, 1 (1953).
35. Heitz, E., and Bauer, H., *Z. Zellforsch. u. mikroskop. Anat.*, **17**, 67 (1933).
36. Hertwig, G., *Z. indukt. Abstamm.-u. Vererblehre*, **70**, 496 (1935).
37. Kaufmann, B. P., and McDonald, M. R., *Cold Spring Harbor Symposia Quant. Biol.*, **21**, 233 (1956).
38. Kaufmann, B. P., McDonald, M., and Gay, H., *J. Cellular Comp. Physiol.*, **38** (Suppl. 1), 71 (1951).
39. Koltzoff, N. K., *Science*, **80**, 312 (1934).
40. ———, *Biol. Zhur.* **7**, 3 (1938).
41. Lafontaine, J. G., and Ris, H., *J. Biophys. Biochem. Cytol.*, **4**, 99 (1958).
42. McClung, C. E., *Biol. Bull.*, **9**, 304 (1905).
43. McMaster-Kaye, R., and Taylor, J. H., *J. Biophys. Biochem. Cytol.*, **4**, 5 (1958).
44. Mechelke, F., *Chromosoma*, **5**, 511 (1953).
45. Metz, C. W., *Cold Spring Harbor Symposia Quant. Biol.*, **9**, 23 (1941).
46. Mirsky, A., and Ris, H., *Nature*, **163**, 666 (1949).
47. ———, and ———, *J. Gen. Physiol.*, **34**, 475 (1951).
48. Nebel, B. R., *Botan. Revs.*, **5**, 563 (1939).
49. Painter, T. S., *Science*, **78**, 585 (1933).
50. ———, *Am. Naturalist*, **73**, 315 (1939).
51. Painter, T. S., and Reindorp, E., *Proc. Natl. Acad. Sci. U. S.*, **26**, 95 (1940).
52. Palade, G., and Siekevitz, P., *J. Biophys. Biochem. Cytol.*, **2**, 671 (1956).
53. Palay, S. L., and Claude, A., *J. Exptl. Med.*, **89**, 431 (1949).
54. Pavan, C., and Breuer, M. E., *J. Hered.*, **43**, 150 (1952).
55. Pelc, S. R., and Howard, A., *Exptl. Cell Research*, Suppl. **2**, 269 (1952).
56. ———, and ———, *Exptl. Cell. Research*, **10**, 549 (1956).
57. Poulson, D. F., and Metz, C. W., *J. Morphol.*, **63**, 363 (1938).
58. Ris, H., *Biol. Bull.*, **89**, 242 (1945).
59. ———, *Genetics*, **37**, 619 (1952).
60. ———, in *Symposium on the Fine Structure of Cells* (Leiden), Intern. Union Biol. Sci. Publ., **B 21**, 121 (1956).
61. ———, in *The Chemical Basis of Heredity* (W. D. McElroy and B. Glass, eds.), p. 23, Johns Hopkins Press, Baltimore (1957).
62. ———, and Crouse, H., *Proc. Natl. Acad. Sci. U. S.*, **31**, 321 (1945).
63. Rückert, J., *Anat. Anz.*, **7**, 107 (1892).
64. Swift, H., *Intern. Rev. Cytol.*, **2**, 1 (1953).
65. ———, in *Current Trends in Molecular Biology* (R. Zirkle, ed.), Univ. Chicago Press, Chicago (1958).
66. ———, and Kleinfeld, R., *Physiol. Zool.*, **26**, 301 (1953).
67. ———, and Rasch, E., data presented in Alfert (2).
68. Taylor, J. H., *Science*, **118**, 555 (1953).
69. ———, McMaster, R. H., and Caluya, M. F., *Exptl. Cell Research*, **9**, 460 (1955).
70. Tomlin, S. G., and Callan, H. G., *Quart. J. microscop. Sci.*, **92**, 221 (**1951**).
71. White, M. J. D., *J. Morphol.*, **78**, 201 (1946).
72. Wischnitzer, S., *Am. J. Anat.*, **101**, 135 (1957).
73. Yasuzumi, G., Odate, Z., and Ota, Y., *Cytologia*, **16**, 233 (1951).

# CHROMOSOMAL REPLICATION AND THE DYNAMICS OF CELLULAR PROLIFERATION—SOME AUTORADIOGRAPHIC OBSERVATIONS WITH TRITIATED THYMIDINE*

## W. L. Hughes

*Medical Department, Brookhaven National Laboratory, Upton, L. I., N. Y.*

### Introduction

In autoradiography, a photographic emulsion is placed in contact with a specimen containing radioactivity and is stored in the dark. The radiation, passing through the emulsion, sensitizes silver bromide granules so that they can be reduced to metallic silver by a photographic developer. One then sees a pattern of silver grains lying above the specimen. (The technique (1) employed in the present studies made use of stripping film which was floated over a histological preparation. In some cases Feulgen staining was performed before autoradiography; in other cases the specimen was stained through the emulsion after photographic development.)

Tritium, $H^3$, with a half-life of 12.3 years, decays to $He^3$. The $\beta$ particle produced in this process is one of the least energetic known, having an average energy of only 5,700 electron volts (Fig. 1). The resulting short range of tritium's $\beta$ particle, while presenting difficulties in many methods for its detection, is of decided advantage in autoradiography because of the high resolution which it permits (2, 3). Even the most energetic $\beta$ particle from tritium can penetrate only 6 microns in water (or tissue) and only 2 microns in photographic emulsion. However, autoradiographic resolution is even better than this range might indicate because most of the $\beta$ rays from tritium have much less energy (Fig. 1), and because, in addition, only the two-dimensional projection of their range is seen. Therefore, essentially all of the silver grains activated by tritium lie clustered within a micron of the labeled locus. This is illustrated in Fig. 2a—an autoradiogram of plant chromosomes labeled with tritium (5).

For comparison, Fig. 2 (right) shows an autoradiogram (6) of similar chromosomes labeled as in Fig. 2 (left) except for the substitution of $C^{14}$ for $H^3$. Note how diffuse the pattern of silver grains has become. The $\beta$ radiation of $C^{14}$, while usually considered weak, has an average energy

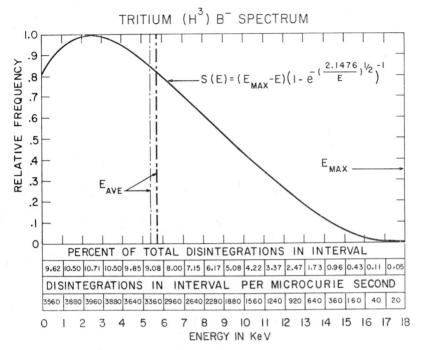

Fig. 1. Tritium (H³) β⁻ spectrum.

of 50,000 electron volts, so that a broad pattern of activated silver grains results.

Hydrogen occurs in almost all metabolites, and methods for its replacement by tritium can usually be devised from published techniques for deuterium exchange (4). When preparing labeled compounds for autoradiography, it is well first to ascertain that specific activities great enough for the problem at hand can be obtained. If one assumes that 20 disintegrations are required on the average to activate one silver grain above a labeled locus, and if the presence of 5 grains is considered sufficient to identify this locus, then the autoradiogram must be exposed long enough for 100 disintegrations to occur in each locus. Since tritium decays at the rate of 0.5 per cent per month, 20,000 tritium atoms must be present in each locus to provide a satisfactory autoradiogram after one month of exposure. If one can label the metabolite which concentrates at the locus intensely enough so that one molecule in fifty is labeled, then $10^6$ molecules of metabolite are sufficient to identify the locus. It is perhaps worth

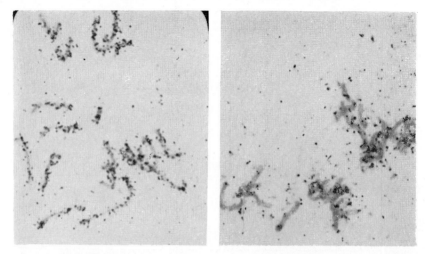

Fig. 2. Autoradiograms of *Bellevalia* chromosomes labeled with tritium (left) vs. *Allium* chromosomes labeled with carbon-14 (right). Note that the black silver grains in 2a lie closely over the chromosomes and are uniformly distributed along the length of the chromosome. However, in 2b the grains are so diffusely scattered that one can only conjecture that the chromosomes are, in fact, labeled.

noting that this labeled metabolite would have a specific activity of 560 curies per mole!

Labeled thymidine with a specific activity in this range has been used in the experiments to be described. The average mammalian nucleus contains $6 \times 10^{-12}$ g. of deoxyribonucleic acid (7), or there are $3 \times 10^9$ thymine units per nucleus. Therefore, mammalian nuclei can be labeled with tritiated thymidine without affecting the size of the precursor pools (8). However, in the studies with *Escherichia coli* (9), which contain only about $10^7$ thymine units per cell, satisfactory labeling has only been obtained with a thymine-requiring mutant which incorporates the labeled thymidine undiluted.

In attempting to label at high specific activities, radiation damage during the labeling process may present real difficulties. Radiation-induced exchange, as developed by Wilzbach, can produce specific activities from 10 to 100 curies per mole (10). However, the coincident radiation damage precludes higher specific activities by this method, and one must turn to the procedures of organic chemistry. Many substances have been labeled by catalyzed exchange with a labeled solvent. However, 10 curies of pure tritium oxide occupy a volume of only 0.003 cm³ and this substance irradiates itself at the rate of $10^9$ rad/day. Therefore, in using tritium oxide as a source of exchangeable tritium, some inert solvent is necessary. An ideal solvent should lack exchangeable hydrogen. However, solvents in this category, such as dioxane or dimethylformamide, proved ineffective in labeling thymidine, and acetic acid was used instead.

Thymidine was labeled (in the position indicated in Fig. 3) by catalytic

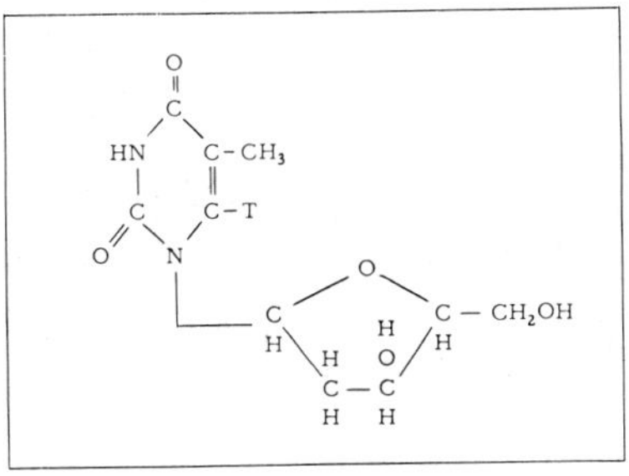

Fig. 3. Tritiated Thymidine.

exchange on reduced palladium black in acetic acid solution containing up to 100 curies of tritium oxide per cm$^3$. This procedure (11), which is analogous to a published method of labeling purines (12), exchanges tritium for the hydrogen bound to carbon in the pyrimidine ring. Tritium was then removed from the labile positions on oxygen or nitrogen, which were simultaneously labeled, by repeated equilibration with water. The product usually contained large amounts of tritiated impurities which were not readily recognized by paper chromatography. These could be detected by crystallization with carrier thymidine. By careful chromatography on anion exchange resins, a fraction was obtained in which all the activity co-crystallized with added thymidine. Cytidine and deoxycytidine have been labeled in a similar manner.

Tritiated thymidine thus prepared does not exchange its radioactivity even in the presence of acid or alkali. Degradation to thymine retains the label in the thymine fraction. However, the assignment of its position to the ring rather than the methyl group is based on the chemical analogy that cytidine, which contains no methyl group, can be similarly labeled.

The remainder of this paper will describe some applications of tritiated thymidine, to indicate the wide range of physiological processes which it has already illuminated.

### Mechanism of Duplication of Deoxyribonucleic Acid

Thymidine is efficiently utilized in the formation of the genetic material, deoxyribonucleic acid (DNA). Current evidence emphasizes the stability of DNA, so that it now stands as a landmark in the shifting currents of physiological processes. While the immutability of the gene has long been recognized, its explanation in terms of a corresponding chemical stability of the genetic material has only recently been forthcoming. Thus radioactive phosphorus has been found preferentially incorporated into the DNA of rapidly dividing tissues (13). However, studies with phosphorus are complicated by its incorporation into many substances besides DNA. Reichard (14) and Friedkin (15) have studied the metabolism of thymidine, whose base, thymine, occurs uniquely in DNA. They have shown that labeled thymidine is utilized almost exclusively for the formation of DNA and that DNA formation occurs in tissues rich in dividing cells. Perhaps the most satisfying evidence for the immutability of DNA comes from studies (5, 9) of the mechanism of chromosomal duplication employing tritium-labeled thymidine. These will now be described in some detail:

*Duplication in Plant Cells* (5)

Seedlings of *Vicia faba* (English broad bean) were grown in a mineral nutrient solution containing 2-3 µg/ml of the radioactive thymidine. This plant was selected because it has 12 large chromosomes, one pair of which is morphologically distinct, and because the length of the division cycle and the time of DNA synthesis in the cycle are known (16). After growth of the seedlings in the isotope solution for the appropriate time, the roots were thoroughly washed with water and the seedlings were transferred to a non-radioactive mineral solution containing colchicine (500 µg/ml) for further growth. At appropriate intervals roots were fixed in ethanol-acetic acid (3:1), hydrolysed 5 minutes in 1 $N$ HCl, stained by the Feulgen reaction, and squashed on microscope slides. Stripping film was applied and exposed one month for autoradiography.

Roots remained in the isotope solution for 8 hours, which is approximately one-third of the division cycle (16). Since about 8 hours intervene between DNA synthesis in interphase and the next anaphase, few, if any, nuclei which had incorporated the labeled thymidine should have passed through a division before the roots were transferred to the colchicine solution. In the presence of colchicine chromosomes contract to the metaphase condition, and the sister chromatids (daughter chromosomes), which ordinarily lie parallel to each other, spread apart. The sister chromatids remain attached at the centromere region for a period of time, but they finally separate completely before transforming into an interphase nucleus. Because colchicine prevents anaphase movement and the formation of daughter cells, but does not prevent chromosomes from duplicating, the number of duplications following exposure to the isotope can be determined for any individual cell by observing the number of chromosomes. Cells without a duplication after transfer to colchicine will have the usual 12 chromosomes at metaphase (c-metaphase), each with the two halves (sister chromatids) spread apart but attached at the centromere. Cells with one intervening duplication will contain 24 chromosomes, and those with two duplications will contain 48 chromosomes.

Two groups of roots were fixed. The first group remained in the colchicine solution 10 hours. The second group remained in the colchicine for 34 hours. In the first group, cells at metaphase had only 12 chromosomes, which indicated that none of these had duplicated more than once during the experiment. The chromosomes in these cells were all labeled, and furthermore, the two sister chromatids of each chromosome were equally and uniformly labeled (Fig. 4, left). The amount of radioactivity in the

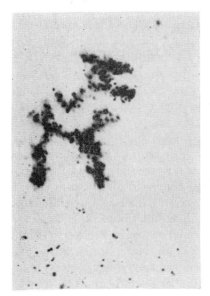 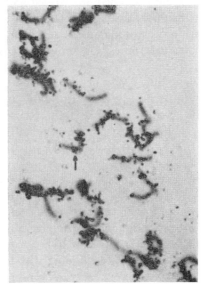

Fig. 4. Autoradiograms of first and second generations of plant chromosomes labeled with tritiated thymidine. *Left,* Duplication with labeled thymidine, showing uniform distribution of label. *Right,* Second duplication of labeled chromosomes in the absence of labeled thymidine, showing labeled and unlabeled chromatids.

chromosomes varied from cell to cell, as would be expected in a non-synchronized population of cells, but within a given cell the label in different chromosomes was remarkably uniform. This is perhaps more clearly demonstrated in Fig. 2 (left), an autoradiogram of similarly labeled chromosomes in *Bellevalia*. (The nucleus shown was slightly further advanced in mitosis so that most of the daughter chromosomes have separated.)

In the second group, cells contained either 12, 24, or 48 chromosomes. Those with 12 chromosomes usually were not labeled, but when labeling occurred, sister chromatids were uniformly labeled as in the first group. In cells with 24 chromosomes, all chromosomes were labeled; however, only one of the two sister chromatids of each was radioactive (Fig. 4, right). Evidently the pool of labeled precursor in the plant had been quickly depleted after the plant was removed from the isotope solution, and these cells with 24 chromosomes had gone through a second duplication in the absence of labeled thymidine.

In the few cells with 48 chromosomes, analysis of all 48 was not possible. However, in several cases where most of the chromosomes were well separated and flattened, approximately one-half of the chromosomes of a complement contained one labeled and one non-labeled chromatid, while the remainder showed no label in either chromatid. The appearance of cells with 48 chromosomes in a 34-hour period in colchicine also indicates that there was some variation in the predicted 24-hour division cycle.

In cells with 24 and 48 chromosomes a few chromatids were labeled along only a part of their length, but in these cases the sister chromatids were labeled in complementary portions (Fig. 4, right, arrow). This can be described as sister chromatid exchange.

These results indicate (1) that the thymidine built into the DNA of a chromosome is part of a physical entity that remains intact during succeeding replications and nuclear divisions, except for an occasional chromatid exchange; (2) that a chromosome is composed of two such entities probably complementary to each other; and (3) that after replication of each to form a chromosome with four entities, the chromosome divides so that each chromatid (daughter chromosome) regularly receives an "original" and a "new" unit. These conclusions are made clearer by the diagrams in Fig. 5. Beginning with two complementary non-labeled strands in a chromosome, the two strands separate with the formation of a complementary labeled strand along each original strand. At metaphase, when the two pairs of strands separate, each chromatid will appear labeled, although it contains both a labeled and a non-labeled strand. After the next replication, in the absence of labeled thymidine, separation of the two pairs at metaphase will reveal a labeled and an unlabeled chromatid. Following another replication only one-half of the chromosomes will contain a labeled chromatid, as demonstrated in those cells with 48 chromosomes.

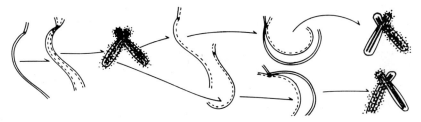

Fig. 5. Diagrammatic representation of mechanism of duplication.

It is immediately apparent that this pattern of replication is analogous to the replicating scheme proposed for DNA by Watson and Crick (17). We cannot be sure, of course, that separation of the two polynucleotide chains in the double helix is involved, for the chromosome is several orders of magnitude larger than the proposed double helix of DNA. However, at least in the case of bacterial DNA, labeled with $N^{15}$, separation of the individual chains does seem to occur (18).

## Duplication in Bacteria

Our own studies with *E. coli* 15 T– labeled with tritiated thymidine (9) seem to indicate a similar mechanism for the duplication of DNA in bacteria.

A log phase culture of *E. coli* 15 T– was added to a glucose-mineral medium containing 2-4 micrograms of tritiated thymidine per milliliter. Following at least 100-fold growth at 37°C in the labeled medium, two types of experiments were performed. In the first, the cells were washed in saline and grown in medium containing excess unlabeled thymidine. Direct cell counts from aliquots were taken at the time of inoculation and every thirty minutes thereafter. Each time a doubling of the population of the culture occurred, as determined by the direct cell count, smears were made. These were fixed by heating and washed with water; autoradiograms were made by the stripping film technique. After various exposure times, varying from 5 to 40 days, the slides were developed and stained with 0.04% crystal violet. In the second type of experiment, labeled cells were isolated micromanipulatively on nutrient agar medium which contained excess unlabeled precursor. Progeny lines were obtained from these isolated cells. Growth was stopped by drying the agar medium, and stripping film autoradiograms were made after fixing in 3:1 ethyl alcohol—glacial acetic acid and a series of alcohol-water solutions.

Both experiments revealed very clearly that the distribution of the incorporated tritium among the daughter cells grown on unlabeled medium was very heterogeneous (Fig. 6). At least as soon as the third generation,

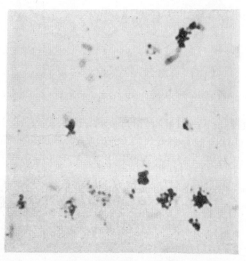

Fig. 6. Autoradiogram of *Escherichia coli* 15 T– labeled by several generations of growth in tritiated thymidine and then grown for about 3 generations in non-radioactive medium. Note that some cells are unlabeled, while in other cases the silver grains are so numerous as to obscure the bacterial cell.

and possibly sooner, one found cells with no label in the presence of those containing considerable label. In the liquid culture experiment, all-or-none splits, where one of an unseparated pair of cells exhibited heavy label (6-14 grains) while the other had no grains above it, were observed occasionally. Such all-or-none splits were also seen in the micromanipulation experiments where the sibling relations in the progeny line were known. These micromanipulation experiments indicate that the heterogeneity observed in bulk culture experiments is not primarily a result of cell death (from radiation or any other cause) or of differences in growth rates for individual bacteria. It should also be noted that at the first division on non-radioactive medium, no case of all-or-nothing distribution has yet been seen, even though attempts were made to select the smallest possible labeled cells as the sources of a progeny line.

The observation that uniform dilution of tritiated thymidine during bacterial growth does not occur would seem to indicate that relatively large amounts of parental DNA can remain functionally associated in transmissions to daughter cells. While the data do not clearly establish a functional two-strandedness for bacterial DNA like that found for the plant chromosomes, the lack of all-or-nothing splits at the first generation and their appearance in later generations are in harmony with this possibility.

### Sequences of Events in Cell Proliferation

In addition to providing information on the mechanism of DNA synthesis, tritiated thymidine is also proving useful in relating the chronology of other events within the cell to DNA synthesis. Thus, if labeled thymidine is available to cells for only a very short interval, the subsequent appearance of labeled nuclei in cells of any recognizable morphological type indicates that those cells were making DNA when the label was available. Hence this interval of the cell's life cycle is measured. Since cells once labeled are presumably labeled forever, subject only to dilution of the label at each cell division, labeled cells may also be followed chronologically through diverse morphological forms. Their migration from their zone of origin to that of function and final death may also be elucidated. Studies along these lines are being performed on mammalian cells in tissue culture and also on intact mammals.

*Tissue Cultures of Mammalian Cells*

In tissue culture the labeling period may be reduced to a few minutes by transferring the cells to an unlabeled medium. Thirty per cent of the cells of a culture of HeLa cells were labeled following a few minutes' exposure (19). Since the cells in such a culture are completely asynchronous, DNA synthesis must require the same fraction (30%) of the generation time. After further incubation in cold medium, labeled mitoses (Fig. 7) began to appear at 3 hours, but did not reach a maximum until after 6 hours. These intervals then measure the minimum and maximum time between the completion of DNA synthesis and the appearance of a mitotic figure (19, 20). The variation in this time must reflect a similar variation among individual cells in the length of this metabolic phase. Extension of these studies to longer times appears to reveal variation in other phases of the generation cycle.

146   THE CHEMICAL BASIS OF DEVELOPMENT

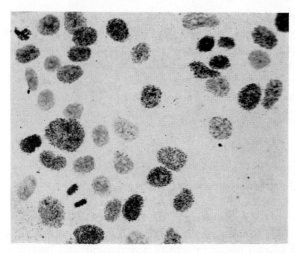

Fig. 7. Autoradiogram of a culture of HeLa cells grown in tritiated thymidine, showing labeled and unlabeled nuclei.

*Intact Mammals*

In mammals (mouse and human) the tritium of labeled thymidine appears to be built into DNA or converted to water during the first hour following injection. Therefore, any cells subsequently found labeled must have been making DNA preparatory to cell division immediately following injection. Autoradiograms of histological sections made a few hours after injection showed a wide variation in the number of labeled cells and thereby permitted the ready identification of the zones where cells are generated (8). The number of labeled cells in young adult mice exceeded that which might have been predicted from previous data on mitotic indices (21). The following section describes some of the findings (8) when young adult male mice were injected intraperitoneally with 0.7 $\mu$c of tritiated thymidine per gram mouse and sacrificed at intervals of from 7 to 48 hours after injection.

Seven hours after injection of the thymidine solution, labeled cells were found in the generative zones of the gut: the basal layer of the forestomach (which is lined with stratified epithelium) (Fig. 8a), the gland necks of the fundus, the crypts of the pyloric region (particularly the lower third, except for the cells at the very bottom of the crypts), the crypts of the small bowel (including the region occupied by the Paneth cells—however, no label was seen in cells containing granules) (Fig. 8b), and the crypts of the colon. The fraction of labeled cells ranged from about one-fifth, in

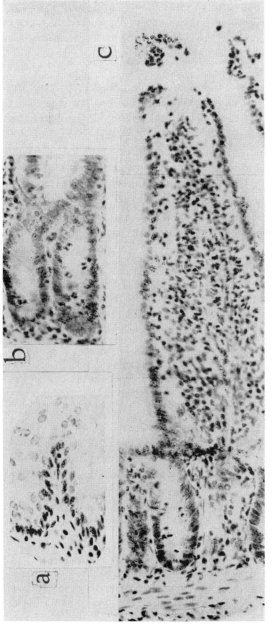

Fig. 8. Autoradiograms of mouse gut after injection of tritiated thymidine.
a. Forestomach 7 hours after injection. Label in basal cell layer.
b. Jejunum 7 hours after injection. Label in crypts only.
c. Jejunum 48 hours after injection. Label diluted in crypts and distributed along entire villus.

the forestomach, through one-third to one-half, in stomach and colon, to about two-thirds, in the jejunum. The amount of label per cell appeared to be fairly uniform. Most mitoses seen at this time were strongly labeled (in a subsequent series, labeled telophases were found as early as 2 1/2 hours after injection).

In preparations taken 14 to 48 hours after injection, labeled cells were seen to have moved from the generative to the functional zones; concomitantly, the number of labeled cells in the generative regions decreased, as did the amount of label per cell (Fig. 8c). An exception seems to be the colon, where a moderate number of heavily labeled cells remained after 48 hours near the bottom of the crypts. The movement of labeled cells was not very regular. This was best seen in the jejunum, with its large percentage of labeled cells: the zone behind the leading edge was characterized by patchy labeling, and the distance to which the label progressed varied considerably among villi and even on the two sides of a single villus.

In fundus, pylorus, jejunum, and colon, the distance from generative zone to surface was covered in less than 2 days, as shown by the appearance in the lumen of desquamated labeled cells 48 hours after injection. In the same 48 hours, in the forestomach, the label moved into the prickle cell layer but did not reach the surface. In the depth of the fundus glands, a few labeled cells were found at this time, but no fully differentiated chief and parietal cells.

The percentage of labeled mitoses was nearly 100 per cent in the preparation taken 7 hours after injection; it was very low 14 hours after injection, and showed irregular variations (between 20% and 60%) in subsequent samples.

In the spleen the most heavily and frequently labeled cell was the normoblast; the label seemed to stay constant for some time, but appeared reduced at 48 hours. Labeled megakaryocytes were found in all preparations, the percentage labeled possibly increasing with time up to about 60 per cent. In the malpighian follicles, a few reticulum cells and about 10 per cent of the lymphocytes were labeled; there were some groups of labeled cells in the germinal centers. The amount of label in lymphocytes was smaller than in the generative cells of the gastro-intestinal tract. During the first day after injection, the percentage of labeled lymphocytes increased and had decreased at 48 hours. A few doughnut-shaped myelocytes were labeled in all specimens; at 48 hours, there were large numbers of labeled mature granulocytes.

In the first series of animals, the pancreas contained label only in some duct cells; in a later series (not fasted) several well-labeled cells and mitoses were found. Occasionally fibroblasts and rarely smooth muscle cells were labeled in all preparations. A labeled liver cell was infrequently found. An occasional labeled cell was found in Brunner's glands. Even neurons, usually considered incapable of division, were occasionally labeled (22, 23).

If a cell population is homogeneous and completely asynchronous, then the percentage of labeled cells present after a brief labeling period measures the percentage of the total generation time during which cells can take up thymidine and use it for DNA synthesis. If the generating zones of the gut are thus considered as homogeneous populations of cells, it follows from our results that the synthetic time must vary from 20 per cent to 70 per cent of the generation time in different regions. If the durations of the periods of synthesis in these different regions are similar, then the rate of cell renewal must vary directly with the fraction of labeled cells. In agreement with this, labeled cells moved more rapidly out of the heavily labeled crypts of the jejunum than out of the more sparsely labeled basal epithelium of the forestomach.

With regard to intensity of labeling, throughout the gastro-intestinal tract, those cells that were labeled at all appeared to be labeled with equal intensity, also suggesting similar synthetic rates. By way of contrast, lymphocytes were weakly labeled. This may possibly be interpreted as indicating a slower synthetic rate for the lymphocyte. However, in addition to the inaccuracies resulting from variation in autoradiographic efficiency, variation in cellular uptake of the isotope or in size of the precursor pool might explain this difference.

The time required for the appearance of labeled mitotic figures following injection measures the interval between completion of DNA synthesis and the beginning of mitosis. Since some mitoses were labeled 2 1/2 hours after injection and all were labeled 7 hours after injection, all cells must proceed into mitosis within 7 hours following the completion of DNA synthesis. In the subsequent intervals studied, the percentage of labeled mitoses decreased to a low value at 14 hours and then varied between 20 per cent and 60 per cent. It is possible that further studies over more frequent time intervals may show a rhythmic variation in labeling which should correspond to the generation time.

Before concluding this discussion it may be well to point to possible complications produced by radiation from the incorporated tritium. In tissue culture, inhibition of cell division has been observed following the uptake of tritiated thymidine (24), but only when the incorporation exceeded many hundredfold that required for autoradiography. More than

$10^7$ tritium atoms per cell are required to inhibit growth, whereas $10^5$ or less will provide satisfactory autoradiograms. However, some effects of radiation, such as chromosome breaks, occur at much lower levels of radiation. These have been observed with $C^{14}$-labeled thymidine (6). Perhaps sister chromatid exchanges observed in the tritium-labeled plant chromosome (5) and the fractional transfer of the tritiated DNA of *E. coli* (9) are also subtle evidence of radiation injury. Further studies are necessary to clarify this situation.

### Conclusions

While the studies reported have been performed on "adult" tissues, their implications in embryology are obvious, and it is hoped that others will find this report useful not only for its contents but also in suggesting new tools for embryology. Thus autoradiography shortly after the injection of tritiated thymidine would permit a quantitative appraisal of rates of cellular proliferation in various parts of the embryo, and subsequent sampling of labeled embryos should help elucidate problems of cell migration and tissue differentiation. The transplantation of labeled cells could provide proof of the viability of transplants and also could permit one to follow the migration of donor cells and thus to determine the origin (donor vs. host) of each cell in the accepted graft.

Finally, it should be noted that other tritiated metabolites should elucidate other biochemical processes at the cellular level, although the situation may be complicated by intracellular turnover, which is lacking in DNA. We are beginning such studies with similarly labeled cytidine as a precursor of RNA and with tritiated arginine as a precursor of protein. Relative to Gall's paper, perhaps most interesting has been the finding that, in *Vicia* roots, cytidine was initially incorporated only in nucleoli (Fig. 9) (25). However, after a few hours the label left the nucleolus and was distributed throughout the cytoplasm, suggesting that RNA may be synthesized in the nucleolus and then transported to functional sites in the cytoplasm.

### Acknowledgments

This paper is a synthesis of reports from several biological and medical research programs utilizing tritium for autoradiography. The author has had the privilege of acting as "liaison" between these groups, which seemed too diverse for a joint paper, and so the author has assumed sole responsibility for this report; but it should be reiterated that credit both for the results and for the concepts expressed belongs equally to his collaborators. Particular appreciation is due Drs. R. B. Painter, J. H. Taylor, and P. S. Woods for their courtesy in making recent, still unpublished, results available.

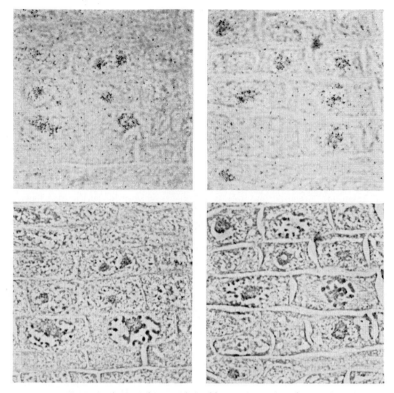

Fig. 9. Autoradiograms (upper photographs) of bean roots grown for one hour in tritiated cytidine. Label is confined almost exclusively to the nucleolus, as can be verified by the lower photographs, focused on the tissue below the autoradiogram.

## REFERENCES

1. Doniach, I. and Pelc, S. R., Autoradiograph Technique, *Brit. J. Radiol.,* 23: 184-192 (1950).
2. Robertson, J. S., and Hughes, W. L., Intranuclear Irradiation with Tritium-Labeled Thymidine, *Proc. Biophysical Society's First Meeting* (in press).
3. Fitzgerald, P. J., Eidinoff, M. L., Knoll, J. E., and Simmel, E. B., Tritium in Radio-autography, *Science,* 14: 494 (1951).
4. Kamen, M., *Isotopic Tracers in Biology,* Academic Press, New York, (1957).
5. Taylor, J. H., Woods, P. S., and Hughes, W. L., The Organization and Duplication of Chromosomes as Revealed by Autoradiographic Studies Using Tritium-Labeled Thymidine, *Proc. Natl. Acad. Sci. U. S.,* 43: 122-128 (1957).
6. McQuade, H. A., Friedkin, M., and Atchison, A., Radiation Effects of Thymidine 2-$C^{14}$, *Exptl. Cell Research,* 11: 249-264 (1956).
7. Vendrely, R., in *The Nucleic Acids,* (Chargaff, E., and Davidson, J. N., eds.) Vol. II, p. 155, Academic Press, New York (1955).

8. Hughes, W. L., Bond, V. P., Brecher, G., Cronkite, E. P., Painter, R. B., Quastler, H., and Sherman, F. G., Cellular Proliferation in the Mouse as Revealed by Autoradiography with Tritiated Thymidine, *Proc. Natl. Acad. Sci. U. S.,* 44:476 (1958).
9. Painter, R. B., Forro, F., and Hughes, W. L., Distribution of Tritium-Labeled Thymidine in *Escherichia coli* During Cell Multiplication, *Nature,* 181: 328-329 (1958).
10. Wilzbach, K. E., Tritium Labeling by Exposure of Organic Compounds to Tritium Gas, *J. Am. Chem. Soc.,* 79: 1013 (1957).
11. Verly, W. G., and Hunebelle, G. have described a similar procedure, *Bull. soc. chim. Belges,* 66: 640 (1957).
12. Eidinoff, M. L. and Knoll, J. E., The Introduction of Isotopic Hydrogen Into Purine Ring Systems by Catalytic Exchange, *J. Am. Chem. Soc.,* 75: 1992 (1953).
13. Hevesy, G. and Ottesen, J., Rate of Formation of Nucleic Acid in the Organs of the Rat, *Acta Physiol. Scand.,* 5: 237-247 (1943).
14. Reichard, P. and Estborn, B., Utilization of Deoxyribosides in the Synthesis of Polynucleotides, *J. Biol. Chem.,* 188: 839-846 (1951).
15. Friedkin, M., Tilson, D., and Roberts, D., Studies of Deoxyribonucleic Acid Biosynthesis in Embryonic Tissues with Thymidine $C^{14}$, *J. Biol. Chem.,* 220: 627 (1956).
16. Howard, A. and Pelc, S. R., Nuclear Incorporation of $P^{32}$ as Demonstrated by Autoradiographs, *Exptl. Cell Research,* 2: 178-187 (1951).
17. Watson, J. D. and Crick, F. H. C., Genetical Implications of the Structure of Deoxyribonucleic Acid, *Nature,* 171: 964 (1953).
18. Meselson, M. and Stahl, F. W., Replication of DNA in *Escherichia coli, Proc. Natl. Acad. Sc. U. S.,* 44:671 (1958).
19. Painter, R. B., Drew, R. M., and Hughes, W. L., unpub.
20. Firket, H. and Verly, W. G., Autoradiographic Visualization of Synthesis of DNA in Tissue Culture with Tritium-Labeled Thymidine, *Nature,* 181: 274-275 (1958).
21. LeBlond, C. P. and Walker, B. E., Renewal of Cell Populations, *Physiol. Revs.,* 36: 255-276 (1956).
22. Kahle, W. has observed labeled nuclei in the neurons of the hippocampus (mouse), pers. commun.
23. Messier, B., LeBlond, C. P., and Smart, I., Presence of DNA Synthesis and Mitosis in the Brain of Young Adult Mice, *Exptl. Cell Research,* 14: 224 (1958).
24. Painter, R. B., Drew, R. M., Hughes, W. L., Inhibition of HeLa Growth by Intranuclear Tritium, *Science,* **127**:1244 (1958).
25. Woods, P. S. and Taylor, J. H., pers. commun.

## DISCUSSION

DR. COWDEN: I would like to ask a two-part question of Dr. Gall. The first part of the question concerns the incorporation of labelled adenine. In its incorporation into the cells, does it go into the chromosomes, or into the nucleolus first?

DR. GALL: These studies have just been done, actually, and I can't tell you the time sequence of the incorporation. There was incorporation into the nucleoli after 8 hours and there was also incorporation into the chromosomes.

DR. COWDEN: The second part is: I think that both Dr. Swift and Dr. Brachet as well as myself have noted that in the metaphase chromosomes basophilia after DNAse treatment seems to reach a peak. How do you reconcile this with your situation where you have more RNA observable in the uncoiled, interphase

# DISCUSSION

condition than in the metaphase meiotic chromosomes of other materials in which RNA is found maximally in metaphase.

Dr. GALL: You do find RNA in metaphase chromosomes but I don't think that proves there are not larger quantities in some interphase chromosomes. I think that the normal interphase cell shows you plenty of RNA and it may in fact be minimal at metaphase. I don't know if there are any quantitative data on that.

Dr. COWDEN: I think that Dr. Swift has some in the 14th Growth Symposium.

Dr. GALL: Is RNA maximal at metaphase?

Dr. SWIFT: Well, we reported it so in our graph, but you will notice that there was a statement below in the paper saying it was impossible to eliminate spindle substance from our measurements, and that metaphase values were thus too high. What I really feel is that RNA is probably maximal at diplotene, and stays high until anaphase when it drops off.

Dr. GALL: Well, the lampbrush chromosomes are prophase chromosomes, they are in the diplotene stage, so these are not interphase chromosomes in the strict sense of the word.

Dr. SPIEGELMAN: I made a mental calculation and perhaps you have done the same with more care; namely, how long would a DNA Watson-Crick model be? Would these lampbrush chromosomes be of approximately the same order of magnitude?

Dr. GALL: Yes, they would be about the same order of magnitude.

Dr. SPIEGELMAN: Perhaps, we can carry this comparison on to compare the DNA content for chromosomes of different sizes.

Dr. GALL: No, I haven't done that, but what I have done is to make the actual estimation for the salamander I have worked on. The DNA length is about 90 centimeters per chromosome. Comparing this with the 5 centimeter length of the chromosome, you must remember that the latter is gotten by measuring only the loops and if there is any amount of strand wound up in the granules, which I think there must be, they may well be of the same order of magnitude. In fact, I don't see any reason for not postulating that the thing is one or two strands, but I know that is a very heretical view.

Dr. SPIEGELMAN: I wonder whether you could then give supporting evidence for this, using the sort of tricks that are used in microbial systems of labeling with very hot $P^{32}$. The suggestion is that you do have a backbone of the same nature as in the Watson-Crick model, and therefore you should see breaks. If you can label with very hot $P^{32}$ and then freeze your material for awhile to allow decay to occur, you should then see the breaks in your preparations.

Dr. GALL: I haven't given any thought to such experiments, but I think that something of the sort should be done now that we know there is DNA throughout the whole chromosome.

Dr. VINCENT: Could I make some comments about nucleoli that might be pertinent if it won't take too long? I have a little evidence here on biochemical analyses of starfish nucleoli which might be pertinent, I think, within the framework of this conference. First of all, take this as the nucleolus: (See Figure 1). It has essentially three compartments that I can find: one is a single

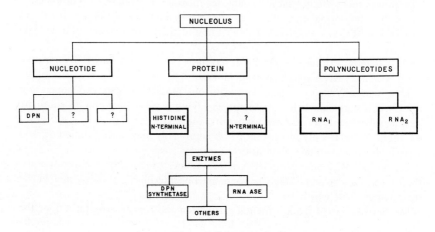

nucleotide and, as indicated earlier, in Dr. Brachet's laboratory, it allows these starfish nucleoli to synthesize DPN. The second compartment is polynucleotide and this is what is called RNA, and I shall call it RNA; and the third one is a protein compartment.

Now we can subdivide two of these compartments at least, and certainly the nucleotide compartment can be subdivided even more to not only DPN but also all of the other mononucleotides which may be significant in the formation of high-energy bonds.

The polynucleotide fraction has been separated by myself in these nucleoli into what I designated $RNA_1$ and $RNA_2$ in which the $RNA_1$ fraction is a fraction which picks up radiophosphate, at least, from an incubation medium very rapidly (this is in the intact egg) and reaches equilibration in a very short time. $RNA_2$ picks up radiophosphate and incorporates it more slowly and it carries on incorporation over a long period of time and so one can plot the specific activities of $RNA_1$ and $RNA_2$ as shown in Figure 2. From curves such as this I think we can conclude that $RNA_1$ did not contribute to $RNA_2$, otherwise the specific activity would never reach this particular level of 2.

Now the protein fraction also has at least two major components. I have just been working on N-terminal end-group analyses of the proteins of starfish nucleoli. One finds two end-groups in starfish nucleoli. I take this to mean that we have just two major species of protein present. One of these proteins is a protein which comes out with $RNA_1$ and it has as its N-terminal end-group histidine. This protein accounts for about 5% to 10% of the total nitrogen of the nucleolus. Thus, the other protein is a very large component of the nu-

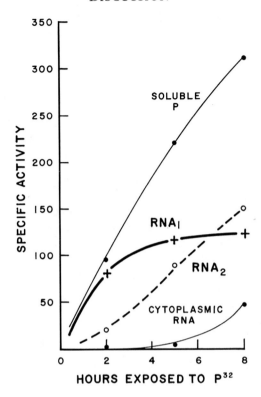

cleolus, in fact some 90% of the nucleolus, and it contains an N-terminal amino acid which I have not yet been able to identify. I might mention in this case it is interesting that it differs from the major cytoplasmic protein of these eggs, the yolk, for the yolk protein has glycine as its N-terminal amino acid. Outside of that, these two proteins appear to resemble each other very much. So, we have then a rather interesting situation in the starfish nucleolus where, if it performs a large number of functions, it must do this with a small number of proteins.

A comment that I would like to make as pertinent to Dr. Mirsky's earlier comment is that we have here a small quantity of protein which is not fractionated any further yet, other than to say it is related to the $RNA_1$ fraction. The $RNA_2$ component—that is this one which is very slow in its incorporation of radiophosphate into RNA—appears to be associated mostly with this protein with the unknown end-group which constitutes 90% of the nucleolar protein. This particular protein is an acid protein, rich in glutamic acid, resembling casein very much in its over-all components.

One might make one comment then in respect to Dr. Allfrey's and Dr. Mirsky's experiments. The combination of these two curves, that is, the curves of incorporation of $RNA_1$ and $RNA_2$ as graphed on the blackboard, gives almost the identical curve that Dr. Allfrey and Dr. Mirsky gave for their "nucleolar RNA." On this basis, then, I would suggest that if the nucleolus that exists in other cells than starfish eggs is similar to the one that I am dealing with here, we might have the nucleolar protein Dr. Mirsky is dealing with possibly containing an extra RNA component similar to this $RNA_1$.

If one takes this RNA component here ($RNA_1$) and analyses it even further one finds, at least the data I have seem best to be interpreted in this manner: that phosphate entering this nucleus, or this nucleolus, is tied up in this $RNA_1$ fraction and later reappears in the acid-soluble nucleotide fraction. This leads me to make the suggestion that only this particular fraction of RNA here ($RNA_1$) acts as a phosphate acceptor for this nucleolus, or for the nucleus, which later then redistributes the phosphate into smaller nucleotide fractions.

Dr. Markert: What is the appearance of these chromosome loops under the electron microscope after RNAse treatment? Are histones present?

Dr. Gall: I haven't looked at those under the electron microscope yet. The ones I showed you were pepsin-treated.

# ORIGIN AND STRUCTURE OF INTERCELLULAR MATRIX[1]

## Mac V. Edds, Jr.

*Biology Department, Brown University,
Providence, Rhode Island*

There are at least two reasons why it is appropriate to consider intercellular matrices in this symposium. First, there are few other materials to which so many significant embryological roles have been assigned. Secondly, numerous recent investigations of these matrices, pursued from the most diverse points of view, have provided new and exciting information which deserves careful evaluation by embryologists.

It is a striking fact that although the importance of the ground substance in embryonic differentiation was pointed out over forty years ago (2), and although embryologists have widely accepted the concepts of Weiss (e.g., 45-47) concerning its role in morphogenesis, we still know almost nothing of how this role is exercised. Grobstein (14) has outlined a provocative hypothesis of the interaction of intercellular matrices during embryonic induction. And Weiss (46), Holtfreter (22), and Moscona (29) have all discussed the possible significance of this material in the affinities and disaffinities of embryonic cells. These various proposals present a compelling challenge. What do we now know about intercellular matrices which could explain their roles in development—and what do we still need to learn?

In trying to answer these questions, it will not be possible even to mention all of the relevant investigations. The intention rather is to gather together a few selected examples which seem to be particularly instructive. As the discussion progresses, it will become apparent that in the present state of our knowledge we are long on speculation and short on explanation. There appear to be opening up, however, fields of research which may in time yield information of crucial significance for our understanding of morphogenesis.

First, a word about definitions. The terms intercellular matrix and ground substance have been used interchangeably with a lack of precision

---

[1] Unpublished research mentioned in this paper was supported by a research grant (A-1755) from the National Institute of Arthritis and Metabolic Diseases, Public Health Service.

158     THE CHEMICAL BASIS OF DEVELOPMENT

not unfamiliar to biologists. In this discussion, "ground substance" designates the optically homogeneous, viscous material which permeates the spaces between the various cells of an organism. "Intercellular matrix" will include the ground substance plus the various fibrous and mineral components which are embedded in it (cf., 9). Obviously, we deal with a motley collection of substances which will have to be carefully dissected in each particular case.

### Ground Substance

The ground substance is a complex mixture composed of water and salts, substances in transit between cells and the vascular system, and such diagnostic compounds as acid mucopolysaccharides, soluble precursors of collagen and elastin, and other proteins. The amount of ground substance varies greatly in different regions and at different ages; it is generally more conspicuous in young tissues and is gradually replaced by fibrous components in later life.

*Origin and Fine Structure*

The structural homogeneity which characterizes the ground substance in ordinary histological preparations is borne out by electron microscopic study. Granular or fibrous structures may be visualized after special treatments (1) or in dried films of extracted materials (8, 43), but the available evidence is against any visibly structured elements in the natural state (e.g., 15, 35, 36).

It has been widely accepted that the ground substance originates as a secretion of mesenchyme cells and some of their derivatives, especially fibroblasts and mast cells. In early stages of development, however, a transparent cell-free jelly occurs between ectoderm and entoderm (3), and the elaboration of ground substance should not be conceived as the exclusive property of cells of mesodermal origin. Moreover, evidence from electron microscopy confirms the earlier views of histologists that small amounts of a homogeneous "cementing" material is laid down between many different kinds of cells. Even in the adult connective tissues, where accumulations of ground substance are particularly conspicuous, specific contributions by other cell types cannot be ruled out. In fact, it would be rather surprising if the special character of the parenchymal cells of various organs was not in this fashion impressed upon the connective tissues with which they are associated.

The most often studied chemical components of the ground substance, the mucopolysaccharides, do, however, appear to arise particularly from fibroblasts and possibly mast cells (26). This view is supported, for example, by the recent studies of Grossfeld et al. (19) on mass cultures of fibroblasts from human embryonic skin and bone. These investigators found that large amounts of mucopolysaccharide, including both chondroitin sulfate and hyaluronic acid, accumulated in the medium. Tissue culture offers a potentially valuable but so far little exploited technique for studying the biosynthesis of polysaccharides (18).

*Mucopolysaccharides in Ground Substance*

Knowledge of the composition of the ground substance is based both on chemical isolations and on histochemical procedures (for reviews, see 9, 21, 28). The histochemical methods, though providing some important information about the localization of certain chemical groups, are generally conceded to lack specificity, even in conjunction with enzymatic digestion. There is, however, little doubt about the presence in ground substance of the acid mucopolysaccharides and carbohydrate-protein complexes described by histochemists.

Chemical analyses have stressed the mucopolysaccharides. Although these compounds are clearly of major importance in the functioning of ground substance, they actually represent only a small fraction of the dry weight of connective tissue (some 0.5 to 5%; 28). The exact amount of mucopolysaccharide in ground substance is unknown; no one has succeeded in isolating ground substance in sufficiently pure form for this type of analysis.

*Types of Mucopolysaccharides*

The number of distinct mucopolysaccharides in connective tissue is also unknown. Due to the pioneering work of Meyer (28) and his associates, it is clear that these substances constitute a family, and the number already isolated is imposing. Hyaluronic acid and its isomer chondroitin represent a class of non-sulfated acid mucopolysaccharides. The sulfated compounds include at least three chondroitin sulfates (A, B, C), in addition to heparin and keratosulfate. All these compounds are high molecular weight polymers; upon hydrolysis they yield hexosamines, and uronic, acetic, and sulfuric acids. In some cases, the chemical description of the monomer units is nearly complete. Hyaluronic acid, for example, is a polymer of hyalobiuronic acid, a disaccharide containing glucosamine and glucuronic acid; its structure has been established by Weissmann (28). The repeating units of other mucopolysaccharides contain such uronic acids as galacturonic, mannuronic, or iduronic acid and such alternate hexosamines as galactosamine.

The occurrence of iduronic acid in chondroitin sulfate B is noteworthy because of the close similarity of this acid to ascorbic acid. Chondroitin sulfate B is one of the most typical components of connective tissue, while ascorbic acid, as is well known, has been implicated in connective tissue maintenance and collagen synthesis (e.g., 13). If ascorbic acid is a precursor of iduronic acid, it would serve not as a vitamin but as an essential nutrient (28).

The various connective tissues contain diverse amounts and proportions of the mucopolysaccharides referred to above. In each case analyzed by Meyer, typically different proportions of hyaluronic acid and one or more of the chondroitin sulfates were encountered. The point is well exemplified by a recent comparative study of the bovine cornea and sclera (32). In the cornea, keratosulfate (lacking a uronic acid) is the major component; chondroitin sulfate A and chondroitin are present in lesser amounts. The sclera contains no keratosulfate but is rich in chondroitin sulfate B and yields lesser amounts of chondroitin sulfates A and C.

*Mucopolysaccharide Molecules*

As macromolecules, the mucopolysaccharides are elongate, flexible particles. They have a high negative charge, and hence a strong affinity for cations and water. Depending on local physical and chemical changes they readily undergo depolymerization and repolymerization, or form reversible linkages with other compounds. Their role in the regulation of water and inorganic ion metabolism has often been acknowledged. Dorfman, for example, has pointed out that "a change in concentration or molecular size of such a substance is obviously of great importance in modifying the capacity of connective tissue to bind water and salts" (9, p. 82). Polatnick, LaTessa, and Katzin suggest that the differences between corneal and scleral mucopolysaccharides are related to the "more closely controlled state of hydration of the cornea" (32, p. 364). This may also be the significance of the varying mucopolysaccharide compositions of other connective tissues. But the possibility is open that constant ratios exist between particular mucopolysaccharides and the protein moieties of some mucoprotein complexes, and that these variations therefore have an altogether different significance. In so far as this matter has been studied, the existence of such a ratio between collagen and mucopolysaccharides is uncertain (cf. 28, 42). But at least in the case of the corneal and scleral analyses considered above, a rather constant ratio between non-collagenous protein and mucopolysaccharides is indicated; the cornea contains about 2.5 times as much mucopolysaccharide as the sclera and also has a proportionately larger amount of non-collagenous soluble protein; the collagen fractions in the two tissues, however, show only minor differences (33).

*Mucopolysaccharide-Protein Complexes*

Although our knowledge of the chemistry of mucopolysaccharides is far from complete, enough information is at hand to suggest that these compounds create more problems for embryologists than they solve. There are, for example, no clear hints that the properties of these substances bear directly on such matters as the specificity of cellular interactions, embryonic induction, or guidance during cell migrations. However, the ground substance is not just a collection of free mucopolysaccharides. On the contrary, it is becoming increasingly apparent that most mucopolysaccharides are complexed with proteins (e.g., 27). This obviously opens possibilities for speculations about specificity, but unfortunately, little of an exact nature is yet known.

Recent analyses of cartilage provide examples of the current status of information about mucoprotein complexes in the ground substance. Bernardi (7) has concluded that about one-third of the chondroitin sulfate (mainly A) of cartilage is linked to a non-collagenous protein, the rest apparently being associated largely with collagen. From analyses of viscosity, sedimentation, and light scattering, Bernardi infers that chondroitin sulfate and polypeptides are joined end-to-end in linear chains.

Partridge and Davis (31) have also shown that at least part of the chondroitin sulfate of cartilage is in firm chemical combination with a non-collagenous protein. This protein has been isolated and its amino acid composition determined. It is altogether different from collagen, being rich in tyrosine and low in both glycine and proline. The protein moiety is apparently influential in forming molecular aggregates, since solutions of the complex rapidly become less viscous upon treatment with proteolytic enzymes. Previous studies (30, 41) had indicated that the ability of these chondroitin-sulfate–protein preparations to combine reversibly with collagen is destroyed by heating or dilute alkali. It therefore seems likely that these linkages with collagen depend not on free chondroitin sulfate but on the mucoprotein complex.

However praiseworthy the analytical work on which these statements are based, it is at once evident that there is no obvious and immediate relation between the known properties of the macromolecules occurring in the ground substance and the developmental events we are seeking to explain. The elongate shape and flexibility of these molecules, their ion-binding characteristics, the ease with which their carbohydrate and protein moieties may associate or dissociate under varying conditions of $p$H or ionic strength, the dynamic equilibria of synthesis and degradation which we may visualize during growth and differentiation—all tell us that we are dealing with a responsive and critical population of substances. But beyond this, the path is dark and sign-posts are lacking.

## Fibrous Components

Collagen is the most abundant and most studied fibrous component of the intercellular matrices. Due to the interest shown in this substance during the past fifteen years by biologists, biochemists, and biophysicists, we now have a remarkably full catalogue of information about it (for reviews, see 4, 5, 20, 34, 40).

For the purposes of this discussion, three important points must be made. First, as Schmitt (37, 38), Gross (16), and others have pointed out, collagen has become a prototype among fibrous proteins—not only because of the intensity with which it has been studied but also because of its significance as a model for studies of morphogenesis at the macromolecular level. Next, collagen participates in the formation of a variety of fibrous fabrics, some with an astonishing complexity of fine structure which also pose intriguing morphogenetic problems. Finally, the often suggested but rarely analyzed relation between collagen and the mineralization of an intercellular matrix is turning out to be even more intimate than previously suspected. Each of these points deserves brief comment.

### Collagen Fibrogenesis

A collagen fibril is an array of adlineated collagen molecules, the tropocollagen particles of Schmitt, Gross, and Highberger (39). The individual molecules are composed of three helically coiled polypeptides strongly linked by hydrogen bonds; they are some 3000 Å long and 14 Å wide. Glycine, proline, and hydroxyproline are conspicuous among the amino acid residues of collagen, the last-named acid being regarded as virtually diagnostic.

The native fibril of adult vertebrate collagen has a cross-banded pattern with an average axial period of about 640 Å. The bands have been attributed to zones of relative molecular disorder resulting from the interactions of long side-chains which repeat regularly along the laterally associated tropocollagen particles (4, 40). In a series of exciting experiments growing out of observations first made by Nageotte, Schmitt and co-workers (40) have demonstrated that collagen which has been dissolved in appropriate solutions can be caused by a number of treatments to reconstitute into fibrils of precisely ordered structure. The treatments include alterations in $p$H or ionic strength, as well as the addition of one of several organic compounds (e.g., chondroitin sulfate, ATP, glycoprotein, or heparin). Depending on the exact conditions employed, the fibrils reconstitute as one of several naturally occurring types with 640 Å, 220 Å, or no axial periods; or as one of at least two unnatural types with periods of 2600 to 3000 Å, that is, approximately the length of the individual tropocollagen particles.

These diverse patterns are interpreted as resulting from the aggregation of elongate molecules which, being polarized, become arranged either in parallel or antiparallel arrays. The molecules are assumed to lie either with their ends in register or staggered by some specific fraction of their length. The important conclusion was reached that the type of fibrillar structure formed depends critically on two factors: one, the built-in properties of the individual collagen macromolecules; and two, the chemical environment which determines the degree of ionization, the interactiveness of specific side chains, and the like. Although it has been claimed that mucopolysaccharides are involved in the biogenesis of collagen fibrils or in stabilizing the linkages between adjacent collagen molecules (e.g., 23), there is no really cogent supporting evidence. Schmitt, Gross, and Highberger have emphasized particularly that "the precipitation of collagen fibrils by various substances is no proof that these substances form an essential part of the structure of native collagen or are involved in its biogenesis" (40, p. 150).

In the intact organism, fibrogenesis presumably involves comparable events. That is, collagen macromolecules which have been synthesized in fibroblasts are deposited in the immediately adjacent ground substance, where they then aggregate into fibrils. Whether any fibril formation occurs intracellularly is still undecided despite several electron microscopic investigations (for literature, see 24). No matter where collagen fibrils first emerge as electron optically visible elements, the main point of the in-vitro work just discussed is that once definitive collagen macromolecules are available, no specific external factors are required to fabricate them into fibrils. The structural order of the fibril will emerge spontaneously. The physical and chemical environment in which the molecules interact imposes limits on the type of fibril which will form. But the detailed specificity of the interaction, and hence the configuration of the final product, depends on the collagen particles themselves.

## Ordered Fibrillar Aggregates

The appearance during development of oriented fibrillar aggregates depends in part on prevailing patterns of tension (e.g., 47), and in part on interactions between individual fibrils and the ground substance in which they are embedded. A developing tendon is a case in point. As Jackson (24) has shown, collagen fibrils each about 80 Å in diameter appear in parallel bundles in the metatarsal tendons of the 8-day chick embryo. The diameter of the fibrils increases progressively with age, an especially rapid increase occurring between the 12th and 18th days. Diameters of some 400 Å are encountered by hatching, 750 Å in the adult bird.

The individual fibrils are embedded in a structurally homogeneous ground substance; they lie in the characteristic hexagonal pattern assumed by closely packed, equidistant cylindrical rods. At any given stage, all fibrils have essentially the same diameter; that is, they are growing laterally at equal rates. As development proceeds, the relative amount of interfibrillar material diminishes; the fibrils thus become ever more closely packed, until in the adult they are separated by distances only a fraction of their diameters.

The growth of the fibrils must therefore be presumed to involve the following types of events. Tropocollagen particles, synthesized in neighboring fibroblasts, continuously permeate the ground substance and become equally distributed around the growing fibrils. Some of these particles adhere to the fibrillar surfaces, aggregate end-to-end and side-to-side with others of their kind, and thus add in an orderly manner to the bulk of the fibrils.

However, some fibrils lie within a few Ångström units of a fibroblast; others near the center of a bundle lie several hundred Ångström units from the nearest source of collagen molecules. Conditions in the interfibrillar region immediately adjacent to the fibrils clearly favor the aggregation of the collagen molecules and their accretion on the fibrils. It is therefore not easy to understand how equal rates of growth are maintained by all fibrils of a given region (35). If the diffusion of collagen molecules were a limiting factor in the growth of the fibrils, then a gradient of diameters—decreasing toward the center of each bundle—would be expected. Since there is no such size gradient, growth control must involve some other mechanism. The simplest hypothesis to account for the facts would assume that collagen molecules en route from the source of supply are in some respect incomplete and hence unable to aggregate on fibrils until "activated." The details of activation are irrelevant; they could include some terminal synthetic event or perhaps the splitting off of a small peptide, as in the activation of fibrinogen (11). Activation would have to occur, however, in the immediate vicinity of the fibrils. And it would have to proceed at a relatively slow rate compared to that at which the collagen molecules were supplied. If these conditions were met, fibrils of uniform diameter would result (44). It is, therefore, of interest that Gross (17) has described evidence implying that in the connective tissues of an actively growing guinea pig the rate of collagen synthesis is greater than the rate of polymerization into fibrils; this rate difference leads to the accumulation of a substantial reserve of collagen monomers in the ground substance.

A similar explanation may be offered for the uniform diameters of fibrils in the intricately structured basement membrane of the amphibian larval epidermis which Weiss and Ferris (50) have studied. Here also the rate of lateral growth of fibrils is the same throughout the membrane, no matter what the distance which separates individual fibrils from the subjacent mesenchyme cells.

There are other morphogenetic problems posed by this membrane, however, which do not seem to have such simple answers. Before discussing them, a brief description of the normal development of the membrane is in order (10). In early tailbud embryos of the frog (about Stage 18), the ectoderm and subjacent mesoderm become firmly adherent. A fine membrane, presumably containing mucopolysaccharides but as yet free of collagen, appears between the two layers and slowly thickens over a period of about two to three days. At this point (Stage 21-22), there are abrupt changes in the character of the union between ectoderm and mesoderm, and in the membrane itself. The two outer germ layers become less firmly bound and the mesoderm forms a typical mesenchyme. The ectoderm can now be stripped from the membrane by considerably lower doses of ultrasonic radiation (6) than are required at earlier stages. The first detectable traces of hydroxyproline appear in the membrane, implying that collagen is being laid down (10). This is supported by electron micrographs of the membrane which show numerous fibrils already beginning to assume the orthogonal order which characterizes the more mature membrane (50). As in older membranes undergoing wound healing (51), orthogonality and layering appear simultaneously. That is, a single layer of fibrils coursing in one direction is found in conjunction with a second layer running at right angles.

In their discussion of the origin of this ordered arrangement during the formation of the basement membrane, Weiss and Ferris (50, 51) called attention to the possible existence in the ground substance of an "underlying lattice system" which would serve to align the fibrils as they formed. However, their argument favored the alternative possibility that the alignment results directly from properties of the interacting fibrils. Such interaction between fibrils might bring their crossbanded regions into register and serve as "nodal points" for the subsequent ordering of new fibrils. The axial periods, fibril diameters, and interfibrillar distances are all of equal dimensions in the mature membrane and fit neatly into a geometric pattern whose fundamental unit is a cube about 500 Å on a side (48, 49). As Weiss and Ferris emphasized, no choice can be made between these alternative explanations. If either of them is to be regarded as correct, some sort of "macrocrystallinity" with a lattice system measured in hundreds of Ångström units must be postulated. How much of the organization of this lattice depends on spontaneous interaction at the macromolecular level, and how much on ordering processes emerging from surrounding cells, is at present impossible to decide. It may be in order, however, to insist that details of the mechanism lie at and must ultimately be sought at the macromolecular level.

## Mineralization of Collagen Fibrils

An equally challenging and even less explored series of developmental problems at the ultrastructural level is posed by the process of mineralization of intercellular matrices. Inorganic salts are normally deposited in various skeletal parts in ordered crystalline patterns. Similar deposits are formed in some pathological states, in tissues which do not usually calcify.

A number of recent electron microscopic and x-ray diffraction studies have revealed that an intimate topographical relation exists between the collagen fibrils in the matrix of bone and the crystals of hydroxyapatite (for literature, see 12, 25). The rod-shaped crystals are of minute dimensions; estimates of width fall mostly between 20 and 50 Å, of length between 200 and 700 Å. All observers agree that these crystals are deposited along the collagen fibrils, either on the main bands or in interband regions. Several speculations have been advanced as to the factors in the organic matrix of bone which might control mineralization. Specific compounds (e.g., chondroitin sulfate, mucoproteins, enzymes), as well as the physical or chemical state of the ground substance (e.g., ion-binding characteristics, degree of polymerization), have been mentioned in this connection.

These speculations, and the descriptive observations on which they have been largely based, must now be reevaluated in the light of an experimental study in vitro of the collagen-hydroxyapatite system by Glimcher, Hodge, and Schmitt (12). According to these authors, "the mechanism of nucleation involves a specific stereochemical configuration resulting from a particular state of aggregation of collagen macromolecules" (12, p. 861). By combining metastable solutions containing calcium and phosphate ions with tropocollagen in solution, or with one of the various types of reconstituted collagen fibrils described above, it was demonstrated that crystallization of hydroxyapatite occurred only on fibrils with the native 640 Å axial period. Neither the individual macromolecules nor fibrillar aggregates of any other type served as nucleation centers. The identity of the inorganic crystals was established by x-ray and electron diffraction and by chemical analysis. It must be assumed, therefore, that when collagen molecules have become so aligned as to form fibrils of the native type, they then present surface groups or sites whose configurations are sterically adapted to the crystal lattice of hydroxyapatite.

The reconstituted fibrils used in this study were prepared with collagen from a number of sources which do not normally become calcified. Thus, whatever the barrier to mineralization of such tissues, it cannot be attributed to the collagen molecules themselves, since despite their origin, they induce the formation of hydroxyapatite if properly aggregated. However, collagen fibrils which are directly isolated from tendon (i.e., *not* reconstituted), do not calcify when exposed to solutions of calcium and phosphate ions even at concentrations where spontaneous precipitation occurs. It is tempting to attribute this difference in the behavior of reconstituted and original fibrils to mucopolysaccharides associated with the latter. Considerable interest will therefore attach to future efforts to dissect this hypothetical inhibitory action of mucopolysaccharides by treatments which remove or modify them.

We thus return to our starting point—the ground substance and the presumably critical but poorly known properties of the mucopolysaccharides and mucoprotein complexes which it contains. In each of the examples which we have considered, these substances apparently play a crucial but tantalizingly obscure role. By turning from the function of the ground substance in events which have often received the attention of embryologists to events in the domain of ultrastructure and macromolecules, we have uncovered a number of fascinating new puzzles, perhaps without greatly clarifying the older problems. But the success already achieved by dealing with morphogenetic events at the new level fosters the hope that some of the less precisely characterizable problems with which embryology has been concerned will ultimately find solutions which can be expressed in at least equally satisfying terms.

## REFERENCES

1. Asadi, A. M., Dougherty, T. F., and Cochran, G. W., *Nature,* **178,** 1061-1062 (1956).
2. Baitsell, G. A., *J. Exptl. Med.,* **21,** 455-479 (1915).
3. ———, *Quart. J. Microscop. Sci.,* **69,** 571-589 (1925).
4. Bear, R. S., *Advances in Protein Chem.,* **7,** 69-160 (1952).
5. ———, *Symposia Soc. Exptl. Biol.,* **9,** 97-114 (1955).
6. Bell, E., *Proc. Biophys. Soc.,* in press (1958).
7. Bernardi, G., *Biochem. Biophys. Acta,* **26,** 47-52 (1957).
8. ———, Cessi, C., and Gotte, L., *Experientia,* **13,** 465-466 (1958).
9. Dorfman, A., in *Connective Tissue in Health and Disease* (G. Asboe-Hansen, ed.), p. 81-96, E. Munksgaard, Copenhagen (1954).
10. Edds, M. V., Jr., *Proc. Natl. Acad. Sci. U. S.,* **44,** 296-305 (1958).
11. Ferry, J. D., Katz, S., and Tinoco, J., *J. Polymer Sci.,* **12,** 500-510 (1954).
12. Glimcher, M. J., Hodge, A. J., and Schmitt, F. O., *Proc. Natl. Acad. Sci. U. S.,* **43,** 860-867 (1957).
13. Gould, B. S., and Woessner, J. F., *J. Biol. Chem.,* **226,** 289-300 (1957).
14. Grobstein, C., in *Aspects of Synthesis and Order in Growth* (D. Rudnick, ed.), p. 233-256, Princeton University Press, Princeton (1955).
15. Gross, J., *Ann. N. Y. Acad. Sci.,* **52,** 964-970 (1950).
16. ———, *J. Biophys. Biochem. Cytol.,* **2** Suppl., 261-274 (1956).
17. ———, *J. Exptl. Med.,* **107,** 265-277 (1958).
18. Grossfeld, H., *Exptl. Cell Research,* **14,** 213-216 (1958).
19. ———, Meyer, K., Godman, G., and Linker, A., *J. Biophys. Biochem. Cytol.,* **3,** 391-396 (1957).
20. Gustavson, K. H., *The Chemistry and Reactivity of Collagen,* Academic Press, New York (1956).
21. Hale, A. J., *Intern. Rev. Cytol.,* **6,** 193-263 (1957).
22. Holtfreter, J., *Ann. N. Y. Acad. Sci.,* **49,** 709-760 (1948).
23. Jackson, D. S. *Biochem. J.,* **56,** 699-703 (1954).
24. Jackson, S. Fitton, *Proc. Roy. Soc. (London),* **B, 144,** 556-572 (1956).
25. McLean, F. C., *Science,* **127,** 451-456 (1958).
26. McManus, J. F. A., in *Connective Tissue in Health and Disease* (G. Asboe-Hansen, ed.), p. 31-53, E. Munksgaard, Copenhagen (1954).
27. Meyer, K., *Discussions Faraday Soc.,* **13,** 271-275 (1953).
28. ———, *Harvey Lectures,* **51,** 88-112 (1957).
29. Moscona, A., remarks at Harvard Cancer Conference (1957).
30. Muir, H., *Biochem. J.,* **62,** 26P (1956).
31. Partridge, S. M., and Davis, H. F., *Biochem. J.,* **68,** 298-305 (1958).
32. Polatnick, J., La Tessa, A. J., and Katzin, H. M., *Biochim. Biophys. Acta,* **26,** 361-364 (1957).
33. ———, ———, and ———, *Biochim. Biophys. Acta,* **26,** 365-369 (1957).
34. Randall, J. T., *Nature and Structure of Collagen,* Academic Press, New York (1953).
35. ———, *J. Cell. Comp. Physiol.,* **49** Suppl., 113-127 (1957).
36. Rothman, S., *Physiology and Biochemistry of the Skin,* University of Chicago Press, Chicago (1954).
37. Schmitt, F. O., *Proc. Natl. Acad. Sci. U. S.,* **42,** 806-810 (1956).
38. ———, *J. Cell. Comp. Physiol.,* **49** Suppl., 85-104 (1957).
39. ———, Gross, J., and Highberger, J. H., *Proc. Natl. Acad. Sci. U. S.,* **39,** 459-470 (1953).
40. ———, ———, and ———, *Symposia Soc. Exptl. Biol.,* **9,** 148-162 (1955).

41. Shatton, J., and Schubert, M., *J. Biol. Chem.*, **211**, 565-573 (1954).
42. Slack, H. G. B., *Biochem. J.*, **65**, 459-464 (1957).
43. Sylvén, B., and Ambrose, E. J., *Biochim. Biophys. Acta*, **18**, 587 (1955).
44. Waugh, D. F., *J. Cell. Comp. Physiol.*, **49** Suppl., 145-164 (1957).
45. Weiss, P., *Amer. Naturalist*, **67**, 322-340 (1933).
46. ———, *J. Exptl. Zool.*, **100**, 353-386 (1945).
47. ———, in *Chemistry and Physiology of Growth* (A. K. Parpart, ed.), p. 135-186, Princeton University Press, Princeton (1949).
48. ———, *Proc. Natl. Acad. Sci., U. S.*, **42**, 819-830 (1956).
49. ———, *J. Cell. Comp. Physiol.*, **49** Suppl., 105-112 (1957).
50. ———, and Ferris, W., *Proc. Natl. Acad. Sci. U. S.*, **40**, 528-540 (1954).
51. ———, and ———, *J. Biophys. Biochem. Cytol.*, **2**, Suppl., 275-282 (1956).

## DISCUSSION

Dr. Steinberg: Dr. Edds, in the earlier part of your discussion you suggested that possibly all cells are capable of producing ground substance. You mentioned in particular the work of Drs. Weiss, Grobstein, and Holtfreter on the role of the so-called ground substance in contact guidance and in induction and possibly in tissue affinities. Do you mean to imply that the intercellular matrix which is studied in connective tissues is present not only in connective tissue spaces, but also between, let us say, epithelial cells of the organism? In other words, do you mean to say that all of the cells in a particular tissue are surrounded by something which we might loosely term intercellular cement?

Dr. Edds: Yes, embryonic cells are imbedded in this ground substance; this is commonly accepted. I was trying to suggest that information from studies of adult connective tissues may provide hints which will help us to understand its role in the embryo.

Dr. Steinberg: I have recently been through some of the literature on the so-called intercellular cement which has been presumed by many to glue the cells of the organism together. It seems to me that the idea of such a ubiquitous cement is a conceptual extension of observations actually made in connective tissue. The presence of such cement between other kinds of cells in different tissues has not really been substantiated. In particular, if one examines electron micrographs of cells in various tissues, one finds that these cells are in very close approximation to one another. The distance in general between the osmiophilic layers of adjacent cell membranes is roughly 200 Ångstroms. This is about the same distance that we find between the two membranes of the nucleus, between the layers of closely packed endoplasmic reticulum, between the membrane pairs constituting the cristae or the outer walls of mitochondria, or between the closely packed layers of myelin, which as we know are Schwann cell membranes. And, of course, this 200 Ångstrom distance between cells is only the maximum possible distance between them, for there may be non-staining layers between the paired osmiophilic ones. Tissue cells seem, therefore, to be as closely approximated as the structure of lipoprotein membranes will permit. Consequently it seems unreasonable to me to assume that there is a proteinaceous, matrix-like cement, distinct from the cell surfaces, between them. Would you care to comment on this?

Dr. Edds: As far as the 200 Ångstroms space is concerned, this is still pretty big when we consider molecular dimensions, even of the macromolecules we are discussing.

Dr. Steinberg: Of course this is only the maximum possible intercellular distance. The presence of a non-osmiophilic component external to the osmiophilic one would reduce or extinguish the apparent "space."

Dr. Edds: Yes, of course. One still does not know what lies within it, however. There is one matter I would like to comment on which I was not able to include in my talk; that is, the situation in the basement membrane of the epidermis of the frog embryo. I cannot see this membrane under the highest resolution of the light microscope until about Shumway stage 19, no matter how I stain it. However, I can find it earlier just by going to the embryo and stripping off the epidermis with ultrasound. I didn't think there was any membrane and it was surprising to find it before it was optically detectable in sections. In fact, this "invisible" basement membrane is thick enough to dissect from the embryo. But it is dissolved by the technique which I was using to study it microscopically. So, I think one has to be very cautious about conclusions concerning the physical existence of these substances until one has explored several different ways of revealing their presence or their absence.

Dr. Afzelius: As an electron microscopist I would like to comment on this 200 Ångstrom distance between cells. This distance might be what actually exists between living cells and then again it might be many, many other things. The possibilities of artifacts are, of course, manifest when one uses the conventional techniques of the day in electron microscopy.

I would now like to comment on the question asked after Dr. Allen's fine paper. Someone asked what happened to those cortical granules of the sea urchin egg that were not extruded during fertilization. I am convinced that the cortical granules which are acid mucopolysaccharides according to the histochemist do not all contribute to the fertilization membrane. Over half of them form the so-called hyaline layer. This layer is homogeneous or slightly fibrous in the electron micrographs. It has been known since 1900 that this layer can be dissolved away, after which the blastomeres will fall apart. If one examines the two-cell stage of the sea urchin egg, the cells seem to be stuck together by a cell cement of acid mucopolysaccharide from the hyaline layer and in the closest places they are only 250 Ångstroms apart. This observation might have some bearing on the problem which has been raised. If one washes away this substance, then the blastomeres fall apart. If they are left alone, however, the blastomeres remain together by this intercellular cement. Therefore, in the case of the cortical granules, you might have a nice opportunity to study materials which may be possible precursors of the ground substance which has been discussed here.

DR. EDDS: That may well be possible. However, I have a feeling that this system may be too minute for such studies. Possibly in vitro culture systems will be useful for studies of biosynthesis. I am afraid that the embryologist has little chance to make a contribution to this field by using cortical granules because the amounts of materials are so very small.

DR. WEISS: I would like to comment briefly on this question raised about intercellular cement. In tissue culture we find that epithelial cells of the same kind form an island which may contain only 4 or 5 cells. We can see that these cells will mill around, will shift their positions relative to one another, and yet the island will remain compact and retain its identity. It seems pretty difficult to explain this in any terms other than assuming that there is an envelope of some kind which surrounds the island. The mode of recognition which keeps cells together cannot be a cement because they are not cemented. They shift constantly relative to one another and yet they retain a common geometrically simple contour. This implies that there must be a ground mat which holds them together. Now Dr. Moscona in our laboratory before he left studied this and in the cases where he could find this envelope under the microscope it stained with periodic acid Schiff. The point I want to make is that the idea of cell glue or cell cement is largely an artifact.

DR. MARKERT: Do you have any evidence concerning the metabolic stability of the intercellular matrix?

DR. EDDS: I am relatively unfamiliar with this aspect of the ground substance but it has been studied extensively, for example by Dorfman. His investigations show that mucopolysaccharides are relatively labile constituents.

DR. ALLEN: I would like to suggest a system for the possible study of the biosynthesis of acid mucopolysaccharides. The sea urchin cortical granules are probably too small, as you suggested, but the fish cortical alveoli, which supposedly contain the same kinds of substances, are extruded from the egg and form the colloid of the perivitelline space. These egg alveoli are large enough and probably concentrated enough to work with. They range from 4 or 5 microns up to 25 or 30 microns in size and show a very intense PAS reaction. They are formed at a very early stage of oogenesis and Dr. Norman Kemp has shown that they come from the primary yolk vesicles.

I would also like to ask a question concerning the remark made about possible false-positive reactions with the PAS stain, and metachromasia. I was wondering if you have any information on what kinds of substances give false-positive reactions.

DR. EDDS: As I understand it, the important step in the PAS reaction is the cleavage of a carbon-carbon bond having adjacent OH or OH and $NH_2$ groups. Since there are a large number of compounds which can be oxidized at such a site, there is always difficulty in interpreting the results of the technic.

DR. LENHOFF: I think that it should be mentioned that embryological studies involving collagens may offer advantages. Since the amino acid hydroxyproline is found almost uniquely and in high concentrations in proteins belonging to the collagen family, it is possible to follow changes in the collagen content of cells and tissues by carrying out the relatively simple and specific chemical assay for hydroxyproline. Thus, cytological changes may be concomitantly followed by the more quantitative chemical assay.

In our own laboratory, Mr. Edward Kline and I have begun to pursue a related problem while studying the differentiation of the nematocysts of *Hydra littoralis*. This problem arose from our observation of radioautographs of $C^{14}$-labeled hydra which demonstrated that specific labeled "loci" in the body tube migrated and concentrated in the batteries of the tentacles. In an attempt to identify the labeled material as the nematocysts, a method was developed for the isolation of these stinging organoids. An analysis of the purified capsules of the previously described "chitinous" nematocysts demonstrated the presence of over 20% hydroxyproline and of a similar amount of proline; these results suggested that the capsule of the nematocyst is an unusual member of the collagen family. Since other studies indicate that hydroxyproline is not present in significant quantities in other hydra proteins, it appears feasible to investigate problems such as nematocyst differentiation and migration, tentacle regeneration, and reconstitution by quantitatively following hydroxyproline.

DR. EDDS: I am glad you gave me an opportunity to comment on hydroxyproline because I have become interested in just this feature of the collagen molecule. Collagen is the only animal protein which we know of that is characterized by a particular amino acid. For this reason, it lends itself to chemical analyses by people who may not be adept at the more complex procedures of protein chemistry. Even if one is not an expert protein chemist it is therefore possible to develop considerable precision in tracing the development of collagen merely by following the one amino acid, hydroxyproline.

DR. SCHEIDERMAN: I should like to remark, Dr. Edds, that F. C. Steward and his colleagues have found that in rapidly growing tissue of many plants, and I believe carrot is one of them, hydroxyproline is found in large quantity in certain proteins. The situation with respect to hydroxyproline here is quite different from that in collagen, but in any case, hydroxyproline is really not unique to collagen.

DR. EDDS: Thank you. Of course, I should have qualified my reply by saying that hydroxyproline is found in a few other restricted places. If anyone can tell the collagen chemists what hydroxyproline is doing in these plant tissues I am sure it would make them very happy.

DR. HADORN: Has anybody examined the ground substance in the jelly fish? That would seem to be a very desirable object to study in this respect.

DR. EDDS: I do not know of any such studies.

DR. BRACHET: I believe that Dr. Bouillon, in Brussels has looked at the jelly fish under the electron microscope. His conclusion was that the mesoglea fibres are not made out of collagen, but that they rather resemble the elastic tissue of the vertebrates.

# CYTOPLASMIC PARTICULATES AND BASOPHILIA[1]

HEWSON SWIFT

*Whitman Laboratory, Department of Zoology
University of Chicago*

INTRODUCTION

Certain particulate components of the cytoplasm (mitochondria, basophilic inclusions, Golgi material) are, with few exceptions, of universal occurrence in all cells. They frequently possess shapes or distributions characteristic of different tissue types, and in many cases change in specific fashion with differentiation and the physiological conditions of the cell. In spite of their wide occurrence, our knowledge of the form and function of cell particulates is still fragmentary, although it has been immeasurably increased in recent years by studies with the electron microscope and from biochemical investigations on isolated cell fractions.

## 1. Mitochondria

Almost 50 years ago Warburg found that certain metabolic enzymes were inextricably bound to cell particulates. In the intervening years of work on isolated mitochondria, begun by Bensley and Hoerr (6) and continued by Claude (17) and his co-workers, the importance of mitochondria as enzyme carriers has been thoroughly established. The enzymes of mitochondria are diverse, functioning in synthesis and hydrolysis, as well as in oxidative metabolism (71). The morphology of mitochondria has also been thoroughly studied with the electron microscope, chiefly after osmium fixation. In the great majority of cells they appear to be limited by a "double membrane" of two dense layers enclosing a less dense space. They contain a number of inner lamellate structures (cristae), or filaments, that have been considered either to be extensions of the inner

---

[1] The author is greatly indebted to Dr. Ellen Rasch for invaluable collaboration in every phase of the work described here. Figs. 2, 5, 11, and 20 were photographed by Dr. Rasch; Fig. 24 was taken by Dr. L. E. Roth; Fig. 29 by Mr. W. Ferris; Fig. 35 by Dr. L. Herman; and Figs. 36 and 37 by Dr. A. Ruthmann. The writer expresses his appreciation to these associates for their kind permission to publish this material. Aided by Grants C-1612 and C-3544 of the U. S. Public Health Service, and the Abbott Memorial Fund.

limiting membrane (48), or as separate structures in contact with it (74). In spite of the wide morphological variation encountered among different tissues, these characters define a surprisingly consistent class of particulates, present in all cells except for those in certain terminal states, such as mature erythrocytes or the cells of outer squamous epidermis.

In most tissues there are also particulates that bear a doubtful relation to typical mitochondria, such as the "microbodies" (68) and "dense bodies" (45) of rat liver, "components I and II" of mouse small intestine epithelium (92), and the "globoid bodies" of HeLa cells (35). Further characterization of such components, and of their enzyme composition if any, must await more subtle isolation procedures, and the further development of enzyme localization techniques for the electron microscope.

Many basic questions concerning mitochondria still remain unanswered. For example, we know next to nothing about mitochondrial origin, although evidence has been presented of their formation from preexisting mitochondria (85, 31), from the cell membrane (31), or from microbodies (68), to name a few current concepts. Is the mitochondrion itself the site of enzyme synthesis, or merely of its aggregation? In this connection the cinematographic observations of Frédéric and Chèvremont (25) on temporary connections between mitochondria and the nuclear membrane are of interest. Mitochondria show morphologies specific for their tissue type. It would be interesting to know how such differences occur in differentiation, whether new types of mitochondria arise, or preexisting embryonic types are modified. These and many other problems await further study.

## 2. Golgi Apparatus

Fine-structure studies have also penetrated somewhat the classical cloud of confusion surrounding the Golgi apparatus far enough to provide a consistent morphological picture. From studies with the light microscope, often employing unpredictable cytological methods, Golgi substance has variously been considered to be fibrous, vacuolar, lamellar, or canalicular, as lipid droplets (3), or as myelin figures produced during fixation (51). In light of this disorder, the constant pattern of Golgi fine structure is most welcome. In the electron microscope after osmium fixation Golgi bodies appear as a series of densely packed parallel double membranes, often slightly expanded at their ends, surrounded by numerous small "vacuoles" or "granules" roughly 50 m$\mu$ in diameter. Since these small vacuoles have limiting membranes similar to the lamellae, it is possible that the two types of structure represent different phases of the same material. In some cases the lamellae may be expanded by large vacuoles of low density, and the whole complex may be surrounded by a dense area of cytoplasmic ground substance (78).

In some tissues, such as mouse epididymis (20), Golgi components are aggregated into characteristic bodies U-shaped in section, and located near the distal nuclear margin. In the small intestine mucosa they exist as several unconnected plate-like structures lying parallel to the cell axis (92), and in rat hepatic cells as somewhat less organized complexes lining the bile canaliculi (23). Similar structures also occur in plant cells, for example, the "dictyosomes" shown by Porter (60). In spite of different arrangements within the cell, the characteristic Golgi morphology is usually readily recognized. An old dispute concerning the Golgi nature of the dictyosomes of certain spermatocytes has thus been adequately resolved, and the acroblast formed from the aggregation of dictyosomes is clearly Golgi-like in structure (12). There are, however, certain membrane complexes showing morphological differences from the usual Golgi pattern, such as the stacked, short lamellar aggregates of flagellates (70), or the smooth-contoured whorls in rat motor neurons (54), whose Golgi nature seems unclear.

A number of cytochemical studies have confirmed the high phospholipid content of Golgi material. Evidence for two different lipid fractions has been presented, one soluble and the other insoluble in 70 per cent alcohol (20). The presence of polysaccharides has been suggested on the basis of the periodic acid-Schiff reaction (30), but lipids have also been considered responsible for this finding (72). The presence of alkaline phosphatase in the Golgi zone of certain epithelial cells has been described on the basis of histochemical tests (22, 46), and also from the isolated Golgi fraction (72). None of these findings are very helpful in elucidating Golgi function. The obvious change in Golgi material with secretion has frequently been demonstrated, for example, by Moore, Price, and Gallagher (44) in rat prostate. Golgi involvement in acrosome synthesis, at least in some spermatids, also seems indicated. However, pilocarpine injection produced no obvious changes in the Golgi apparatus in pancreas, although in acinar cells a series in size and density could be arranged from small Golgi vacuoles to fully formed zymogen granules (78). Except for a poorly defined role in secretion, possibly in condensing or partitioning products synthesized elsewhere, we are almost completely ignorant of Golgi function.

## 3. Cell Membranes

It is evident from a number of electron microscope studies that cell membranes are not merely passive boundaries. They are strikingly convoluted in many tissues (55, 65), and may extend into "microvilli" at free cell surfaces (90, 92), affording good morphological support to physiological concepts of active transport and absorption. Such findings suggest that cell membranes frequently invaginate to form cytoplasmic channels and intracellular vesicles, taking fluids or large particulate materials across cell membranes by pinocytosis, phagocytosis, or extrusion. Certain cell inclusions thus may continually arise from internal blebbing of cell membranes (31, 50, 90). Where such processes occur between cells it seems possible that large chunks of cytoplasm may freely be exchanged. The importance of such processes in cell interaction, for example in embryonic induction, is as yet completely unexplored.

### 4. Basophilic Components

Our knowledge of the RNA-containing structures of the cytoplasm has emerged spasmodically over a long period. A number of early studies indicated that basophilic cell components (i.e., those stainable with basic dyes) must somehow be associated with synthetic processes. For example, in his studies on gland cells Garnier (28) considered that the filamentous, lamellate, or whorled structures comprising the "ergastoplasm" were important in the process of secretion. In subsequent studies the term ergastoplasm was also applied to filamentous or flocculent inclusions found in a variety of other animal and plant cells, including starfish oocytes and embryo sacs of lily (63), where the "ergastoplasm" was thought to be engaged in the elaboration and transformation of various cell substances. In his studies on annelid oocytes, Lillie (40) suggested that the highly basophilic particles, which he called "microsomes" after Hanstein (34), were in some way associated with yolk synthesis. Similar conclusions were reached for other cell types by numerous investigators, although in many cases the "microsomes" were inadequately differentiated from mitochondria and other cytoplasmic materials. Finely granular or filamentous components of the cytoplasm have also been called "chromidia" (36), because their staining properties were similar to certain components of the nucleus. Goldschmidt (32), among others, considered that the "chromidial apparatus" was formed from basophilic components extruded from the nucleus, and that it possessed a synthetic or trophic function. Several later workers refuted this theory, chiefly on the grounds that chromidia were probably mitochondria.

It has been known for many years (43) that the basophilia of certain cell structures was due to their nucleic acid content. In addition, van Herwerden (87) showed that cytoplasmic basophilia could be depressed if tissues were first treated with ribonuclease. Our more recent interest in the function of RNA, however, dates from the findings of Caspersson and Schultz (14) from ultraviolet absorption studies, and from the studies of Brachet (10) with basic dyes and ribonuclease. These authors have repeatedly stressed the presence of high concentrations of RNA in cells engaged in active secretion or growth. Further evidence for the involvement of RNA in protein synthesis has come largely from the studies on isolated RNA-containing components, begun by Claude (17). These studies have now expanded into a highly important analytic approach to biochemical mechanisms of protein synthesis.

The RNA-containing components of the cytoplasm occur in a variety of forms. They have been isolated from cell homogenates by a number of different methods that affect in various ways their composition and structure. The "small granule" fraction isolated from rat liver by Claude (18), was considered to contain particles more or less homogeneous in size, averaging about 200 m$\mu$ in diameter, and containing a high concentration of lipoproteins and RNA. To these particles Claude applied the name "microsomes," using a term frequent in the older cytological literature, but for structures he felt were below the level of light microscope visibility. Later more extensive fractionation studies demonstrated that the microsomal fraction was far from homogeneous in composition, but could be divided into subfractions of varying RNA and lipoprotein content (15, 47). Ribonucleoproteins were also isolated as very much smaller particles, 5 to 24 m$\mu$ in diameter, of high density, essentially free of lipids. These fine particles were called "ultramicrosomes" by Barnum and Huseby (4), and "macromolecules" by Petermann and Hamilton (56), who distinguished different classes of particles on the basis of density and charge. With desoxycholate the fine particles were readily recovered from the microsomal fraction, leaving behind a membranous complex of lipoproteins (91). It is of interest to note, however, that considerable RNA may remain within the membranous component, after the fine particles have been removed (16).

That labeled amino acids were most rapidly incorporated into the proteins of the "microsomal" fraction was shown by Borsook (9) and later by many others, as a clear indication of the importance of these particulates as sites of protein synthesis. Ability to incorporate amino acids was later found to reside mainly in the fine particle fraction, and only slightly in the lipoprotein membranes (41). This incorporation was demonstrable in vitro, where it was rapidly depressed by ribonuclease, and was thus dependent on the integrity of RNA (41, 91). It is also interesting to note that cholesterol synthesis in the lipoprotein membranes, as determined by labeled acetate incorporation, was independent of the presence of the fine particle components, and unaffected by ribonuclease (11). Thus although nucleoprotein particles and lipoprotein membranes are in intimate association in the microsomal fraction, protein synthesis in the particles and cholesterol synthesis in the membranes seem largely independent.

Basophilic components of the cytoplasm, as seen in the electron microscope, possess a varied and complex morphology. Disagreement in the current literature concerning matters of their classification and interpretation attest to our present state of ignorance. In studies on whole cells, grown in tissue culture, Porter, Claude, and Fullam (61) described strings of small vesicles and tubules that were disposed as a network throughout the cytoplasm, but were absent from the outermost extensions of the cell. This vesicular system was later called the "endoplasmic reticulum" (62), a descriptive term obviously quite appropriate to this component as found in cultured cells. The development of adequate sectioning techniques precipitated a flood of observations on tissue sections. In many tissues investigators described vesicles, tubules, lamellae, and membranes arranged in a variety of ways (21, 8, 75). In a study comparing sections with whole cultured cells, Palade and Porter (52) identified a series of membrane-lined vesicles with the endoplasmic reticulum. Although in sections the vesicles usually appeared as isolated entities, analysis of serial sections showed that, at least in some cases, vesicles were interconnected and the "reticular" nature of the system was still evident. Tissue sections also demonstrated that vesicles frequently occurred in the form of large flattened sacs or "cisternae," and that in some cells, for example, in acinar cells of pancreas or in the parotid gland (58, 89), the cisternae were stacked to form complex arrays of parallel lamellae. In describing these structures, Porter (60) and Palade (50) have emphasized the occasional interconnections visible between parallel cisternae. They thus felt that "the main feature of the system" is that it forms a reticulum permeating the entire cytoplasm. In addition, these authors (50) continue to use the name "endoplasmic reticulum" because they have not been able to find a better term, although admittedly the concept of an endoplasm is inappropriate to most tissue sections.

The concept of the endoplasmic reticulum has been gradually expanded in the light of additional findings. The correspondence between the distribution of cell basophilia and the lamellar or vesicular systems was recognized, and the endoplasmic reticulum was thus considered as a component responsible for cytoplasmic basophilia (19, 58, 89). In many cases, the membranes of the endoplasmic reticulum were further seen to be associated with fine particles 10 to 15 m$\mu$ in diameter, with a distribution characteristic of the tissue, and lining the outer surface of the cisternae or vesicles (49, 77). In addition, particularly in embryonic and tumor cells, these particles occurred freely distributed in the cytoplasm, and not associated with the membranes. It was suggested that the RNA responsible for cell basophilia occurred in this particulate component rather than in the membranes (49), and that the particles were probably homologous with the "ultramicrosomes" of Barnum and Huseby (4).

In other investigations it was evident that cytoplasmic vesicles also existed to which no granules were attached. Their lining membranes were thus smooth contoured, as seen in sections, as opposed to the particle-lined "rough" membranes. In most cases these smooth structures were clearly distinguishable as elements of the Golgi material (20), but in some cases single cisternae, for example in rat liver (50) or rat motor neurons (54), appeared to be particle-lined in some places and smooth in others. In addition, membrane complexes of the endoplasmic reticulum have been described as being in continuity with infoldings of the cell membrane at the periphery (50) and also with the nuclear membrane (88). The endoplasmic reticulum has thus been considered as "a continuous net-work of membrane-bound cavities permeating the entire cytoplasm from the cell membrane to the nucleus" (50). As such, the system includes both basophilic "rough" membranes, and "smooth" elements of the Golgi apparatus, the nuclear membrane, and infoldings of the cell membrane. All are considered to be interconnecting or "reticular," but differentiated into recognizably different components in specific areas of the cell.

This is a sweeping concept, postulating that all the membrane systems of the cell are combined into one integrated system. If the conclusion is sound that all of these membranous components are essentially interconnecting, then it is certainly of value to stress this aspect with a single generic term. The concept of the endoplasmic reticulum has been criticized, however, on several grounds. First, the name has been considered unsuitable, since the structures included are "not reticular, and there is little reason to talk about endoplasm in tissue cells" (76). Second, there have been objections to the use of a generic name for several different cell components that in the great majority of cases can be clearly distinguished by their fine structure. Third, several investigators have failed to find the reported interconnections between the membrane systems of different types (76, 77, 92). Partly because of these criticisms, other investigators have used different terms in describing these components. The name "ergastoplasm" has been used for the stacks of lamellae characteristic of acinar pancreas, salivary gland, and hepatic parenchymal cells, since these structures are obviously comparable with the basophilic bodies originally studied by Garnier (8, 89). However, use of the term "ergastoplasm" was criticized by Sjöstrand and Hanzon (77), since it implies a function as yet unverified. Sjöstrand (76) and Zetterqvist (92) have preferred to use the completely neutral name α-*cytomembranes* for the particle-lined components, β-*cytomembranes* for membranes connected with or derived from the cell membrane, and γ-*cytomembranes* for the lamellar or vesicular membranes of the Golgi complex. These membrane components were considered to be readily distinguishable by their morphology and characteristic disposition in the cell; no connections between them were reported.

The electron microscope has helped to characterize the "microsomes" of Claude (18). As isolated from rat liver homogenates, the microsomal fraction appeared to contain a collection of particle-lined vesicles, occurring singly or in strings (79). It thus seemed likely that microsomes were largely the fragmented portions of the particle-lined membrane system, although considerable contamination from fragments of mitochondria and Golgi vesicles was also found (7). Similar conclusions have been reached by other investigators (53). The word "microsome," as now pre-empted by biochemists, is an entity created in the centrifuge and defined by its chemistry. In this sense it should obviously not be employed for structures in intact cells (7).

In spite of the complex morphology it has demonstrated, the electron microscope has so far added little to our concepts of the function of basophilic components. Postulated functions for the endoplasmic reticulum include: (1) providing a membrane surface for enzyme distribution; (2) providing compartments in the cytoplasm for segregation of certain metabolites; and (3) acting as an interconnected system for rapid diffusion (60). It has also been assigned a function in conduction of the excitatory impulse in muscle (59). Other suggestions have been somewhat more in line with the cytochemical evidence relating microsomes with protein synthesis. Sjöstrand and Hanzon (77) have assumed that the membranes and particles of the exocrine pancreas cells are important in enzyme synthesis, and that the Golgi apparatus, mitochondria, zymogen granules, and the cell membrane are primarily involved in the secretion process. In general, the vagueness of all such theories serves to emphasize the wide gap that still exists between our concepts of function at the biochemical level and our knowledge of cell morphology.

## Results

Although the problem is discussed with growing frequency, we still are largely ignorant concerning the role played by the nucleus in the origin of cytoplasmic RNA. Some observations are reported here on two aspects of the problem: cytochemical studies have been made on the RNA of amphibian liver with microphotometry and radioautography; and the morphology of basophilic components of amphibian liver and pancreas cells has been investigated with the electron microscope.

### 1. Cytochemical Observations

RNA occurs inside the nucleus as a component of the nucleolus, but also in association with the chromosomal material. As mentioned by Dr. Gall, in the polytene chromosomes of *Drosophila* and *Chironomus* it forms in the expanded puffs (73, 82), and it occurs in the lateral loops of the lampbrush chromosomes of amphibian oocytes (76). In some cells RNA also occurs in the nuclear sap. The chromosomes in amphibian oocyte nuclei may be centrifuged to one side of the nucleus, so as to leave a diffuse material that stains faintly for RNA in the non-chromosomal portion. Many workers have suggested that RNA is first formed at these nuclear sites, and later extruded into the cytoplasm (33, 37, 42).

As shown by several investigators (39), when rats are starved or fed low protein diets, the RNA of the cytoplasm is depleted, and upon refeeding normal levels are restored. In an attempt to analyze the role of the nucleus in this restoration process, we have measured by microphotometry the amount of RNA in nuclear and cytoplasmic fractions of single liver cells. *Ambystoma tigrinum* larvae, which were used for these studies, were reared under constant conditions from a single batch of eggs. They were starved for 3 weeks prior to refeeding. RNA was estimated on tissue sections by azure B binding after the removal of DNA by deoxyribonuclease (Worthington, 0.02% in 0.003 $M$ magnesium sulfate, $pH$ 6.5, at 25°C for 1 hour). The cytoplasmic RNA was measured by the two-wavelength method to avoid error due to non-random dye distribution. Details of the methods and calculations used have been presented elsewhere (83, 84). A total of 6 larvae were studied.

After refeeding, the cells increased in size and total protein content, and the amounts of stainable RNA in nucleus and cytoplasm showed marked changes. Microphotometric determinations on three different liver cell components—nucleoli, nuclei (excluding nucleoli), and cytoplasm—are graphed in Fig. 1. At maximum stimulation (36 hours) the relative amounts in nucleolus, nucleus, and cytoplasm were 1, 4.7, and 82 respectively. Thus both of the nuclear RNA fractions we have measured showed roughly parallel changes during the recovery process. We can also say that the nucleolus contributed only about one-fifth of the total stainable RNA of these nuclei.

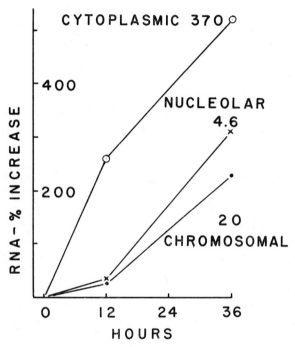

Fig. 1. RNA levels, plotted as percent increase over control values, of refed *Ambystoma tigrinum* larvae after starvation for 3 weeks. Larvae were all about 3 cm. in length. Each point represents the mean of from 10 to 20 measurements. RNA was determined by microphotometry of tissue sections stained with azure B (84). The total RNA, in relative photometric units, is also indicated for the 36-hour determinations.

To characterize the nuclear RNA fractions further, incorporation of adenine-$C^{14}$ into the RNA of salamander tissues was studied by radioautography (Fig. 2). Two adult *Triturus viridescens dorsalis* were injected with 2.5 μc of adenine-8-$C^{14}$ (1 mc/mM) in Holtfreter's solution, and were sacrificed 5 hours later. Tissues were fixed in acetic alcohol (1:3), and paraffin sections were cut at 4μ. Deparaffinized tissues were treated with 5% trichloroacetic acid at 3°C for 5 minutes, thoroughly washed and

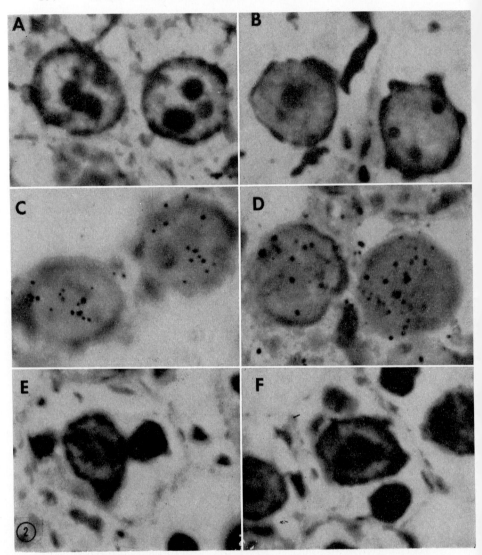

Fig. 2. Adult *Triturus* liver. Magnification 1,800 ×. A. Cells stained with azure B for RNA and DNA. B. Same, except the DNA has been removed with deoxyribonuclease before staining. Note diffuse RNA staining of nuclei. Central clumps of chromatin are practically unstained, but are faintly outlined by the RNA. C and D. Radioautographs showing incorporation of adenine-$C^{14}$ into nuclear RNA. E and F. Sections of the same tissue shown in A, but after treatment with Holtfreter's solution for 3 hours. Nuclei are shrunken, and basophilic inclusions in the cytoplasm are enlarged.

air-dried before covering with Kodak, Ltd. stripping film. Control sections were also treated with ribonuclease (Armour, 0.1% at $p$H 6.5 for 1 hour at 25°C) before covering. Slides were stored for 6 weeks before development. RNA was estimated on adjacent sections by azure B binding as before; DNA was removed with deoxyribonuclease (Worthington, 0.02% at $p$H 6.5 for 90 minutes at 25°C) before staining. Results of these determinations are summarized in Table 1. Photographs of the stained cells are shown in Fig. 2.

TABLE 1

RNA AMOUNTS AND RELATIVE SPECIFIC ACTIVITY IN TRITURUS LIVER AND PANCREAS CELLS[1]

| Tissue | RNA-azure B (photometric units) | Relative specific activity |
| --- | --- | --- |
| Liver (animal #1) | | |
| Nucleolus | 2.0 | 0.81 |
| Nucleus | 10.0 | 4.0 |
| Cytoplasm | 124 | 0.21 |
| Liver (animal #2) | | |
| Nucleolus | 1.8 | 1.1 |
| Nucleus | 13.5 | 2.7 |
| Cytoplasm | 90 | <0.2 |
| Pancreas (animal #2) | | |
| Nucleolus | 3.9 | 0.73 |
| Nucleus | 15.0 | 5.0 |
| Cytoplasm | 221 | 0.37 |

[1] Slides were measured at a wavelength of 550 m$\mu$ for nuclei and nucleoli; cytoplasm was measured with the two-wavelength method at 510 and 550 m$\mu$. Photometric units equal extinction per micron section thickness multiplied by the volume in cubic microns (83). RNA values represent means of 15 measurements. Relative specific activities are given as grains per $\mu^2$ per unit of RNA-azure B concentration, computed as extinction per micron section thickness, as determined on adjacent sections. Each value represents the mean of 30 to 40 counts on single nucleoli, nuclei, or on random areas 6.6 $\mu$ in diameter taken in the cytoplasm. These values assume stoichiometry of the RNA-dye complex.

In adult *Triturus* liver and pancreas, as in *Ambystoma* larval liver, nuclear RNA occurred outside of the nucleolus, associated with the chromosomes. The relative specific activity of this fraction, determined by relating absorption measurements on dye concentration to the number of autograph grains per $\mu^2$, was also considerably higher than nucleolar RNA. These studies indicate that the fraction we have called "chromosomal RNA" comprises a large part of the RNA of the nucleus and shows a relatively high rate of incorporation. It should be emphasized, however, that the computations of relative specific activity assume stoichiometry of

186    THE CHEMICAL BASIS OF DEVELOPMENT

dye-RNA binding. We feel that this is justified, since absorption curve analysis has shown Beer's law to hold for this method (24, 80).

Estimates of RNA and DNA per nucleus were also made using ultraviolet light at 260 m$\mu$ (Figure 3). Measurements were performed on untreated and deoxyribonuclease-treated slides to obtain DNA absorption by difference. Slides were further treated with 5% trichloroacetic acid at 90°C for 15 minutes to remove RNA, and the RNA was also computed by difference. These determinations indicated a DNA/RNA ratio of 1.5 for nuclei (excluding nucleoli). Comparative DNA-Feulgen measurements on diploid rat liver and *Triturus* liver nuclei showed that the *Triturus*

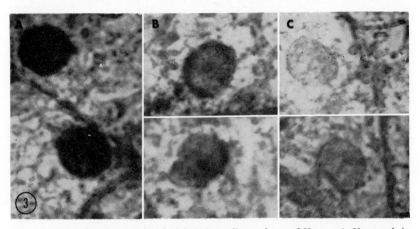

Fig. 3. Ultraviolet photographs of adult *Triturus* liver, taken at 260 m$\mu$. A. Untreated tissues. B. An adjacent section after treatment with deoxyribonuclease. C. An adjacent section after treatment with trichloroacetic acid (5% at 90°C for 15 minutes) to remove both nucleic acids. The decrease in density between B and C represents absorption due to RNA.

diploid value was 16.4 times that of the rat. Assuming the rat contains $6 \times 10^{-9}$ mg. DNA per diploid nucleus (86), then *Triturus* contains $9.8 \times 10^{-8}$ mg. DNA per nucleus, and the RNA per nucleus of this animal (excluding the nucleolus) was $6.8 \times 10^{-8}$ mg. Our DNA figure is slightly higher than that given by Gall (26), the difference being largely attributable to the different reference value used. Our values are also for a different subspecies of *Triturus viridescens*.

From visual observations on slides stained for either DNA or RNA, it was obvious that distributions were different within the nucleus. DNA was aggregated into several very dense clumps in the nucleus center, but also occurred diffusely around these clumps, and showed smaller concentrations near the nuclear membrane. RNA, in addition to the 1 to 4 nucleoli, was largely absent from the dense clumps, and was most evident in areas where the DNA concentration was low (Fig. 2, B). RNA occurred in the cytoplasm in a few dense lamellar or more rarely spherical bodies. These frequently occurred against the nuclear membrane and scattered through the cytoplasm, or less often along the cell membrane (Fig. 2,B).

In the process of studying incorporation into liver explants, it was apparent that in Holtfreter's solution these basophilic inclusions of the cytoplasm often increased in size (Fig. 2, D, E). This process was considered to be a condensation from the faint diffuse basophilia of the background cytoplasm, and apparently did not involve an increase in the total RNA of the cell. In hypotonic medium (0.3% Holtfreter's solution) the cells increased in volume with imbibition of water, and the inclusions became vacuolate and disperse, although the cell did not return completely to its original appearance. Somewhat similar aggregates have been described as forming around the nucleus of sea urchin eggs in hypertonic media (68a), and in guinea pig spermatocyte cytoplasm in Tyrode solution (36a).

## 2. Electron Microscopy

So far, the electron microscope has added little to our understanding of the mechanisms of nuclear control of cell processes. The detail presented in the nuclear membrane region is largely too complex and too poorly understood for functional interpretation. A few specific observations, however, have been considered of possible importance in nuclear-cytoplasmic interaction. Fine particles have been reported in the nucleolus (58), nuclear sap (27), and along the inner edge of the nuclear membrane (88), and bear at least a superficial resemblance to particles in the cytoplasm. The nuclear membrane, in a wide variety of tissues, has been shown to contain discontinuities, which some have interpreted as "pores" (88), although others have felt that the presence of actual holes in the nuclear membrane has not been convincingly demonstrated (76). Projections from the nuclear membrane have also been mentioned as possibly a sign of nuclear-cytoplasmic exchange (57). A morphological similarity between the nuclear membrane and the cisternae of the cytoplasm has been demonstrated. This has been interpreted as indicating either that the nuclear membrane is essentially cytoplasmic in origin, being merely another modification of the endoplasmic reticulum (50, 88), or that cisternae and cytoplasmic vesicles may originate from the nuclear membrane by a process of blebbing or delamination (5, 29).

Many of the observations reported here have been made on amphibian liver or acinar cells of the pancreas, although other material has been added for comparative purposes. In these cell types the nuclei were variable in appearance, probably influenced by the physiological state of the cell, but also by its fixation in osmic acid and other aspects of preparation. Nucleoli were usually the most dense intranuclear structures. In both tissues the nucleoli frequently contained peripheral areas having in the electron microscope a coarse thread-like appearance surrounding a more dense and roughly circular center. Adjacent sections stained for nucleic acids (Figs. 4-6) indicated that RNA occurred in both regions but was more concentrated in the thread-like areas. Vesicular regions of lower electron density occasionally occurred adjacent to the nucleoli. These contained only protein and no measurable nucleic acid. There was no evidence for nucleolus-associated chromatin, such as is present in many mammalian cells.

In many cells the nucleus contained chromatin areas, usually of uniform density, surrounded by less dense and less regular interchromatin regions. In addition, some nuclei showed very dense "heterochromatic" clumps around the nuclear membrane. These clumps closely resembled the thread-like nucleolar substance in fine structure, but when studied on adjacent stained sections showed the presence of DNA and no RNA. Nuclear sap areas contained thread-like extensions from the chromatin, and also irregular clusters of particles varying in diameter up to about 50 m$\mu$ (Figs. 6, 7, 14, 15). These particles were frequently distributed around clumps of chromatin, so as to resemble the distribution of "chromosomal RNA" stained for the light microscope. For this reason we have suggested that they might be the site of this RNA fraction (84). Further studies have shown that the largest of these particles are no longer apparent when thin sections are treated for 3 hours with 3% hydrogen peroxide, a treatment that removes certain lipid inclusions from osmium-fixed tissue, but little or no RNA. It thus seems likely that some of these particles contain an osmophilic component that is possibly lipid in nature (Figs. 10, 18).

The nuclear membrane appeared somewhat variable from cell to cell. In many cases it was clearly double with the inner and outer components separated by an area of low density. The inner membrane often appeared denser, probably because of chromatin fibrils that were packed against it. In some cells this gave a double appearance to the inner membrane (Fig. 8). In sections of sufficient thinness the nuclear membrane was visibly interrupted at certain places, where inner and outer membranes were joined (Figs. 7-9, 21, 22, 29).

In oblique cuts through the nuclear membrane the discontinuities were

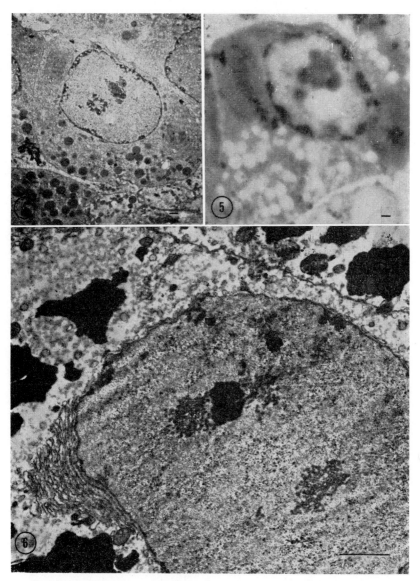

Figs. 4, 5. Larval *Ambystoma tigrinum* pancreas, showing the same cell in adjacent sections as seen with the electron microscope, and with the nucelic acids stained for the light microscope with azure B. Note that whorled structures at either end of the nucleus are basophilic, while the secretory granules do not stain. Fig. 6. Larval *Ambystoma* liver cell. Note the heterogeneous nucleolus, the interchromatin areas of the nucleus containing dense particles, and the lamellar basophilic body against the nuclear membrane. The irregular black areas in the cytoplasm are probably lipid inclusions.

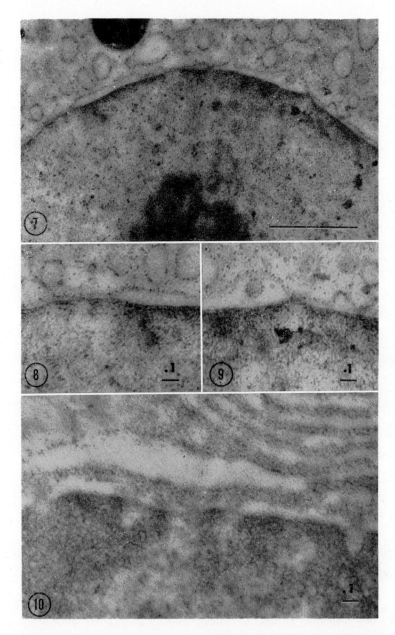

Figs. 7-9. Larval *Ambystoma* pancreas. The nuclear membrane contains a darkly outlined inner component, which appears double in Fig. 8, probably because of filamentous material lying in contact with it. Fig. 9 shows a section through an annulus. The inner and outer nuclear membranes appear to join, and there is a characteristic region of reduced density in the nucleus adjacent to the annulus. Fig. 10. The same tissue as in Figs. 7-9, except that the section has been treated for 3 hours with 3% hydrogen peroxide. Obvious particles in the interchromatin areas are absent. The section shows portions of five annuli. Note their association with less dense areas of the nucleus and their contact with cytoplasmic cisternae.

seen to be surrounded by dense rings of material, or "annuli," as reported by many authors for a wide variety of tissues (1, 13, 38). The structure of annuli in the nuclear membrane of several tissue types is shown in Figs. 22 to 33, A. Several points are of interest. Annuli were usually present where interchromatin areas were against the nuclear membrane, and much less often when the membrane was in contact with clumps of chromatin, a relation also pointed out by Porter (60). When annuli occurred amid chromatin material, small areas of lower density were almost always present in the adjacent chromatin (Figs. 9, 10). The annulus center frequently contained a small dense central body, as described by Afzelius (1) (Fig. 33). The annulus wall in oblique section presented a variable morphology. In some cases it apparently contained small rings or "subannuli" about 20 m$\mu$ across. These might be interpreted as cross sections through small filaments possessing relatively electron-light centers (Figs. 23; 33,A).

From comparisons with other cell types, additional characteristics of annuli were evident. The annulus in some tissues extended as a tube or rod a considerable way down inside the nucleus, and out into the cytoplasm (1, 81). In some cases, as in the snail and spider oocytes shown in Figs. 22, 23, 25-28, the "annular tubule" apparently widened out in the cytoplasm. When the nuclear membrane was cut obliquely, annular tubules were transected at different levels, and thus their organization above and below the nuclear membrane was evident. Annuli thus cut at different levels possessed different morphologies. This was particularly evident in the onion telophase nucleus section shown in Fig. 30. Here the annuli were most evident at the level of the nuclear membrane, as opaque circular structures, usually with a dense center. The annular tubule apparently extended down into the nucleus, between chromatin areas where it usually became smaller in cross-section and less distinct in organization. In the cytoplasm it became paler and more expanded, and again less distinct. In the spider oocyte the extensions sometimes appeared as an amorphous cloud of material, as described by André and Rouiller (2).

In some tissues, namely, the *Triturus* pancreas and spider oocyte, the annular tubule had a wall that closely resembled in structure certain small vesicles or tubules occurring in the cytoplasm (Figs. 11, 25, 26).

Annuli in some cases were associated with fine particles (27, 88), although the relation between such particles and the annular tubule was often difficult to interpret. In the pancreas section shown in Fig. 11, the projections, that were probably annular tubules, clearly contained particles on the outer surface. In some cases there was a suggestion of a spiral row of particles around the projection. In rat liver and onion root oblique cuts

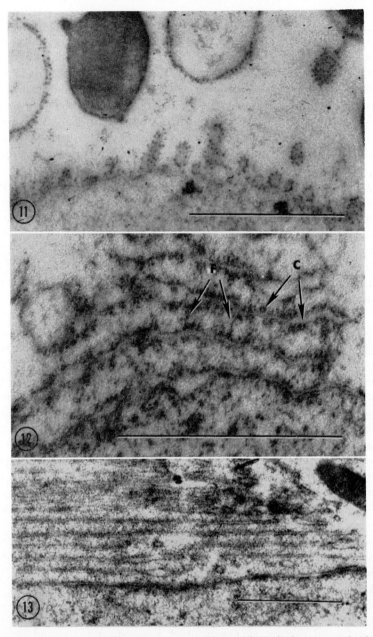

Fig. 11. Nuclear margin of larval *Ambystoma* pancreas. The projections are associated with small particles. Note the cytoplasmic vesicles also with particle-lined membranes. The dark body is a secretory granule with a characteristic smooth limiting membrane. Fig. 12. A small lamellar aggregate on the margin of a larval *Ambystoma* liver nucleus. Note the rather poorly defined filamentous component F, connecting the membranes perpendicular to them. Arrows at C indicate small circles, possibly filaments in cross section. Fig. 13. A similar lamellar body from the pancreas of the same animal, seen at lower magnification.

through nuclei frequently demonstrated rings of particles of a similar diameter to the annulus, as pointed out by Watson (88) (Figs. 30, 33). Particles were also arranged in semicircles or spirals. These arrangements, in both rat liver and onion root, appeared more commonly near the nucleus, and rarely further out in the cytoplasm, where particles usually showed less regular groupings. Although in some cases these rings, semicircles, or spirals were in contact with cisternal membranes, in other cases no such attachment was apparent, and they rather appeared to lie detached in the cytoplasm, possibly surrounding projecting annular tubules.

Annuli apparently form in telophase, as the nuclear membrane forms, as shown in both onion root and embryo chick somites (Figs. 29, 30). Short annular tubules are evident in the chick annuli shown in Fig. 29. The chick nuclear membrane arises apparently by the coalescence of a series of cytoplasmic vesicles. Telophase nuclei were, as might be expected, more compact than interphase nuclei, and the interchromosomal areas were practically absent. Where annuli occurred, however, interchromosomal areas were often evident.

The relation between the nuclear membrane region and adjoining areas of cytoplasmic basophilia was complex, variable, and difficult to analyze. In amphibian liver most nuclei contained orderly nuclear membranes, but in a number of cells this region was irregular and the nuclear membrane was interrupted for wide areas, particularly in relation to the interchromatin regions (Fig. 14). The cytoplasm surrounding the nuclei contained numerous small vesicles, in some cases grouped into somewhat irregular lamellar aggregates. Although the actual point of the nuclear margin was difficult to recognize, some vesicles apparently occurred inside the nucleus (Fig. 15). Since nuclei of this type were interspersed with cells where nuclear membranes were intact, it seems unlikely that disruption of the membrane can be attributable to poor fixation or other artifacts.

In other cells, stacks of cisternae frequently occurred against the nucleus. These regions were intensely basophilic when studied on adjacent sections, stained for the light microscope, and are homologous to the dense perinuclear bodies shown in Fig. 2. Similar aggregates also occurred against the cell membrane, but these were usually smaller and less frequent. In most instances the membranes were not obviously particle-lined, although at high magnification they may appear finely granular (Fig. 17). In many cases they occurred in relation with a more or less amorphous electron-dense material that occurred in the spaces between cisternae (Fig. 16). In other cases there was a suggestion that filaments, approximately 150 m$\mu$ in diameter, ran through the cisternal membranes at right angles to them

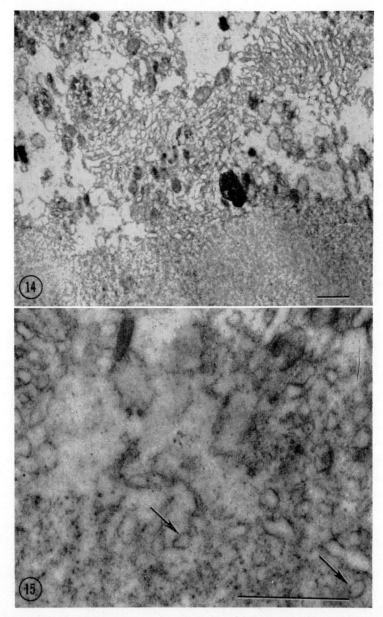

Figs. 14, 15. Nuclear margin from the liver of a starved *Ambystoma* larva 24 hours after refeeding. The nuclear membrane area is poorly defined. Small vesicles (Fig. 15, arrows) occur in association with the interchromatin areas, apparently lying slightly inside the nucleus.

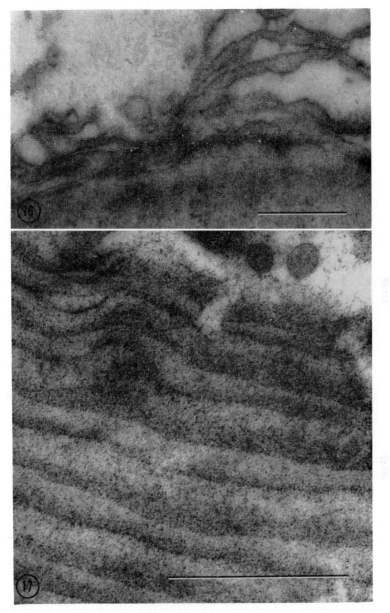

Figs. 16, 17. Lamellar aggregates from the liver of a starved *Ambystoma* larva, 24 hours after refeeding. Membranes are cut transversely in Fig. 16, and obliquely in Fig. 17. Lamellae are associated with a dense, faintly granular material, but contain no definite particles. These lamellae appear basophilic, as seen in adjacent sections.

(Figs. 12, 13). In a few favorable cases such fine filaments also appeared to penetrate the nuclear membrane and similar structures were also visible inside the nucleus. Where filaments were in contact with membranes, areas of greater electron density usually surrounded the filament. Where such an area was cut obliquely a ring or oval of dense material was evident in contact with the membrane. These could easily be interpreted as particles, but at least in some cases may be the filamentous component cut transversely (Fig. 12).

Somewhat similar stacks of cisternae were also seen against the nuclear membrane in *Triturus* pancreas (Fig. 18). Such lamellar arrays usually appeared to contain particles between adjacent cisternae. In the cytoplasm of the same cells whorls of membranes also occurred, lined with typical particles, about 15 m$\mu$ in diameter, as found in guinea pig pancreas by Sjöstrand and Hanzon (77) and many others. In a few regions, however, these "particles" also appeared to stretch in a linear fashion between adjacent membranes, so as to suggest the presence of short filaments rather than a regular series of particles (Fig. 20).

We were anxious to see the distribution of RNA in relation to these particles, and consequently a number of thin sections were prepared from which the osmium was removed by treatment with 3% hydrogen peroxide for 3 to 12 hours (Fig. 10), followed by "staining" for RNA with 10% ferric chloride in 0.1 $N$ HCl. After treatment from 1 to 3 hours, sections were differentiated for 3 to 10 minutes in 0.1 $N$ HCl to remove excess ferric chloride (83a). Results of this treatment are shown in Figs. 18, 19, 21, and 31. These micrographs must be interpreted with caution, since it is possible that the position of RNA may be altered during the staining process. In general, the stainable material was distributed as the particulate component, but appeared more finely dispersed in fine dots and lines between the cisternae. This work is still preliminary.

In other tissues, as in amphibian liver and pancreas, filamentous components were also observed in association with basophilic inclusions. This was particularly evident in lamellar bodies found in snail oocytes, where poorly defined filaments, about 15 m$\mu$ in diameter, were seen oriented at right angles to lamellae (Fig. 24). These filamentous components showed some resemblances to the projecting walls of the annuli of the same cell, as shown in Fig. 22. As we have reported earlier (64, 81), the lamellae in these cells also contain annuli. A quite different arrangement of filamentous material was seen in the lamellae of amphibian (*Ambystoma*) thyroid, where filaments about 20 m$\mu$ in diameter appeared in intimate association with basophilic lamellar structures, running along the membranes

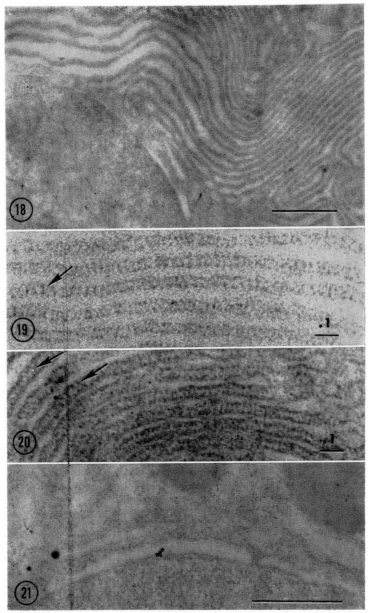

Figs. 18, 19, 21. Adult *Triturus* pancreas, treated with hydrogen peroxide to remove osmic acid, and then stained with 10% ferric chloride to visualize nucleic acids. Fig. 20. An untreated control section. As visible in Figs. 19 and 21, densities after staining are finely distributed as irregular lines and dots, following the distribution of the cisternae. In some places dense material appears as short filaments between cisternal membranes (arrows).

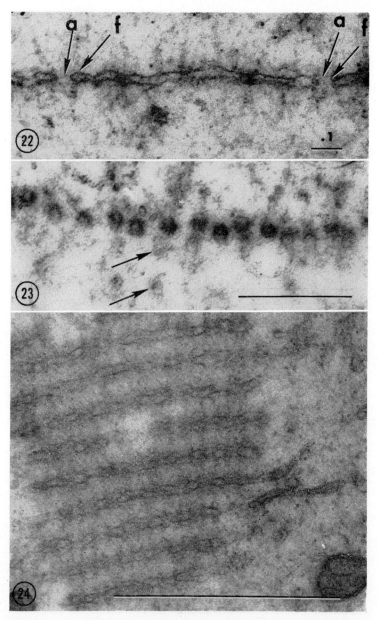

Figs. 22-24. Oocytes of the helicid snail *Otala lactea*.

Fig. 22. Cross section of the nuclear membrane showing a section through two annuli (a). Note the faint filamentous material (f) at either side of the annulus, and also at other places along the membrane. Cytoplasm above. Fig. 23. An oblique cut through the nuclear membrane, showing annuli. Annular tubules have been cut across as they project into the nucleus (arrows). Fig. 24. Lamellar inclusion in oocyte cytoplasm, showing interconnecting filamentous structures. (Micrograph by L. E. Roth).

and apparently forming a part of them (Fig. 35). These filaments appeared to be outlined by very fine dense particles arranged along the membrane, with a diameter that was variable from cell to cell, in Fig. 35 averaging roughly 5 m$\mu$. Another interesting filament membrane relationship, also highly basophilic, has been found in crayfish (*Cambarus*) spermatocytes (69). Here the filaments occurred in a lattice-like array, traversed by lamellae at right angles to them. In some cases the lamellar and filamentous structures merged into one another, suggesting that they may be two different types of aggregation of the same components, or at least have certain constituents in common. The lamellar component was further associated with cytoplasmic vesicles of two types, either single-walled or doubled-walled. In double-walled vesicles the two membranes were separated by the width of one filament, and in fact when such membranes were cut across, rings and ovals of dense material in the membranes suggest that filament integrity is still maintained inside the membrane (Figs. 36, 37). No particulate component was evident in these inclusions.

In tissues where fine particles are present they may occur both in association with membranes and away from them. In the case of regenerating muscle cisternae (Fig. 34), the particles were arranged along poorly defined filaments that project at right angles to the membrane. The arrangement of particles in linear clusters, also suggestive of some filamentous orientation, occurred in a wide variety of tissues, such as onion root cells, although in this tissue membranous components were not apparent in the cytoplasm (Fig. 30). In rat liver, when cisternal membranes were cut in tangential section, rows of fine particles were often evident (Fig. 32), somewhat similar to those seen in the amphibian thyroid. This orientation suggests that the particles may be outlining a convoluted filamentous component lying in the membrane. In addition, small rings of particles were frequent between and on the lamellae, as a characteristic of this tissue. It is at present extremely difficult to reconstruct a three-dimensional conception of particle arrangements from thin sections through this complex structure. It seems possible, however, that these small rings of particles could represent sections through convoluted filaments lying between cisternal membranes (Fig. 32). In sections stained for RNA with ferric chloride the distribution of stain was similar to the distribution of fine particles, except that convoluted lines of dense material were present instead of rows of particles, and the cisternal membranes themselves seemed to bind the ferric ions. Mitochondria were completely unstained (Fig. 31). This distribution may be caused in part by changes in RNA distribution during staining. It suggests, however, that RNA may occur in areas where obvi-

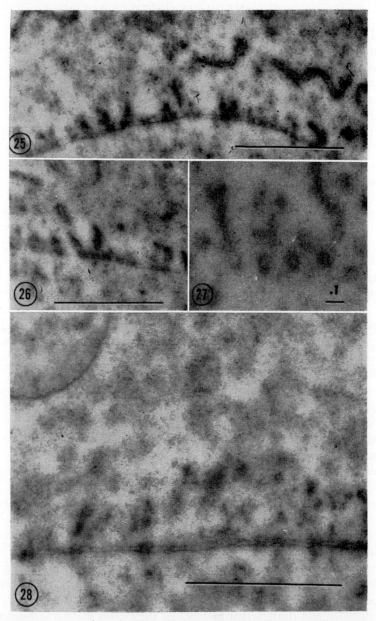

Figs. 25-28. Oocytes from the spider *Theridion tepidariorum*, showing projecting annular tubules. The nucleus is below and the cytoplasm above. In Figs. 25 and 26 note the similarity between some of the projections and vesicles in the cytoplasm.

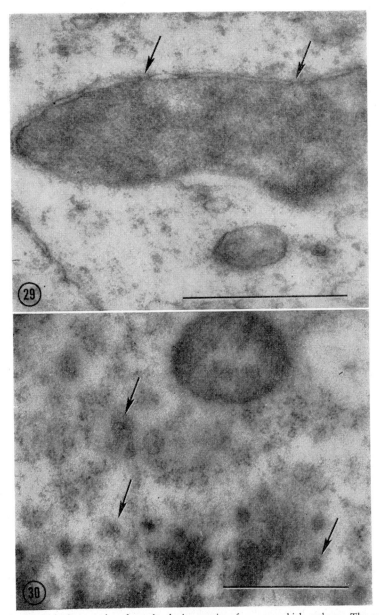

Fig. 29. Telophase nucleus from developing somite of a young chick embryo. The membrane has not completely formed below but annuli are apparent at upper surface (arrows). Notice the less dense nuclear areas in association with the annuli. (Micrograph by W. Ferris).

Fig. 30. Oblique section through a late telophase nucleus of an onion root. Annuli are most evident at the level of the nuclear membrane, but annulus-like structures occur in the nucleus (below) and in the cytoplasm a considerable distance from the nuclear margin (arrows).

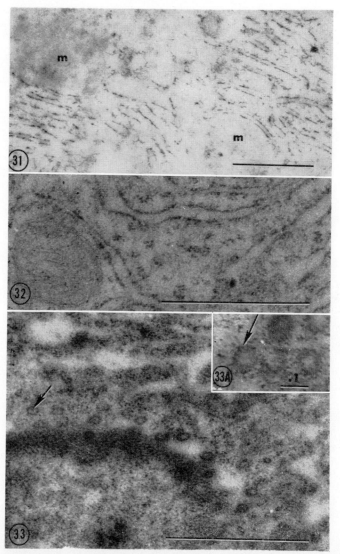

Figs. 31 to 33. Rat hepatic cells. Fig. 31. Section treated with hydrogen peroxide to remove osmic acid and then stained with ferric chloride to visualize nucleic acids. The distribution of stain is similar to the distribution of fine particles, except that convoluted lines of dense material are present instead of rows of particles, and cisternal membranes themselves seem to bind the ferric ion. Note that the mitochondria (m) are completely unstained.

Fig. 32. Note the small rings of particles which could represent sections through convoluted filaments lying between cisternal membranes, and rows of particles at lower right.

Fig. 33. Oblique section through nuclear membrane, showing rings of particles of similar diameter to annuli. Note that the annular center contains a small dark central body (arrow).

Fig. 33A. Annuli in cross section showing subannuli (arrow).

ous particles are lacking. It is interesting to note that Chaveau, Moulé, and Rouiller described RNA as being present in the membranous component of rat liver homogenates (16).

## Discussion

From the observations presented, several conclusions can be drawn. In *Triturus* liver and pancreas there is a large RNA fraction associated with the chromosomes. This component surrounds the clumped chromatin and fills the interchromatin regions. It also actively incorporates adenine-$C^{14}$. Where interchromatin regions lie in contact with the nuclear membrane, small dense rings or annuli are most evident.

Annuli, at least in some tissues, arise in telophase as the nuclear membrane forms, and persist as a characteristic component of the nuclear membrane throughout interphase. They are complex structures, frequently with visible extensions (annular tubules) projecting into the nucleus and out into the cytoplasm. Telophase nuclei may appear compact, with the interchromatin areas absent. As the nucleus grows in volume the interchromatin areas become apparent. This suggests that annuli, at least during the telophase through early interphase period, may be sites for the entrance of materials from the cytoplasm. In spite of this, annuli are not simple pores in the nuclear membrane. They often contain a small "central granule" visible in cross sections of the annular tubule, and in some tissues, such as onion root, have dense centers.

In some tissues (salamander pancreas, spider oocytes) there is a resemblance between the fine structure of vesicles or tubules in the cytoplasm and the projecting annular tubules of the nuclear membrane. It thus seems possible that some cytoplasmic components may originate from annular tubule material by a process of extension of these structures. Strings of cytoplasmic vesicles, as in the "endoplasmic reticulum" of whole cultured cells, could arise in this way. We have reported earlier (81) that annulate lamellae, in which structures resembling annular tubules are incorporated, apparently break down into cytoplasmic vesicles formed in part by the tubule wall.

The annular tubule occasionally appears made up of a circle of filaments, appearing in cross section as small rings (subannuli). Above and below the annulus these filaments become vague in outline and irregularly arranged. Filamentous structures also form a part of several types of cytoplasmic basophilic inclusions. In snail oocytes, and in lamellae lying against the nuclear membrane in salamander liver and pancreas, filaments

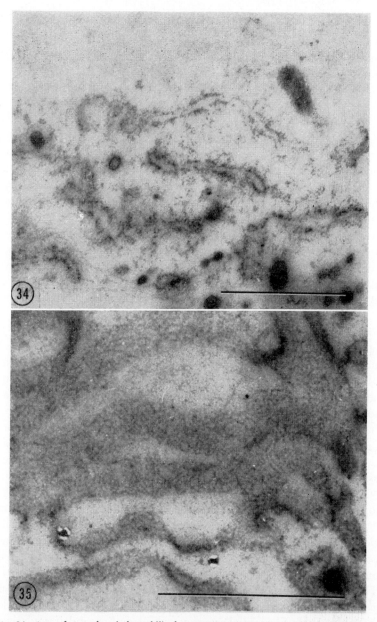

Fig. 34. Area of sarcoplasmic basophilia from regenerating mouse muscle. Cisternae show particles arranged along poorly defined filaments that project at right angles to the membranes.

Fig. 35. Filaments in association with basophilic lamellar structures of the cytoplasm in adult *Ambystoma tigrinum* thyroid treated with 0.1 mg. of thyroid-stimulating hormone for 10 days. Here the filaments appear outlined by fine dense particles arranged along the membranes. (Micrograph by L. Herman).

run perpendicular to the plane of the lamellae. They occasionally appear to penetrate lamellae, and to traverse the nuclear membrane. A similar basophilic structure occurs in crayfish spermatocytes, except that the filaments are regularly packed in a lattice-like arrangement, traversed by evenly spaced double membranes (69). In amphibian thyroid lamellae the filamentous structures, instead of running perpendicular to the lamellae, appear to lie within the lamellar membrane in a convoluted pattern. In some respects these lamellar membranes thus appear to be composed of fused filamentous components. In crayfish spermatocytes the continuity between filaments and lamellae lends support to this interpretation.

The relation between these filamentous structures and the fine particles often considered responsible for cell basophilia is not clear. In many cases the membranes of cytoplasmic vesicles are clearly particle-lined (as in Fig. 11). Where cisternae are closely opposed, however, and the membranes properly oriented with the section plane, the "particles" lining adjacent cisternae may appear to merge into short filaments. This linear orientation of short filaments perpendicular to the lamellae may merely be a secondary ordering, influenced by association with the membranes. On the other hand, since filaments are occasionally demonstrable in the perinuclear lamellae of the same tissue, it may indicate that the "particulate" material in the closely packed cisternal aggregates is also predominantly filamentous in nature. Sjöstrand and Hanzon (77), in a schematic diagram of the α-membranes, have shown the particles with an elongate shape. We feel that this configuration is equally well interpreted as an oblique section through a filament, oriented perpendicular to the membrane. Although a great deal more work is needed before the fine structure of these components, and their manner of change, becomes clear, we may conclude that in certain tissue types the "rough profiles" of "particle-lined" membranes are possibly equally well considered to be membrane-filament complexes.

In many basophilic cell types, areas of diffuse basophilia appear to be associated with more or less disordered areas of dense material. These show a variety of forms, from the largely amorphous clouds seen in spider oocytes, to the particles in onion root. This dense material in the cytoplasm never appears completely random in distribution. Particles are frequently aggregated into small strings or clusters. In a few cases, e.g. in liver (Fig. 32) or muscle (Fig. 34), particles appear in rings or short rows, as if outlining a filamentous component of lower electron density.

It is easy to conclude from these few examples that the concept that cytoplasmic RNA occurs only as regular particles is inadequate. In the

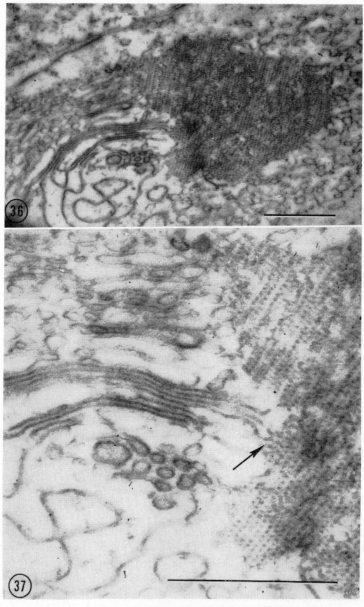

Fig. 36. Basophilic body in cytoplasm of crayfish (*Cambarus virilis*) spermatocyte. Note the lattice-like arrangement of filamentous units in cross section.

Fig. 37. Same, but at higher magnification. Note also the transition of lamellae and filamentous components (arrow). (Micrographs by A. Ruthmann).

various basophilic components described briefly above, some contain no resolvable particles at all, and in others the particles are extremely small, or absent from some basophilic regions of one cell and present in others. Particle size may be in part a matter of fixation, but this possibility can probably be ruled out when variations occur consistently within single cells. One probably should further differentiate between large particles (14 to 20 m$\mu$ in diameter), which we have suggested may represent filaments in section, and smaller particles (usually below 10 m$\mu$ in diameter), which apparently lie along the filaments, as for example in amphibian thyroid.

In studies of rat liver during refeeding after starvation, Fawcett (23) described the formation of membranous components that at first lacked the associated fine particles. For this reason he has suggested that the "membranous framework" of the basophilic inclusion may arise before they acquire RNA. Somewhat similar observations have been made by Rouiller and Bernhard (68). In the light of our findings that the lamellae of amphibian liver are in some cases smooth-contoured but nevertheless basophilic, it seems possible that rather than a change from nonbasophilic to basophilic with accumulation of particles, the change may somehow be associated with the function of these components in protein synthesis. In the whorled structures found in neurone cytoplasm by Palay and Palade (54), particles were present in some areas and absent in others. This is possibly another instance of the change from nonparticulate to particulate membranes. Such structures could be the basis for considering the "rough profiles" and Golgi membranes to be part of one interconnecting system, the endoplasmic reticulum (50). In our material the Golgi membranes were almost invariably distinguishable from the membranes associated with basophilic inclusions, whether or not they contain obvious particles.

We had hoped that the RNA staining for electron microscopy might indicate that filamentous RNA material might be associated with filament centers, as suggested for chromosomal structures by Ris (66). The stained sections are difficult to interpret, but in rat liver and amphibian pancreas no evidence was found for this hypothesis.

Whether or not the RNA-filament association emphasized here is important in nuclear-cytoplasmic exchange is unclear. Certain processes of the cell involve the active extension of filamentous structures, for example, the formation of chromosomal fibers between centrosome and centromere, or the fibrillar systems of ciliate protozoa (67). It is possible that similar extensions of filaments from nucleus to cytoplasm could act in the transfer of information. In this connection the filamentous material apparently radiating from nucleoli may be of interest (82).

SUMMARY

Studies on RNA of amphibian liver and pancreas, with microphotometry and radioautography, have indicated a significant and active nuclear RNA fraction outside of the nucleoli, in association with the chromosomes. Electron microscope studies on these and other tissues have indicated that RNA occurs in association with filamentous and lamellar structures which may or may not contain fine particles. It is suggested that extensions from the nuclear membrane (annular tubules), and filamentous structures, may function in nuclear-cytoplasmic exchange.

REFERENCES

1. Afzelius, B. A., *Exptl. Cell Research*, **8**, 147 (1955).
2. André, J., and Rouiller, C., in *Electron Microscopy, Proceedings of the Stockholm Conference, September 1956* (Sjöstrand, F. S. and J. Rhodin, eds.), p. 162, Academic Press, New York (1957).
3. Baker, J. R., *Quart. J. Micros. Sci.*, **90**, 293 (1949).
4. Barnum, C. P., and Huseby, R. A., *Arch. Biochem. and Biophys.*, **19**, 17 (1948).
5. Bennett, H. S., *J. Biophys. Biochem. Cytol.*, **2** (Suppl.), 99 (1956).
6. Bensley, R. R., and Hoerr, N. L., *Anat. Rec.*, **60**, 449 (1934).
7. Bernhard, W., Gautier, A., and Rouiller, C., *Arch. anat. microscop. morphol. exptl.*, **43**, 236 (1954).
8. ———, Haguenau, F., Gautier, A., and Oberling, C., *Z. Zellforsch. u. mikroskop. Anat.*, **37**, 281 (1952).
9. Borsook, H., Deasy, C. L., Haagen-Smit, A. J., Keighley, G., and Long, P. H., *J. Biol. Chem.*, **184**, 529 (1950).
10. Brachet, J., *Compt. rend. soc. biol.*, **133**, 88 (1940).
11. Bucher, N. L. R., and McGarrahan, K., *J. Biol. Chem.*, **222**, 1 (1956).
12. Burgos, M. H., and Fawcett, D. W., *J. Biophys. Biochem. Cytol.*, **1**, 287 (1955).
13. Callan, H. G., and Tomlin, S. G., *Proc. Roy. Soc. (London)*, **B**, **137**, 367 (1950).
14. Caspersson, T., and Schultz, J., *Proc. Natl. Acad. Sci. U. S.*, **26**, 507 (1940).
15. Chantrenne, H., *Biochim. et Biophys. Acta*, **1**, 437 (1947).
16. Chaveau, J., Moulé, Y., and Rouiller, C., *Exptl. Cell Research*, **13**, 398 (1957).
17. Claude, A., *Cold Spring Harbor Symposia Quant. Biol.*, **9**, 263 (1941).
18. ———, *Biol. Symposia*, **10**, 3 (1943).
19. Dalton, A. J., *Am. J. Anat.*, **89**, 109 (1951).
20. ———, and Felix, M., *Amer. J. Anat.*, **92**, 277 (1953).
21. ———, Kahler, H., Striebich, M. J., and Lloyd, B., *J. Natl. Cancer Inst.*, **11**, 439 (1950).
22. Deane, H. W., and Demsey, E. W., *Anat. Rec.*, **93**, 401 (1945).
23. Fawcett, D. W., *J. Natl. Cancer Inst.*, **15** (Suppl.), 1475 (1955).
24. Flax, M., and Himes, M., *Physiol. Zool.*, **25**, 297 (1952).
25. Frédéric, J., and Chèvremont, M., *Arch. Biol. (Liége)*, **63**, 109 (1953).
26. Gall, J., *Brookhaven Symposia*, **8**, 17 (1955).
27. ———, *J. Biophys. Biochem. Cytol.*, **2**, (Suppl.), 393 (1956).

28. Garnier, C., Medical Thesis, Nancy (1899).
29. Gay, H., *Proc. Natl. Acad. Sci. U. S.*, **41**, 370 (1955).
30. Gersh, I., *Arch. Pathol.*, **47**, 99 (1949).
31. Gey, G., *Harvey Lectures*, **50**, 154 (1956).
32. Goldschmidt, R., *Zool. Jahrb.*, **21**, 49 (1904).
33. Goldstein, L., and Plaut, W., *Proc. Natl. Acad. Sci. U. S.*, **41**, 874 (1955).
34. Hanstein, J., *Das Protoplasma*, Heidelberg (1880).
35. Harford, C. G., Hamlin, A., Parker, E., and van Ravenswaay, T., *J. Biophys. Biochem. Cytol.*, **2** (Suppl.), 347 (1956).
36. Hertwig, R., *Arch. Protistenk.*, **1**, 1 (1902).
36a. Ito, S., and Revel, J. P., *Anat. Rec.*, **130**, 319 (1958) (Abstr.).
37. Jeener, R., and Szafarz, D., *Arch. Biochem.*, **26**, 54 (1950).
38. Kautz, J., and deMarsh, Q. B., *Exptl. Cell Research*, **8**, 394 (1955).
39. Lagerstedt, S., *Acta Anat., Suppl.*, **9**, 1 (1949).
40. Lillie, R. R., *Wilhelm Roux' Arch. Entwicklungsmech. Organ.*, **14**, 477 (1906).
41. Littlefield, W., Keller, E. B., Gross, J., and Zamecnik, P. C., *J. Biol. Chem.*, **217**, 111 (1955).
42. Marshak, A., and Calvet, F., *J. Cell. Comp. Physiol.*, **34**, 451 (1949).
43. Matthews, A., *Amer. J. Physiol.*, **1**, 445 (1898).
44. Moore, C. L., Price, D., and Gallagher, T. F., *Amer. J. Anat.*, **45**, 71 (1930).
45. Novikoff, A. B., Beaufay, H., and deDuve, C., *J. Biophys. Biochem. Cytol.*, **2** (Suppl.), 179 (1956).
46. ———, Korson, L., and Spater, H. W., *Exptl. Cell Research*, **3**, 617 (1952).
47. ———, Podber, E., Ryan, J., and Noe, E., *J. Histochem. and Cytochem.*, **1**, 27 (1953).
48. Palade, G. E., *J. Histochem. and Cytochem.*, **1**, 188 (1953).
49. ———, *J. Biophys. Biochem. Cytol.*, **1**, 59 (1955).
50. ———, *J. Biophys. Biochem. Cytol.*, **2** (Suppl.), 85 (1956).
51. ———, and Claude, A., *J. Morphol.*, **85**, 35 (1949).
52. ———, and Porter, K. R., *J. Exptl. Med.*, **100**, 641 (1954).
53. ———, and Siekevitz, P., *J. Biophys. Biochem. Cytol.*, **2**, 171 (1956).
54. Palay, S., and Palade, G. E., *J. Biophys. Biochem. Cytol.*, **1**, 69 (1955).
55. Pease, D. C., *J. Biophys. Biochem. Cytol.*, **2** (Suppl.), 203 (1956).
56. Petermann, M. L., and Hamilton, M. G., *Cancer Research*, **12**, 373 (1952).
57. Pollister, A. W., Gettner, M., and Ward, R., *Science*, **120**, 789 (1954). (Abstr.).
58. Porter, K. R., *J. Histochem. and Cytochem.*, **2**, 346 (1954).
59. ———, *J. Biophys. Biochem. Cytol.*, **2** (Suppl.), 163 (1956).
60. ———, *Harvey Lectures*, **51**, 175 (1957).
61. ———, Claude, A., and Fullam, E. F., *J. Exptl. Med.*, **81**, 233 (1945).
62. ———, and Kallman, F. L., *Ann. N. Y. Acad. Sci.*, **54**, 882 (1952).
63. Prenant, A., Bouin, P., and Maillard, L., *Traité d'histologie*, vol. **5**, Masson & Co., Paris (1905).
64. Rebhun, L., *J. Biophys. Biochem. Cytol.*, **2**, 93 (1956).
65. Rhodin, J., Correlation of ultrastructural organization and function in normal and experimentally changed proximal convoluted tubule cells of the mouse kidney, Thesis, Karolinska Institutet, Stockholm (1954).
66. Ris, H., in *The Chemical Basis of Heredity* (McElroy, W. D. and B. Glass, eds.), p. 23, Johns Hopkins Press, Baltimore (1957).
67. Roth, L. E., Ph. D. Thesis, Univ. of Chicago (1957).
68. Rouiller, C., and Bernhard, W., *J. Biophys. Biochem. Cytol.*, **2** (Suppl.), 355 (1956).
68a. Runnström, J., *Exptl. Cell Research*, **8**, 49 (1955).
69. Ruthmann, A., *J. Biophys. Biochem. Cytol.*, **4**, 267 (1958).
70. Sager, R., and Palade, G. E., *J. Biophys. Biochem. Cytol.*, **3**, 463 (1957).
71. Schneider, W. C., *J. Histochem. and Cytochem.*, **1**, 212 (1953).

72. ———, and Kuff, E. L., *Anat. Rec.*, **92,** 209 (1954).
73. Schurin, M., Ph. D. Thesis, Univ. of Chicago (1957).
74. Sjöstrand, F. S., *Nature,* **171,** 30 (1953).
75. ———, *Nature,* **171,** 31 (1953).
76. ———, *Intern. Rev. Cytol.*, **5,** 455 (1956).
77. ———, and Hanzon, V., *Exptl. Cell Research*, **7,** 393 (1954).
78. ———, and Hanzon, V., *Exptl. Cell Research*, **7,** 415 (1954).
79. Slautterback, D. B., *Exptl. Cell Research,* **5,** 173 (1955).
80. Swift, H., in *The Nucleic Acids* (E. Chargaff and J. N. Davidson, eds.), Vol. 2, p. 51, Academic Press, New York (1955).
81. ———, *J. Biophys. Biochem. Cytol.*, **2** (Suppl.), 415 (1956).
82. ———, in *Current Activities in Molecular Biology* (R. E. Zirkle, ed.), Univ. Chicago Press, Chicago (in press).
83. ———, and Rasch, E., in *Physical Techniques in Biological Research* (Oster, G. and A. W. Pollister, eds.), **3,** 353, Academic Press, New York (1956).
83a. ———, and ———, *RCA Scient. Instr. News,* **3,** 1 (1958).
84. ———, Rebhun, L., Rasch, E., and Woodard, J., in *Cellular Mechanisms in Differentiation and Growth* (D. Rudnick, ed.), Princeton Univ. Press, Princeton (1956).
85. Tahmisian, T. N., Powers, E. L., and Devine, R. L., *J. Biophys. Biochem. Cytol.,* **2** (Suppl.), 325 (1956).
86. Thomson, R. Y., Heagy, F. C., Hutchinson, W. C., and Davidson, J. N., *Biochem. J. (London)*, **53,** 460 (1953).
87. Van Herwerden, M. A., *Arch. Zellforsch.*, **10,** 431 (1913).
88. Watson, M. L., *J. Biophys. Biochem. Cytol.*, **1,** 257 (1955).
89. Weiss, J. M., *J. Exptl. Med.*, **98,** 607 (1953).
90. Yamada, E., *J. Biophys. Biochem. Cytol.*, **1,** 445 (1955).
91. Zamecnik, P. C., and Keller, E. B., *J. Biol. Chem.*, **209,** 337 (1954).
92. Zetterqvist, H., *Ultrastructural Organization of the Columnar Absorbing Cells of the Mouse Jejunum,* Thesis, Karolinska Institutet, Stockholm (1956).

## DISCUSSION

Dr. Brachet: I should like to thank Dr. Swift for his excellent presentation of such a stimulating paper, and for showing us these most excellent electron photomicrographs.

Dr. Lehmann: Bairati and I have shown for instance that in amoeba the fine structure of fibrillar and chromidia-like formations are very poorly preserved by buffered osmium tetroxide (Bairati and Lehmann, *Protoplasma, 45,* 525, 1956). The same is true for the plasmalemma. In *Tubifex* astral and spindle structures disappear practically after osmium fixation. On the contrary the lipid-containing mitochondria or the nuclear membrane in amoeba is very well preserved by buffered osmium tetroxide. Here again the problem of the association of unsaturated lipids with certain cell structures emerges. This leads to the question whether osmium tetroxide is only a good histochemical reagent to unsaturated lipids and whether other cell structures like spindles and asters require better adapted fixatives. Dr. Swift, are you aware of this problem?

Dr. Swift: I would like to talk with you about some of these things later. There are a couple of things, though, I might say now. Osmium, as you know, in addition to lipid binding, is also bound by proteins. In fact, you can greatly enhance the electron density of proteins in sections by floating them on an osmic acid solution. Therefore I think we are showing more with osmic acid fixation than just the lipid, though unsaturated lipid inclusions stain very darkly of course. Concerning the electron density of basophilic components, the fine particles appear dark even after acetic-alcohol fixation, so some of this electron density is due to the components of the particle itself. The rest may well be due to osmic acid binding by protein. It is my opinion, and I am no authority on this, that once you isolate the fine particulate component from liver or pancreas microsomes, as Zamecnik and associates have done, that these contain relatively little lipid. Is that not right, Dr. Mirsky? I think several workers have found nearly equal amounts of RNA and protein and virtually no lipid in the fine particle component. Most people, I think, would tend to believe that the lipids were in the membranes with which the fine particles are often associated, and not in the particles themselves. We have no indication that these particles, or other components poor in lipids, are not adequately fixed by osmic acid.

Dr. Novikoff: It is unfortunate that time limitation kept you from discussing a topic of great interest to embryologists: the origin of mitochondria.

I would like to ask two specific questions: (1) Did the "microbodies" show areas of high electron density in sections not treated with ferric chloride? and, (2) did you mean to imply that mitochondria possess no RNA?

Dr. Swift: In the first place one has to interpret our nucleic acid staining for the electron microscope carefully. It is messy at this particular level, since the osmic acid fixative penetrates tissue blocks very poorly and very slowly, and you have a fixation gradient. This means that different regions of the same block stain differently. We take a normally fixed tissue block and section it for the electron microscope. These serial sections are then treated in various ways. A control section is untreated, another is floated on 3% hydrogen peroxide for 3 to 12 hours, which removes most or all of the osmic acid. After this treatment the density of rat liver microbodies is greatly reduced, but is increased again if sections are stained with ferric chloride. On the other hand, if sections are treated with perchloric acid before staining, to remove RNA, the microbodies don't stain. On the basis of this purely operational definition we tentatively suggest that microbodies contain RNA.

Dr. Novikoff: No, quite the contrary. We have put ourselves out on a thin limb, by suggesting that mitochondria of rat liver have little or no RNA. This was based on parallel decreases in RNA and esterase activity (used as a microsomal marker) when the isolated mitochondrial fraction was repeatedly washed. However, similarly indirect evidence has led investigators in at least three other laboratories to conclude that the low level of RNA in washed fractions was not due to microsomal contamination but was intrinsic to mitochondria.

Dr. Swift: Oh yes, I see what you mean. We have seen in our liver sections the very dark particulates you and deDuve have called dense bodies. In our material these seem to be different from the microbodies, if one uses the term as applied by Rouiller and Bernhard. Microbodies usually do not contain the fine dense particles you have suggested might be ferritin.

On the other hand, typical liver mitochondria, in our sections, contain no stainable components in them at all, and this seems to indicate they have no RNA. But they are very intimately associated with RNA on the membranes which are often wrapped around them. Do you disagree with this?

DR. NOVIKOFF: Oh, I am very happy with this. We had been way out on a limb. As we purified our mitochondrial preparations the RNA content got to the point where I considered it to be insignificant. However, data from other laboratories indicated that this isn't so, so I am very pleased to hear your data.

DR. SWIFT: If you consider a microbody to be an incipient mitochondrion, then RNA may be associated with mitochondrial formation.

DR. MIRSKY: I want to say one thing about the nucleolus. In particular I would like to speak about the nucleolar organizer. You remember the beautiful work of Barbara McClintock showed that the material for the nucleolus probably came from all over the chromosomes to the nucleolus organizer site where it formed the definitive nucleolus.

DR. SWIFT: The matrix theory as presented by McClintock has been refuted rather adequately by several recent studies.

DR. MIRSKY: I have read the papers by the South American workers criticizing McClintock, but her figures show that nucleolar material probably comes from the chromosomes. When the nucleolar organizer is damaged, one can see material accumulated on the chromosomes.

DR. SWIFT: Dr. McClintock's is a very beautiful and important paper, but I would like to point out that there is now considerable evidence against her suggestion that the nucleolar locus "organizes" chromosomal matrix material to form the nucleolus. For example, lagging chromosomes in the monosomic wheat strains studied by Dr. Crosby-Longwell (*Amer. J. Bot.* 44:813, 1957) often produce small separate nuclei. In some cases such small nuclei form nucleoli of their own, but with nuclei formed from some chromosomes no nucleoli at all are formed. It thus seems likely that the matrix is not involved in nucleolus formation, but rather that nucleoli are the synthetic products of particular chromosome loci. The fact that multiple nucleoli arise in cells containing a damaged "organizer" locus, seems better interpreted as the result of competition between potential nucleolus-forming sites. The fact that potential loci compete in nucleolus formation has been shown by McClintock and several others. That is, the same chromosome may normally form a large nucleolus, but when it is combined with chromosomes containing a "stronger" locus, for example through hybridization, it may form only a small nucleolus or none at all.

DR. ALLEN: I have one very short comment on the lamellar bodies that you showed in oocytes. The egg of *Spisula* has a very high viscosity as measured by the centrifuge method. If you centrifuge one of these eggs at about 200,000 times g for 5 minutes, with some difficulty you can stratify the cytoplasmic components. If you now look at one of these oocytes with the polarizing microscope, you see that there is a very intensely birefringent ring around the nucleus. When I saw this some years ago, I attributed it to strain birefringence on the advice of Dr. Inoue. From some recent electron micrographs by one of your former students, Dr. L. Rebhun, it has become apparent that these are the lamellar bodies that have been aligned due to centrifugal force, probably due to collisions with various cytoplasmic components among which these structures are lying. It struck me that this might be an interesting way to test for the presence of these lamellar bodies in living cells without having to go to the bother of electron microscopy.

Dr. Swift: Yes, that is a very nice idea.

Dr. Brachet: Well, I think that this session has been a very successful one and that certainly developmental cytology has been making a lot of progress and is going to make much more. And we all have been working very well and very hard, especially the speakers, and I wish to thank them and all those who took part in the discussion.

# Part II

*CELLULAR AND TISSUE INTERACTIONS IN DEVELOPMENT*

# EMBRYONIC INDUCTION [1]

Tuneo Yamada

*Biological Institute, Faculty of Science*
*Nagoya University,*
*Nagoya, Japan*

## Introduction

At the present time a vivid revival of interest in the mechanism of embryonic induction may be witnessed. New approaches are being made from different angles. According to the method adopted, the following groups of investigations may be distinguished in this field: (a) study of exchange of substances between the inducing and reacting systems by isotope technique; (b) elucidation of the physical nature of the agent mediating induction by the use of a filter with known porosity; (c) demonstration of the diffusible agent of the organizer by culture in vitro of the embryonic cells; (d) morphological study of the cells in inductive interaction by electron microscopy; and (e) isolation and characterization of the substances responsible for regional induction by adult tissues. In the present paper some of the recent progress made in the last-mentioned group of experiments will be reviewed and discussed together with some of the data obtained in other groups of experiments.

Holtfreter (13) was first to show that some differentiated tissues of various animals give strong inductive effects which are either neural or mesodermal, when implanted in the *Triturus* gastrula. Experiments subsequently performed along this line by Chuang, Toivonen, and their followers have revealed that the inductive effects of many of these tissues are regionally specific. However, no general relationship was found to exist between the regional effect of an organ and its embryological derivation. The first attempt at chemical characterization of the factor responsible for regional effects was made by Toivonen, who chose the liver and kidney tissue of the guinea pig as source materials. It had been shown by

[1] The research reported in this paper was supported by a science research fund of the Education Ministry and by a grant from the Rockefeller Foundation.
The writer wishes to express his thanks to Dr. E. Babbott for her kind help in preparing the manuscript.

TABLE 1

REGIONAL INDUCTIVE EFFECTS

|   | Regionality (Lehmann) | (Dalcq) | Structures |
|---|---|---|---|
| A | Archencephalic | Acrogenetic | Forebrain, eye, nose |
| D | Deuterencephalic | Deutogenetic | Midbrain, hindbrain, ear vesicle |
| SC | Spino-Caudal | Tritogenetic | Spinal cord, tail-notochord, tail-somites |
| TM | Trunk-Mesodermal[1] | — | Trunk-notochord, trunk-somites, pronephros, blood islands |

[1] According to Yamada (40).

Toivonen (28) himself that the former tissue induces forebrain, eye, nose, and lens (archencephalic (19) or acrogenetic (4) structures), while the latter tissue induces spinal cord, tail-notochord, and tail-somites (spino-caudal (19) or tritogenetic (4) structures), or midbrain, hindbrain, and ear vesicle (deuterencephalic (19) or deutogenetic (4) structures). According to Toivonen's earlier papers (29), the factor responsible for archencephalic induction is relatively thermostable, soluble in petroleum ether, accompanies the nucleoproteins in fractionation, endures protracted ethanol treatment and is readily dialysable, while the factor responsible for spino-caudal induction is thermolabile, not extractable with petroleum ether, non-dialysable, and is inactivated by formalin and proteolytic enzymes. The work was continued by Kuusi (16, 17, cf. 1) in a series of extensive experiments. She tested inductive effects of cell components separated by differential centrifugation, nucleohistone samples, saline extracts, tissues treated with formalin, and homogenates treated with ribonuclease by implanting the samples into the blastocoel of the early amphibian gastrula. After formalin treatment both tissues lost their spino-caudal and deuterencephalic effects and showed a low frequency of archencephalic effects. On the other hand, ribonuclease failed to produce clear suppression of any type of induction. Kuusi observed a large difference between the effects of the kidney tissues and those of the various fractions extracted from the latter. The tissue or the residue of centrifugation was very often characterized by the spino-caudal and deuterencephalic effects, while the fractions separated from the kidney tissue indicated mostly archencephalic and deuterencephalic tendencies. This led her to conclude that centrifugal separation of the spino-caudal factor from the tissue is more difficult than that of the archencephalic factor. Yet she succeeded in obtaining from the supernatant fraction of the kidney spino-caudal effects as well as archencephalic and deuterencephalic ones. Considering various data, she maintained that archencephalic and spino-caudal factors are chemically distinct, and that the former is represented by granules containing pentose nucleic acid (PNA) and the latter by a protein. In a later paper she tested the inductive effects of the homogenates of the kidney, which were tested with some reagents for suppressing certain active groups of protein contained therein (17). In recent papers Kuusi (18) has reported her efforts toward getting into solution the factors responsible for the mesoderm-inducing effects of guinea pig bone marrow. We shall discuss these results later.

Using the extract of the chick embryo as source material, Tiedemann and Tiedemann (26) succeeded in isolating various protein fractions with either archencephalic or spino-caudal and deuterencephalic effects. Extraction with phenol, pyridine, and ethanol, precipitation with trichloroacetic acid and ammonium sulfate and high speed centrifugation were the chief techniques adopted by these German workers for isolation of the fractions.

In our own laboratory, research is being carried out to separate and characterize the factors responsible for regional induction by liver, kidney, and bone marrow of the guinea pig. The results to be presented here in some detail have been obtained with the collaboration of Dr. Y. Hayashi, Mr. K. Takata and Miss C. Takata.

### The Testing of Inductive Effects

In contrast to the experiments mentioned above, in which the sample was introduced into the blastocoel of a whole gastrula, in the present study the sample was tested on the isolated ectoderm of the early gastrula cultured in vitro. This procedure excluded possible interference with the regional effect by the host, which has been demonstrated to occur upon implantation of an inducer in the whole embryo by Chuang (2, 3) and Vahs (34). Other complications which might be caused by the presence of endoderm, mesoderm, and blastocoel fluid in the direct vicinity of the sample, and the difficulty of controlling the direct contact of the ectoderm with the sample were also avoided in our experiments. The samples were always tested in solid form. The solutions were precipitated with ethanol and the precipitate was collected, washed in Holtfreter's solution, and tested. A piece of the sample measuring about 0.4 mm. in diameter was enclosed between two pieces of the isolated ectoderm each measuring about 1 mm. in diameter. The ectoderm pieces fused to give rise to a vesicle. The explants were cultured in Holtfreter's solution adjusted to $pH$ 7.3 for 10 to 14 days at 15°C. All explants were fixed and studied in sections. The technique allows a reliable morphological evaluation of the inductive effects. Its drawback consists in the quantitative assay, although the total frequency of induction gives in most cases a measure for the inductive ability of the sample. A technique for quantitative assay is now being worked out. As the donor of the ectoderm, early gastrulae of *Triturus pyrrhogaster* were used exclusively. In Holtfreter's solution the ectoderm of the species differentiated only epidermal cells clustered irregularly (38, 10).

### Morphogenetic Effects of Kidney Pentose Nucleoprotein

Preliminary experiments (41) indicated that a piece of ethanol-treated guinea pig kidney evoked spino-caudal and deuterencephalic structures in the isolated ectoderm, as it had in the insertion experiments of Toivonen (28). The same regional effects were obtained by an ethanol precipitate of the 0.14 $M$ NaCl extract. Moreover, an acid-precipitable fraction of the extract containing the bulk of the pentose nucleic acid (PNA) present in the extract revealed the same effect. Comparing the inducing ability of the cytoplasmic fractions of the tissue prepared by differential centrifugation, the "small microsome" fraction containing the highest amount of PNA was found most effective in inducing spino-caudal structures (39). These results were taken to suggest that the pentose nucleoprotein (PNP) of the original extract might be responsible for the induction. This led us to prepare from the saline extract a sample of PNP by precipitating with streptomycin sulfate and subsequent washing and dialysis. The ethanol precipitate of this sample induced spino-caudal and deuterencephalic structures in the isolated ectoderm at a high frequency (43). The sample, which indicated a typical nucleoprotein absorption curve, was found to contain a considerable amount of PNA (245 PNA-P $\mu$g. per mg. protein-N) and to be free of DNA (cf. also 25). In electrophoresis, the sample indicated a single boundary whose mobilities were calculated by K. Takata to be $-6.1 \sim -6.3$ for the ascending limb and $-5.3 \sim -5.4$ for the descending limb (in cm$^2$/sec./volt $\times 10^{-5}$, $p$H 7.5, borate or phosphate buffer, $\mu = 0.2$, 4°C).

The sample was further purified by centrifuging at 100,000 g for 1 hr., and the supernatant containing a higher amount of PNA was precipitated with ethanol and tested (Table 2). An intensive inducing effect was registered, in which the deuterencephalic tendency was dominating. A moderate tendency of the spino-caudal type and a weak tendency of the archencephalic type were also noted (see p. 232).

In an earlier paper, Hayashi reported that a ribonuclease treatment of a phosphate extract of the kidney removed a large part of the PNA present in the extract, but did not appreciably reduce the spino-caudal and deuterencephalic effects of the extract (10). This was in accordance with the

### TABLE 2

MORPHOGENETIC EFFECTS OF SUBFRACTIONS OF PENTOSE NUCLEOPROTEIN
PREPARED BY ULTRACENTRIFUGATION FROM THE KIDNEY
(Yamada, Hayashi, and Takata)
100,000 × g, 1 hr.

|  | Supernatant | Sediment |
|---|---|---|
| PNA content (PNA-P µg./mg. protein-N) | 358.7 | 124.6 |
| No. of available explants | 85 | 87 |
| Total induction | 84 (99%) | 86 (99%) |
| Spino-caudal induction | 25 (29%) | 16 (18%) |
| Deuterencephalic induction | 56 (66%) | 72 (83%) |
| Archencephalic induction | 9 (11%) | 24 (28%) |
| Non-regional induction | 6 (7%) | 1 (1%) |

results that Kuusi (16) obtained after a ribonuclease treatment of the homogenate of the tissue. On the other hand, trypsin and chymotrypsin suppressed the inducing ability of the kidney tissue and its extracts (42). These experiments harmonize with each other in suggesting that the spino-caudal and deuterencephalic effects of the kidney extract are due not to its PNA but to its protein component. This conclusion was further supported by the experiments in which PNA isolated from the kidney of the guinea pig was found to possess only weak morphogenetic effects on the isolated ectoderm (44). In no case was a spino-caudal or deuterencephalic effect observed (cf. also 27).

### MORPHOGENETIC EFFECTS OF LIVER PENTOSE NUCLEOPROTEIN

Hayashi (11) fractionated the 0.14 $M$ NaCl extract of guinea pig liver, which showed considerable inductive effect of the archencephalic type alone, into two fractions by adding streptomycin sulfate. The pentose nucleoprotein fraction precipitable with streptomycin sulfate caused archencephalic structures at a higher frequency than the original extract, while the non-precipitable fraction induced these structures at a lower frequency. In a later series of experiments a liver PNP sample induced not only archencephalic but also deuterencephalic structures (cf. 30). In any case, we have here an interesting situation, that from the liver and kidney of the guinea pig with an identical technique PNP samples were isolated, which have distinctly different morphogenetic effects corresponding to the effects of the source materials. Also in the case of liver PNP, the electrophoretic pattern indicated a single boundary. The mobilities of the sample[2] as calculated by K. Takata did not significantly differ from the above reported

---

[2] −6.0 ∼ −6.1 for the ascending limb, and −5.1 ∼ −5.3 for the descending limb. For further data see p. 220.

values for kidney PNP. However, as in the sedimentation diagram some heterogeneity was observed, the sample was separated into the supernatant and sediment by ultracentrifugation. The supernatant, containing a large amount of PNA, showed deuterencephalic and archencephalic effects in a very high percentage of cases, while the sediment, containing a smaller amount of PNA, gave archencephalic and non-regional neural effects in a lower percentage of cases (Table 3). Probably the supernatant represents the bulk of PNP, and the sediment contains some protein contaminating the original sample of PNP.

TABLE 3

The Inductive Effect of Subfractions of a Sample of Liver Pentose Nucleoprotein Separated by Means of Ultracentrifugation (Hayashi and Takata)

|  | 100,000 × g precipitate | 100,000 × g supernatant | 173,490 × g precipitate | 173,490 × g supernatant |
|---|---|---|---|---|
| PNA-P μg./mg. N | 10.2 | 92.7 | 21 | 197.5 |
| No. of available explants | 53 | 59 | 31 | 37 |
| Neural induction | 38 (72%) | 59 (100%) | 21 (68%) | 36 (96%) |
| Non-neural induction | 12 (23%) | 0 | 5 (16%) | 1 |
| Archencephalic | 29 (55%) | 20 (34%) | 17 (55%) | 9 (24%) |
| Deuterencephalic | 5 (9%) | 42 (71%) | 9 (29%) | 27 (73%) |
| Induction without regional character | 20 (38%) | 5 (8%) | 5 (16%) | 5 (14%) |

The obvious question to be asked was whether the induction by the nucleoprotein sample is due to its nucleic acid or its protein component. In order to answer the question, effects of proteolytic enzymes and ribonuclease on the inductive ability of the sample were studied by Hayashi. He succeeded in removing the bulk of PNA present in the PNP sample with ribonuclease in the following two series of experiments. In the first series an ethanol-precipitated sample of liver PNP was used as the substrate (series A). As shown in Table 4, no significant difference was observed

TABLE 4

Effect of Ribonuclease on the Inductive Ability of Pentose Nucleoprotein from the Liver (Hayashi)

|  | Series A | | Series B | |
|---|---|---|---|---|
|  | Control | Treated | Control | Treated |
| PNA content (PNA-P μg./mg. protein N) | 86.6 | 3.5 | 148.6 | 0.84 |
| No. of available explants | 25 | 33 | 49 | 52 |
| Neural induction | 25 (100%) | 33 (100%) | 49 (100%) | 52 (100%) |
| Archencephalic induction | 24 (96%) | 27 (82%) | 7 (14%) | 8 (15%) |
| Deuterencephalic induction | 1 (4%) | 3 (9%) | 40 (82%) | 40 (77%) |
| Non-regional induction | 1 (4%) | 4 (12%) | 8 (18%) | 9 (17%) |

in the inducing ability of the sample. A better removal of PNA was attained by treating a liver PNP sample purified by ultracentrifugation with ribonuclease without previous ethanol-precipitation (series B). Even in this sample practically free of PNA, no reduction of inducing ability was detected as compared with the control sample.

The results were in accord with Hayashi's earlier work, in which an extract of guinea pig liver showed, after incubation with ribonuclease and deoxyribonuclease, an inducing frequency comparable to that of the untreated sample (10). That the treatment with ribonuclease does not affect the inductive power of tissues and their homogenates has been repeatedly observed in experiments with whole embryos (16, 5, 35).

Thus no indication was obtained for an active participation of PNA in the induction by the liver. On the other hand, the importance of protein in induction was clearly evidenced in the experiments with pepsin and trypsin (12). Hayashi incubated a liver PNP sample with 0.1% pepsin at $pH$ 4.0 and 28°C for 30, 60, and 120 minutes. After incubation, the $pH$ was adjusted to 7.5 and ethanol was added to the mixture (Table 5). The

TABLE 5

Effects of Pepsin on the Inductive Ability of Liver Pentose Nucleoprotein (Hayashi)

Enzyme 0.1%, $pH$ 4.0, 28°C.

|  | Control | Pepsin 30 min. | Pepsin 60 min. | Pepsin 120 min. |
|---|---|---|---|---|
| No. of available explants | 70 | 51 | 31 | 47 |
| Neural induction | 66 (94%) | 45 (88%) | 17 (55%) | 20 (43%) |
| Archencephalic induction | 53 (76%) | 25 (49%) | 9 (29%) | 6 (13%) |
| Deuterencephalic induction | 9 (13%) | 2 (4%) | 0 | 0 |
| Non-regional induction | 13 (19%) | 21 (41%) | 19 (61%) | 34 (72%) |

precipitate was tested on the isolated ectoderm and the effects were compared with those of the control sample, to which enzyme in inactive condition was added after a sham incubation. The result indicated a progressive suppression of the regional inductive ability of the sample, although even after 120 minutes no complete suppression was achieved.

In the case of the trypsin series, incubation was done at $pH$ 7.6 and 28°C for 30, 60, and 120 min. with 0.01% enzyme. After each incubation period, ethanol was added and the precipitate thus obtained was heated in 95% ethanol for 5 min. at 85°C, in order to effect inactivation of the enzyme. In the control series after incubation without the enzyme, trypsin was added and immediately followed by ethanol. The precipitate thus obtained was heated in ethanol as in the experimental series. As shown in

TABLE 6

EFFECTS OF TRYPSIN ON THE INDUCTIVE ABILITY OF PENTOSE
NUCLEOPROTEIN FROM THE LIVER (HAYASHI)
Enzyme 0.01%, $p$H 7.6, 28°C.

|  | 30 min. | | 60 min. | | 120 min. |
| --- | --- | --- | --- | --- | --- |
|  | Control | Exper. | Control | Exper. | Exper. |
| No. of available explants | 45 | 35 | 50 | 45 | 53 |
| Neural induction | 45 (100%) | 16 (46%) | 46 (92%) | 5 (11%) | 3 (6%) |
| Archencephalic induction | 43 (96%) | 12 (34%) | 42 (84%) | 3 (7%) | 0 |
| Deuterencephalic induction | 2 (4%) | 0 | 0 | 0 | 0 |
| Non-regional induction | 0 | 10 (29%) | 8 (16%) | 12 (27%) | 11 (21%) |

the table, a progressive inactivation of the inducing ability was clearly demonstrated. In this series, after 120 minutes of incubation, the regional effects of the sample had completely vanished.

## THE SUBSTANCE RESPONSIBLE FOR THE MESODERM-INDUCTION BY THE BONE MARROW

As stated above, Toivonen (31) demonstrated that guinea pig bone marrow induces almost exclusively mesodermal structures from the presumptive ectoderm. Recently Kuusi (18) has published papers concerning the chemical nature of the bone marrow factor. In her attempt to get the factor in the extract, Kuusi obtained induction of mesodermal structures by a phosphate extract of the tissue. However, a saline extract of the tissue proved to be very weak in morphogenetic effects. Without knowing the work of Kuusi, we started in 1956 a series of experiments aiming at the isolation and characterization of the bone marrow factor. Before going further it may be convenient to point out that most of the mesodermal structures evoked by a freshly prepared sample of guinea pig bone marrow are those characteristic for the trunk-region, and are easily distinguished from the mesodermal structures induced, for instance, by guinea pig kidney, which are caudal in nature. Hence the present writer (40) proposed that the typical structures induced by the bone marrow be called the "trunk-mesodermal structures" and be distinguished from the caudal mesodermal structures which are included in the spino-caudal type (Table 1). Probably the trunk-mesodermal effects are further given by the bone marrow of the rat (24) and mouse (14, 15), and also by the skin of the adult frog (7, 8, 21).

Contrary to Kuusi's results (18), the 0.14 $M$ NaCl extract of the bone marrow of the guinea pig gave in our case a high percentage of trunk-mesodermal induction in the isolated ectoderm after ethanol-precipitation (see Fig. 5). Two series of fractionations were done with the saline extract as the starting sample. In the first series, streptomycin sulfate was added to the extract in order to precipitate PNP. After washing, dialysis, and ethanol-precipitation the PNP sample, containing a large quantity of PNA (399 µg. PNA-P/mg. protein-N), was tested on the isolated ectoderm. Contrary to our expectations, the sample proved to be a very weak inducer, giving rise only infrequently to small neural or mesodermal structures (Fig. 1). On the other hand, the fraction not precipitable with streptomy-

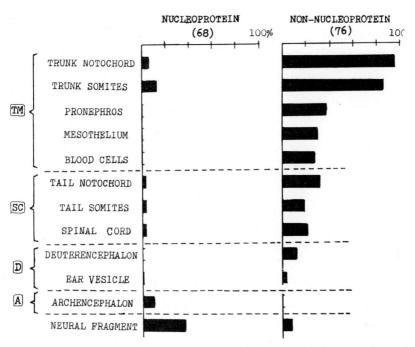

Fig. 1. Inductive effects of the pentose nucleoprotein fraction and the non-nucleoprotein fraction of the bone marrow extract. The frequency is expressed as the percentage of explants carrying a particular structure among the total number of explants of the respective series (in parentheses). This explanation applies also to Figs. 3, 5, and 7.

cin sulfate and possessing only a small quantity of PNA (12 µg. PNA-P/ mg. protein-N) revealed an intensive activity of mesodermization on the isolated ectoderm. The results, repeatedly confirmed, imply that the bone marrow factor is not represented by the streptomycin-precipitable PNP fraction.

In the second series of fractionations, the sediment and supernatant were prepared from the saline extract according to the scheme of Fig. 2. The

226  THE CHEMICAL BASIS OF DEVELOPMENT

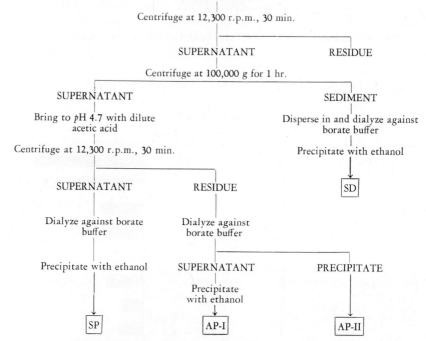

Fig. 2. The scheme of fractionation of the bone marrow extract. SD, final sediment; SP, final supernatant; AP, acid-precipitable fraction.

former fraction, containing probably the PNP fraction of the last-mentioned series, exerted only a weak morphogenetic effect on the ectoderm. The supernatant fraction was further fractionated by carefully acidifying with acetic acid to $pH$ 4.7. The final supernatant showed almost no morphogenetic effects on the ectoderm. The fraction precipitated at $pH$ 4.7 was suspended in a borate buffer and dialyzed. After the dialysis the part of the sample which was in solution was separated from the sediment found in the dialysis bag. The former was called acid-precipitable 1 (AP-I) and the latter acid-precipitable 2 (AP-II). Both samples caused very intensive induction of the trunk-mesodermal type when tested after ethanol-treatment (Fig. 3). Thus the active substance present in the original saline extract was recovered in the acid-precipitable and non-dialyzable fraction of the supernatant obtained by centrifuging at 100,000 × g for 1 hour. The data of chemical analysis (Table 7) imply that only about one-eighteenth

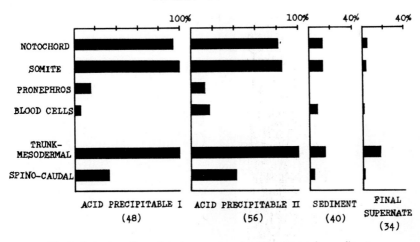

Fig. 3. Inductive effects of the bone marrow fraction, separated according to the scheme of Fig. 2.

of the protein-N present in the original saline extract is accounted for by the acid-precipitable fraction.

In the sedimentation diagram of sample AP-1, a major component with $S_{20} = 5$ was found accompanied by a minor component with $S_{20} = 11$ (Fig. 4). A sample of AP-1 was fractionated into the supernatant SII and sediment SI by centrifuging at 220,000 × g for 1 hour, and both subfractions were tested for morphogenetic effects on the isolated ectoderm. The data shown in Fig. 5 indicate strong trunk-mesodermal effects by both subfractions.

A sample of AP-I was studied in a Tiselius apparatus (Fig. 6). Five components were recognized whose mobilities are given in Fig. 6. The slower-moving components (D and E) were separated as one fraction

TABLE 7

Analytical Data on the Fractions of the Bone Marrow Extract

| Samples | PNA-P µg. per mg. protein-N | Total protein N (mg.) | Total PNA-P (µg.) |
|---|---|---|---|
| 0.14 M NaCl extract | 19.1 | 88 | 1680 |
| Supernatant fraction | 9.7 | — | — |
| Acid non-precipitable of supernatant (BM-SP) | 6.0 | — | — |
| Acid precipitable of supernatant (BM-API) | 12.4 | 5 | 62 |

228  THE CHEMICAL BASIS OF DEVELOPMENT

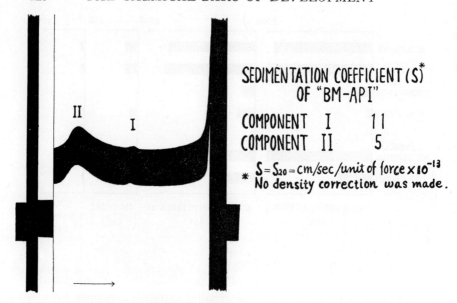

after 16 min. at 56,100 r.p.m.

Fig. 4. Sedimentation pattern of bone marrow AP-I.

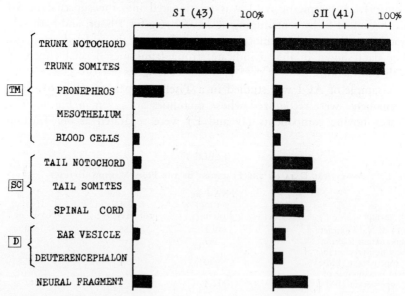

Fig. 5. Inductive effects of the subfractions of AP-I separated by ultracentrifugation.

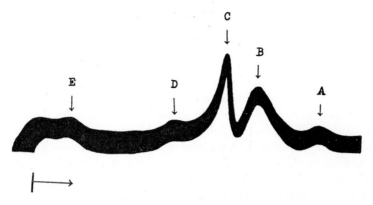

Fig. 6. Electrophoretic pattern (ascending) of BM–AP-I. 0.14 μ borate buffer at pH 7.5 after 70 min. in a field of 9.5 volts per cm. Ascending Mobilities of the Components of "BM–AP-I," in cm./sec./volt $\times 10^{-5}$.

| A | B | C | D | E |
|---|---|---|---|---|
| −9.2 | −7.4 | −6.1 | −4.4 | −1.9 |

D and E bracketed as S-E.

from the descending limb after ethanol-precipitation. This sample (S-E) caused intensive mesodermization of the isolated ectoderm (Fig. 7). We have not been successful so far in testing the morphogenetic effects of the fast-moving components.

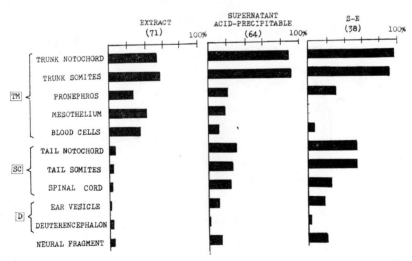

Fig. 7. Inductive effects of the original extract, fraction AP-I and S-E, an electrophoretically separated subfraction of the latter.

Under the influence of the acid-precipitable fraction not only were practically all the explants induced, but the main part of the isolated ectoderm was converted into mesodermal structures. In most of the explants the only cells retaining their ectodermal nature were those of the thin epidermis covering the surface of the explant. In a smaller number of cases the whole mass of the explanted ectoderm was turned into mesodermal structures which were exposed directly to the medium. Besides the mesodermal structures listed in Fig. 7, limb buds, blood vessels, and mesenchyme could also be recognized. Associated with these trunk-mesodermal structures, spino-caudal ones such as the tail notochord, tail somites, and spinal cord appeared in the series of the bone marrow fractions. In most cases both types of structures were continuous, giving rise to a more or less organized axis with cephalocaudal polarity (Fig. 8, B).

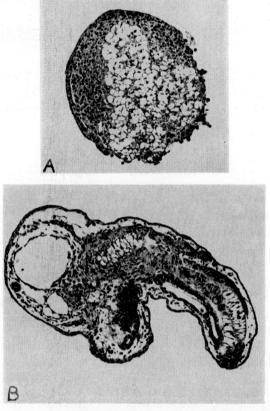

Fig. 8. A. Induction of notochord and mesoderm by fraction $S_1$. B. Induction of notochord, myotome, and spinal cord by fraction S-E.

From the beginning of our work on bone marrow induction, I had been attracted by occurrence of a group of yolk-rich cells within the induced mesodermal tissue. The problem is now under investigation with the collaboration of Miss C. Takata. Recently we have obtained convincing

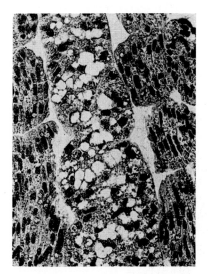 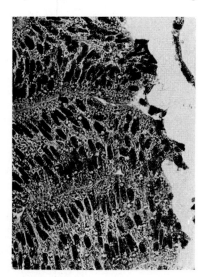

Fig. 9. Induced myotomes *left* and notochord with somites *right* in the explants of series S-E.

data showing induction of endodermal tissues from the ectoderm by the bone marrow fractions. In favorable cases differentiation of the pharynx and stomach was indicated in the explant containing a piece of acid-precipitable fraction. Those endodermal structures appear always closely associated with the trunk-mesodermal tissues but not with other types of induced structures. In her recent paper Kuusi (18) pointed out that the bone marrow extract enhanced differentiation of the host endoderm at the site of implantation. It remains to be decided whether this acceleration has something to do with the induction of endodermal tissues observed in our experiments.

### Progressive Change in Regional Effects

Chuang (2, 3) was first to show a progressive change in the regional effects of the adult tissue caused by heat treatment. According to him the mesodermal induction was the first to vanish, while the induction of brain, nose, and eye not only withstood heating but showed an increase in frequency. Recently Vahs (33) has studied the effects of heating on induction by the mouse kidney. Classifying the effects in terms of regionality, he reached the conclusion that heat suppresses the deuterencephalic effects but increases the archencephalic effects under his experimental conditions. It was shown further that during a long ethanol-treatment of mouse kidney (35, 6), guinea pig kidney, and parotid gland (5), a quick decrease of spino-caudal effects and a later decrease of deuterencephalic effects were evident. A gradual increase in archencephalic effects kept the total induction frequency at the original level.

A comparable shift of regionality by various treatments occurs not only on the level of the tissue but also on the level of the extract (41) or of a fairly homogeneous sample. For instance, in the case of our sample of kidney PNP, the spino-caudal effects were easily suppressed and the deuterencephalic effects enhanced, if the sample was left in solution for a longer time or kept more than several days in ethanol before testing. This fact helps to explain why the fractions separated from the kidney of the guinea pig after a longer procedure often showed a lower frequency of spino-caudal effects and a larger frequency of deuterencephalic effects (cf. Table 2 and p. 220). Even archencephalic effects could appear. A complete suppression of spino-caudal effects and a concomitant appearance of archencephalic effects were noted upon treating the sample of kidney PNP with acetic anhydride, according to the method suggested by Hugh for acetylating protein.[3] Hayashi found that a similar change in regional effect can be evoked in the same sample by treating it with pepsin. Upon addition of the enzyme to a neutral solution of kidney PNP, suppression of the spino-caudal tendency, increase of the deuterencephalic tendency, and appearance of the archencephalic tendency were demonstrated. When the sample was incubated with pepsin at $p$H 4.0, a quick disappearance of the spino-caudal and slower disappearance of the deuterencephalic tendency were noted. In the course of incubation the archencephalic effects also diminished. The results suggest that the presence of pepsin under suboptimum conditions causes a change in the protein part of PNP, which is reflected in the change of its regional effects. The enzymatic activity of pepsin seems to suppress all types of regional induction, by acting strongly on the spino-caudal one and weakly on the archencephalic one. Using a sample of liver PNP with deuterencephalic effects, Hayashi (12) observed the disappearance of deuterencephalic effects after a short heat treatment. A similar effect of pepsin has been reported by Tiedemann and Tiedemann

[3] However, Kuusi (17) observed no clear shift of regionality when she treated the kidney homogenate in the same way.

(26) for protein fractions of chick embryos, which induce primarily spino-caudal structures.

According to Toivonen (32) and Kuusi (18), the bone marrow of the guinea pig is different from the above-mentioned tissues and their fractions in that it loses its inductive effects altogether when heated, without converting its inductive quality. This claim is based on the experiments of Toivonen (32) that ethanol-treated bone marrow tissue lost all its mesodermal effects and induced only insignificant epidermal structures after immersion in 80°-90°C water for 10 minutes. However, by steaming a thin layer of bone marrow tissue, the present writer (40) could confirm that during the first 150 seconds of treatment the tissue quickly changes its regional effects, and induces trunk-mesodermal, spino-caudal, deuterencephalic, and archencephalic structures in sequence (Fig. 10), before the

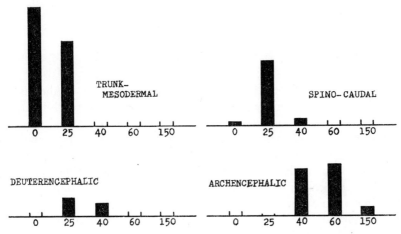

Fig. 10. Regional effects of bone marrow tissue steamed for 0, 25, 40, 60, and 150 seconds respectively (Yamada).

total induction frequency drops to a very low level. Probably Toivonen observed in his last-mentioned experiment only the very end of this sequence of events. Nonetheless we have to admit that the bone marrow factor is more heat-sensitive than the factors of other tissues studied so far. A similar conversion of the regional effect of an inducer of the trunk-mesodermal type was demonstrated earlier by Okada (21), when he observed the induction of various neural structures by the heat-treated frog skin.

When the acid-precipitable fraction of the supernatant which carries the inducing activity of the bone marrow (p. 226) was steamed for 40 seconds and the inductive effects were compared with those of the untreated sample, the spino-caudal effects and deuterencephalic effects were found to be increased at the expense of the trunk-mesodermal effects. Thus also in the case of the bone marrow it is likely that the progressive change in regional effects observed in the tissue is based on the comparable change in its active component. An extensive study on this point is being planned.

Summarizing the results reviewed above, various treatments convert the regional effects according to the sequence of regional types represented in Fig. 11. In many cases the progressive change in regional effects pro-

$$\text{TRUNK-MESODERMAL} \downarrow \text{SPINO-CAUDAL} \downarrow \text{DEUTERENCEPHALIC} \downarrow \text{ARCHENCEPHALIC}$$

Fig. 11. The sequence of regional types of induction.

ceeds without any change in the total induction frequency. In the case of steaming the bone marrow or of pepsin-treatment of kidney PNP, a progressive decrease in the total induction frequency accompanied the progressive regional change. Many results presented above speak for the possibility suggested by us earlier (41, 42) that these progressive changes in regional effects are due to comparable changes of the active component of the tissue, which is probably a protein. It is tempting to assume that the protein molecule responsible for induction goes through a series of changes in configuration or in any other physical property, which are reflected in the regional effects.

It may be pointed out that on the surface of an early gastrula of the urodele, the area having the prospective significance of archencephalic structures is located near the animal pole, followed by the prospective areas of the deuterencephalic, spino-caudal, and mesodermal structures in sequence toward the vegetal pole. That the sequence of the regional types observed in various inducing samples upon various treatments follows the vegetal-animal sequence of the corresponding areas can hardly be a mere coincidence. It may be added that even the induction of the endodermal structures by the bone marrow factor fits in well with this sequence.

## The Specificity of the Inductive Phenomena

As to the specificity of the induction studied in the present paper, the following remarks may be pertinent. First, in all types of inducing sample discussed above, the effect observed is region-specific but not organ-specific or tissue-specific. Also, during the course of step-wise purification of the samples, no indication was obtained to speak for any concomitant segregation of the factor inducing one specific organ. Among the large number of induction experiments with various adult tissues hitherto reported, no reliable case of organ-specific induction has been found. The claim that the thymus of the guinea pig is a specific inducer of the lens (28) could not withstand later critical experiments (23). Also in the extensive experiments of Kuusi (16, 17, 18) and of Tiedemann and Tiedemann (26), in which more or less purified samples with inductive ability have been obtained, the effects which have been reported are region-specific but not organ-specific. Induction of several distinct organs or tissues belonging to one or two regionalities by a fairly homogeneous sample could be interpreted by assuming a secondary induction of some of these structures by one structure which alone is primarily induced by the sample. This interpretation is however at least not always applicable, because in all regional types of heterogeneous induction most of these structures have been induced directly from the inducing sample, independently from other structures. Thus we are forced to admit that the inducing sample initiates in the ectoderm a morphogenetic activity which is characteristic for a particular region, but not for an individual organ or tissue. Thus the segregation of the whole explant induced by a sample into a number of organs and tissues belonging to the same regionality may be most readily understood by assuming a gradient of morphogenetic activity, which is established within the explant according to the topographical relation to the sample, and which brings forth a different morphogenesis on its different levels.

This idea is supported by observations on the relative frequencies of various structures of one regional type in a series of induction experiments with the samples containing various concentrations of the active substance. Observations by Hayashi (11) suggest that samples with a higher concentration of the active substance of the archencephalic type induce the eye and forebrain more readily, whereas samples with a weaker concentration tend to induce the mesenchyme and melanophores more frequently. These observations confirm the earlier results of Yamada (38). A comparable situation is suggested also for the tissue of the trunk-mesodermal type: by a sample of lower concentration of the active substance (for instance, the extract of Fig. 7) the blood islands and mesothelium are more readily induced than by a sample of higher concentration (for instance, S-E of Fig. 7); while the opposite holds for the notochord and somites. Furthermore, the topographical relationship between various mesodermal rudiments induced within an explant often reminds us of the normal dorso-ventral distribution of the rudiments, as Okada (21) pointed out in his experiments with frog skin. The same is true for all other types of regionality (38, 11). These facts accord with the assumption of a gradient of morphogenetic activity.

Secondly, the specificity of induction caused by our sample is limited by the phenomenon which has been called the shift of the regional type. This tends to widen the range of the morphogenetic effects of a particular sample. In the case of the bone marrow factor, in particular, almost all the principal organ rudiments belonging to three germ layers may be induced from the presumptive ectoderm by this factor if it is applied under fresh and modified conditions.

### Concluding Remarks

The results reported in this paper concerning the bone marrow factor seem to raise the question whether in the normal organizer cells a similar substance is present at a higher concentration and directs differentiation and mediates induction.[4] The demonstration of Niu (20) that a substance or substances are discharged from the developing organizer cells into the medium and cause the ectoderm cells cultured in the medium to differentiate in a neural, mesectodermal, or mesodermal direction, may be interpreted as suggesting such a possibility. The recent work of Grobstein (9) indicates that the tubule-inducing effects of the spinal cord on the metanephrogenic mesenchyme of the mouse embryo can be transmitted without any cytoplasmic contact between the two tissues, and suggests that some substance responsible for the induction and separable from the cell body is transferred from the acting system into the reacting system. A transfer of protein from the organizer into the ectoderm cells is also made probable by some of the recent isotope experiments (cf. 37). However, facts are known which indicate that the hypothetical factor involved in the normal organizer, or other normal inducers if present, must be chemically different from the bone marrow factor—for the morphogenetic substances found in the adult tissues are stable toward ethanol-treatment, but the amphibian

---

[4] As the bone marrow factor is known to be active as an antigen (36, and oral communication of Dr. Kawakami), it may be interesting to study the effect of the antiserum against the bone marrow factor on the differentiation of the normal organizer.

organizer loses its characteristic effects when treated with ethanol. It was shown by Okazaki (22) that a piece of the organizer loses its inducing effect when devitalized by a mild method, such as treatment with subzero ethanol or freezing-drying, and that the archencephalic inducing ability can be secondarily evoked in such a piece by a subsequent treatment with 5°C ethanol. Thus the inducing ability exhibited by a "dead organizer" may have nothing to do with the original inducing ability. Further, it must be pointed out that also in the specific induction of the later embryonic period the inducer loses its effect upon devitalization, while the adult tissues are capable of "specific" induction after devitalization. In any case, the question raised above needs more coordinated information from the different fields suggested in the Introduction.

## REFERENCES

1. Brachet, J., Kuusi, T., and Gothié, S., *Arch. Biol.*, **63**, 429 (1952).
2. Chuang, H. H., *Wilhelm Roux' Arch. Entwicklungsmech. Organ.*, **139**, 556 (1939).
3. ———, *Wilhelm Roux' Arch. Entwicklungsmech. Organ.*, **140**, 25 (1940).
4. Dalcq, A. M., *Acta Anat.*, **30**, 242 (1957).
5. Engländer, H., and Johnen, A. G., *J. Embryol. Exptl. Morphol.*, **5**, 1 (1957).
6. ———, ———, and Vahs, W., *Experientia*, **9**, 100 (1953).
7. Fujii, T., *J. Facul. Sci. Imp. Univ. Tokyo*, Sec. IV, **5**, 425 (1941).
8. ———, *J. Facul. Sci. Imp. Univ. Tokyo*, Sec. IV, **6**, 451 (1944).
9. Grobstein, C., *Exptl. Cell Research*, **13**, 575 (1957).
10. Hayashi, Y., *Embryologia*, **2**, 145 (1955).
11. ———, *Embryologia*, **3**, 57 (1956).
12. ———, *Embryologia*, **4**, 33 (1958).
13. Holtfreter, J., *Wilhelm Roux' Arch. Entwicklungsmech. Organ.*, **132**, 307 (1934).
14. Kawakami, I., and Mifune, S., *Mem. Fac. Sci. Kyushu Univ.*, Ser. E, **2**, 21 (1955).
15. ———, and ———, *Mem. Fac. Sci. Kyushu Univ.*, Ser. E, **2**, 141 (1957).
16. Kuusi, T., *Ann. Zool. Soc. Zool. Botan. Fennicae Vanamo*, **14**, 1 (1951).
17. ———, *Arch. Biol.*, **64**, 189 (1953).
18. ———, *Arch. Soc. Zool. Botan. Fennicae Vanamo*, **11** (2), 136 (1957); **12** (1), 73 (1957).
19. Lehmann, F. E., *Einführung in die physiologische Embryologie*, Verlag Birkhäuser, Basel (1945).
20. Niu, M. C., in *Cellular Metabolism in Differentiation and Growth* (D. Rudnick, ed.), p. 155 (1956).
21. Okada, Y. K., *Proc. Japan Acad.*, **24**, 22 (1948).
22. Okazaki, R., *Exptl. Cell Research*, **9**, 579 (1955).
23. Rotmann, E., *Naturwiss.*, **30**, 60 (1942).
24. Saxén, L., and Toivonen, S., *Ann. Med. Exptl. et Biol. Fennicae (Helsinki)*, **34**, 235 (1956).
25. Takata, K., and Osawa, S., *Biochem. et Biophys. Acta*, **24**, 207 (1957).
26. Tiedemann, H., and Tiedemann, H., *Hoppe-Seyler's Z. physiol. Chem.*, **306**, 7 (1956).
27. ———, and ———, *Hoppe-Seyler's Z. physiol. Chem.*, **306**, 132 (1957).
28. Toivonen, S., *Ann. Acad. Sci. Fennicae*, Ser. A, **55**, 3 (1940).
29. ———, *Rev. Suisse Zool.*, **57**, Fasc. Suppl. 41 (1950).

30. ———, *Experientia,* **8,** 120 (1952).
31. ———, *J. Embryol. Exptl. Morphol.,* **1,** 97 (1953).
32. ———, *J. Embryol. Exptl. Morphol.,* **2,** 239 (1954).
33. Vahs, W., *Z. Naturforsch.,* **10,** 412 (1955).
34. ———, *Biol. Zentr.,* **75,** 360 (1956).
35. ———, *Wilhelm Roux' Arch. Entwicklungsmech. Organ.,* **149,** 339 (1957).
36. Vainio, T., *Ann. Acad. Sci. Fennicae, Ser. A,* **35,** 7 (1957).
37. Waddington, C. H., and Mulherkar, L., *Proc. Zool. Soc. Bengal,* Mookerjee Memor. vol., 141 (1957).
38. Yamada, T., *Embryologia,* **1,** 1 (1950).
39. ———, *Symposia Soc. Cellular Chem.,* **4,** 1 (Jap.) and 147 (Eng. résumé) (1956).
40. ———, *Experientia,* **14,** 81 (1958).
41. ———, and Takata, K., *J. Exptl. Zool.,* **128,** 291 (1955).
42. ———, and ———, *Exptl. Cell Research,* **3,** (Suppl.), 402 (1955).
43. ———, and ———, *Embryologia,* **3,** 69 (1956).
44. ———, ———, and Osawa, S., *Embryologia,* **2,** 123 (1954).

# INDUCTIVE SPECIFICITY IN THE ORIGIN OF INTEGUMENTARY DERIVATIVES IN THE FOWL

JOHN W. SAUNDERS, JR.[1]

*Department of Biology, Marquette University, Milwaukee, Wisconsin*

### INTRODUCTION

The domestic fowl bears a variety of regionally distinctive integumentary derivatives. The body is covered with feathers, the head is adorned with the comb, beak, and wattles, and the feet are armed with scales and claws; a small claw tips Digit II of the wing. These skin derivatives are of dual origin; the comb and wattles comprise thickened and highly vascular modifications of the dermis, with a thin epidermal covering; the feathers, scales, and claws, on the other hand, are chiefly epidermal structures, each associated with a more obscure dermal portion.

The feathers are of particular interest. They are arranged in tracts (pterylae) separated by featherless areas (apterylae). Each tract contains feathers of a distinctive range of structural and physiological characteristics, ordered in parallel rows long both longitudinal and diagonal coordinates (Fig. 1). Within each tract there are graded variations in structure and physiological features (size, degree of asymmetry, hormone thresholds, etc.) from feather to feather along each row. Each feather is thus an unique component of an orderly pattern (1, 3).

The origin of this pattern presents a challenging problem. More than two decades ago, Professor Willier and his colleagues sought to analyze the embryonic origin of spatial distribution and tract-specificity in feathers by grafting skin ectoderm from the head to the wing bud in embryos of genetically different breeds of fowl. The grafts contributed little or no epidermis to the host, however, and hence did not affect the regional characteristics of feather structure at the graft site. On the other hand, they frequently "infected" the host tissues with melanoblasts, prospective pig-

---

[1] This research has been supported in part by research grants (C1481-C5, C6) from the National Cancer Institute of the National Institutes of Health, by a grant (G2253) from the National Science Foundation, and by grants from the American Cancer Society, Milwaukee Division.

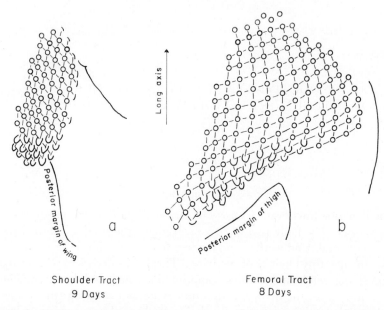

Fig. 1. "Grid" patterns showing the arrangement of feather germs in *a*, the *humeral* and *b*, the *femoral* feather tracts of the chick, traced from camera-lucida drawings. The feather germs form a pattern of intersecting coordinates.

ment cells, of donor origin, and accordingly brought about the formation of donor-colored feathers in the region of the implant. This observation led to a most successful analysis of the role of the migratory pigment cell in the origin of color patterns in the plumage of birds (11-16, 24, 26-30), but further work on the origin of the distribution pattern and tract-specificity of feather germs was deferred.

This being so, it would seem particularly appropriate that, in a Symposium Volume honoring Professor Willier, some attempt should be made to pursue the analysis which he originally projected. This report, based chiefly on work initiated in his laboratory and originally under his direction, takes up this analysis as an aspect of the general problem of the origin of regional specificity in the epidermis and its derivatives. It treats chiefly of (1) ectoderm-mesoderm interactions in the locus of origin of the feather germ; (2) the order in which tract-specific feather characteristics are established; and (3), the regionally specific inductive action of the mesoderm in the origin of distinctive epidermal derivatives, i.e., kinds of feathers, scales, claws.

## Induction of Tract-Specific Feather Characteristics

### Ectoderm-Mesoderm Relationships in the Mature Feather

The familiar contour feather is a bilaterally organized epidermal structure which develops in intimate association with a vascular mesodermal papilla located at the base of a pit in the skin, the feather *follicle*. When the feather of an adult bird is plucked or moulted, the denuded dermal papilla becomes encircled by a ring of epidermal cells, the *collar*, originating from the adjoining lining of the follicle. From this epidermal collar is regenerated a new feather similar in size, structure, and physiological characteristics to the one that was lost.

The dermal papilla permanently resides at the base of the follicle and is essential to feather formation; a follicle from which it has been removed will not regenerate a feather. Apparently, therefore, the dermal papilla induces feather formation in the epidermis. Furthermore, it determines the bilateral organization of the feather vane, for rotation of the dermal papilla in the base of the follicle brings about a corresponding rotation of the regenerating feather which it induces (10).

On the other hand, when a dermal papilla from a follicle of the distinctive "saddle" feather tract is grafted to a follicle of the "breast" tract from which the papilla was previously removed, a breast feather is formed. Conversely, grafting a dermal papilla from the breast tract to an empty follicle in the saddle tract results in the formation of a saddle feather. Clearly, therefore, the inductive action of the dermal papilla is generalized, or nonspecific, and the tract-specific characteristics of the feather are the property of the follicular epidermis (23).

### Embryonic Origin of the Feather Germs

The spatial distribution of feather germs, the inductive activities of the dermal papillae, and the definitive response capacities of the follicular ectoderm might reasonably be expected to originate during embryonic development. Accordingly, it is appropriate to describe briefly the embryonic origin of the feathers.

The feather germs first appear in the back skin as early as 6½ days of incubation, and by the tenth or eleventh day the tract patterns of the future plumage are essentially complete. Within each feather tract, the first primordia appear as a row of mesenchymal condensations, usually oriented longitudinally with respect to the embryonic axis. On each side of this "row of origin" (7) secondary rows of mesodermal condensations appear, each occupying a position corresponding to the space between condensations in the primary row; thus diagonally intersecting coordinates are formed (cf. Fig. 1). Additional rows are added in a similar manner, extending the tract until its pattern is complete.

Meanwhile the ectoderm overlying each dermal condensation becomes thicker. Proliferation of both the ectoderm and the dermal condensation forms the conical feather papilla which projects above the level of the skin, usually sloping somewhat posteriorly in the direction of orientation of the definitive feather. Subsequently, each primordium sinks into the skin so that its insertion is at the base of a pit, the follicle. There the dermal condensation forms the dermal papilla and pulp of the down feather; the proliferating ectoderm of the germ forms the feather vane.

The down feather is superficially a radially symmetrical structure, but the manner of its origin forecasts the future bilateral organization of the juvenile and adult feathers. The first barb-vane ridges arise on the side of the feather papilla which forms the outer obtuse angle of its incidence with the skin (8, 9, 24). This is the *dorsal* side, the site of origin of the rachis of the definitive feather. Further, the melanophores first become aligned in rows along the dorsal side (24) of the germ, and alkaline phosphatase in the feather pulp shows a greater initial activity dorsally (6, 9).

Details of the formation of the down and its replacement by the juvenile and adult plumage are well known (2, 4, 5, 8, 24). They are not particularly pertinent to this report and will not be described.

*Factors Determining the Origin of Feather-Germ Loci*

The intimate association of epidermal and mesodermal elements in the origin of the feather suggests that mutual interactions between these layers are responsible for determining the feather-germ loci. This has been shown experimentally. Feather formation is not initiated by either ectodermal or mesodermal elements of the embryonic skin cultured alone in vitro on media which will support the differentiation of feather germs in intact skin (22). Also, mesoderm of the dorsal surface of the wing bud in vivo will not organize a typical corium or dermal papillae in the absence of the covering ectoderm (21, 25).

On the other hand, since the first signs of feather-germ formation are seen in the mesoderm, it is reasonable to suggest that mesodermal factors may be responsible initially for determining where feather germs will form. This has also been demonstrated (18). Rectangular blocks of tissue (ectoderm plus subjacent mesoderm) were excised from the dorsal side of the wing bud in the 3½-day chick and replaced by mesodermal isolates of similar size and shape taken from the prospective thigh region of the leg bud. Ectoderm from the regions of the wing bud adjoining the graft then healed over the implant, and the resulting wings frequently showed feather papillae in the normally apterous region adjacent to the distinctive shoulder tract (Fig. 2). It is clear, therefore, that feather formation is induced

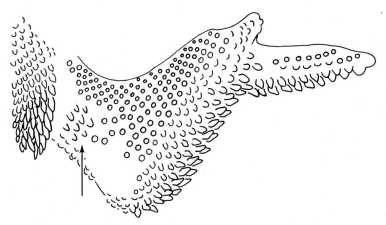

Fig. 2. The induction of supernumerary feather germs (arrow) in epidermis of the normally apterous region of the upper arm by an implant of prospective thigh mesoderm. Camera lucida drawing of the 10-day embryonic wing.

by mesoderm for, in these cases, feather germs developed in ectoderm which normally would not form feathers. Hence, the locus of feather-germ formation is probably controlled initially by mesodermal factors.

Under some conditions, however, the ectoderm may also affect the site of feather-germ formation. When epidermal wounds are made on the dorsal side of the wing bud distal to the shoulder tract, the coordinate pattern of feather germs in the tract is frequently distorted in the direction of the wound (Fig. 3). Presumably this effect results from shifting of the ectodermal layer with respect to the underlying mesoderm of the prospective feather tract during wound-healing movements. The embryonic ectoderm is thus an active partner in feather-germ formation (21).

*The Progressive Organization of a Feather Tract*

Since mesodermal factors apparently bear the initial responsibility for the origin of feather-germ loci, it follows that the number and spatial distribution of feather germs in the various tracts are under mesodermal con-

244    THE CHEMICAL BASIS OF DEVELOPMENT

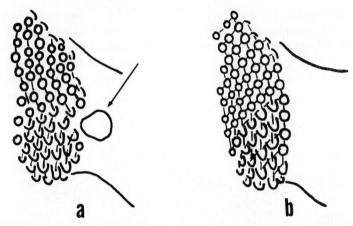

Fig. 3. Effects of a persisting epidermal wound (arrow) on the orderly arrangement of feathers in the humeral tract. *a*, 10-day embryonic wing; epidermis stripped lateral to the shoulder tract at stage 24+ (ca. 4½ days); note distortion of the antero-posterior coordinates of the humeral-tract pattern. *b*, the normal feather arrangement in the shoulder tract. Camera lucida drawings of the specimens.

trol. It is also of interest to learn whether other aspects of feather-tract organization, such as the structure, orientation, size relationships, and physiological characteristics of the feathers are determined by mesodermal factors and, further, to examine the sequence in which characteristics of the feather tracts are fixed.

The humeral (shoulder) feather tract, which originates from materials located proximally on the wing bud, was chosen for this analysis. This is a well-defined tract (cf. Fig. 1) which has been thoroughly studied (1), and its prospective materials are readily accessible to experimental manipulation during the young wing-bud stages. In embryos of stages 17 to 24 (3 to 4½ days of incubation), rectangular blocks of mesoderm (with or without the covering ectoderm), comprising chiefly materials of the postero-lateral quadrant of the shoulder tract and distally adjoining apterium, were excised from the dorsal side of the right wing bud to a depth of approximately 0.1 mm., rotated 180° in the horizontal plane, and replaced in situ. The operation thus displaced materials of the future humeral tract distally and in reversed antero-posterior and medio-lateral orientation (Fig. 4). Simultaneously, it shifted materials of the prospective apterium into the region of the humeral tract.

The results of these experiments were analyzed in a large number of adult and embryonic specimens (20). Many of them showed completely

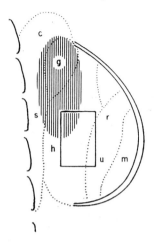

Fig. 4. Fate map for wing bud in stage 21 (adapted from Saunders, 17). The rectangular outline indicates the wing parts affected by the reorientation operation. The approximate location of the materials of the humeral tract is designated by vertical hatching. Mapped skeletal areas are enclosed by dotted lines: *c*, coracoid; *g*, glenoid region; *h*, humerus; *m*, manus; *r*, radius; *s*, scapula; *u*, ulna.

normal patterns of feather distribution. In others (*Type I*), the humeral tract was defective, usually lacking from 2 to 25 feather germs in the postero-lateral quadrant, but the feathers of more distal wing parts were normally arranged. Approximately one-half of the specimens likewise had defective humeral tracts but, in addition, bore distinctive feathers, structurally like those of the humeral tract, in the apterous region distal to it (*Type II*). These feathers, presumably of humeral-tract origin and displaced distally by the operation, emerged simultaneously with feathers of the humeral tract in the juvenile plumage and were directed posteriorly or postero-distally. They showed normal dorso-ventral relationships; i.e., the dorsal aspect of each feather was dorsally oriented. In some favorable cases it could be shown that the normal antero-posterior order of feather sizes in the displaced group of humeral-tract feathers was reversed: larger feathers, normally found at the posterior edge of the tract, were located anteriorly in the displaced group. Finally, a few specimens (*Type III*) showed defective humeral tracts and displaced feathers, as described above, but among the displaced feathers the apices of at least some were directed anteriorly, conforming to the original axial relationships of the reoriented isolate. In the only bird of this kind which hatched and achieved the adult plumage, dorso-ventral axiation of the displaced feathers was appropriate to the feather slope (i.e., dorsal sides dorsally oriented).

The results of one group of experiments, ordered with respect to the developmental stages of the embryos at the time of operation, are summarized in Fig. 5. Analysis of the frequency with which the various feather patterns were found, as a function of the stage of operation, shows: (1) a

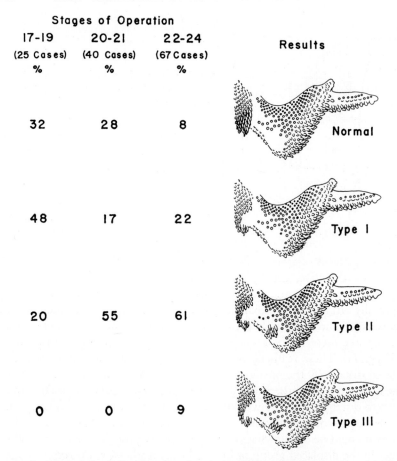

Fig. 5. Effects of the reorientation operation on the distribution and orientation of feather germs in the wing bud. Atypical feather patterns designated as Types I, II, and III, were redrawn, with suitable modifications, from a camera-lucida tracing of the normal pattern.

progressive loss of the ability of the humeral tract to reorganize its typical pattern after the reorientation operation; (2) a progressive increase in the ability of displaced parts of the prospective humeral tract to differentiate the characteristic structure, spatial order, and time of emergence of its feathers in a new location; and (3), in the latest stages treated, a loss of the capacity to regulate the initial direction of feather outgrowth.

It is noteworthy that the dorso-ventral organization of the individual feather-germ does not appear to be correlated necessarily with the slope of the newly-emerged papilla. This is suggested by the fact that in embryonic specimens which showed some feather papillae of reversed slope after the reorientation operation, there were also feather papillae which were apparently forced into a more normal orientation through mechanical factors of crowding. Also, among cases hitherto unreported, in which blocks of tissue entirely within the humeral tract were reoriented 180°, rosettes of feather papillae, variously oriented, were observed; but, in adult specimens, dorso-ventrality always corresponded to feather slope regardless of direction (i.e., dorsal sides dorsally oriented). It follows, therefore, that any causal connection between feather slope and dorso-ventrality is effective at the time the direction of feather slope is finally fixed, or later, and not during the initial outgrowth of the feather papilla. [This raises the question whether other evidences of dorso-ventrality (e.g., order of origin of barb-vane ridges, or distribution of melanophores) in the newly-formed feather papilla are necessarily related to the future dorso-ventral organization of the feather germ. An answer to this question is now being sought in appropriate experiments.]

These results suggest that the properties of the prospective humeral tract are progressively established in the following sequence: (1) the delimitation of the tract area; (2) the structural properties, space-size distribution, and time of emergence of the prospective feathers; (3) the orientation of their initial outgrowth with respect to the body axis; and (4) the establishment of their dorso-ventral organization. Clearly, the differentiation of these properties occurs first in the embryonic mesoderm, for the effects of the reorientation operation are the same regardless of whether the ectoderm is rotated along with the mesoderm. The mesoderm subsequently induces in the ectoderm the appropriate tract-specific response properties.

*Regionally Specific Inductive Action of Feather-Tract Mesoderm*

Since, in the adult bird, each feather generation arises from the epidermis of the follicle as a consequence of the non-specific inductive action of the dermal papilla, and since each feather is an unique component of an orderly pattern, it follows that there are differences from follicle to follicle in the response properties of the epidermis. These epidermal response patterns arose, as shown above, through the inductive action of the mesoderm during early developmental stages; therefore, it must further follow that during the origin of the feather tract the mesoderm associated with each prospective feather germ gives different inductive cues to its overlying epidermis. (Presumably these inductive cues also determine the sequence of structural changes which characterize the successive feather generations of the juvenile plumage; that is, the embryonic induction establishes a pattern of changing epidermal responses to non-specific inductive action of the dermal papilla, and the response pattern becomes stabilized only in the adult feather follicle. There is a possibility, however, that specific influences of the dermal papilla in the juvenile bird affect the characteristics of feathers regenerated during juvenile moults. This should be studied.)

This conclusion has been examined further by means of experiments in which a mesodermal isolate excised from the prospective thigh feather tract in the 3- to 4-day chick was grafted to the dorsal surface of the wing bud from which an isolate of similar size and shape was previously removed (1). The thigh mesoderm became covered with wing ectoderm after the operation and induced in it the formation of typical thigh feathers, in some cases clearly replicating the particular gradient pattern of change from feather to feather characteristic of a portion of the normal thigh tract. This result occurred at all levels of the wing from shoulder to wrist. It is apparent, therefore, that mesoderm of the prospective thigh feather tract exercises a regionally specific inductive action on its covering ectoderm. Likewise, there is a graded variation in this inductive specificity from one prospective feather locus to another, such that the epidermal component of each feather follicle in the adult bird forms a distinctive feather vane which is a component of an orderly pattern of structural variation within the feather tract.

The regional response properties of the epidermis are thus entirely dependent on the nature of the mesodermal stimulus. It is important, however, to recognize that these response properties are subject to genetic limitations. For example, when an isolate of mesoderm from a normal White Leghorn embryo is grafted to the wing bud of a host homozygous for silky (a gene affecting feather structure), the wing epidermis responds to the mesodermal induction by forming feathers of the silky type (1).

*Time of Inductive Action*

Results of the foregoing experiments clearly indicate that the potentiality of mesoderm to give specific inductive cues to the epidermis of the prospective feather tract is differentiated long before there are morphological signs of incipient feather formation. But, they do not provide evidence of the time at which these potentialities are exercised or the manner in which the inductive action is carried out. It seems possible, however, that specification of the future epidermal response may be induced as soon as the dermal condensations are formed and epidermal outgrowth of the feather papilla begins. At this time there occurs a complex interplay of inductive reactions between mesodermal and ectodermal components of the feather germ, as demonstrated by experiments in vitro. These reactions are of such interest as to merit further discussion.

The mesodermal condensations of the first feather germs of the back skin appear in embryos of stage 30 (6½ days), or slightly before. Small isolates of back skin explanted to tissue culture media at this time form typical feather papillae but differentiate poorly, if at all, when explanted at earlier stages. Furthermore, when ectoderm from undifferentiated skin is explanted in combination with skin-mesoderm in which the dermal condensations have just formed, feather germs develop. It would appear, therefore, that the initial outgrowth of feather germs may be induced by the dermal condensations in vitro (22).

This initial inductive action of the mesodermal condensations on the overlying epidermis is apparently of short duration. When isolates of mesoderm from embryonic skin in which the epidermal response has begun are associated in vitro with epidermis (from a younger donor) which has not been induced, feather germs are not formed. But, epidermis which has initiated feather formation is itself inductive, for when combined with mesoderm which has not yet formed dermal condensations, it induces their origin and typical feather germs arise (22). It is possible, of course, that these relationships do not occur in vivo and that, although the mesodermal component of the feather germ may lose its inductive capacity shortly after it is formed, it subsequently regains it, for, as noted above, the dermal papilla, a derivative of the original mesodermal condensation, induces the formation of successive feather generations in each follicle after hatching.

## Specific Induction of Scales and Claws

Whereas the foregoing sections show that regionally specific differences in the character of the plumage in the fowl have their origin in specific inductive actions of the mesoderm, the question now arises whether other integumentary derivatives likewise arise in response to regionally specific mesodermal influences. An answer to this question has been sought in experiments in which mesodermal tissues of the prospective foot, typically associated with the origin of scales and large claws, have been combined with ectoderm of the wing, which does not normally form these structures.

In the 3- to 4-day embryo, mesoderm of the apical region of the leg bud, which forms the foot, was stripped of ectoderm, excised, and grafted to the apex of the wing bud subjacent to the apical ectodermal ridge (1). The graft was thus initially supplied with wing ectoderm of the apical ridge, and subsequently became covered with ectoderm regenerating from the proximal portions of the wing bud. In most of the resulting embryos, the wings terminated in abnormal toes bearing scales and large claws characteristic of the foot.

In other experiments grafts of apical leg-bud mesoderm from older (ca. 4½ days) donor embryos were implanted in the shoulder region of the wing bud. In these cases, also, wing-bud ectoderm covered the graft, and scales and large claws formed on toe-like structures which developed proximally on the wing. Since epidermis of the wing normally does not form scales and long claws, it is clear that these structures differentiated in response to a specific inductive action of the leg mesoderm.

### Origin of Regional Differences in Inductive Specificity

It is now appropriate to give particular attention to the differences observed when mesoderm from different proximo-distal levels of the leg bud is grafted to the wing bud in association with ectoderm of wing-bud origin. As shown in the foregoing sections, mesoderm originating from the proximal portion of the leg bud (prospective thigh) induces typical thigh feathers in the wing ectoderm at all levels of the wing from shoulder to wrist. Mesoderm of the distal portion of the leg bud, however, in suitable grafts to the shoulder or the wing-bud apex, induces scales and claws in the wing ectoderm.

These results emphasize, perhaps more dramatically than others reported here, the striking differences in inductive specificity which are exercised by closely associated regions of the embryonic mesoderm. The origin of these differences is a problem which currently offers a better opportunity for analysis than those more subtle differences which result in the induction of tract-specific and follicle-specific feather characteristics which have been treated in a considerable part of this report.

It should be noted, however, that no great progress has been made as yet in this analysis, and the only evidence contributing appreciably to the problem has been obtained from experiments on the developmental role of the apical ectodermal ridge (17). This structure, which caps the apical zone of the young limb bud, acts as the inductor of limb outgrowth (19, 31-33); it elicits the sequential formation of materials for the prospective limb parts in proximo-distal sequence in the apical mesoblast. During the outgrowth of the limb bud, therefore, the apical zone is a region of intensive morphogenetic activity and might reasonably be expected to modify the inductive specificities of tissues implanted in it.

This has proved to be the case. Grafts of prospective thigh mesoderm, which would normally induce thigh feathers in the covering ectoderm, were implanted at the apex of the wing bud subjacent to the apical ecto-

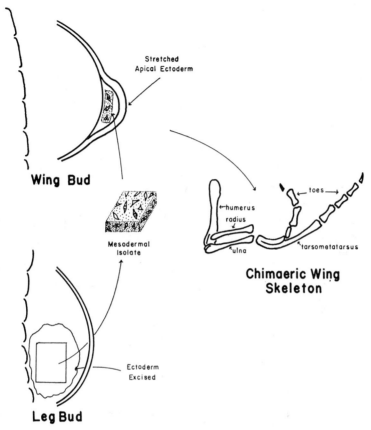

Fig. 6. The effect of grafting an isolate of prospective thigh mesoderm subjacent to the apical ectodermal ridge of the wing bud. The skeleton of the chimaeric appendage was traced from an enlarged x-ray photograph of the disarticulated wing shortly after hatching.

dermal ridge (Fig. 6). In the resulting chimaeric wings the grafts differentiated typical foot parts in the region of the manus and induced the formation of warty and overlapping scales and large claws in the wing ectoderm. This occurred, however, only when the implant developed in intimate association with the apical ectodermal ridge. When a small amount of mesoderm was interposed between the graft and the apical ridge, foot parts were never formed; instead, the implant developed at more proximal wing levels and usually induced the formation of typical thigh feathers in the wing ectoderm. From these results it is apparent that suitable implants of prospective thigh mesoderm may retain their ability to differentiate leg-specific qualities in the wing environment. In the special circumstance of contact with the apical ridge, however, the thigh-specific regional potentialities of the implant are not realized; instead, the graft differentiates foot parts and the inductive specificities which provide for their appropriate armor of scales and claws. Thus the origin of regional differences in inductive specificity in the integumentary mesoderm is related to the morphological differentiation of its underlying skeletal parts. The latter, however, differentiate under the influence of the apical ectodermal ridge.

### Conclusions

From these studies one may draw the general conclusion that the regionally distinctive integumentary derivatives in the fowl arise as the consequence of a complex inductive interplay between their epidermal and mesodermal components during early embryonic development. The cooperation of these components is schematized in Fig. 7. The ectoderm is an

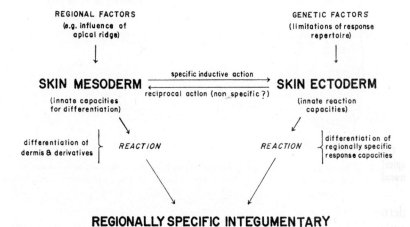

Fig. 7. Scheme of the interaction between embryonic ectoderm and mesoderm in the origin of regionally specific integumentary derivatives in the fowl.

active partner in the process, but the potentialities for qualitative regional differences develop initially in the mesoderm, which then induces a specific epidermal response.

It remains a challenging problem for the future to determine the factors responsible for the origin of inductive specificities in the mesoderm, and the mechanisms whereby the mutual influences of the epidermal and mesodermal components of the skin are exercised in the differentiation of the complex integumentary system. These problems possibly will be attacked successfully through application of the newer methods of biochemical and biophysical analysis.

## REFERENCES

1. Cairns, J. M., and Saunders, J. W., *J. Exptl. Zool.*, **127**, 221-248 (1954).
2. Davies, H. R., *Morphol. Jahrb.*, **15**, 560-645 (1889).
3. Fraps, R. M., and Juhn, M., *Physiol. Zool.*, **9**, 378-406 (1936).
4. Goff, R. A., *J. Morphol.*, **85**, 443-482 (1949).
5. Hamilton, H. L. (ed.), *Lillie's Development of the Chick*, 3rd edition, Henry Holt & Co., New York (1952).
6. ———, and Koning, A. L., *Am. J. Anat.*, **99**, 53-80 (1956).
7. Holmes, A., *Am. J. Anat.*, **56**, 513-535 (1935).
8. Jeffries, J. A., *Proc. Boston Soc. Nat. Hist.*, **22**, 203-241 (1883).
9. Koning, A. L., and Hamilton, H. L., *Am. J. Anat.*, **95**, 75-180 (1954).
10. Lillie, F. R., and Wang, H., *Physiol. Zool.*, **14**, 103-135 (1941).
11. Rawles, M. E., *J. Genet.*, **38**, 517-532 (1939).
12. ———, *Proc. Natl. Acad. Sci. U. S.*, **26**, 86-94 (1940).
13. ———, *Physiol. Zool.*, **17**, 167-183 (1944).
14. ———, *Physiol. Zool.*, **18**, 1-16 (1945).
15. ———, *Physiol. Zool.*, **20**, 248-269 (1947).
16. Ris, H., *Physiol. Zool.*, **14**, 48-66 (1941).
17. Saunders, J. W., *J. Exptl. Zool.*, **108**, 363-404 (1948).
18. ———, *Anat. Rec.*, **111**, 450 (1951).
19. ———, Cairns, J. M., and Gasseling, M. T., *J. Morphol.*, **101**, 57-88 (1957).
20. ———, and Gasseling, M. T., *J. Exptl. Zool.*, **135**, 503-528 (1957).
21. ———, and Weiss, P., *Anat. Rec.*, **108**, 581 (1950).
22. Sengel, P., *Experientia*, **13**, 177-182 (1957).
23. Wang, H., *Physiol. Zool.*, **16**, 325-354 (1943).
24. Watterson, R., *Physiol. Zool.*, **15**, 234-259 (1942).
25. Weiss, P., and Matolsty, A. G., *Nature*, **180**, 854 (1957).
26. Willier, B. H., and Rawles, M. E., *Proc. Natl. Acad. Sci. U. S.*, **24**, 446-452 (1938).
27. ———, and ———, *Physiol. Zool.*, **13**, 177-199 (1940).
28. ———, and ———, *Yale J. Biol. and Med.*, **17**, 321-340 (1944).
29. ———, and ———, *Genetics*, **29**, 309-330 (1944).
30. ———, ———, and Hadorn, E., *Proc. Natl. Acad. Sci. U. S.*, **23**, 542-546 (1937).
31. Zwilling, E., *J. Exptl. Zool.*, **128**, 423-441 (1955).
32. ———, *J. Exptl. Zool.*, **132**, 151-171 (1956).
33. ———, *J. Exptl. Zool.*, **132**, 173-187 (1956).

## DISCUSSION

Dr. Holtfreter: I want to congratulate you, Dr. Yamada, on your beautiful results. They are novel and most revealing. At the same time they are puzzling, leaving many of the old problems still unanswered. As an old-timer in this field I may be permitted to put your findings into a wider context.

As many of you know, the quest for the chemical nature of the inductive agents operating in *normal* amphibian development (and this, I may insist, is our goal) has been tortuous and disheartening. Some two decades ago this inquiry started out with a dazzling bang. Embryologists at various places discovered almost simultaneously that a frightening diversity of artificial unrelated stimuli can imitate the effects of the living inductive tissues of the embryo. In most instances, however—and this is important for my subsequent remarks—the stimuli, such as tissue extracts or pure chemicals, would produce only the neuralizing, brain-inducing ("archencephalic") effects of the natural inductors.

The story became more involved. Very early two interconnected observations were made. (1) Devitalization of the natural inductors by heating, freezing, or various chemicals did not readily inactivate their neuralizing power but quickly destroyed their notochord—somite-inducing or "mesodermizing" power. (2) A great variety of tissues from all sorts of adult animals, including man and guinea pig, when grafted into an amphibian gastrula, could perfectly replicate the effects of either the archencephalic or the mesodermizing inductors of the embryo. The latter activity vanished in the adult tissues about as quickly by heat treatment as it did in the natural spino-caudal inductor.

From these observations it has been concluded by some authors that the multitude of effective adult tissues contain (for no apparent morphogenetic reason) the same inductive agents as are operating in the embryo. The search for the chemical nature of the agents went off on a new tangent. From now on, except for some abortive attempts at characterizing the inductors of the embryo proper, most researchers took to model experiments. They tested the activity of various extracts or fractions from readily available mammalian tissues, such as liver or kidney of the guinea pig. It was thought implicitly that if these chemically identifiable preparations can act like living inductors they would provide clues to the chemical nature of the latter.

Let us be aware of this sleight-of-hand procedure. From all we know about phenocopies, parthenogenesis, carcinogenesis, and, more fundamentally, about the potential side-tracking of enzymatic reactions, there is no *a priori* reason to assume that the effective agents present in adult tissues are identical with, or even related to those acting in normal induction.

This sceptical attitude became strengthened when it turned out that a great variety of biologically alien, so-called unspecific chemical agents can elicit normal conversion of gastrula ectoderm. Such agents were alkali or acids, versene or distilled water, alcohol, detergents, or what not. Who would dare attempting an identification of any one of these foreign agents with those employed by the embryo? And yet, what is the sense of making experiments if we cannot relate them rationally to the mechanisms of the embryo?

Other methodological difficulties have been pointed out in previous reviews. They offer further serious obstacles in interpreting our experimental results. Due to these difficulties many embryologists have given up in despair. But you, Dr. Yamada, and a few others have courageously followed up this line of approach, trying to specify the chemical compounds engaged in amphibian induction by resorting to experiments with adult mammalian tissues.

The saving grace of your endeavors lies, I believe, in the fact that your various preparations not only neuralize the ectoderm explants—as many slightly damaging shock treatments would do—but that some of them bring about mesodermal and even entodermal transformations. For one thing, your results support the view that embryonic induction is not simply due to a trigger mechanism, or represents an alternative choice, say between epidermal and neural differentiation. Whatever the chemical nature of your inductive preparations may be, they produce qualitatively different results. One may infer, indeed, that in normal development, too, the different structures are induced by different inductive agents. Comparable to the ubiquitous auxins, the very substances of the embryonic inductors may indeed occur throughout the organic world.

Dr. Cohn: I would like to ask Dr. Yamada what was the precision of the assay used to fractionate the various extracts that he made.

Dr. Yamada: In the case of the acid-precipitable fraction of the bone marrow-supernatant both the light and heavy fractions were found to be active in our test. However, it might be that one is the real inducer and the others are just contaminated by small amounts of the first component. We need a quantitative assay technique for determning this and other similar questions. We are now trying hard to work out such a method. We realize that the test of morphogenetic effects used in the present study is suited for morphological evaluation of the inductive effects, but not for quantitative study.

Dr. Grant: Most of your fractionation began with nucleoprotein, is that right? In other words, you have a salt extract with protein.

Dr. Yamada: Yes. In the case of liver and kidney an electrophoretically homogeneous sample of ribonucleoprotein has the inducing effect.

Dr. Grant: Do you have any indication that protein other than nucleoprotein will serve as inducing agent?

Dr. Yamada: Yes. The ribonucleoprotein fraction has been found to be effective only in the liver and kidney. In the bone marrow the same fraction of nucleoprotein has nothing to do with induction. Even in the case of kidney and liver it might be that there are other fractions which have a similar effect. For instance, we have obtained very clear spino-caudal effects by the kidney nucleus fraction separated by ordinary differential centrifugation. But as the sample was contaminated with whole cells, we don't use this result as a basis of argument.

Dr. Novikoff: I would like to ask how the $0.14\ M$ salt extract was prepared. Was it a soluble extract?

Dr. Yamada: This is soluble. We use different methods of preparation but ordinarily we use unbuffered sodium chloride, centrifuging it at about 3,000 g.

Dr. Novikoff: The reason I ask is, I think if we start with soluble material, it may be questionable whether we should subsequently use the word "microsomes" for a particulate fraction. I don't wish to argue with this. This is a beautiful piece of work and has nothing to do with what we are calling a particular fraction, but I remember Hewson Swift's use of the word "microsomes" and it is bad enough when we get out of the cells certain materials which are in the cells if we call those microsomes but we confound the problem, I think, if we start with a soluble extract of particulates, presumably of the cell, and we treat them subsequently in the centrifuge and call the resulting pellet "microsomes." Now the word "microsomes," as Swift pointed out, had a past which is both venerable and confusing, and was resurrected by Claude.

Dr. Yamada: What we know is only that it contains a large amount of RNA. What I wanted to indicate with the designation was just to give you an idea of the magnitude of the centrifugal force we use for sedimenting this fraction. I agree with your suggestion.

Dr. Novikoff: Would it not be better that a pellet obtained from a salt extract be called "sediment" or "precipitate"? The term "microsome" is confusing enough, even for cell particulates isolated from homogenates.

Dr. Yamada: Yes. Perhaps better to call it a sediment.

Dr. Niu: I want to congratulate Dr. Yamada for contributing so much to the problem of embryonic induction. We are also doing some simple experiments on the subject. First of all let me mention our purpose. As an embryologist, I am a little more interested in the physiological functioning of nucleoproteins or nucleic acids. Of course we have been approaching this problem in a different way. Many of you know of the technique we are using. To begin with, we have been interested in "ribonucleoprotein" and so we tried to isolate "ribonucleoprotein" as Dr. Yamada did. I use the term ribonucleoprotein in quotation marks because I don't know how pure it is. But I do know the following properties. It has an ultraviolet absorption spectrum typical of nucleoprotein. Of course, this curve varies somewhat from sample to sample. A second property is its ratio of nitrogen to phosphorus. According to our experiments this ratio varies from 5 up to 50. The content of RNA in these samples is about 5-20 percent. We want to know which part of this preparation is active. To my mind there are three ways to study this. The first method, and one which Dr. Yamada employed, is to use an enzyme, either a proteolytic enzyme or ribonuclease to hydrolyze the substrate. We have followed Dr. Yamada's work closely along these lines and can say that we have confirmed what he has obtained, but with some precautions. For instance, trypsin treatment in our system removes all of the activity. However, this enzyme is still in the system. When we added trypsin inhibitor, the activity returned. Following treatment with ribonuclease we still found activity but that activity was lower than in the control. Therefore it is clear that RNA-ase has some effects on the sample but

## DISCUSSION

this effect is not complete. In other words, after RNA-ase treatment some RNA is still present in the sample. Since RNA-ase is destructive to living cells and lower concentration has no inducing effect, it appears that the loss of activity is due to the loss of RNA by treatment with RNA-ase. As to what is the active factor, you may draw your own conclusion. It immediately occurred to us that we should isolate some RNA by a newer method. It is the Kirby method which is published in the *Biochemical Journal*. With this RNA we had some specific activity but the activity was higher in the preparation of ribonucleoprotein. In the RNP we found 70% of the activity and only 30% residing in the RNA. The effects of these two were qualitatively the same but quantitatively different. If the isolated RNA is combined with a foreign protein, such as serum albumin, or a plant protein, then the activity is increased considerably. I don't know the mechanism involved here, but do think that pinocytosis may have something to do with it. It should be mentioned that if you prepare RNA you have to really be careful that you don't dry it or the activity of the fraction is lost. Finally, the concentration of RNA is extremely important. At the optimal concentration RNA initiates ectodermal differentiation. A factor of 5 or 10 will change the picture completely. Higher concentration tends to produce inhibitory effect. I certainly would appreciate hearing Dr. Yamada's comments on our work.

Dr. Yamada: Do you know about the work by Tiedemann and Tiedemann who tested nucleic acids prepared with phenol and came to the conclusion that they are not responsible for regional induction? Were the techniques the same or were they different?

Dr. Niu: Yes. It is interesting to note that they obtain low activity in their dried RNA. Using phenol, we also extracted protein from 9-day chick embryos. The solubility of this extract is too low to be tested in our assay system.

Dr. Yamada: I think your result is extremely important. Up until now we have tested no nucleic acid which was prepared in your way. What we used was the guanidine-hydrochloride method. This is the only one we have tested carefully enough and so it may be that the difference in preparation might be important in this case. How do you explain your result with the protease series?

Dr. Niu: I think Dr. Hayashi might be better able to answer that question. However, we compared ectodermal explants explanted into medium containing ribonucleoprotein treated with trypsin with explants placed in similar medium to which had also been added soy bean inhibitor. After two days, the explants in medium containing only trypsin became free-swimming vesicles. With inhibitor, however, the explants had attached to the glass and later gave rise to outgrowths which may contain special structure, neural tissue, or both.

Dr. Yamada: I was told by Tiedemann that he tried also soy bean inhibitor in his experiments using trypsin. Trypsin alone suppressed the inducing ability of his sample. In the presence of soy bean inhibitor this effect of trypsin was inhibited. He concluded that the enzymatic activity of trypsin was responsible for suppressing the inducing ability.

Dr. Niu: That pleases me very much.

Dr. Yamada: Of course, this hasn't been published yet.

Dr. Brachet: I would like to make the point that it is really important to know what trypsin and its inhibitors are really doing. I wonder if Dr. Yamada, or any of the other people here, has considered this question by looking at the liberation of amino acids or nucleotides from the material which was treated with these enzymes. In other words, which materials are being released by the action of these enzymes? I would also like to say that I think it would be valuable to study the combined action of ribonuclease and trypsin on the explants. I think that it is very difficult to remove all of these enzymes from the solid preparations you are using, and it then becomes very important to know what the enzyme itself is doing. Finally, I would like to make briefly a comment on something Dr. Niu mentioned, that is that he gets an increased activity of RNA when he adds a non-specific protein to the preparation. This might be due to pinocytosis induced by the nonspecific protein. In our own experience with the amoebae, we have found that if you add RNA to the medium, say, for instance, labeled RNA, you get very little uptake whatsoever. However, if you add some protein to the RNA in the medium, this increases the pinocytosis of the amoebae and the uptake becomes considerable. So I would just like to offer as a suggestion that the effect of the protein in your system, Dr. Niu, is to increase pinocytosis, which may in turn increase the uptake of RNA by a factor of three or four.

Dr. Yamada: Thank you very much Dr. Brachet for the very important suggestion. I would like to comment on your question about the effect of the enzymes on the cells in the explants. For a control we always used a sample to which the enzyme was added after a sham incubation. The enzyme in these particular cases was always treated with heat. Compared with the sample without added heat-treated enzyme there was no significant difference in the outcome of induction.

Dr. Brachet: However, these enzymes can be heated to a great degree and still retain their activity, say trypsin, for instance. They are both very thermostable enzymes.

Dr. Yamada: We compared the heat-treated and non-heat-treated series of trypsin-incubation of the liver ribonucleoprotein sample for inductive ability. In the not heat-treated sample we get a complete suppression of induction. But if the enzyme is heat-treated after the incubation period, there is still a clear suppression of induction probably due to enzyme action on the sample during incubation. On the other hand, addition of heat-treated enzyme to the reaction mixture after a sham incubation does not cause a clear change in induction, when compared with the original untreated sample. This suggests a heat inactivation of the enzyme even if it is not complete.

Dr. Niu: Dr. Brachet, I am very glad you made the point on the use of non-specific proteins to increase the rate of pinocytosis by cells. We were very fortunate to have Dr. Holtfreter with us for a few days. I was happy to hear of his work on induced pinocytosis in amoeba. It is highly possible that this may well be what is going on.

Dr. Weiss: The organism is to us still very much of a black box which responds to an experimental input Y with a signal X without revealing the complete network of connections and the route that has led from Y to X. Most attention in the problem of induction has been given to the identification of various Y's. I would like to get back to the other side of this question. Now we have been concerned mostly with what kind of substance produces a given "X" reaction. We have administered these substances to the black box, ignorant of what goes on inside, and then we observed the end result. With one kind of substance one got one particular type of reaction, with another substance one got another particular type of reaction. I think it is very significant that Dr. Yamada has shown us today the same substance, put through a series of treatments, can produce in this black box any particular reaction, depending upon the previous treatment. This removes the stigma of specificity from the chemical quality of the substance. In addition, if Dr. Niu is right, that the reaction of these materials differs according to the method of preparation, then again this contradicts specificity in the sense of a one-to-one matching reaction between the substance and the reacting system. Now where does this leave us? I would like to ask a question and it is this: what do we know about what happens between the time when we apply the substance and the time when we see the results of its action? Naturally, it will make a considerable difference just how fast the cell picks up this material, and just how fast it gets through the three or four or more reacting steps it may have to go through after it has gotten into the cell. All of these things must be taken into consideration. Now, to what extent can the different results we heard about be explained by differences in the penetration reaction kinetics, and so on, of the agent Y—for instance, extracts from bone marrow or some other source—and by the different stage of development of the reacting system—our black box—at the time when it is actually "hit" by Y, when the crucial event takes place? What do we know about progressive changes in our reacting systems irrespective of what agent we might apply to them? What effect do these materials you have extracted have on ectoderm taken from different stages, say from early and late gastrulae? How does this affect the type of results you get? The reacting system is usually treated as if it were stagnating, as if it were dead in a way. I think Dr. Holtfreter has shown that the ectoderm which is in these explants, although it may not form neural tissue or undergo specific differentiation types, is changing in competence—meaning constitution—and the reacting system is therefore different when taken from different stages.

Dr. Yamada: We don't have any experiments on this point but I would like to mention the work of Toivonen and his group in this respect. They put bone marrow on their explants of some stage for different lengths of time. They found that the results obtained were qualitatively the same irrespective of the length of time the bone marrow was applied to the explant. They left the bone marrow on from several hours to several days and the effect was always of the trunk-mesodermal type. What did change, however, was the frequency with which the effect was noted; increasing time of application gave a greater frequency of effect.

Dr. Weiss: When you say frequency, do you mean that it was different types or the same type increasing in frequency. It seems to me that I remember that Toivonen showed that there were some admixture of types here.

Dr. Yamada: Yes, but I think that this was insignificant. What was quite apparent to me was that, in general, the time of application did not affect the quality of the induction noted. Another example is that studied by Johnen in Cologne. She studied the effect of kidney on gastrula ectoderm for different durations of time. When there is only a short period of contact with ectoderm she gets archencephalic effects, but if she leaves the sample on for a longer time she gets spino-caudal effects. Thus the effect seems to be qualitatively changing with different lengths of time but she feels that this may only be a quantitative difference and she interprets this along the lines which Nieuwkoop has suggested.

Dr. Holtfreter: I think it would be most desirable if we could reduce some of these troubles to changing properties of the reacting system in time, but I am not sure that this is the complete story.

Dr. Weber: I should like to ask a question concerning heat treatment of the bone marrow extract. I think it would be of interest to find out what happens to the extract after different lengths of heat treatment in biochemical terms. Are there progressive changes in these proteins or peptides, for instance, which could be correlated with the regional inductions obtained? Have you any information about this point?

Dr. Yamada: No, I am afraid that has not been done, but of course it should be. Our approach has been to use different treatments and it was interesting that a number of different treatments such as high salt concentration, longer ethanol-treatment, addition of pepsin at $pH$ 7 and so on gave rise to the same kind of effects and we feel that something "spontaneous" is happening to the proteins. Not only tissues or extracts but fairly purified samples show this sequential change in regional effects after various treatments. We are planning to study the change of the sample, which is responsible for the change in regional effects, after we get a reasonably purified sample of the bone marrow factor.

Dr. Hayashi: Going back to the earlier discussion I would like to make a few comments to the questions raised by Dr. Niu. As Dr. Yamada mentioned we always added the same amount of enzyme to the control samples after incubation as that used in the experimental series. Only in the case of experiments with trypsin both the control and experimental samples were heated after incubation (and after the addition of trypsin to the control samples) in order to inactivate the enzyme. In preliminary experiments the control sample, to which trypsin was added and which was used without inactivating treatment, was found to lose its inductive effect almost completely. This might suggest either that trypsin absorbed to the sample acted directly on the reacting ectoderm or that the enzyme digested the implanted sample after implantation. Anyway, the control sample which was treated with heat for 5 minutes in 95% ethanol after the addition of trypsin retained its inductive effect. We also have controls on the effect of simple heat-treatment of the sample without the enzyme. This heat-treatment suppressed the deuterencephalic effect of the sample and shifted the regional inductive effect toward archencephalic. Thus, as far as the heat-treatment was involved we could only test the effect of trypsin on the archencephalic activity of the sample. Unfortunately, we could not use soyabean inhibitor because it is so expensive for us.

In the cases of ribonuclease-treatment and pepsin-treatment, no enzyme-inactivating treatment was used because the control samples of liver ribonucleoprotein, to which the enzymes were added after incubation, retained their inductive activities almost unchanged. However, in the case of kidney ribonucleoprotein we found that the mere addition of pepsin to the sample suppressed to some extent its spino-caudal effect, shifting the regional effect toward deuterencephalic. This suggests a more labile nature of the spino-caudal factor than the archencephalic factor of the kidney nucleoprotein.

As to the question of the effective concentration of RNA, I should like to say that we tested many ribonucleoprotein samples containing varying amounts of RNA and got results indicating that the inductive effect of those samples is largely dependent upon the nature of the protein-component and not upon the presence or amount of RNA in the sample. For example, we got almost the same regional effect at a comparable frequency with protein samples which were extensively reduced in their RNA content by ribonuclease-treatment as with the original ribonucleoprotein samples having high RNA content. Also in the case of pepsin-treatment, experimental samples which contained increased amounts of RNA (relative to the protein-moiety) showed only weak inductive effect. This fact indicates that the proteolytic enzyme did degrade the protein-moiety of the ribonucleoprotein and thus suppressed the inductive effect of the sample. We cannot completely exclude the possibility that RNA had something to do in the process of induction, but even if it participates it seems evident that some integral structure of protein is indispensable for the induction of regional structures in the ectoderm at least under the experimental conditions we adopted.

I also tested the inductive effect of protein samples, which were derived from the liver ribonucleoprotein after removing RNA by phenol-treatments (Kirby's method and the method of Westphal et al. as described by Tiedemann and Tiedemann). With those protein samples I could not get any induction at all. The loss of inductive effect seems to me to be caused by severe denaturation of the sample and not to be caused by removal of RNA.

DR. NIU: I realize now that our discussion is coming to quantitative aspect of inducers. If so, it would be wise to discuss the problem on a quantitative basis. Unfortunately, time does not permit us to do so.

DR. STREHLER: How did you determine what concentration of protein or other substances to use? Did you use the same amount of protein in each one of these cases or was it the concentration you ended up with in a particular fraction?

DR. YAMADA: This type of experiment must be very curious for biochemists. For testing morphogenetic effects we have to put a solid sample into the ectodermal explant. When a supernatant fraction was to be tested, cold 90% ethanol was added to make 70% and the precipitate developed was gathered by centrifugation and applied (undiluted) to the ectoderm, after careful washing in the culture medium. In all series roughly 0.01 mg. (in dry weight) of such a sample was given to an ectodermal explant of approximately 1 mm. diameter.

DR. HOLTFRETER: Dr. Yamada, I wonder if you get any results using non-precipitated whole fractions?

DR. YAMADA: We have not yet obtained regional induction by putting fluid sample in the culture medium. But there are some experiments of von Woellwarth in Germany and Kawakami in Japan speaking for regional inductive effects of chick embryonal extract added to the culture medium.

DR. COHEN: I am still not clear as to the quantitative methods which you use, Dr. Yamada. Were there increments of solid put on the reacting system and did the differences in quantity reveal any different effects? Just how can you quantitate these effects in your assay system?

DR. YAMADA: Well, we tried to put approximately the same amount of solid material onto the explant. However, a very serious difficulty in even becoming semi-quantitative in this system is the difficulty in obtaining the reacting explants in uniform size. Even though we take them from the same region of the gastrulae it is very difficult to precisely control how much material we have. So we are aware that this test is not quantitative enough, although it allows a reliable identification of the induced structures. We are working out another test technique which gives us quantitative informations.

DR. COHEN: What I am thinking of is the concentration in the medium.

# DISCUSSION

Dr. Yamada: I think I told you that we do not add the material in solution but that we add it as a solid material to the reacting system.

Dr. Cohen: Well, do you quantitate it by taking one, two, or three times a given amount of solution and then adding all the precipitate obtained from those samples?

Dr. Yamada: No, we did not do that, because from earlier experiments we know that that type of test does not work in this method. After the technique of quantitative assay will be worked out, we are planning to do that type of experiment.

Dr. Cohen: I would like to say that however exciting it may seem to be to look for stimulating compounds which produce these effects, it seems to me that the same kind of results could be explained on the basis of specific inhibitors which act upon the right system at the right time. Thus when you add something to a system of this complexity the effect may be due to an inhibitor or a stimulating factor, and both of these possibilities must be taken into account when you interpret your results.

Dr. Racker: I would like to re-emphasize a point brought up by Drs. Cohn, Strehler and Cohen, namely, the quantitative evaluation of inducers. I have recently read some of the literature on the purification of inducers and noted that although serious attempts are being made to grade the effects in a quantitative manner, which is undoubtedly difficult, few attempts are made to express these data in terms of dry weight or any other suitable unit of material. One is reminded of the earlier days in the field of enzymology when emphasis was placed on concentration rather than purification, and the meaning of specific activity was not generally recognized.

Dr. Niu: Well, Dr. Racker, in our work we know precisely the amount of RNA we used, the RNA content of nucleotprotein and particularly the specific structure induced by the freshly isolated substances.

# A DEVELOPMENTAL ANALYSIS OF CELLULAR SLIME MOLD AGGREGATION [1]

MAURICE SUSSMAN[2]
*Northwestern University,*
*Evanston, Illinois*

## INTRODUCTION

Aggregation is a stage in the developmental cycle of the cellular slime molds (Acrasiales). It is a device by which heretofore independent myxamoebae are brought together into a compact mass, a single organized, multicellular entity, whose terminal expression is a fruiting body. The process is illustrated by the time lapse series of photomicrographs shown in Fig. 1. The entire developmental cycle is shown schematically in Fig. 2.

This system is considered by the writer to exemplify cellular differentiation and to possess attributes that permit a rigorous, quantitative analysis thereof. Because the writer comes to the subject of developmental biology by the unorthodox route of microbial genetics, it may be of interest to examine his rationale for these statements.

We start with the observation that morphogenesis in multicellular systems is invariably accompanied by a sequential and progressive emergence of many new cell types. These specialized cells may persist as constituents of the terminal structure or may appear only transiently, play a specific role in the developmental sequence, and then disappear. Any such phenotypic modifications are adjudged to be within the domain of cellular differentiation. Whether the new phenotypes are readily reversible or, on the contrary, are irreversible by any presently known means is not considered to be of significance in defining this domain. Inclusion is based solely upon whether or not the new cell type can be shown to play a causal role in the construction or functioning of the terminal entity.

What kind of information, then, could provide a complete description of

---

[1] The investigations summarized in this paper were supported by grants from the National Cancer Institute and the Office of Naval Research.
[2] Present address: Department of Biology; Brandeis University, Waltham, Mass.

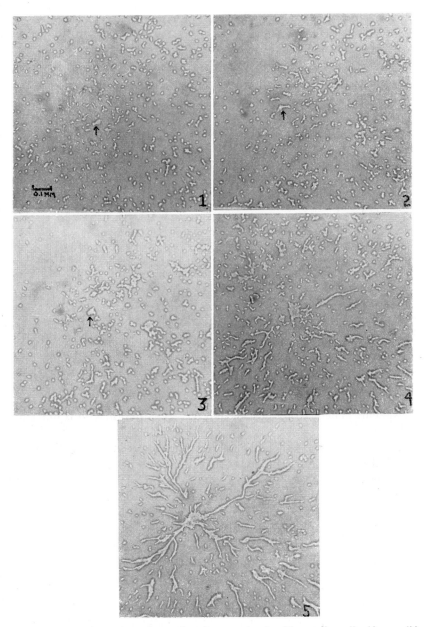

Fig. 1. Time lapse photomicrographs of aggregation by *Dictyostelium discoideum* wild type. The arrows point to the cell at which the aggregative center formed. The myxamoebae had been incubated on washed agar for periods of 9.5, 10.4, 10.75, 11.5, and 12.25 hours, respectively. (Sussman, Ennis, and Sussman, 1958).

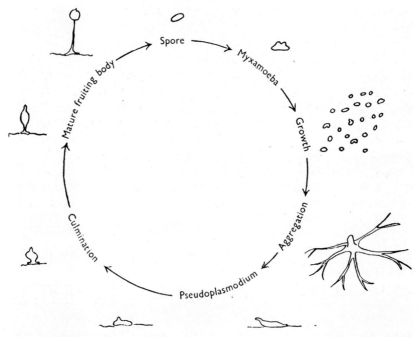

Fig. 2. A schematic diagram of the developmental cycle of *D. discoideum*. (Sussman, 1954).

the system? The need for at least three classes of information is immediately apparent:

1. We would have to describe the new phenotypes in quantitative, biochemical terms. This is necessary since ultimately we must explain the physiological role played by each in the developmental sequence and the manner in which it was derived from its parental variety.

2. We would have to learn whether each novel phenotype proceeds from a transient, purely physiological alteration or involves changes in the genetic apparatus of the cell and, further, to describe the nature of these changes. At present, there are three ways of doing this. One is by recombination analysis of matings between cells. Another is by nuclear transplants. Where these are impossible, one must infer the nature of the change by studying the rate of appearance and disappearance of the phenotype in relation to the growth of the total population, a practice employed with great precision in microbial genetics (13).

3. We would have to specify the cell interactions and other environmental factors which may induce many of the phenomic modifications and

which certainly must regulate the numbers of cells of each type to be found in the total population.

Clearly the complexity of the task demands that the total morphogenetic cycle be broken up into a serial array of smaller sequences during each of which the emergence of only a limited number of new phenotypes need be accounted for. Further, the system should be examined in a rigorously defined chemical and physical environment. The differential growth and death rates of the cell types must be calculable by growth of clones and concomitant viability determinations. The number of cells taking part in the development and their population density must be controlled.

Happily, for consistency's sake, we can report that cellular slime mold aggregation, as well as its other developmental stages, satisfies these criteria. At present only two cell types have been recognized to play a causal role during aggregation. The normal process can be carried out by myxamoebae free of exogenous nutrients and dispensed on washed agar. No detectable changes in cell number need occur during the process, and growth rates prior to it have been determined by clonal platings. The dependence upon cell number and population density has been extensively investigated. While the current description of the system is by no means complete, the broad outlines are, I think, now clear. The contributions to it that have emerged from our laboratory and that will be discussed in this paper were in large part due to the efforts of Drs. Raquel R. Sussman, H. L. Ennis, and S. G. Bradley, Mr. N. S. Kerr, Mrs. Frances Lee Fu, and Miss Elizabeth Noel.

## The Position of Aggregation in the Developmental Cycle

The developmental cycle begins with spore germination. Myxamoebae emerge, feed on bacteria, and increase exponentially. With the onset of the stationary phase, morphogenesis begins. The randomly distributed cells elongate, orient radially, and move toward central collecting points, forming ramified streams as they do so. As the cells arrive at the center, a conical mound develops, and this is transformed into an organized, slug-shaped pseudoplasmodium. Ultimately, by a series of complex morphogenetic movements, the slug is transformed into a terminal fruiting body with a sorus or spore mass, a parenchymatous, cellulose-enclosed stalk, and a basal disk. In one genus this basic structure is modified by the appearance of whorled branches, each possessing an apical spore mass.

The developmental fate of the myxamoebae is determined by the order in which they entered the aggregate (15, 2). The first contingent constitutes the lower stalk of the fruit; the next arrivals become the upper stalk.

Cells arriving later comprise the spore assembly, and the last ones form the basal disk.

## THE DEPENDENCE OF CENTER FORMATION ON POPULATION DENSITY

The ability of population density to determine the number of aggregative centers was originally demonstrated with *Dictyostelium discoideum* wild type (24). Aliquots containing a constant number of washed myxamoebae were dispensed on washed agar at population densities between 80 and 750 cells/mm$^2$. Counts of aggregative centers were then made after suitable incubation. At very low density, no centers were formed. At very high density, the number of centers was depressed, presumably by competition between rival potential center-forming agencies for a common audience. At an intermediate value (200 cells/mm$^2$), center formation was maximal.

The dependence on population density thus revealed was found to hold true for 6 other stocks representing 3 species and 2 genera. As Table 1

TABLE 1
DEPENDENCE OF CENTER FORMATION ON POPULATION DENSITY AND CELL NUMBER

| Species | Strain | Threshold density | Optimal density | Center:Cell ratio |
|---|---|---|---|---|
| *D. discoideum* | wild type | 80 | 200 | 1:2200 |
| | Fruity(*Fty-1*) | 40 | 200 | 1:24 |
| | Fruity(*Fty-2*) | 40 | 200 | 1:43 |
| | Bushy(*Bu-1*) | 70 | 350 | 1:1810 |
| *D. purpureum* | wild type | 20 | 100 | 1:330 |
| | Bushy(*Bu-1*) | 20 | 20 | ca. 1:100 |
| *P. violaceum* | wild type | 50 | 250 | ca. 1:180 |

Density values are given as no. cells/mm$^2$, determined by direct count and by calculation from total number of cells in sample and total area covered. Data for the first five strains were published previously (Sussman and Noel, 1952; Sussman, 1955). The results for the last two are from preliminary experiments and represent rough approximations.

shows, great differences were encountered in the values of threshold and optimal densities.

## THE DEPENDENCE OF CENTER FORMATION ON CELL NUMBER

The data also indicate that, at optimal density, the number of centers was proportional to the number of cells present. Thus, $10^5$ *D. discoideum* wild type cells produced about 50 centers; $5 \times 10^4$ cells, about 25 centers; $2.5 \times 10^4$, about 12.5 centers. This corresponded to a distribution of one center among approximately 2000 cells. As the result of many such determinations in the range between $5 \times 10^3$ and $10^5$ cells, a value of 1:2200 for the center: cell ratio at optimal density was obtained.

The proportional relationship also held for the other six strains. However, as shown in the last column of Table 1, a wide variety of center : cell ratios was encountered. The lowest was that of *D. discoideum* wild type (1:2200) and the highest, that of the fruity mutant, *Fty-1,* of *D. discoideum* (1:24).

Apropos of the latter, *Fty-1* fruits illustrate some of the remarkable properties of this morphogenetic system. Consider that, at optimal density, the aggregates contain approximately 24 cells and in some cases as few as 10 to 12. The fruits which these cells construct are exquisite miniatures of the relatively enormous wild-type fruits such as emerge on growth plates at high population density. Photomicrographs of the mutant structures and considerations of the import that they bring to the problem of cellular interactions have been published elsewhere (33, 28).

Returning to the proportional relationship between centers and cells, this suggested that single myxamoebae were initiating the centers by inducing their neighbors to aggregate. This conclusion followed since, were more than one cell required, the relationship would have been exponential, not linear. The question remained, however, as to whether only special cells in the population could do this or, alternatively, whether any cell could ultimately attain initiative capacity. The latter would imply that the center : cell ratio reflected only the proportion of cells that had attained initiative capacity first.

The point was settled by examining the distribution of aggregates among very small, replicate population samples. Cell assemblies containing 250, 500, 900, 1025, and 2100 myxamoebae were dispensed on washed agar with a machined loop at densities high enough so that if they could have aggregated, they would have. Yet, as seen in Table 2, only a relatively small

TABLE 2

Distribution of Aggregates Among Small Population Samples
(Sussman and Noel, 1952; Ennis and Sussman, 1958b)

| No. cells per sample | Total no. samples | Proportion of samples with: | | | | Center: Cell ratio |
|---|---|---|---|---|---|---|
| | | 0 centers | 1 center | 2 centers | 3 centers | |
| 2100 | 85 | 0.42 | 0.50 | 0.07 | 0.012 | 1:2420 |
| 1025 | 91 | 0.63 | 0.32 | 0.044 | 0.011 | 1:2230 |
| 900 | 123 | 0.69 | 0.29 | 0.025 | 0.0 | 1:2430 |
| 500 | 99 | 0.80 | 0.19 | 0.01 | 0.0 | 1:2170 |
| 250 | 306 | 0.88 | 0.12 | 0.0 | 0.0 | 1:2020 |
| | | | | | Mean | 1:2250 |

proportion of the samples aggregated. Had a large number of the myxamoebae been capable of initiating centers, this result could not have been obtained. Instead, all of the samples would sooner or later have aggregated. This datum compelled acceptance of the conclusion that a small number of myxamoebae are uniquely constituted to be initiator cells and that these are distributed at random among their neighbors, the responder cells.

Confirmation was obtained by comparing the proportions of samples containing 0, 1, 2, and 3 aggregative centers with the Poisson distribution which holds only for systems in which the probability of an event is small and its distribution completely random. The general terms of this expansion are $P_k = \dfrac{m^k e^{-m}}{k'}$, where $P_k$ is (in this case) the probability that a sample will contain $k$ aggregates, and $m$ is the average number of aggregates per sample. Comparison of the calculated and observed proportions of samples with 0, 1, 2, and 3 aggregates by Chi square test revealed a very good statistical agreement. Further, the Poisson expression was used to calculate the value of $m$ from the observed $P$ values. The term, $m$/no. cells per sample, yielded the center : cell ratio. The last column of Table 2 lists the calculated ratios, which agree closely with the ratio previously arrived at for *D. discoideum* wild type by examination of large populations.

### The Identification of the Initiator Cell

The populational analysis of aggregation had provided a firm basis for belief in the existence of a specialized initiator cell. Therefore, in 1955, an attempt was begun in collaboration with Dr. Gaylen Bradley to determine if in *D. discoideum* wild type the initiator cell could be distinguished from its neighbors by morphological criteria prior to the onset of aggregation. A unique cell type (termed the I-cell) was indeed detected, and preliminary experiments encouraged the belief that it might be the initiator cell; but it was not until the past year, largely due to the efforts of Dr. Herbert Ennis, that micromanipulative and other techniques could be evolved to enable a crucial test of the possibility (8).

The I-cell is two to three-fold larger than the responder cells in diameter, three to ten-fold larger in area. It is much flatter, so as to acquire a bluish tint under appropriate illumination. It is highly motile in random directions before aggregation has begun. Enveloping lobopodia and great numbers of filopodia protrude constantly and explosively. Finally, it is more heavily granulated and vacuolated and its nucleus is considerably larger. Fig. 3 shows photomicrographs of I-cells and Fig. 4 presents histograms to illustrate the size difference between I-cells and responders.

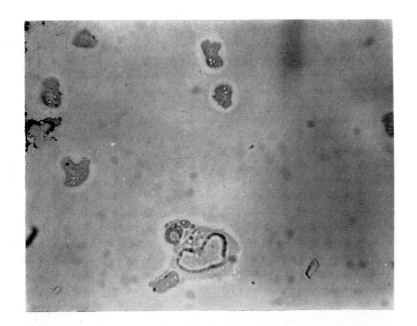
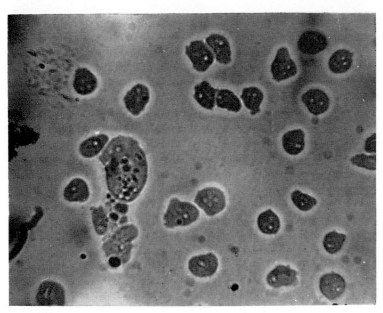

Fig. 3. Photomicrographs of two I-cells. 460 × magnification. (Ennis and Sussman, 1958b).

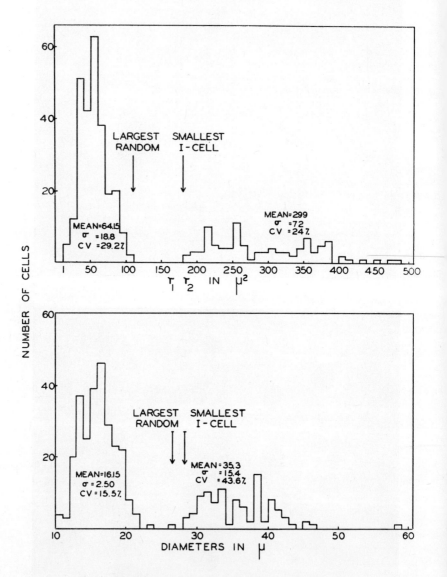

Fig. 4. Size differences between I-cells and randomly chosen cells after incubation on washed agar (Sussman et al., 1958).

The following evidence supports the contention that the I-cell is in fact the initiator:

## 1. Agreement Between Number of I-Cells and Number of Centers

Differential counts of myxamoebae during the preaggregative period have revealed that the ratio I-cells : total cells is 1:1940 or 15 per cent in excess of the center : cell ratio (1:2200) for *D. discoideum*. Table 3 presents the data.

TABLE 3

The Proportion of I-cells in the Population
(Ennis and Sussman, 1958b)

| Incubation period | No. cells examined | No. I-cells | Ratio I-cells: total cells |
|---|---|---|---|
| 3.5 hours | 17,160 | 9 | 1:1910 |
| 5 hours | 18,720 | 9 | 1:2090 |
| 10 hours | 114,300 | 59 | 1:1940 ± 107$\sigma$ |

Washed myxamoebae were dispensed on washed agar at optimal density. They were incubated for the periods noted in the first column before inspection. The 10-hour determination was performed in four separate experiments.

## 2. Correlation Between Positions of I-Cells and of Centers

Myxamoebae were dispensed on washed agar at optimal density. Two to three hours before the onset of aggregation, plates were fixed on microscope stages and low-power fields were chosen at random. Fields containing I-cells were culled from these. At approximately 15-minute intervals, camera lucida drawings were made to record the positions of I-cells and their nearest neighbors. Aggregates appeared in almost 80 per cent of these fields, whereas in fields randomly chosen and not inspected for I-cells, aggregates occurred in 25 per cent. Moreover, in 42 per cent of the I-cell fields, a center formed at the precise position of the I-cell. (Since the center : cell ratio is 1:2200, the random chance of predicting that a center will form precisely at a particular cell would be 0.05%). In 36 per cent, the center occurred near but not at the I-cell. Table 4 summarizes the data.

The camera lucida drawings showed that, immediately prior to aggregation, the small cells nearest the I-cell began to nestle about it and form a small clump. Thereafter clumps arose concentrically about, and at a distance up to a few hundred micra from, the I-cell. Clumping never occurs prior to this time. Then cell elongation became apparent and the clumps transformed into streams. Usually the streams moved toward the I-cell, but when one of the concentric clumps was close to the I-cell (usually

TABLE 4

CORRELATION BETWEEN POSITIONS OF I-CELLS AND OF CENTERS
(Ennis and Sussman, 1958b)

| Category | Result | | No. fields | % of total |
|---|---|---|---|---|
| A | Center formed at I-cell | | 29 | 41.5 |
| | Center formed near I-cell | | 25 | 35.7 |
| | | Total | 54 | 77.2 |
| | No center in field | | 16 | 22.8 |
| B | Center in field | | 7 | 25 |
| | No center in field | | 21 | 75 |

Category A: Fields chosen which contained I-cells.
Category B: Fields chosen at random without inspection for the presence of I-cells.

about 50-100 $\mu$ distant) and extremely large, the cells and streams would gravitate toward it, and the center would slowly take form at the exceptional clump rather than at the I-cell. In this case the I-cell would rarely enter the center along with its neighbors but generally would remain solitary and unaffected outside. Figs. 5 and 6 are time-lapse drawings which illustrate the two modes of aggregation.

## 3. Correlation Between the Incidence of I-Cells and of Centers in Small Population Samples

Replicate samples of 500 myxamoebae were dispensed at a density of 300 cells/mm$^2$ on washed agar with a machined platinum loop. After 8 hours incubation, the samples were examined and predictions were made as to whether or not they would form aggregates, on the basis of whether or not they contained an I-cell. In the best experimental series (A), 13 samples were chosen which certainly contained I-cells and 13 which certainly did not. All of the samples containing I-cells formed aggregates; none of the samples not containing I-cells did so. In another series of three experiments (B), 99 samples were inspected and predictions were made for all without discarding doubtful choices. Of these, 67 samples, it was decided, contained no I-cells. Two of the 67 aggregated. The remaining 32 samples were considered to contain at least one I-cell and 18 of these aggregated. Table 5 shows the data and a statistical analysis of the distribution. Actually, subjective bias probably reduced the impressiveness of the correlation. At high density it becomes increasingly difficult to distinguish true I-cells from groups of two or three responder cells so closely appressed as to appear as a single cell. Unfortunately, for technical reasons, the experiment had to be carried out at high density, and hence in about 20 per cent of the

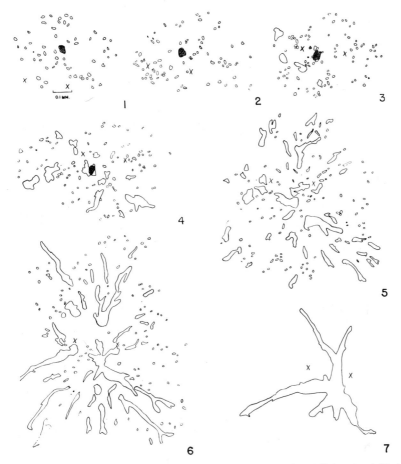

Fig. 5. Time-lapse camera lucida drawings of aggregation. The I-cell is colored black. The large uncolored forms are cell clumps. The X marks are points of reference. Note that the center formed at the final position of the I-cell. The drawings were made after incubation periods of 10.8, 11.25, 11.7, 11.9, 12.4, 13.0, and 14.4 hours respectively. (Sussman et al., 1958).

32 samples the presence of I-cells was considered doubtful. Our bias entered here, since these were classed with the I-cell samples.

### 4. Inhibition of Center Formation by Prior Removal of I-Cells

Replicate samples of 500 washed myxamoebae, maintained in the cold before use, were dispensed on washed agar with a machined loop. Within

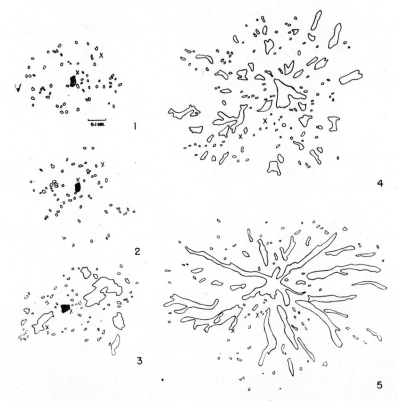

Fig. 6. Time-lapse camera lucida drawings of an aggregation. See Fig. 5 for details. Note that in this instance the center formed not at the I-cell but at the large cell clump nearby. The drawings were made after incubation periods of 12, 12.5, 13.0, 13.4, and 14 hours respectively. (Sussman et al., 1958).

2 minutes after deposition, each drop was inspected and, wherever an I-cell appeared, it was removed by micromanipulation. These and control samples from which I-cells had not been removed were incubated and the distribution of centers recorded.

Of 18 experimental samples only one aggregated. Of 16 control samples, 10 aggregated.

These determinations had been preceded by a number of less successful attempts. At first, we waited 8 hours before removing I-cells and obtained the disconcerting result that center formation remained unaffected. On the supposition that the I-cells had completed their mission by 8 hours, the time of removal was pushed back to four hours and then to one hour, but

### TABLE 5

**Correlation between the Incidence of I-cells and Centers in Small Population Samples**
(Ennis and Sussman, 1958b)

Series A: *Samples with I-cells*
    No. with centers    13
    No. without centers    0
  *Samples without I-cells*
    No. with centers    0
    No. without centers    13

| | Experiments | | | |
|---|---|---|---|---|
| | 1 | 2 | 3 | Total |
| Series B: *Samples with I-cells* | | | | |
| No. with centers | 7 | 8 | 3 | 18 |
| No. without centers | 5 | 6 | 3 | 14 |
| *Samples without I-cells* | | | | |
| No. with centers | 1 | 1 | 0 | 2 |
| No. without centers | 18 | 28 | 19 | 65 |

Total number of samples
in Series B = 99            $x^2 = 37.27$
No. with centers = 20          d.f. = 3
Ratio centers:cells = 1:2170[1]     $p < .001$

[1] Calculated from the $P_0$ term of the Poisson series.

still without effect. A bit frantically, we removed I-cells at 20 minutes after deposition and were then rewarded with a 30 per cent decrease in the frequency of centers; and after 5 minutes, with a 50 per cent drop. It would appear, therefore, that the I-cell can complete its part of the aggregative process within a few minutes of contact with its neighbors.

### 5. *Induction of Centers by Addition of I-Cells*

The data obtained in these experiments are shown in Table 6. I-cells were moved with a micromanipulator to 1 mm² test areas containing an average of 250 myxamoebae placed there previously. Since the center : cell ratio is 1:2200, one would expect that about 11 per cent of such squares would aggregate spontaneously. The value actually obtained in control runs was 12.1 per cent. When randomly chosen cells were moved to the test squares, the frequency of center formation remained the same; but when I-cells were added, the frequency rose to 61 per cent. The highest figure in a single experiment was 72 per cent. This is even more impressive when it is considered that the micromanipulation killed about 25 per cent of the I-cells and possibly damaged others. Ten per cent of the I-cells, but none of the responders, lysed in transit. In viability trials of 20 cells of

## TABLE 6

### Induction of Aggregates by I-cells
(Ennis and Sussman, 1958b)

| | I-cells | | | Random cells | | | Background | | |
|---|---|---|---|---|---|---|---|---|---|
| Exp. | Squares | Centers | % | Squares | Centers | % | Squares | Centers | % |
| 1 | 10 | 6 | 60 | | | | 124 | 14 | 11 |
| 2 | 7 | 5 | 72 | | | | 53 | 6 | 11 |
| 3 | 15 | 8 | 53 | | | | 32 | 2 | 6 |
| 4 | 11 | 7 | 64 | | | | 18 | 4 | 22 |
| | 43 | 26 | 60.5 | | | | 227 | 26 | 11.4 |
| 5 | | | | 12 | 2 | 17 | 31 | 4 | 13 |
| 6 | | | | 18 | 1 | 6 | 24 | 3 | 13 |
| 7 | | | | 23 | 4 | 17 | 24 | 4 | 17 |
| | | | | 53 | 7 | 13.2 | 79 | 11 | 13.9 |
| 8 | 6 | 2 | 33 | 13 | 0 | 0 | 30 | 1 | 3.5 |
| 9 | 10 | 3 | 30 | 16 | 1 | 7 | 64 | 1 | 2.6 |
| 10 | 7 | 3 | 43 | 10 | 0 | 0 | 65 | 1 | 1.5 |
| | 23 | 8 | 34.7 | 39 | 1 | 2.5 | 159 | 3 | 2.0 |

each type, 75 per cent of the I-cells could produce clones when moved to nutrient agar and when bacteria were added, whereas 100 per cent of the responder cells produced clones.

In three experiments shown in Table 6, too few cells were inadvertently delivered to the test areas. The number was about 150 per square, and the density was accordingly well below optimal. As a result the background frequency of aggregation was 2 per cent. Addition of randomly chosen cells again did not affect the frequency, but addition of I-cells raised it to 34.7 per cent, or 17-fold greater than background.

It appears therefore that a clear phenomic difference exists between the I-cells and the responder cells (R-cells) with respect to initiative capacity. Taken together with the data previously described, we can see no alternative to the conclusion that the I-cells are in fact the initiator cells for slime mold aggregation, and that if a different route to aggregation does exist, its contribution is insignificant under these conditions.

### The Persistence of the I-Cell Phenotype

The data at hand, though fragmentary, indicate strongly that the difference between I-cells and R-cells is genetically based and that I-cells do not appear, de novo, immediately before or during aggregation, but exist throughout the life cycle.

In a previous investigation (31), *D. discoideum* myxamoebae were grown under conditions where the duration of the lag and log phases, the growth rate, and the ultimate cell crop in the stationary phase could be controlled. Cells were harvested at late lag phase, at early, middle, and late log phases, and at the stationary phase, and were dispensed on washed agar in order to determine their aggregative capacities. No differences were detectable. That is, all displayed a threshold density of 80 and an optimal density of 200 cells/mm$^2$. The center : cell ratio at optimal density was 1:2200 in every case, subject to a slight sampling and experimental error. This implied that I-cells must be present during all of the growth phases and in a constant proportion. Subsequent microscopic examination did in fact reveal the presence of I-cells and in the predicted proportion. These I-cells were also tested for initiative capacity and 70 to 80 per cent could induce centers when added to test cells in the manner previously described, whereas randomly chosen cells at the same stage had no effect (35). There is apparently no time during the life cycle of the myxamoeboid population when I-cells are not present.

A preliminary study had been made of clones derived from I-cells. Thus far we have examined the aggregative performances of 12 of these. Ten gave a normal picture with respect to the density parameters and the distribution of centers. Two were decidedly anomalous. The optimal densities of both were below 100 cells/mm$^2$ and the center : cell ratios were significantly greater than 1:600. Microscopic examination showed the presence of correspondingly greater numbers of I-cells. Much more work will be necessary before it will become profitable to propose hypotheses to account for these data, but certainly the thesis that I-cells and R-cells differ genetically seems to be justified.

It should also be noted that most, if not all, spores from the terminal fruit are R-cells, and yet clones derived from them aggregate in normal fashion and display the usual distribution of centers and incidence of I-cells. We may therefore tentatively infer that (1) I-cells arise during the growth of the R-cell population; (2) the transformation is reversible during growth of I-cell clones; and (3) the steady state I-cell : R-cell ratio in *D. discoideum* is maintained because of the kinetics of these transformations and the interactions of the two cell types with the environment and each other.

## Is Aggregation a Sexual Process?

Wilson (39) and Wilson and Ross (40) examined fixed, stained smears of early aggregates and encountered, at low frequency, cells which had engulfed or were in the act of engulfing their neighbors, usually at the fringe of the aggregate. These, it was concluded, were zygotes resulting from syngamy via engulfment. Wilson proposed a scheme which made syngamy a mandatory accompaniment to aggregation. He specified that all cells must undergo syngamy immediately before entering the aggregate and that the first-formed zygotes become the aggregative center. Subsequently Wilson and Ross amended the view that syngamy bears a causal relation to aggregation and denied that syngamy need be universal. They suggested instead that aggregation might occur without the inferred zygote formation, which might require special environmental conditions for its appearance.

Our own view coincides with this and springs from three considerations. First, were zygosis universal and the first zygotes the centers of aggregation, it would have been impossible to obtain a Poisson distribution in small population samples. Instead, all the samples should have aggregated. Second, did syngamy (and meiosis) mandatorily accompany aggregation (and fruiting), the population would be expected to undergo drastic changes in cell number. Yet R. R. Sussman has made direct counts at intervals during the aggregation and fruiting of the *D. discoideum Fty-1* mutant (whose aggregates are tiny enough to permit accurate cell counts) and found that the same number of cells which had entered an aggregate could be detected at all of the developmental stages thereafter, including the terminal fruit. Moreover, no nuclear activity could be detected in hematoxylin or aceto-orcein stained preparations, nor did cells fixed in situ on glass while streaming toward early aggregates show evidence of nuclear activity or of having engulfed their neighbors (34). Finally, a large number of attempted crosses with both spontaneous and induced mutants provided no evidence of recombination.

The question remains as to whether the I-cell might not be a zygote. These cells very frequently are seen to contain what appears to be an engulfed myxamoeba and (rarely) as many as five. Previously, Wilson reported the occurrence of abnormally large cells with inclusions resembling myxamoebae in his stained preparations, and these may well have been I-cells. We do not feel that the occurrence of engulfed cells is in itself compelling evidence of syngamy. The data at hand are at present not sufficient to distinguish among the possible explanations, of which zygosis may be one. Perhaps a study of I-cell clones, now in progress, will provide such data.

Cellular Interactions during Aggregation

*1. The Chemotactic Complex*

Earlier investigators of the slime molds had surmised that chemotaxis might be the modus operandi of aggregation (14, 1). The first evidence to support this view was offered by Runyon (19), who showed that cells placed above and below a semipermeable membrane aggregated in coincidence, center for center and stream for stream. This effect was later confirmed with agar membranes of controllable thickness over distances up to 200 $\mu$ (29). In 1947, Bonner (3) extended Runyon's findings by showing that preestablished centers under salt solution could direct the establishment of a center by unaggregated myxamoebae at distances up to 800 $\mu$. When centers were permitted to form in a current of salt solution, Bonner found that only cells downstream from the center were attracted to it, and never cells upstream. In a subsequent paper (4), it was shown that migrating pseudoplasmodia of *D. discoideum* could also attract myxamoebae and therefore presumably produced the chemotactic agent, which was given the generic name acrasin.

In 1953, Shaffer (20) utilized the chemotactic capacity of the pseudoplasmodia to demonstrate that myxamoebae could be attracted by cell-free washings. He cut a small square of agar supporting an early aggregate or portion thereof and inverted this over a drop of water on a slide. The cells came off the agar, sank through the thin film of water, and adhered to the glass. Left untreated, they quickly reaggregated to form the characteristic wheel-like pattern of streams. If however, washings from pseudoplasmodia were periodically applied to one edge of the block, the test cells elongated, became oriented in the direction of that edge, and moved toward it, forming streams as they did so. Fig. 7 illustrates a modification of the Shaffer assay.

Shaffer found that the washings were genus-specific. Material from *D. discoideum* did not attract myxamoebae of *Polysphondylium violaceum*. Washings from early *Polysphondylium* aggregates attracted only homologous myxamoebae. Washings from later aggregates attracted both kinds and from very late *Polysphondylium* centers attracted only heterologous test cells, not homologous ones. This confirmed the earlier findings of Raper and Thom (16), who had shown that intergeneric mixtures of the myxamoebae did not enter into communal aggregates. Instead, separate aggregates were formed without any sign of interaction.

Shaffer also reported that the chemotactic agent was highly unstable

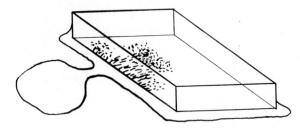

Fig. 7. A modification of the Shaffer assay for chemotactic activity. A narrow bridge is made between a drop of material to be tested and the small amount of water surrounding the agar block. The test solution actively seeps under the block. If active, the test cells elongate, orient toward the point of application, and move toward it in streams.

and lost activity within minutes at room temperature. This property is not particularly surprising. To sense direction in response to a diffusible agent, a cell must detect a concentration gradient. Maximum sensitivity would be attained if the agent disappeared at a high rate and thereby reduced the noise level at the rear of the sensing cell. It would therefore be to the advantage of such cells if they were to secrete an enzyme in order to insure destruction of the agent (30, 22). In simultaneous publications, Shaffer (21) and Sussman, Lee, and Kerr (32) reported the successful application of this hypothesis. Although the methods were different, both involved the extraction of acrasin under conditions which promoted protein denaturation, and both yielded completely stable preparations. Moreover, the crude washings were found to contain a heat-labile, ammonium sulfate precipitable fraction which inactivated the chemotactic agent.

The crude chemotactic preparation was chromatographed on paper with 70 per cent ethanol. Two fractions, termed A and B, were eluted which were, themselves, inactive or inhibitory but which, when mixed in the proper proportions, displayed the full activity of the original (32). Mixtures of A and B, treated with the inactivating enzyme, were reactivated by addition of fresh B-component. Accordingly the inactivating enzyme was tentatively termed a "B-ase."

One would suppose that the system thus revealed would be confusing enough. Unfortunately a recent collaborative effort by Dr. R. R. Sussman, Mrs. Fu, and myself has succeeded in complicating the situation even more. Additional purification of the inactive A-fraction by countercurrent extraction yielded a component which was optimally chemotactic. Reconstitution of the original from the separate countercurrent fractions once again resulted in an inactive preparation which could as before be reactivated by addition of the B-component. The intimation that there were three components in the complex, and not two, was confirmed by careful elution of paper chromatograms of the crude acrasin. Two tightbands (B and A) with $R_f$ values of 0.14 and 0.33, and a region (C) extending from above the A-band to the solvent front, were eluted separately. Mixtures of the three components were assayed in coded series by the Shaffer procedure to yield the data given in Table 7. Each determination was made in replicate by at least two operators.

TABLE 7

Possible Components of the *D. discoideum* Chemotactic Complex

The number of plus signs refers to arbitrarily chosen but reproducible levels of response intensity; minuses mean that cells become rounded up and would neither respond to the agent nor form their own center; zeros mean that cells formed their own center. Concentrations are given in activity equivalents per ml. (a.e./ml.) of the crude acrasin from which the components were prepared.

A. Relation between A and C:

| A a.e./ml. | C a.e./ml. 0 | 1.7 | 3.3 | 5.0 | 6.7 |
|---|---|---|---|---|---|
| 1.7 | ++++ | ++++ | ± | ± |  |
| 3.3 | ++++ | ++++ | ++++ | ± | ± |
| 6.7 | ++++ | ++++ | ++++ | ++++ | +± |

B. Relation between B and C:

| A a.e./ml. | C a.e./ml. | B a.e./ml. 0 | 1.7 | 3.3 | 6.7 |
|---|---|---|---|---|---|
| 1.7 | 3.3 | ± | ++++ | ++++ |  |
|  | 5.0 | ± | ± | ++++ |  |
| 3.3 | 5.0 | ± | +0 | +± | ++++ |
|  | 6.7 | ± | ± | +± | ++++ |

C. Relation between A and B:

| A a.e./ml. | C a.e./ml. | B a.e./ml. 0 | 1.7 | 3.3 | 6.7 |
|---|---|---|---|---|---|
| 1.7 | 5.0 | ± | ± | ++++ |  |
| 3.3 |  | ± | +0 | ++++ |  |
| 3.3 | 6.7 | ± | ± | +± | ++++ |
| 6.7 |  | +± | +++ | ++++ | ++++ |

The A-component alone was titrated in serial two-fold dilutions and was minimally active at 0.83 a.e./ml. (150 µg./ml.). As seen in Table 7, some mixtures of A and C were inactive. The higher the concentration of A in the mixture, the more C was required to inactivate. The addition of the B-component reactivated these. The higher the concentration of C, the more B was required to reactivate. Finally, in the presence of a constant amount of C, components A and B appeared to spare each other.

The question arises, of course, as to whether the B and C-components might be artifacts and not involved in the chemotactic system. We as yet have no results which provide a clear answer. It is true that the inactivating enzyme seems to operate by destruction of B and this suggests physiological meaningfulness. Further, fractionation of *Polysphondylium* acrasin has revealed the presence of three functionally identical components.

The aforementioned countercurrent extraction procedure was applied to both the A-component (obtained by chromatography) and the crude acrasin. Both yielded preparations of A, minimally active at between 2 and 5 μg./ml. and representing a purification of the crude material to the extent of about 1000-fold. Fig. 8 illustrates one such run starting with crude ma-

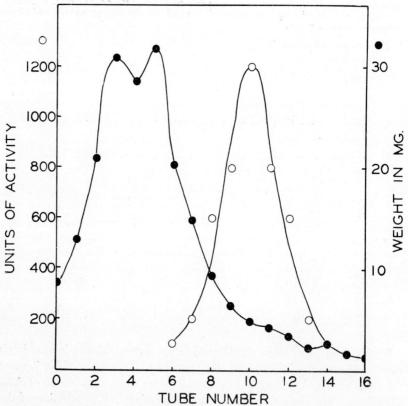

Fig. 8. The distribution of chemotactic activity (open circles) and weight (black circles) after countercurrent extraction of crude acrasin. See text for details. The term "units of activity" refers to the number of ml. in which a fraction could be diluted to yield minimal activity.

terial. The solvent pair was n-butanol and 0.01 $N$ $NH_4OH$. After 158 transfers, the activity peak was located in the tenth tube, indicating a partition coefficient of 0.06. The weight distribution and determinations of fluorescence and spectral absorption all agreed in suggesting that the active material was still grossly impure. At the moment, we have no information regarding the chemical nature of this component, but Dr. Barbara Wright-Kalckar will describe some exciting results that bear on this subject.

Having scarcely driven a wedge into the problem of the chemotactic complex, it would of course be premature to think seriously at present about the sensing apparatus of the myxamoebae or about the integrated actions of the cell assembly within this morphogenetic field. In this connection, Bonner has made the extremely important observation that not only the center but also the radial cell streams can produce the chemotactic agent (3, 4). His data suggest that a cell becomes sensitized to and excited by the incoming agent and then actively produces it (or perhaps some of its components).

## 2. Cell Adhesion During Aggregation

When one attempts to manipulate well-formed streams and centers, it is immediately obvious that the cells are no longer loosely massed but possess active, adhesive properties. Thus, segments of the aggregate can be removed intact from the substratum, and even prolonged trituration yields large cell clumps. In fact, the only way by which we have been able to prepare homogeneous, unclumped suspensions of cells in this stage for clonal plating, etc., has been treatment with trypsin at high $p$H, in accordance with procedures employed by embryologists for tissue dispersal (36).

Observations by many workers indicate that this adhesive property augments and complicates the chemotactic process (1, 16, 5, 23). For example, streams converging upon a center need not coalesce even though they are extremely close together. The streams often approach a center tangentially over a highly curved path, a response not easily interpretable as a purely chemotactic one. When the distal segment of a stream is separated from the proximal, it will often stop moving and form a satellite center, so as to suggest that chemotaxis and adhesion must operate in conjunction to maintain stream integrity.

Some of the adhesive properties may conceivably be explained in terms of the production of an extracellular cement. A precedent for this has been established by Raper and Fennell (18), who showed that the pseudoplasmodium of *D. discoideum* is held together by a slime envelope which is spun out as a cylindrical ribbon behind the advancing slug. However, earlier observations of Raper and Thom (17) are not so easily explainable in these terms. They found that intergeneric mixtures of myxamoebae formed completely separate aggregative centers, the streams of which crossed each other and became intimately interlaced without disturbing the integrity of either. In mixtures of *D. discoideum* and *D. mucoroides,* common radiate aggregation patterns were formed and, by vital staining, it could be shown that both species were contributing to a common center. However, the cells were reshuffled upon reaching the center and two separate pseudoplasmodia appeared, each characteristic of one species. Chimeric hybrids were never encountered and replating of the spores provided no evidence that both types of cells had entered a single pseudoplasmodium. It is difficult to understand these actions without invoking some sort of specific adhesion.

To us, the most promising attempt at interpretation is that of J. H. Gregg (9, 10), who is proceeding upon the supposition that forces similar to those operating in antigen-antibody reactions can account for the specific adhesion. Gregg has followed the appearance of surface antigens by titration with rabbit antisera. His data indicate the presence of more than one kind of antigen and a regulated time sequence of appearance. He has also begun to study antigenic properties of a collection of mutants. The ultimate expectation is to correlate adhesive properties and morphogenetic potencies with antigenic topography. As Gregg has pointed out (9), this approach is predicated upon the hypotheses advanced by Tyler (37) and by Weiss (38) to explain the adhesion of metazoan and metaphytic tissues.

## 3. Interactions During Synergistic Aggregations by Wild Type and Mutant Cells

That the initiator cell must by its actions and its position affect the developmental fate of its neighbors is obvious, since whether they aggregate, and the order in which they enter the aggregate, determines their place in the terminal fruit (i.e. whether they are to become spores, stalk cells, or basal disk cells). That the responder cells might in turn affect the initiators is not immediately obvious but has been revealed through a study of mixed aggregations (7, 25, 26, 27).

The ability of potential initiator cells to induce their neighbors to aggregate has been found to depend in part upon the nature of the audience which they must excite (28). This was demonstrated by comparing the ability of potential initiator cells of the fruity mutant *Fty-1* to induce centers among a homologous audience, with their ability when faced with an

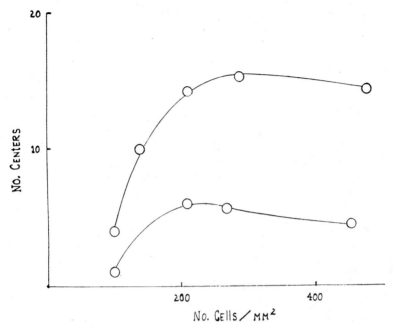

Fig. 9. The relation between center formation and population density. The lower curve was obtained with populations of 10,000 *D. discoideum* wild-type cells; the upper with mixtures of 10,000 wild-type and 500 *Fty-1* mutant myxamoebae. (Sussman, 1955).

audience of wild type. Mixtures containing 500 *Fty-1* cells and 10,000 wild type were dispensed on washed agar over a range of densities. Fig. 9 records the center production by these cells as compared with equivalent populations composed only of wild type. Subtraction of the second from the first gave a measure of the number of centers attributable to the mutant cells in the mixture. At higher densities a plateau value was reached, and this was found to be directly proportional to the number of *Fty-1* cells present. Thus one could be sure that single mutant cells were inducing the extra centers. Because the mutant cells were numerically insignificant, one could also be sure that it was the wild type which supplied the bulk of the responder cells. Table 8 shows the plateau ratios for mixtures of 10,000 wild type with 500, 750, and 1000 mutant cells. The ratio of centers induced by the mutant cells to the number of mutants present was found to be 1:58. But when *Fty-1* cells were permitted to aggregate alone, the center : cell ratio at optimal density was 1:24 (see Table 1). This would imply that when potential *Fty-1* initiators are faced with a homologous

### TABLE 8

**THE ABILITY OF *Fty-1* CELLS TO INDUCE CENTERS AMONGST WILD-TYPE RESPONDERS**

| No. *Fty-1* cells | No. centers at plateau densities | Ratio |
|---|---|---|
| 1000 | 15.8 | 1:63 |
|  | 15.6 | 1:64 |
| 750 | 11.9 | 1:63 |
|  | 14.6 | 1:50 |
| 500 | 8.8 | 1:57 |
|  | 9.8 | 1:51 |
|  | Mean | 1:58 |

The term ratio refers to the ratio of centers induced by the mutant to the number of mutant cells present.

audience, more than twice as many can actually induce a center and be scored as initiators than when faced with a wild-type audience.

Further examples of interactions emerged from a study of aggregations by mixtures of *D. discoideum* wild-type and aggregateless mutants. In these mixtures, small numbers of wild type were present at densities so low that, alone, they could form no centers at all or, at most, a small fraction of the number that they could have produced had they been at optimal density. The addition of aggregateless cells permitted a progressive increase in center formation until, in the presence of excess mutants, a plateau value was attained. Fig. 10 shows the results for mixtures of wild-type and two mutant stocks, Agg-70 and Agg-204.

The plateau values were again directly proportional to the number of wild type present. Fig. 11 demonstrates this relationship in greater detail for mixtures with two mutant stocks, *Agg-53* and *Agg-53A*. As before, the linear response suggested that single (wild-type) cells had induced the formation of centers, but note that the number of wild type that could do so depended very drastically upon the nature of the audience. Thus with *Agg-53A*, one out of 1780 wild type apparently could induce a center; with *Agg-53*, one out of 980; with *Agg-204*, one out of 76! Table 9 is a summary of the center : wild-type cell ratios encountered in the presence of excess numbers of nine aggregateless strains, and gives further evidence of the degree of variation. Finally, it is noteworthy that the wild-type potential initiators appeared capable of inducing many more centers in the presence

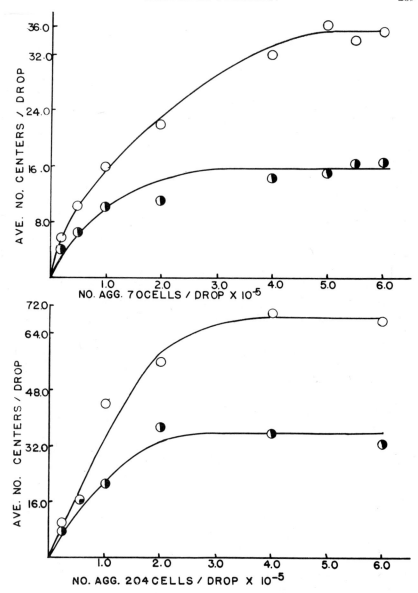

Fig. 10. Center formation by mixtures of *D. discoideum* wild type and two aggregateless mutant stocks. The upper curves were obtained with mixtures containing 5000 wild type (open circles); the lower, 2500 (half-shaded circles). (Ennis and Sussman, 1958a).

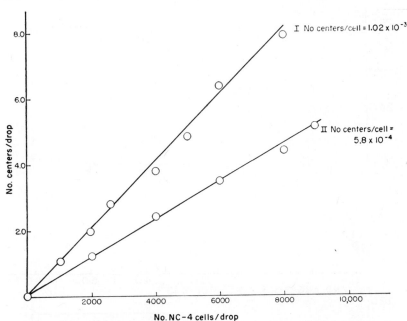

Fig. 11. Formation of aggregative centers by mixtures of wild type NC-4 with excess numbers of (I) NC-4 *Agg-53*, or (II) NC-4 *Agg-53A*. (Sussman, 1952).

of these mutant cells (though not among all the strains studied) than they could when faced with an exclusively wild-type audience.

The key to the two problems, that of variation in the number of centers induced in the presence of different morphogenetically deficient partners and that of the multiplicity of centers, was found in the data given in Fig. 11. It will be noted that the center : wild-type cell proportionality held for

TABLE 9

THE ABILITY OF WILD-TYPE CELLS TO INDUCE CENTERS
AMONGST AGGREGATELESS MUTANT RESPONDERS

| Mutant | Plateau ratio of centers : wild-type cells |
|---|---|
| 53 | 1:980 |
| 53-A | 1:1780 |
| 70 | 1:148 |
| 77 | 1:395 |
| 204 | 1:76 |
| 85 | 1:300 |
| 89 | 1:210 |
| 91 | 1:118 |
| 83 | 1:84 |

mixtures containing as many as 8000 to 9000 wild-type cells. This is remarkable, since all of the mixtures had been dispensed on washed agar in identical fluid volumes and consequently covered the same surface area. This means that while the lower numbers of wild type (500-6000) were present in the mixtures at population densities below optimal, the higher numbers (6000-9000) were at a density above optimal (200 cells/mm$^2$) and thus could have aggregated perfectly well, even had there been no aggregateless cells. Yet here too a multiplicity of centers was encountered, far more than the wild type alone could have produced. We were forced to conclude that the synergism can operate under two entirely different circumstances.

These were shown to be experimentally separable when the mutants and wild type were physically separated by a thin agar membrane. Excess mutant cells were dispensed on one side, and on the other side were placed wild-type cells either at a density below the threshold for any center formation or alternatively at a density above optimal, thereby simulating the circumstances alluded to above. At low density the wild type produced no centers at all, indicating that when the wild type is too sparse to aggregate alone, the mutant cells must contribute their physical presence if synergism is to operate. The mere supply of diffusible material is not enough.

In contrast, when the wild type was present at high density and was opposed by excess mutant cells, far more centers could be produced than in the absence of the latter. Table 10 gives the data for several pairings and

TABLE 10

Action of Aggregateless Cells Upon Wild Type Across Agar Membranes

| Mutant | Center:wild-type cell ratio in mixtures | No. centers formed by 20,000 wild-type cells on agar membranes | | |
|---|---|---|---|---|
| | | Experimental | Control | Fold increase |
| Agg-204 | 1:76 | 34.2 | 3.6 | 9.5 |
| Agg-70 (3 × 10$^5$ cells) | 1:148 | 30.1 | 3.5 | 8.6 |
| Agg-70 (1.5 × 10$^5$ cells) | | 19.1 | 3.5 | 5.4 |
| Agg-53 | 1:980 | 5.3 | 3.2 | 1.7 |
| Agg-72 | 1:00[1] | 13.8 | — | 3.8 |

[1] The wild type can form no centers at all, even when mixed with small numbers of Agg-72 cells and even when the wild type is at a density high enough for it to aggregate alone. This mutant actively inhibits the process, but, as seen in the data, only when physically present.

shows that the intensity of the effect depended on the number of mutant cells which had been placed on the membrane, and moreover that it was strain-specific.

This result forced the conclusion that, in mixtures where the wild type is dense enough to aggregate alone, the mutants need only supply diffusible material to effect the synergistic multiplicity of centers and need not be physically present. Confirmation of this view was obtained in synergistic aggregations of wild type with *Agg-91*, a strain whose myxamoebae and (synergistically produced) spores are dwarf-sized and can be distinguished by inspection. In mixtures containing wild type at low density, the *Agg-91* cells comprised the major component of the fruit; but in mixtures containing wild type at optimal density or above, the *Agg-91* cells could not be found. Yet far more centers could be formed by the wild type in this circumstance than when alone.

The question arose as to whether cell-free preparations of mutant cells might not be active in affecting initiative capacity. As Table 11 indicates,

TABLE 11

EFFECT OF CELL-FREE EXTRACTS ON WILD-TYPE
AGGREGATION AT OPTIMAL DENSITY

| Source of extract | Fold activity[1] |
|---|---|
| *Agg-70* | 2.6 |
| *Agg-72* | 2.5 |
| *Fty-1* | 11.4 |
| Wild type | 7.8 |

[1] Fold activity refers to the ratio of centers produced by 20,000 wild-type cells at optimal density on extract agar to that on control agar.

active preparations could be obtained not only from the aggregateless mutants but from the fruity (*Fty-1*) stock and from the wild type itself. This result becomes especially noteworthy when it is considered that no other treatment among trials of many kinds has ever succeeded in influencing initiative capacity.

To summarize, the foregoing indicates that aggregateless cells perform at least two functions in synergistic mixtures. First, they play an active role as responder cells, enter the aggregate, and become part of the fruit. As such, they can offset the lack of their wild-type counterparts where this is needed. Second, they produce diffusible material which can affect the initiative capacity of the wild type. This is the only function that is needed when adequate numbers of wild type are present and, in at least the case of *Agg-91*, it is the only function actually performed.

The implication of these findings for the relationship between wild-type initiator and responder cells is obvious, in view of the fact that cell-free material from the wild type can also affect initiative capacity. One may ask why the mutant cells can exert the effect in mixtures when wild-type responders cannot. The answer is probably contained in the fact that the mutant cells could be packed around the wild type at densities as high as 20,000 cells/mm$^2$ in these mixtures. This cannot be done with exclusively wild-type populations because packing the cells more densely than 200 cells/mm$^2$ brings the initiator cells close enough together that they compete for a common audience and center formation is then depressed, as Fig. 3 shows.

The most interesting question at present concerns the bearing of these phenomena upon the I-cell. Unfortunately, the extract experiments were performed prior to the I-cell investigation. We mean to repeat the incubation of wild type on the extract agar and determine if the I-cell count rises in proportion to the increase in center formation.

A final and most interesting complex of interactions was uncovered in mixtures of wild type with *Agg-91*. When small numbers of the former (500-2000) were mixed with an excess of the latter, the ratio center : wild-type cells remained constant at 1:118. When larger numbers of wild type (2500-8000) were added, the ratio progressively decreased to 1:442. Because these mixtures had all been dispensed on washed agar in 0.01 ml. drops and consequently covered the same surface area, the increase in the wild-type contingent from 500 to 8000 meant a 16-fold increase in the density of the latter. When, however, the wild-type density was reduced by dispensing them on washed agar in larger-sized drops (while still maintaining an excess density of *Agg-91*), the larger numbers of wild type could induce proportionately as many centers as the smaller, and the ratio rose to 1:118.

This meant to us that the capacity of larger numbers of the wild type to induce centers in the presence of the mutant had originally been hampered by the wild type itself, since reducing the density of the wild type while maintaining a constant density of the mutant relieved the hindrance. The situation is analogous to that which obtains when wild type aggregates alone. That is, at densities greater than optimal the potential center-forming agencies are brought closer together, compete with each other, and thereby decrease the number of centers. The difference is that this depression is observed at densities greater than 200 cells/mm$^2$, whereas when excess *Agg-91* cells are present the depression in center formation becomes significant at 38 cells/mm$^2$!

Clearly the mutant cells must act upon the wild type in order to bring this about. But as mentioned before, the reduction in center formation ultimately involves interactions between the wild-type cells themselves. In short, a hierarchical system operates in which one cell class acts upon another and causes the second to interact within itself. Coupled with these, it must be remembered, is the diffusible material contributed by the mutant, which at the very least permitted the wild type to induce many more centers (1:442) than they could when alone (1:2200). It would seem that the description of aggregation will not be complete unless the interactions cited above can be defined in chemical terms.

## Environmental Effects

It is probable that the onset of aggregation is itself the direct result of an environmental influence, namely, the exhaustion of nutrients and the entrance of the population into the stationary phase. Raper (16) found that when aggregates or pseudoplasmodia were disrupted, the cells immediately reaggregated and completed the morphogenetic sequence. If, however, bacteria were added at this time, the myxamoebae reverted to the vegetative state until the fresh nutrient supply was exhausted. As mentioned previously, myxamoebae harvested from the lag and log phases and washed free of bacteria could aggregate normally (31). Yet left undisturbed, on the growth plate, these cells would have continued their vegetative state.

Hirschberg et al. (11, 12) have examined the effects of a wide variety of enzyme poisons, carcinogenic agents, and known cell constituents upon aggregation. Some of these were highly effective in inhibiting aggregation, but since viability determinations could not be performed at that time, it was not determined if the active compounds inhibited aggregation directly or did so only secondarily, as a consequence of having killed the cells. Hirschberg has since indicated the potentialities of this system as a screen for carcinogenic agents (12).

A subsequent examination was made of a large number of substances in which changes in the relation of center formation to cell number and density were employed as criteria of interference with aggregation or encouragement of it (6). Two interesting groups of compounds were encountered. The first of these consisted of adenine and guanine, which depressed center formation and greatly increased the threshold and optimal densities without killing the cells. The second included histidine and imidazole, which did not alter initiative capacity but which increased the sensitivity of the cells so that they could aggregate at phenomenally low densities. Studies of histidine uptake revealed that the cells became acutely sensitive simultaneously with entrance of histidine into the free amino acid pool, and that removal of exogenous histidine after that time did not reduce the sensitivity. The free intracellular histidine was subsequently dissipated, some by entrance into protein, the rest by destruction and/or transformation of the imidazole nucleus.

## REFERENCES

1. Arndt, A., *Wilhelm Roux' Arch. Entwicklungsmech. Organ.,* **136,** 681-747 (1937).
2. Bonner, J. T., *Am. J. Botan.,* **31,** 175-182 (1944).
3. ———, *J. Exptl. Zool.,* **106,** 1-26 (1947).
4. ———, *J. Exptl. Zool.,* **110,** 259-272 (1949).
5. ———, *Biol. Bull.,* **99,** 143-151 (1950).
6. Bradley, S. G., Sussman, M., and Ennis, H. L., *J. Protozool.,* **3,** 60-66 (1956).
7. Ennis, H. L., and Sussman, M., *J. Gen. Microbiol.,* **18,** 241-260 (1958).
8. ———, and ———, *Proc. Natl. Acad. Sci. U. S.,* (in press).
9. Gregg, J. H., *J. Gen. Physiol.,* **39,** 813-820 (1956).
10. ———, *Anat. Rec.,* **128,** 558 (1957).
11. Hirschberg, E., and Rusch, H. P., *J. Cellular Comp. Physiol.,* **37,** 323-336 (1951).
12. ———, *Bacteriol. Revs.,* **19,** 65-78 (1955).
13. Luria, S. E., and Delbrück, M., *Genetics,* **28,** 491-511 (1943).
14. Potts, G., *Flora* (Ger.), **91,** 291-307 (1902).
15. Raper, K. B., *J. Elisha Mitchell Scient. Soc.,* **56,** 241-282 (1940).
16. ———, *Growth* (3rd Symposium), **5,** 41-75 (1941).
17. ———, and Thom, C., *Am. J. Botan.,* **28,** 69-78 (1941).
18. ———, and Fennell, D., *Bull. Torrey Botan. Club,* **79,** 25-51.
19. Runyon, E. H., *Collecting Net, Woods Hole,* **17,** 88 (1942).
20. Shaffer, B. M., *Nature,* **17,** 975 (1953).
21. ———, *Science,* **123,** 1173 (1956).
22. ———, *J. Exptl. Biol.,* **33,** 645-657 (1956).
23. ———, *Quart. J. Microscop. Sci.,* **98,** 377-392 (1957).
24. Sussman, M., and Noel, E., *Biol. Bull.,* **103,** 259-268 (1952).
25. ———, *Biol. Bull.,* **103,** 446-457 (1952).
26. ———, and Sussman, R. R., *Ann. N. Y. Acad. Sci.,* **56,** 949-960 (1953).
27. ———, *J. Gen. Microbiol.,* **10,** 110-121 (1954).
28. ———, *J. Gen. Microbiol.,* **13,** 295-309 (1955).
29. ———, and Lee, F., *Proc. Natl. Acad. Sci. U. S.,* **41,** 70-78 (1955).
30. ———, *Ann. Rev. Microbiol.,* **10,** 21-50 (1956).
31. ———, *Biol. Bull.,* **110,** 91-95 (1956).
32. ———, Lee, F., and Kerr, N. S., *Science,* **123,** 1172 (1956).
33. ———, and Sussman, R. R., in *Cellular Mechanisms in Differentiation and Growth* (D. Rudnick, ed.), p. 125-153, Princeton Univ. Press, Princeton (1956).
34. Sussman, R. R., (unpub.).
35. ———, Ennis, H. L., and Sussman, M., (unpub.).
36. Trinkaus, J. P., and Groves, P. W., *Proc. Natl. Acad. Sci. U. S.,* **41,** 787-795 (1955).
37. Tyler, A., *Growth,* **10** (suppl.), 7-19 (1946).
38. Weiss, P. A., *Yale J. Biol. and Med.,* **19,** 235 (1947).
39. Wilson, C. M., *Am. J. Botan.,* **40,** 714-718 (1953).
40. ———, and Ross, I., *Am. J. Botan.,* **44,** 345-350 (1957).

# ENZYME PATTERNS DURING DIFFERENTIATION IN THE SLIME MOLD

BARBARA E. WRIGHT AND MINNIE L. ANDERSON

*Laboratory of Cellular Physiology and Metabolism, Enzyme Section, National Heart Institute, National Institutes of Health, Public Health Service, U. S. Department of Health, Education, and Welfare, Bethesda, Maryland.*

## INTRODUCTION

One of the simplest, and hence one of the most appealing model systems for a biochemical approach to differentiation is found in the slime molds. The unique life cycle of these microbes includes a vegetative, independent single cell as well as a multicellular stage. Multicellular differentiation involves only two cell types, and proceeds in the absence of growth. Thus any biochemical changes occurring during differentiation are neither confused by nor interdependent on biochemical changes accompanying growth.

Some hours after vegetative myxamoebae exhaust their food supply and have been washed free of any nutrient material (i.e., dead bacteria) and placed on purified agar, they aggregate in response to a diffusible, chemotactic substance called acrasin (25, 1) to form a multicellular pseudoplasmodium composed of about 150,000 cells. It should be emphasized that, although a multicellular unit is formed, the component cells do not arise as progeny from the growth of a single cell. The unit forms as a result of the aggregation of many single myxamoebae in the absence of cell division.

Once the pseudoplasmodium has formed, gradual specialization into two cell types begins (23, 2). The anterior third of the pseudoplasmodium contains the presumptive stalk cells of the eventual fruiting body, which is composed of a cellulose-rich stalk supporting a mass of spores. The presumptive stalk cells gradually acquire the ability to synthesize and deposit cellulose extracellularly; after so doing they are incorporated into the stalk, become highly vacuolated, and eventually die. They are the somatic cells of the organism. The presumptive spore cells in the posterior two-thirds of the pseudoplasmodium become transformed to spores by the gradual accumulation of polysaccharides and the extracellular deposition of a cellulose husk (5, 24). In view of this very simple life cycle, it would not be surprising to find that the main biosynthetic activities during differentiation are associated with polysaccharide synthesis.

Gregg has shown that the dry weight of the slime mold drops only 6 per cent during differentiation, and that total reducing substances increase about 45 per cent based on dry weight (15, 16). A simultaneous decrease in total protein nitrogen occurs, accompanied by ammonia formation. These changes occur in the absence of external nutrients, and all the evidence supports the conclusion that endogenous protein is the major source of energy and building blocks for carbohydrate synthesis (16, 17).

Biochemically, how are these changes in metabolism brought about? This being an exclusively endogenous system, the new emphasis on carbohydrate synthesis must stem from internally induced stimulants and regulators of metabolism. By far the greatest protein utilization and carbohydrate synthesis occurs *after* the formation of the migrating pseudoplasmodium. Thus the new pattern of substrates, enzymes, and products must fully establish itself by the time that the pseudoplasmodium is formed.

Such a life cycle offers some rather unique experimental possibilities, and poses interesting questions about the metabolism of slime molds during differentiation.

A useful definition of differentiation is "the gradual production of differences between the descendents of the same ancestral type." This concept includes differentiation on a cellular level—phenomena such as enzyme induction in response to a substrate, as well as gradual changes in the metabolic pattern of a cell. The end product of "biochemical differentiation" in the slime mold is mainly represented by the accumulation of polysaccharide—predominantly cellulose. What is the gradual sequence of events by which the rate of carbohydrate synthesis is increased? Which specific biochemical pathways become activated, in what order, and how do they become activated? By the availability of essential cofactors? Or by the removal of specific inhibitors? Or by the sequential formation of substrates which induce the synthesis of their specific enzymes? It is clear that the answers to such questions depend upon a detailed knowledge of the particular metabolic characteristics of slime molds: the presence, concentration, and activity of relevant substrates, enzymes, and products. Even fragmentary information of this type about a critical biosynthetic pathway might supply some of the necessary keys for an understanding of the mechanisms controlling biochemical differentiation.

## Methods

The myxamoebae of the slime mold *Dictyostelium discoideum* are grown on complex media with either live or dead bacteria (3, 7). They are then washed three times with cold distilled water and spread on washed agar plates, under conditions where no growth occurs (30). By freezing and thawing five times, cell-free preparations were made of cells harvested at succeeding stages of differentiation up to the migrating slug. With later stages, samples were also ground with alumina. The following enzymatic activities were followed spectrophotometrically at 340 m$\mu$, by the increase or decrease in reduced pyridine nucleotide: isocitric dehydrogenase (TPN- and DPN-linked [1]), malic dehydrogenase (TPN and DPN), glucose-6-phosphate dehydrogenase (TPN), 6-phosphogluconic dehydrogenase (TPN), glutamic dehydrogenase (DPN), glyceraldehyde-3-phosphate dehydrogenase (DPN), and lactic acid dehydrogenase (DPN). G6P-phosphatase activity was followed by measuring orthophosphate liberation.

The enzyme forming uridine diphosphoglucose (UDPG) from uridine triphosphate (UTP) and glucose-1-phosphate is most conveniently followed by measuring glucose-1-phosphate (G1P), which is formed in the reaction: UDPG + PP → UTP + G1P (21). TPN, 1,6-phosphoglucomutase, and glucose-6-phosphate (G6P) dehydrogenase were present during the experiment in order to measure the G1P formed. To demonstrate the actual accumulation of G1P in the system, the incubation was carried out in the absence of added TPN, mutase, and G6P dehydrogenase. Following heat inactivation, these components were added to measure the G1P present. Glucose was determined in the hexokinase assay, and was further identified by analysis with glucose oxidase and glucose dehydrogenase. Routine quantitative determinations were made on deproteinized samples by means of the hexokinase assay (commercial zwischenferment was used as a source of both G6P dehydrogenase and hexokinase). A specific chemical test (20) was used to detect small amounts of fructose which might be present, since this compound was also active in the hexokinase assay. About 10-15 per cent of the hexose present was fructose.

The various enzymatic assays employed in these studies will be presented in greater detail elsewhere (33).

---

[1] Ammonium sulfate fractionation has demonstrated the separation of these two enzymes.

Experimental Results

*Nature of the Chemotactic Substance*

Shaffer developed a test system by which sensitive cells undergoing aggregation could be used to demonstrate the presence of the chemotactic agent (26). From the nature of its action, it appeared possible that acrasin might be a known hormone. Using the Shaffer test, crude hormone preparations were examined. The urine of a pregnant woman was extremely active, and consequently a number of steroids were tested for activity. Active compounds were: estradiol; estrone sulfate; 17(a)-methyl $\Delta^5$ androstene-3($\beta$), 17($\beta$)-diol; $\Delta^4$ androstene-3,17-dione; progesterone; testosterone; and 17-ethynyl-19-nor-testosterone. Inactive compounds included: cholesterol; androstane-diol, 3,17; $\Delta^1$ androstene-3,17-dione; androstane-3,17-dione; diethyl stilbestrol; 17$\alpha$ methyl-19-nor-testosterone; pregnanediol; $\Delta^1$ hydrocortisone; hydrocortisone; and somatotropin. With the exception of estrone sulfate, the exact potency of most of the active steroids could not be tested due to their extreme insolubility. However, three different preparations of estrone sulfate were found to be active at $<0.05$ $\gamma$/ml. Such activity is clearly in the range of physiological significance. Stability studies on acrasin are not incompatible with its being a steroid (27, 28, 29). However, the response of the amoebae to estrone sulfate is not as striking as to "pregnant urine" or to crude acrasin. Estrone sulfate may therefore be closely related to, but not identical with, acrasin (34).

*The Acquisition of a More Aerobic Metabolism*

Perhaps the most general and biochemically relevant description to be made of the slime molds is that their acquisition of a multicellular life is accompanied by the transition to a more aerobic metabolism. Gregg has shown that oxygen consumption increases strikingly from the amoeba stage until the stage of multicellular differentiation just prior to stalk formation; it then decreases, and is finally undetectable in the mature fruit (14). We have extended Gregg's data by showing that in fact the amoebae can grow as well "anaerobically" as aerobically, and have confirmed his observation that multicellular differentiation cannot proceed anaerobically. By growing the myxamoebae on dead bacteria, it could be shown that after 48 hours they had multiplied as rapidly under essentially anaerobic conditions (in an evacuated dessicator) as they had in the presence of air (Table 1). After removing a section of each plate to count the amoebae, incubation was continued, with the results shown. The food supply was exhausted on both plates, and under aerobic conditions multicellular differentiation proceeded normally. It is quite possible that under completely anaerobic conditions the multicellular stages of differentiation would have been stopped altogether.

TABLE 1

Oxygen Requirement of Differentiation

|  | Condition of plate | | |
| --- | --- | --- | --- |
| Hours | 48 | 72 | 96 |
| Aerobic | $2.5 \times 10^7$ amoebae | all slugs ca. 800 | fruits |
| Limited $O_2$ | $2.0 \times 10^7$ amoebae | no slugs some aggreg. | 10% slugs 90% aggreg. |

A number of inhibitors have been found which, at a given concentration, do not interfere with amoebae growth but do not inhibit pseudoplasmodium formation to various degrees. These are listed in Table 2.[2] In the case of malonate and iodacetate, no pseudoplasmodia had formed by the time that fruiting had occurred on the control plate. The inhibition by malonate and dicoumerol is of particular interest, since it suggests a new importance for Krebs cycle enzymes and oxidative phosphorylation at this stage of differentiation. The inhibition by 8-azaguanine will be discussed below. Hirschberg has studied the effect of 2,4-dinitrophenol on aggregation of the myxamoebae, and has shown that at $10^{-4}$ $M$ aggregation is reversibly inhibited (i.e., transferring the cells to agar without inhibitor allows aggregation to proceed). Furthermore, concomitantly with progressive weakening of the power of aggregation, the concentration of inorganic phosphate in the myxamoebae is increased; it reaches about 250 per cent of the initial value by the time the cells can no longer aggregate (19).

TABLE 2

Inhibitors of Multicellular Differentiation

| Inhibitor | Concentration ($M$) |
| --- | --- |
| 8-azaguanine | $2 \times 10^{-3}$ |
| iodoacetate | $2 \times 10^{-3}$ |
| dicoumerol | $2 \times 10^{-3}$ |
| malonate | $5 \times 10^{-3}$ |

[2] Had these inhibitors been in direct contact with the amoebae instead of being in the agar, their effective concentration would probably have been a good deal lower.

All of these data demonstrate that oxidative metabolism is required for multicellular differentiation in slime molds, and suggest that oxidative phosphorylation is also occurring and may be necessary for morphogenesis.

## Enzymatic Changes During Differentiation

As an initial approach to understanding the over-all metabolism of slime molds during differentiation, a number of well-known, easily assayed enzymes were investigated. During the formation of the pseudoplasmodium, prior to migration, enzymes involved in aerobic metabolism, as well as an enzyme required for cellulose synthesis (uridine diphosphoglucose synthetase)[3] change from undetectable or very low activity in the amoebae to strikingly enhanced activity in the pseudoplasmodium during a period of a few hours at 25°C. Preliminary experiments consisted of preparing cell-free extracts from cells harvested at three stages during differentiation: the amoebae; the pseudoplasmodium (or "slug"); and the final fruit. The experiments shown in Figs. 1 and 2 were done with comparable amounts of protein from the three kinds of extracts.

It can be seen that amoebae extracts are very low in isocitric dehydrogenase and UDPG synthetase, whereas the pseudoplasmodium extracts are extremely active. In the absence of added TPN, mutase, and G6P dehydrogenase, G1P should accumulate from UDPG and PP. Such an accumulation occurred, and is illustrated by the data in Table 3.

### TABLE 3
#### Uridine Diphosphoglucose Synthetase

I. UDPG + PP → G1P + UTP.
II. Heat inactivate.
III. Measure G1P.

| Enzyme source | Omitted | $\mu$M G1P accumulated |
|---|---|---|
| amoebae | — | 0.004 |
| slug | — | 0.266 |
| slug + amoebae | — | 0.216 |
| slug | UDPG | 0.011 |
| slug | PP | 0.016 |

In the case of isocitric dehydrogenase and UDPG synthetase, as well as other enzymes to be described, no indication was found of (1) any inhibition of one extract by another, or of (2) any difference in relative activities of cell-free extracts as compared to homogenates (33). The increased enzymatic activities appear to be genuine, in the sense that they are not due

---

[3] Uridine diphosphoglucose has recently been implicated as the activated form of glucose for the synthesis of cellulose (13).

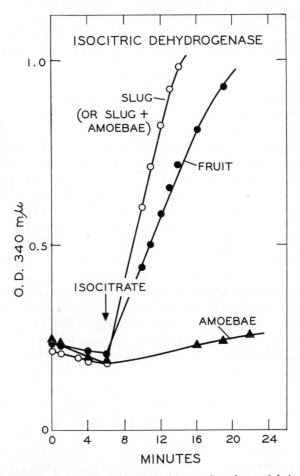

Fig. 1. Isocitric dehydrogenase activity measured in amoebae, slug, and fruit preparations. The amoebae were grown on dead bacteria; differentiation occurred at 20°C. Specific activities ($\mu M$ TPNH per 10 min. per mg. protein) of the three preparations were: amoebae, 0.4; slug, 21.8; fruit, 15.1.

to the removal of a simple inhibitor or to differences in enzyme solubilities at various stages of differentiation. It has also been demonstrated that the changes in enzymatic activities are the same, whether the amoebae are grown on live or dead bacteria prior to washing them and allowing differentiation to proceed on washed agar (33).

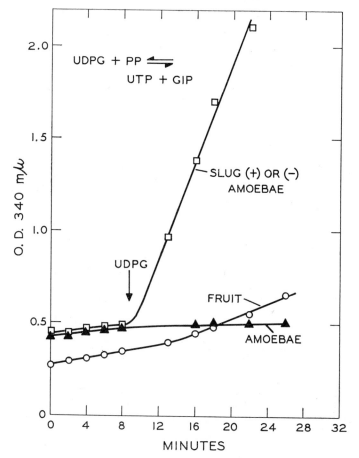

Fig. 2. UDPG synthetase activity. Conditions as for Fig. 1, except that the amoebae were grown on live bacteria. The specific activities were: amoebae, 0.00; slug, 33.2; fruit, 5.0.

*Enzyme Localization*

One essential characteristic of multicellular differentiation is localization of specific enzymatic activities. It is known that the chemotactic substance, acrasin, is produced predominantly by the presumptive stalk cells (4). Histochemically, studies have shown that these cells are also rich in an alkaline phosphatase activity, compared to the presumptive spore cells (5).

With a micropipette, about 50 migrating slugs were divided, their anterior and posterior portions pooled, and cell-free enzyme preparations were made by freezing and thawing. Isocitric dehydrogenase and UDPG synthetase activities were examined in such extracts, with the results shown in Table 4. The relative specific activities are based on protein. Apparently the extract prepared from the posterior portion of the slugs is the most active with respect to these two enzymes. This was an unexpected result, and may represent an exception to Bonner's generalization that the anterior cells are enzymatically the most active.

TABLE 4

LOCALIZATION OF ENZYMES

| Enzyme | Relative specific activity | |
|---|---|---|
| | Anteriors | Posteriors |
| Isocitric dehydrogenase | 1.40 | 1.81 |
| UDPG synthetase | 1.21 | 1.73 |

*Changes in Enzyme Activities at Progressive Stages of Differentiation*

In order to attempt a correlation of enzymatic activities with the morphological changes (e.g., stalk formation) occurring during the life cycle, enzyme extracts were prepared at more frequent intervals from the time prior to aggregation until the final fruit had formed. Since the changes are extremely rapid at room temperature, most of these experiments were carried out at 20° or 13°C, in order to obtain more values at critical times. Fig. 3 shows the changes in relative specific activities of a number of enzymes, calculated on the basis of protein. The highest value obtained was given a figure of 100, and the others were calculated relative to this value. The enzymes examined in this experiment were UDPG synthetase, isocitric dehydrogenase (TPN- and DPN-specific), malic dehydrogenase (DPN), glucose-6-phosphate dehydrogenase (TPN), 6-phosphogluconic dehydrogenase (TPN), and glutamic dehydrogenase (DPN).

It can be seen that those enzymes which are involved in the Krebs cycle, the hexose monophosphate shunt, and UDPG synthesis increase strikingly in activity. The greatest increases in activity occur during the formation of the pseudoplasmodium. All of the enzymes examined thus far are listed in Table 5; they are grouped according to their maximum change in relative specific activity. It should be emphasized that these enzyme systems are being examined in crude extracts, and that the assays employed may not be definitive. Although it is clear that striking changes in enzymatic activities do occur, it is conceivable that some of these effects are secondary.

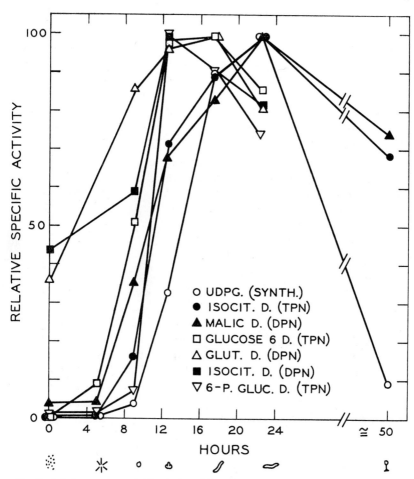

Fig. 3. Relative enzyme activities during differentiation on washed agar at 20°C; activities are based on protein. The amoebae were grown on dead bacteria. Zero time is arbitrary, and prior to aggregation. The enzymes examined and the stages of development at which samples are taken are indicated.

TABLE 5

Enzyme Summary

| 1.5 to 3-fold increase | >10-fold increase |
|---|---|
| 1. Lactic acid dehydrogenase | 1. Glucose 6-PO$_4$ dehydrogenase |
| 2. Glutamic dehydrogenase | 2. 6-phosphogluconic dehydrogenase |
| 3. Isocitric dehydrogenase (DPN) | 3. UDPG synthetase |
| 4. Malic dehydrogenase (TPN) | 4. Isocitric dehydrogenase (TPN) |
| 5. Glyceraldehyde phosphate dehydrogenase | 5. Malic dehydrogenase (DPN) |
| 6. Glucose-6-phosphatase | |

*Glucose Accumulation*

During an attempt to assay for the presence of hexokinase in the slime mold extracts, an endogenous substrate was found there which required ATP and G6P dehydrogenase in order for TPN to be reduced. This endogenous material, which accumulated during differentiation, was identified as glucose. Fig. 4 gives the relative changes in glucose con-

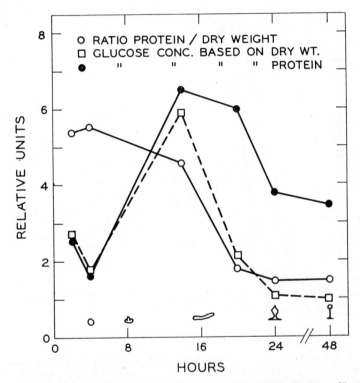

Fig. 4. Glucose concentration at various stages of differentiation in "starved" cells at 20°C. The amoebae were initially grown on live bacteria. The concentration is expressed on the basis of dry weight and of protein; the ratio of protein to dry weight during the experiment is also plotted. See text.

centration based on protein and on dry weight. Starved cells were used in this experiment, in the hope of observing maximum changes in endogenous glucose concentration during stalk formation. Starvation was encouraged by keeping the amoebae anaerobic for 48 hours at 13°C, so as to prevent multicellular differentiation. The experiment was initiated by placing the plates aerobically at 20°C; within a few hours formation of the pseudoplasmodium began. It can be seen that the "metabolism of differentiation" in the absence of growth results in a net accumulation of glucose during formation of the slug. The glucose concentration then drops while the cellulose stalk is being formed. Polysaccharide synthesis is occurring prior to the stalk formation, in so far as the "slime sheath" is being deposited (24) and the presumptive spore cells in the posterior two-thirds of the slug are accumulating non-starch polysaccharides (5). Thus the accumulation of carbohydrate (glucose) represents the amount in excess of that necessary for husk formation. Only during more active polysaccharide synthesis (stalk formation) is there a net decrease in glucose concentration. Since protein is the main endogenous material used, and since the dry weight remains fairly constant, the concentration of glucose (as well as enzymatic activities) is best calculated on a dry weight basis. Based on protein, the concentration is abnormally high, particularly in the later stages of differentiation when the most striking decrease in protein occurs (15). The ratio of protein to dry weight is also shown in Fig. 4, and is a direct expression of the fact that the extractable protein drops relative to the dry weight during the fruiting process.

Fig. 5 shows the course of glucose accumulation and depletion in unstarved cells on the basis of dry weight. The relative activities of two enzymes are also indicated. As might be expected, the glucose level does not fall as far during stalk formation in unstarved cells as it did in the case of the starved cells. Although it is not plotted in this experiment, glutamic dehydrogenase again (see Fig. 3) increased two-fold; its decrease was very similar to that shown for isocitric dehydrogenase. Since the first experiment of this type was based on protein instead of dry weight (Fig. 3), the value for isocitric dehydrogenase in the final fruit was higher than in the present experiment, which based the relative activities on dry weight. Gregg's data on oxygen consumption at various stages of development have also been plotted (14). As with the enzymatic activities, the highest value (at preculmination) was taken as 100, and the others are given relative values.

Table 6 presents data for the absolute amounts of glucose and total reducing sugar present in slime molds. On a dry weight basis, the accumulation of free hexose is similar to that found in plants; this value is almost one-hundred-fold higher than that for mammalian tissue. In relating Gregg's values for total reducing sugar to the amount of glucose accumulated in unstarved slime molds, it appears as though most of the reducing sugar in the pseudoplasmodium can be accounted for as free

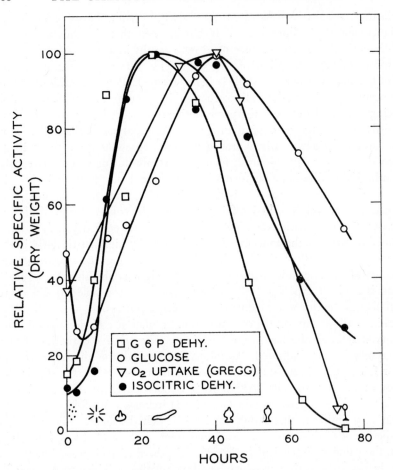

Fig. 5. Glucose concentration and enzyme activities in unstarved cells at 13°C; activities are based on dry weight. The amoebae were grown on live bacteria. Gregg's data on $O_2$ uptake during differentiation are plotted, as well as enzyme activities for glucose-6-phosphate dehydrogenase and isocitric dehydrogenase.

hexose. The amount of glucose found in the cells is the same whether or not they are washed once during their harvest; this suggests the presence of some permeability barrier in the cell membrane.

Fig. 6 presents more data on enzymatic changes during multicellular development. UDPG synthetase shows the usual pattern, followed also by the Krebs cycle and shunt enzymes. G6P-phosphatase was examined since it might be involved in the accumulation of glucose. Glyceraldehyde

TABLE 6

GLUCOSE AND TOTAL REDUCING SUGAR VALUES

| Myxamoebae | | mg. glucose / mg. dry wt. | mg. total[2] reducing sugar / mg. dry wt. |
|---|---|---|---|
| Starved | amoebae | 0.016 | — |
| | pseudopl. | 0.058[1] | — |
| | fruit | 0.010 | — |
| Not starved | amoebae | 0.025 | 0.085 |
| | pseudopl. | 0.088[1] | 0.103 |
| | fruit | 0.046 | 0.125 |

[1] About 10-15% of the hexose accumulation is fructose.
[2] After Gregg (15).

phosphate dehydrogenase and lactic acid dehydrogenase increase relatively little in activity during pseudoplasmodium formation, although their decrease in activity is similar to that of UDPG synthetase. Apparently, G6P-phosphatase is unique among the enzymes tested thus far, in that no striking decrease in activity was observed during formation of the final fruit. Similarly, very little inhibition was observed for phosphatase action on 1,6-fructose diphosphate or $\beta$-glycerol phosphate. The ratio of protein to dry weight decreased by about 50 per cent during this experiment.

## Discussion

In a sense, the slime molds appear to be facultative aerobes, even though the multicellular stage of their life cycle cannot proceed anaerobically. There is a precedent, to be found in other facultative microbes, for the enzymatic changes observed in the slime molds. The investigations of Ephrussi and Slonimski (11), Hirsch (18), Chantrenne (8), and Englesberg and Levy (10) are all relevant to this discussion. The last-mentioned work is perhaps in best analogy to the situation in the slime molds. Englesberg and Levy examined a number of enzymes in extracts of *Pasteurella pestis* which had been (a) grown anaerobically, (b) grown aerobically, or (c) grown anaerobically, and then adapted to aerobic metabolism in the absence of growth. Glucose was present in all cases. Under conditions (b) and (c), they found enzymes involved in aerobic metabolism to be more active than in cells grown anaerobically. The most striking example was isocitric dehydrogenase (60-fold). Typical glycolytic enzymes, on the other hand, were as active or more active in anaerobically-grown cells.

In the slime molds, in contrast to *Pastuerella pestis*, it is the organism

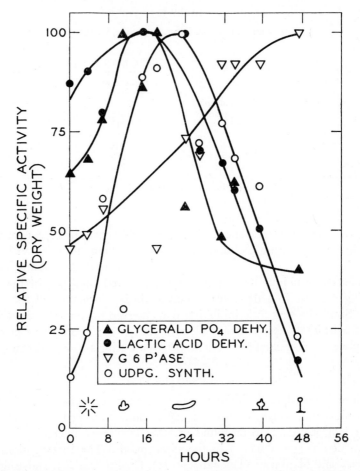

Fig. 6. Relative enzyme activities during morphogenesis in unstarved cells at 13°C; activities are expressed on a dry weight basis. Amoebae were grown on live bacteria. The enzymes examined are glyceraldehyde phosphate dehydrogenase, lactic acid dehydrogenase, glucose-6-phosphate phosphatase, and UDPG synthetase.

rather than the investigator which induces anaerobic or aerobic metabolism. The mechanism for this control is as yet completely unknown. It may be of general interest in understanding differentiating systems, since other cases are known in which more complex development is accompanied by a requirement for more aerobic metabolism (e.g., 6). In the slime molds, activity of the Krebs cycle may afford the energy necessary for migration and, in particular, polysaccharide synthesis.

One of the most immediate problems is to understand the mechanism underlying the observed enzymatic changes. Are we dealing with activation of already formed protein, or with substrate-induced synthesis of enzymes? The latter alternative is the most appealing one at the moment, although definitive experiments are not yet available. Preliminary results indicate that isocitrate can induce the formation of isocitric dehydrogenase, as well as some other enzymes to a lesser degree. If enzyme induction operates normally during differentiation, substrates inducing Krebs cycle enzymes might well arise via the amino acids liberated from protein breakdown.

Since there is a rise and fall of enzymatic activity in the absence of cell multiplication, it is possible that we are dealing with a dynamic state of enzyme synthesis and degradation. In other words, there may be a very high protein turnover, and a measured enzymatic activity at any time would represent the resultant of rapid synthetic and degradative reactions.[4] A number of factors could conceivably contribute to the general decrease in enzymatic activities observed. In view of the data on the inhibition of enzyme induction by glucose (9, 12, 22), it is tempting to attribute the decline in activity of various enzymes to an accumulation of glucose, which in turn could inhibit the continued induction of an enzyme by its substrate. In this connection, it is possibly significant that the initial drop in glucose concentration coincides with the time of the most rapid rise in enzymatic activities (Figs. 4 and 5). Although the data are not particularly good, it is clear that G6P-phosphatase does not decrease in activity during the later stages of differentiation (Fig. 6). This situation may be related to the hypothesis of Neidhardt and Magasanick, that hydrolytic types of enzymes are not inhibited by glucose, since their activity does not result in the accumulation of products similar to those formed by the metabolism of glucose (22).

The decline in enzymatic activities may also be a reflection of the general utilization and eventual depletion of protein material as a source of energy and building blocks for cellulose synthesis. Yet another factor to consider is the gradual exhaustion of substrates responsible for the presumed enzyme inductions. The observed pattern of decline in certain enzymatic activities may play a role in allowing the terminal steps of cellulose synthesis to proceed more rapidly. For example, the loss of G6P dehydrogenase would facilitate G1P accumulation (Fig. 5). One might expect to find many control mechanisms operating, interacting, and inter-

[4] The inhibition by 8-azaguanine referred to earlier might reflect a requirement for RNA synthesis during this period of active protein synthesis.

dependent in order to account for the stability and reproducibility of morphogenesis.

One of the most fascinating problems concerns the role of the chemotactic agent (acrasin) responsible for the initial aggregation of the myxamoebae into the multicellular unit. This is, after all, one of the more primary causes of differentiation, in that the amoebae are delegated at this time to the anterior or posterior region of the future pseudoplasmodium. Moreover, acrasin continues to be produced in the anterior portion of the pseudoplasmodium. Starvation of the amoebae apparently initiates a change in metabolism such that the cells produce the chemotactic hormone, become sensitized to it, or both. One appealing site of action for acrasin would be in the stimulation of endogenous protein breakdown during starvation. Activation of proteolytic enzymes could lead to the accumulation of critical substrates, and thus initiate a series of sequential enzyme inductions.

Evidence was presented earlier that acrasin may be a steroid-like hormone. Apparently there are a number of factors involved in the aggregation phenomenon, which is a very complicated process (31). The recent contribution of Talalay and Williams-Ashman (32) on the role of steroid hormones as coenzymes for transhydrogenase may be of particular interest in this connection. It remains to be seen whether acrasin may exert its effect on metabolism in some similar manner.

### Summary

When the slime molds begin their multicellular life in the absence of external nutrition, aerobic metabolism is acquired. Closely correlated with an increase in oxygen uptake are striking increases in the activities of a number of enzymes involved in aerobic metabolism. An enzyme active in cellulose synthesis, UDPG synthetase, also increases markedly during pseudoplasmodium formation. These enzymatic changes occur *prior* to migration and cellulose synthesis. This observation is not incompatible with the hypothesis that the enhanced enzymatic activities are, in fact, necessary in order for these stages of differentiation to proceed. More specifically, these enzymes are at maximum activity prior to the period of major protein utilization and carbohydrate synthesis which occurs during morphogenesis. It is reasonable to assume that their activity may in part cause these gross metabolic changes.

### Acknowledgment

We are grateful to Dr. E. R. Stadtman for stimulating criticism concerning this work. Drs. M. and R. Sussman introduced one of us (B.E.W.) to the slime molds, and we are indebted to them for their continued helpfulness. We would like to thank Dr. W. Albers for generous aid and advice concerning microtechniques. Our thanks are also due to Drs. R. W. Bates, D. M. Bergenstal, and M. Horning for samples of various hormone and steroid preparations.

## REFERENCES

1. Bonner, J. T., *J. Exptl. Zool.*, **106,** 1 (1947).
2. ———, *Amer. J. Bot.*, **31,** 175 (1944).
3. ———, *J. Exptl. Zool.*, **106,** 1 (1947).
4. ———, *J. Exptl. Zool.*, **110,** 259 (1949).
5. ———, Chiquoine, A. D., and Kolderie, M. Q., *J. Exptl. Zool.*, **130,** 133 (1955).
6. Brachet, J., *Arch. biol.* (Liége), **45,** 611 (1934).
7. Bradley, S. G., and Sussman, M., *Arch. Biochem. and Biophys.*, **39,** 462 (1952).
8. Chantrenne, H., and Courtais, C., *Biochim. et Biophys. Acta*, **14,** 397 (1954).
9. Cohn, M., in *Enzymes: Units of Biological Structure and Function* (Gaebler, O. H., ed.), p. 41, Academic Press, New York (1956).
10. Englesberg, E., and Levy, J. B., *J. Bact.*, **69,** 418 (1955).
11. Ephrussi, B., and Slonimski, P., *Biochim. et Biophys. Acta,* **6,** 256 (1950).
12. Gale, E. F., *Bacteriol. Revs.*, **7,** 139 (1943).
13. Glaser, L., *Biochim. et Biophys. Acta*, **25,** 436 (1957).
14. Gregg, J. H., *J. Exptl. Zool.*, **114,** 173 (1950).
15. ———, and Bronsweig, R. D., *J. Cellular Comp. Physiol.*, **48,** 293 (1956).
16. ———, and ———, *J. Cellular Comp. Physiol.*, **47,** 483 (1956).
17. ———, Hackney, A. L., and Krivanek, J. O., *Biol. Bull.*, **107,** 226 (1954).
18. Hirsch, H. M., *Biochim. et Biophys. Acta*, **9,** 674 (1952).
19. Hirschberg, E., and Rusch, H. P., *J. Cellular Comp. Physiol.*, **37,** 323 (1951).
20. *Methods in Enzymology,* (Colowick, S. P. and N. O. Kaplan, eds.), III, p. 76, Academic Press, New York (1957).
21. Munch-Petersen, A., Kalckar, H., Cutolo, E., and Smith, E. E. B., *Nature,* **172,** 1036 (1953).
22. Neidhardt, F. C., Magasanik, B., *Nature,* **178,** 801 (1956).
23. Raper, K. B., *J. Elisha Mitchell Scient. Soc.*, **56,** 241 (1940).
24. ———, and Fennell, D. I., *Bull. Torrey Botan. Club,* **79,** 25 (1952).
25. Runyan, E. H., *Collecting Net,* **17,** 88 (1942).
26. Shaffer, B. M., *Nature,* **171,** 957 (1953).
27. ———, *Science,* **123,** 1172 (1956).
28. ———, *J. Exptl. Biol.*, **33,** 645 (1956).
29. Sussman, M., et al., *Science,* **123,** 1171 (1956).
30. ———, pers. commun.
31. Sussman, R. R., Sussman, M., and Lee Fu, F., *Bacteriol. Proc.*, p. 32 (1958).
32. Talalay, P., and Williams-Ashman, H. G., *Proc. Natl. Acad. Sci. U. S.*, **44,** 15 (1958).
33. Wright, B. E., and Anderson, M. L., *Biochim. et Biophys. Acta,* (in press).
34. ———, *Bacteriol. Proc.*, p. 115 (1958).

## DISCUSSION

DR. HADORN: I would like to ask Dr. Sussman what the nature of the mutants are with which he deals. Are these true so-called Mendelian mutants?

DR. SUSSMAN: Dr. Racquel Sussman has worked extremely hard trying to demonstrate a Mendelian segregation of characteristics but we have been very much hampered by the fact that we can demonstrate no sexual process taking place here.

Dr. Hadorn: In your cultures which have mixed cell populations, that is, different mutants, do you obtain true chimaeric aggregates? And if you do obtain such aggregates, is the size a reflection of the genetic properties of the $I$ or of the $R$ cells.

Dr. Sussman: Yes, we do obtain chimaeric aggregates and we have looked at this situation by using a particular mutant called *Agg-91*. These cells are much smaller and give rise to small, non-viable spores in synergistic fruits. You will remember that there are two circumstances: if the wild type were present in low numbers a mutant had to contribute both cells and materials, but if the wild type were present in high numbers, then all that the mutant had to contribute was materials. Now in the system to which I am referring when the wild type was present in low numbers, the largest contingent in the aggregates and in the fruits, were the *Agg-91* cells but at the point where the concentration of wild type was high enough in the mixture so that they could aggregate alone, you found no mutant cells in the fruits. In this system the aggregateless mutant actually performs two functions with low concentrations of wild type and only the one with the high concentration of wild type.

Dr. Markert: Is it true that $I$ cells appear to be unresponsive to acrasin?

Dr. Sussman: I only know that in the great majority of cases when the center of the aggregate does not form exactly at the position of the $I$ cell, the $I$ cell is eventually left outside the aggregate.

Dr. Markert: I wonder if one of the differences between the $I$ and $R$ cells might not be the inability of the $I$ cells to destroy acrasin. If this were so, the $I$ cell would fail to detect any differences in the concentration of acrasin, because it would be continuously surrounded by abundant acrasin produced by itself.

Dr. Sussman: Yes, of course the simplest hypothesis is that the $I$ cell is the superior producer of acrasin and it is the initiator of the streaming process and that this is due to the lack of an enzyme which inactivates acrasin or some other similar phenomenon. I would hesitate at the moment to attribute the ability of the $I$ cell to the production of a co-factor merely on the basis of economy.

Dr. Brown: What did you mean by a milliequivalent of crude powder?

Dr. Sussman: Well, a solution of crude powder could be diluted serially and have a minimal activity at about 0.8 mg./ml. Now we could chromatograph 10 milligrams of this crude extract on paper to get out a component, let's say the A component, which was then dissolved in 1 cc. This amount of a component was then called a 10 milliequivalent ml. solution.

Dr. Brown: I had the impression that you actually knew the molecular (or equivalent) weight of the compound when you used this expression.

Dr. Sussman: No, of course not, we were merely referring it to the total activity of the original crude extract in order to standardize our procedures and see that we are not losing any large amount of activity in any one chemical step.

Dr. Pappenheimer: One thing was not clear to me. In these "starved" amoebae there seems to be a lot of protein synthesis going on because the organisms are synthesizing new enzymes. I was wondering where the nitrogen and the amino acids come from to serve this function.

Dr. Sussman: The word "starved" is an unfortunate one in this particular instance. In the normal sequence of events, even on a growth plate as Dr. Raper has shown, these cells stop growing, enter morphogenesis, and undergo their transformations. They are really not starved. I might mention, in addition, that we have obtained a defined medium for these organisms. A curious bacterial protein which is a specific growth factor is required and in addition one must add either proteose-peptone or gelatin or casein which can be used as a sole source of carbon and energy. Now these organisms have a tremendous endogenous supply of nutrient materials. You can blow them for 40 hours in a Warburg flask without substrate and they still continue to respire on their endogenous materials. Secondly, on washed agar they can go ahead and aggregate, form a fruit, the fruit will fall over and the spores hit the agar. The spores regerminate and produce amoebae which will then reaggregate and form another fruit. So they are not starved under any sense of the word.

Dr. Cohn: By growth do you mean division?

Dr. Sussman: No, by growth I mean growth. That is, an increase in cell mass. These cells are washed completely free of bacteria and put on agar which has been washed between 6 to 15 times depending upon the experiment and so there is nothing exogenous upon which they can increase their mass and all materials must come from within the cell.

Dr. Spiegelman: I am afraid I am confused now. Dr. Wright pointed out that these cells were breaking down their protein at a very high rate and therefore there was plenty of supply to form new enzymes from the catabolism taking place within the cells.

Dr. Cohn: Is this an experiment or a hypothesis?

Dr. Wright: No, these are the results as obtained by Gregg. The amount of endogenous protein goes down strikingly and the amount of total reducing sugar increases a great deal. On a dry weight basis, there is 10 times as much protein broken down as carbohydrate synthesized. Ammonia is also formed in very large quantities, so they are breaking down their protein into amino acids and these are the sources of energy and the building blocks for new specific proteins and for carbohydrate synthesis. There is a very high endogenous metabolism which goes on during differentiation.

Dr. Wilde: I am a little confused about growth and differentiation in this system. As I understand it, the $I$ and the $R$ cells coexist during the logarithmic phase of growth. When are the $I$ and $R$ cells formed in the clone?

Dr. Sussman: Since they are present at all stages they must arise during the growth of the population. I think the confusion which exists here is one which exists in much of embryology. Namely, cellular differentiation does not refer to the differentiated state of the cell but refers to the process which gave rise to it. Aggregation is simply an expression or a scoring, if you will, of the two cell types. The problem of cellular differentiation in this organism, and certainly in any morphogenetic system, is before the expression of the differentiation.

Dr. Wilde: But then one can no longer maintain that, in the case of the cellular slime molds, growth is distinct and different from differentiation.

Dr. Sussman: I didn't say that. Growth is certainly separate from morphogenesis.

Dr. Wilde: Yes, that's so but certainly not from the origin of different cell types.

Dr. Weiss: It is very easy to disaggregate the slug and I wonder what happens when you do this. Do they revert to a former state?

Dr. Wright: Yes, you can take an early slug and disaggregate it. And the cells will sit down in the liquid and the enzyme activities will not increase. In fact, the enzyme activities will drop a little. However, if you add isocitrate then the level of isocitric dehydrogenase will increase as it would have if the cells had remained together in the slug. So by disrupting the cells they do dedifferentiate enzymatically at a particular stage when, had they remained together, the enzyme activity would have been rising rapidly.

Dr. Niu: I would like to ask you to comment about the alkaline phosphatase and the formation of the fruiting body.

Dr. Wright: The alkaline phosphatase activity was shown by Dr. Bonner to be localized in the pseudoplasmodium and it is predominantly in the presumptive stalk cells, that is, the anterior one-third of the slug. The enzyme was localized with the substrate beta-glycerol phosphate, and the presumptive spore cells did not have much activity. We studied a number of phosphatases, including alkaline phosphatase, in the whole homogenate, and found that they continued to rise in activity during slug and fruit formation and were not inhibited like the other enzymes. One might say then that if Dr. Magasanik's ideas on the glucose inhibition of the formation of enzymes is right, the phosphatases do not fall into the class of "synthetic type" enzymes which should be inhibited by glucose. Phosphatase activity would probably be necessary for glucose formation, for if glucose is formed from the breakdown of amino acid which then moved through the reversal of the Krebs cycle and up through the glycolytic pathway, then glucose-6-phosphate must be dephosphorylated by the phosphatase in order for glucose to accumulate.

Dr. Grant: I am just wondering if you considered the possible role of carbon dioxide in the differentiation of the plasmodium. I say this because of the work of Cantino on the effect of $CO_2$ on differentiation in the water mold. In the slime mold, there might be extreme differences in the local concentration of $CO_2$ in the plasmodium which would not be evident in the amoeboid state.

Dr. WRIGHT: No, I haven't considered that possibility. I only know that high concentrations of $CO_2$ will inhibit differentiation. Isn't that true, Dr. Sussman?

Dr. SUSSMAN: Well, Cohn did this, but he used fantastic amounts of carbon dioxide. It was completely unphysiological.

Dr. WRIGHT: $CO_2$ is produced along with ammonia during differentiation. However, I don't know any specific effect of it.

Dr. SWIFT: What is the evidence for or against the $I$ cells being polyploid and possibly arising by nondisjunction?

Dr. SUSSMAN: There is really no evidence for this but there are several possibilities. Wilson has suggested from his cytological studies that zygote formation occurs during aggregation. In fact, he infers that it is a mandatory accompaniment of this process. From his pictures I suspect that what he has seen are $I$ cells. He later amended his views and said that although it had no relationship to aggregation, that zygotes did occur. Now, these $I$ cells are frequently much larger than these other cells, their nuclei are larger, and you often find them engulfing other myxamoebae. They might be zygotes by engulfment but I think I would prefer to call this cannibalism rather than mating. We hope to determine if the cells are polyploid by doing DNA determinations, but this has not been done as yet.

Dr. GALL: It would be very interesting to try a low concentration of colchicine to see if an increase in tetraploid cells would lead to an increase in number of centers.

# EXCHANGES BETWEEN CELLS

M. ABERCROMBIE

*Department of Anatomy and Embryology, University College,*
*Gower Street, London, W. C. 1.*

MORPHOGENESIS IN both embryo and adult obviously involves a great deal of interaction between the parts of the developing system; though it is true, since many of the causal connections are trigger actions initiating complex responses, that limited aspects of the process for limited periods in limited regions are "self-differentiating." In analysing this pervasive nexus of interactions we can choose our units, the parts between which our hypotheses assert that interactions occur, from any of several levels of the hierarchy of separable parts making up the developing system. Classical transplantation embryology has for technical reasons worked with somewhat ill-defined multicellular regions of the embryo, and has been able to state many useful hypotheses, such as those concerned with induction, in terms of interactions between these units. There is a tendency now to jump a stage in the hierarchy, straight to the molecular level. It is not, however, possible to take full account of the organization of developing systems without pressing much further forward an intermediate level of analysis, the cellular. Metazoan organization is such that the cell is an indispensable unit for an important part of the analysis of growth (by mitosis), of spatial shifts of material (by cell locomotion), and of differentiation (by intracellular specialization). Much that the chemical level of analysis must ultimately deal with remains to be discovered by those who make cells their units of investigation.

Though the subject has been decisively put on the agenda by Holtfreter, Weiss, and others, what goes on between cells during morphogenesis is largely unknown. For that matter, what goes on between cells during the normal working of an adult is little known too. But interactions of two general kinds between adult cells are usually distinguished. One is by cell-to-cell contact, exemplified in the working of the nervous system. The other is by diffusion of substances (more or less assisted by flow of body fluids) through the tissues independently of mutual cell contacts, as exemplified in the activities of the endocrine glands. These two general kinds of intercellular communication, highly specialized in their best known manifestations in the adult, are represented in the embryo in more primitive forms. No doubt one should not try to enforce a rigid segregation of the two types in morphogenesis, if only because "contact" will surely prove to be a relative term to the electron microscopist. But the distinction may be a useful approximation. As in the adult, very close cellular contact is likely to provide a refined and discriminating influence, and it might too ensure a more immediate "feed-back" of responding cell on stimulating cell; diffusion through the tissues is likely to be engaged in more generalized, less finely adjusted, responses.

Diffusion through the tissues naturally means the transfer of some substance from stimulating to reacting cells, though not necessarily to the inside of the reacting cells; it may act directly on their surfaces, for instance by blocking groups exposed there—the experiments of Spiegel (20) provide a model. An influence through cell contacts may on the contrary involve no material transfer, but merely a transfer of pattern. It was Needham (14) who first pointed out the possibility of such an effect, and the suggestion has been much developed by Weiss (26 and later). But contact action does not imply a mere pattern transfer. The necessity for intimate contact and the degree of localization of effect would for instance be almost the same if a stimulatory substance, but an unstable one, was produced by the stimulating cell.

## LOCOMOTION OF FIBROBLASTS

The control of cell locomotion provides a clear instance of cellular interaction through mutual contact, which can best be considered by contrasting the behavior of normal and malignant fibroblasts in tissue culture. When cultured on a plane surface, normal fibroblasts show a kind of mutual behavior which we have called contact inhibition (5). Contact inhibition means that when a fibroblast travelling in a given direction makes contact with another fibroblast in its path its continued movement in that direction is obstructed. We are not here concerned with the results of this behavior; though it is worth mentioning that as they display themselves in a large population these results are probably not trivial. In fact, in conjunction with Weiss's contact guidance (25) perhaps all the directional aspects of the control of movement in a fibroblast colony in vitro can be explained. The radial nature of cell movement from an explant, the circular form taken up by fibroblast cultures, or the longitudinal migration on a cylindrical surface (Weiss) can for instance all be derived from contact inhibition alone. There has always been an inexplicable impulse amongst tissue culturists to explain fibroblast behavior by first trying to make something else, a diffusion gradient or interfacial layer, take up by its internal interactions the required distribution and orientation, which it then imposes on the cells. But the cells by their direct interaction through contact inhibition can do all that is required.

The point for discussion here is not so much the results of contact inhibition as what exactly the interaction is. The following account of what happens when one fibroblast makes contact with another is chiefly based on an analysis of films of chick heart fibroblasts made with an interference microscope (2). Any moving fibroblast has its leading edge (if this is free from contact with other fibroblasts) in the form of a ruffled membrane extended over the substrate surface. There may also be other small undulating membranes elsewhere around the cell. When the leading membrane touches an obstructing fibroblast it appears to fuse with it. Although the junction is somewhat changeable and may partly break down and re-form with considerable rapidity, where the junction exists no border can be seen with a light microscope between the two cells. These are the adhesions described by Lewis (12) and investigated by Kredel (11) by means of microdissection, and found to require some force to break. This is also shown when adhering cells move away from each other. The point of adhesion then distorts the cells, and when it breaks the cells give the impression of snapping apart (4). The ruffling of the membrane that makes the contact is suppressed by the adhesion and with it the apparent pinocytosis that can often be seen to accompany the undulations. If the ruffled membrane was expanding in area at the time of the contact its expansion ceases. The general impression is in fact as if the membrane has become frozen into immobility. Sometimes it gives the impression of coming under tension. Soon after the adhesion has formed the inhibition of movement occurs. The fibroblast ceases to move in its original direction if it has run into a stationary cell, though this cessation may be preceded by a slight acceleration of movement as the contact is formed, perhaps connected with the apparent tension in the contacting membrane: such acceleration was detected statistically (4), but has not since been investigated. Exactly how the freezing of the membrane is related to the checking of the movement of the cell cannot profitably be discussed without more knowledge of the precise nature of the locomotory mechanism than is at present available.

When a contact inhibition has occurred in the way described there is a common but not invariable consequence. A new membrane, or more often a growth of an already existing minor membrane, starts up and, if it can, leads the cell off in a new direction. We interpret the redirection of the cell's movement by a new membrane in terms of competition between membranes, a hypothesis already developed for behavior during contact guidance by Weiss and Garber (28). The new membrane cannot always redirect the cell, because it may immediately run into another obstructing cell and be itself inhibited. If a cell is completely surrounded by stationary cells it too must be stationary, apart from minor oscillations.

What is it possible to say about the transaction between the two cells which brings contact inhibition about? In the first place, it depends entirely on contact: there is no effect until the cells touch. It is not, however, the non-specific result of mere adhesion, since it must be remembered that the cell is adhering fairly firmly to its substrate. Adhesion is a prerequisite, but it is adhesion to a particular surface, that of another fibroblast, that counts. When such an adhesion occurs, it is hardly conceivable that no transfer of material between the cells happens. But it is also hardly conceivable that such a transfer of material is responsible for the events described. The contact is not only between two similar cells but may occur between similar parts of similar cells: two leading ruffled membranes can inhibit each other. There seems little possibility of one cell gaining something from the other that it has not already got. Nevertheless, there seems to be transmission of the effect within the inhibited cell. Events in other parts of the cell are altered by the contact of the membrane, in that the original movement is stopped, a new leading membrane is established, and locomotion in a new direction occurs. In perhaps an elementary way the events of contact inhibition seem to be a model of an entirely superficial interaction between cells, such as Weiss has so vigorously advocated in opposition to ideas of material transfer between cells.

This simple reaction between fibroblasts is not apparently highly specific. It can occur between fibroblasts from different organs and between fibroblasts from chick and mouse (6). Whether it is completely non-specific is not known, since only a limited range of the possible variants have been tested, and since most of them have not been tested in a way which would reveal small quantitative differences.

Even with contacts between similar cells the inhibition is not invariable. In a proportion of cases cells, which visible observation strongly suggests meet on a plane surface, nevertheless disregard each other, one cell moving over the other unobstructed by the contact. The impression that film shots of such an event give is that a cell momentarily loses some essential part of the contact inhibitory mechanism, perhaps its relevant surface properties, its "fibroblast surface markers," or perhaps a link in the chain connecting specific adhesion with inhibition of movement; and this may last an appreciable time, since one cell may miss inhibition twice or more in successive encounters. It has proved possible to induce loss of the ability to undergo contact inhibition by applying very high concentrations of embryo juice to the cultures (Curtis, unpub.) and it is conceivable that this effect may represent a saturation of the surface markers, though there are several other possible hypotheses. Fibroblasts also seem to lose their surface markers when they enter mitosis since they then apparently cease to inhibit fibroblasts which run into them (Karthauser, unpub.).

## Locomotion of Sarcoma Cells

A more striking case of loss of the properties responsible for contact inhibition occurs in malignant fibroblasts. So far as we have tested sarcomata (mouse sarcomata 37, 180, MC1M, Klein M2C), they at least contain many cells that are not inhibited by contact with normal fibroblasts (6) (Abercrombie and Karthauser, unpub.). Nor are fibroblasts inhibited by contact with S37 or S180 cells, though their reaction to fibrosarcoma cells is more complex. If the sarcoma is a solid form, that is to say is not one specially adapted to ascitic life, its cells in culture stick well to glass and to the surfaces of normal fibroblasts, on both of which they move. Malignancy therefore does not involve a completely non-specific loss of adhesiveness, though it may be true that some degree of diminution of general adhesiveness occurs.

A surface property which sarcoma cells have not lost is that responsible for contact guidance. In our experience sarcoma cells are highly susceptible to the classical situation of Weiss (25) where the cells are caused to move in or on orientated fibrin. They show Weiss's "two-center" phenomenon, when explants are placed near together, very strongly. It is perhaps stretching the topic of exchanges between cells unduly to include such behavior. But because of their deficient contact inhibition and strong propensity to be contact-guided, sarcoma cells become capable of a form of cell interaction that is forbidden to normal fibroblasts: they can be guided by orientated fibroblasts. This behavior proved at first very confusing. We found that sarcoma cells in vitro showed a remarkably direct movement towards a nearby fibroblast explant, travelling over the surface of the fibroblast outgrowth towards it, a direct movement performed by single sarcoma cells isolated from their kind, so that we at first interpreted it as chemotaxis. We soon found however that their movement was equally direct *away* from the fibroblast explant if that is where they were made to start from. It is one of the consequences of mutual contact inhibition that fibroblasts move radially from an explant, in doing so tending to take up a polarized form; and it seems that this provides an effective contact guidance for sarcoma cells. Such a movement along cell surfaces is closely comparable to the movement of the amoeboid tips of nerve fibres along Schwann cells. In both cases the cells are so strongly polarized that they rarely turn back. It is a different transaction between cells from that involved in contact inhibition, but is again a surface relationship probably involving competition between pseudopodial membranes. In theory at least it may have high specificity (Weiss's *selective* contact guidance), but its main importance is probably between cell and extracellular structure rather than between cell and cell.

## Aggregation

When cells in vitro are not moving actively their surface properties concerned with adhesion may come into play, with results somewhat different from the contact inhibition between fibroblasts I have discussed. It seems that their tendency to adhere to each other and to extend these mutual adhesions comes to exceed their ability to adhere to and move on the substrate. The result is aggregation. It has been perhaps most beautifully shown by Twitty (21) in amphibian melanoblasts. It does not occur in ordinary cultures of fibroblasts, which adhere well to the substrate, though it is probably expressed in the fusion of two fragments of chick heart (18) and of course in the reaggregation of disaggregated chick cells (13). Fibrosarcoma cells (mouse MC1M tumor) also become clumped late in the life of a culture, and the explant tends to become spherical. Numerous other examples of such aggregation can be found. The precise relation of the adhesion involved to that of contact inhibition is unknown, but the same argument can be applied, though with less assurance because there is less information: these are surface interactions between similar cells which seem hardly likely to be mediated by exchange of substances.

## Cell Locomotion: Conclusions

Directed locomotion, and the groupings produced by adhesions and manifested in vitro by aggregation, are, it is now generally agreed, of great importance in establishing the pattern of cells in vertebrate morphogenesis, embryonic or adult. Contact reactions between cell and cell and between cell and extracellular structures are probably therefore a major source of the precise control of cell pattern. Contact inhibition between fibroblasts, for example, can provide a simple self-regulatory mechanism by which any space deprived of fibroblasts becomes exactly filled. It would be unwise, though, to dismiss diffused substances as an orientating influence on cell locomotion, particularly in view of the work of Twitty and Niu (22). The Weiss and Garber (28) hypothesis of pseudopodial competition in fibroblast movement makes it easy to see how chemotaxis could work in a fibroblast, by differential inhibition or encouragement of the pseudopodia at different parts of the cell surface exposed to different concentrations of some substance. We have indeed found a possible case of this, though an artificial one, in that chick fibroblasts are slowed in their movement towards cells of mouse sarcoma MC1M when several hundred micra away, and when very near turn aside. But the effectiveness of known contact reactions is potentially so considerable that we are justified at present in assuming that *directional* control of movement by diffusing substances, when it occurs, is either supplementary or a special case. Where substances diffusing independently of cell contacts have their main role in controlling cell movement is, I have recently argued (1), in influencing the general level of activation of the amoeboid mechanism. This influence on locomotion is closely connected with the control of mitotic rate.

## Mitosis

In cultures of chick fibroblasts, where contact effects on locomotion are so striking, we have been unable to detect any relation between amount of cell contact and number of mitoses. The overwhelming influence on mitosis in such cultures seems to be one of inhibition in the neighborhood of the explant, an inhibition which is particularly extensive in the direction of another nearby explant. This suggests that a diffusing substance is concerned, and may well represent competition between the cells for some nutrient in the medium. These results reflect in miniature the wider picture of the control of mitosis. Contact effects, though they may be reasonably suspected in the embryonic central nervous system (see 10), are not yet demonstrated in vertebrates, as far as I am aware. Mitotic stimulation or inhibition by diffusing substances is on the other hand well known in adult morphogenesis, as in the thyroid-pituitary system (some instances are discussed in ref. 1). If such diffuse interactions exist in the embryo too, it is not surprising. Since the precise control of pattern is already provided by cell locomotion, exact location of mitosis is unnecessary.

## Induction

I have considered in detail a contact reaction between cells, contact inhibition, that most probably involves no transfer of material; and I have mentioned the existence of interchanges controlling mitosis that operate by diffusion of material through a distance. The other possible mechanism, by which material is transferred through cell contacts, is less easy to exemplify with confidence. But morphogenetic interaction by this means has long been accepted as the mechanism of many cases of induction in embryonic development. Neural induction, which normally involves close contact between the presumptive chorda–mesoderm and the overlying presumptive neural plate, is a suitable case for brief discussion. Thanks largely to the substantial measure of agreement reached between the schools of Toivonen, Yamada, and Nieuwkoop, there is now good foundation for supposing that in urodeles neural induction is due to two influences from the chorda–mesoderm, present in progressively different proportions along the cranio-caudal axis. One influence corresponds to the neural evocation long ago distinguished by Waddington (see 23) and produces complex archencephalic structures when acting alone. The other influence is the work of the mesodermal inductor, modifying the neural evocation in a more caudal sense. The nature of the neural evocation, now called the archencephalic inductor, is a notorious problem of many years standing, because the natural agent has so many successful imitators, while it appears itself to be highly labile (16). There is, however, so far no conclusive evidence that the mesodermal inductor is not a unique substance. It appears to be a protein, though it is found associated with pentose nucleic acid (29). It can be obtained in solution in saline, and Niu's results (15) suggest that it can induce mesoderm in such solution. There is therefore good evidence that a transferable substance *may* act in normal induction, at least by the mesodermal inductor; though it would still be legitimate to think that the inductor exerts its effects while remaining localized at the interface between chorda–mesoderm and ectoderm. Evidence as to whether the inductor leaves the chorda–mesoderm in vivo and whether it enters the ectoderm would be helpful. The former has been sought by interposing porous membranes, the latter chiefly by tracer experiments.

Induction has not yet been shown to traverse a gap separating chorda–mesoderm from ectoderm (7); but unfortunately while a positive result would be relatively unequivocal, a negative is not. Instability of the substance (at least in the interior of the embryo) might for instance make it impossible to transfer sufficient material across a gap; some such restriction on diffusion would of course improve the precision of the normal induction. The tracer experiments have not been much more helpful. Experimental juxtaposition of radioactively marked inducers and unmarked competent tissue (3, 8, 17, 19, 24) has merely indicated that there is no large-scale transfer of macromolecules. But sensitivity to drugs and hormones suggests that a very small transfer might suffice for an induction. If on the contrary cells are reacting in exceedingly close contact, then even though the induction is a mere transfer of pattern, their surfaces are unlikely to be so stable that no passage of the surface macromolecular material of one into the cytoplasm of the other cell occurs. Some transfer of substance seems then to be compatible with hypotheses asserting that the inducing stimulus is either a pattern-transfer or a material-transfer, and further tracer studies of increased sensitivity seem unlikely to be decisive.

The problem of the exact transaction between inducing chorda–mesoderm and responding ectoderm is not therefore satisfactorily resolved, but the example must serve as the nearest we can get to interaction by transfer of material through contact, which at least is the general opinion of the process. Contact, with its accompanying precision of action, is certainly involved as it is in some other inductions. But yet others may be different. Grobstein (9) has shown how one inductor (though an "abnormal" one) can act across an extracellular space to produce metanephric tubules. Induction in such systems, perhaps even more so in the induction of the considerable mass of mesonephric tissue by the nearby Wolffian duct, may reasonably be by the shot-gun method of extracellular diffusion. The possibility should not be completely disregarded that such diffusion might be from induced to inductor, i.e., might be the consumption of an inhibitor.

### Regulation

It would be idle to suppose that clear-cut induction, as exemplified by the neural reaction, is a very important form of cellular exchange during the process of determination in vertebrate embryos. The main problem, because it is commoner and so obscure in its mechanism, is the exchange that goes on during the development of a region capable of regulation. Although nothing precise can be said about the nature of this exchange, no opportunity should be lost of calling attention to the problem.

When a region capable of regulation, say a limb-field, is halved, or two are fused, and a single normally proportioned organ develops from the half or double region, then substantially all the cells do not develop as they would normally have done. Such a thorough reorganization of cellular presumptive fates precludes the possibility that there persists, undisturbed by the operation, some controlling group of cells. But something must in a sense persist. In some essential respect the new system must have the same spatial organization as the original, since the pattern of the limb that emerges is the same. It might be thought that if the original pattern were a system of inherent axial gradients, half the region would possess the same pattern, and the problem of continuity of organization is solved. It is not solved, however, because the problem is not merely to keep the same crucial organization after the operation, but to reallocate the cells within the organization so that their individual presumptive fates are recast to fit the reduced or increased size of the organ-forming region. Persistent axial gradients are merely a secondary influence: they provide for persistent polarities, though they are not even the only way of doing that. It seems to me that there is only one form of relationship that will provide both continuity of pattern and reorganization of cellular presumptive fate. This is the inside-outside relation based on the difference between cells within the regulatory region (halved, double or single in origin) and cells outside it, which in turn means a difference between the peripheral and central parts of the region. Weiss (27) has considered how such a simple inside-outside relation in a mass of tissue can be translated into complex structure. The point in the present context is that it cannot be so translated, regardless (within limits) of differences in the size of the mass of tissue produced by operation, without extensive mutual influences between the cells. Directly or indirectly it seems that all cells must influence each other, so that each is informed of its relative position in the system. There is at present no indication of what form such interactions could take, and it may be that there is a role here for a so far unsuspected kind of conduction through cell contacts.

## Conclusion

This discussion has been concerned with only a few of the large variety of cellular exchanges that take part in morphogenesis. I have concentrated on influences which trigger a complex response. The real task of the stimulus of complex cellular behavior, like that of a releaser in animal behavior, is simply to be potent and unequivocal: it is not required to contribute energy or material or organization to the ensuing response, though it may. The response goes its own way, once the stimulus threshold is past. Its complexity may be very great, as in neural induction, where the archencephalic inductor sets in train a further extensive set of interactions between ectodermal cells; or it may be relatively small, as in contact inhibition. It is of course precisely its repertoire of complex responses to simple stimuli that makes the cell an indispensable unit of interaction in the analysis of development.

## REFERENCES

1. Abercrombie, M., *Symposia Soc. Exptl. Biol.*, **9**, 235-254 (1957).
2. ———, and Ambrose, E. J., *Exptl. Cell Research* (in press).
3. ———, and Causey, G., *Nature*, **166**, 229 (1950).
4. ———, and Heaysman, J. E. M., *Exptl. Cell Research*, **5**, 111-131 (1953).
5. ———, and ———, *Exptl. Cell Research*, **6**, 293-306 (1954).
6. ———, ———, and Karthauser, H. M., *Exptl. Cell Research*, **13**, 276-292 (1957).
7. Brachet, J., *Rev. Suisse Zool.*, **57**, 57-70 (1950).
8. Ficq, A., *J. Embryol. Exptl. Morphol.*, **2**, 204-215 (1954).
9. Grobstein, C., *Exptl. Cell Research*, **13**, 575-587 (1957).
10. Hamburger, V., in *Cellular Mechanisms in Differentiation and Growth* (D. Rudnick, ed.), p. 191-212, Princeton Univ. Press, Princeton (1956).
11. Kredel, F., *Bull. Johns Hopkins Hosp.*, **40**, 216-227 (1927).
12. Lewis, W. H., *Anat. Rec.*, **23**, 387-392 (1922).
13. Moscona, A., and Moscona, H., *J. Anat.*, **86**, 278-286 (1952).
14. Needham, J., *Biochemistry and Morphogenesis*, Cambridge, at the University Press (1956).
15. Niu, M. C., in *Cellular Mechanisms in Differentiation and Growth* (D. Rudnick, ed.), p. 155-172, Princeton Univ. Press, Princeton (1956).
16. Okazaki, R., *Exptl. Cell Research*, **9**, 579-582 (1955).
17. Pantelouris, E. M., and Mulherkar, L., *J. Embryol. Exptl. Morphol.*, **5**, 51-59 (1957).
18. Sigurdson, B., *Proc. Soc. Exptl. Biol. Med.*, **50**, 62-66 (1942).
19. Sirlin, J. L., Brahma, S. K., and Waddington, C. H., *J. Embryol. Exptl. Morphol.*, **4**, 248-253 (1956).
20. Spiegel, M., *Biol. Bull.*, **107**, 149-155 (1954).
21. Twitty, V. C., *J. Exptl. Zool.*, **100**, 141-178 (1945).
22. ———, and Niu, M. C., *J. Exptl. Zool.*, **125**, 541-574 (1954).
23. Waddington, C. H., *Principles of Embryology*, Allen & Unwin, London (1956).
24. ———, and Mulherkar, L., *Proc. Zool. Soc. Bengal*, Mookerjee Memor. Vol., 141-147 (1957).
25. Weiss, P., *J. Exptl. Zool.*, **68**, 392-448 (1934).
26. ———, *Yale J. Biol. and Med.*, **19**, 235-278 (1947).
27. ———, *J. Embryol. Exptl. Morphol.*, **1**, 181-211 (1953).
28. ———, and Garber, B., *Proc. Natl. Acad. Sci. U. S.*, **38**, 264-280 (1952).
29. Yamada, T., and Takata, K., *Embryologia*, **3**, 66-79 (1956).

# SOME PROBLEMS OF PROTEIN FORMATION IN THE SCLERA AND CORNEA OF THE CHICK EMBRYO

## Heinz Herrmann

FOR SOME TIME work has been in progress at our laboratory on protein formation in embryonic tissues, for the purpose of obtaining information about the properties of the protein-forming system in developing cells. This system may be regarded as the final target of nuclear and cytoplasmic factors which control the formation of proteins as chemically measurable indices of differentiation.

While continuing the exploration of protein formation in developing muscle tissue and in explanted chick embryos we have recently extended our analyses to the sclera and cornea of the developing chick. The consideration which led us to this line of work was the following. According to Laguesse (6) and Redslob (10) the corneal area of the chick eye contains only a few scattered mesenchymal cells up to the sixth day of development. Following this stage, mesodermal cells begin to migrate from the scleral portion of the head mesenchyme into the corneal area, and the divergent trends in the development of the corneal and scleral cells become apparent. The cells which remain in the sclera develop a cartilaginous matrix, whereas in the corneal area a non-cartilaginous material is formed which contributes to the transparency of the corneal stroma.

Characterizing the developments of these two tissues in chemical terms, both are found to produce collagen as the main protein component of their intercellular substance; but in the case of the sclera chondroitin sulfuric acid is formed as the intercellular mucopolysaccharide (MPS) constituent, while in the cornea keratosulfate appears as the additional MPS component (9). This must mean that before the sixth day of development the synthetic apparatus has not become stabilized in all scleral cells and that MPS synthesis in the cells which migrate into the corneal area is directed in the new environment toward synthesis of keratosulfate. Alternatively, one would have to assume that the scleral mesenchyme represents a heterogeneous population in respect to MPS synthetic capacities, and that the cells which have the synthetic apparatus for corneal MPS production migrate selectively to the corneal area. According to Weiss and Amprino (12), the cartilage-forming capacity of the chick sclera becomes determined on the fourth day of development, and therefore synthetic functions must begin to become stabilized at least in a large number of scleral cells at that stage of development.

In approaching the question of the stabilization of synthetic capacities in scleral and corneal cells we compared first the formation of collagen, the protein component of the intercellular matrix of the sclera and the cornea. Thus far we have found that the analytically determined accumulation of collagen in the corneal area and in the sclera follow a similar course. However, a striking difference in the protein-forming system of the two cell types is indicated by tests in vitro which show that incorporation of tracer amino acids into the collagen fraction of the cornea is dependent to a large extent upon the presence of the ectodermal component of this tissue, while in the sclera the incorporation of the tracer into scleral collagen seems to be independent of other adjacent tissues (4). The experiments which led us to this conclusion are discussed in the following section of this paper.

## Experimental Results

The first set of graphs shows the changes in the DNA content of the sclera and the cornea respectively (Fig. 1). This parameter is regarded as an index of the magnitude of the cell population in these tissues during the examined period of development (7). It can be seen that the increase in the DNA content of the sclera is a discontinuous process. During the periods from the 8th to 12th day, 16th to 18th day, and 22nd to 28th day, no statistically significant change in the DNA of this tissue could be observed, while the increases in the intervening periods were found to be statistically significant. This discontinuity may be indicative of alternating periods of proliferative activity and more stationary states, or it could suggest periodic phases of cell degeneration balancing continuous cell proliferation. In the cornea, periods of statistically significant increases of DNA were observed to continue up to the 16th day of development. After this period no statistically significant increase of the DNA content of the cornea could be established with 15, 14, and 9 determinations for the 16th to 18th day and the 22nd to 28th day periods, respectively. To what extent migration of scleral cells into the cornea or proliferation of the corneal cells themselves contribute to the rise in the corneal DNA content has not been determined.

The accumulation of collagen fractions in the cornea and in the sclera

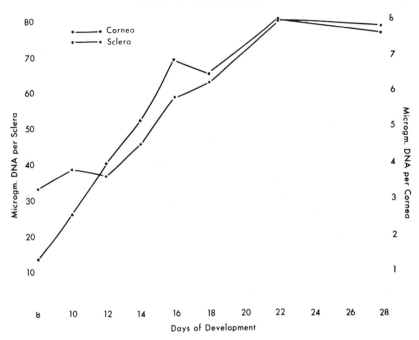

Fig. 1. Content of DNA in the sclera and cornea of the developing chick embryo.

is given in Fig. 2. Determined as hydroxyproline in hydrolysates of hot trichloroacetic acid extracts of the respective tissues, the curves for the collagen content of the sclera and of the cornea follow two similar, sigmoid curves with the inflection point at about the 18th day. In order to obtain comparable values for the protein-forming capacity for the scleral and corneal cells respectively, the ratios of total hydroxyproline/total DNA (HP/DNA) for the two tissues were calculated. The plots for these values give the collagen content in each tissue per unit of DNA (Fig. 3) or, for the equivalent, constant number of cells (7). These HP/DNA curves are also similar in respect to the overall acceleration as well as the total increase in the HP/DNA ratio. However, certain divergent aspects should be pointed out. It can be noticed that on the 8th day the HP/DNA ratio is markedly higher in the cornea (0.40) than in the sclera (0.076). Attempts to find determinable amounts of HP in the corneal area on the 6th day have failed so far, a result indicating that only negligible amounts of collagen can be present in the corneal area before arrival of mesenchymal cells. Another difference between the two tissues can be seen in compar-

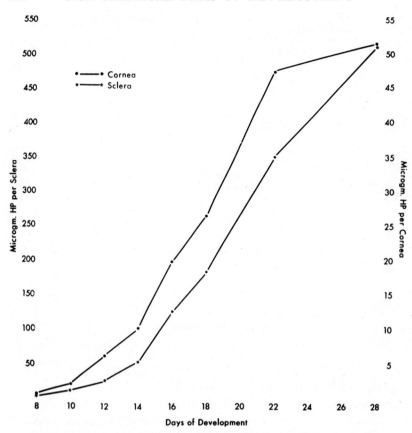

Fig. 2. Content of collagen, determined as hydroxyproline (HP), in the sclera and cornea of the developing chick embryo.

ing the curves for the HP/DNA ratios and for the ratio of non-collagenous proteins NCP/DNA (Fig. 3). In case of the cornea the HP/DNA values increase faster than the NCP/DNA values for the period examined except during the 14th-16th day interval, during which both parameters increase at the same rapid rate. In the case of the sclera, the two curves run parallel until the 22nd day, after which the HP/DNA curve maintains a steeper slope.

The above values for collagen accumulation were compared to the rates of tracer incorporation into the collagen fractions of the sclera and cornea. Using glycine-1-$C^{14}$ the uptake of the tracer was observed for 1- to 4-hour incubation periods during which the tissues were suspended in two milli-

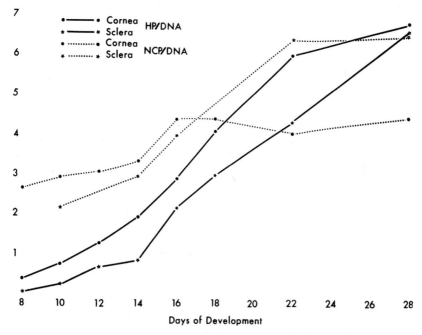

Fig. 3. Ratios of hydroxyproline (HP) per DNA in the sclera and cornea of the developing chick embryo.

liters of a nutrient solution to which was added 1 μc. of the labelled amino acid. To make these data comparable with the figures for accumulation, the tracer incorporation was expressed as cts./min./DNA of the respective tissues. Comparing the incorporation rates for the 9th-10th day period and for the 17th day of development (Table 1), a marked increase in the tracer uptake (cts./min./DNA) was observed which corresponds to the enhanced rate of increase of the HP/DNA ratio comparing the 8th-10th day period with the 16th-18th day period. The data show also that the values for incorporation of glycine-1-$C^{14}$ into the collagen fraction of the two cell types are very similar on the 17th day of development. On the 10th day, however, the cornea shows a markedly higher incorporation rate under our experimental conditions than the sclera. This means, apparently, that on the ninth day of development collagen is formed more rapidly in the cornea than in the sclera, while the rates of collagen formation are of the same order of magnitude during later phases of development. These values were obtained for the collagen fraction formed in the stroma of the two tissues during incubation of the mesodermal portion together with the adjacent

TABLE 1

Uptake of Glycine-1-C$^{14}$ into Collagen Fraction of Sclera and Cornea at Different Stages of Development

| Age in days | cts./min./µg. DNA | |
|---|---|---|
| | Sclera | Cornea |
| 9 | 5, 5, 9, 14 (1) | 41, 41 (4) |
| 9-10 | 14, 14, 18 (1) | 66, 67, 70 (4) |
| 17 | 193, 171, 216 175, 218, 218 (1) | 231, 224 237, 157 (2) |

The figures in parentheses indicate number of scleras or corneas, respectively, for each determination.

ectodermal components; with the retinal pigment epithelium or, if possible, with the entire retina in case of the sclera and with the epithelium in the case of the cornea. The ectodermal cells were removed after incubation, and the collagen fraction was prepared from the remaining mesodermal layer. Incubation of the tissues after removal of the ectodermal components did not lead to significant differences in the case of glycine incorporation into the collagen fraction from the scleral mesoderm. In the case of the cornea, however, a sharp drop in the incorporation was observed when the mesoderm was incubated in the absence of the corneal ectoderm (Table 2). The drop in incorporation was found to be proportional to the decrease in the stroma surface covered by the epithelium; metabolic inhibitors like dinitrophenol or iodoacetate markedly diminished the incorporation rate, and the presence of glucose in a salt solution enhanced the tracer uptake.

## Discussion

In introducing these experiments, it was suggested that differences in the chemical make-up of the mesodermal cells of the cornea and of the sclera may be due to a selective migration of one cell strain from a mixed cell population in the sclera, or that the cells derived from a relatively homogeneous population change their characteristics in their new environment. Although the first possibility cannot be ruled out, for the time being we are pursuing the alternative of a change in the scleral cells in the course of their interaction with the new corneal environment. This would mean that scleral cells, originally autonomous in respect to protein formation, become dependent for a rapid accumulation of the collagen fraction upon some factor contributed by the corneal epithelium. Since the corneal

TABLE 2

Uptake of Glycine-1-$C^{14}$ into a Collagen Fraction of the Corneal Stroma of the Developing Chick (4)

The data are given as means of cts./sec./mg. N for the number of determinations indicated by the figures in parentheses.

| 10-day chick embryo | | Stroma with epithelium | Stroma without epithelium | |
|---|---|---|---|---|
| Complete medium | | | | |
| Incubation time | 1 hr. | 453 (8) | 77 (6) | 17%[1] |
| Incubation time | 2 hr. | 1403 (2) | | |
| Incubation time | 4 hr. | 3182 (4) | 574 (2) | 18% |
| Incubation time | 6 hr. | 5680 (2) | | |
| **3- to 8-day hatched chick** | | | | |
| Complete medium | | | | |
| Incubation time | 1 hr. | 106 (16) | 6 (12) | 6% |
| Incubation time | 4 hr. | 386 (4) | 15 (4) | 4% |
| Incubation time | 1 hr. | | | |
| ½ epithelium removed | | 56 (4) | | |
| ¾ epithelium removed | | 36 (4) | | |
| Complete medium | | | | |
| Incubation time | 1 hr. | | | |
| Dinitrophenol | | | | |
| Final concentration | $3 \times 10^{-5}$ M. | 84 (3) | | |
| | $1 \times 10^{-4}$ M. | 56 (3) | | |
| | $1 \times 10^{-3}$ M. | 25 (3) | | |
| Iodoacetate | | | | |
| Final concentration | $1 \times 10^{-3}$ M. | 3 (2) | | |
| P. C. solution | | | | |
| Incubation time | 1 hr. | 75 (10) | | |
| Incubation time | 4 hr. | 258 (2) | | |
| P. C. solution with glucose | | | | |
| Incubation time | 1 hr. | 99 (10) | | |
| Incubation time | 4 hr. | 422 (2) | | |

[1] The percentage values refer to the fraction of tracer incorporated after removal of the epithelium.

stroma cells, especially in later stages of development, are at quite some distance from the epithelial layer, these factors would have to be assumed to be quite readily diffusible or readily propagated in some other fashion. At least in several tissues composed of an ectodermal and mesodermal component, the ectodermal layer seems to be the site of rather active oxidative metabolic reactions while the mesodermal component is poor in such oxidative enzymatic activity (2, 3). A similar situation seems to exist in the cornea at least of the fully developed beef eye (5). It is therefore suggested, as a tentative interpretation of our results, that in migrating to the corneal area the mesodermal cells become adapted to the utilization of the oxidative energy produced in the epithelium for their own synthetic activities, in particular for the formation of proteins. This adaptation leads apparently to a more rapid initial production of collagen in the cornea, while the increase in the rate of protein formation in the sclera is somewhat slower and must be based on an energy supply provided by the scleral mesodermal cells themselves.

In this connection it must be remembered that other eye primordia (optic cup, lens) are held responsible for the induction of the cornea. To what extent these primordia participate in the control of the development of the corneal stroma remains to be investigated.

The interpretation of the different incorporation rates in the sclera and cornea has yet to be established conclusively by observations in vivo of the incorporation rates in the collagen fraction of the sclera and cornea, and by tests in vitro with a wider range of concentrations of the tracer, and a more detailed study of the time course of the incorporation in order to obtain information about the significance of the amino acid pool and possible collagen precursors in the final incorporation data. Also, it has to be demonstrated that the purified collagen contains the same amount of hydroxyproline at different stages of development. Finally, more evidence as to the metabolic pathways involved in this activation will be necessary to substantiate our hypothesis. These questions are being studied in our laboratory at the present time.

Whatever its mechanism may be, the fact of an ectoderm-dependent formation of protein in the corneal mesoderm has led us to approach experimentally the two following problems. First, it should be of interest to find out whether interaction of the corneal mesodermal cells with the corneal epithelium changes the rate of collagen production not only quantitatively but also gives rise to a qualitative change in the MPS-producing enzymes such as to form in the corneal stroma keratosulfate instead of chondroitin sulfuric acid, as in the sclera. Secondly, recent work by Saunders (10) and Zwilling (13) has shown that in the case of the limb bud the ectodermal component is essential for the normal development of the mesoderm of this primordium. In preliminary experiments with limb buds we have not yet been able to demonstrate an ectoderm-dependent incorporation of glycine into the protein moiety or into the nucleic-acid-containing trichloroacetic acid extract of the limb bud mesoderm. However, tests with a wider range of conditions in attempting to demonstrate such a dependence in the developing limb bud are in progress. It is hoped that observations in more numerous instances of a dependence of protein formation in one embryonic cell type upon the metabolic activities or other processes in its cellular environment should prove to be useful in the investigation of mechanisms of cellular interactions and of their roles in embryonic development.

## Materials and Methods Used

Fertilized eggs from Hi-Line hens were obtained from a commercial source and incubated at the laboratory at 38°C. For preparation of the sclera the eyes were dissected at the desired stages of development and the looser connective tissue was removed. The eyes were opened from the posterior pole with four meridional incisions to the rim of the cornea. The lens, vitreous body, cornea, and iris were discarded, to leave the four separate scleral sections with the retina attached. The corneas were excised from the eye remaining in situ, particular care being taken not to injure the epithelium. For incubation with the tracer, the scleras were used with or without attached whole retina or retinal pigment epithelium, and the corneas were also incubated with or without epithelium. The incubation was carried out in a Dubnoff shaker with nutrient solutions of 1 or 2 ml. volume and to which was added 1 $\mu$c. of tracer glycine-1-C$^{14}$ (Nuclear, Chicago). If present during the incubation period, the epithelial components were removed after termination of the incubation. Preparation of the collagen fraction and counting procedure used in the tracer experiments were carried out as described previously (4). Collagen contents were determined using the methods of Fitch, Harkness and Harkness (1), and Martin and Axelrod (8). The quantities of collagen obtained by the two procedures used in the tracer experiments and in the content determinations respectively were in close agreement, and the ratios of hydroxyproline/total protein nitrogen were only slightly lower in case of the trichloroacetic acid extraction.

### Acknowledgments

The author is indebted to Mr. R. D. Salerno and Mrs. M. Cooper for their competent technical assistance in carrying out the experiments reported in this paper.

The original work was supported by Grant B-549 from the National Institute of Neurological Diseases and Blindness of the National Institutes of Health, United States Public Health Service.

## REFERENCES

1. Fitch, S. M., Harkness, M. L. M., and Harkness, R. D., *Nature,* **176,** 163-164 (1955).
2. Flexner, L. B., and Stiehler, R. D., *J. Biol. Chem.,* **126,** 619-626 (1938).
3. Friedenwald, J. S., Buschke, W., and Michel, H. O., *Arch. Ophthalmol.,* **29,** 535-570 (1943).

4. Herrmann, H., *Proc. Natl. Acad. Sci. U. S.,* **43,** 1007-1011 (1957).
5. ———, and Hickman, F. H., *Bull. Johns Hopkins Hosp.,* **82,** 225-250 (1948).
6. Laguesse, E., *Arch. anat. micros.,* **22,** 216-265 (1926).
7. Leslie, I., and Davidson, J. N., *Biochim. et Biophys. Acta,* **7,** 413-428 (1951).
8. Martin, C. J., and Axelrod, A. E., *Proc. Soc. Exptl. Biol. Med.,* **83,** 461-462 (1953).
9. Meyer, K., Linker, A., Davidson, E. A., and Weissmann, B., *J. Biol. Chem.,* **205,** 611-616 (1953).
10. Redslob, E., *Arch. anat. histol. embryol., Strasbourg,* **19,** 135-229 (1935).
11. Saunders, J. W., Cairns, J. M., and Gasseling, M. T., *J. Morphol.,* **101,** 57-87 (1957).
12. Weiss, P., and Amprino, R., *Growth,* **4,** 245-258 (1940).
13. Zwilling, E., *J. Exptl. Zool.,* **128,** 423-441 (1955).

# CELL MIGRATION AND MORPHOGENETIC MOVEMENTS

Robert L. DeHaan

*Department of Embryology,
Carnegie Institution of Washington,
Baltimore, Maryland*

## Introduction

For purposes of experimental investigation or theoretical consideration the subject at hand can conveniently be divided into several more or less separable subtopics, each referable to specific questions. A primary question is that of cell adhesion: what is the manner in which the cells of tissues, organs, or embryonic germ layers are bound together? Since cells seem to adhere to one another as if glued by some intercellular "cement" substance, what is the mechanism whereby this adhesiveness is reversible, allowing cells to move about on one another? Of equal importance is the problem of the motive force behind cell movements. It is clear that some cells, such as those derived from neural crest, move as individuals, by inherent ameboid activity. In these circumstances, what are the triggering forces that instigate and maintain such movements? In other situations, such as those obtaining during the processes of gastrulation, neurulation, or of gut formation, entire layers may be involved in mass cellular migrations. In these cases, what are the relative roles (1) of changes in size and shape of cells of the layer, (2) of ameboid activity of individual cells of the layer, (3) of localized bursts of mitotic activity, or (4) of contraction of intercellular connections? Questions may be posed also regarding the guidance of cell movements. How are cells directed along their routes of migration to their ultimate destination? What are the means (surely diverse) whereby intricately timed movements of masses of cells are coordinated into the reproducible patterns that we see in every developing organism?

I propose to examine some of these questions utilizing information derived both from the literature relevant to these topics, and to a lesser extent from the results of some experiments of my own. No attempt will be made to review the literature fully. References, for the most part, will be limited to investigations which emphasize the *chemical basis* of cell migration and morphogenetic movements, as suggested by the title of this symposium.

## Movements of Individual Ameboid Cells

### Chromatophore Migration

Perhaps the simplest manifestation of cell migration is to be seen in the movements of single ameboid cells. From the paired ridges of the vertebrate neural folds the neural crest develops and gives rise to an array of at first undifferentiated ameboid cells which migrate ventrolaterally and develop into cells of such diverse nature as chromatophores, chondrogenic and ganglionic cells, Schwann cells, ectomesenchyme, etc. (71). The motions of these ubiquitous cells have been studied in detail, especially the movements of those that later develop into chromatophores (for reviews of the pigmentation literature, see 110, 112).

It has frequently been observed that propigment cells will migrate most readily into an area which is not already occupied by melanoblasts. If a piece of melanocyte-free chicken skin is grafted to a pigmented host chick, the melanoblasts of the host migrate freely into the graft (108). Migration also takes place into skin grafts partially populated with melanoblasts, but not into skin grafts containing a full complement of them (27, 109). Evidence for similar preferential migration of propigment cells from a pigmented to a less pigmented area has been provided in amphibia by Twitty (139), in mammals by Rawles (111), and in the teleost by Goodrich and coworkers (47, 48). The last-named investigators have also demonstrated conclusively that the ability of an area to allow penetration by propigment cells is not necessarily a function of factors permitting later differentiation of chromatophores in that area. Scales from a nonpigmented zone, transplanted to a melanized area, were soon invaded by chromatophores from the adjoining dark stripe. On the other hand, into a patterned area freed of chromatophores by freezing, melanoblasts first migrate in a widely dispersed manner, and are only secondarily localized to conform with the pattern, by retention and differentiation in the appropriately dark areas, and destruction in the nonpigmented areas.

A secondary localization of pigment cells also occurs in the urodele, though by a somewhat different mechanism. As migratory activity of the propigment cells begins to decline following emigration from their site of origin, the cytoplasmic processes of the differentiating melanophores become increasingly elongate and ramified, and characteristically develop firm adhesions with the processes of neighboring cells. This dispersed meshwork of melanoblasts soon condenses into a dense black band of pigmented cells on the dorsolateral aspect of the animal. During this aggregation the interconnecting processes appear to undergo a shortening or retraction, which draws the cells into close proximity (138, 139).

The observation that melanoblasts will invade a nonpigmented area, but not one already fully populated by propigment cells has been interpreted as representing the establishment of an equilibrium (chemically mediated, presumably) between the melanoblasts and the skin tissues. It has been suggested that a constant ratio is maintained between the number of melanoblasts and the cells of the skin (109, 156). Several investigators have suggested various forms of chemical gradients which might account for this phenomenon. In one of his early studies, Twitty (137) tentatively proposed that carbon dioxide produced by the migratory cells themselves, probably through its effect on the $p$H of the medium, influences cell migration. Flickinger, also with some reservations, has adopted this view and provides evidence in its support (40).

From more recent investigations of chromatoblast migration in the urodele, Twitty and his collaborators have derived an hypothesis which may account for much of the migratory behavior of crest derivatives. When pieces of neural ridge and crest are explanted into glass capillary tubes filled with physiological saline or coelomic fluid, the migration of the propigment cells from the explant can be observed and to some extent manipulated. Migration is not influenced by products of the explant itself, since no effect is seen upon removal or change in position of the explant (137). However, the extent of migration by individual cells is proportional to the size and density of the cell population. Moreover, migration is more extensive under conditions which favor retention in the environmental medium of diffusible substances released by the cells. Thus it is more active in capillary tubes of small than of large diameter, and in deeper positions within the tube than near the mouth. This was shown more directly in experiments in which the explants were cultured with a modified hanging drop technique. The conditions were so arranged that part of the peripheral growth zone spread, as usual, on the coverslip, while the remainder advanced into a confined space under a tiny roof of glass supported close to the coverslip. Upon entering this restricted volume, the cells began to migrate more actively and became more widely spaced than those spreading on the open coverslip. Those cells which continued on to the other side of the confining glass plate regained their original slower rate of movement and again assumed a crowded arrangement. Twitty concludes from these observations that the initiation and direction of migration are referable to mutually repulsive influences exerted by the chromatophores themselves, through the action of diffusible substances (139, 140).

More recent tests lend further support for such an hypothesis. Propigment cells may be isolated singly or in small groups in capillary tubes. Single cells remain essentially stationary, although undergoing active changes in shape, and differentiating normally. On the other hand, cells isolated in pairs move away from one another, consistently, and often for considerable distances. Three or more cells behave in the same manner, and, in general, the more cells in the tube, the greater is the distance over which they distribute themselves (141). On the basis of these observations, one would predict that in a population of pigment cells situated as close together as their mutual repulsions would allow, entry of new melanoblasts would be prevented by the concentration of repellent substance maintained in the area.

Of some importance is the suggestion that this hypothesis is generalizable to widely diverse types of cells. Cultures of chick heart fibroblasts and of subcutaneous mouse tumor are said to respond to growth into confined spaces in a manner identical with neural crest cells (141). Such an idea is difficult to reconcile with observations of Abercrombie and his colleagues suggesting that chick heart and mouse muscle fibroblasts obstruct each other's movements upon contact. Both Holtfreter (67) and Weiss (148) have noted a tendency of cells, in culture, to restrict their movements when coming into contact with their neighbors. Abercrombie and coworkers have extended these observations and postulated a coherent theory of "contact inhibition" (3, 4, 5). When chick heart fibroblasts, cultured on glass in a liquid medium, are observed by means of time-lapse cinemicrography, an inverse relationship can be seen between the rate of displacement of a cell and the number of other cells with which it is in contact. If the number of contacts is experimentally increased by placing two explants close together on the coverslip and allowing their outgrowths to fuse, the rate of cell movement after junction in the inter-explant area is much lower than before junction. Although culture outgrowth is thus substantially prevented in the area of fusion, lateral outwandering of cells from the explants continues unabated. Moreover, even after junction, the cells between the two explants tend to remain as a "monolayer," rarely moving over one another's surfaces. Such inhibition of movement is seen upon contact of even widely divergent types of fibroblasts, for example, cells from mouse muscle and chick heart. Mouse sarcoma cells, on the other hand, are not noticeably affected by coming in contact with normal fibroblasts, and migrate over them with apparent ease. Abercrombie (2) has suggested that contact inhibition might play an important role in morphogenesis by providing a mechanism for the cessation of cell movements, and the restriction of such movements to noninhibited areas.

*Orientation of Ameboid Cells*

One possible source of direction for cell movements is to be found in the layer on which the cells are migrating. Weiss has postulated the theory of "selective contact guidance" (146, 148, 150) to account for the directed movements of mesenchyme and Schwann cells, and of axon outgrowths, from explanted tissue fragments. According to this theory, cells or cell processes follow lines of ultrastructural organization in the substratum, produced by stress or other orienting forces acting on the micellar components of the material. At least for some systems, both in vitro (147) and under certain circumstances in vivo (149), such a guiding mechanism can be shown to be operative. (For a recent review of this theory, especially as applied to neuron outgrowth, see ref. 151.) To account for the apparent fact that cells are capable of choosing "the right track" out of a multiply aligned system, Weiss suggests that the contact substrata have different specific properties which act as cues. He depicts such guide structures, situated in the intercellular matrix or in the surfaces of neighboring cells, as oriented lines of specific molecular groups, to which the cell can form temporary linkages by means of complementary groups in its own surface. A cell following such an oriented substratum would be exemplifying, according to Weiss's terminology, "selective conduction" (148). Extending these ideas to the movements of cells which are not guided by their substratum, "selective fixation" would account for any eventual pattern of localization. Cells would continue migrating until complementary groups in the substrate were encountered, at which time linkages would form and further movement cease. (This seems not unreasonable as a mechanism for Abercrombie's contact inhibition.) Conversely, cells might be stimulated to leave an area, or be destroyed there, a process termed "selective elimination."

Various other proposals have been made concerning the directive influences on single cells. Rosin (117) has suggested that propigment cells might be drawn ventrally toward an epidermal implant in response to a true chemotactic attraction exerted by the latter. Holtfreter has also described instances of pigment cell behavior indicative of their responsiveness to positive chemical attraction by lipoid substances (67). As mentioned above, an even less specific agent, carbon dioxide, has been tentatively implicated by Twitty (137) and by Flickinger (40) in the direction of cell migration, and in a totally different system Harris (59) has provided clear evidence that mammalian granulocytes, in culture, respond to chemical gradients. Chemotaxis is also known to occur in the Myxomycetes or slime molds, a group which has been rigorously investigated by Bonner (17, 18) and by Sussman and associates (131 and this Symposium). Another orientative influence mediated by diffusible substances is to be seen in the direction of polarity in the egg of Fucus (153, 73). It should be noted, incidentally, that such chemical attractions "at a distance" between cells must be translated, operationally, into microgradients at the cell surface, which can affect the pseudopodial activity or adhesiveness of the plasma membrane.

Although Holtfreter has recognized the need for an "inside-outside chemical gradient" to understand fully the migratory regulation of amphibian cell masses, he has placed greatest emphasis, for many years, on the surface interactions of cells in explaining their movements (65, 66, 133). He has postulated a mechanism of cell guidance, which, like that of Weiss, is based on selective cell surface compatibilities. According to this theory, embryonic cells have specific positive affinities toward certain of their neighbors, while lacking such affinities toward others. Continued migration over the latter, and adhesion to the former, could eventually localize cells in their definitive positions. The control of movement of the propigment cells, for example, could be dependent upon their specific interactions or contact relationships with the cells of the epidermis. Holtfreter (65) has demonstrated such a positive affinity between chromatophores and epidermis, and Twitty and Niu (140) have confirmed and extended this observation. The latter authors were able to show that chromatophores in contact with epidermis migrate upon it whether or not a layer of mesoderm underlies it. On the other hand, chromatophores in contact only with somite mesoderm do not migrate. Finally, when epidermis is stripped from the underlying mesoderm subsequent to emigration of the propigment cells from the neural crest, all of the melanoblasts adhere to the epidermis instead of the mesoderm.

Further evidence in favor of selective association of cells of specific types was obtained by Weiss and Andres (152). They injected dissociated embryonic chick cells, including melanoblasts, from potentially pigmented donor strains, into the blood stream of host chick embryos of unpigmented breeds. These injected cells became scattered throughout the body of the host, but the donor melanoblasts proliferated and synthesized melanin only in locations in the host equivalent to those in which they normally would have developed pigment in the donor. Never were donor melanocytes found in unusual cell associations. Caution must be employed however, in attributing such localization to properties of the cell surfaces alone. Ebert (34) has recently demonstrated that subcellular fractions (microsomal and supernatant) of spleen or kidney homogenates are differentially localized in the homologous host organ when injected into 9-day chick embryos.

In spite of the evidence that the adherence and migration of melanoblasts on different substrata are influenced by specific cell affinities, and that various chemotactic and chemotropic responses may be involved in the initiation and maintenance of such movements, it should be emphasized that there is no evidence of specific predetermined routes in the embryo, along which the chromatoblasts are unerringly guided. In normal development, pigment cells originate dorsally and descend in a ventrolateral direction. If, however, the crest is removed from its normal position along the neural tube, and reimplanted in a more ventral site, it has been shown by Twitty and Niu (140) that the chromatoblasts will migrate as readily in a dorsad direction as they normally do ventrally. In the words of these investigators, "We may best state the situation merely by saying that the cells spread from their source, wherever this may be situated on the embryo, and in so doing behave quite independently of any directional pulls or pathways associated with the axes of the organism" (p. 426). Control of the final position taken up by these cells would then represent an example of "selective fixation" rather than "selective conduction" (148).[1]

## Mass Cell Movements

Much of the activity which constitutes true morphogenesis consists of the movements of interconnected masses or continuous layers of cells, rather than of cells as individuals. The invaginations and evaginations occurring during gastrulation, neurulation, and gut formation, or the migrations of organ primordia to their definitive sites of differentiation may be cited among the many examples of such mass movements. Whereas isolated cells can move only as a result of their own activity (excluding for the moment passive carriage by other tissues or along vascular routes), members of such layers or masses are clearly subject to other motive forces as well. Growth of a cell layer by mitotic division will cause peripheral areas of the layer to spread centrifugally, or, if these areas are anchored

---

[1] However, an interesting experimental study by W. M. Reams (*J. Morphol.*, **99**, 513-548, 1956), suggests that melanoblasts may indeed follow dorso-ventral lines of orientation in their invasion of the coelomic lining of the chick embryo.

firmly to the substrate, buckling, warping, or thickening of the layer must occur. Even without mitosis, if each member of a layer of cells alters its shape by an active process, as for example by flattening and expanding its area of contact with the substratum, the proportions of the entire layer will also change; in this case, again, expanding or buckling. This process is clearly exemplified in the blastodermal epiboly of fish embryos, in which expansion is caused by flattening and migration of the underlying syncytial periblast layer (90, 134, 135).

Another motive force influential in morphogenesis is that imparted to cells by the contraction of filopodial intercellular connections. The extent of such "filose" activity (9) of embryonic cells is apparently much greater than is generally recognized (101). Even before the turn of the century, intercellular connections were revealed between the cells of various invertebrate embryos (8, 9). These were considered not only as a means of cellular cohesion by which cells could be drawn together, but also as true "physiological connections" whereby the exchange of living protoplasm could take place.[2] Much more recently Dan and Okazaki (26) and Gustafson and Kinnander (56) have described the important role played by filopodial contraction during gastrulation in the sea urchin. Both primary and secondary mesenchyme cells spin out long thin pseudopodia, which attach to the ectoderm and draw the cell body, or the archenteron tip to which it adheres, toward the point of pseudopodial attachment.

In vertebrate tissues, similar intercellular connections have been reported. The contraction of chromatophore cell-processes in amphibia, as described by Twitty (138, 139), was noted above. In the chick embryo, Spratt (128) has seen fibers in living specimens, connecting adjacent and nonadjacent cells within and between germ layers. Many are elastic, and some appear to be spatially oriented in certain important growth regions such as the node area and neural plate. Even in the human embryo, extensive fibrillization between cells of the cardiogenic plate has been described in fixed material, although Davis (28) considers such structures largely artifactitious.

Most of the earlier references on intercellular connections are provided in Fisher's extensive discussion of the topic (38, especially chapters 2 and

---

[2] In 1897, Andrews (9) was able to make the following statement, which, I think, could well be taken as a standard for aesthetic and (in the light of modern knowledge) prophetic scientific prose: "If we look upon the cell wall as a part of the machinery of the embryo . . . , that is, as an organ of the mass of living substance, just as is the nuclear membrane, or any other local modification for physiological purposes, we shall probably be nearer the truth than if we keep to an earlier conception, and hold that, for the living substance, cell walls a prison make" (p. 389).

3). He also takes up the question of whether such connections represent true protoplasmic bridges. Although opinions of workers in the field are divided, Fisher seems to conclude that in many tissues, and in cultured cells, cytoplasmic continuity can occur. On the other hand, Caesar, Edwards, and Ruska (22) have recently examined many adult tissues with the electron microscope, hoping to answer this question. They are able, in every case, to see continuous membranes between cells, even in tissues previously accepted as syncytial. They tentatively conclude that "continuity of protoplasm from cell to cell [does] not exist in vertebrate tissues, and perhaps not in any animal tissue" (p. 873). Whether or not such a generalization is valid, and whether or not it is equally applicable to embryonic cells, remains to be seen.

In spite of the evidence for such intercellular forces, there are many situations in which movements of cell masses are brought about by ameboid activity of individual cells or small groups of cells constituting the mass. Spratt (127, 129) has recently shown this to occur in the process of notochord elongation in chick embryos. He has found that in the head-process embryo there is a fairly well-defined group of cells, lying beneath the primitive pit and extending anteriorly to the edge of Hensen's node, which constitutes a single, median notochord center or "chorda bulb." Removal of the chorda center during regression of the primitive streak results in cessation of further notochord formation posterior to the site of operation, but has essentially no effect on continued formation of somites. Elongation of the notochord is accomplished partly by mitotic division of chordal cells, but to a greater extent by addition of cells originally posterior to the chorda bulb. Apparently most of the cells of the chorda center migrate actively in a posterior direction, at a rate which is much faster than the regression of the postnodal mesoderm. Since it is only if the chorda bulb cells accomplish this movement through the postnodal mesoderm that the latter are transformed into notochord, it is felt that chorda center cells exhibit an inductive influence upon the cells with which they come in contact during their migration.

What forces, it may be asked, maintain these wandering cells in their (literally) straight and narrow path down the middle of the embryonic area? Unfortunately, no experimental evidence is forthcoming from Spratt's studies which will shed light on this question, and we must thus look to analogous situations elsewhere in development. One of these is seen in the migration of myoblasts during the formation of discrete muscle bundles. It has been demonstrated in amphibian embryonic muscle cells in culture (68) and in chick myoblasts both in vitro and in freshly teased preparations (70, 80) that motility is a function of pseudopodial activity concentrated mainly at the leading tip of the cell. Therefore, the selective contact guidance theory of Weiss and the specific cell affinity theory of Holtfreter may be applicable here. Hall (57) has studied myoblast movements in the forming head musculature of the urodele. He began his investigation with the working hypothesis that the pre-muscle masses elongate and migrate toward their points of origin and insertion in response to specific influences emanating from these points of attachment. At early stages of development, well before the muscle mesoderm had begun to segment, the prospective points of attachment or route of growth of the levator and depressor mandibulae muscles were modified by various operations. The results of these experiments emphasized the lack of importance, for differentiation and orientation of the muscles, of the structures which afford their origin and insertion. Excellent development was found in muscles lacking one or the other of these structures, and even in parts of muscles which lacked both. On the other hand, profound modifications of muscle structure, orientation, and position were obtained after neural crest extirpations had resulted in deficiencies in certain head cartilages. These modifications, it should be emphasized, often occurred in the presence of a well-developed origin and insertion. Hall concluded, therefore, that it is the mesectodermal elements which orient, and exert a trophic influence upon, the developing muscle primordia. He suggested that this might occur during elongation and differentiation of the muscles, at which time the neural crest cells have obtained their definitive positions and are differentiating into cartilages. It may also have occurred at earlier stages during which the neural crest material was migrating ventrally and surrounding the muscle mesoderm.

A similar directive influence of mesenchymal tissue has been described by Holtfreter in the development of the pronephric system (66). He was able to show that the caudal extension of the Wolffian duct is oriented by cells in the environment. The tendency of the duct to follow its normal course along the lower border of the somites seems to be due to a selective adhesion between the nephric cells and the endothelium of blood vessels, especially the posterior cardinal veins. In experimental animals in which accessory somites and blood vessels had been induced from implanted cells, the host duct was frequently seen to forsake its normal surroundings and follow, instead, the alternate route along the accessory structures. It is difficult to interpret these examples, as well as others which might be mentioned (see, e.g., 148) as representing anything other than an orientation of movements resulting from characteristics of the cells or substratum on or through which the moving cells travel.

Townes and Holtfreter (133) have recently published an experimental analysis of this problem, the results of which add notably to our information regarding cell surface interactions. Tissues of known prospective fate were excised from amphibian embryos between gastrula and late neurula stages. The tissues were either explanted together in various combinations, or they were first disaggregated (with high $p$H), the individual cells of the tissues intermingled, and the "compound" tissues then allowed to reaggregate. Within a matter of hours the individual cells of such an aggregate begin to perform "directed movements" corresponding to those of the tissue fragments from which they were derived. Cells from the neural plate or mesoderm which had been combined with either epidermal or endodermal cells exhibited inward movements. (A clear example of "unipolar ingression," as claimed for the yolky endoderm cells of the early blastula, 122; but see 11.) The epidermal and, to a lesser extent, endodermal cells manifested a tendency to migrate centrifugally. Having reached the periphery of the aggregate, these cells spread out to form a flattened epithelium. As a consequence of these movements, the different cell types sort out into distinct homogeneous layers, corresponding to the normal germ-layer arrangement after gastrulation. The tissue segregation is then completed by an emergence of further selectivity of cell adhesion, homologous cells remaining in contact, nonhomologous cells being separated, either by continued migration, or by the formation of fluid-filled spaces between the cells.

Gregg and his associates (52, 53, 102) also have attempted to analyze the relation between tissue interaction and gastrulatory movements, using tissue combinations in vitro. A piece of presumptive head endoderm, notochord mesoderm, or ventral ectoderm from a stage 10 frog gastrula is explanted on a "base" of yolk endoderm in such a way that distinct types of cell movements can be observed separately. By applying various metabolic poisons and by using tissues from nongastrulating hybrid embryos, they have been able to demonstrate characteristic movements of specific cell masses, and to relate failure of these movements to developmental defects in intact embryos. Although their results to date are incomplete, much is to be expected from this type of attack.

The principle of specific cell affinities may also be applied in interpreting some of the observations on embryonic grafts in which donor tissue is found in contact only with homologous host tissues, often at sites distant from the point of implantation. Such a situation has been described in endodermal implants in both amphibia (66) and fish (100); similarly, the histotypic development of organized tissues in chimeric aggregates can thus be depicted (95, 96, 97). Even those movements which cannot be accounted for wholly by autonomous activity of the cells, designated as "correlative movements" by Schechtman (123), are influenced by cellular adhesive properties, since traction forces set up in one area of an embryo would have to be transmitted to another part through cohesive chains or layers.

From the above examples it would appear that those movements of gastrulation, neurulation, and organ formation which involve active ameboid behavior of the individual cells of a mass are to a large extent controlled by the characteristics of the cell surfaces, and by the resultant "migration tendencies" (133) inherent in each cell type at a given stage in its differentiation and in response to a given contact surface. Changes in the membrane properties would result in alterations of both the migration tendencies and the specific contact affinities of cells.

## Coordinated Movements of Cell Groups and Layers

Obviously, morphogenetic movements consist, not only of the motions of individual cells sliding past or over one another, but also of the movements of entire layers of cells in which each individual member maintains a relatively stable juxtaposition with its neighbors. A prime example of this is provided by the process of foregut development in the chick. By carefully marking the endoderm of early chicks with carbon particles, Bellairs (12, 12a) has determined the presumptive foregut areas and has followed the movements of the endodermal layer during development. She distinguishes "two-dimensional" and "three-dimensional" movements, the former occurring in the original plane of the endoderm, the latter folding certain regions medioventrally to form the floor of the foregut (Fig. 1). The anterior end of the gut is formed first as a pocket between two sets of opposing movements in the endoderm, a forward two-dimensional movement beneath the head process, and an obliquely mesiocaudad, three-dimensional movement on either side. These oblique movements result in a U-shaped ridge in the endoderm bordering the anterior intestinal portal. As the movements continue progressively backwards, the two limbs of the ridge are brought together in the mid-line, where they fuse and thus gradually close off the cavity of the foregut from that of the yolk-sac.

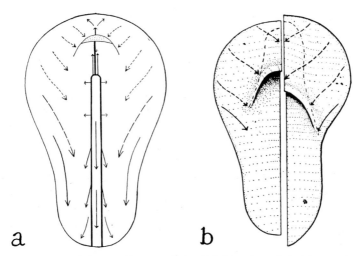

Fig. 1. Diagrams of the ventral aspect of the chick blastoderm during foregut formation and elongation. (After Bellairs, 12). Solid arrows show endodermal movements in the plane of the drawing. Broken arrows represent three-dimensional movements folding the endoderm medioventrally.
  a. Initiation of head fold.
  b. Stages in elongation of gut.

In seeking a mechanism for these mass movements, Bellairs has examined two possibilities. Spatially limited bursts of mitotic activity would be expected to cause locally variable changes in area and shape of the layer. Mitotic counts of the area pellucida endoderm, however, yielded no evidence suggesting that differences in mitotic rate were prerequisite for the observed morphogenetic movements (14). Furthermore, in a study of the anomalies produced by treatment of chicks with a mitotic inhibitor (menadiol diphosphate), it was concluded that failure of closure of the foregut, and other alterations of normal cell movements, were not attributable solely to interference with cell division (13). These findings are supported by the results of many recent studies, which largely eliminate cell division as a major causal factor in mass cell movements (133, 42, 144). Nonetheless, it is not to be concluded that mitosis plays no part whatsoever here, since both Bellairs (15) and Abercrombie (1) have recently suggested a secondary role of cell division in stimulating cell movement or ameboid activity.

A second candidate for the major motive force underlying the endodermal movements noted above is the phenomenon of change in shape of individual cells of the layer. This is an idea which dates back to Rhumbler's studies of gastrulation on model systems (114). He conceived that the observed changes in cell shape might be causally, rather than consequentially, related to the invagination of the blastular wall. More recent workers have obtained much evidence supporting this concept (78, 79, 16, 21, 29, 42, 81, 144). Bellairs describes in detail such changes occurring in the cells of the endoderm: the thinning of the presumptive foregut roof, and thickening of the floor and lateral walls prior to and during gut formation (12, 12a). These observations and those in her more recent work (15) suggest that it is the locally varying patterns of spreading or contraction of the cells of the endoderm which produce the observed folds and movements of that layer.

Holtfreter has suggested that ameboid behavior similar to that seen in isolated cells (67) might be active in the folding of cell layers. Forces within each cell, probably related to some sol-gel transformation mechanism (61), and thus perhaps to a calcium-binding by cytoplasmic proteins (85, 88, 113, 116), could lead to autonomous cell shrinkage or extension. A mechanism early suggested to account for such shape-changes was the differential intake of water into cells. Glaser (44, 45) obtained evidence supporting this idea of differential imbibition, and Giersberg (41) and Hobson (64) were able to reverse the closure of the neural folds with strongly hypertonic media. Unfortunately for this hypothesis, however, Brown, Hamburger, and Schmitt (20) have been unable to demonstrate changes of density of more than five to ten per cent in the respective neural tissue of amphibia during gastrulation, indicating no large uptake of water. Moreover, the criticisms of Gillette (42), bringing into question the validity of the interpretation of experiments such as those of Giersberg and of Hobson, leave little support for any hypothesis demanding extensive cell swelling.

To account for such changes in shape without invoking water uptake, Waddington (144) has postulated the formation of fibrous structures in the internal cytoplasm. In the neck of the endodermal flask cells which form in the amphibian blastopore region, he has shown double refraction, indicating the presence of such fibrous structures. In most cells, however, there is little evidence that the bulk of the internal cytoplasm also contains oriented material of this kind (143). "It is therefore probable," admits Waddington, "that intracellular fibers play at best a very minor role in amphibian early morphogenesis" (144, p. 439).

Recent studies of morphogenetic mechanisms have attached great importance to the properties of cell membranes. As cited above, Holtfreter has emphasized the importance of cell surface characteristics to his concept of specific cell affinities. Similarly, changes in cell adhesiveness could be postulated which would produce just those changes of cell shape observed in the endoderm and neural ectoderm. An increased tendency of a cell for adhesion to its neighbors would increase the area of contact, resulting in a transition from squamous to columnar forms, and decreasing the area of the layer constituted by these cells (20). Waddington (144) feels that it is quite probable that such forces arising from adhesiveness of the cell membranes play a part in the elongation and narrowing of the dorsal material, and the compensating expansion of the more ventral parts of the amphibian embryo during gastrulation and neurulation.

It has been demonstrated clearly by Gillette (42), in his detailed analysis of the size and shape of cells of the medullary plate during neurulation, that in the process of folding the cells become flask-shaped by shrinkage of that part of their surface facing the exterior, resulting in a shift of the nucleus and the bulk of the cytoplasm toward the internal side of the ectoderm. Lewis (81) attributes these form changes to contraction of the superficial gel layers of the cells. However, increased adhesiveness of the cells along their upper sides, resulting in a greater area of intercellular contact on that side, could bring about a similar effect. Sol-gel transformations involving calcium transfer within different parts of the cell (as mentioned above), coordinated with these surface changes, might also be involved.

Another process of development which has recently been examined from the point of view of changes in cell shape is the formation of the notochord in the mid-line of the invaginated amphibian mesoderm (93). It is quite clear here that the area of contact between cells does increase considerably as the notochord forms. Beginning as a group of loosely connected, rounded cells, the notochord grows progressively more compact, the cells taking on an arrangement comparable "to a pile of coins" (p. 402), each cell becoming a more or less flat disc with the maximum possible contact with other similar cells. Pending further evidence, we might conclude from these examples, as does Waddington (144), that "increase in cell contact is clearly one of the basic phenomena of morphogenesis, and it seems not unreasonable to accept it as a causal explanation of the events. Forces arising from the cell membranes may well be the prime cause of the changes of tissue configuration during the whole process of gastrulation and neurulation" (p. 451).

## The Structure of the Cell Membrane and the Mechanism of Adhesion

We are brought back, then, to the first question asked in the introduction to this paper: what is the mechanism of cellular adhesiveness in the embryo? We must begin by considering first where such adhesiveness takes place. That is, we must recognize that a cell adheres to its neighbor or its substratum by means of its plasma membrane, and that it will be the properties and structure of this membrane, the properties and structure of the cell surface, to which we must look for mechanisms of cell adhesion. (For a recent discussion of specialized adhesive surface structures in adult cells, see 36.)

A detailed analysis of the various theories concerning the physico-chemical structure of the plasma membrane would be redundant, since several such reviews have been published in the last few years, examining the problem from one point of view or another (69, 91, 30, 60, 118, 106, 19). It is sufficient for our purpose here to note only that the protoplasmic membrane behaves, under numerous circumstances, like a highly condensed double phosphatide layer, complexed with proteins and probably with cholesterol (19, 30). Such lipo-protein membranes exhibit evidence of rather complex patterns of oriented molecules and even some suggestion of paracrystalline lattice structures (145, 124). The phosphatides most frequently noted are cephalin, lecithin, and possibly phosphatidic acid (19). At least in parts of the membrane, protein of both enzymic and antigenic nature is found at the surface (118, 125, 126), probably associated with nucleic acids (77, 119, 25).

As mentioned earlier, linkages formed between specific molecular groups in contact surfaces might be expected to bind the cells in a manner analogous to an enzyme-substrate or antigen-antibody system. Tyler has expanded upon this idea in elaborating his auto-antibody concept (see, e.g., 142). Weiss (146, 148, 150) has suggested a similar hypothesis wherein tissue affinities would be based on the presence in the contact surfaces of reciprocally identical, or complementary, protein configurations, interlocking and binding by "molecular attraction forces" (146, p. 188). Different degrees of adhesion might be explained by differences in the spacing of the molecules in each membrane.

These theories, based largely on analogy, are supported by three major classes of observation: (1) the marked specificity of cell adhesions, which can be accounted for readily in terms of the specificities of the postulated molecular groupings; (2) the disaggregation which is effected by proteolytic enzymes, such as trypsin or papain (152, 97, 37), indicating that peptide linkages are indeed involved somewhere in the adhesive mechanism; and (3) the inhibition of reaggregation of cells by antisera prepared against their surface antigens (125, 126).

As formidable as this evidence may appear, it must be pointed out that neither individually nor in the aggregate do the data form more than a suggestive basis for such a concept. Surface specificity can indeed be accounted for by any system incorporating patterned chemical bonding sites (see below), but, as suggested by Pauling (105), this might include configurations of ionic bonds, covalent, metallic, or hydrogen bonds, or London-van der Waals forces. Specific contact affinities might be provided by spatially patterned groups of any of these nonspecific interatomic forces, thus retaining none but the most general similarity to that observed in antigen-antibody reactions. Disaggregation by proteolytic enzymes, though practiced routinely in several laboratories, does not always produce a reversible separation. Feldman (36) disaggregated early amphibian neurula tissues with trypsin, papain, pepsin, or calcium-binding agents. Only with the last-named, in the absence of enzymes, was he able to get satisfactory reaggregation. Townes (132) has obtained similar findings. As to the inhibition of such reaggregation by antibodies to surface antigens, this would be expected whether or not those antigens represented the sites of cell binding. Indeed, Spiegel (125) recognizes the possibility that each of his antibody molecules simply "overhangs or somehow interferes with the activity of the adhesion site" (p. 144), while actually binding to an uninvolved antigen nearby. Finally, and perhaps crucially, the "lock and key" theory of cell adhesion does not take into consideration the voluminous evidence implicating divalent cations, especially calcium, in the adhesive mechanism.

As early as 1890, Ringer (115) was able to show that traces of calcium ions are required to prevent dispersion of the cells of mussel gills, and in the next few years the importance of calcium in maintaining tissue integrity was confirmed and extended by Herbst (62), Overton (103), and Lillie (82). Largely on the basis of the work of these authors, it was generally agreed that the cohesive material binding cells together was probably a reversible calcium salt of a weak acid (158). Other investigators have visualized an intercellular "cement substance" consisting of mucoprotein (50), calcium proteinate (158), histone (124), or other viscous material (see e.g. 23, 50). With the more recent use, not only of calcium-free solutions, but of various calcium-complexing agents, it has been possible to confirm that cells of many embryonic and adult tissues depend on the presence of free divalent cations for their adherence to one another (although, see 63). Townes (132) and Feldman (37) have used citrate, oxalate, and glycine (an excellent calcium-binder) to produce reversible disaggregation of embryonic amphibian cells. Zwilling (159) introduced the use of a commercially available chelating agent, ethylene diamine tetraacetic acid ("Versene") for the dissociation of chick embryonic cells, and this same substance, along with other calcium complexers, has been found useful in separating the cells of adult rat liver (7). Furthermore, we know that calcium and other divalent cations are bound at the cell surface (77, 86, 92, 118, 6), and also would be expected to form an integral part of any mucoprotein matrix (76, 50, 55) if such exists in the embryo (see the discussion of intercellular matrices by Edds, in this Symposium).

In view of our present knowledge of the chemical composition of the cell membrane and its ability to bind ions, it seems most parsimonious to assume that the "weak acid" referred to by earlier workers, which was supposed, as a calcium salt, to form the intercellular adhesive material, actually is part of the membrane. The carboxyl groups of the membrane protein or—more likely—the orthophosphate or polyphosphate groups of phosphatides and nucleic acids would fill this role. Calcium ions might then form "bridges" between two cells, one valency of each ion binding to one of the contact surfaces. Enzymatic conversion between lecithin, cephalin, and phosphatidic acid, initiated from within the cell (19), could provide a means for altering the dissociation constant of the calcium complex, and thereby the degree of adhesiveness of any part of the cell membrane. A "zipper mechanism" analogous to that proposed by Schmitt (124) to account for changes in cell shape is thus made possible, utilizing only substances known to exist at the cell surface.

On the basis of observations on disaggregated amphibian cells, Steinberg (130) has reached a similar view. He has, in fact, devised a theory which accounts for both cell adhesiveness and specific cell affinities, in terms of a lattice array of calcium-binding sites on the cell surface. The degree of adhesiveness would be a function of the degree of congruency of such sites, and changes of cohesion during differentiation might be associated with alterations in the geometry of the lattice pattern.

Heretofore, very few workers (67, 150, 130) have recognized that a cell can adhere to its neighbor or to its substratum with more or less firmness, depending on the state of the membrane and the environment in which the cell finds itself. There has been a tendency on the part of most investigators in this field to treat cell dissociation as an all-or-none phenomenon. I should like to question this attitude, and emphasize that embryonic cells can adhere to each other with varying strengths. Moreover, I propose not only that calcium is required for cell adhesion, but that it is an important factor in controlling the strength of that adhesion; i.e., that there is a correlation between the amount or "activity" (in the physical chemist's sense) of calcium in the surrounding medium, and the extent or firmness of cellular contact. Furthermore, it should be noted that the phenomenon of cell movement is in turn a function of cell adhesion. We have seen earlier how this is the case in cell layers. Although it has been said that "in morphogenesis the forces controlling directed movements must overcome those of cell adhesion" (133, p. 110), it is clear that migratory cells cannot move at all except in firm contact with a cellular or noncellular substrate (146, 151, 2).

In view of the above considerations, it should be possible, by means of limited removal of calcium from a developing embryo, to produce disturbances of intercellular or cell-substrate adhesion of just sufficient severity to cause abnormal morphogenetic movements, yet not yield disaggregation.

## The Effects of Limited Cell Dissociation on Cardiac Development

In the normal development of the chick, heart-forming capacity can be traced back to a pair of mesodermal anlagen situated bilaterally on either side of Hensen's node in the head-process stage embryo (107). With the splitting of the mesoderm into somatic and splanchnic layers, the amniocardiac vesicles are formed. The presumptive cardiac cells remain in the splanchnic layer, closely applied to the endoderm (121). As the foregut is formed, the ventrocaudal folding of the endoderm (12a) carries the heart primordia toward the mid-line, where further mesiad migration brings the newly formed endocardial "gutters" together at their rostral ends to form the conus arteriosus. The familiar process of "zipping" together of the two sides over the regressing arch of the anterior intestinal portal then occurs, allowing progressive fusion of the bulbus cordis, the ventricle, the atrium, and finally the sinus venosus of the primitive tubular heart. This process has been described in detail in several vertebrates (28, 104, 120, 49, 155), and can readily be observed in the living chick embryo, explanted ventral side up, according to the method of New (99). Under these circumstances, with the embryo resting in its normal position on the vitelline membrane, development progresses satisfactorily for two to three days after explantation at the primitive streak stage. The heart and endodermal structures can be visualized clearly, especially during early phases, and manipulation or treatment of the chicks can be accomplished from the endodermal surface.

If a small crystal of acetylcholine (Ach) is allowed to dissolve on the endoderm of a primitive streak stage embryo, the folding of that layer and the migration of the presumptive cardiac mesoderm toward the mid-line will frequently be prevented or modified, producing a "cardia bifida" monster with two bilaterally situated beating hearts (31). An attempt to investigate the mechanism of this disruption of morphogenetic movements, and to relate the Ach effect with the hypothesis concerning cell adhesion and cell movements set forth above, will constitute the remainder of the present paper.

To confirm and extend the previous observations, a total of 250 embryos was divided into groups of about twenty, and each group was treated with 10 microliters of a solution of Ach in calcium-free Howard-Ringer (72). Control chicks were treated with 10 microliters of the same solution, lacking Ach, in which the $pH$ had been adjusted to various levels with hydrochloric acid. Figs. 2-5 show a control and three treated embryos, fixed at 44-48 hours of incubation, the latter three having developed from the primitive streak stage in the presence of approximately molar Ach. This series demonstrates nicely the range of effects, from only a slight indentation in the anterior intestinal portal and improper fusion of the ventricles (Fig. 3), through complete separation of the right and left heart (Fig. 4), and finally to the stage having wide separation of the hearts, complete lack of a foregut, and open neural tube (Fig. 5). In each group treated with concentrations of $0.89\ M$ Ach or greater, some of the embryos were completely disaggregated. Within a few minutes after addition of the Ach solution, separation of the two halves of the primitive streak occurred along the mid-line, the edges of the hole being drawn back toward the area opaca by tension in the blastoderm. Within 15-20 minutes, the space previously occupied by the area pellucida was empty except for a few isolated clumps of cells. By this time, usually, the area opaca had become reticular and "patchy," cells separating into strands, again migrating toward the periphery. By the next morning, each of these disaggregated embryos was represented only by a ring of vesicular tissue around the edge of the glass ring, the vitelline membrane being essentially empty.

In view of this continuum of effects, a system of grading was devised for the experimental embryos, using the following criteria:

Excellent—Heart, somites, and nervous system conform well with those of controls and the Hamburger-Hamilton normal stages (58); heart beating (if stage ten or older); tissues distinct and translucent.

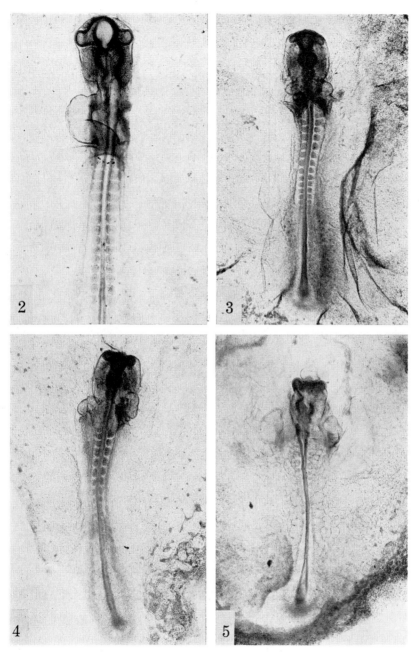

Fig. 2. Control chick, stage 11, explanted and treated at stage 4 with 10 microliters of calcium-free Howard-Ringer. Fixed in 10% neutral formalin. Stained in alcoholic cochineal. Photographed from ventral view. × 20.

Figs. 3-5. Cardia bifida (CB) embryos produced by treatment with acetylcholine (Ach) at stage 4. Mounted and photographed as with control.

Fig. 3. Stage 11, 1.0 $M$ Ach, Partial CB.

Fig. 4. Stage 10+, 1.0 $M$ Ach.

Fig. 5. Badly deformed stage 10+, 1.1 $M$ Ach. Note indistinct somites and paired hearts; open, flattened neural tube, complete lack of foregut, and blistered appearance of blastoderm.

360  THE CHEMICAL BASIS OF DEVELOPMENT

Good—Only minor variations from "excellent," e.g., no heart beat, abnormal flexures, etc.

Fair—Heart, brain, and foregut recognizable, and somites countable; permissible are stunted central nervous system structures, heart not beating, heart present in abnormal position as in situs inversus, but with normal structure; tissues more or less opaque.

Poor—Embryo noticeably disorganized, somites usually lacking or indistinct, tissues opaque; definitely not good enough to be called "fair."

Disaggregate—Area pellucida or entire blastoderm dissociated, leaving clear vitelline membrane (extreme care has been exercised not to confuse this situation with that in which the embryo has been pulled over to one side of the ring, due to separation of the blastoderm from the vitelline membrane).

Cardia bifida—Any of the first three grades (excellent, good, or fair) having double hearts, which are either beating spontaneously or respond to gentle mechanical stimulation.

The over-all concentration relations are shown in Fig. 6, in which is plotted increasing concentrations of Ach against developmental effect. The open circles represent the percentage of each group of chicks in which de-

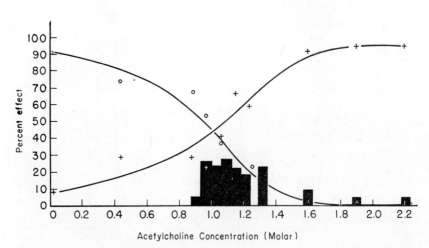

Fig. 6. Acetylcholine dose-response curves. Open circles, embryos graded as "excellent," "good," or "fair"; plus symbols, disaggregated or "poor"; histogram bars, percentage cardia bifida (CB).

velopment was normal (i.e. excellent, good, or fair). The plus symbols show the percentage which were disaggregated or poor; frequently these embryos were partly disaggregated and partly disorganized. The solid bars represent cardia bifida embryos, and have not been counted in either of the other two groups. It can be seen that at concentrations above 0.9 $M$, 20 to 30 per cent of the chicks form double hearts. Above 1.3 $M$, none of the chicks develop normally, most are disaggregated, and the small remainder are cardia bifida.

The anomalies of cardiac development produced in these experiments, if taken alone, would hardly suggest that poor cellular adhesion might be involved in their causation. However, the total range of effects, from partial inhibition of foregut fusion to complete cell dissociation, points more strongly in that direction.

At present there are only three known methods for chemically disaggregating tissues: (1) the use of proteolytic enzyme preparations; (2) the use of highly basic or acidic solutions; and (3) removal of divalent cations from the tissue. The question may be asked, then: does Ach, in the present circumstances, produce its effects by decreasing cellular adhesion, and if so, is its mode of action related to one of the above three possible mechanisms?

Acetylcholine bromide (Eastman Organic Chemicals), as used in these experiments, is a crystalline compound with strictly controlled, and very low, permissible limits of organic impurities, having no known activating or unmasking effect on proteolytic enzymes. A molar solution in Howard-Ringer (non-buffered) has a $p$H of about 5.4. Ten microliters of such a solution, added to a chick resting on a pool of about one milliliter of egg albumen ($p$H 7.8-8.2), a powerful buffer, can hardly be expected to shift the tissue $p$H markedly. To test this possibility, however, control chicks were treated with ten microliters of Howard-Ringer at $p$H levels of from eight to three. No adverse effects were seen. Thus, the first two possibilities mentioned above may be discarded.

Acetylcholine, as an organic cation, however, might be expected to have an effect on cell adhesion, if such adhesion is indeed dependent upon calcium-binding at the cell membrane. Because of its ion-binding capacities, referred to above, the cell surface, with its negative carboxyl and phosphoryl groups, has been analogized to an ion-exchange resin (61). It is well known that such resins will bind organic cations, and that these will be in equilibrium with metallic ions (154, 10). For example, choline, acting as a potent base, replaces some or most of the metal ions bound to the resin "Permutit" (154). Moreover, in unpublished experiments, it has been possible to show that Ach (or choline) will compete with $Sr^{89}$ for binding sites on a sulfonate resin IR-120 (Amberlite). The amount of $Sr^{89}$ bound is in inverse proportion to the concentration of Ach, at the levels tested (0 to 0.5 $M$).[3]

In a related system, Gross (54) has suggested that competition for binding sites on cytoplasmic colloids, between organic cations and calcium, might be involved in the mechanism of sol-gel transformation. Moreover, a physiological cation-like effect of Ach has been recognized and traced to its charge reversal effects on phosphatides (19). In fact, Heilbrunn (61) has put forth the idea that the neuronal effects of Ach, and other "stimulants," may be explained in terms of calcium displacement. Thus, it seems not unreasonable to propose that Ach might produce its effects on chick embryos by displacing calcium-ion bridges from their sites of binding between two cell surfaces.

It should be emphasized here that these experiments should not necessarily be taken as support for the theory that Ach acts as a neurohumor simply through its ability to substitute for metallic ions. The concentrations required in the present work make it unlikely that the effects were within the "physiological" range.

In view of this high concentration needed to produce cardia bifida and disaggregation, the possibility that sheer hypertonicity might have played an important role cannot be completely ruled out. Furthermore, observations such as those of Hobson (64) that hypertonic solutions of sodium chloride or glycerol can prevent or reverse fusion of the neural folds in chick embryos tend to support this view. Several points about Hobson's work should be noted, however: (1) the solutions used were much more concentrated than those in the present experiments, being 20% (3.45 $M$) NaCl and 50% or 100% (6.8 or 13.7 $M$) glycerol; (2) blastoderms were completely immersed in a relatively large volume of one of these solutions; (3) with 20% NaCl some degree of cell dissociation is not surprising, since one would expect that high enough concentrations of any univalent cation might tend to displace calcium from the cell surface; and (4) Hobson clearly demonstrated (p. 126) that only osmotic balance, and not cellular cohesion, was affected by his glycerol solutions.

To test further the effects of hypertonicity on chicks under experimental conditions comparable to those in which cellular dissociation was produced, forty-one embryos were explanted as before and treated with 10 microliters each of 1.0 or 1.5 $M$ sucrose. It was assumed that this substance would have no tendency to displace calcium ions from the cell surface. Most of

---

[3] I wish to thank Dr. Laurence S. Maynard, of the Department of Anatomy, Johns Hopkins School of Medicine, for performing these experiments with radiostrontium.

these animals developed quite normally, not differing from saline controls. Development in six of them was affected to the extent that spatial relations were distorted, nervous structures stunted, and tissues were more or less opaque. None developed double hearts, although one did have a heart the two halves of which had fused improperly over the anterior intestinal portal, but which were beating synchronously (one of the several hundred untreated or saline control embryos also developed such an improperly fused heart). Ten of these embryos, however, exhibited degenerative changes similar to disaggregation, in which the area pellucida was empty upon examination after 24 hours of culture.

Accepting that this destruction of the embryos with sucrose was the same as the disaggregation noted above when organic cations were used, one must conclude either that the degree of osmotic imbalance imposed by any 1 $M$ solution is sufficiently deleterious to disperse 20-25 per cent of the embryos, or that sucrose itself has physico-chemical actions other than its influence on tissue osmolarity. In view of the known effects of monohydric alcohols on condensed phosphatide layers, as to their ion-binding properties and state of coacervation (19), it seems not unreasonable that a polyhydric alcohol (sucrose) would have related effects. For the present, all that can be said is that no substance applied to chick embryos in concentrations of 1 $M$ or higher was without deleterious effect. It does not appear, however, that this is sufficient to account for the specific anomalies produced with Ach. Moreover, we shall see that other substances can yield effects similar to those of Ach, in concentrations far too low to influence tissue osmolarity. Hypertonicity, then, cannot be excluded from involvement, but it is clearly neither necessary nor sufficient as an agent in producing limited cell dissociation.

If the foregoing ideas are at all valid, one might generalize that removal of divalent cations from between cells, by displacement with univalent ions or by any other means, should decrease cellular adhesiveness, producing a spectrum of morphogenetic anomalies and disaggregation. In my first report of the production of cardia bifida with Ach (31), I noted that crystals of sodium acetate had a similar effect, and that this substance also produced much disaggregation in the embryos. It was this observation and the knowledge that acetate can form weak complexes with polyvalent cations (43) which, in fact, suggested the entire hypothesis outlined above. As pointed out by Martell and Calvin (84, p. 477), however, the stability constants for metal complexes with acetate are exceedingly low. Thus, this argument, alone, is relatively weak. More significant evidence was provided by a limited number of chicks treated with sodium citrate, in which a few cardia bifida embryos and several disaggregates were found. The effective concentration of this substance was less than 0.1 $M$.

Therefore, a series of experiments was set up to determine whether a powerful chelating agent, at carefully controlled concentrations, would produce effects similar to those of Ach, presumably by removal rather than displacement of calcium from between cells.

A total of 545 chicks was explanted as before, in groups generally of 20 to 30 each. Each chick was treated with 10 microliters of Versene[4] at concentrations of 0-10 m$M$, in Howard-Ringer. The droplet of solution was placed on the endoderm with a constriction pipette, and was generally sufficient to flood the area pellucida and some of the area opaca. The chicks were treated in a warm box, under the dissecting microscope, and then removed to a 38°C incubator overnight.

Figs. 7 and 8 show Versene-produced cardia bifida embryos. Their similarity to those formed in the presence of Ach is apparent. It should be noted that these effects could not have resulted from an osmotic imbalance, since the concentration of Versene used was only 5-6 m$M$. The chicks shown in Figs. 9 and 10, treated with 4.8 and 6.5 m$M$ Versene, respectively, are relatively unusual examples of partial dissociation. In these embryos, the primitive streak split and separated when Versene was applied. Instead of leading to total disaggregation, however, the process stopped before Hensen's node was damaged, and development could continue. The more anterior structures, apparently being more resistant, were disturbed only enough to cause cardia bifida and abnormalities of the nervous system.

The dose-response curves of Versene in calcium-free solution are given in Fig. 11. Cardia bifida was produced in the critical concentration range of 4.5-5.4 m$M$, and the concentration at which 50 per cent of the embryos were damaged ("$D_{50}$") is 4.3 m$M$. Below that range, little damage was done, and above it essentially all of the chicks were disaggregated or totally disorganized. To test whether calcium is actually the ion which was involved in these effects, a second dose-response curve was obtained (Fig. 12) with Versene dissolved in Howard-Ringer containing 1.52 m$M$ calcium chloride. Under these circumstances the cardia bifida range is seen to be 5.3-6.7 m$M$ and the $D_{50}$ is 5.8 m$M$. Thus a close agreement is noted between the amount of calcium available and the concentration of Versene required to produce its effects. Moreover, concentrations of Versene as high as eight millimolar (which yield 100 per cent disaggregation in calcium-free solution) have no effect whatsoever when added to embryos in solutions containing excess calcium (15 m$M$). Of greater significance is

[4] Disodium salt of ethylenediamine tetraacetic acid, Bersworth Company.

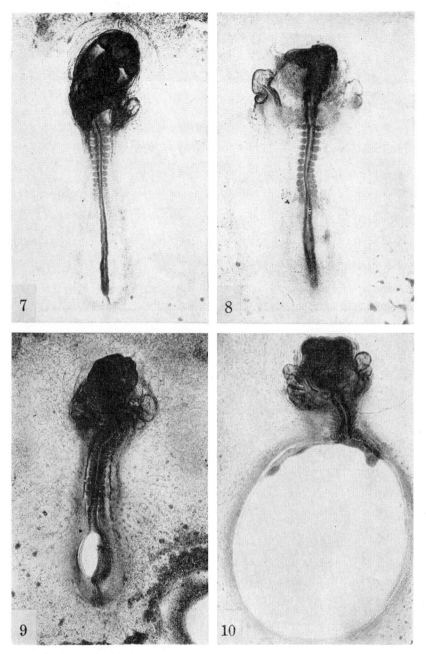

Figs. 7-10. Cardia bifida embryos produced by treatment with Versene. Compare with Figs. 2-5. × 20.

Fig. 7. Stage 11+, 6.2 mM Versene in Howard-Ringer.

Fig. 8. Stage 10, 5.65 mM Versene in Howard-Ringer.

Fig. 9. Stage 12, 4.8 mM Versene in calcium-free Howard-Ringer. Note opacity of tissues and arrested disaggregation which began in early primitive streak and extra-embryonic areas in the blastoderm.

Fig. 10. Approx. stage 11, 6.5 mM Versene in Howard-Ringer. Arrested disaggregation.

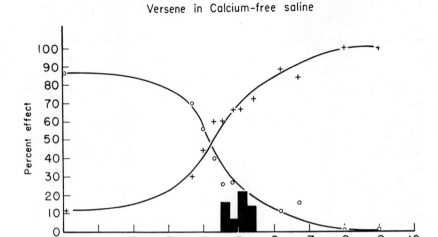

Fig. 11. Dose-response curve of Versene in calcium-free Howard-Ringer Symbols as in Fig. 6.

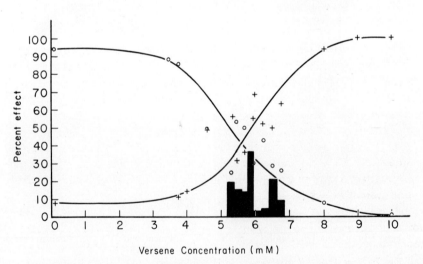

Fig. 12. Dose-response curve of Versene in Howard-Ringer containing 1.52 m$M$ calcium chloride. Compare with Fig. 11. Note that critical responses, i.e., range of CB and $D_{50}$, are displaced about 1.5 units to the right along the abscissa.

the finding that equimolar concentrations of Versene and calcium chloride (4.5 m$M$), when added to chicks together, have no deleterious effects. Magnesium also will prevent Versene dissociation of chick tissues; however, it is less potent in this protective role than is calcium. In a group of chicks treated with 4.5 m$M$ Versene in calcium-free Ringer containing 4.5 m$M$ magnesium chloride, 22 per cent were cardia bifida and an equal number were disaggregated or disorganized. Versene (4.5 m$M$) in the presence of a ten-fold excess (45.0 m$M$) of magnesium produces no cardia bifida or disaggregation. However, neither the experimental animals nor the controls (treated with 45 m$M$ magnesium but no Versene) develop quite normally, probably due to the high concentration of the metal.

The combination of Versene and the alkaline earth metals is low on the scale of stability constants when compared with other di- and tri-valent cations. Calcium is bound somewhat more strongly than magnesium, but manganese, iron, cobalt, etc., are much more tightly chelated (84). Thus, if Versene were producing its effects by removing one of these heavy metals from the cell (thereby perhaps disturbing an enzymatic function), neither calcium nor magnesium should offset the effect. If removal of magnesium by the chelating agent were the primary cause of cell dissociation, addition of Versene and calcium in equimolar concentrations should have a protective effect, but the amount of magnesium required to prevent cardia bifida and disaggregation should be equal to or less than that of calcium. This, as we have seen, is not the case, and we must conclude, therefore, that calcium is, in fact, the ion involved.

These experiments tend to support the hypothesis set forth earlier, stating that morphogenetic movements are dependent upon proper cellular adhesion, and that cell adhesiveness is in turn dependent upon calcium bound at cell surfaces. There are other possible interpretations of this work, however. One is that the developmental anomalies produced resulted from more general toxic effects of the agents used. It is difficult to predict the influences of molar concentrations of a quaternary ammonium compound on various aspects of cell physiology or structure. However, we have already seen that Ach markedly alters the colloidal behavior of phosphatides (19), and it has also been suggested that this substance can modify the configuration of proteins (98, 32). Its effect on enzyme systems is unknown. Versene, too, is known to have effects on cells, unrelated to the chelation of surface-bound calcium ions. Fabiny and Hamilton (35) have reported an inactivation of phosphatase by this compound, resulting in decreased synthesis of ribonucleic acids, and Mazia (87), Kaufmann, and co-workers (74, 89) and Kihlman (75) have described chromosomal aberrations in animal and plant cells resulting from treatment with Versene.

Another idea to account for the anomalies observed above stems from the report of Dornfeld and Owczarzak (33) that Versene causes marked surface "bubbling" and ameboid activity of chick heart fibroblast cells. Holtfreter (69) has described a similar response of amphibian cells to calcium-free or highly alkaline solutions. Thus, as Steinberg suggests, under circumstances of markedly increased cell surface activity and bleb-formation, a fabric of cells would tend to dissociate simply because the cell membranes were irregularly deformed (130). Irrespective of whether one envisions cell adhesion as involving calcium bridges or complementary protein configurations, some stable pattern is required for its maintenance.

Still a third possible mechanism hinges on a disturbance of normal inductive processes. Flickinger (40a) has reported inhibition of melanoblast differentiation with amino acids, and suggests that normal development of these cells is dependent upon factors contributed by solubilized yolk proteins. This solubilization normally requires calcium, and, therefore, is prevented when calcium is chelated by the amino acids. Bellairs (15) has recently suggested that the folding of the foregut results from induction by the splanchnic mesoderm. Thus, it is conceivable that normal folding of the endoderm is dependent upon the availability of some substance requiring divalent cations for its solubilization or entry into the cells. The action of complexing agents such as Versene or citrate would thus be predicted. The effects of Ach, however, would have to be accounted for in some other manner.

Since none of these last interpretations of the experimental results has more than suggestive evidence supporting it, further discussion would seem unprofitable for the present.

### Recapitulation and Conclusions

I might conclude by reiterating briefly those factors which, at the present time, seem most likely to be involved in the initiation and organization of morphogenetic movements, in cases of directed movements of single ameboid cells, such as those derived from the neural crest. (1) these cells are probably dispersed from their site of origin by a chemically mediated mutual cell repulsion, each cell migrating away from its fellows: (2) there is as yet no evidence that this migration, in vivo, is limited to predetermined routes or directions dictated by chemical attraction to distant sites or organs, or by specific pathways built into the substratum upon which the cells are migrating. There is, however, suggestive evidence of influences by apparently nonspecific chemical gradients such as $CO_2$, oxygen, or lipids. (3) Some crest derivatives, though probably not directionally oriented by their substrate, do indeed choose a given cell-type upon which to migrate. Thus, the importance of the cell-to-substrate relationship is reaffirmed. (4) The final pattern of cell locations is evidently determined by specific cell affinities and disaffinities. Secondary localizing factors such as contraction of originally dispersed cell groups into dense areas, or destruction of cells in noncompatible sites, also occurs.

In the case of mass cell movements, in which groups or aggregates of cells move by means of their own pseudopodial or gliding activity, two factors have been implicated: (a) relatively nonspecific chemical gradients; and (b) specific cell affinities and disaffinities. Evidence that embryonic cells (other than neural crest derivatives) may be influenced in their movements by chemical gradients stems from a limited number of in vitro or abnormal circumstances. Whether such gradients play a role in normal embryogenesis is not clear at the moment. On the other hand, observations are plentiful supporting the idea that cell masses are guided in their movements by the neighboring cells and structures over which they migrate. Moreover, the fact that cells can "recognize" and adhere, or fail to adhere, to their neighbors, depending on specific surface characteristics of the latter, is well substantiated. It seems reasonable to assume that such surface recognition would be mediated through patterned molecular groupings in the cell membrane. It must be assumed here that embryonic cells continue to migrate until a sufficient number or a special type of group is encountered at the contact surface so that firm linkages may be formed and further movement inhibited.

In epithelial layers, where each cell maintains a more or less fixed position in relation to its adjacent neighbors, the folding, invagination, spreading, or thickening that is seen must be explained in different (though related) terms. Alterations in adhesiveness of different parts of the cell surface, and the resultant change in area of cell contact, seem sufficient in general to account for such movements. Increased cell cohesion of one side of a layer would tend to cause the cells to contract or taper on that side, and shift the main mass of cytoplasm toward the other side, thus buckling or evaginating the stratum. Shortening of intercellular connections of various sorts, which can often be observed in embryos, is recognized as a probable contributing factor. It should be noted that when a filopodial process makes contact with, and sticks to, another cell, this represents merely a special case of the general problem of cell adhesion.

Of the three major hypotheses put forth as a mechanism for cellular adhesiveness (viz., complementary protein linkages, viscous intercellular cement, and calcium bridges), evidence has been marshalled favoring that invoking the binding of calcium ions between cell surfaces. The production of limited cell dissociation with high concentrations of organic cations or with calcium-complexing agents, resulting in a range of effects from minor disturbances of neural or cardiac fusion to disaggregation, lends support to this idea, although other interpretations are not ruled out.

Such an hypothesis immediately suggests several lines of experimental investigation to test its validity further. For example, is it possible to demonstrate differential adsorption of calcium at the surface of cells which appear to be increasing in adhesiveness during differentiation? Such a question might well be answered by histological studies of embryological material, using refined histochemical or biophysical techniques for ions (46, 83, 157). What is the effect of limited dissociation of cells, when studied under circumstances such as those employed by Twitty and Niu (141) or by Gregg and associates (52, 53, 102)? What is its effect on "contact inhibition" or on "contact guidance"? The abolition by Versene of Fetuin-induced adhesiveness of cells cultured on glass has been described recently (39). Further analysis of this phenomenon with carefully controlled removal of divalent cations might prove fruitful. An especially interesting question arises as to the effects of limited dissociation on a system such as that used by Chiakulas (24) to demonstrate specificity of cell adhesion. Are the adhesive sites identical with those providing specificity, as implied by Steinberg's lattice hypothesis (130), or is this latter property indeed provided by large molecular configurations, leaving cohesion, but not the specificity thereof, to be a function of calcium?

Professor B. H. Willier has, over the years, ended his graduate seminars with a statement to the effect that the value of any investigation or discussion is to be measured less by the number of answers it provides than by the number of new questions, and lines of experimental attack on these questions, that it stimulates. Perhaps, in this sense, the present discussion may represent a small contribution.

### Acknowledgments

I wish to acknowledge my debt to Dr. James D. Ebert and Prof. Johannes Holtfreter for their critical reading of my manuscript, and to Drs. Malcolm S. Steinberg and Mary E. Rawles for several stimulating discussions of various aspects of this paper. I would also like to thank Mrs. Nancy C. Fait for her superb technical assistance with the original work described herein.

# REFERENCES

1. Abercrombie, M., *Symposia Soc. Exptl. Biol.*, **11**, 235-254 (1957).
2. ———, *Proc. Zool. Soc. Bengal*, Mookerjee Memor. Vol., 129-140 (1957).
3. ———, and Heaysman, J. E. M., *Exptl. Cell Research*, **5**, 111-131 (1953).
4. ———, and ———, *Exptl. Cell Research*, **6**, 293-306 (1954).
5. ———, ———, and Karthauser, H. M., *Exptl. Cell Research*, **13**, 276-291 (1957).
6. Alexander, A. E., Teorell, T., and Aborg, C. G., *Trans. Faraday Soc.*, **35**, 1200-1205 (1939).
7. Anderson, N. G., *Science*, **117**, 627-628 (1953).
8. Andrews, E. A., *Zool. Bull.*, **2**, 1-13 (1898).
9. Andrews, G. F., *J. Morphol.*, **12**, 367-389 (1897).
10. Applezweig, N., *Ann. N. Y. Acad. Sci.*, **49**, 295-314 (1947).
11. Ballard, W. W., *J. Exptl. Zool.*, **129**, 77-98 (1955).
12. Bellairs, R., *J. Embryol. Exptl. Morphol.*, **1**, 115-124 (1953).
12a. ———, *J. Embryol. Exptl. Morphol.*, **1**, 369-385 (1953).
13. ———, *Brit. J. Cancer*, **8**, 685-692 (1954).
14. ———, *J. Embryol. Exptl. Morphol.*, **3**, 242-250 (1955).
15. ———, *J. Embryol. Exptl. Morphol.*, **5**, 340-350 (1957).
16. Boerema, I., *Wilhelm Roux' Arch. Entwicklungsmech. Organ.*, **115**, 601-615 (1929).
17. Bonner, J. T., *J. Exptl. Zool.*, **106**, 1-26 (1947).
18. ———, *J. Exptl. Zool.*, **110**, 259-272 (1949).
19. Booij, H. L., and Bungenberg de Jong, H. G., *Protoplasmatologia*, **1**, 1-162 (1956).
20. Brown, M. G., Hamburger, V., and Schmitt, F. O., *J. Exptl. Zool.*, **88**, 353-372 (1941).
21. Burt, A. S., *Biol. Bull.*, **85**, 103-115 (1943).
22. Caesar, R., Edwards, G. A., and Ruska, H., *J. Biophys. Biochem. Cytol.*, **3**, 867-878 (1957).
23. Chambers, R., *Cold Spring Harbor Symposia Quant. Biol.*, **8**, 144-153 (1940).
24. Chiakulas, J. J., *J. Exptl. Zool.*, **121**, 383-417 (1952).
25. Curtis, A. S. G., *Nature*, **181**, 185 (1958).
26. Dan, K., and Okazaki, K., *Biol. Bull.*, **110**, 29-42 (1956).
27. Danforth, C. H., and Foster, F., *J. Exptl. Zool.*, **52**, 443-470 (1929).
28. Davis, C. L., *Contribs. Embryol. Carnegie Inst.*, **19**, 245-284, #107 (1927).
29. Davis, J. O., *Biol. Bull.*, **87**, 73-95 (1944).
30. Davson, H., and Danielli, J. F., *The Permeability of Natural Membranes*, Cambridge, at the Univ. Press (1952).
31. DeHaan, R. L., *Proc. Natl. Acad. Sci. U. S.*, **44**, 32-37 (1958).
32. Demin, N. N., *Biokhimiya*, **20**, 317-327 (1955).
33. Dornfeld, E. J., and Owczarzak, A., *Anat. Rec.*, **128**, 541 (1957).
34. Ebert, J. D., *Carnegie Inst. Wash. Year Book*, 327-329 (1957).
35. Fabiny, R. J., and Hamilton, H. L., *Anat. Rec.*, **128**, 545 (1957).
36. Fawcett, D. W., and Selby, C. C., *J. Biophys. Biochem. Cytol.*, **4**, 63-72 (1958).
37. Feldman, M., *J. Embryol. Exptl. Morphol.*, **3**, 251-255 (1955).
38. Fisher, A., *Biology of Tissue Cells*, Gyldendalske Boghandel, Copenhagen (1946).
39. Fisher, H. W., Puck, T. T., and Sato, G., *Proc. Natl. Acad. Sci. U. S.*, **44**, 4-10 (1958).
40. Flickinger, R. A., *J. Exptl. Zool.*, **119**, 1-22 (1952).
40a. ———, *Am. Naturalist*, **41**, 373-380 (1957).
41. Giersberg, H., *Wilhelm Roux' Arch. Entwicklungsmech. Organ.*, **103**, 387-424 (1924).
42. Gillette, R., *J. Exptl. Zool.*, **96**, 201-222 (1944).
43. Gilman, H., and Jones, R. G., *Chem. Revs.*, **54**, 835-890 (1954).

44. Glaser, O. C., *Anat. Rec.*, **8**, 525-551 (1914).
45. ———, *Science*, **44**, 505-509 (1916).
46. Gomori, G., *Microscopic Histochemistry*, Univ. Chicago Press, Chicago (1952).
47. Goodrich, H. B., Hines, R. L., and Reynolds, J., *J. Exptl. Zool.*, **114**, 603-626 (1950).
48. ———, Marzullo, C. M., and Bronson, W. R., *J. Exptl. Zool.*, **125**, 487-506 (1954).
49. Goss, C. M., *Amer. J. Physiol.*, **137**, 146-152 (1942).
50. Gray, J., *Brit. J. Exptl. Biol.*, **3**, 167-187 (1926).
51. ———, *Experimental Cytology*, Cambridge, at the Univ. Press (1931).
52. Gregg, J. R., and Klein, D., *Biol. Bull.*, **109**, 265-270 (1955).
53. ———, and Ornstein, N., *Biol. Bull.*, **105**, 466-476 (1953).
54. Gross, P. R., *Trans. N. Y. Acad. Sci.*, **20**, (ser. II), 154-172 (1957).
55. Grossfeld, H., Meyer, K., Godman, G., and Linker, A., *J. Biophys. Biochem. Cytol.*, **3**, 391-396 (1957).
56. Gustafson, T., and Kinnander, H., *Exptl. Cell Research*, **11**, 36-51 (1956).
57. Hall, E. K., *J. Exptl. Zool.*, **113**, 355-378 (1950).
58. Hamburger, V., and Hamilton, H. L., *J. Morphol.*, **88**, 49-92 (1951).
59. Harris, H., *J. Pathol. Bacteriol.*, **66**, 135-146 (1953).
60. Harvey, E. N., *Protoplasmatologia*, **2**, 1-30 (1954).
61. Heilbrunn, L. V., *Dynamics of Living Protoplasm*, Academic Press, New York (1956).
62. Herbst, C., *Wilhelm Roux' Arch. Entwicklungsmech. Organ.*, **9**, 424-463 (1900).
63. Herrman, H., and Hickman, F. H., *Bull. Johns Hopkins Hosp.*, **82**, 182-207 (1948).
64. Hobson, L. B., *J. Exptl. Zool.*, **88**, 107-134 (1941).
65. Holtfreter, J., *Arch. exptl. Zellforsch. Gewebezücht.*, **23**, 169-209 (1939).
66. ———, *Rev. Can. biol.*, **3**, 220-250 (1944).
67. ———, *J. Morphol.*, **80**, 25-55 (1947).
68. ———, *J. Morphol.*, **80**, 57-91 (1947).
69. ———, *Ann. N. Y. Acad. Sci.*, **49**, 709-760 (1948).
70. Holtzer, H., Marshall, J. M., and Fink, H., *J. Biophys. Biochem. Cytol.*, **3**, 705-724 (1957).
71. Hörstadius, S., *The Neural Crest*, Oxford Univ. Press, Oxford (1950).
72. Howard, E., *J. Cellular Comp. Physiol.*, **41**, 237-260 (1953).
73. Jaffe, L., *Proc. Natl. Acad. Sci. U. S.*, **41**, 267-270 (1955).
74. Kaufman, B. P., and McDonald, M. R., *Proc. Natl. Acad. Sci. U. S.*, **43**, 262-270 (1957).
75. Kihlman, B. A., *J. Biophys. Biochem. Cytol.*, **3**, 363-380 (1957).
76. Kwart, H., and Shashoua, V. E., *Trans. N. Y. Acad. Sci.*, **19**, 595-612 (1957).
77. Lansing, A. I., and Rosenthal, T. B., *J. Cellular Comp. Physiol.*, **40**, 337-345 (1952).
78. Lehmann, F. E., *Wilhelm Roux' Arch. Entwicklungsmech. Organ.*, **108**, 243-282 (1926).
79. ———, *Wilhelm Roux' Arch. Entwicklungsmech. Organ.*, **113**, 123-171 (1928).
80. Lewis, W. H., *Contribs. Embryol. Carnegie Inst.*, **18**, 1-22 (1926).
81. ———, *Anat. Rec.*, **97**, 139-156 (1947).
82. Lillie, R. S., *Am. J. Physiol.*, **17**, 89-141 (1906).
83. Lindstrom, B., *Acta Radiol.* (Suppl.), **125**, 1-206 (1955).
84. Martell, A. E., and Calvin, M., *Chemistry of the Metal Chelate Compounds*, Prentice-Hall, New York (1952).
85. Mast, S. O., and Fowler, C., *Biol. Bull.*, **74**, 297-305 (1938).
86. Mazia, D., *Cold Spring Harbor Symposia Quant. Biol.*, **8**, 195-203 (1940).
87. ———, *Proc. Natl. Acad. Sci. U. S.*, **40**, 521-527 (1954).
88. McCutcheon, M., and Lucké, B., *J. Gen. Physiol.*, **12**, 129-138 (1928).
89. McDonald, M. R., and Kaufman, B. P., *Exptl. Cell Research*, **12**, 415-417 (1957).
90. Milkman, R., and Trinkaus, J. P., *Anat. Rec.*, **117**, 558-559 (1953).
91. Mitchison, J. M., *Symposia Soc. Exptl. Biol.*, **6**, 105-127 (1952).

92. Monné, L., *Experientia*, **2**, 153-196 (1946).
93. Mookerjee, S., Deuchar, E. M., and Waddington, C. H., *J. Embryol. Exptl. Morphol.*, **1**, 399-409 (1953).
94. Moore, A. R., *J. Exptl. Zool.*, **87**, 101-111 (1941).
95. Moscona, A., *Proc. Soc. Exptl. Biol. Med.*, **92**, 410-416 (1956).
96. ———, *Proc. Natl. Acad. Sci. U. S.*, **43**, 184-194 (1957).
97. ———, and Moscona, H., *J. Anat.*, **86**, 287-301 (1952).
98. Nachmansohn, D., and Wilson, I. B., in *Electrochemistry in Biology and Medicine*, (T. Shedlovsky, ed.), p. 167-186, John Wiley & Sons, New York (1955).
99. New, D. A. T., *J. Embryol. Exptl. Morphol.*, **3**, 326-331 (1955).
100. Oppenheimer, J. M., *J. Exptl. Zool.*, **128**, 525-560 (1955).
101. ———, *Surv. Biol. Progr.*, **3**, 1-46 (1957).
102. Ornstein, N., and Gregg, J. R., *Biol. Bull.*, **103**, 407-420 (1952).
103. Overton, E., *Arch. ges. Physiol., Pflüger's*, **105**, 176-290 (1904).
104. Patten, B. M., *Physiol. Revs.*, **29**, 31-47 (1949).
105. Pauling, L., in *Specificity of Serological Reactions* (K. Landsteiner), rev. ed., p. 275-293, Harvard Univ. Press, Cambridge (1945).
106. Ponder, E., *Protoplasmatologia*, **10**, 1-123 (1955).
107. Rawles, M. E., *Physiol. Zool.*, **16**, 22-45 (1943).
108. ———, *Physiol. Zool.*, **17**, 167-183 (1944).
109. ———, *Physiol. Zool.*, **18**, 1-16 (1945).
110. ———, *Physiol. Revs.*, **28**, 383-408 (1948).
111. ———, *Am. J. Anat.*, **97**, 79-128 (1955).
112. ———, in *Analysis of Development* (B. H. Willier, P. Weiss, and V. Hamburger, eds.), p. 499-519, W. B. Saunders Co., Philadelphia (1955).
113. Reid, M. E., *Physiol. Revs.*, **23**, 76-99 (1943).
114. Rhumbler, L., in *Handbuch der biologischen Arbeitsmethoden* (E. Abderhalden, ed.), **5** (3A), 219-440, Berlin (1923).
115. Ringer, S., *J. Physiol.*, **11**, 79-84 (1890).
116. Robertson, J. D., *Biol. Revs. Cambridge Phil. Soc.*, **16**, 106-133 (1941).
117. Rosin, S., *Rev. Suisse Zool.*, **50**, 485-578 (1943).
118. Rothstein, A., *Protoplasmatologia*, **2**, 1-86 (1954).
119. ———, and Meier, R., *J. Cellular Comp. Physiol.*, **38**, 245-270 (1951).
120. Sabin, F. R., *Contribs. Embryol. Carnegie Inst.*, **6**, 63-124 (1917).
121. ———, *Contribs. Embryol. Carnegie Inst.*, **9**, 213-262, (1920).
122. Schechtman, A. M., *Univ. Calif. Publ. Zool.*, **39**, 303-310 (1934).
123. ———, *Univ. Calif. Publ. Zool.*, **51**, 1-40 (1942).
124. Schmitt, F. O., *Growth*, **5** (Suppl.), 1-20 (1941).
125. Spiegel, M., *Biol. Bull.*, **107**, 130-148 (1954).
126. ———, *Biol. Bull.*, **107**, 149-155 (1954).
127. Spratt, N. T., *J. Exptl. Zool.*, **128**, 121-164 (1955).
128. ———, *Anat. Rec.*, **124**, 364 (1956).
129. ———, *J. Exptl. Zool.*, **134**, 577-612 (1957).
130. Steinberg, M. S., *Am. Naturalist*, **92**, 65-82 (1958).
131. Sussman, M., and Sussman, R. R., in *Cellular Mechanisms in Differentiation and Growth* (D. Rudnick, ed.) p. 125-154, Princeton Univ. Press, Princeton (1956).
132. Townes, P. L., *Exptl. Cell Research*, **4**, 96-101 (1953).
133. ———, and Holtfreter, J., *J. Exptl. Zool.*, **128**, 53-120 (1955).
134. Trinkaus, J. P., *J. Exptl. Zool.*, **118**, 269-319 (1951).
135. ———, and Drake, J. W., *J. Exptl. Zool.*, **132**, 311-348 (1956).
136. ———, and Groves, W., *Proc. Natl. Acad. Sci. U. S.*, **41**, 787-795 (1955).
137. Twitty, V. C., *J. Exptl. Zool.*, **95**, 259-290 (1944).

138. ———, *J. Exptl. Zool.*, **100**, 141-178 (1945).
139. ———, *Growth*, **13** (Suppl.), 133-161 (1949).
140. ———, and Niu, M. C., *J. Exptl. Zool.*, **108**, 405-438 (1948).
141. ———, and ———, *J. Exptl. Zool.*, **125**, 541-574 (1954).
142. Tyler, A., *Growth*, **10** (Suppl.), 7-19 (1946).
143. Waddington, C. H., *J. Exptl. Biol.*, **19**, 284-293 (1942).
144. ———, *Principles of Embryology*, Macmillan Co., New York (1956).
145. Waugh, D. F., and Schmitt, F. O., *Cold Spring Harbor Symposia Quant. Biol.*, **8**, 233-241 (1940).
146. Weiss, P., *Growth*, **5** (Suppl.), 163-203 (1941).
147. ———, *J. Exptl. Zool.*, **100**, 353-386 (1945).
148. ———, *Yale J. Biol. and Med.*, **19**, 235-278 (1947).
149. ———, *J. Exptl. Zool.*, **113**, 397-461 (1950).
150. ———, *Quart. Rev. Biol.*, **25**, 177-198 (1950).
151. ———, in *Analysis of Development* (B. H. Willier, P. Weiss, and V. Hamburger, eds.), p. 346-401, W. B. Saunders Co., Philadelphia (1955).
152. ———, and Andres, G., *J. Exptl. Zool.*, **121**, 449-487 (1952).
153. Whitaker, D. M., *Growth* (Suppl., Second Symposium), 75-90 (1940).
154. Whitehorn, J. C., *J. Biol. Chem.*, **56**, 751-764 (1923).
155. Wilens, S., *J. Exptl. Zool.*, **129**, 579-606 (1955).
156. Willier, B. H., *Bios*, **23**, 109-125 (1952).
157. Zeitz, L., and Baez, A. V., *X-ray Microscopy and Microradiography* (V. E. Cosslett, A. Engström, and H. H. Pattee, Jr. eds.) p. 417-434, Academic Press, New York (1957).
158. Zweifach, B. W., *Cold Spring Harbor Symposia Quant. Biol.*, **8**, 216-223 (1940).
159. Zwilling, E., *Science*, **120**, 219 (1954).

## DISCUSSION

DR. HOLTFRETER: I wonder to what extent your observations on the chick embryo can be related with the general topic of your paper. The method of disaggregating embryonic tissues by various means, of mixing the isolated cells together in different permutation and allowing them to reshuffle themselves so as to form well-segregated tissues, has become quite fashionable. You did not do this sort of experiment. You applied some chemicals to the prospective heart region of the chick embryo. The result was anatomic malformation, especially of the heart.

There are several ways of connecting causally the intervening chemical agent with the final morphological manifestation. At face value it seems that the heart malformations observed cannot be readily connected with the principles of cellular locomotion or cell affinity which are your general topic. The chemicals you used may have simply acted as toxic or semi-toxic agents. Some of your results, such as two lateral hearts can be achieved by surgical interferences.

All that matters here is to prevent a junction of the lateral heart primordia in the ventral midline. It might be expected that by killing or damaging the midventral juncture heart region by chemical means would produce such an obstacle. We know that embryonic cells can be killed by an excess of oxalate, citrate, Versene; what high concentrations of acetylcholine would do to the cells I don't know. Have you studied microscopically the immediate and successive reactions of the individual cells subjected to your treatments? My suspicion is that the treatments either killed the area affected outrightly or incapacitated it in such a way that it became a barrier to the fusion of the heart primordia.

Dr. DeHaan: Yes, I have checked these preparations under the high powered microscope. I don't know that I haven't killed cells but I can say that most of the cells appear to be alive. I say this not only on the basis of their appearance under the microscope, but I also wish to refer you to the appearance of the partial disaggregates which I referred to during my discussion. In these cases disaggregation apparently occurred and then the reaggregation occurred around the periphery of the glass circle after 24 hours time.

Dr. Zwilling: Have you tried a solution of acetylcholine on early chick embryos to see if it will disaggregate their tissues?

Dr. DeHaan: No, as a matter of fact I have not checked this with isolated tissues.

Dr. Zwilling: This happens very readily with Versene. The other thing is whether you have considered another possibility. As you know in our laboratory a great deal of evidence has been accumulated by Dr. Landauer indicating that various metabolites are involved in teratology. In these systems the coenzymes seem to be implicated very strongly. We know that many of the coenzymes need metal activation to perform their functions, say with iron or magnesium for instance. Virtually all of the work which you have presented could be explained on the basis of the inhibition of coenzyme function which has resulted from the chelation of metal-activating agents.

Dr. DeHaan: I think you have a good point there and it is one that I have considered. I can only say that all of the agents which do produce the effect are, in some way, related to calcium. They can either displace it from a surface, or bind it. Of course, these agents can do this to other ions as well. It is just that when you talk about coenzymes it seems rather unlikely that acetylcholine would be displacing magnesium or iron from an enzyme site in the cell. This is especially so since Nachmansohn and others have pointed out that acetylcholine does not penetrate into cells very well. Of course the one molar concentration which we have used here might indeed get into cells. Another point is that I am basing a large part of this hypothesis on the continuum idea, that is, that we see a gradation of effects which merge imperceptibly into one another when we increase the concentration of the agent in the medium, culminating in disaggregation. It seems to me rather unlikely that such a continuum would be caused by displacement of increasing amounts of an ion from a coenzyme, since I know of no evidence linking coenzyme function and cell adhesion. Also, I would refer you to my discussion of the stability constants of the chelate compounds formed between Versene and various polyvalent metals, suggesting that calcium is indeed the ion involved in these effects, and not those metals known to act as coenzymes such as iron, magnesium, manganese, cobalt, etc.

Dr. Zwilling: But here I think the point which Dr. Holtfreter made is very important. The role of cell death has to be evaluated carefully in this situation. Cell separation of some sort is a frequent symptom of early moribundity.

Dr. DeHaan: I would certainly agree with you that this may be an important factor.

Dr. Boell: I should like to ask if you observe abnormalities in the nervous system. You would expect that the neural folds would not fuse if your explanation is a valid one.

Dr. DeHaan: Indeed you do observe this effect. In two of the slides which I presented this morning, even though seen through the ventral side of the embryo it is quite apparent that the folds of the neural tube have not fused completely. This again can occur in varying degrees of severity. I think that one of the reasons why we get the teratogenic effects on the ventral side of the embryo more frequently than on the dorsal side is simply because that is where I am directly adding the agent. Therefore the agent has to diffuse through three layers of cells before it can get to the nervous system in the dorsal side of the embryo.

Dr. Boell: But aren't these hearts themselves tubular?

Dr. DeHaan: Yes, that too is a very good point, but I just don't know what happens there. I do get beautiful tubular structures in the double hearts in the case of *cardia bifida*. In some cases, if I let them go long enough, circulation is established so that I know the tubular vascular system is intact. I am afraid I just can't explain that. I can only suggest that since the original tubules of the heart form from vaso-formative mesenchymal tissue as it is folding under with the endoderm possibly we have two different mechanisms; one for the folding of epithelial sheets or tissues which is dependent upon calcium, as I have suggested, and another mechanism which is involved in the hollowing out of mesenchymal tissues. I just don't know.

Dr. Trinkaus: In our work on *Fundulus* gastrulation we have found support for the hypothesis that coordinated movements of masses of cells may be due in part to a change in the character of the individual constituent cells. In this study we dissociated blastoderms of *Fundulus heteroclitus* in late cleavage, blastula and early gastrula stages with Versene and cultured suspensions of individual cells in standing drops on a glass substratum. The medium consisted of 0.2% proteose peptone and 0.5% glucose in double strength Holtfreter's solution buffered with phosphate and bicarbonate to $p$H 7.4. Under these conditions cells from late cleavage stages and from blastulae (up to stage 10) are semispherical and actively protrude lobopodia during the first few hours in culture. In contrast, cells from eggs in which epiboly has already begun (stage 11 to 12) become attached to and actively spread over the substratum during the first 30 minutes after culturing. In this fashion they rapidly achieve a fibroblast-like shape with abundant filopodia and contact each other to form a flattened lacework of cells. Cells from advanced blastulae (just prior to the onset of epiboly, i.e. stage $10\frac{1}{2}$) are typically spherical with lobopodia during the first hour in culture, but during the next hour they spread on the substratum. At this time intact control eggs have commenced epiboly (stage $11\frac{1}{2}$).

We conclude from these observations that the spreading of the whole blastoderm in epiboly is due in part to the acquisition by individual cells of the capacity to spread over the substratum. Moreover, this spreading character is relatively stable under conditions in vitro.

# Part III

*PROBLEMS OF SPECIFICITY IN GROWTH AND DEVELOPMENT*

# COMPARATIVE CYTOCHEMISTRY OF THE FERTILIZED EGG

J. J. PASTEELS

*Embryological Laboratories of the Faculty of Medicine, Université libre de Bruxelles, Belgium*

INTRODUCTION

The time is past when the relations between the morphologists and the biochemists consisted only of a "dialogue between deaf men." Refinements of techniques have brought them both to the same level of ultrastructure, where spatial organization and enzymatic mechanisms are necessarily complementary. The different methods of study are now often employed in the same laboratory, often by the same man, to achieve the essential goal of the Biologist: the integration of the structure and the function of the cytoplasm.

The correlation between structure and biochemical mechanisms has been studied mainly in adult cells, particularly the hepatic cells of mammals. We propose to examine the way in which this correlation may be considered in the case of the egg, limiting ourselves to the fertilized, unsegmented egg. There is no doubt that more problems will be proposed than will be solved. The fact that a problem is presented is nevertheless of considerable interest, and its subsequent study can be of importance in helping solve the main problem under consideration.

The egg should be considered from two points of view. (1) The egg as a cell: this presents a paradox, since the egg is undifferentiated on the one hand, and highly specialized on the other. The question arises as to whether this particular case of the egg-cell also has particularities of structure and chemistry. (2) The egg as a future embryo: one can forget too easily that the principal function of an egg is to form an embryo. This ultimate morphogenesis must essentially be preenvisaged by structural particularities. Thus, it is important to be familiar with the relationship between chronology and topography. The correlation between the structure and function of the egg will only be considered as definitely established when the results obtained by different techniques agree. This agreement has been obtained only rarely in the particular case of the egg. It is important to consider the value of these various methods, always keeping in mind limitations in order not to overestimate the results.

Before attempting to get at the base of these two problems, it might therefore be a good idea to review briefly the methods of study:

(a) *The isolation of cytoplasmic particles by fractional centrifugation of cellular homogenates.* This method is to a large extent responsible for the recent rapid development of cytology. Nevertheless, there does exist a great danger of misinterpretation due to artifacts. Fortunately, the source of the artifacts produced by this method is known today. Thus, it appears that the microsomes formerly considered as fundamental organelles of the cytoplasm and controlling protein synthesis, have no real value as such but are the result of an artifact consisting of a fragmentation of the ergastoplasm (11). The method of homogenization in electrolytic media has been criticized (61) on the basis that it can provoke elutions, contamination, or agglomeration of the particles. Thus, most of the results of studies made on eggs have been obtained by this method, and the current improvements in the technique, such as the homogenization of the material in solutions containing glucose, have been utilized only in very recent studies (90, 91, 101, 149).

(b) *The analysis of enzymatic or respiratory activity of fragments of eggs obtained by centrifugation of eggs in the living state* (57). This method permits consideration of these activities with respect to the presence or absence of specific cytoplasmic particles (62, 64, 65). It should be pointed out that although the danger of an artifact is much more limited in this than in the preceding technique, it is not totally excluded, since the sedimentation of the particulates in an area can modify their enzymatic and respiratory activities (17, 155). The results obtained by this method can be justly evaluated only after a correlation with the results obtained by cytochemical and microscopical investigation, particularly with the electron microscope. Indeed, we shall see that the presence or absence of mitochondria can be confirmed only after a study of the ultrastructure made with ultrathin sections (cf. pp. 386-387).

(c) *The study of the respiratory or enzymatic activities of isolated blastomeres or egg fragments without previous centrifugation* (8, 9, 10, 67). This method gives us information on the subsequent linkage between these activities and the morphogenetic tendencies of the territories. The application of this method is, for reasons easy to understand, rather limited. In addition to this, the method is not exempt from artifacts: the activity of an isolated fragment can be different, following a morphogenetic regulation, from that of the corresponding territory in an intact egg.

(d) *Cytochemical studies of sections.* One disadvantage of this technique is that it can only be applied to a limited number of substances. This method gives us important topographical information and has an advantage over the others in that it permits us to establish more easily the relationship between morphology and chemistry, although it is very rarely quantitative. The field of cytochemistry is constantly growing. Even at the present time it is possible to study the cytochemistry of the living egg by using metachromatic stains (34, 35, 39, 122). Another recent aspect of cytochemical study gives promise of a bright future. This involves the use of radioactive tracers in sections (1, 18, 49, 158). Up to now, however, these particular methods have practically not been used for the study of the first stages of development. It would seem that a combination of these techniques with microsurgical techniques might greatly increase the scope of their application (1).

(e) *Studies with the electron microscope.* The analysis of the fine structure of the egg with the electron microscope is still in its infancy. This method seems indispensable in permitting the determination of the nature of the constituents, as revealed by the other techniques. The weakness of this method lies in the difficulty of choosing an adequate fixative. It cannot be overemphasized that the osmic fixatives which are excellent for the preservation of all lipoproteic structures preserve the other structures very poorly or not at all (86, 87).

### The Egg as a Cell

*A. The Cytoplasm*

1. *Ribonucleic acid.* It is because of its decisive role in protein synthesis, in embryonic cells as well as in adult cells, that the RNA is considered to be a fundamental constituent of the egg. Cytochemical tests have provided valuable information on its topographical localization. It is paradoxical to state, however, that its structural localization (that is, its relationship with the ultrastructure of the cytoplasm) has been until recently entirely ignored. It is generally accepted that the RNA is more loosely combined in the egg than in the adult cell; that is, it is associated with much smaller particles, possibly even as isolated molecules. The technique of high-speed, fractional centrifugation of homogenates shows that in amphibians (14, 19) and sea urchins (53, 170), the RNA is found in the supernatant. In *Arbacia* eggs, it has been shown that the RNA is found in the particles of which only 25 per cent sediment at a centrifugal force of 25,000 g, the remainder sedimenting at 100,000 g. The latter have been estimated to

have a diameter of $3 \times 10^{-6}$ cm. (300 Å), corresponding to a molecular weight of 10,000,000 (170). It should be pointed out, however, that these conclusions result from experiments with homogenates in solutions containing KCl. The authors have observed that the solubilization of RNA increases with the concentration of KCl. A more recent study has confirmed these conclusions by showing that the sedimentation of these granules is accelerated in the presence of $Ca^{++}$ ions (53).

These conclusions seem to be corroborated by the results of cytochemical studies. In *adult cells* centrifuged in the living state, the ergastoplasm is displaced toward the centrifugal pole (11, 134), whereas in the case of the centrifuged *eggs*, the RNA is more often found centripetally, in the hyaloplasm just beneath the lipid pole (19).

Several remarks can be made in regard to this point. (1) It is very difficult to compare the relative sedimentation conditions of an adult cell and of an egg which is very rich in heavy vitelline platelets. (2) In addition to the RNA in the hyaloplasm, it has been possible to demonstrate the presence of RNA in heavy particles which are displaced toward the centrifugal pole. This has been shown in *Limnaea* eggs (134), and, as we will show later, in *Paracentrotus* eggs. (3) Interesting observations made with *Cyclops* show that during the phases of the approach of the pronuclei and of the first mitotic divisions, there is a cyclic evolution of RNA, sometimes dispersed in the hyaloplasm, sometimes accumulating in large and heavy granules which sediment more rapidly. (4) Finally, studies made with invertebrates such as *Ascaris, Acanthoscelides* (a bruchid beetle), and *Cyclops* (117, 103, 164) have shown that a relationship exists between the richness of the granular system of RNA and the origin of the gonocytes, a relation explaining also the basophilia which is so unique in genital determinants. It thus seems that RNA can be linked to these different cytoplasmic organites and is modified during the cellular cycle.

The study of the ultrastructure of the cytoplasm of the ovule, particularly its relationship with its basophilia, is much less advanced than that of adult cells. The few studies which have appeared up to now concern the polar plasm of *Tubifex* eggs (86, 87, 173), the unfertilized egg of *Psammechinus* (2), and oocytes of the mollusks *Otala* and *Spisula* (138, 139, 168).

Although the cytochemistry of the oocyte will not be considered here, some space will be devoted to the unexpected discovery of the new concept of the ergastoplasm as consisting of annulate membranes (membranae fenestratae, periodic lamellae). These have a structure which is typical of that of a nuclear membrane, with its double lamella. This can be shown by osmic fixatives to be perforated regularly by pores, which are more or less blocked by a mass of granules. In the unfertilized ovum of *Paracentrotus* (2) these membranes surround heavy bodies which are very rich in RNA and which resemble nucleoli. These have been considered to originate from the membrane of the germinal vesicle. In the oocytes of *Otala* and *Spisula* such laminations form very basophilic masses in the midst of which appear vitelline granules. At certain points these lamellae are dilated to form vesicles which form a typical ergastoplasm containing Palade's granules. These groupings of lamellae are found at first in contact with the nuclear membrane, of which they are a sort of replica (138, 139). These new organites have also been found in *Ambystoma* pancreas and in rat spermatids (168), as well as in cancerous cells (154, 174).

A study now in press (123) permits us to visualize the relationships between the ultrastructure and the cytochemistry of the fertile and unsegmented egg of *Paracentrotus*. The centrifuged egg of this species sediments in four layers which are, beginning with the centripetal region: (1) lipid; (2) hyaloplasmic; (3) vitelline; and (4) mitochondrial. RNA (demonstrated by the Unna-Brachet reaction, with an RNAase control) is found in layers 2, 3, and 4, but in distinctly different quantities. In the hyaloplasm, staining with pyronine is uniform but moderate; in the vitelline zone, the coloration is found with the same intensity only in the spaces between the platelets; on the other hand, the reaction is very pronounced in the mitochondrial zone, where round bodies with smooth contours can be distinguished in a more diffuse background.

The study with the electron microscope of ultrathin serial sections of eggs fixed in Palade's fluid has shown in the hyaline zone and in the spaces between the platelets an atypical ergastoplasm consisting of vesicles sometimes in continuity with flattened cylinders, the walls of which are always surrounded by many Palade granules. Furthermore, in the vitelline zone, numerous annulate lamellae are found. Those organites are here free (i.e., are not associated with heavy bodies); they are continuous with a typical ergastoplasm (double lamellae with Palade granules). We have confirmed the conclusions which also show the ergastoplasmic nature of these structures in mollusk oocytes (139, 168).

The basophilia, shown by the Unna-Brachet reaction, is by far most pronounced in the centrifugal cap containing the mitochondria. However, after osmic fixation, no ergastoplasmic structure can be discerned. Instead, we find heavy bodies (2), which are dense granular masses 1-3 $\mu$ in diameter. On the other hand, in the fertilized egg they are only rarely surrounded by annulate lamellae. These bodies are nevertheless widely dispersed and cannot be considered as being responsible for the intense basophilia of structures apparently poorly fixed by osmic acid, but which nevertheless show neither lamellae nor Palade granules when observed in the limit of the resolution of the microscope used. If one tries to associate the intense basophilia of this zone with a poorly fixed and unknown structure which is found between the mitochondria or which may be the mitochondria themselves, uncertainty remains. The observations seem to be more in agreement with the latter hypothesis, especially those made with centrifuged *Arbacia* eggs, in which the mitochondrial zone above the vitellus is found to be equally rich in RNA (105).

Thus the results of this study show that the RNA in the *Paracentrotus* egg is linked to four different structures. (1) A vesicular ergastoplasm, atypical but rich in Palade particles. This type of ergastoplasm seems to be the same as the osmiophilic vesicles described in the polar plasm of *Tubifex* (172) as well as that which has recently been described in the mammary adenocarcinoma of the rat (154). (2) Annulate membranes, a peculiar type of ergastoplasm the structure of which seems to indicate a nuclear origin. (3) Large bodies and dense heavy bodies, the structure of which is reminiscent of that of nucleoli. (4) An undefined structure, which could be linked to the mitochondria but which is composed of the most dense material in the egg.

With what information we have at present, we can only compare the results obtained from homogenates of *Arbacia* to those of cytochemical and electron microscopical studies of *Paracentrotus*.

The conclusions, though only provisionary, which can be drawn from these studies are as follows. (1) One finds in the egg of the sea urchin two ergastoplasmic modalities, which although different from those described in most adult cells can both be found in a certain tumor cell (the mammary adenocarcinoma of the rat). (2) A large percentage of the ovular RNA is found to be linked to non-ergastoplasmic structures, which although not yet definite, may be the mitochondria. (3) The slow sedimentation of these RNA-carrying particles in the homogenate seems to be due to the great fragility of these structures in a solution of KCl.

2. *The enzymes.* From the point of view of the enzymes as well, numerous authors have not hesitated to make a clear distinction between the egg and the adult cell (54, 63, 76, 163). According to one thesis, the enzymes, including those concerned with aerobic respiration, are not linked to the mitochondria but are dispersed in the cytoplasm of the egg; their incorporation into the mitochondrial structures would correspond progressively with differentiation. Such a thesis, however, lacks a very strong basis.

It is true that homogenization in a saline (KCl) medium has shown that in *Arbacia* cytochrome oxidase is not found specifically in the sediment but rather in the supernatant (71). The same method has also shown that the oxidative phosphorylation occurs in the supernatant containing these particles but lacking the mitochondria (75). However, the results obtained with homogenates of amphibian eggs lead to different conclusions. In *Rana* oocytes, 70 per cent of the cytochrome oxidase is found in the mitochondrial fraction (140). Such is also the case in the unfertilized egg of *Xenopus,* where the mitochondria exhibit a typical cytochrome oxidase activity. This activity increases strongly during development, but the increase is too great to be attributed to the eventual increase in the number of mitochondria. One should also consider the possibility of a true differentiation of the mitochondria, either in the sense that they increase in enzyme content, or that the conditions which regulate the activity of the enzyme become more favorable (173). It should be pointed out, however, that according to this concept cytochrome oxidase would be linked to the mitochondria of the unsegmented egg.

The theory that the enzymes are not linked to the particles has been reenforced particularly by the results of centrifugation experiments with the living egg, in which the "clear" fragments could be separated from the "granular" fragments (57-59). A comparison of the activity of centrifugal and centripetal fragments, accompanied by a computation of the respective volumes of the hyaloplasm, has led to the conclusion that dipeptidase and catalase are found to be distributed in this "ground cytoplasm" (62, 64). Finally, since it has been well shown that a clear fragment which was supposed to be completely deprived of mitochondria is capable of undergoing segmentation and developing into a complete larva (57-59), the obvious conclusion was that all of the enzymes necessary for development are found in the hyaloplasm, and that the mitochondria which are found in the cells of the developing larva are formed de novo.

Studies with the electron microscope have recently shown the weakness of this argument, which at first seemed infallible. The resolution of the light microscope does not suffice to show conclusively the absence of mitochondria in the centrifuged fragments. In centrifuged *Arbacia* (80) and *Paracentrotus* (123) eggs mitochondria settle out in two separate layers, one of which, contrary to what might be expected, remains mixed with the lipid fraction at the centripetal cap. Our own studies have shown that this association involves a physical connection. In the normal (non-centrifuged) egg, each lipid droplet is completely surrounded by a wreath of mitochondria. This thus constitutes a very good argument, although a morphological one, to explain the role of mitochondria in the lipid metabolism of the egg.

The conclusion drawn from the results of E. B. Harvey's elegant method has been somewhat premature. However, hers is still an excellent method for studying the significance, the activity, and eventually the origin of the cytoplasmic particles, upon condition that the method is controlled by complementary studies made with the electron microscope.

On the other hand, cytochemical studies favor the existence of some link between aerobic respiratory enzymes and mitochondria. There are excellent cytochemical reactions for the demonstration of cytochrome oxidase, such as the Janus green B reaction (81) and the Nadi reaction, which should be used in conjunction with cytochrome oxidase inhibitors as controls. All of the researchers who have utilized these techniques, whether with the eggs of ctenophores (129, 130, 144), annelids (83, 84, 85, 130, 131, 147), mollusks (5, 6, 92, 134, 135, 136, 150, 151), ascidians (142, 143, 146, 152, 171), or the centrifuged eggs of the sea urchin *Sphaerechinus* (93), have found the reaction to be positive in the mitochondrial mass. Such is also the case for succinic dehydrogenase as demonstrated by the neotetrazolium technique in the *Tubifex* egg (172).

A remark should be made at this point. The Nadi reaction has its limitations in the facility with which the indophenol blue can be visualized. The coloration by the dye is very definite when the mitochondria are densely packed, either in the natural "yellow crescent" state of ascidians, or as the result of centrifugation. However, in dorsal blastomeres of ascidians, where the mitochondria are scattered, the reaction is very faint. There is no reason to doubt the existence of cytochrome oxidase in these blastomeres. The cytochrome oxidase activity has been studied in homogenates of ventral and dorsal blastomeres of *Ciona* (9), and has been found to be in a ratio of 2.7:1, respectively. Nevertheless, it has been demonstrated that the consumption of oxygen by isolated dorsal and ventral blastomeres is identical. This fact shows that the *total* enzymatic activity shown after homogenization or by cytochemical tests does not necessarily operate in the *actual activities* of the egg during the first stages of development (as we shall see later). There is thus evidence for the existence of cytochrome oxidase in the mitochondria of the eggs of all invertebrates which have been studied in this respect; but the actual methods of study do not permit us to conclude its existence *only* in mitochondria. What we can confirm, however, is that the theory that enzymes are not linked to cytoplasmic particles at the beginning of the egg's development is certainly exaggerated and is based at least partly on errors in technique.

As for the localization of RNA, the results obtained by homogenization in media containing electrolytes should be considered with suspicion, since recent studies (90, 91, 101, 149), whereby a sucrose solution was utilized for the homogenates, have led to different conclusions. Proteases acting in the same $pH$ range have been displaced in the supernatant as well as in the mitochondrial layer of *Paracentrotus* material. The respective activities of these two enzymatic fractions vary in a different manner after fertilization. The activity increases in the supernatant and diminishes in the mitochondrial fraction (90). ATPase has been shown to be fixed to the mitochondria of the unfertilized or recently fertilized egg of the same sea urchin (101). It has been shown that there are two levels of ATPase activity, which are activated by $Mg^{++}$ ions, but at different $pH$ optima, and which are linked to two kinds of particles, both probably mitochondrial, but sedimenting differently and differently pigmented (149). The cytochrome system has been localized in one of these mitochondrial fractions, the most pigmented one (91). These new studies point out the diversity of the granular enzyme-carrying systems in the egg. This point seems to be one of the present concerns of cytologists studying the adult cell.

Recent works (3, 4, 46, 47, 51, 107, 108) have shown the existence in hepatic cells of a new cytoplasmic particle smaller than the mitochondria and containing enzymes such as acid phosphatase, ribonuclease, deoxyribonuclease, cathepsin, and $\beta$-glycuronidase. These "lysosomes" have peculiar properties, in that they are surrounded by a membrane isolating the enzymes from their substrates.

It is difficult to decide at the present time if particles analogous to the lysosomes exist in the cystoplasm of the egg. It is possible, however, to confirm that acid phosphatase is localized in particular systems different from the mitochondria and that the linkage of the acid phosphatase to mucopolysaccharides can be demonstrated by vital metachromatic staining.

The significance of the metachromasia of the toluidin blue reaction in vivo has been linked to the presence of mucopolysaccharides in the eggs of mammals (33-37, 41). Later, it was shown that the production of these granules is linked to nucleolar activity and that they are incorporated in an area close to the nucleus and rich in plasmalogen. The histochemical reactions in these same rat eggs show a progressive activity of acid phosphatase beginning at fertilization, progressing during the stages of the pronuclei, and accentuating during the first phases of segmentation (104). This enzymatic activity appears in two systems: (1) in the large metachromatic perinuclear granules, and (2) in small, diffuse granules. The appearance of the latter coincides with a high acid phosphatase activity in the nucleolar region of the pronuclei. The author of these observations (104) has tried to assimilate the large juxtanuclear granules to the mitochondria, whereas the small diffuse granules would be analogous to the lysosomes. It is hoped that these hypotheses will be tested with the electron microscope.

However that may be, the linking of acid phosphatase to granules rich in mucopolysaccharides and staining metachromatically with dyes such as toluidin blue, brilliant cresyl blue, and azure B is considered to be a general phenomenon, since it has been demonstrated in the eggs of two mollusks (122), three sea urchins (105, 119, 120), one annelid (120), and one ascidian (39). In the eggs of invertebrates studied thus far, the metachromatic stains are fixed selectively on the "$\alpha$ granules" which could not be distinguished from the vitelline platelets except by their particular density after centrifugation and their selective staining by alcian blue at a $p$H of 0.2 after oxidation with permanganate (120). Ultimately, and parallel with the spreading of the asters (spermaster or asters of segmentation), the stain leaves the $\alpha$-granules to become fixed to a new type of granule, $\beta$, which is lighter and more strongly metachromatic. Future studies should try to define or clarify the structure of these granules, and the conditions which determine their affinity for toluidin blue. It is no less true that the $\beta$ granules always show a strong acid phosphatase activity, and observations made with *Paracentrotus* discredit their mitochondrial nature. In centrifuged eggs of this species these $\beta$ granules separate in the superior vitelline layer. It is precisely in this zone that there appear (1) the alcian blue reaction at $p$H 0.2, demonstrable without preoxidation (hence indicating very acid mucopolysaccharides); and (2) the very strong acid phosphatase activity, which can already be demonstrated after 10 to 20 minutes incubation. No definite structure has been able to be demonstrated in this zone with the electron microscope, unless a great number of Golgi vesicles. Future studies will show whether or not this is only a coincidence.

In *Arbacia* and *Chaetopterus* these granules sediment in the hyaline layer. They give a positive acid phosphatase reaction and a reaction with alcian blue (105). Thus, in all of the kinds of eggs which have been examined up till now (sea urchin, mollusk, annelid, ascidian, and rodent) the phosphatase activity appears on the particles which are rich in mucopolysaccharides and which are metachromatic in vivo. This result thus indicates the existence of a new cellular organelle of unknown function, but presumably important because of its general properties. Its relationship to the lysosomes of the hepatic cells should be considered.

*In summation:* (1) it is at least exaggerated to pretend that the ovular enzymes are not linked to the cytoplasmic particles, and become linked only during the course of development; (2) enzymes such as cytochrome oxidase, succinic dehydrogenase, ATPase, and acidic phosphatase have been localized on organelles in the ovular cytoplasm; and (3) certain cytoplasmic particles other than mitochondria can be considered as "platforms" for enzymes.

## B. The Nucleus

In the egg, where the nucleus is considerably enlarged, the measurement of DNA, which is diluted, presents many difficulties. The illusion that the nucleus of the unfertilized egg contains no DNA is due to this very dilution (95, 96). However, the Feulgen technique has shown the presence of DNA in the form of a thin crust just within the nuclear membrane (18, 21). Aside from certain divergent opinions (89, 121), the majority of authors agree that from the beginning the nuclei of the two gametes contain an equal amount of DNA, equivalent to half that of the diploid adult cells (98, 167, 169). Without attempting to discuss this point here, it seems that the idea of an absolute constancy of nuclear DNA, expressed in all its rigidity, is excessive.

Apart from such reservations, two facts seem well established. (1) The quantity of DNA doubles in each of the pronuclei during its swelling (42). (2) In the first mitoses of segmentation, the synthesis of DNA is brought about at the very beginning of interkinesis, whether this stage be short, as in the egg of the sea urchin (98), or whether it be exceptionally prolonged, as in the case of rodents (42).

The mitotic apparatus during the first stages is characterized by its great richness in RNA. The results of a study made on the segmentation of *Cyclops* (165) show that the spindle RNA originates from the nuclear sap. The same author has also shown (166) that the interkinetic nuclei and the mitotic apparatus of segmenting eggs in various organisms (*Sabellaria, Cyclops, Clupea harengus,* and *Gadus morrhua*) are very rich in polysaccharides, as demonstrated by the periodic acid Schiff (PAS) reaction, whereas this reaction is negative in the nuclei of cells which have a low mitotic activity. The nuclei of sea urchin eggs are also rich in polysaccharides (99) and cytochemical observations have shown a cyclic variation during the first division (74) of these substances on the asters as well as on the chromosomes.

The fixation of polysaccharides can present real difficulties, and these observations should be verified. Furthermore, these observations have only a descriptive value, and their relationship with the mitotic mechanism does not yet seem clear. On the other hand, the great richness of the spindles in —SH groups (132) has been correlated with the structure of a specific protein extracted from the achromatic apparatus of the egg of the sea urchin (97).

In most of the eggs which have very short interkinesis phases, the nucleolar activity is easily passed unnoticed. In the case of the eggs of mammals, however, and particularly of rodents, the very long interkinesis phase has permitted the observation of a cyclic activity in the nucleus with active intervention by the nucleolar mechanism, by means of cytochemical methods (31, 36-38). It also appears that the number of nucleoli and their content of RNA and of polysaccharides vary during the phases of swelling of the pronuclei and the first interkinesis. This activity has been correlated with the elaboration of juxtanucleic cytoplasmic bodies, particularly those complexes which are rich in mucopolysaccharides and in plasmalogene (38).

Finally, the cytochemical studies made with *Ascaris* eggs have shown the formation of an appreciable quantity of RNA at the area of contact with the male pronucleus (109-111, 117). The chemical reactions in the nucleus during the first stages of development are still not very well understood. The results so far obtained indicate that a continuation of the work would be well worthwhile.

### The Egg as a Future Embryo

The length of this paper prohibits any complete examination of all of the relationships between structure and morphogenesis of the unsegmented egg. Such an analysis would require the scope of a full-length book. The present purpose is only to examine certain key problems. Can we show that in the unsegmented egg there is a cytochemical pattern capable of explaining, completely or in part, its ultimate morphogenesis? In what way are the morphogenetic properties of the diverse ovular territories linked to this kind of pattern?

With the actual techniques available today one cannot demonstrate that in the egg of the sea urchin there is any certain pattern related to its morphogenetic properties. RNA, platelets rich in polysaccharides, lipids, and mitochondria are dispersed in a rather uniform manner. The only topographical differentiation which appears in such an egg is a perinuclear area containing fewer platelets and more ergastoplasm. However, there doesn't seem to be any indication of *bilateral symmetry* or *polarity*. It is in later phases of development that there appear some cytochemical manifestations in relationship with the gradients shown by experimental analysis. The results of studies of differential reduction with $rH$ indicator dyes can be cited here (68, 69, 70); also, modifications of the nucleus appear after the reaction of Hale once the eggs have been placed in sea water without sulfate (73). An unequal distribution of mitochondria along the animal-vegetal axis has also been described. The latter observations have however been the object of serious criticism (156, 157).

In numerous other cases, however, it has been possible to show a direct relationship between a morphogenetic activity and a cytochemical peculiarity in a certain region of the unsegmented egg.

Certain authors are tempted to believe that the first case involves "regulative" eggs, and the second case "mosaic" eggs. Such a distinction does not seem too likely, however (29). It implies too easily, and wrongly, an eventual qualitative difference between two types of development. In effect, all transitions can be found from eggs which display partial development following merogony or a separation of the first blastomeres to those which give evidence of a widespread aptitude toward regulation. Moreover, eggs which are very regulative, such as those of amphibians, show a direct relationship between a cytochemical pattern and morphogenesis.

## Polarity

The cytochemistry of the egg in relationship to its polarity has been the object of numerous studies. Particularly, it has been shown with annelids, mollusks, and fishes that the polarity can be only faintly demonstrated in the unfertilized egg and that it is defined and accentuated during maturation, especially at the moment when the two pronuclei fuse. It is this which has previously been referred to as "bipolar differentiation" (133, 150, 151, 160-162), or "ooplasmic segregation" (27, 28).

There is a question as to whether there is a cause and effect relationship between this cytochemical polarity of the fertilized egg and the polarity of the future organism. This relationship has often been denied, since in many cases the distribution of inclusions and cytoplasmic particles has been altered by centrifugation without changing the future embryo. One should not attempt to overgeneralize this idea, since in at least one case, that of the amphibians, a direct cause and effect relationship could be shown between the distribution of particles and inclusions according to the polarity of the egg, and the polarity of the adult organism.

In the unsegmented amphibian egg, there exists a gradient in the distribution of RNA with a maximum at the animal pole (13, 15, 16). On the other hand, there is a decreasing gradient both in dimension and number of vitelline platelets from the marginal zone to the animal pole (118). This distribution of cytoplasmic particles can be modified by gravity. Turning an egg down and centrifuging it lightly can invert the orientation of the gradients of RNA and vitellus. Such an egg will give a larva with an inverted cephalo-caudal polarity. The head will always form in the area which is richest in RNA, whether that be at the animal pole in the normal egg, or at the vegetal pole in the inverted egg.

Maintaining the egg in this inverted position but without centrifugation will produce partial alterations in the distribution of the inclusions. Double, and even triple, embryos can be formed from the "marginal zones," that is, from the territories situated at the limit of the vitelline mass (114, 125-127). Cytochemical analysis always shows a mixture of granules rich in RNA and abundant vitelline platelets, analogous to the distribution found in the normal marginal zone (118).

Finally, by centrifuging the unsegmented egg in the normal position, it is possible to dissociate the ribonucleic granules from the vitelline platelets. The former accumulate in the centripetal or animal pole, while the latter accumulate at the centrifugal or vegetal pole (19, 118). Such a manipulation leads to hypomorphic embryos which, in spite of a sufficient gastrulean invagination, are incapable of differentiation in the middle layer (115), with a correlative lack of induction of the outer layer.

This series of experiments tends to imply that, in the case of amphibians, the differentiation of the chordomesoblast and its cephalo-caudal polarization depend directly on the gradient between the animal and vegetal poles of cytoplasmic particles in the unsegmented egg.

*Bilateral Symmetry*

In many cases, cytochemical studies have revealed a bilateral symmetry in the fertilized egg, sometimes even in the unfertilized egg. We will cite only the two best examples of such a bilateral pattern: rodent eggs and ascidian eggs.

In the oocyte and in the fertilized egg of the rat (30, 36, 37, 43), two kinds of granules rich in RNA can be observed: small granules forming a "cloud" in the center of the cytoplasm, and large granules distributed under the cortex. These two categories of particles clearly visible with the light microscope are rich in phospholipids and have been considered provisionally as mitochondria. Their exact nature is being studied at present

(40). The cortical "mitochondria" are clearly dominant on one side of the egg, whether it be fertilized or unfertilized, while the cytoplasm on the opposite side has an alveolar structure. The ultimate evolution is clearly different in these two ovular territories. The side which is rich in cortical RNA granules leads to blastomeres which are smaller, but which, beginning with the 8- or 16-cell stages, become intensely basophilic following the secondary appearance of fine granules surrounding the nucleus. These are the blastomeres which are destined to form the ectophyll (ectoblast + chordomesoblast) of the embryo. The other side of the egg gives larger blastomeres which are characterized by the appearance of granules which are rich in mucopolysaccharides and acid phosphatase (cf. pp. 386, 387), and which stain metachromatically in vivo. These large blastomeres envelop the smaller blastomeres and form the trophoblast as well as the endophyll.

It is not excluded that the bilateral symmetry of mammalian eggs, like those of other vertebrates, might be modified at the moment of fertilization. However, no indication of such an alternative has appeared up to now. Certain observations made on cheiropteran eggs give strong indication of a continuity between the bilateral symmetry of the oocyte and that of the fertilized egg (159).

The evolution of the cytochemical pattern of the mammalian egg thus shows a primary bilateral symmetry in the oocyte which, after fertilization, seems to maintain a continual relationship with the diversification of the embryonic and extra-embryonic parts of the germ.

With the present techniques for the cultivation of mammalian eggs, it has not been possible to determine whether an experimental modification in the distribution of the cytoplasmic particles would provoke an alteration of the ultimate morphogenesis.

Such an investigation has been made in the case of ascidian eggs (142, 143, 146, 152, 171). The classical observations of Conklin have elucidated the signification of the ventral "yellow crescent," that is, the manner in which the crescent organizes after fertilization, and the fact that centrifugation of this egg, contrary to most cases, provokes a profound disturbance in the morphogenesis, particularly in the formation of the myoblasts. It has been shown (142, 143, 146, 152, 171): (1) that the "yellow crescent" contains an accumulation of mitochondria rich in cytochrome oxidase and in benzidine peroxidase; (2) that the mitochondria are found in the myoblast; and (3) that the ectopic differentiation of the myoblast in the embryos deriving from centrifuged eggs can be related to the displacement of the mitochondria.

In the particular case of ascidian eggs, a mitochondrial pattern with its specific enzymatic components has been able to be directly related to an important differentiation of the embryo.

### Signification of Mitochondrial Patterns

The Nadi reaction and the Janus green B reaction have permitted the study of mitochondrial patterns in numerous marine eggs. Accumulations of mitochondria with their respiratory enzymes thus appear in the cortical plasm of ctenophores (129, 130, 144). They have also been followed from the unsegmented egg stage through the cell linkage which is so particular in annelids and mollusks (5, 6, 83, 84, 85, 130, 131, 147). The results of all of these works indicate that a mitochondrial mass selectively prepares the differentiation of locomotor organs such as muscles, in the case of ascidians, ciliated bands in the case of ctenophores, or in the larva of mollusks. It has been concluded (144) that the mitochondria could be considered as "organ-forming substances," in the sense that they afford a "supply of energy." If in effect this is so, the accumulation of respiratory enzymes in a selective territory of the unsegmented egg does not necessarily have an actual signification, that is, they do not necessarily alter the metabolism in that territory of the unsegmented egg (see p. 387).

Parallel observations made with ascidians show that such a position is well founded. Microchemical methods have confirmed the results of cytochemical studies. After homogenization of the ventral blastomeres, it has been possible to show a cytochrome oxidase activity 2.7 times greater than that of the dorsal blastomeres (9). However, the consumption of oxygen by the isolated dorsal and ventral blastomeres has been shown to be identical (67).

The cytochemical localizations of the unsegmented egg are thus not necessarily linked to the actual life of this unsegmented egg. Such a conception can explain the contradictions which seem to arise too frequently between the study of the cytochemistry of the egg and its relation with morphogenesis.

### The Case of the Polar Lobe of Spiralia

The preceding observations made on the eggs of amphibians, mammals, and ascidians tend toward an optimistic view: there exists an easily explainable relationship between these patterns and morphogenesis. This relationship seems to be so direct that at first hand one would be tempted to consider the mitochondrial mass as the true "germinal localizations." It is to be noted, however, that such a situation is really exceptional. In many cases, a comparative study of different species having the same type of morphogenesis makes the existence of a common cytochemical pattern difficult to conceive. The polar lobe and the polar plasma of annelids and mollusks can be cited as examples of this.

We know that these plasms, which are eventually united in a polar lobe, have essential morphogenetic properties corresponding to the formation of the somatoblast (*2d* and *4d*). The cytoplasm of the polar lobe of the mollusk *Ilyanassa* (102), and the annelid *Sabellaria* (137) does not show any apparent difference from the rest of the vegetative portion of the egg. The polar plasmas of *Tubifex* (86, 172) are characterized by a great abundance of Nadi-positive mitochondria and a vesiculated ergastoplasm. In the aberrant annelid *Myzostoma,* one also finds in the polar lobe a "green plasma" which has been shown to be an accumulation of Nadi-positive mitochondria (130). This "plasma" is found at the vegetative pole from the time when the egg is laid. In these forms, displacement by centrifugation does not affect segmentation, especially the formation of the lobe and the inequality of the first two blastomeres (113). Similarly, in *Ilyanassa* the polar lobe forms normally regardless of the abnormal stratification produced by centrifugation (102). In the mollusk lamellibranch *Gryphaea,* one finds in the polar lobe (and in the corresponding vegetative territories of the egg from the beginning of maturation) an accumulation of RNA, while the vitelline platelets are much less abundant (122). In another lamellibranch, *Mytilus,* cytological examination does not show any notable particularity in the lobe (8). Once the latter has been isolated, it is characterized by a consumption of oxygen which is 25 per cent less (per unit volume) than that of the rest of the egg. In addition to this, the respiration of the blastomere *CD* (which contains the polar lobe) is less than that of *AB* (10). The authors of these experiments insist on the fact that this last notation shows that the decreased respiration of the polar lobe is in fact real and is not due to the fact that the isolated polar lobe is an anuclear fragment. However, it has been demonstrated that the incorporation of labeled amino acids is less in a polar lobe that has been isolated than in one which has been left in situ (1).

## Discussion

The results of the experiments performed using the usual cytochemical techniques thus tend to lead to conclusions which are contradictory. In certain cases (polarity of the amphibians; bilateral symmetry of the ascidians) a direct link can be established between a cytochemical pattern and development, and a topographical alteration of this pattern produces a disturbance which can be correlated with development. In other cases, however, one cannot define a general pattern for all forms having the same type of morphogenesis. What is even more striking is that an alteration of pattern has no effect on the ultimate morphogenesis.

Two remarks can be made at this point. The first concerns the significance of centrifugation experiments. It is practically a classical statement to say that one can easily centrifuge an egg without altering its morphogenesis and by consequence state that the "inclusions," taken in the broadest sense of the word, do not play an essential role in this morphogenesis. In that form, such a conclusion is certainly false. We have already seen two examples where centrifugation of an unsegmented egg, although moderate, can cause profound disturbances in development, and it would be easy to find other examples (112). On the other hand, the real effects of centrifugation on a living egg, especially if negative, should be most carefully scrutinized before one can derive valuable conclusions, as has already been pointed out (136). The sedimentation caused by centrifugal force can return to the normal position more rapidly than had been formerly perceived. The presence of intercellular membranes does not always present an obstruction to such rearrangement (136). Finally, as we have seen previously (p. 386), the exact distribution of sedimented materials cannot be considered as being known until after a profound examination, including one with the electron microscope. Each experiment concerning centrifugation of the egg should be studied minutely and systematically before coming to formal conclusions.

The second remark concerns the methods used up to now for the study of these patterns. In such studies as those which have a topographical nature, only cytochemical methods could be utilized up to now. What has been demonstrated consists of a limited number of enzymes (cytochrome oxidase and phosphatases), ribonucleic acid, and polysaccharides. In other words, we are better informed on the tools than on the skilled worker. These "tools" can be put into place earlier or later according to the different types of eggs, but sometimes even according to the species having the same type of development, as in the case of the polar lobe. A disturbance of the topography can lead to irreversible consequences, as after the centrifugation of eggs of ascidians, amphibians, and annelids (*Tubifex*). Very often, however, the fundamental structure of the egg can possess sufficient resources to repair the disturbed mechanism before it is too late.

Besides morphogenetic regulation, one should consider structural regulation, the study of which could present one of the most fascinating of all embryological problems. Thus, by studying the mechanisms and methods of repair of the machinery, one will get to know the engineer. But this engineer should be studied also by direct means.

The main problem remains open: to study the protein organization of the fundamental cytoplasm, particularly in the ovular cortex. This could be done by applying on a topographical plan the most refined cytochemical studies such as the incorporation of labelled isotopes or immunological methods. We are convinced that a complete cytochemical study of the protein structure, as well as the energy-producing mechanisms, seems to be the best future plan of organization for the study of development in direct correlation with the structure of the unsegmented egg.

## Conclusion

The comparative cytochemical study of the unsegmented egg has been considered under two aspects. The first consists of a comparison of an ovular structure to that of the adult cell. It seems that the differences in organization between these two types of cells have been overestimated. In particular, the most recent studies made with reliable techniques show that the enzymes which are linked to cytoplasmic particles in the adult cell are likewise so linked in the egg. As in the adult cells, there appears to be a number of enzyme-carrying systems in the egg. The question regarding the eventual association of non-mitochondrial acid-phosphatase-carrying granules with the lysosomes in hepatic cells is still open. Although the RNA is linked, in part at least, to the ergastoplasmic systems, the latter can, however, be of very atypical structure, as in certain tumor cells.

The second aspect consists of a correlation between the cytochemical pattern and the morphogenetic properties of an unsegmented egg. A synthetic view appears here more difficult, and a comparison made with different animal types has led to contradictory conclusions. It is reasonable here to refer to our insufficient information on this subject. The topographical cytochemistry, with the methods utilized used up to now, informs us more about the "machinery" than regarding the "direction" of the morphogenetic undertaking. It is hoped that a topographical cytochemistry of the proteins will soon fill this gap. It is no less true that the direct intervention of localized accumulations of mitochondria or of ergastoplasm have been very nicely demonstrated in certain types of morphogenesis.

## REFERENCES

1. Abd-El-Wahab, A., and Pantelouris, M., *Exptl. Cell Research,* **13**, 78-82 (1957).
2. Afzelius, B. A., *Z. Zellforsch u mikroskop. Anat.,* **45**, 660-675 (1957).

3. Appelmans, F., and Duve, C. de, *Biochem. J. (London)*, **59**, 426-433 (1955).
4. ———, Wattiaux, R., and Duve, C. de, *Biochem. J. (London)*, **59**, 438-455 (1955).
5. Attardo, C., *Ricerca sci.*, **25**, 2797-2800 (1955).
6. ———, *Acta embryol. et morphol. exptl.*, **1**, 65-70 (1957).
7. Ballentine, R., *Biol. Bull.*, **77**, 328 (1939).
8. Berg, W. E., *Cell Biology* (Proc. 15th Ann. Biol. Colloq., Oregon State Coll., Corvallis) 30-34 (1954).
9. ———, *Biol. Bull.*, **110**, 1-7 (1956).
10. ———, and Kutsky, Ph. B., *Biol. Bull.*, **101**, 47-61 (1951).
11. Bernhard, W., Gauthier, A., and Roullier, C., *Arch. anat. microscop. morphol. exptl.*, **43**, 236-275 (1954).
12. Boell, E. J., in *Analysis of Development* (Willier, B. H., Weiss, P. A., and Hamburger, V., eds.), p. 520-555. W. B. Saunders, Philadelphia (1955).
13. Brachet, J., *Embryologie chimique*. Desoer, Liége; Masson & Cie., Paris (1944).
14. ———, *Experientia*, **3**, 329 (1947).
15. ———, *Compt. rend. Soc. Biol.*, **142**, 1241-1254 (1948).
16. ———, *Chemical Embryology*, Interscience, New York (1950).
17. ———, *Biochemical Cytology*, Academic Press, New York (1957).
18. ———, and Ficq, A., *Arch biol. (Liége)*, **67**, 431-446 (1956).
19. ———, and Pasteels, J., unpub., cited in 17.
20. ———, and Shaver, J. R., *Experientia*, **5**, 204-205 (1949).
21. Burgos, M. H., *Exptl. Cell Research*, **9**, 360-363 (1955).
22. Carrano, F., *Ricerca sci.*, **25**, 3049-3052 (1955).
23. ———, *Atti accad. nazl. Lincei, Rend., Classe sci. fis. mat. e nat.* [8] **22**, 216-219 (1957).
24. ———, *Ricerca sci.*, **27**, 1121-1124 (1957).
25. ———, and Palazzo, F., *Riv. biol. (Perugia)*, n. s., **47**, 193-201 (1954).
26. Clement, A. C., and Lehmann, F. E., *Naturwiss.*, **43**, 478-479 (1956).
27. Costello, D. P., *J. Elisha Mitchell Sci. Soc.*, **61**, 277-289 (1945).
28. ———, *Ann. N. Y. Acad. Sci.*, **49**, 663-684 (1948).
29. Dalcq, A., *L'oeuf et son dynamisme organisateur*. Albin Michel, Paris (1941).
30. ———, *Koninkl. Ned. Akad. Wetenschap., Proc., Ser. C*, **54**, 351-372; 469-477 (1951).
31. ———, *Compt. rend. assoc. anat.*, **38e** Réun. (Nancy), 345-361 (1951).
32. ———, *Arch. anat. et embryol. (Vol. Jub. Prof. Dubreuil)*, **34**, 157-169 (1951).
33. ———, *Biol. Jaarboek (Gand)*, **19**, 52-59 (1952).
34. ———, *Compt. rend. soc. biol.*, **146**, 1408-1411 (1952).
35. ———, *Compt. rend. assoc. anat.*, **39e**, Réun. (Nancy), 513-516 (1952).
36. ———, *Compt. rend. soc. biol.*, **148**, 1332-1373 (1954).
37. ———, *Stud. on Fertility* (Blackwell Scientific Publications, Oxford) **7**, 113-122 (1955).
38. ———, *Exptl. Cell Research*, **10**, 99-119 (1956).
39. ———, *Bull. soc. zool. franc.*, **82**, 296-316 (1957).
40. ———, pers. commun.
41. ———, and Massart, L., *Compt. rend. soc. biol.*, **146**, 1436-1439 (1952).
42. ———, and Pasteels, J., *Exptl. Cell Research*, suppl. 3, 72-97 (1955).
43. ———, and Seaton-Jones, A., *Compt. rend. assoc. anat.*, **36e** Réun. (Nancy), 170-175 (1949).
44. Dalton, A. J., and Felix, M. D., *Symposia Soc. Exptl. Biol.*, **10**, 148-156 (1957).
45. Deutsch, H., and Gustafson, T., *Arkiv Kemi*, **4**, 221-231 (1952).
46. Duve, C. de, *J. de physiol. (Paris)*, **49**, 113-115 (1957).
47. ———, *Symposia Soc. exptl. Biol.*, **10**, 50-61 (1957).
48. Evola-Maltese, C., *Acta embryol. et morphol. exptl.*, **1**, 99-104 (1957).

49. Ficq, A., *Exptl. Cell Research*, **9**, 286-293 (1955).
50. Gersch, M., *Arch. exptl. Zellforsch.*, **22**, 549-564 (1939).
51. Gianetto, R., and Duve, C. de, *Biochem. J. (London)*, **59**, 433-438 (1955).
52. Green, D. E., *Symposia Soc. exptl. Biol.*, **10**, 30-49 (1957).
53. Gross, P. R., *J. Cellular Comp. Physiol.*, **47**, 429-447 (1956).
54. Gustafson, T., and Hasselberg, I., *Exptl. Cell Research*, **2**, 642-672 (1951).
55. ———, and Lenicque, P., *Exptl. Cell Research*, **2**, 642-672 (1952).
56. ———, and ———, *Exptl. Cell Research*, **8**, 114-117 (1955).
57. Harvey, E. B., *Biol. Bull.*, **64**, 125-148 (1933).
58. ———, *Biol. Bull.*, **81**, 114-118 (1941).
59. Harvey, E. N., *Arch. exptl. Zellforsch.*, **22**, 463-476 (1939).
60. ———, and Lavin, G. L., *Biol. Bull.*, **86**, 163-168 (1944).
61. Hogeboom, G. H., Schneider, W. C., and Palade, G. E., *Proc. Soc. Exptl. Biol. Med.*, **65**, 320-322 (1947).
62. Holter, H., *Arch. exptl. Zellforsch.*, **19**, 232-237 (1937).
63. ———, *Advances in Enzymol.*, **13**, 1-19 (1952).
64. ———, *J. Cellular Comp. Physiol.*, **8**, 179-200 (1936).
65. ———, Lanz, H., Jr., Linderstrøm-Lang, K., *J. Cellular Comp. Physiol.*, **12**, 112-127 (1938).
66. ———, Lehmann, F. E., and Linderstrøm-Lang, K., *Compt. rend. trav. lab. Carlsberg (Ser. chim.)*, **21**, 259-262 (1938).
67. ———, and Zeuthen, E., *Compt. rend. trav. lab. Carlsberg, (Ser. chim.)*, **25**, 33-65 (1944).
68. Hörstadius, S., *J. exptl. Zool.*, **120**, 421-436 (1952).
69. ———, *J. Embryol. exptl. Morphol.*, **1**, 257-259 (1953).
70. ———, *J. exptl. Zool.*, **129**, 249-256 (1955).
71. Howatson, A. F., and Ham, A. W., *Cancer Research*, **15**, 62, 69 (1955).
72. Hutchens, J. O., Kopac, M. J., and Krahl, M. E., *J. Cellular Comp. Physiol.*, **20**, 113-118 (1942).
73. Immers, J., *Exptl. Cell Research*, **10**, 546-548 (1956).
74. ———, *Exptl. Cell Research*, **12**, 145-153 (1957).
75. Keltch, A. K., Strittmatter, C. F., Walters, C. P., Clowes, G. F., *J. Gen. Physiol.*, **33**, 547-553 (1949).
76. Krahl, M. E., *Biol. Bull.*, **98**, 175-217 (1950).
77. ———, *Biochim. et Biophys. Acta*, **20**, 27-32 (1956).
78. ———, Keltch, A. K., Neubeck, C. E., Clowes, G. H. A., *J. Gen. Physiol.*, **24**, 597-617 (1941).
79. Krugelis, E. J., *Biol. Bull.*, **93**, 209 (1947).
80. Lansing, A. I., *J. Histochem. Cytochem.*, **1**, 266 (1953).
81. Lazarow, A., and Cooperstein, S. J., *J. Histochem. Cytochem.*, **1**, 234-241 (1953).
82. Lehmann, F. E., *Naturwiss.*, **29**, 101 (1941).
83. ———, *Naturwiss.*, **29**, 101 (1941).
84. ———, *Folia. Biotheor.*, **3**, 7-24 (1948).
85. ———, *Rev. suisse zool.*, **55**, 1-43 (1948).
86. ———, *Naturwiss.*, 289-296 (1956).
87. ———, and Wahli, H. R., *Z. Zellforsch.*, **39**, 618-629 (1954).
88. Lindahl, P. E., and Holter, H., *Compt. rend. lab. Carlsberg, (Ser. chim.)*, **23**, 249-255 (1940).
89. Lison, L., and Pasteels, J., *Arch. biol. (Liége)*, **62**, 1-43 (1951).
90. Maggio, R., *J. Cellular Comp. Physiol.*, (in press).
91. ———, and Ghiretti, A., *Exptl. Cell Research*, (in press, quoted by personal communication, courtesy A. Monroy).
92. Mancuso, V., *Ricerca sci.*, **24**, 1886-1888 (1954).

93. ———, *Atti Accad. nazl. Lincei, Rend.*, (*Classe sci. fis. mat. e nat.*) [8] **21,** 504-506 (1956).
94. ———, *Ricerca sci.*, **26,** 2756-2760 (1956).
95. Marshak, A., and Marshak, C., *Exptl. Cell Research*, **5,** 288-300 (1953).
96. ———, and ———, *Exptl. Cell Research*, **8,** 126-146 (1955).
97. Mazia, D., and Dan, K., *Proc. Natl. Acad. Sci. U. S.*, **38,** 826-838 (1952).
98. McMaster, R., *J. Exptl. Zool.*, **130,** 1-23 (1955).
99. Monné, L., and Slautterbach, D. B., *Exptl. Cell Research*, **1,** 477-491 (1950).
100. ———, and ———, *Arkiv. Zool.* (ser. 2), **3,** 349-356 (1952).
101. Monroy, A., *J. Cellular Comp. Physiol.* (in press).
102. Morgan, T. H., *J. Exptl. Zool.*, **64,** 433-467 (1933).
103. Mulnard, J., *Bull. classe sci., Acad. roy. Belg.*, **36,** 767-778 (1950).
104. ———, *Arch. biol.* (*Liége*), **66,** 525-685 (1955).
105. ———, pers. commun.
106. ———, and Dalcq, A., *Compt. rend. Soc. Biol.*, **149,** 536-839 (1955).
107. Novikoff, A. B., *Symposia Soc. Exptl. Biol.*, **10,** 92-109 (1957).
108. ———, Beaufay, H., Duve, C. de, *J. Biophys. Biochem. Cytol.*, suppl. **2,** 179-184 (1956).
109. Panijel, J., *Bull. soc. chim. biol.*, **29,** 1098-1106 (1947).
110. ———, Thèses Fac. Sci. Univ. Paris (Ser. A 2326), n°3198, 1-331 (1951).
111. ———, and Pasteels, J., *Arch. biol.* (*Liége*), **62,** 354-369 (1951).
112. Parseval, M. von, *Roux' Arch. Entwicklungsmech. Organ.*, **51,** 468-497 (1922).
113. Pasteels, J., *Arch. Anat. microscop.*, **30,** 161-197 (1934).
114. ———, *Arch. biol.* (*Liége*), **49,** 629-667 (1938).
115. ———, *Arch. biol.* (*Liége*), **51,** 335-386 (1940).
116. ———, *Arch. biol.* (*Liége*), **52,** 321-339 (1941).
117. ———, *Arch. biol.* (*Liége*), **59,** 405-446 (1948).
118. ———, *Bull. Soc. zool. Fr.*, **76,** 231-270 (1953).
119. ———, *Bull. classe sci. Acad. roy. Belg.* (5e série), **41,** 760-768 (1955).
120. ———, *Arch. biol.* (*Liége*) (in press).
121. ———, and Lison, L., *Nature*, **167,** 948 (1951).
122. ———, and Mulnard, J., *Arch. biol.* (*Liége*), **68,** 115-163 (1957).
123. ———, Castiaux, P., and Vandermeerssche, G., *J. Biophys. Biochem. Cytol.* (in press).
124. Peltrera, A., *Pubbl. staz. zool. Napoli*, **18,** 20-49 (1940).
125. Penners, A., *Z. wiss. Zoöl.*, **148,** 189-220 (1936).
126. ———, and Schleip, W., *Z. wiss. Zoöl.*, **130,** 365-454 (1928).
127. ———, and ———, *Z. wiss. Zoöl.*, **131,** 1-156 (1928).
128. Petrucci, D., *Acta embryol. et morphol. exptl.* **1,** 105-117 (1957).
129. Pittoti, M., *Pubbl. staz. zool. Napoli*, **18,** 250-272 (1940).
130. ———, *Pubbl. staz. zool. Napoli*, **21,** 93-100 (1947).
131. Pollicita, M., *Ricerca sci.*, **25,** 3114-3115 (1955).
132. Rapkine, L., *Ann. physiol. physicochim. biol.*, **7,** 382-405 (1931).
133. Raven, Chr. P., *Acta Neerl. Morphol. norm. pathol.*, **1,** 337-357 (1938).
134. ———, *Arch. néerl. zool.*, **7,** 91-119 (1945).
135. ———, *Biol. Revs. Cambridge Phil. Soc.*, **23,** 333-368 (1948).
136. ———, and Bretschneider, L. H., *Arch. néerl. zool.*, **6,** 255-278 (1942).
137. Raven, P., Van Brink, J. M., Van de Kamer, J. C., *Verhandel. Koninkl. Ned. Akad. Wetenschap., Afdel. Natuurk.* (Sect. II), **47,** 1-48 (1950).
138. Rebhun, L. I., *J. Biophys. Biochem. Cytol.*, **2,** 93-104 (1956).
139. ———, *J. Biophys. Biochem. Cytol.*, **2,** 159-170 (1956).
140. Recknagel, R. O., *J. Cellular Comp. Physiol.*, **35,** 111-129 (1950).
141. Reverberi, G., *Pubbl. staz. zool. Napoli*, **18,** 129-139 (1940).

142. ———, *Experientia,* **12,** 55-61 (1956).
143. ———, *Pubbl. staz. zool. Napoli,* **29,** 187-212 (1957).
144. ———, *Arch. embryol. morphol. exptl.,* **1,** 134-142 (1957).
145. ———, *Scientia (Milan),* (ser. 6), **51,** 1-6 (1957).
146. ———, and Pittoti, M., *Comment. pontif. Acad. Sci.,* **3,** 469-488 (1939).
147. ———, and ———, *Pubbl. staz. zool. Napoli,* **18,** 256-263 (1940).
148. ———, and ———, *Pubbl. staz. zool. Napoli,* **21,** 237-258 (1947).
149. Ricotta, C. M. B., *Experientia,* **13,** 491-2 (1957).
150. Ries, E., *Pubbl. staz. zool. Napoli,* **16,** 364-399 (1937).
151. ———, *Arch. exp. Zellforsch. Gewebezücht.,* **22,** 569-584 (1939).
152. ———, *Arch. exp. Zellforsch. Gewebezücht.,* **23,** 95-119 (1939).
153. Schleip, W., *Die Determination der Primitiventwicklung.* Akademische Verlagsgesellschaft, Leipzig (1929).
154. Schultz, H., *Oncologia,* **10,** 307-329 (1957).
155. Shapiro, H., *J. Cellular Comp. Physiol.,* **6,** 101-116 (1935).
156. Shaver, J. R., *Exptl. Cell Research,* **11,** 548-559 (1956).
157. ———, in *The Beginnings of Embryonic Development* (A. Tyler, R. C. von Borstel, and C. B. Metz, eds.), p. 263-290, Am. Assoc. Advance. of Sci., Washington, D. C. (1957).
158. Sirlin, J. L., *Experientia,* **11,** 112 (1955).
159. Skreb, S., *Arch. biol. (Liége),* **68,** 381-428 (1957).
160. Spek, J., *Protoplasma,* **18,** 497-545 (1933).
161. ———, *Roux' Arch. Entwicklungsmech. Organ.,* **131,** 362-372 (1934).
162. ———, *Protoplasma,* **21,** 394-405 (1934).
163. Steinbach, H. B., and Moog, F., in *Analysis of Development* (B. H. Willier, P. A. Weiss, and V. Hamburger, eds.), p. 70-90. W. B. Saunders, Philadelphia (1955).
164. Stich, H., *Roux' Arch. Entwicklungsmech. Organ.,* **144,** 364-380 (1950).
165. ———, *Z. Naturforsch.,* Pt. b., **6,** 259-261 (1951).
166. ———, *Chromosoma,* **4,** 429-438 (1951).
167. Swift, H. H., *Physiol. Zoöl.,* **23,** 169-198 (1950).
168. ———, *J. Biophys. Biochem. Cytol.,* **2** (suppl.), 415-418 (1956).
169. ———, and Kleinfeld, R., *Physiol. Zoöl.,* **26,** 301-311 (1953).
170. Tsuboi, K. K., De Terra, N., and Hudson, P. B., *Exptl. Cell Research,* **7,** 32-43 (1954).
171. Urbani, E., and Urbani, Mistruzzi L., *Pubbl. staz. zool. Napoli,* **21,** 69-82 (1947).
172. Weber, R., *Rev. suisse zool.,* **63,** 277-288 (1956).
173. ———, and Boell, E. J., *Rev. suisse zool.,* **62,** 260-268 (1956).
174. Wessel, W., and Bernhard, W., *Z. Krebsforsch.,* **62,** 140-162 (1957).
175. Wicklund, E., *Nature,* **161,** 556 (1948).
176. Woodward, A. A., *Biol. Bull.,* **99,** 367 (1950).
177. Yčas, M., *J. exptl. Biol.,* **31,** 208-217 (1954).

# A CYTOCHEMICAL STUDY OF THE GROWTH OF THE SLUG OOCYTE

RONALD R. COWDEN[1,2]

*Biology Division, Oak Ridge National Laboratory*[3]
*Oak Ridge, Tennessee*

### INTRODUCTION

The development of the germ cells of gastropods has been the subject of a number of classical cytological studies. The main object of this investigation, however, was to consider the changes in the morphological relationships of a number of the intracellular substances, as demonstrated by cytochemical methods, during the growth of the slug oocyte. In an earlier study, Bridgeford and Pelluet (6) described some of the chemical relations between the primary nucleolus and the one or more secondary nucleoli that appear during development of the slug oocyte. In view of the general relation considered to exist between the nucleolus and protein and ribonucleic acid (RNA) synthesis (7), an investigation of the unusual nucleolar system found in the slugs is of particular interest and is the focal point of this investigation.

The slugs used in this investigation were collected on the grounds of the Biology Division of the Oak Ridge National Laboratory and were subsequently classified as *Deroceras reticulatum*. Some other unclassified species of the genus *Deroceras* were also used. The animals were decapitated, the viscera exposed, and the ovotestes quickly removed. These were fixed in one of three fixatives: acetic-acid–alcohol (1:3), 10 per cent buffered formalin, or Baker's calcium formal. These were embedded in Tissuemat and sectioned at 15 $\mu$.

For the study of each class of nucleic acid, a number of methods were employed. For deoxyribonucleic acid (DNA) the Feulgen reaction and Einarson's (11) gallocyanin-chromalum after ribonuclease (RNAase) di-

---

[1] This work was performed while the author was a U. S. Public Health Postdoctoral Fellow (HF-6176).

[2] Present address: Department of Biology, The Johns Hopkins University, Baltimore 18, Maryland.

[3] Operated by the Union Carbide Corporation for the U. S. Atomic Energy Commission.

gestion were used. For RNA, Kurnick's (13) modification of the methyl-green-pyronin method, the azure B bromide method of Flax and Himes (12), and the previously mentioned gallocyanin-chromalum method of Einarson (11) were used. The material stained for RNA received prior incubation in deoxyribonuclease (DNAase). Both DNAase and RNAase, Worthington products, were used as suggested by Swift (17). Acetic-acid-alcohol (1:3) fixed material was used for all studies on nucleic acids.

Two procedures were used for the demonstration of proteins: the fast green method of Alfert and Geschwind (1) for basic nucleoproteins (histones) and the naphthol yellow S method of Deitch (9) for the dibasic amino acid residues of lysine, arginine, and histidine. The latter was coupled with the Feulgen reaction as suggested by Deitch (10). When used in this manner, the stain is still qualitatively specific, but there is some reduction in the intensity of naphthol yellow S binding resulting from the hydrolysis in 1 $N$ HCl.

The mucopolysaccharides were demonstrated by the periodic acid Schiff (PAS) reaction, in some cases preceded by incubation in 1 per cent malt diastase at 37°C for 4 hours to remove glycogen. Lison's alcian blue method (15) was used to stain mucin and other acid mucopolysaccharides. Prior incubation at 37°C in 1 per cent bovine testis hyaluronidase had no effect on the staining in this material. Both formalin-fixed and acetic-acid–alcohol-fixed material were used for this purpose.

Phospholipids were demonstrated by the alcoholic alcian blue method of Pearse (16), and control sections were extracted in pyridine at 60°C for 8 hours. Bradley's method (5), which involves squashing in acetocarmine and then in 45 per cent acetic acid saturated with Sudan black B, was used for the demonstration of lipids in general.

An ocular micrometer was used to measure the diameters of the two classes of nucleoli in all the uncut nucleoli of two sets of serial sections stained by the azure B bromide method.

OBSERVATIONS

*The Chromosomes*

In young oocytes, the chromosomes at leptotene were observed as distinct elongated threads filling the space around the nucleolus (Fig. 1). At pachytene (Fig. 2) their "lamp-brush" nature could be discerned. Likewise, the diplotene stage was distinctly observable (Fig. 3), in which the much shortened and thickened chromosomes stained deeply with gallocyanin-chromalum. The nucleolus persisted throughout this development, but diakinesis

406　THE CHEMICAL BASIS OF DEVELOPMENT

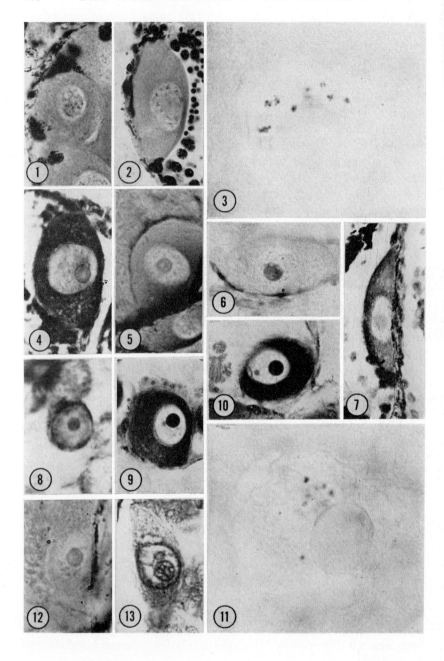

seemed to be the latest stage of meiosis reached in the ovotestis. The chromosomes appeared to be stained much more readily by gallocyanin-chromalum after RNAase digestion than by the Feulgen reaction. Some metachromatic binding of azure B bromide was noted in the chromosomes of the oocytes after DNAase treatment. Fast green was also faintly bound by the chromosomes of the oocytes—so faintly that it was not possible to make adequate photographs. In other types of nuclei, e.g., nurse cells, spermatocytes, the specific binding of fast green in only the chromosomes and nuclei was evident.

*The Cytoplasm*

The cytoplasm of all sizes of oocytes contained a demonstrably large amount of RNA. In the smaller cells, the binding of various stains was more intense than in the larger ones; this could be caused by a greater increase in the volume of the cell than of new RNA. The same relation held for the dibasic amino acid residues of lysine, arginine, or histidine, and for phospholipids. The cytoplasmic glycogen also appeared to be more intensely concentrated in the cytoplasm of the smaller oocytes. Alcian blue

---

Fig. 1. The leptotene chromosomes stained with gallocyanin-chromalum after RNAase treatment. 450 ×.
Fig. 2. Pachytene chromosomes stained with gallocyanin-chromalum after RNAase treatment. 450 ×.
Fig. 3. Diakinesis chromosomes stained with gallocyanin-chromalum after RNAase treatment. 970 ×.
Fig. 4. Oocyte stained for phospholipids by Pearse's method, nucleolus counterstained with aqueous neutral red. 450 ×.
Fig. 5. Oocyte stained by the PAS reaction. 450 ×.
Fig. 6. Oocyte stained by the PAS reaction after digestion in malt diastase. 450 ×.
Fig. 7. Oocyte stained in 0.1 per cent fast green after 5 per cent TCA extraction at 90°C for 15 min. 450 ×.
Fig. 8. Youngest oocyte stained with azure B bromide after DNAase treatment showing a single nucleolus. 970 ×.
Fig. 9. A later oocyte, stained as in Fig. 8, and showing a small secondary nucleolus lying next to the primary nucleolus. 450 ×.
Fig. 10. A fully developed oocyte with a primary and secondary nucleolus, stained as in Fig. 8. 450 ×.
Fig. 11. The nucleolar membrane shows some binding of gallocyanin-chromalum after treatment with RNAase, probably representing nucleolus-associated chromatin in the diplotene oocyte. 970 ×.
Fig. 12. Oocyte stained with alcian blue for acid mucopolysaccharides; the nucleolar membrane binds this dye. 450 ×.
Fig. 13. Oocyte stained with naphthol yellow S for dibasic amino acid residues; both classes of nucleoli bind this dye. 450 ×.
(Photomicrographs by Mr. John Spurbeck, Illustration Division, Johns Hopkins University).

did not stain the cytoplasm to any degree, but some sudanophilia was observed. Fast green at $p$H 8.0 stained the cytoplasm of the oocytes to a considerable degree, but not that of any other cell types. An example of the distribution of the above-mentioned materials in the largest class of oocytes is given in Figs. 4-7.

*The Nucleoli*

As mentioned earlier, the unusual nucleolar system in slugs, previously described by Bridgeford and Pelluet (6), was the object of principal interest in this investigation. According to their description the primary nucleolus, which was present in the youngest and smallest oocytes, underwent cyclical changes from acidophilic to basophilic and back to acidophilic; the secondary nucleoli were never basophilic, and seemed to arise from the primary nucleoli during growth and development. Although no direct evidence of the origin of secondary nucleoli from the primary nucleolus was observed, the fact that the secondary nucleolus is always seen in direct contact with the primary nucleolus during the early part of its growth would tend to support their interpretation. In Figs. 8-10, the origin of the secondary nucleolus and the separation of the nucleoli in young oocytes are shown. In these pictures, the disparity in azure B bromide binding is also apparent between the two classes of nucleoli. The differences in volume do not seem great enough to account for the differences observed in the intensity of staining. Just as the cytoplasm appears to have a higher concentration of RNA in the smaller oocytes, there are some changes in intensity of staining in the very large primary nucleoli that suggest once more that the volume has increased at a higher rate than RNA. In the large nucleoli, some inhomogeneity in the distribution of RNA was observed. In most instances, the secondary nucleoli were less intensely stained, indicating a generally lower concentration of RNA; there was a greater variability in the intensity of basophilia in secondary nucleoli of a given size class than in primary nucleoli of the same size class. In material stained with the Feulgen reaction or with gallocyanin-chromalum after RNAase treatment, there was a perinucleolar binding of dye (Fig. 11) that was removable by prior incubation in DNAase. The periphery of the primary nucleolus also stained with fast green at $p$H 8.0, indicating the presence of histones (Fig. 7). The secondary nucleoli gave no evidence of such binding. Furthermore, only the primary nucleoli appeared to possess a mucopolysaccharide-containing envelope that stained with either the PAS reaction or alcian blue, and which was not removable with either 1 per cent malt diastase or 1 per cent hyaluronidase at 37°C for 4 hours (Figs. 6, 12). Both the classes of nucleoli stained with the naphthol yellow S method of Deitch (10) for basic amino acids, and with apparently equal intensity (Fig. 13).

The primary nucleolus has a ring of DNA around it that may be chromatin associated with the nucleolus. The secondary nucleoli do not, since they are unable to bind dyes specific for DNA and histones. Furthermore, it is possible that the primary nucleoli possess some form of nucleolar membrane containing mucopolysaccharides that is not found in the secondary nucleoli. In view of the disparity in the concentration of RNA between the two types of nucleoli, it is particularly interesting that both seem to bind naphthol yellow S equally, so as to indicate the presence of dibasic amino acids, i.e. lysine, arginine, or histidine.

The cytochemical data gathered from these investigations are summarized in Table 1.

The diameters of all the nucleoli of both classes that were undamaged by sectioning were measured and the volumes of the two classes of nucleoli were compared in two ways. In Fig. 14 the volumes of the secondary

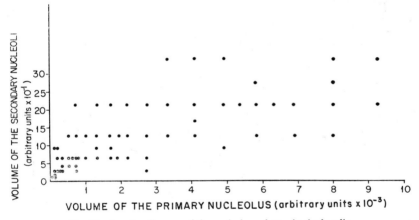

Fig. 14. A scatter diagram of the nucleolar volumes in single cells.

nucleoli were compared with the volumes of the primary nucleoli in each individual cell. The resulting scatter diagram indicated that there probably is no linear relationship between the sizes of the two classes of nucleoli. In Fig. 15 a frequency distribution is given of the ratio of volumes of primary to secondary nucleoli. Although the most common ratio observed was eight, i.e., the diameter of the secondary nucleolus is one-half that of the primary nucleolus, there was enough variation to invalidate any general statement about the parallel growth of the two classes of nucleoli.

TABLE 1

A Summary of the Cytochemical Findings on Slug Oocytes

| Staining procedure | Stains | Oocyte cytoplasm | Oocyte chromosomes | Primary nucleolus | Secondary nucleolus |
|---|---|---|---|---|---|
| Feulgen<br>after DNAase | DNA<br>Removes DNA | −<br>Pink | +<br>− | Periphery only<br>Pink | −<br>− |
| Gallocyanin-chromalum<br>after RNAase<br>after DNAase | DNA<br>RNA | −<br>+ | +<br>− | Periphery only<br>+ | −<br>+ |
| Azure B bromide<br>after RNAase<br>after DNAase | DNA<br>RNA | −<br>+ | + light<br>+ | −<br>+ | −<br>+ light |
| PAS<br>after malt diastase | Mucopolysaccharides<br>Removes glycogen | +<br>− | −<br>− | ++<br>+ | −<br>− |
| Alcian blue<br>after hyaluronidase | Acid mucopolysaccharides<br>Removes some ground substance | −<br>− | −<br>− | +<br>+ | −<br>. |
| Naphthol yellow S | Basic proteins: dibasic amino acid residues of lysine, arginine, and histidine | + | − | + | + |
| Fast green | Histones | + | + | Periphery only | − |
| Pearse phospholipid<br>after pyridine | Phospholipids<br>Removes phospholipids | +<br>− | −<br>− | Nucleolus stained red<br>Nucleolus stained red | −<br>− |
| Sudan black B | Neutral fats | + | − | − | − |

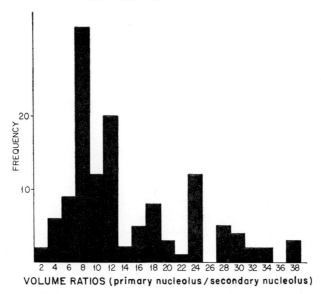

Fig. 15. The frequency distribution of classes of volume ratios between the primary and secondary nucleoli in the same cell.

## Discussion

The chromosomes do not represent an unusual situation in any respect in this slug material. Characteristic of oogenesis, the "lamp-brush" stage was clearly observed, particularly in the material stained by gallocyanin-chromalum after incubation in RNAase. By diplotene the chromosomes had undergone considerable contraction, occupying only a small fraction of the nucleus.

The employment of gallocyanin-chromalum as a stain for DNA has vastly improved the quality of the staining of the chromosomes in oocyte material over that generally found in Feulgen-stained material. Although the Feulgen method has given excellent results in other types of tissues where the chromosomes were more condensed, in oocyte material it has left a great deal to be desired. This is particularly interesting since both these methods are supposedly qualitatively and quantitatively specific for DNA. Perhaps Wood's report (19) that the acid hydrolysis, which is an integral part of the Feulgen reaction, tends to remove a portion of the DNA accounts for this difference in intensity of staining. The difference in stain intensity may lie in the different mechanisms by which the two dyes attach to DNA.

The data concerning the distribution of cytoplasmic materials are about what one would expect from the previous descriptions of Brachet (4) and his school. The lack of nuclear phospholipids might be explained by the recent evidence introduced by Chayen, LaCour, and Gahan (8) that calcium formal does not fix nuclear phospholipids, but that Lewitsky's fixative (equal parts 10 per cent formalin and 1 per cent chromic acid) will fix this material so that it may be demonstrated by Berg's (3) benzpyrine-caffeine method. Otherwise the data are as would be expected, including the observed decrease in staining intensity of the cytoplasm of the larger oocytes. None of these staining methods was subjected to semiquantitation with a microphotometer; however, the visual differences were great enough to make this unnecessary in most cases. The most notable exception to this is the lysine, arginine, and histidine method where differences in the intensity of naphthol yellow S binding are difficult to assess visually. Fast green at $p$H 8.0, which normally stains basic nucleoproteins exclusively, was found to stain the cytoplasm of the oocytes. Alfert and Goldstein (2) observed cytoplasmic binding of this dye in a protozoan. In a personal communication, Alfert suggested that such cytoplasmic binding *might* be caused by accumulations of oxidative enzymes of the proper isoelectric point to bind fast green at $p$H 8.0. This possibility underlines the ease with which erroneous interpretations may be drawn from supposedly specific cytochemical reactions that depend on only the interaction of a basic or acidic dye at a particular $p$H with the material in question.

Most cytologists today accept the general conclusions of Caspersson (7), that the nucleolus is the principal site of RNA and protein synthesis, and that the function of the nucleolus is in some way influenced by the "nucleolus-associated chromatin" or nucleolar organizer. Recent evidence to support this has come from Lin (14), who found a linear increase in RNA per nucleolus in maize microsporocytes with each extra nucleolus organizer inserted in the genome. Taylor, McMaster, and Caluya (18) found that $P^{32}$ was incorporated first into the nucleolar organizer, next into the nucleolus, and last into the chromosomes of *Drosophila* salivary gland cells. In the present investigation, the most obvious differences between the two classes of nucleoli concern the lack of DNA or histone associated with the secondary nucleoli. Since the secondary nucleolus seems to arise during the course of development from the primary nucleolus, it follows that the initial synthesis of RNA and protein probably was regulated by the primary nucleolus, and possibly under the control of the DNA and histone associated with the primary nucleolus. Since the secondary nucleoli seem to grow independently of primary nucleoli, and since the rate of growth does not seem to be specifically controlled with respect to the rate of growth of the primary nucleolus, it is possible that the secondary nucleoli represent an independent center of RNA and protein synthesis.

SUMMARY

1. The growth of the slug oocyte was studied cytochemically with respect to DNA, RNA, histones, basic proteins, mucopolysaccharides, and phospholipids.

2. It was discovered that the two classes of nucleoli found in the oocyte nuclei of some slug species are of different chemical composition.

3. A primary nucleolus is present from the beginning of oocyte growth throughout the life of the cell in the ovotestis. It is characterized by perinucleolar DNA, histone, and a mucopolysaccharide-containing nucleolar envelope; in common with the secondary nucleolus, it contains RNA and dibasic amino acid residues of lysine, arginine, and histidine.

4. Secondary nucleoli do not appear until later in the development of the oocyte and appear to arise from the primary nucleoli; while these nucleoli contain both RNA and the basic amino acids, lysine, arginine, and histidine, the concentration of the former is lower and much more variable in secondary nucleoli as compared to primary nucleoli of the same size class.

5. No consistent volume ratio relationship was observed to exist between the two classes of nucleoli.

6. From these observations, it is suggested that secondary nucleoli may represent independent centers of protein and RNA synthesis produced during the course of oocyte growth.

REFERENCES

1. Alfert, M., and Geschwind, I. I., *Proc. Natl. Acad. Sci. U. S.*, **39**, 991 (1953).
2. ———, and Goldstein, N. O., *J. Exptl. Zool.*, **130**, 403 (1955).
3. Berg, N. O., *Acta Pathol. Microbiol. Scand.*, Suppl., **90** (1951).
4. Brachet, J., *Chemical Embryology*. Interscience Publishers, New York (1950).
5. Bradley, M. V., *Stain Technol.*, **32**, 85 (1957).
6. Bridgeford, H. B., and Pelluet, D., *Can. J. Zool.*, **30**, 323 (1952).
7. Caspersson, T., *Cell Growth and Cell Function*, Van Nostrand, New York (1950).
8. Chayen, J., La Cour, L. F., and Gahan, P. B., *Nature*, **180**, 652 (1957).
9. Deitch, A., *Lab. Invest.*, **4**, 324 (1955).
10. ———, *Anat. Record*, **117**, 583 (1953).
11. Einarson, L., *Acta Pathol. Microbiol. Scand.*, **28**, 82 (1951).
12. Flax, M. H., and Himes, M. H., *Physiol. Zool.*, **25**, 297 (1951).
13. Kurnick, N. B., *Stain Technol.*, **30**, 213 (1955).
14. Lin, M., *Chromosoma*, **7**, 340 (1955).

15. Lison, L., *Stain Technol.*, **29**, 131 (1954).
16. Pearse, A. G. E., *J. Pathol. Bacteriol.*, **70**, 554 (1955).
17. Swift, H., in *The Nucleic Acids,* Vol. II (E. Chargaff, ed.), p. 51. Academic Press, New York (1955).
18. Taylor, J. H., McMasters, R. D., and Caluya, M. F., *Exptl. Cell Research,* **9**, 460 (1955).
19. Woods, P. S., *J. Biophys. Biochem. Cytol.*, **3**, 71 (1957).

## DISCUSSION

DR. ALLEN: I would like to introduce a word of warning about the interpretation of staining properties underneath the fat cap of centrifuged eggs. It is seen that in the electron microscope studies, many of the heavy particles which it seems ought to be centrifuged down, simply don't move down in this area. So, if yolk granules are in this area, and you felt yolk granules give you artifactual results, they might give you false staining reactions.

DR. PASTEELS: Yes, but in the case of *Paracentrotus* you can see that there is no collection of granules under the oil cap. A very peculiar fact is that there is a correlation between the different zones of stratification of different eggs.

DR. WEBER: It bothers me a little bit that the localization of acid phosphatase differs in *Paracentrotus* and *Arbacia.* Did you try to detect the activity of this enzyme in particle fractions, isolated from homogenates?

DR. PASTEELS: No, we haven't done that. It will be done later, of course. We have only studied the Gomori reaction.

DR. SWIFT: I wondered about the localization of the metachromatic granules under the electron microscope and also where the annulate lamellae are in your scheme of things.

DR. PASTEELS: As for the annulate lamellae, in the case of *Arbacia,* they are probably just above the yolk region. Sometime ago there was some work done with the electron microscope by McCullough that showed lamellae in this region. However, I did not see anything of the metachromatic granules under the electron microscope. Perhaps they are too easily confused with the yolk platelets, or possibly they are not fixed with osmic acid. In the case of *Arbacia* these metachromatic granules are above the yolk and could be distinguished from the yolk. Dr. Mulnard has already made fixations for detecting these granules and I hope in a few weeks we will have the results of his studies.

DR. NOVIKOFF: Where is the Golgi material located in the striated egg?

DR. PASTEELS: The Golgi material is in the upper part of the yolk layer.

DR. NOVIKOFF: It is interesting that Golgi material and acid phosphatase activity are found in the same zone.

Our electron microscopic study, with Dr. de Duve, of the "lysosome" fraction isolated by him from liver led us to suggest the possibility that acid phosphatase (and the other hydrolytic enzymes) were localized in the peribiliary "dense bodies." In attempting to test this suggestion by the staining method for acid phosphatase, we have been impressed with the similar locations of acid phos-

## DISCUSSION

phatase-rich granules and Golgi apparatus. In three experimental situations we have studied, when the granules move away from the bile canaliculus region so does the Golgi apparatus.

I wonder, Dr. Pasteels, if you would comment on the fact that in your photographs the stain for acid phosphatase activity seemed diffuse, whereas in liver (and kidney, too) we always find it associated with particles about the size of small mitochondria.

DR. PASTEELS: Yes, but perhaps this is a matter of fixation techniques. We fix these in cold calcium saline formol.

DR. NOVIKOFF: We fix them in the same medium.

DR. FLEXNER: I wonder if Dr. Pasteels wouldn't like to add the fluorescent antibody technique to his array of cytochemical techniques. You gave us a very adequate coverage of the pitfalls and advantages of cytochemical techniques, Dr. Pasteels, and I wondered if you would not like to add the fluorescent antibody technique to your arsenal.

DR. PASTEELS: Yes, we certainly would like to try this technique in the future.

# CHANGES IN ENZYMATIC PATTERNS DURING DEVELOPMENT

S. C. SHEN *

*Osborn Zoological Laboratory, Yale University*
*New Haven, Connecticut*

THE CONCEPT of a chemical basis of embryonic development derives, though not its very origin, a major source of inspiration from the discovery of the phenomenon of embryonic induction. For over thirty years, a vigorous search has been sustained for some chemically identifiable substance or substances that might act as a causal agent in initiating early embryogenesis. The implication was that the chemical nature of such an agent, if and when discovered and understood, would provide a powerful clue to the chain of events it initiates or catalyzes, first in chemical terms, then translated into biological terms. The outcome of many years' extensive investigation with this objective in mind, unfortunately, has not thus far yielded information as illuminating as it was optimistically anticipated. In the light of our newer knowledge of protein synthesis, and aided by technical advances in chemistry and physical optics, the problem of embryonic induction is being investigated with renewed vigor and sophistication in the hands of some particularly valiant and capable embryologists. Rewarding results are beginning to emerge from these investigations (29). Encouraged by these and perhaps more significantly by the enviably successful work of the geneticists, particularly the microbial geneticists, there has been a phenomenal trend in the past decade toward a mass transformation of experimental embryologists into what are now known as molecular biologists. This extraordinarily rapid advance of the frontier of embryology from study of the whole embryo to one of submicroscopic dimensions of life inevitably leaves behind it a vast territory in which a prodigious number of mopping up operations remain to be carried out if some sort of tangible communication between the front and the rear echelons is to be established.

It is this scarcely explored land that is assigned to the cytochemists as their base of operation. Their task is a challenging one: to reorient the

* Present address: Department of Anatomy, College of Physicians and Surgeons, Columbia University, New York, N. Y.

extracted macromolecules in their proper perspective or, conversely, to identify their orientation in vivo. It is surely too elementary a thought to suggest that a particular species of macromolecule, were it to give rise to a specific biological structure or function, must not only be present in quantitative adequacy but also with spatial specificity at a given time of the developmental processes. Enzymes, as species of specific protein molecules, must also conform to such requirements whether they act as causative agents or are present merely as end-products of protein synthesis. Evidences from contemporary enzymology strongly suggest that most, if not all, intracellular enzymes are structurally bound in the living system. It is a moot question whether such enzymes may be regarded as structural proteins even if they can be solubilized or crystallized by very drastic means.

The present discussion therefore attempts to focus our attention on the change of enzyme patterns not as patterns of abstract curves of enzyme content or because of their metabolic implications, but rather as structural patterns formed on a histological or cytological level during embryogenesis. There are a number of recent reviews summarizing and evaluating a considerable body of observations on changes in the enzyme content of embryo or tissue homogenates (1, 20).

## Distribution of Enzymes

In recent years, embryologists interested in developmental enzymology have become increasingly aware of the possibility that an unequal distribution of enzymes among different cell types, or within a given cell at different stages of development, may illuminate the chemical basis of cellular differentiation. With sufficient refinement, techniques for enzyme assays applicable to embryo or tissue homogenates can also be applied, with comparable quantitative precision in some instances, to a single cell. The chief difficulty is, however, in the isolation of a single cell from a multicellular system without radically altering its structure and physiological properties. Spratt (28) has shown that an isolated piece of chick blastoderm exhibits a very different cytochemical behavior from that of the same piece in situ. It is quite apparent that practically all the inherent weaknesses in enzyme assay techniques of homogenates are present in cytochemical staining of tissue sections. The chief advantage of the latter technique is that it does preserve to a large extent structural orientations of individual cells as well as cell-to-cell relations. Under suitable conditions, enzyme localization may be resolved to subcellular dimensions. It thus enables one to see, as it were, both the topography of a forest and the characteristics of individual trees. Such information would provide, in part, a useful background for conceiving the process of embryogenesis basically as one of cellular and ultimately molecular ecology.

The major criticism against the present-day techniques of enzyme cytochemistry is their lack of quantitative precision as compared with direct chemical analysis. No one would wish to contest such a criticism; there is however no need for undue pessimism that available techniques cannot be so improved upon as to yield meaningful quantitative data. Indeed, one should be much encouraged by the successful photometric measurement of cytochemical reactions of nucleotides in the cell nucleus. To discredit the contributions of cytochemistry because of the present inadequacy in quantitative precision is like discrediting the anatomical and cytological descriptions upon which rest the validity of experimental biologists. At present, enzyme cytochemists are more preoccupied with the following problems: (1) to devise methods for detecting and localizing enzymes that have hitherto eluded cytochemical procedures; (2) to minimize structural and chemical damage to cells by cytochemical manipulations; and (3) to increase the resolution of enzyme localization by minimizing diffusion artifacts and by reducing the dimensions of the visualized products. To determine the precise distributional patterns of enzyme molecules on biological fine structures must surely be among the fondest hopes of cytochemists, electron-microscopists and, possibly, molecular embryologists.

## MITOCHONDRIAL ENZYMES IN EMBRYOGENESIS

The cell nucleus is generally regarded as the exclusive domain of the geneticists. Embryologists who are interested in cellular differentiation must content themselves with the remainder of the cell: the cytoplasm and the cell membrane. Fortunately, the cytoplasm alone appears to be sufficiently complex and intriguing to divert the embryologists. Within the resolving power of the optical microscope, mitochondrial particles are perhaps the most unique among cytoplasmic inclusions. Embryologists have long suspected that mitochondria have an important role in cytodifferentiation and embryogenesis. For example, mitochondria in myoblasts were believed to be the actual precursors of myofibrils (11). Independently, chemical embryologists, not particularly impressed by the fine points of cytology, were largely concerned with cellular respiration and the oxidative enzymes as clues to developmental processes. The postulation of an oxidation-reduction gradient or field in developing embryos has long dominated the thinking in causal embryology. The remarkable discovery in recent years of the fact that mitochondria are virtually the exclusive loci of oxidative and phosphorylating enzymes, among others, has profoundly impressed embryologists of divergent discipline and interest. It is only natural that investigations on mitochondria during cellular differentiation have been undertaken with renewed enthusiasm and perhaps a fresh viewpoint. It is not the intention of the present paper to make a comprehensive survey of the rather extensive literature, except to mention briefly the highlights of some of these investigations.

Most of these studies were made on annelids (18), ascidians (22) and echinoderms (14). The annelid and ascidian eggs have in common the embryological characteristic of being of the mosaic type. That is to say, at very early stages in the cleavage of a fertilized egg, the developmental potency or fate of each daughter cell is more or less determined. It follows that if mitochondria and the associated enzymes do play an important role in cellular differentiation, the different but already determined fate of the daughter cells may be reflected by their mitochondrial contents, quantitatively or qualitatively. Without the encumbrance of experimental details, the observations of these investigations may be generalized as follows. (1) The mitochondrial particles in these cells may be identified either by their cytological structure or by their complement of oxidative enzymes, as demonstrated by cytochemical means. (2) In unfertilized eggs or in fertilized eggs prior to the first cleavage, mitochondria are present in abundance but show a relatively uniform distribution with respect to the egg's axis or plane of cleavage, although characteristic concentrations around the nuclear and cell surfaces are generally observed. (3) A significantly unequal distribution of the mitochondria population in the daughter cells occurs at the first and subsequent cleavages. (4) The mitochondria are much more concentrated in those cells that give rise to neural and somatic structures than they are in the endodermal cells. In the ascidian eggs, by treatment at the beginning of development with a mitochondrial stain, the distributional pattern of the particles may be followed to the tadpole stage. There is a conspicuous shift of the mitochondrial concentration toward the tail-forming region and finally a localization in the muscle cells of the tail.

To give further support to the thesis that mitochondrial enzymes are causally related to cellular differentiation and embryogenesis, attempts were made to block these enzymes by treating the eggs with more or less specific inhibitors. When a sufficient degree of inactivation of these enzymes is achieved, such eggs generally result in embryos with defective structures which are normally rich in, and hence presumably more dependent for their differentiation on, the mitochondrial enzymes. In contrast to mosaic eggs, the regulative eggs, such as those of amphibians, do not show any marked degree of topographic segregation of mitochondria in early cleavages. When such eggs were treated with various inhibitors of mitochondrial enzymes, they generally resulted in an early arrest of development or in abnormalities of a non-specific nature. In the chick embryo, Spratt (28) noted a high concentration of mitochondrial enzymes, as indicated by the intense cytochemical reactions at the nodal region, which has been shown to be the center of differentiation of the axial musculature.

From the few examples cited above, it appears suggestive that mitochondrial enzymes constitute at least in part an essential chemical machinery for cellular differentiation and probably morphogenesis. Two crucial questions, however, remain unanswered. (1) What is the mechanism by which mitochondria are differentially segregated at the early stages of cleavage, particularly in the mosaic eggs? (2) In what way do mitochondria and their enzymes contribute to the differentiation processes? To answer the first question, at least three possibilities may be considered: (a) the cortical layer and the nuclear apparatus, to both of which mitochondria appear to have a relatively great affinity at the time of cell division, may play a decisive role; (b) there may be active migration of the mitochondria during cell cleavage; or there may be a (c) differential rate of mitochondria proliferation and/or internal differentiation, structurally and enzymatically, after their segregation. Although crucial evidence is lacking, it is suggestive that populations of mitochondria isolated from cells at different stages of differentiation show very different enzyme content on a per unit protein nitrogen basis (4, 25). The problem is greatly complicated by the high probability that within a given cell type there are structurally and functionally diversified species of mitochondria. A further possibility deserves serious consideration: a given mitochondrion of a specific enzyme makeup may perform very differently in a different environment within a cell.

Attempts to offer any plausible explanation of the chemical contribution of mitochondria to cellular differentiation can go no further than sheer speculation at present. Impressive progress in the biochemistry of protein biosynthesis, of an unspecific nature, overwhelmingly implicates the involvement of the microsomal fraction. In in-vitro systems, incorporation of amino acids as criteria of protein synthesis by the microsomal RNA can apparently be brought about without the direct participation of mitochondria, but with an externally supplied source of energy. It seems quite probable, however, that in vivo mitochondria constitute that source of energy. Furthermore, recent investigations provide evidence that the mitochondrial RNA, from muscle cells, is also capable of incorporating labeled amino acids.

### Distribution of Cholinesterase in Cellular Differentiation

Cholinesterase (ChE) may be chosen as an example of developmental enzymology for the following reasons. (1) It is a relatively simple enzyme system that can readily be characterized and assayed. (2) A number of satisfactory techniques have been devised for its cytochemical localization. (3) Extensive physiological and pharmacological evidence implicates the enzyme as having a specific function in synaptic transmission. (4) Unlike the mitochondrial enzymes, cholinesterase is probably not directly, if at all, involved in the general metabolism of cells. It is not present in a detectable amount in early ontogeny. Hence it is perhaps a more suitable enzyme for the investigation of its relation to cellular differentiation of a specific nature.

*Cholinesterase in the Central Nervous System*

Since the enzyme cholinesterase is believed to be primarily concerned with neural function, its distributional pattern has been studied largely in the vertebrate central nervous system. On a wet or dry weight basis, the adult central nervous system as a whole has the highest cholinesterase content relative to other tissues. Enzyme assays on homogenates of crude subdivisions of the central nervous system indicated a rather characteristic pattern of distribution of cholinesterase, in contrast to that of the oxidative enzymes, which are generally uniformly distributed among the diverse cellular elements in the central nervous system. Within the limitation of experimental error, and with a few unexplained exceptions, the distributional pattern of cholinesterase bears a striking resemblance to that of acetylcholine and cholinacetylase in the central nervous system (5). However, the interpretation that the cholinesterase-rich regions therefore represent regions of cholinergic synapses is not much more than an inference based on rather dubious and perhaps circular arguments.

Similar investigations on developing embryos further revealed that structural and functional maturation of the central nervous system as a whole is accompanied by an unusually high rate of cholinesterase synthesis in contrast to that of the oxidative enzymes (25). Furthermore, within the central nervous system, a demonstrable amount of the enzyme in the various regions appears at different stages of neurogenesis. The subsequent rates of enzyme accumulation in these regions are also quite different. In general, these chemical events may be correlated well with the events of morphological and functional differentiation within the central nervous system (3). All these investigations were made on tissue homogenates. The immensely complex interneuronal organization and the unique structure and function of the individual neurons in the central nervous system immediately set a serious limitation to any critical evaluation based on such crude, though initially desirable, analysis. Within the limitations of the available methods, and there are many of these, cytochemical visualization of the enzyme in tissue sections offers certain advantages over the homogenate approach.

Cytochemical localization of cholinesterase, based on the method devised by Koelle (16), has been studied in this laboratory largely on amphibian material. In the adult central nervous system, the pattern of cholinesterase distribution agrees well with that found by direct assays on homogenates. The cytochemical pictures, however, further reveal that the enzyme is concentrated in well-defined neurological structures identified as synaptic centers in which the chief cytological components are the interneuronal synapses. During normal development, the first appearance of the enzyme and its subsequent accumulation in a given area correspond closely to the cytological differentiation of synapses in that area. It may be mentioned as an example that the formation of synaptic fields in the optic tectum, as the optic fibers extend into that region, proceeds from the cephalic towards the caudal pole of the optic lobe. This is accompanied by a similar sequence of cholinesterase distribution.

The close correlation between synaptic formation and cholinesterase synthesis is further borne out by the fact that experimental interference with the normal differentiation of synaptic centers similarly interferes with the appearance of the enzyme at these centers. For example, experimental interruption of the optic fibers would result in structural defects in the chain of synaptic stations along the primary optic pathway, with concomitant non-formation or, once formed, disappearance of the enzyme in the affected centers (2).

*Cholinesterase in the Neuroretina*

Perhaps a better example of the enzyme-synapse relationship is afforded by the vertebrate neuroretina, embryologically an integral part of the central nervous system. It is well known from the classical descriptions of the vertebrate neuroretina that primary and secondary synapses in the retina are topographically well separated from the mass of cell bodies. In the fully differentiated retina, these synapses are located in two well-defined zones: the external and internal plexiform layers, respectively. Furthermore, in the retina of certain species, notably avian, the secondary synapses in the inner plexiform layer are stratified into distinct zones within a dense network of fibrous processes. If cholinesterase is associated exclusively with the synaptic terminals, its distributional pattern should correspond precisely to the synaptic stratifications. That this is indeed the case is clearly illustrated in the chick retina (27). In the inner plexiform layer of a fully differentiated chick retina, there are four distinct zones of synapses identifiable cytologically by means of silver impregnation as well as cytochemically by cholinesterase localization. In addition to these four synaptic zones, cholinesterase is also present in considerable quantity on the surface of the ganglion cells and of some of the amacrine cells. It is cytologically demonstrable that the axonal processes of some of the bipolar cells synapse directly on the bodies of the ganglion and amacrine cells. On the other hand, the enzyme is conspicuously absent in the external plexiform layer, where the visual elements enter into synaptic relations with the dendrites of the bipolar cells. It has been postulated (12) that the optic pathway consists of a chain of alternately cholinergic and non-cholinergic neurons. Direct evidence in support of this postulation is, however, lacking. It is of interest to note that in the amphibian retina, cholinesterase is found in both external and internal plexiform layers.

Based on these observations, the distribution pattern of cholinesterase was followed in the chick retina during retinal differentiation. As anticipated, the morphological event of synaptic differentiation is concomitant with the chemical event of cholinesterase synthesis and accumulation at the site of synapses. When the inner plexiform layer is first formed, the poorly aligned synaptic terminals are reflected by a diffuse zone of enzyme concentration. With progressive stratification of the synaptic structures, the enzyme-rich zone resolves into two increasingly distinct layers within the plexiform layer. With further differentiation, two additional zones of cholinesterase concentration corresponding to synaptic terminals appear between the two initial zones, to result in a cytological and cytochemical pattern virtually identical to that of an adult retina (Fig. 1).

It may be of considerable interest to note that during early histogenesis of the retina, the first trace of cholinesterase is found not in the plexiform layer, but on the surface of the ganglion cells, the first neuronal elements to undergo differentiation. At that time, the prospective bipolar cells remain undifferentiated. It is highly improbable that structural synapses have been established on the ganglion cells. It seems suggestive that the enzyme is initially formed and localized on the post-synaptic surface prior to and

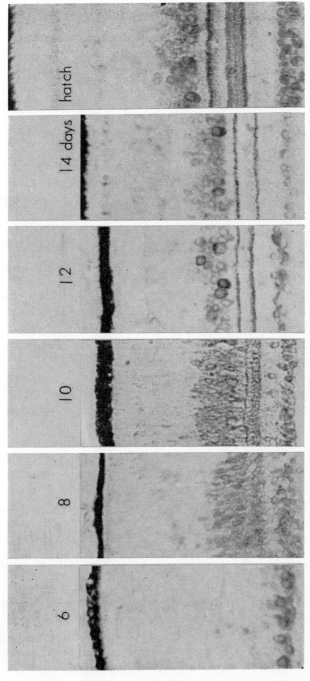

Fig. 1. Histochemical localization of cholinesterase in the chick retina from embryos of varying age up to hatching. Cross section through the fundus region. 250 ×. The ganglion-cell layers are at the bottom of each picture. The dark band at the top of the pictures is the pigment epithelium and does not represent enzyme concentration.

independent of the incoming presynaptic terminals. Even at the time the enzyme first appears in the inner plexiform layer, it is doubtful if the axonal terminals of the bipolar cells have reached that region. It is quite probable that the enzyme localization in that region represents the dendritic processes of the ganglion cells or, in other words, the post-synaptic elements of the yet incompletely formed synapse. The initial and independent formation of the enzyme on the post-synaptic surface will be better illustrated in the neuromuscular differentiation to be described presently.

No serious or successful attempts have yet been made to alter experimentally the neurogenesis of the retina. Explants of very young optic vesicles were made in the coelomic cavities. Such explants, unfortunately, were invariably resorbed by the host tissue before becoming fully differentiated. Those explants that were recovered at the peak of their differentiation, however, showed fairly extensive histogenesis, but with a greatly reduced number of ganglion cells. Histochemically, the enzyme is present in the few ganglion cells and the two well-stratified zones in the inner plexiform layer. No detailed cytological examination of the neuronal organization of the explants has yet been made. One fact seems quite clear, however, that retinal differentiation accompanied by cholinesterase synthesis can proceed to a considerable extent without central connections with the central nervous system. It is not certain if the few well-differentiated ganglion cells in the explants represent the total number or the survivors from a larger number of differentiated cells.

*Cholinesterase in the Myoneural Junctions*

A synaptic structure, defined as a cytological entity according to the neuron theory, consists of at least two components: the pre-synaptic and the post-synaptic membranes. The above-described observations on cholinesterase distribution in the central nervous system and the neuroretina, while strongly suggesting a close association of the enzyme with the synapses, do not distinguish the two components of a synapse. Without the benefit of a greatly improved method of enzyme localization on fine structures, direct evidences are as yet not available. Experimental alteration of the pre- or post-synaptic neurons of a complex synapse does not necessarily provide an unequivocal answer. The immediate effect may conceivably be on both pre-synaptic and post-synaptic surfaces, considering the extreme proximity, structurally and functionally, of the two. The actual situation is particularly complicated in the central nervous system, in which most, and probably all, neurons are in synaptic relations with more than one other neuron. An ideal structure for such investigation is therefore one in which (1) the synapse consists of a single pair of pre- and post-synaptic members; (2) where the developmental history of the two members can be followed separately in the course of normal or experimentally modified differentiation; and, of course, (3) where the structural and functional differentiation of the synapse is characteristically associated with cholinesterase synthesis. The vertebrate myoneural junction, while not exactly an idealized synapse, does offer many more advantages for experimental analysis than do the central synapses.

Although the transmission of motor impulses across the vertebrate motor endplate via a chemical mediator, acetylcholine, has been postulated many years ago (8), convincing evidence has been obtained only recently by a brilliant series of experiments (9). The normally rapid recovery from an endplate excitation demands a similarly rapid destruction of the liberated acetylcholine by its specific enzyme, cholinesterase. The low mobility of the enzyme molecules demands their presence at an adequate concentration at the immediate neighborhood, if not the very site of liberation of acetylcholine. The local concentration of cholinesterase at the motor endplate has been convincingly demonstrated. However, the exclusive association of the enzyme with the post-synaptic membrane was an implication rather than experimental demonstration. It appears that developmental analysis of the initial formation and the subsequent pattern of cytological distribution of the enzyme in relation to endplate differentiation may shed some additional light on the question.

While there is a considerable body of literature on myogenesis, descriptions of the cytodifferentiation of the motor endplate are relatively scanty and have yielded conflicting interpretations. Some investigators (7, 21) maintain that the soleplate is derived entirely from Schwann cells, and hence is of a neural origin; others (10) consider it as of exclusively muscular origin. Still others (6) believe that the endplate is derived from both sources. The fine structure of a motor endplate, as revealed by electron microscopy, has been published only very recently (23). The myoneural junction in lizard skeletal muscle appears to consist of a complex membrane with extensive foldings. The synaptic membrane itself seems to be a complex of several membranes of both neural and muscular derivation. Although nuclei, both neural and muscle, mitochondria, and cytoplasmic granules are more concentrated around the endplate, they do not however appear to contribute directly to the synaptic membranes. Cytochemical localization of cholinesterase on the endplate, while bearing a striking resemblance to the synaptic structure as a whole as shown by the electron microscope, does not resolve the question of the pre- or post-synaptic localization of the enzyme.

In following the cytochemical distribution of cholinesterase in rat muscle during myogenesis, Kupfer and Koelle (17) concluded that the enzyme is initially formed in some specialized muscle nuclei which later aggregate to form the endplate. It is not clear whether these nuclei give rise directly to the synaptic membranes. Unfortunately, their method of enzyme localization is open to question. They used a technique characterized by nuclear staining which Koelle attributed to enzyme diffusion artifact. When such diffusion is minimized by means of an improved method (16), nuclear staining is completely absent. This observation has been confirmed in our own experience with amphibian materials.

Recently, we have made some observations on the cholinesterase distribution in relation to the differentiation of the myoneural junction in the amphibian axial muscle. In the adult, motor innervation of these muscles is restricted to the ends of each muscle fiber. No innervation, motor or sensory, has ever been observed elsewhere along the muscle fiber. Similarly, cholinesterase localization on these muscles coincides precisely with the point of motor innervation. During normal development, cholinesterase first appears at the time and at the site of motor termination on the muscles. Physiologically, this also coincides with the first elicitable behavior of a neurogenic response, although a myogenic response may be observed prior to that. The innervation of these muscles by their corresponding motor neurons, during development, proceeds in an orderly cephalo-caudal direction. The appearance of cholinesterase on these segmental muscles follows exactly the same sequence. At an early free-swimming stage, the patterns of motor innervation and cholinesterase distribution are virtually identical to those in an adult, especially in urodeles (Figs. 2-10). Based on these observations, it is difficult to escape the conclusion that cholinesterase syntheses and myoneural differentiation are an integral system.

The following question immediately arises: is the initial formation of the motor endplate, as a specialized apparatus of these muscles, a direct consequence of the motor innervation? This may readily be tested in amphibian embryos. It is a simple procedure to remove a part of or the entire presumptive neural material in an amphibian embryo, which will then continue to develop into a partially or totally nerveless tadpole. Harrison (15) has shown convincingly that normal muscle differentiation proceeds to an extensive extent in the total absence of motor and sensory innervation. Sawyer (24), using the cholinesterase content of muscle homogenate as a

428  THE CHEMICAL BASIS OF DEVELOPMENT

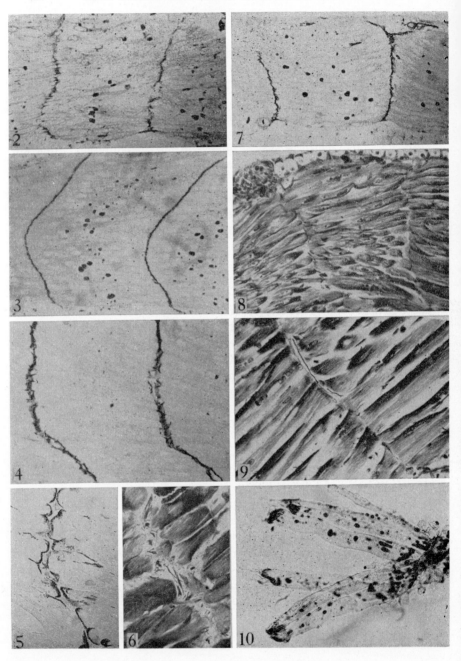

criterion, observed however that in the nerveless embryos there is no appreciable synthesis of muscle cholinesterase as compared to the normal muscles. He therefore concluded that the formation of muscle cholinesterase is dependent on and contributed directly by the neural elements, presumably the Schwann cells that accompany the incoming motor nerves.

Using histochemical techniques, we have found, however, that the initial formation and localization of cholinesterase in the nerveless muscles are exactly the same as in normal muscle. There is no detectable difference in the cytological appearance of the two, other than the total absence of nerve terminals in the nerveless ones (see Figs. 7-8). Furthermore, quantitative measurement shows that up to a stage, which interestingly enough corresponds to the youngest embryos Sawyer investigated, the cholinesterase content per unit muscle protein and the rate of its increase are identical in the nerveless and the control muscles. After that stage, while the control muscle continues to accumulate cholinesterase, the nerveless one ceases to do so and finally declines in its enzyme content, accompanied by extensive muscular dystrophy (Fig. 11).

A further evidence that cholinesterase synthesis, at least its initial phase, is intrinsic with the mesodermal cells is provided by tissue culture technique. Under suitable conditions, the isolated presumptive mesoderm may undergo extensive muscular differentiation in vitro in the complete absence of other cell types. This differentiation is accompanied by a substantial accumulation of cholinesterase. Due to the disorientation of the muscle cells under such conditions, however, it is difficult to establish a clear-cut localization of the cholinesterase on the muscle cell. In fact, our

---

PLATE I (opposite). Cholinesterase localization and innervation patterns of the amphibian axial muscles during myogenesis. Longitudinal sections through several adjacent segments. 100 ×. Enzyme localization is at the ends of each muscle fiber. The dark granules along the muscle fibers are pigment granules which disappear during myogenesis.

Fig. 2. Normal *Amblystoma* punctatum, Harrison Stage 36. ChE localization.
Fig. 3. Normal *Amblystoma*, Stage 38. ChE localization.
Fig. 4. Normal *Amblystoma*, Stage 45. ChE localization.
Fig. 5. Adult *Triturus*. ChE localization.
Fig. 6. Adult *Triturus*. Silver preparation showing motor innervation restricted to the inter-segmental zone.
Fig. 7. Nerveless *Amblystoma*, Stage 37. ChE localization. Compare with Fig. 2.
Fig. 8. Nerveless *Amblystoma*, Stage 45. Silver preparation showing absence of nerve.
Fig. 9. Normal *Amblystoma*, Stage 45. Silver preparation showing normal pattern of innervation at the inter-segmental zone. 250 ×.
Fig. 10. Frog tail muscles. Fresh teased preparation. 250 ×. Note enzyme concentration at the terminals of each muscle fiber. Dark spots along the muscle fibers are pigment granules.

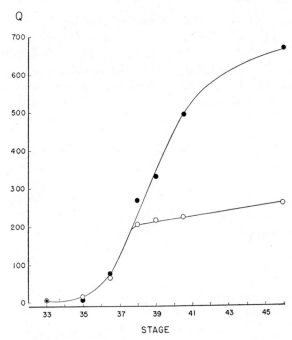

Fig. 11. Cholinesterase accumulation at the motor endplates of control (closed circles) and nerveless (open circles) *Amblystoma* during development. Q, cholinesterase activity per µg. protein nitrogen per hour. Developmental stages according to Harrison.

preliminary observations tend to suggest that the enzyme is initially synthesized not by the muscle cells, but by the connective fibers that later become a part of the specialized region of the sarcolemma. The dual origin of the sarcolemma, a complex of the modified plasma membrane of the muscle cells and an overlying fibrous structure derived from extracellular material, must however remain as a postulate that awaits further investigation.

The implications from these observations are the following. (1) Muscle cells, or possibly a specialized portion of their surface membranes, are capable of cholinesterase synthesis without benefit of neural elements. (2) In normal development, it is highly probable that synthesis of cholinesterase occurs prior to innervation, and hence may be a determining factor in the site of motor termination on the muscle. (3) The continued accumulation of the enzyme, together with cytological differentiation of the junctional structure, may well require the participation of the neural, or presynaptic, elements. With some stretch of imagination, the "trophic" action of the motor terminal may be conceived as possibly a process akin to enzyme induction, in which the enzyme is provided with a suitable atmosphere of its substrate. While no one should be encouraged to dwell for long on speculations, these above-mentioned points may serve as an example of enzyme-structure-function relationships in the analysis of a chemical basis not only of cellular differentiation but also of intercellular reactions.

## Summary

Developmental processes involve growth and differentiation. Chemically, the two aspects of the same process imply the synthesis of additional molecules of the preexisting type and also of new ones. Furthermore, such molecules wherever they are being synthesized must ultimately be so aggregated and oriented as to give rise to recognizable biological forms. The development of a concept that enzymes, as a specific species of proteins, are directly involved in the emergence of specific cellular structures should be a significant step toward closing the vast gap between biological and chemical entities, levels of organization, and the laws that govern the behavior of these entities. The change in the enzyme pattern of cholinesterase during cellular differentiation is cited as one example illustrating some specific relation between an enzyme and cytological structures. Further investigations made on this and similar systems would undoubtedly contribute to our understanding of a chemical basis of development.

## REFERENCES

1. Boell, E. J., in *Analysis of Development* (B. H. Willier, Weiss, P. A., and Hamburger, V., eds.), p. 520-555, W. B. Saunders, Philadelphia (1955).
2. ———, Greenfield, P., and Shen, S. C., *J. Exptl. Zool.*, **129**, 415-452 (1955).
3. ———, and Shen, S. C., *J. Exptl. Zool.*, **113**, 583-600 (1950).
4. ———, and Weber, R., *Exptl. Cell Research*, **9**, 559-567 (1955).
5. Burgen, A. S. V., and Chipman, L. M., *J. Physiol.*, **114**, 296-305 (1951).
6. Couteaux, R., *Intern. Rev. Cytol.*, **4**, 335-375 (1955).
7. Cuajunco, F., *Contrib. Embryol. Carnegie Inst. Wash.*, **30**, 127-152 (1942).
8. Dale, H. H., *Harvey Lectures*, **32**, 229-245 (1937).
9. Del Castillo, J., and Katz, B., *Proc. Roy. Soc. (London)*, B, **128**, 339-381 (1957).
10. Del Rio-Hortega, P., *Compt. Rend. soc. biol., Paris*, **92**, 3 (cited by Couteaux) (1925).
11. Duesberg, J., *Verhandl. anat. Ges. Anat. Anz.*, **34**, 123-127 (1909).
12. Feldberg, W., and Vogt, M., *J. Physiol.*, **107**, 372-381 (1948).
13. Gaebler, O. H. (ed.), *Enzymes: Units of Biological Structure and Function*, Academic Press, New York (1956).
14. Gustafson, T., and Lenecque, P., *Exptl. Cell Research*, **8**, 114-117 (1955).
15. Harrison, R. G., *Am. J. Anat.*, **3**, 197-220 (1904).
16. Koeller, G. B., *J. Pharmacol. Exptl. Therap.*, **103**, 153-171 (1951).
17. Kupfer, C., and Koeller, G. B., *J. Exptl. Zool.*, **116**, 397-413 (1951).
18. Lehmann, F. E., and Wahli, H. R., *Z. Zellforsch. microskop. Anat.*, **39**, 618-629 (1954).

19. McElroy, W. D., and Glass, B. (eds.), *Chemical Basis of Heredity*, Johns Hopkins Press, Baltimore (1957).
20. Moog, F., *Intern. Symposium on Cytodifferentiation*, Chicago Univ. Press, Chicago (in press).
21. Noel, R., and Pommé, B., *Rev. Neurol.*, **57**, 589-611 (1932).
22. Reverberi, G., in *The Beginnings of Embryonic Development* (A. Tyler, R. C. von Borstel, and C. B. Metz, eds.), p. 319-340, Am. Assoc. Advance. Sci., Washington, D. C. (1957).
23. Robertson, J. D., *J. Biophys. Biochem. Cytol.*, **2**, 381-393 (1956).
24. Sawyer, C. H., *J. Exptl. Zool.*, **92**, 1-47 (1943).
25. Shen, S. C., in *Biological Specificity and Growth* (E. G. Butler, ed.), p. 73-92, Princeton Univ. Press, Princeton (1955).
26. ———, Greenfield, P., and Boell, E. J., *J. Comp. Neurol.*, **102**, 717-744 (1955).
27. ———, ———, and ——— *J. Comp. Neurol.*, **106**, 433-462 (1956).
28. Spratt, N., this Symposium.
29. Yamada, Y., this Symposium.

# BIOCHEMICAL PATTERNS AND AUTOLYTIC ARTIFACT

J. Lee Kavanau

*Department of Zoology, University of California*

## Introduction

Methodology is all-important in studies of the chemical basis of development. Great precautions are required to maximize the information obtained about the living system and to minimize the background of artifact introduced by autolysis and other disruptive factors that may come into play during the handling, processing, and assaying of the material.

Autolysis *per se* has been little studied. Often, the investigator employing aqueous extractions depends upon the use of low temperature and non-physiological $pH$ for protection from autolytic artifacts. If proteins need not be maintained in the native state, some degree of protection against autolysis can be obtained by preliminary enzyme denaturation. Where enzyme activity is the object of study, aqueous homogenates are often employed and protection against autolysis is usually at a minimum.

The author has recently been engaged in a study of extraction methods suitable for the fractionation of components of living systems, where it is desired to preserve for assay a wide spectrum of substances from a single batch of experimental material. With this objective, the avoidance of artifact is more difficult than when merely a particular component is to be studied to the exclusion of others. The goal sought was an extraction procedure attaining the following objectives: (a) complete avoidance of autolysis; (b) attainment of sharp quantitative separations; and (c) minimum destruction of labile substances or interconversion of structurally related molecules, as well as minimum intermingling of identical molecular species which, as they occur in the cell, may be partitioned by physicochemical barriers.

In the course of this study it was found that, unless enzymes are first inactivated, extensive and serious autolytic changes take place in aqueous cellular homogenates (of marine invertebrate embryos). This occurs at both high and low dilutions, regardless of the precautions of maintenance of low temperature and non-physiological $pH$.

As a result of this finding, it seemed desirable to sacrifice the information that might be obtained by fractionation and individual assay of different protein components, in order to ensure maximum protection against autolysis. To this end the proteins are denatured and precipitated *in toto* in the first extraction steps; the subsequent assay of this fraction gives only the "lump sum" of proteins.

To denature enzymes thoroughly and yet avoid extensive destruction or modification of other labile substances, mild physicochemical treatments known to denature proteins are employed. These treatments are applied before contact with aqueous media (except boiling water), and contact with water is subsequently limited to the shortest practicable duration.

One of the ultimate objectives of these investigations has been to examine the patterns of chemical changes during the first few cleavages of marine invertebrate embryos. Information concerning the quantitative changes in such substances as free amino acids, vitamins, coenzymes, nucleic acid precursors, etc., during the division cycle could lead to a deeper understanding of the mechanism of cell division. In a more general way, by following in close detail the time course of these changes, it should be possible to gain further insight into the cellular mechanisms of control of many vital processes, i.e., how such processes are initiated, how their rates are regulated, and how they are terminated. Thus, studies of the correlations between changes in these substances may yield information as to whether the supply or withdrawal of substrate, coenzyme, enzyme, or primary energy-yielding compounds is rate-limiting in any given instance. For example, it might be found that the amount of a vitamin decreased simultaneously with the increase in the amount of a coenzyme derived from it, and that the utilization of some substrate or the appearance of some end-product began only after the concentration of the coenzyme had begun to rise. This would suggest that the coenzyme concentration was the limiting factor for the process in question. The control mechanism would then be placed at least as far back as the determination of the conversion of the vitamin to the coenzyme.

The rates of many intermediary steps of metabolism have been shown to be limited mainly by the concentration of substrates, the enzyme systems being present in excess of their need. In other cases, reaction rates are believed to depend on other factors than the amounts of enzyme or substrate. However, the assignment of rate limitation to a single factor, be it enzyme, coenzyme, substrate, inhibitor, primary energy source, or other, is only a simplification and merely the first step in the determination of control mechanisms. It remains to be shown, for example, by what physicochemical means the interposition and removal of constraints is mediated in the cell. Although we may be far from our goal of elucidating the ultimate control mechanisms—which must be presumed to be exceedingly complex and perhaps inextricably interwoven with the structure of cells—it should, at least, be possible to construct sequential networks of chemical events underlying morphological changes. Greater familiarity with such networks may lead to the recognition of feedback and other interrelationships which, when the network reactions are once set into motion, normally "automatically" lead to the attainment of certain critical conditions which may "trigger" the next sequence of changes, etc.

It seems most appropriate in investigations with these objectives to concentrate one's attention on cleaving embryos, which at first are relatively simply constituted—adult organelles being absent. Moreover, during early cleavages the metabolic apparatus of the cell appears to be primarily engaged in the business of cell division, this phase being a "silent period" for developmental processes which come into prominence in later stages. The surf-clam, *Spisula solidissima,* is highly favorable experimental material for a number of reasons:

(1) In a batch of many millions of fertilized eggs, almost perfect synchrony of division can be obtained, at least during the early cleavage period.

(2) Collecting, fertilizing, and culturing of the material is relatively easy.

(3) Sufficient eggs for a developmental series can be obtained from a single female, thus ensuring near-maximum homogeneity of the starting material.

(4) The eggs and embryos contain relatively small quantities of free amino acids, so that changes in free amino acids in the course of development are not swamped in a large pool.

## Material and Methods

The experimental material consisted of eggs and embryos of *Spisula solidissima,* collected, lyophilized, and processed according to techniques already described (1, 4, 5), except as noted below. The collecting was carried out at the Marine Biological Laboratory, Woods Hole, Massachusetts in the summer of 1955. In the new extraction procedure, the lyophilized material was soaked for weekly periods (at room temperature) in petroleum-ether–ethanol (2:1) and acetone–ethanol (1:1) followed by a 10-minute exposure to boiling water. The material was then homogenized and extracted repeatedly in ice-cold 0.022 $M$ acetic acid. Each extraction consisted of rapid freeze-thaw cycling between the limits 0°C and −20°C, the suspensions not being allowed to remain for more than a few minutes in the liquid state during each cycle. The residue (insoluble protein fraction) from this extraction, containing over 99 per cent of the total protein, was sedimented by centrifugation and subsequently hydrolyzed, preparative to microbiological assay for amino acids. The combined extracts were lyophilized and extracted repeatedly in 80% ethanol at −20°C. The material insoluble in 80% ethanol (soluble protein fraction) was also sedimented by centrifugation and hydrolyzed. The soluble constituents were recovered in aqueous solution and were assayed microbiologically for various low molecular weight components.

In any given developmental series, each stage collected was represented by the same number of embryos ($\pm \sim \frac{1}{2}\%$) and was processed quantitatively, utilizing such safeguards as to absolutely preclude loss of material.

The microbiological assay methods used for amino acids were essentially the same as those employed previously (4, 5). For assays of vitamins, coenzymes, and other growth factors, the appropriate Bacto Vitamin Assay Media were used. These assays were essentially standard, except for certain published modifications (2, 6, 7) and other refinements to be published in detail elsewhere, or indicated in the text below.

## Results

A preliminary comparison of the results obtained using an aqueous saline extraction method employed previously (5) with those obtained using the new "non-aqueous" method led to the following general findings concerning the inadequacy of the aqueous saline method:

1. Autolytic degradation of proteins to free amino acids and peptides occurs, to such an extent that the true amino acid patterns might be obliterated.

2. Autolytic processes lead to varying degrees of destruction of certain coenzymes and vitamins, as well as to lesser chemical modifications and possibly, in certain cases, to freeing of bound forms.

In view of these findings, a very cautious approach to the interpretation of results obtained with non-denatured aqueous homogenates is in order, particularly so when such homogenized material is incubated, dialyzed, or extracted with aqueous solutions for long periods.

On the other hand, these studies suggest that the results obtained with material subjected to controlled autolysis (i.e., autolysis under such conditions that all extracts are given exactly the same chance to autolyze) may be very useful for interpreting the normal patterns of chemical changes during development, once these normal patterns have been well established. Thus, for example, differences from the normal patterns detected after controlled autolysis of different stages would be important indications of qualitative or quantitative differences in the extracts of these stages, and would reflect developmental changes in the experimental material.

Included in the experimental material was a developmental series consisting of 29 stages of *Spisula* embryos (819,000 embryos per stage $\simeq$ 30 mg. dry weight). These were collected at 5-minute intervals, beginning one minute after fertilization and continuing through the period of formation of the polar bodies and the first three cleavages. Freeze-drying of this particular batch of material was inadvertently interrupted too soon, and a small core of undried material remained undetected in 9 of the 29 collecting tubes. The sealed tubes were subsequently exposed to ambient summer temperature for six hours during transport and were stored thereafter for one year at about $-15°C$. Moisture was detected in the 9 tubes of this series preliminary to extraction of the experimental material. Since this was the only series on hand representing the early post-fertilization period, it was decided to proceed with the investigation of the series, letting the 9 incompletely lyophilized stages serve as pre-autolyzed reference stages. It should be pointed out that the autolysis which took place was not controlled autolysis, inasmuch as different amounts of moisture were present in the different tubes, and that the extent of autolysis which did take place was far less than occurs under the conditions of the aqueous saline extraction employed previously.

The study of this "mixed" series proved to be profitable. The general pattern of post-fertilization changes was obtained and serves as a control against which to compare the effects of relatively limited autolysis. The autolytic artifacts proved to be qualitatively the same as those occurring under controlled conditions, and they are helpful for the interpretation of the normal patterns.

The results obtained for valine are given in Fig. 1. Similar patterns and distributions were found for tyrosine, the only other amino acid investigated in this series. It will be noted that less than 0.5% of the total valine present one minute after fertilization exists in the free form, and that this fraction increases gradually to a level 60 per cent higher during polar body formation and the first cleavages. There is no evident correlation between the small fluctuations in the free valine curve and cell division cycles, although such a correlation cannot be definitely ruled out. The solid black circles above the curve represent the determinations for the pre-autolyzed

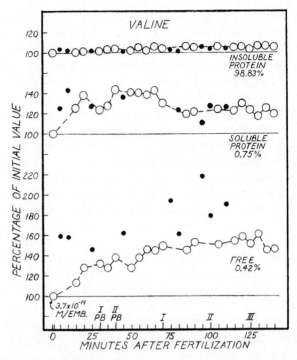

Fig. 1. Changes in valine fractions during early cleavage of *Spisula solidissima* embryos. The large open circles represent the results obtained with material free of autolytic artifact, the small solid black circles results obtained with pre-autolyzed material. The *time of completion* of stages is given by Roman numerals along the abscissa: I PB, first polar body; II PB, second polar body; I, first cleavage; II, second cleavage; III, third cleavage.

stages. In all stages, autolytic changes lead to the liberation of free valine, doubtless at the expense of protein valine. It is evident that artifacts introduced by even a relatively minor degree of autolysis would probably obliterate the true pattern of changes in free valine. More extensive protein degradation occurs during the course of autolysis in aqueous homogenates, and leads to the liberation of far greater amounts of free amino acids.

The curve for soluble protein valine represents a protein fraction, probably of low molecular weight, which remains soluble in 0.022 $M$ acetic acid through the denaturation treatments to which it is subjected. The biological significance of this fraction is unknown. From the evidence at hand it is difficult to decide whether it is affected by mild autolysis of the original experimental material.

The insoluble protein valine comprises almost 99 per cent of the total valine present (one minute after fertilization). The curve deviates by only about ±2% from being smooth and regular, indicating that the experimental error in the techniques upon which the determination of this curve is based is at most of that magnitude. The "artifact stages" deviate only slightly from the curve, showing that the amount of protein degraded autolytically is small relative to the total amount present.

Curves for determinations of the nicotinic acid group are shown in Fig.

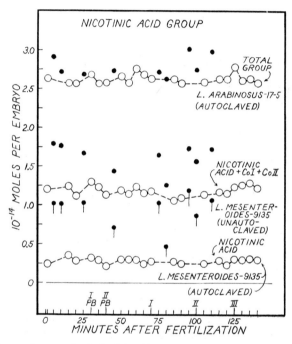

Fig. 2. Changes in unresolved nicotinic-acid group activity. The curves give the actual assay results based upon the activity relative to the common standard, nicotinic acid. The organisms used for the assays are given below the respective curves. Other details as in Fig. 1.

2. The total activity of the group is given by the upper curve, obtained by assay with *Lactobacillus arabinosus* 17-5 (ATCC 8014). Nicotinic acid, nicotinamide, Co-I (DPN), Co-II (TPN) and some other derivatives are more or less equally active (on a molar basis) in the nutrition of this microorganism. The breakdown products formed by autoclaving the coenzymes are also active in this assay. The middle curve represents assays with *Leuconostoc mesenteroides* (ATCC 9135), which responds only to nico-

tinic acid, Co-I and Co-II, other derivatives having negligible activity (8). In this assay, the material is unautoclaved to avoid breakdown of the coenzymes, for the breakdown products are inactive for *L. mesenteroides*. The lower curve was obtained by the same assay carried out on extracts in which the coenzymes were first destroyed by autoclaving (20 minutes at 15 lbs. pressure). This latter curve is thus essentially representative of changes in nicotinic acid alone. In all assays with *L. mesenteroides* (ATCC 9135), the assay medium was supplemented with supra-physiological amounts of nicotinamide to eliminate the possibility of even minor interference by this form of the vitamin.

Inspection of these curves (Fig. 2) shows that, in agreement with accepted findings concerning the biological occurrence of free nicotinic acid, only relatively small amounts are present in *Spisula* embryos. On the other hand, as shown by the artifact stages, a relatively large amount of nicotinic acid is released from other members of this group of compounds by mild autolysis. The significance of the findings is clearer if the "difference curves" for the resolvable components of this group are examined (Fig. 3). In calculating these difference curves, corrections have been applied for the destruction of the coenzymes by 10-minute treatment with boiling water (34.5%) and 20-minute autoclaving (91.9%), as determined with controls.

Referring to Fig. 3, it can be seen that the coenzymes are responsible for over 50 per cent of the total activity of the group. Autolysis leads to destruction of the coenzymes, the ultimate active degradation product being nicotinic acid. There appear to be changes in the total amounts of Co-I plus Co-II that are present during early development, with some indication of a correlation of these changes with cycles of cell division. The curve for members of the group exclusive of nicotinic acid and the coenzymes (probably mostly nicotinamide) also shows evidence of changes which might be synchronized with cleavage. The consistently lower position, relative to this curve, of points representing the artifact stages, indicates a possible autolytic degradation of these other derivatives to nicotinic acid.

In Fig. 4, the findings for the pantothenic-acid–CoA group are shown. Pantothenic acid is present in roughly two hundred times the amount of coenzyme A. The curve for pantothenic acid is based on assays (unautoclaved) with *Lactobacillus fermenti*-36 (ATCC 9338), which does not respond to CoA, pantethine, or pantetheine (3), with which the assay medium is heavily supplemented. There appears to be a considerable drop in pantothenic acid after fertilization (the decline in the first minute is not

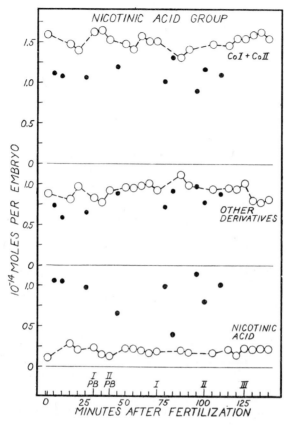

Fig. 3. Changes in Co-I plus Co-II, nicotinic acid, and other derivatives in terms of equivalents of nicotinic acid as common standard. The curves have been computed from the results of the three assays represented in Fig. 2, correcting for the known destruction of the coenzymes during processing (see text). Other details as in Fig. 1.

recorded in this series), followed by a gradual decline in the next $2\frac{1}{2}$ hours. Deviations from this pattern are small and may not be of developmental significance. It is interesting to note that mild autolysis does not result in measurable destruction of pantothenic acid, as shown by the artifact stages.

The findings represented in the lower curve for the sum of CoA and pantethine (or pantetheine) are of unusual interest. The assay was carried out on unautoclaved extracts with *Lactobacillus acidophilus* (ATCC 4355), which does not respond to free pantothenic acid (3) under the assay conditions employed. The assay medium was heavily supplemented with pantothenic acid to eliminate the possibility of synergistic interference.

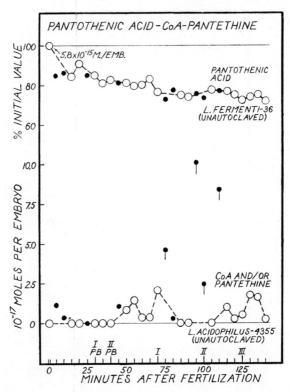

Fig. 4. Changes in pantothenic acid and unresolved CoA and/or pantethine. Pantothenic acid was used as standard for the upper curve, pantethine for the lower. Other details as in Fig. 1.

No detectible amount of CoA or pantethine is found at certain stages of development. At other stages small amounts of CoA and/or pantethine are found, and the changes may well correlate with cell division. The origin of this activity could be either synthesis de novo or liberation from a bound or otherwise inactive condition. The findings for the artifact stages are variable. In certain cases a large increase in CoA and/or pantethine occurs. In others, the change from the control curve is small or negligible. It is difficult to assess the significance of these differences without further experimental data. One possibility is that at certain stages, but not at others, autolytic release of CoA or pantethine from a bound condition takes place. In the bound form the activity would not normally be detected because it would pass into the insoluble protein residue. The differences produced at different stages by this mechanism could depend on numerous factors. Another possibility is that the autolytic artifact is small, the findings for the autolyzed stages giving a true picture of the changes in CoA and/or pantethine.

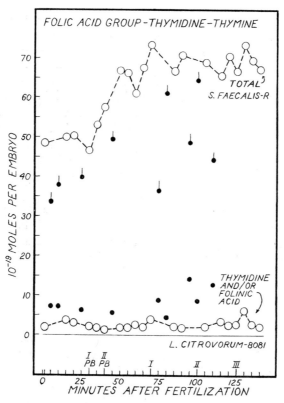

Fig. 5. Changes in total folic-acid-replacing activity for *S. faecalis*-R and folinic-acid-replacing activity for *L. citrovorum*-8081 (see text). The ordinate gives the activity in equivalents of folinic acid as the common standard (see text). Other details as in Fig. 1.

Fig. 5 shows the findings for the group of metabolically interrelated substances including folic acid, folinic acid (citrovorum factor), thymidine, and thymine. The lower curve was obtained by assay with *Leuconostoc citrovorum* (ATCC 8081), which responds only to thymidine and folinic acid (6). Apparently there is very little free thymidine or folinic acid present in *Spisula* embryos. On the other hand, the artifact stages show that autolysis leads to the appearance of much greater activity. The most reasonable interpretation, which would accord with the findings of Lansford and Shive (9), is that this increased activity is due to thymidine, liberated from DNA by autolysis. The upper curve represents an assay with *Streptococcus faecalis*-R (ATCC 8043). This microorganism responds to folic acid and folinic acid (and to their higher glutamic acid derivatives) as well as to both thymine and thymidine. Since it is already established that folinic acid and thymidine are normally present in only minute amounts, the activity observed must be due primarily to thymine and/or folic acid. It seems clear that significant changes take place in one or the other (or both) of these substances, and there is definitely indication of a correlation with cell division. Autolysis leads to a marked decrease in total activity, notwithstanding the (probable) increase in thymidine. This decrease most probably represents autolytic degradation of folic acid (or of its higher glutamic acid derivatives), and suggests the likelihood that this vitamin is responsible for a major fraction of the activity represented by the curve.

The common reference standard for the quantitative comparison of the two assays for this group, as plotted in Fig. 5, is synthetic folinic acid. Referred to thymidine as a common standard, the curves are only slightly modified but the ordinate must be increased by a factor of roughly 5000.

Discussion

*Extraction Method*

Judging from the results presented and those of other comparative studies of extraction methods (unpub.), the new "non-aqueous" extraction procedure appears to be free of autolytic artifact and promises to be a valuable tool for the investigation of the chemical basis of development. However, assays of free and microbiologically active forms of growth factors and other substances in extracts prepared by a single extraction technique give only a partial representation of the patterns of chemical changes. A molecular species of given function(s) may be partitioned among several states within the cell, for example, in the free form, in a bound form, or in the form of a free but metabolically inactive derivative. Consequently, treatments altering the normal patterns of partitioning may be required to render bound or inactive forms accessible to microassay. Controlled autolysis of homogenates may prove to be useful for this purpose, inasmuch as it may be possible to bring about differential destruction, liberation, or modification of certain forms of metabolites by this means.

## Microbiological Assay

Because it is necessary to follow small quantitative changes of metabolites occurring over short intervals of development, only analytical methods of high relative accuracy, which permit simultaneous multisample determinations, are suitable. Microbiological assay methods are ideal for this purpose.

Assays carried out in liquid media have the greatest sensitivity. In some cases, for example, the assay of pantothenic acid with *L. fermenti*-36, assays in liquid media appear to be highly specific. In other cases several substances can replace a single nutrient deficiency of the assay medium. For this reason it was not possible to make definite assignments for the activity represented by several of the assay curves. However, this is not necessarily a limitation upon assays utilizing microorganisms having a multiple-response spectrum; in one way it is advantageous. Thus, if chromatographic techniques are utilized for the preliminary resolution of different members of, say, a vitamin group, each member can then be assayed individually (but simultaneously) by using an organism with a multiple-response spectrum. This is accomplished in a single bioautographical assay of the activity of a paper chromatogram (the paper is layered over an agar-solidified assay medium seeded with the appropriate organism). This is an elegant method for detecting simultaneous changes in various members of a metabolically interrelated group of substances. For example, cycles of interconversions of members of the thiamine group (probably phosphate derivatives) have been detected during sea urchin development (6). Assays in liquid media and utilizing microorganisms having a multiple-response spectrum should be regarded primarily as convenient tools for a preliminary survey, to be supplemented with bioautographic assay techniques giving more specific information, wherever the findings appear to warrant a more detailed analysis.

## General Considerations

The results presented above are part of an exploratory program to develop satisfactory techniques for the investigation of the time course of quantitative changes in substances involved in specific metabolic and developmental events. The techniques now at hand are eminently suitable for such investigations and their application to specific biological problems is underway. The chemical background of cell division is both an ideal and a logical starting point. The exploratory results presented for the early post-fertilization development of *Spisula* embryos suggest that quantitative changes in certain substances may be synchronized with and characteristic of different phases of cell division. A more detailed study of cleaving marine invertebrate eggs utilizing the same experimental techniques and incorporating improvements instituted since the time of the *Spisula* work would probably lay bare certain aspects of the underlying chemical changes. One of the indicated modifications to facilitate this achievement is the collecting of stages at even more frequent intervals of the mitotic cycle, spanning at most two cell-division cycles.

One rather definite indication of the *Spisula* findings, which is of particular interest, is that no marked increase or decrease in free valine (or tyrosine) occurs in the course of division cycles. If the formation and breakdown of the mitotic apparatus involved synthesis and degradation of proteins from and to free amino acids, cyclic variations in the free amino acid pool might be expected to occur during early cleavage. Although other interpretations are possible, the failure to detect such changes is in agreement with Mazia's (10) polymerization theory of the origin of the mitotic apparatus, according to which realignments in protein disulfide links are at the basis of spindle formation and breakdown.

### Summary and Conclusions

A new technique for the extraction of biological materials, and one which appears to be free of autolytic artifact, has been developed. This extraction method is designed to preserve unaltered a wide spectrum of low molecular weight components of living systems. Early post-fertilization stages of the surf-clam *Spisula solidissima*, including several stages known to have been exposed to conditions favoring autolysis, were extracted by the new method and were assayed microbiologically for their content of certain metabolites. In certain cases, for example, in regard to CoA (or pantethine) and folic acid (or thymine), it appears that there may be correlations between cycles of cell division and changes in the quantity or state of low molecular weight cellular components. In other cases, for example, those of free nicotinic acid and free valine, there is little or no evidence of correlated changes. More detailed studies along these lines will probably lead to a greater understanding of the mechanisms of cell division and of biological control mechanisms in general.

Autolysis appears to be virtually unavoidable in aqueous cellular homogenates unless enzymes are first inactivated. The results obtained for autolyzed stages, when compared with those for stages protected from autolysis, show that even relatively mild autolysis can lead to serious artifacts. These artifacts are of such magnitude that if present they could obliterate the true patterns of chemical changes during development. It is clear from these findings that avoidance of even mild autolysis is imperative in the extraction and processing of biological materials for many quantitative chemical studies.

ACKNOWLEDGMENTS

The author is greatly indebted to Professor Paul Weiss for his continued interest in this work and for the generous provision of the facilities of his laboratory, in which these studies were carried out. He is also indebted to Dr. O. D. Bird, of Parke, Davis & Company for the gratuitous supply of pantethine.

## REFERENCES

1. Allen, R. D., *Biol. Bull.*, **105**, 213-239 (1953).
2. Banhidi, Z. G., and Kavanau, J. L., *Exptl. Cell Research*, **10**, 405-414 (1956).
3. Craig, J. A., and Snell, E. E., *J. Bacteriol.*, **61**, 283-291 (1951).
4. Kavanau, J. L., *J. Exptl. Zool.*, **122**, 285-337 (1953).
5. ———, *Exptl. Cell Research*, **7**, 530-557 (1954).
6. ———, *Proc. Intern. Congr. Developmental Biol.*, 1956 (in press).
7. ———, and Banhidi, Z. G., *Exptl. Cell Research*, **10**, 415-423 (1956).
8. Koser, S. A., and Kasai, G. J., *J. Infectious Diseases*, **83**, 271-278 (1948).
9. Lansford, E. M., and Shive, W., in *Antimetabolites and Cancer*, p. 33-45, A.A.A.S., Washington, D. C. (1955).
10. Mazia, D., *Symposia Soc. Exptl. Biol.*, **9**, 335-357 (1955).

# ENZYME INDUCTION IN DISSOCIATED EMBRYONIC CELLS[1]

RICHARD N. STEARNS AND ADELE B. KOSTELLOW[2]
*Department of Physiology, Albert Einstein College of Medicine,
New York*

## INTRODUCTION

Histogenesis may be described as a process of development and stabilization, by the differentiating cells, of a complex of specialized synthetic properties, upon which the subsequent structural and functional characteristics of the adult cell depend. Although the end-product of this process, the differentiated cell, cannot yet be completely described in biochemical terms, the enzymatically catalysed activities carried out by certain specialized cells are considered to be definitive characteristics of many types of tissue. We therefore felt that it would be interesting to investigate the conditions under which a specific instance of such a definitive activity could be initiated, and maintained, by the embryonic cell.

Much of our present knowledge about the mechanism of enzyme synthesis comes from the field of microbiology, especially from work with adaptive enzymes. The latter phenomenon has been of particular interest to embryologists, in that the initiation of synthesis of a specific protein is shown to occur only in those members of a cell population which are exposed to a relatively simple and specific environmental change.

Analysis of microbial systems can be carried out on populations of cells, each cell being equally exposed to the controlled environment of a carefully defined medium. This situation obviously cannot be simulated with the intact vertebrate embryo, in which groups of cells are subjected to the inductive effects of their immediate neighbors. Therefore it seemed most desirable to conduct our experiments with cultures of dissociated embryonic cells, maintained under conditions which would permit cytological differentiation.

[1] This work was supported by grants from the National Science Foundation, the U. S. Public Health Service, and the Jon Miles Davidson Fund, Inc.
[2] Research fellow, Damon Runyon Memorial Fund.

## Dissociated Cell System

Dr. Stearns had developed a culture medium, having the same ionic composition as frog serum, plus 0.5% human serum albumin. This medium supports the differentiation of completely dissociated cells from *R. pipiens* embryos (9). (Table 1.)

TABLE 1

Stearns' Solution Grams/Liter

| NaCl | 4.36 | NaHCO$_3$ | 2.13 |
|---|---|---|---|
| KCl | 0.18 | Na$_2$SO$_4$ | 0.27 |
| Na$_2$HPO$_4$ | 0.089 | CaCl$_2$ | 0.22 |
| KH$_2$PO$_4$ | 0.019 | MgCl$_2$·6H$_2$O | 0.263 |
| Human serum albumin (Fraction V) dialyzed 12 hrs. | | | 5.0[1] |

[1] The albumin was donated by the Red Cross Blood Program.

Methods have been developed by means of which several thousand *R. pipiens* eggs can be dissociated and cultured as a population of cells. The jelly and vitelline membrane are removed using a modification of the technique described by Spiegel (8). The eggs are dissociated with $0.002\,M$ Versene, in culture medium from which the calcium and magnesium salts, and the albumin, have been omitted. The cells settle onto the agar-coated bottom of the dissociation vessel, and changes of media are accomplished by siphoning off the supernatant fluid. The dissociated cells are rinsed thoroughly with culture medium from which only the albumin is omitted, and are finally rinsed with the complete culture medium. For the substrate induction experiments, the inducer is dissolved in the complete medium in which the cells are incubated. All of the procedures described, after the initial dejellying, are carried out under sterile conditions, and at a temperature of 16°C.

It is possible to separate the dissociated cells with respect to the germ layer of their origin, by utilizing the variation in cell density from animal to vegetal pole which results from the gradation in yolk distribution (described by Barth, 1). A suspension of dissociated cells is floated over a series of layers of gum arabic having graded densities, and is centrifuged gently. The surface cells, presumptive ectoderm, presumptive mesoderm, and presumptive endoderm cells each settle at a different level of the series, and can be withdrawn and cultured separately. Populations of cells with different presumptive fates can thus be compared in their capacity to respond to a given environmental stimulus.

## Substrate Induction

One of the first enzymes with which we obtained positive results was tryptophan peroxidase (10). Knox and Mehler (7) have shown that in mammals this enzyme is found only in the liver, and that its specific activity can be increased tenfold by injecting tryptophan into the animals a few hours before assaying. Using their assay methods for formed kynurenine (6), we found detectable levels of tryptophan peroxidase activity only in liver homogenates of the adult *R. pipiens,* and were able to obtain a four-fold increase in activity after the injection of tryptophan. No tryptophan peroxidase activity could be detected in the embryo before hatching.

*R. pipiens* embryos were dissociated into cell cultures at a stage just prior to gastrulation. L-Tryptophan (0.01 $M$) was added to the culture medium. After four or five hours of incubation at 16°C, tryptophan peroxidase activity could be demonstrated in homogenates of the cells that had been exposed to the tryptophan (Fig. 1). Activity continued to increase in the incubated cells, reaching a maximum level between eight and ten hours. No activity was detected in the control cultures. Intact embryos kept at this temperature had just begun to gastrulate.

These experiments were repeated, using cell cultures that had been separated before the onset of gastrulation into populations of presumptive endoderm cells, and populations of ectoderm plus mesoderm cells. There was no significant difference in the amount of tryptophan peroxidase activity per milligram of protein between homogenates of the two types of culture. Although tissues derived from the presumptive ectoderm and mesoderm layers would not, in the ordinary course of development, ever make appreciable amounts of the enzyme, their capacity to respond to induction seemed equal to that of the germ layer from which liver tissue is normally derived.

Induction was then attempted with cells obtained from embryos which had completed gastrulation. At this stage, the density gradient layers employed separated the cells into three main populations: ectoderm and most of the mesoderm cells; mesoderm and lighter endoderm cells (from the prospective gut-forming region); and heavy endoderm cells (those which still remain laden with yolk in the belly region of the hatched tadpole). It was found that only in cultures containing the presumptive gut cells could tryptophan peroxidase activity be induced which was comparable to that shown by early gastrula cells. The ectoderm plus mesoderm cells showed very slight activity; the heavy endoderm cells showed none.

The data are not yet complete on post-gastrula stages. So far, however,

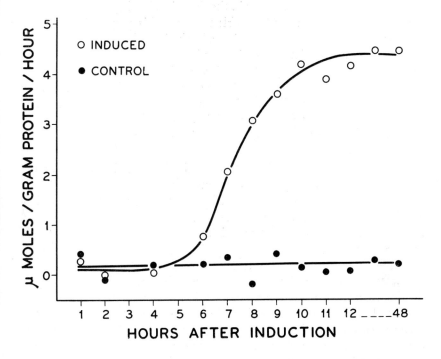

Figure 1. Tryptophane peroxidase activity in induced and control cultures of early gastrula cells. The enzyme was assayed by measuring the rate of conversion of tryptophan to kynurenine.

they are consistent with a pattern of progressive restriction in the ability of non-endoderm cells to respond to this particular inductive stimulus.

It had been observed that cell populations which had been induced before gastrulation continued to show maximum enzyme activity after 48 hours, in the presence of the inducer. Experiments were conducted to determine how long this activity could be maintained after removal of the inducer. It was found that an hour's exposure of early gastrula cells to tryptophan was sufficient for induction, full activity appearing eight to ten hours after the exogenous supply of tryptophan had been washed off. Again, the activity was maintained for a period of at least 48 hours.

Positive results have also been obtained by using niphegal and lactose to induce β-galactosidase activity in the dissociated cell system. Up to this time, however, we have not been able to induce this enzyme in early gastrula cells. Induction will occur in cells obtained from embryos which have completed gastrulation, and in populations of younger cells after they have been cultured for a period of time equivalent to that required for gastrulation of the intact embryo. Since spontaneous galactosidase activity could not be detected until some time after hatching, induction in the cell cultures does represent premature activity. The pattern of distribution of this activity among the various presumptive cell layers has not yet been determined.

Preliminary experiments indicate that histidase activity, found in the liver of the adult *R. pipiens,* can be induced by incubation of pre-gastrula cells with histidine. It should perhaps be added that incubation with histidine does not induce tryptophan peroxidase activity, nor does incubation with tryptophan induce histidase activity.

## Discussion

These experiments have been concerned with the phenomenon of induction in a more restricted sense than is usually inferred by the term, "embryonic induction." Alterations during development in the ability of a given type of embryonic cell to respond to a specific inducer with a single, chemically defined change in protein synthesis, may help to solve a fundamental problem in the field, which may be outlined as follows.

A. Mechanisms for the synthesis of the structural and enzymatic substances which characterize all of the various mature cell types may already be present in each undifferentiated cell. The process of determination would then activate a given complex of these mechanisms, while presumably inactivating all of the others. Determination of cell fate would actually consist of a progressive restriction of synthetic abilities.

Our success in inducing formation of the enzyme tryptophan peroxidase indicates that at least one such synthetic mechanism is present in the cells of all germ layers of the early embryo (unless it is assumed that the substrate has been able to activate directly the gene or genes concerned). After gastrulation is complete, we have shown that all of the germ layers can no longer be induced.

B. Pregastrular "totipotent" cells may, on the other hand, possess very few of the special synthetic mechanisms which characterize the adult cell. Their potential ability to develop any given complex of these mechanisms may reside only in their possession of the appropriate genetic factors. The process of determination would then consist of the activation or segregation of the necessary complex of genes, and, again, presumably the inactivation or elimination of genes which control synthetic processes specific to cells of other developmental fates. Briggs and King (3) have presented evidence that a change in nuclear potentialities actually does take place after gastrulation in the frog embryo. Our results concerning the induction of lactase indicate that this synthetic mechanism may not be present in undetermined embryonic cells. Induction by substrate was obtained, however, at a stage considerably earlier than that at which this enzyme would normally be detectable. Possibly in this case the enzyme-synthesizing site developed or became available for activation only after gastrulation. Further experiments will reveal whether or not this active site is present in cells of all germ layers.

Experiments by other investigators (2, 4, 5) indicate that chick embryonic cells are also responsive to substrate induction. We know of no other investigations, however, in which undetermined cells have been induced to make enzymes characteristic of a particular differentiated adult tissue.

Although these experiments were primarily designed to reveal the state of the adaptive mechanism in the undetermined embryonic cell, a positive response must alter the cell to the extent that at least a part of its protein-synthesizing activity is diverted towards the formation of an enzyme which it would not normally be required to produce. In the case of tryptophan peroxidase induction, there is preliminary evidence that the sequential induction of formylase occurs in the system.

It will also be of interest to see whether or not two enzymes, each specific to a particular type of differentiated tissue, can be induced in the same embryonic cell; and, finally, whether the evocation or activation of these mechanisms can limit or alter the cell's developmental fate.

## REFERENCES

1. Barth, L. G., *Physiol. Zool.*, **15**, 30 (1942).
2. Boell, E. J., *Anat. Record*, **96** (Suppl.), 91 (1946).
3. Briggs, R., and King, T. J., in *Biological Specificity and Growth* (E. Butler, ed.), p. 207. Princeton Univ. Press, Princeton (1955).
4. Burkhalter, A., Jones, M., and Featherstone, R. M., *Proc. Soc. Exptl. Biol. Med.*, **96**, 747 (1957).
5. Gorden, M. W., and Roder, M., *J. Biol. Chem.*, **200**, 859 (1953).
6. Knox, W. E., and Mehler, A. H., *J. Biol. Chem.*, **187**, 419 (1950).
7. ———, and ———, *Science*, **113**, 237 (1951).
8. Spiegel, M., *Anat. Record*, **111**, 544 (1951).
9. Stearns, R. N., *Anat. Record*, **122**, 444 (1955).

## DISCUSSION

Dr. Markert: Does the cholinesterase in muscle disappear completely or just go down to a very low level?

Dr. Shen: We haven't followed it very far because these animals die because of lack of feeding. There is no way to feed these nerveless animals and I don't know how far we could carry them on.

Dr. Markert: How far were you able to carry them on?

Dr. Shen: At normal temperatures, and this animal is *Amblystoma,* these animals can develop from Harrison stage 33 to stage 46 in about two weeks time. I wish we could carry this on longer but they simply won't survive.

Dr. Spratt: It might be interesting to graft some of these denervated muscles to normal hosts and thus check to see whether or not there is a reappearance of the cholinesterase and how this might be correlated with secondary innervation of the transplant.

Dr. Shen: That is what we had hoped to do; perhaps we can even induce the nerves to enter the transplant at a specific point.

Dr. Coulombre: Dr. Shen, you have evidence that synthesis of cholinesterase takes place in the post-junctional side of the myoneural junction, and that it is essentially nerve-independent. In the case of the neuronal synapse is cholinesterase localized on the dendritic side, and is it independent of the formation of a fully constituted synapse?

Dr. Shen: Thank you for your question, for I didn't go into that. I think that the case in the chick retina is very convincing. I don't know whether I mentioned that or not, but in the earlier stages we only have ganglion cells staining and there are unlikely any nerve processes coming into them. So, all the presynaptic members have not formed yet, but nevertheless there is still enzyme on the neuron which would indicate, just as in the case of muscle, that the enzyme is synthesized at the postsynaptic membrane independent of the presence of incoming presynaptic terminals.

Dr. Weiss: I would like to know just how sure we can be that these are the sites where the enzyme is synthesized rather than merely places where it is accumulated?

Dr. Shen: That is a very good question. I think we need some further work here utilizing tissue culture techniques because we have some interesting observations that would indicate to us that perhaps the enzyme is not actually synthesized on the muscle cell. There are a lot of fibers coming in and finally in the tissue culture we can see at certain stages that the enzyme is indeed right on these fibers. We don't know just what their origins are. We have a suspicion that the enzyme on the motor end plate represents a specialized region of the sarcolemma. There are some theories as to a possible dual origin of the sarcolemma; it might be partly a modified membrane of the muscle cells with, in all probability, another layer of fibrous structure on top of that, which we presume is contributed by these before-mentioned fibrous elements, and a portion of this then becomes specialized with the enzyme right on it.

DR. LIPMANN: You mentioned several times the liberation of acetylcholine, meaning, I suppose, the not too well defined release of bound acetylcholine. One might, however, somewhat more hopefully inquire about the related synthesis of acetylcholine, and I wonder if you might have something to say about the localization of the synthetic reaction and its relation to the location of hydrolysis.

DR. SHEN: Yes, it would be one of our fondest hopes that someone would be able to devise a fine cytochemical technique to localize the choline acetylase, that has been found to be specific for the synthesis of acetylcholine from choline and acetate. This occurs presumably, and again there is no direct evidence on the presynaptic nerve. For that matter we don't know exactly where acetylcholine comes from either, whether from the strictly presynaptic cell, or again from somewhere else. There have been a lot of vesicles found in the presynaptic membrane in the electron microscope, and it is presumed again that this is where the acetylcholine is found. Later on there is some confusing evidence which shows that these vesicles are also present in the postsynaptic neuron as well. Therefore, we really don't know anything about the localization of the enzymes responsible for the synthesis of acetylcholine. The precise localization of these enzymes would be a very great contribution to knowledge of biochemical pathways.

DR. COWDEN: I wanted to add a comment here. I have recently been studying the RNA distribution in amphibian muscle cells and apparently there is a very distinct perinuclear region of RNA and also some at the tip of the muscle cells. This may indicate that here at the tip is where some sort of synthesis is going on, perhaps the synthesis of these enzymes.

DR. SHEN: Yes, that would be a very interesting thing for us to do but, before that, we would like to know something about the cytology of the muscle at the tip. In fact, in my search of the literature I have found no description of that except in the latest edition of the textbook of histology of Maximow and Bloom showing a nice picture of Keith Porter's and the picture is labeled as *Amblystoma* myoblast; from the picture it's very obvious that it's a piece of trunk musculature and this is exactly the type of picture we want to see. The interesting part to us is that the myofibrils come down and suddenly stop a short distance before the end of the sarcolemma between the muscle fibers and the tendon. There is no direct connection between the muscle fibers and the tendon. There seems to be a great deal of infolding at the very tip of the sarcolemma, and we believe that that might be where the enzyme is located.

DR. RUDNICK: Some preliminary results have been obtained on the question of distribution of glutamotransferase (Waelsch, H., Adv. in *Enzymology,* 13, 237-319, 1952) in the retina of the chick eye at the time when this tissue is undergoing a spectacular increase in specific activity in that tissue (Rudnick, D.

and Waelsch, H., *J. Exptl. Zool.*, **129**, 309-326, 1955). It was believed, in view of the very high specific activity manifested by the neural retina in chicks at hatching, that it might be possible to locate the activity within tissue layers of the retina at this time, using microdissection methods. This presupposes that the activity is associated with definite cell types, as might possibly be indicated by the fact that during this period the retina is maturing physiologically and that specific structural changes have just been completed in the visual cells.

Whole eyeballs of chicks ranging from 18 days' incubation to 3 days after hatching were frozen in dry ice; sections of these cut at $0°C$ in a cryostat were desiccated at the same temperature and stored in this state in the deep-freeze for 1-7 days. The neural retina was dissected from the sections at room temperature under a binocular microscope. Samples to be assayed were prepared as follows: a) the neural retina including all layers; b) the visual layer separated from the overlying layers by a cut estimated to pass a little inside the external limiting membrane; c) the remaining retinal layers after the removal of b), including bipolar and ganglionic cells.

The frozen-dried tissues were homogenized in small amounts (0.08-0.1 ml.) of phosphate buffer (0.02 $M$, $pH$ 7.1) in suitable micro-homogenizers, and assayed for glutamotransferase as described in Rudnick, D., Mela, P. and Waelsch, H., *J. Exptl. Zool.*, **126**, 297-322, 1954, except that the system was cut to 10% of the volume used in that investigation. Protein was determined by the method of Lowry (Lowry, O. H., Rosebrough, N. J., Farr, A. L. and Randall, R. J., *J. Biol. Chem.* **193**, 265-275, 1951). The microsystem gives results comparable to those obtained with the larger system if total protein concentration is kept within an optimum range. Specific activity of the entire neuroretina (sample a) rises from 0.7 on the 18th day of incubation to 6.2 two days after hatching.

In most experiments, though not in all, the specific activity of the visual layer (sample b) was found to be lower than that of the entire neuroretina; a reduction of 40-50% was observed in most cases. The bipolar-ganglionic layer (sample c) tended to have values much closer to those of the complete tissue (10% maximum reduction). Some of the values obtained may require revision because of too wide variation in protein concentration; nevertheless it is perfectly clear that the activity is not exclusively or even preponderantly associated with the visual cell layer.

Experiments previously reported (Rudnick, D., *Anat. Rec.*, **124**, 356 abstract, 1956) have shown, by a quite different method, that activity is not concentrated appreciably in the ganglion cell layer. By transplanting chick eyes to heterotopic sites on the 2nd day of incubation, retinas are obtained, at hatching age, in which the ganglionic layer is reduced to a mere vestige, the other cell layers appearing normal histologically. Such retinas, in some cases though not uniformly, show a somewhat reduced (as much as 50%) specific activity as compared with their individual controls. Even these reduced values are, however, not outside the range of normal variability.

It thus remains a possibility that the ganglionic cell layer has a slightly higher activity than do other retinal layers, but it is clear that all 3 of the major cellular layers of the retina participate in the sudden rise in transferase activity just preceding hatching; and that this activity is very far from being associated exclusively or even preponderantly with any one layer.

# ON THE DIFFERENTIATION OF A POPULATION OF *ESCHERICHIA COLI* WITH RESPECT TO β-GALACTOSIDASE FORMATION[1]

MELVIN COHN

*Department of Microbiology,*
*Washington University School of Medicine,*
*St. Louis, Missouri*

BACTERIAL POPULATIONS show changes in properties which superficially, at least, appear similar to the phenomenon of differentiation in animal cells. The purpose of this paper is to describe one case of such an induced differentiation in bacteria. The system referred to is the β-galactosidase of *Escherichia coli,* which in recent years has been so extensively studied that the parameters of its induction are well understood.

One of the ways to study the induction of β-galactosidase is to add the inducer to an exponentially growing culture and to follow the formation of enzyme as a function of increase in bacterial mass (10). This relationship between enzyme formed and total bacterial mass synthesized has been termed the differential rate of synthesis (10). If such an experiment is carried out under proper conditions (see Fig. 1, open circles) a beautifully simple relationship emerges. The formation of enzyme begins virtually upon addition of the inducer, and the increase in enzyme formed is a linear function of bacterial mass. The addition of inducer has produced a new population of cells which have differentiated from the old ones with respect to a character that is related to the "β-galactosidase system." However, stated this way, the analogy to differentiation in higher organisms is not very probing. The possession of the "β-galactosidase system" actually confers upon the cell some unusual new properties, in particular in its response to low concentrations of inducer and to the inhibition of induction by glucose.

If a non-induced culture is placed either (1) into glucose plus inducer (Fig. 1A, black circles) or (2) into a properly chosen low inducer concentration (Fig. 1B, black circles) it will not synthesize β-galactosidase. If a

[1] This work has been supported by grants from The American Cancer Society, National Science Foundation, and National Institutes of Health.

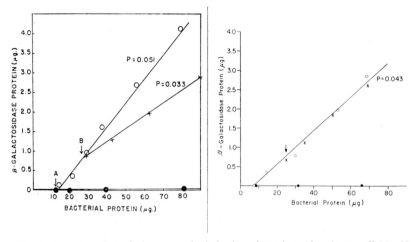

Fig. 1. *Left,* the effect of glucose on the induction of β-galactosidase in *E. coli* ML 30 (lac +). At a concentration of cells indicated at the arrow marked "A," a non-induced culture is placed into methyl-β-D-thiogalactoside ($10^{-4}$ M), curve with open circles, or into inducer plus glucose ($10^{-3}$ M), curve with black circles. After more than doubling (at arrow marked "B"), glucose is added to the preinduced culture, which continues to produce enzyme (curve with crosses).

*Right,* the effect of low concentrations on the induction of β-galactosidase in *E. coli* ML 30 (lac+). A non-induced culture is placed into a low concentration of inducer ($8 \times 10^{-6}$ M), curve with black circles, or into a high concentration of inducer ($10^{-4}$ M), curve with open circles. The preinduced culture is centrifuged and resuspended in inducer ($6 \times 10^{-6}$ M) at a slightly lower concentration than the non-induced culture, to allow for trapped inducer. The continued synthesis of β-galactosidase follows the curve marked by crosses.

preinduced culture is placed under the same conditions into glucose plus inducer (Fig. 1, Left, crosses) or into low inducer concentration (Fig. 1, Right, crosses), enzyme will continue to be made. Such an experiment is illustrated in Fig. 1, where a culture is preinduced for one to two divisions and is then placed either under conditions of glucose inhibition or of low inducer concentration. This result shows that one can establish two distinct populations of identical genotype in an identical environment, the one producing enzyme and the other not. This situation has been termed maintenance by Novick and Weiner (11) and preinduction by Monod (7).

If maintenance in bacteria is to enjoy the dignity of being compared to differentiation in higher organisms, then it should persist over many generations; and in fact it does (Table 1). Either by a proper choice of conditions of glucose inhibition or of low inducer concentration, two genotypically identical cultures which differ phenotypically by the possession of the "β-galactosidase system," can be maintained for over 150 generations

TABLE 1

The Duration of Maintenance in E. coli (lac+)

|  | Number of generations | $P\left(\dfrac{\text{enzyme protein}}{\text{bacterial protein}}\right) \times 10^2$ | |
| --- | --- | --- | --- |
|  |  | Glucose $10^{-3}$ M + TMG $10^{-4}$ M | TMG[1] $5 \times 10^{-6}$ M |
| Induced | — | 3.3 | 2.2 |
| Non-induced |  | <0.01 | <0.01 |
| Induced | 150 | 3.1 | 2.2 |
| Non-induced |  | <0.01 | <0.01 |

[1] Data calculated from Novick and Weiner (11).

in the same medium. The culture producing enzyme has somehow recorded that 150 generations previously it had been preinduced for a period of a generation.

These two steady-state conditions, synthesis of enzyme vs. non-synthesis of enzyme, are delicately poised and one can shock the system to go from one state to the other. The most interesting example, in the case of the glucose inhibition, is anaerobic shock (Fig. 2). Consider a glucose-inhib-

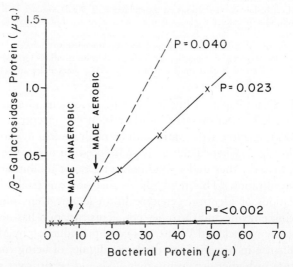

Fig. 2. The effect of anaerobic shock on induction of $\beta$-galactosidase in the presence of glucose. A culture of E. coli ML (lac+) which is not producing $\beta$-galactosidase aerobically in the presence of glucose (crosses) is made anaerobic, and when growth resumes enzyme begins to be formed. After one division anaerobically it is restored to aerobic conditions, whereupon it continues to make enzyme. The culture left aerobic throughout the experiment is illustrated by the curve with black circles.

ited culture placed abruptly under anaerobic conditions. As soon as growth resumes enzyme synthesis begins. If after an hour of induction in anaerobiosis, the culture is restored to aerobiosis, then enzyme synthesis continues in the presence of glucose. Anaerobic shock is equivalent to preinduction. I point this situation out, since as a model for cellular differentiation it illustrates how a "non-specific" environmental change such as anaerobiosis can affect a "specific" character which seems unrelated to oxygen.

Thus a new character, that of synthesizing an enzyme, tends to become stabilized after it appears in the cell. In order to understand this phenomenon we must develop in more detail the kinetics of induced $\beta$-galactosidase synthesis. A linear differential rate of synthesis which begins upon addition of inducer is found under two conditions: (1) when saturating concentrations of inducer are given to a growing culture of the *E. coli* mutant which can utilize lactose as a carbon and energy source (lactose-positive) (Fig. 3, Left); or (2) when any concentration of inducer sufficient to provoke measurable enzyme formation is put into the growth medium of the *E. coli* strain which cannot utilize lactose (lactose-negative cryptic) (Fig. 3, Right). In the case of the lactose-positive strain, if the concentration of inducer is lowered from that which gives a linear response, then an acceleration phase appears in the differential rate of synthesis which gradually reaches constancy (Fig. 3, Left). This behavior of the *E. coli* lac+ strain, discovered by Buttin, Cohen-Bazire, and Monod (1) is in sharp contrast with that of the *E. coli* lac— cryptic strain, which responds immediately at all effective concentrations of inducer to give a linear differential rate of synthesis (5).

What is the difference between the *E. coli* cryptic type investigated by Herzenberg and Monod and the *E. coli* "common" or "normal" type studied by Buttin, Cohen-Bazire, and Monod? These two strains were derived via the classical mutation of *E. coli* mutabile from a type which cannot utilize lactose to one which can. It might be recalled that the utilization of lactose was once thought to require only the presence of $\beta$-galactosidase itself. The lac-negative cryptic strain was believed to lack this enzyme. However, when Lederberg et al. (6) showed that $\beta$-galactosidase formation could be provoked in the cryptic strain if alkyl $\beta$-D-galactosides were used, then it became probable that this mutation specifically affected the permeability of the cell to galactosides. In fact, just such a specific induced permeability factor was discovered by Rickenberg, Cohen, Buttin, and Monod (14), and this observation clarified a host of unpublished experiments which had filled notebooks over the years. This permeability factor was termed "permease" and given the symbol "Y." As we know today

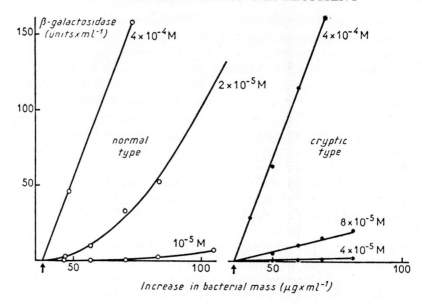

Fig. 3. *Left*, the induction of β-galactosidase by methyl-β-D-thiogalactoside in the lactose-positive strain. Data of Buttin, Cohen-Bazire, and Monod (1). *Right*, the induction of β-galactosidase in the lactose-negative cryptic strain. Data of Herzenberg and Monod (5).

(2), there is a large number of permeases specific for various amino acids, carbohydrates, organic acids, etc. It is the galactoside permease which is absent in the lac— cryptic strain and is inducible in the lac+ strain.

The data on the induction of the lac+ strain best fit the following hypothesis (Fig. 4). The external inducer, Ie, enters the cell via the Y system, where as internal inducer, Ii, it induces both galactoside permease (Y) and β-galactosidase (E) formation (14). The inducer, Ii, leaves the cell via another route to become once again Ie. A steady state is set up, and Ii can accumulate in the cell to levels over 100 times that in the medium. Although the cell is virtually impermeable to lactose in the absence of galactoside permease, alkyl galactosides such as methyl β-D-thiogalactoside (Fig. 5) can enter by diffusion across the cell wall, but cannot be concentrated by this reaction. Therefore, in the lac— cryptic strain, induction of enzyme (E) formation can take place if a high enough concentration of an alkyl galactoside inducer, Ie, is put into the medium. It should be noted though that at low concentrations of Ie, the building of an effective concentration of Ii in the cell would depend upon the presence of Y, the very system being induced. E mirrors this induction, since both Y and E are induced by Ii.

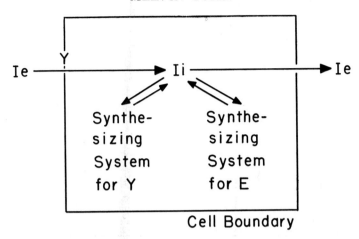

Fig. 4. Schema to illustrate the pathway of induction of $\beta$-galactosidase in *E. coli*. Y = galactoside permease; E = $\beta$-galactosidase.

How does this permease model account for the preinduction or maintenance effect of Monod (7) and of Novick and Weiner (11)? Clearly, a cell which has a high level of Y can capture an external inducer Ie more efficiently than a cell which does not have Y. The internal concentration of Ii depends, then, on both the external concentration of Ie and the level of Y. If the Ie concentration is made so low that the probability that a non-induced cell will become induced is negligible, and yet high enough so that the preinduced cell can capture Ie and concentrate it to an effective internal inducer, Ii, then, as we have just shown, one can maintain steady-state conditions of enzyme synthesis or non-synthesis in a given medium.

Glucose enters the cell by a distinctly different constitutive permease, Y' (2), to become the inhibitor, symbolized as glucose$_i$ (Fig. 6). It is the ratio

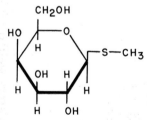

Fig. 5. The structure of a commonly employed inducer methyl $\beta$-D-thiogalactoside. Other alkyl thiogalactosides have been studied; isopropyl $\beta$-D-thiogalactoside is somewhat more active.

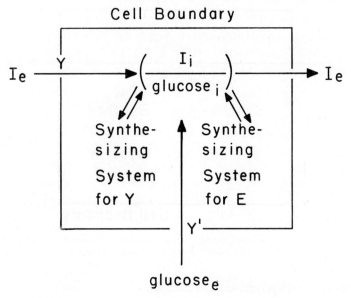

Fig. 6. Schema to illustrate the maintenance effect in the presence of glucose. (See text).

of internal inducer to internal inhibitor ($I_i/glucose_i$) which determines the induction of both E and Y. Non-induced cells, which have little or no Y, will evidently have a low ratio of $I_i/glucose_i$, and consequently will be inhibited by glucose for the production of both Y and E. Induced cells with high levels of Y will have high internal concentrations of $I_i$, a higher ratio of $I_i/glucose_i$, and consequently the glucose inhibition will be reduced with the maintenance of a steady state of induction. In our picture, then, glucose is equivalent to decreasing the effective concentration of inducer and preinduction is equivalent to increasing it.

If the permease model is applicable and the increased "sensitivity" of a preinduced lac+ culture is actually due to the Y system, then the lac— cryptic strain, which lacks Y, should not show the maintenance effect. And in fact it does not, being equally inhibited by glucose and identically responsive to inducer whether or not it is preinduced. The $\beta$-galactosidase system is then the first analyzed example of the type of steady-state model proposed by Delbrück (3) some ten years ago to explain the appearance of a stable "new" character in cells without invoking genotypic changes.

The comparison of maintenance with the development of persistent changes in cells during differentiation is only one part of the analogy.

Actually we know that only a part of the embryo's undifferentiated cellular population becomes converted to a given specialized tissue. Therefore, for the analogy to be important, conditions should be found under which the response of the *E. coli* population to induction would be heterogeneous, i.e., with some cells producing enzyme and others not, and this heterogeneity should be of long duration, covering many cell generations. The most obvious cause of heterogeneity of the type referred to here is mutation and selection, but this model of differentiation does not seem to have met with much success. Actually in the $\beta$-galactosidase system, even if mutation and selection are eliminated, a heterogeneous response to induction is generated whenever the non-induced lac+ strain is placed at low inducer concentration. We owe the analysis of this induced discontinuity in enzyme synthesis to Novick and Weiner (11) who, by applying continuous culture techniques, were able to carry out a brilliant analysis of the kinetics of induction at low inducer concentration. Using the maintenance phenomenon as a tool, they showed that the acceleration phase in the induction of the $\beta$-galactosidase of *E. coli* lac+ at low inducer concentration (Fig. 3, Left) was due to the heterogeneity of the population, some cells producing enzyme and other cells nothing.

The experiment described in Fig. 7 illustrates the Novick-Weiner effect. As has been shown (see Fig. 1), a preinduced culture responds to low inducer concentration without lag and with a linear relationship between enzyme produced and increase in bacterial mass. The non-induced culture shows a marked acceleration phase in induction (see Figs. 1 and 3, Left). The problem now is to determine the state of induction of individual cells.

Any population can be analyzed for induced and non-induced cells by use of the maintenance phenomenon. If an induced cell is placed in glucose plus inducer or in a low inducer concentration, the resultant clone will contain enzyme, whereas the non-induced cell will give a clone which contains no enzyme.

All of the single cells derived from the control preinduced culture (see Fig. 7) and placed in maintenance, give rise to clones which are induced. Therefore, this population is homogeneous. The non-induced culture yields only non-induced clones. It, too, must be homogeneous. However, as induction proceeds at low inducer concentration, the proportion of cells making enzyme increases. The population becomes heterogeneous, some cells making enzyme and the others not. It should be recalled that the maintenance and the heterogeneity also applies to the permease, whose presence the $\beta$-galactosidase merely mirrors in these experiments. Thus,

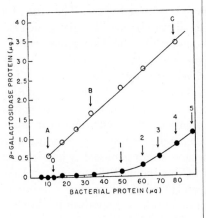

| Sample | Starting Culture | % Positives |
|---|---|---|
| A | Induced | >99 |
| B | Induced | >99 |
| C | Induced | >99 |
| 0 | Noninduced | <1 |
| 1 | Noninduced | 3 |
| 2 | Noninduced | 11 |
| 3 | Noninduced | 18 |
| 4 | Noninduced | 24 |
| 5 | Noninduced | 31 |

Fig. 7. The heterogeneity of response of *E. coli* ML (lac+) to induction at low inducer concentration ($1.5 \times 10^{-5}$ M).

*Left*, the kinetics of induction of a preinduced culture (open circles), and of a non-induced culture (black circles).

*Right*, the analysis of the population for induced and non-induced cells. The culture was diluted to low multiplicity into glucose $10^{-3}$ M + inducer $10^{-4}$ M. The "% positives" refers to the percentage of resultant clones possessing enzyme. (See text).

from a uniform non-induced population, a fraction of the cells take on a new character in much the same way that a fraction of the population of embryonic cells differentiate.

Novick and Weiner (11) point out that "the induced synthesis of β-galactosidase at low concentrations of inducer bears a close resemblance to the phenomenon of mutation. . . . All the offspring of a mutant bacterium are mutant; all the progeny of an induced bacterium are induced, as long as maintenance inducer is present. Bacteria undergo mutation as the result of some random single event; likewise uninduced bacteria make the transition to the induced state as the result of a random single event.

"However, unlike mutation, the induction system requires the continued presence of a low concentration of inducer to maintain the distinction between the induced and uninduced states. At high concentrations of inducer, all the bacteria become induced, while in the absence of inducer the entire population becomes uninduced."

How are we to understand the "pseudogenetic" behavior of this system?

A general model to account for just such a situation was developed by Monod (9) before the permease had been discovered. He posed the following situation, which I shall reorient in the light of our present knowledge about this system. Suppose that the probability that any cell will synthesize permease (and also enzyme) in a generation time is (1) infinitesimal in the absence of inducer, (2) small in the presence of inducer for those cells which do not possess one unit of permease, and (3) large in the presence of inducer for cells which already possess one unit of permease. Under these conditions the synthesis of permease (and also of enzyme) at the cellular level would not only be discontinuous, but the heterogeneity would be perpetuated during a certain time in the descendants. This model predicted, as was found (11), a situation in which the capacity to synthesize permease as well as enzyme becomes a clonal property appearing like a mutation.

Herzenberg and Monod (5) eliminated the advantage that preinduced cells have in being further induced by using a lac— strain which lacks the galactoside permease. With this organism no acceleration phase is demonstrable in the kinetics of induction, very probably because the response to induction is essentially homogeneous. However, it might be pointed out that by lacking the Y system, this strain is incapable of differentiating.

The induction of $\beta$-galactosidase by a low inducer concentration is in all likelihood another example of the type of system beautifully studied by Spiegelman and his colleagues (16) under the name of long-term adaptation of yeast to galactose. The suggestion of Douglas and Condie (4) that yeast do in fact possess a permease for galactose has been confirmed by Robichon-Schulmajster (15); and it seems reasonable that both *E. coli* and yeast are reacting similarly as the result of possessing an identical mechanism.

The "permease" model has been extended by Monod (8) and by Pappenheimer (12) to the problem of antibody synthesis. The general problem of deadaptation when external inducer is removed has been recently compared by Pollack (13) with the phenomenon of dedifferentiation in animal cells. Both of these interesting problems are beyond the scope of this paper, which has had as its goal simply to show how a clone of cells differing by a specifically inducible property can arise from a population without invoking any preceding change in genotype.

## REFERENCES

1. Buttin, G., Cohen, G., Bazire, G., and Monod, J., cited in ref. 7.
2. Cohen, G., and Monod, J., *Bacteriol. Revs.*, **21**, 169 (1957).

3. Delbrück, M., in *Unités Biologiques Douées de Continuité Génétique, Colloq. intern. centre natl. recherche sci. (Paris),* **8,** p. 33 (1949).
4. Douglas, H. C., and Condie, F., *J. Bacteriol.,* **68,** 662 (1954).
5. Herzenberg, L., and Monod, J., cited in ref. 7.
6. Lederberg, J., Lederberg, E. M., Zinder, N. D., and Lively, E. R., *Cold Spring Harbor Symposia Quant. Biol.,* **16,** 413 (1951).
7. Monod, J., in *Enzymes: Units of Biological Structure and Function* (O. Gaebler, ed.), p. 7. Academic Press, New York (1956).
8. ———, *Trans. N. Y. Acad. Sci.* (in press).
9. ———, and Cohn, M., *Advances in Enzymol.,* **13,** 67 (1952).
10. ———, Pappenheimer, A. M., and Cohen-Bajire, G., *Biochim. et Biophys. Acta,* **9,** 648 (1952).
11. Novick, A., and Weiner, M., *Proc. Natl. Acad. Sci. U. S.,* **43,** 553 (1957).
12. Pappenheimer, A. M., Jr., *Ford Symposium 1958* (in press).
13. Pollock, M., *Brit. J. Exptl. Pathol.* (in press).
14. Rickenberg, H., Cohen, G., Buttin, G., and Monod, J., *Ann. inst. Pasteur,* **91,** 829 (1956).
15. Robinchon-Schulmajster, H., pers. commun.
16. Spiegelman, S., *Cold Spring Harbor Symposia Quant. Biol.,* **16,** 87 (1951).

# FEED-BACK CONTROL OF THE FORMATION OF BIOSYNTHETIC ENZYMES[1]

LUIGI GORINI

*Department of Bacteriology and Immunology,
Harvard Medical School,
Boston, Massachusetts.*

and

WERNER K. MAAS[2]

*Department of Microbiology,
New York University College of Medicine,
New York, New York.*

IN THE PAST, a good deal has been learned about the nature of the biochemical pathways involved in intermediary metabolism but relatively little about the regulatory mechanisms governing the flow of biochemical reactions in the living cell. Yet it is the understanding of these regulatory mechanisms, particularly those controlling protein synthesis, which is of importance for attacking the problems of differentiation on a biochemical level. In this paper we shall consider one such mechanism, the control of enzyme formation by feed-back inhibition.

Most of our information on control of enzyme production comes from studies on microorganisms, because it has been much easier to study enzyme synthesis in microbial cells than in cells of higher forms. Inducible enzymes in particular lend themselves to such studies, and until quite recently most of the work in this field was done on them, especially on β-galactosidase. During the last few years, however, it has become equally feasible to study the formation of other enzymes, mainly through the discovery that certain metabolites inhibit the formation of enzymes involved in their own biosynthesis.

The metabolic pathways in which such negative feed-back control was found include the formation of amino acids, purines, and pyrimidines. It has been known for some time that in certain microbial species, such as

---

[1] This work was supported by grants No. G-3344 and G-2492 from the National Science Foundation and by grant No. E-2011 from the United States Public Health Service.

[2] Senior Research Fellow, United States Public Health Service, SF-129.

*E. coli,* addition of these metabolites to the growth medium inhibits their endogenous formation. This inhibition is specific, intermediates in the production of the metabolites or other related substances not producing such an effect. More recently, the inhibition was analyzed on an enzymatic level. It was found that in some cases the end-product inhibited the activity of enzymes leading to its formation; in others, it interfered with the production of such enzymes; and in still others, both types of inhibition were observed. Although the present paper is concerned mainly with inhibition of enzyme formation, it should be pointed out that for the problem of differentiation, feed-back inhibition of enzyme activity may be equally as significant as feed-back control of enzyme production.

End-product inhibition of enzyme formation was first demonstrated in 1953 by Monod and his co-workers. They showed that methionine inhibits the production of methionine synthase (1) and that tryptophan inhibits the production of tryptophan desmolase (9). Later, Vogel found that arginine interferes with the production of acetylornithinase (13), and we showed inhibition by arginine of the formation of two more enzymes in the arginine pathway, ornithine transcarbamylase (4) and argininosuccinase (3). Similar inhibitions of enzyme formation were demonstrated by Yates and Pardee for the formation of pyrimidines (14), by Magasanik and his co-workers for the synthesis of purines (7) and histidine (10), and by Umbarger for the synthesis of isoleucine and valine (12). Thus, the phenomenon of feed-back inhibition of enzyme formation is widespread among biosynthetic pathways.

Our own studies have dealt chiefly with the arginine pathway, particularly with the formation of ornithine transcarbamylase. Fig. 1 is an overall picture of arginine biosynthesis in *E. coli,* including its connection with pyrimidine biosynthesis, with carbamyl phosphate as a common precursor. Fig. 2 shows the ornithine transcarbamylase reaction in detail: the car-

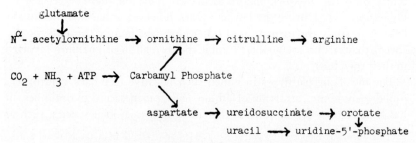

Fig. 1. Abbreviated scheme of the biosynthesis of arginine in *E. coli.*

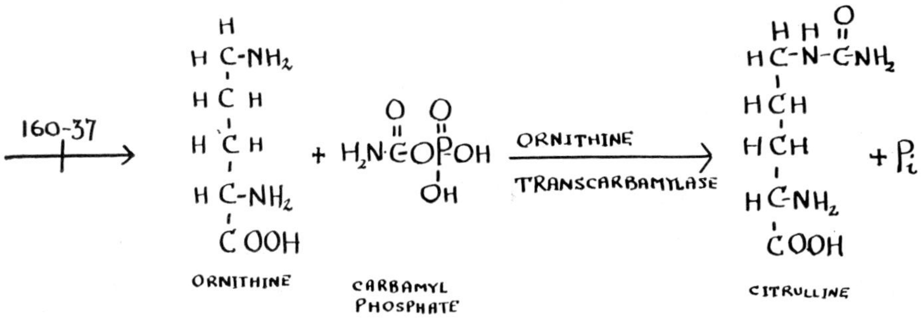

$$\text{160-37} \longrightarrow \begin{array}{c} \text{H} \\ \text{H C-NH}_2 \\ | \\ \text{H C H} \\ | \\ \text{H C H} \\ | \\ \text{H C-NH}_2 \\ | \\ \text{COOH} \\ \text{ORNITHINE} \end{array} + \begin{array}{c} \text{O O} \\ \| \ \| \\ \text{H}_2\text{N-C-O-P-OH} \\ | \\ \text{OH} \\ \text{CARBAMYL} \\ \text{PHOSPHATE} \end{array} \xrightarrow[\text{TRANSCARBAMYLASE}]{\text{ORNITHINE}} \begin{array}{c} \text{H H O} \\ | \ | \ \| \\ \text{H C-N-C-NH}_2 \\ | \\ \text{H C H} \\ | \\ \text{H C H} \\ | \\ \text{H C-NH}_2 \\ | \\ \text{COOH} \\ \text{CITRULLINE} \end{array} + P_i$$

Fig. 2. The ornithine transcarbamylase reaction.

bamyl group is transferred from carbamyl phosphate to ornithine to form citrulline and inorganic phosphate.

When arginine is added to a growing culture of *E. coli*, the synthesis of ornithine transcarbamylase stops immediately, and the enzyme present in the cells is diluted out during the subsequent growth. Thus, after overnight growth on arginine, the concentration of enzyme per cell is about one-hundredth that of cells grown in the absence of arginine. If now the arginine-grown cells are washed and inoculated into fresh medium without arginine, enzyme synthesis is resumed. This system provides a method for studying the kinetics of enzyme formation, just as the addition of an inducer to non-induced cells provides a means for studying the kinetics of formation of inducible enzymes.

For the latter case, extensive studies, particularly on β-galactosidase formation, had established that upon the addition of an inducer to non-induced, growing cells, enzyme synthesis starts immediately and at a constant rate per cell (8). This means that the maximal level of enzyme per cell is attained only after several generations.[3] In the absence of information on other enzyme systems, it had been assumed that in exponentially growing cultures, all enzymes are formed according to these constant-rate kinetics.

The kinetics found for transcarbamylase formation, illustrated in Fig. 3, were therefore most surprising. Here the enzyme level reached a maximum very rapidly, after half a division, and then declined during the remainder of the growth phase. As a working hypothesis to explain these

---

[3] The increase in enzyme is described by the following equation (4): $\% \ E_{max} = 100 \ (1 - e^{-at})$, where $E_{max}$ is the maximum final level of enzyme per cell and $a$ is the rate constant of the equation for exponential growth, $dN/dt = aN$ ($N$ = number of bacteria per unit volume).

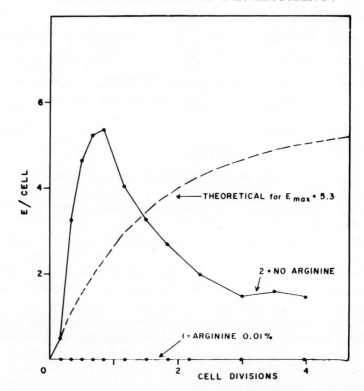

Fig. 3. Enzyme synthesis in flask cultures of the wild-type. Washed cells of *E. coli* strain W taken from exponentially growing cultures maintained on arginine were inoculated into minimal medium A (2) + lactate 0.5%, supplemented as indicated. Incubation was at 37°C with shaking. Enzyme activity was determined in toluenized cells, according to the method of Jones et al. (5). One enzyme unit = amount of enzyme which synthesizes 1 μmole of citrulline per hour; lower limit of method = 0.01 enzyme units. E/cell = units of enzyme per mg. dry weight bacteria. Cell division time was 60 minutes. Theoretical curve was calculated according to the equation described in footnote 3.

unexpected results, we postulated that not only exogenously added arginine but also endogenously formed arginine can inhibit enzyme formation. Initially, after overnight growth on arginine and washing, the cells contained very little arginine and therefore ornithine transcarbamylase (and presumably other enzymes in the arginine pathway), are synthesized very rapidly. As a result of this rapid synthesis of enzymes arginine is also formed rapidly and accumulates sufficiently to slow further enzyme formation.

To test this hypothesis conditions were provided which produced a lower level of intracellular arginine than that present in the wild-type strain growing on minimal medium. The first method to achieve this end was to grow an arginine auxotroph blocked before the transcarbamylase reaction (160-37-h, see Fig. 2) in a chemostat (11), growth being limited by the supply of arginine. It is a reasonable assumption, though it has not been proven, that when the rate of protein synthesis is limited by the concentration of arginine, the intracellular level of arginine is lower than under ordinary growth conditions of the wild-type strain in minimal medium, where arginine is not limiting. Strain 160-37-h has an additional block in histidine biosynthesis, so that in control experiments growth can be limited by another metabolite besides arginine.

The results of the chemostat experiments are described in Fig. 4. The cells were pre-grown on arginine, as in the experiment in Fig. 3, washed, and transferred to the chemostat. During subsequent growth, samples were removed and tested for enzyme activity, as described in the legend of Fig. 3. With arginine limiting the growth rate (Fig. 4, Curve 1B) one would expect the same rapid initial enzyme formation as described in Fig. 3, but now the cells should continue to form enzyme at a rapid rate since the intracellular level of arginine remains depressed. This was actually observed; the enzyme is formed at a constant rate and reaches a level about 25 times as high as in the wild-type culture growing on minimal medium (Fig. 3). The other two curves in Fig. 4 are control experiments. In Curve 1A, growth of strain 160-37-h is limited on histidine, and arginine is supplied at a concentration of 1 μg. per ml. in excess of that required for growth. This slight excess is sufficient to prevent enzyme formation completely. In addition, with a strain having a block in histidine but not in arginine synthesis (Curve 2) the kinetics of enzyme formation are similar to those of the wild-type grown on minimal medium (Fig. 3), as expected.

In the second method for lowering the intracellular level of arginine a mutant (160-37-L3) was used with a partial block between $N^{\alpha}$-acetylornithine and ornithine, derived by Richard Novick in our laboratory from strain 160-37 which has a complete block at this step. This "leaky" mutant can grow slowly on minimal medium, with a division time of 150 minutes as compared to 60 minutes for the wild-type strain. Upon addition of arginine, growth of the mutant is accelerated to the wild-type rate, thus showing that the supply of arginine is limiting the growth rate. In such a mutant growing on minimal medium, one would expect a similar high level of ornithine transcarbamylase as in the chemostat with arginine limiting growth. This expectation was fulfilled. As shown in Fig. 5, during growth on minimal medium the enzyme rises to about the same level as

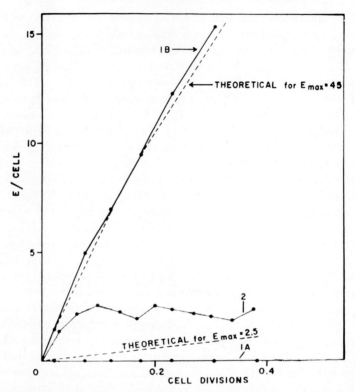

Fig. 4. Enzyme synthesis in the chemostat. The minimal medium and the method for the enzyme determinations were the same as those described in the legend to Fig. 2. The histidine arginine double auxotroph 160-37-h was grown with (μg./ml.): histidine 1 + arginine 6 (Curve 1A); histidine 10 + arginine 5 (Curve 1B). The histidine auxotroph was grown with histidine 1 (Curve 2). The level of growth with histidine limiting at 1 μg./ml. was the same as that with arginine limiting at 5 μg./ml. Inocula were washed cells taken from cultures growing exponentially in excess of arginine. Cell division time was 460 minutes. Theoretical curves were calculated from the constant E/cell values, reached for both curves 1B and 2 after 4 cell divisions.

in the chemostat experiment. In this experiment, in contrast to the experiments with the chemostat, arginine is not supplied exogenously, and thus any possible explanation for the rise in the enzyme level having something to do with the uptake of arginine from the medium is ruled out.

However, since the release of feed-back inhibition has been observed so far only in mutants, it would be desirable to demonstrate the same phenomenon in the wild-type strain in order to rule out any possible effect of a mutation per se on the rise in the level of the enzyme. To do this, it is

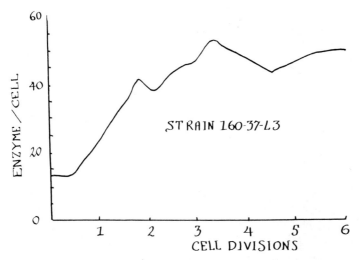

Fig. 5. Enzyme synthesis in a partially blocked arginine auxotroph. Cells of strain 160-37-L3 (see text) were taken from a culture fully grown on minimal medium A + 0.5% lactate + arginine 20 μg. per ml. (arginine was exhausted at the end of growth) and inoculated into fresh minimal medium A + 0.5% lactate. Incubation was at 37°C with shaking. During the subsequent growth, samples for the determination of ornithine transcarbamylase were removed at hourly intervals. The cells were growing exponentially during the time shown in the figure, with a doubling time of 150 minutes. Enzyme units per cell were determined as described in the legend to Fig. 3.

necessary to find conditions for this strain which result in a lowering of the intracellular level of arginine. Two such circumstances have been found under which the supply of arginine for growth becomes sufficiently low to result in an increase in the level of ornithine transcarbamylase. Under both conditions, it appears to be a limitation in the supply of carbamyl phosphate, which in turn limits the formation of arginine.

The first of these conditions is anaerobiosis. When the bacteria are grown in an $N_2$ atmosphere, in the absence of added $CO_2$, the level of ornithine transcarbamylase rises to 45 enzyme units per mg. of dry weight bacteria, as compared to a level of about 3 units during aerobic growth. The addition of 5 per cent $CO_2$ to the $N_2$ atmosphere results in a lowering of the enzyme level to 32 units per mg. dry weight. Since it has been shown that in bacteria carbamyl phosphate is formed from $CO_2$, $NH_3$, and ATP (6), this effect of $CO_2$ on the level of enzyme indicates that under anaerobic conditions carbamyl phosphate is the limiting factor for arginine synthesis. However, since added $CO_2$ lowers the enzyme level only slightly it is presumably the limited availability of ATP that is chiefly responsible for the shortage of carbamyl phosphate. That the feed-back control by arginine itself is not disturbed by anaerobiosis is shown by the observation that exogenous arginine inhibits the formation of ornithine transcarbamylase as well under anaerobic conditions as it does under aerobic conditions.

The second condition for obtaining a rise in the level of ornithine transcarbamylase involves the use of an arginine-free growth medium, enriched with vitamins, purines, pyrimidines, and other amino acids. In such a medium, the wild-type strain will grow about twice as fast as in the minimal medium consisting of only inorganic salts and lactate or glucose as a carbon source. Richard Novick has studied the kinetics of enzyme formation in bacteria after transfer from minimal to the enriched medium. As shown in Fig. 6, the enzyme level rises, though not quite as high as under the conditions of the experiments described before. If the limited supply of carbamyl phosphate is responsible for the rise in the level of enzyme, one would expect that the addition of either arginine or citrulline should cause a depression of the enzyme level, whereas the addition of ornithine should not do so. As shown in Fig. 6, this was found to be the case. Apparently, the rate of carbamyl phosphate synthesis after transfer to the enriched medium does not increase to the same extent as the growth rate. It should be noted in Fig. 6 that in the cultures containing either ornithine or no additional supplement there is a second abrupt rise in the level of enzyme shortly before the end of growth. Since such a rise is not observed in the cultures containing added citrulline or arginine, it appears that at that stage of growth carbamyl phosphate becomes even more limiting for arginine formation, presumably because the supply of ATP for carbamyl phosphate synthesis is becoming exhausted.

If the supposition that the rate of carbamyl phosphate formation is limiting the supply of arginine is correct, one might expect that the level of the enzymes concerned with carbamyl phosphate formation is not raised when the intracellular level of arginine is lowered. James Schwartz in our laboratory has measured the level of the enzyme that forms carbamyl phosphate from ATP and ammonium carbamate (6) and has found that its concentration in the bacteria is not increased under conditions of release of feedback inhibition of ornithine transcarbamylase. For instance, when strain 160-37 is grown in the chemostat with arginine limiting the growth, the level of the carbamyl-phosphate-synthesizing enzyme is the same as that of the wild-type strain growing on minimal medium. Moreover, the addition of arginine to the minimal medium does not depress the level of this enzyme. In similar experiments with uracil, it was found that this substance also had no effect on carbamyl phosphate synthesis. Yates and

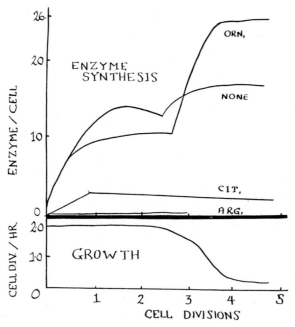

Fig. 6. Enzyme synthesis in the wild-type growing on arginine-free enriched medium. Wild-type cells, pregrown on minimal medum A + 0.5% lactate + arginine 100 μg. per ml., were washed and inoculated into enriched, arginine-free medium (see text), supplemented with the amino acids indicated, each at a concentration of 100 μg. per ml. Incubation was at 37°C with shaking. The lower part of the figure shows the growth rates, which were identical for the 4 cultures. The upper portion shows the concentrations of ornithine transcarbamylase, determined from samples taken every 30 minutes. Enzyme units per cell were determined as described in the legend to Fig. 3.

Pardee have shown that the later steps in pyrimidine synthesis are under feed-back control by uracil (14). It thus appears that in the case of this sequence of reactions feed-back control does not extend beyond the branch point introduced as a result of the two pathways having a common precursor. Since there are many common precursors in the biosynthetic machinery of the cell it will be of interest to look for feed-back regulation of their formation in other systems.

The results described in this paper lead to the conclusion that the intracellular level of endogenously formed arginine or a derivative of arginine controls the rate of ornithine transcarbamylase formation. Such a control is also suggested in the other previously mentioned systems in which the end-product inhibits enzyme formation in its own pathway. This type of feed-back inhibition thus appears to be a general physiological mechanism for controlling enzyme synthesis. One of the striking aspects of these findings is the high potential of the cells to synthesize ornithine transcarbamylase, the final possible level being much greater than what is needed for growth. This high enzyme-forming potential enables the cell to conserve its enzyme-forming machinery and yet to adapt rapidly from an environment in which the end-product is present to one in which it is absent. Negative feed-back control coupled with a high enzyme-forming potential is thus one of the mechanisms responsible for the adaptability of microorganisms.

In relation to the problem of development, the most significant feature of the control mechanism described here is the ability of metabolites of low molecular weight to regulate the enzymatic constitution of the cell. In this regard, negative feed-back control resembles enzyme induction. A change in the enzymatic constitution resulting from either feed-back inhibition or enzyme induction could be the step initiating the series of reactions necessary to produce a differentiated cell.

## REFERENCES

1. Cohn, M., et al., *Compt. rend. Acad. Sci. Paris*, **236**, 746 (1953).
2. Davis, B. D., and Mingioli, E. S., *J. Bacteriol.*, **60**, 17 (1950).
3. Gorini, L., unpub.
4. ———, and Maas, W. K., *Biochim. et Biophys. Acta*, **25**, 208 (1957).
5. Jones, M. E., et al., *J. Am. Chem. Soc.*, **77**, 819 (1955).
6. ———, *Proc. 3rd Intern. Congr. Biochem., Brussels*, p. 278 (1955).
7. Magasanik, B., *Ann. Rev. Microbiol.*, **11**, 221 (1957).
8. Monod, J., et al., *Biochim. et Biophys. Acta*, **9**, 648 (1952).
9. ———, and Cohen-Bazire, G., *Compt. rend. Acad. Sci. Paris*, **236**, 530 (1953).
10. Moyed, H. S., *Federation Proc.*, **17**, 279 (1958).
11. Novick, A., and Szilard, L., *Science*, **112**, 715 (1950).
12. Umbarger, H. E., pers. commun.
13. Vogel, H. J., in *The Chemical Basis of Heredity* (W. D. McElroy and B. Glass, eds.), p. 276. Johns Hopkins Press, Baltimore (1957).
14. Yates, R. A., and Pardee, A. B., *J. Biol. Chem.*, **227**, 677 (1957).

# COMMENT ON THE POSSIBLE ROLES OF REPRESSERS AND INDUCERS OF ENZYME FORMATION IN DEVELOPMENT[1]

HENRY J. VOGEL

*Institute of Microbiology,
Rutgers, The State University,
New Brunswick, New Jersey*

IN THEIR EFFORTS to gain a better understanding of the molecular basis of development, many investigators have devoted particular attention to the process of enzyme formation. For example, Spiegelman (24) in a succinct, although perhaps oversimplified, phrase characterized differentiation as the controlled production of unique enzymatic patterns. Inevitably, two related control mechanisms of enzyme biogenesis have been considered as factors in development: induction of enzyme formation, and more recently, repression of enzyme formation. Both of these mechanisms involve effects of specific *small molecules* on the synthesis of macromolecules (namely enzymes).

## INDUCTION OF ENZYME FORMATION

Our knowledge of induced enzyme formation has been gathered primarily through experiments with microorganisms, particularly with bacteria. However, induced enzyme formation is by no means confined to microorganisms; perhaps the best-analyzed instance of this phenomenon in higher organisms is the case of tryptophan peroxidase-oxidase studied by Knox (16). A number of detailed reviews of induced enzyme formation have been compiled in recent years (7, 21, 22, 25, 32). For the terminology of induced enzyme formation see Cohn et al. (9).

## REPRESSION OF ENZYME FORMATION

The repression of enzyme formation, like enzyme induction, is a widespread cellular control mechanism. Enzyme repression has been defined (30) as "a relative decrease, resulting from the exposure of cells to a given

---

[1] Aided by grants from the Damon Runyon Memorial Fund for Cancer Research and the National Science Foundation.

substance, in the rate of synthesis of a particular apoenzyme." In a number of instances, it has been found that the "end-products" of various biosynthetic pathways are efficient repressers of one or more enzymes within the respective pathways. Such repression by end-products constitutes a regulatory device of a "feed-back" nature which presumably was positively selected in the course of evolution (30, 31).[2] The term enzyme repression, however, embraces not only this type of feed-back event, but also decreases in the rate of enzyme synthesis occasioned by normal or abnormal metabolites other than end-products or by artificial substances. Repression and induction of enzyme formation are considered to be related in terms of the actual molecular mechanisms involved (see below).

The rather general occurrence of enzyme repression is attested to by demonstrations of this phenomenon in pathways leading to a variety of amino acids and nucleic acid components. For instance, one or more repressible enzymes have been found in pathways leading to ornithine (28, 30), citrulline (14), proline (36), methionine (8, 34), valine (1), tryptophan (20), pyrimidine compounds (35), and purine compounds (19). As in the case of enzyme induction, the bulk of the work on enzyme repression has been carried out with microorganisms. Recently, however, DeMars (10) reported an instance of enzyme repression involving a mammalian glutamine-synthesizing enzyme which was studied with the aid of tissue culture methods.

### Possible Relationships between Specific Small Molecules and the Nucleus

Enzyme induction as well as enzyme repression occur against a constant genetic background (with respect to the genic control of the potential or actual formation of the enzyme involved). The consequences of induction or repression will, nevertheless, continue as long as the environmental (small-molecule) stimulus persists. A protracted induction effect which depends in part on the *previous* environmental history of the cells involved has been revealed by studies of an inducible enzyme whose inducers can, under appropriate conditions, be concentrated in the cells through a separate system that itself is elicitable by the same inducers (see M. Cohn, this volume). It is thus clear that specific small molecules, which include normal metabolites such as amino acids and others, can have pronounced

---

[2] It should be emphasized that this type of feed-back mechanism depends on an antagonism to enzyme *formation* and is distinct from another widespread type of feed-back mechanism which involves enzyme *function*. The two types of mechanisms may, however, coact (30).

effects on cellular enzyme patterns without the necessity for any genetic changes (cf. Ephrussi, 12). There is, however, substantial evidence (15, 18, 23; cf. Lederberg and Iino, 17) for the occurrence of certain nuclear changes in development; such changes could be (but are not necessarily) of a genetic nature. Accordingly, considerable importance attaches to the molecular mechanisms underlying these nuclear alterations; of particular interest in this connection are possible roles of cytoplasmic substances.

One possibility (which has been discussed repeatedly at this symposium) is that some nuclear alterations in development are occasioned by *macromolecules* that originated in or passed through the cytoplasm. Such an influence of macromolecules would depend on their ability to gain access to the nucleus and, presumably, to relevant intranuclear regions; however, at least some types of macromolecules appear to be excluded from some nuclei (cf. Mirsky and Allfrey, this volume). A second, highly intriguing possibility (which does not preclude the first) is that nuclear alterations are somehow related to the activity of specific small molecules. In the following, three such relationships will be briefly considered.

(a) The role of specific small molecules (as environmental factors) in the selection of spontaneous mutants in cell populations is well known from studies with microorganisms (2). In view of the regularity with which certain mutant cell types become predominant within a cell population under a given set of selective conditions, it is conceivable that the occurrence of spontaneous, undirected mutations may play a contributory role in development, especially when large cell populations are involved (3). More generally, however, if we are to envisage mutation-like processes in development, it seems necessary to contemplate orderly, directed gene mutations (11). Clearly, what is known about gene mutation in microorganisms does not encourage any hypothesis of directed gene changes (see, for example, Braun and Vogel, 4). On the other hand, the available evidence does not eliminate the possibility of the occurrence, under suitable conditions, of something like directed mutations. In the following, it will be attempted to outline, in molecular terms, a conceivable mechanism that could bring about this type of genetic change.

(b) Let us first consider the previously proposed (31) mechanism of induction or repression of enzyme formation. Enzyme inducers and enzyme repressers are thought to act, respectively, by accelerating or retarding the separation of a newly formed template product (nascent enzyme protein) from its template.[3] Inducers and repressers generally are chemically

---

[3] To characterize a template existing under conditions of enzyme repression, Dr. Werner Braun has suggested the term "clogged template" or, more metaphorically, "stuck zipper"; under conditions of enzyme induction we would have a "greased zipper."

related to the *substrates* of the corresponding enzymes. The action of inducers and repressers may thus well depend on their affinity for the "dynamic site" of the nascent enzyme (which is similar to but not identical with the finally shaped enzyme).[4] In line with recent assumptions, enzyme induction and repression are considered to involve secondary, cytoplasmic templates.[5] It will be noted that the secondary template carries the *"information"* for the structure of the enzyme, but that *affinity* for the inducer or represser is attributed to the corresponding nascent enzyme protein. Now, considering the primary gene template, the latter would again have the information and, therefore, a protein produced at the site of the gene with the aid of this information may have affinity for inducers or repressers. That such genic proteins are formed appears to be a reasonable assumption (see Mirsky and Allfrey, this volume). Accordingly, the possibility exists that inducers or repressers, which are involved in the *phenotypic* phenomena of enzyme induction and enzyme repression, have affinity for the very portions of the *genome* that control the formation of the corresponding enzymes. It is further conceivable that the presence of inducers or repressers (particularly of repressers) in the regions of the corresponding genes may produce permanent genetic changes either in the course of replication or in some other way.

(c) If specific small molecules can penetrate to and have affinity for specific genic sites in the manner contemplated above, it is perhaps not too far-fetched to imagine that such molecules can affect the functioning (cf. Glass, 13) or "state" (cf. Lederberg and Iino, 17; Stern, 26) of genes without actually causing mutation-like events.

### Heritable Lowering of an Enzyme Level Observed upon Continuous Cell Cultivation in the Presence of a Represser

Recently, a *heritable* lowering of the level of the enzyme, acetylornithinase, was noted (6) when *Escherichia coli* (ATCC # 9637) was grown

---

[4] Alternatively, their affinity may be directed at a protein that resembles the nascent enzyme but forms part of the template. A chemical similarity of inducers or repressers to the corresponding substrates is not considered to be an intrinsic feature of enzyme induction or enzyme repression. Thus, it is conceivable that the formation of some enzymes may be induced or repressed (in the same general manner as previously proposed, 31) through the agency of substances (for example, certain hormones) that are not chemically related to the corresponding substrates. If so, hormone-like inducers or repressers would presumably act at a site other than the "dynamic site." If induction results from the neutralization of a repressive effect (cf. Vogel, 31), the inducer involved could exert its antagonizing action either at the "dynamic site" of the nascent enzyme or in the synthesis (or uptake) of the represser.

[5] A possible involvement of nuclear templates in enzyme induction or repression does not, however, seem to be excluded by the available evidence.

at about 30°C, without forced aeration, in a continuous cultivation device, on a glucose-salts (minimal) medium with or without a supplement of L-ornithine hydrochloride (20 mg. per liter). The cultivation device used was a slight modification of the turbidostat of Bryson (5). Acetylornithinase (33) mediates the last step in the synthesis of ornithine in *E. coli* (27, 29); ornithine, in turn, is a precursor of arginine. Both ornithine and arginine are repressers of acetylornithinase formation.

In the continuous cultivation experiments, cultures were sampled at daily intervals and streaked to yield single colonies on unsupplemented solid medium. From such colonies, stock cultures in liquid minimal medium containing casein hydrolysate (0.2%) were derived. For assays of acetylornithinase levels, inocula from each stock culture, after preliminary growth on liquid minimal medium, were again cultivated on minimal medium. Enzyme levels (in units per mg. protein) were determined, as previously described (30), in extracts of sonically disrupted organisms.

Between the third and seventh day of continuous cultivation in the presence of ornithine, isolates were obtained that exhibited acetylornithinase levels ranging downward to 10-15% of those of the inoculum used. In contrast, isolates from continuous cultures grown without ornithine showed no substantial decreases in acetylornithinase levels, but often showed increases.

These results indicate that upon continuous cultivation of cells in the presence of a represser, which can *phenotypically* antagonize the formation of a particular enzyme, organisms are readily isolable that exhibit a *heritably* reduced level of the same enzyme. While the available data do not permit any definite conclusion as to the manner in which the observed phenomenon arises, they are consistent with a hypothesis based on undirected mutation and selection. However, in view of the foregoing discussion, the possibility of the involvement of something resembling a directed gene change is being seriously considered.

## REFERENCES

1. Adelberg, E. A., and Umbarger, H. E., *J. Biol. Chem.*, **205**, 475 (1953).
2. Braun, W., *Bacterial Genetics*. W. B. Saunders Company, Philadelphia (1953).
3. ———, in *Symposium on Genetic Approaches to Somatic Cell Variation*, Gatlinburg, *J. Cellular Comp. Physiol.* (in press).
4. ———, and Vogel, H. J., in *Bacterial and Mycotic Infections of Man* (R. Dubos, ed.), Chapter 3, p. 28. J. B. Lippincott Company, Philadelphia (1958).
5. Bryson, V., *Science*, **116**, 48 (1952).
6. Cocito, C., and Vogel, H. J., *Proc. 10th Intern. Congr. Genet.* (Montreal), p. 55 (1958).
7. Cohn, M., *Bacteriol. Revs.*, **21**, 140 (1957).

8. ———, Cohen, G. N., and Monod, J., *Compt. rend. Acad. Sci. Paris*, **236**, 746 (1953).
9. ———, Monod, J., Pollock, M. R., Spiegelman, S., and Stanier, R. Y., *Nature*, **172**, 1096 (1953).
10. DeMars, R., *Biochim. et Biophys. Acta*, **25**, 208 (1957).
11. Ephrussi, B., *Nucleo-cytoplasmic Relations in Microorganisms*. Oxford University Press, London (1953).
12. ———, in *Enzymes: Units of Biological Structure and Function* (O. H. Gaebler, ed.), p. 29. Academic Press, New York (1956).
13. Glass, B., *Science*, **126**, 683 (1957).
14. Gorini, L., and Maas, W. K., *Biochim. et Biophys. Acta*, **25**, 208 (1957).
15. King, T. J., and Briggs, R., *Proc. Natl. Acad. Sci. U. S.*, **41**, 321 (1955).
16. Knox, W. E., in *Physiological Adaptation* (C. L. Prosser, ed.), American Physiological Society, Washington (in press).
17. Lederberg, J., and Iino, T., *Genetics*, **41**, 743 (1956).
18. Lehman, H. E., in *The Beginnings of Embryonic Development* (A. Tyler, R. C. von Borstel, and C. B. Metz, eds.), p. 201. American Association for the Advancement of Science, Washington (1957).
19. Magasanik, B., *Ann. Rev. Microbiol.*, **11**, 221 (1957).
20. Monod, J., and Cohen-Bazire, G., *Compt. rend. Acad. Sci. Paris*, **236**, 530 (1953).
21. ———, and Cohn, M., *Advances in Enzymol.*, **13**, 67 (1952).
22. Pollock, M. R., *Symposium Soc. Gen. Microbiol.* (*Adaptation in Microorganisms*), p. 150 (1953).
23. Sonneborn, T. M., *Proc. 9th Intern. Congr. Genet.* (Bellagio), p. 307 (1954).
24. Spiegelman, S., *Symposia Soc. Exptl. Biol.*, **2**, 286 (1948).
25. ———, in *Enzymes: Units of Biological Structure and Function* (O. H. Gaebler, ed.), p. 67. Academic Press, New York (1956).
26. Stern, C., in *Analysis of Development* (B. H. Willier, P. A. Weiss, and V. Hamburger, eds.), p. 151. W. B. Saunders Company, Philadelphia (1955).
27. Vogel, H. J., *Proc. Natl. Acad. Sci. U. S.*, **39**, 578 (1953).
28. ———, *Proc. 6th Intern. Congr. Microbiol.* (*Rome*), **1**, 269 (1953).
29. ———, in *Amino Acid Metabolism* (W. D. McElroy and B. Glass, eds.), p. 335. The Johns Hopkins Press, Baltimore (1955).
30. ———, in *The Chemical Basis of Heredity* (W. D. McElroy and B. Glass, eds.), p. 276. The Johns Hopkins Press, Baltimore (1957).
31. ———, *Proc. Natl. Acad. Sci. U. S.*, **43**, 491 (1957).
32. ———, in *Handbuch der Pflanzenphysiologie* (W. Ruhland, ed.), Vol. 11, Springer-Verlag, Heidelberg (in press).
33. ———, and Bonner, D. M., *J. Biol. Chem.*, **218**, 97 (1956).
34. Wijesundera, S., and Woods, D. D., *Biochem. J., (London)*, 55, viii (1953).
35. Yates, R. A., and Pardee, A. B., *J. Biol. Chem.*, **227**, 677 (1957).
36. Yura, T., and Vogel, H. J., *J. Biol. Chem.* (in press).

# THE METABOLIC REGULATION OF PURINE INTERCONVERSIONS AND OF HISTIDINE BIOSYNTHESIS [1]

BORIS MAGASANIK

*Department of Bacteriology and Immunology,
Harvard Medical School, Boston, Massachusetts*

THE ENZYMATIC REACTIONS responsible for the biosynthesis of inosine 5'-phosphate (IMP) de novo and for its conversion to adenosine 5'-phosphate (AMP) and guanosine 5'-phosphate (GMP) have been elucidated in recent years. The latter part of this biosynthetic pathway is shown in some detail in Fig. 1, and in schematic form in Fig. 2. The experiments whose results are described in this essay were carried out with Enterobacteriaceae (*Escherichia coli, Aerobacter aerogenes, Salmonella typhimurium*). The pathway of purine nucleotide synthesis in these organisms and in animal cells is essentially the same.

IMP is converted to AMP in two steps (9): the first of these is the condensation of IMP with aspartic acid to give adenylosuccinic acid (AMP-S), and it requires guanosine triphosphate (GTP) as the source of energy; the second step is the hydrolysis of AMP-S to AMP and fumaric acid. Mutants lacking one of the two enzymes catalyzing these reactions require adenine for growth (6); this base is converted to AMP by reaction with phosphoribosylpyrophosphate (PRPP) (7).

The conversion of IMP to GMP also occurs in two steps: the first is the oxidation of IMP to xanthosine 5'-phosphate (XMP) with diphosphopyridine nucleotide ($DPN^+$) as the hydrogen acceptor (12); the second step is the amination of XMP by $NH_3$ to give GMP, and this step requires adenosine triphosphate (ATP) as the source of energy (16). Mutants lacking the enzyme catalyzing the first of these reactions, IMP-dehydrogenase, require xanthine or guanine for growth, while those lacking the enzyme catalyzing the second step, XMP-aminase, have a specific requirement for guanine. These purine bases, too, are apparently converted to the corresponding nucleotides by reaction with PRPP.

---

[1] The experimental work described in this essay has been supported by research grants from the U. S. Public Health Service (C-2864) and from the National Science Foundation (N.S.F.-G1295).

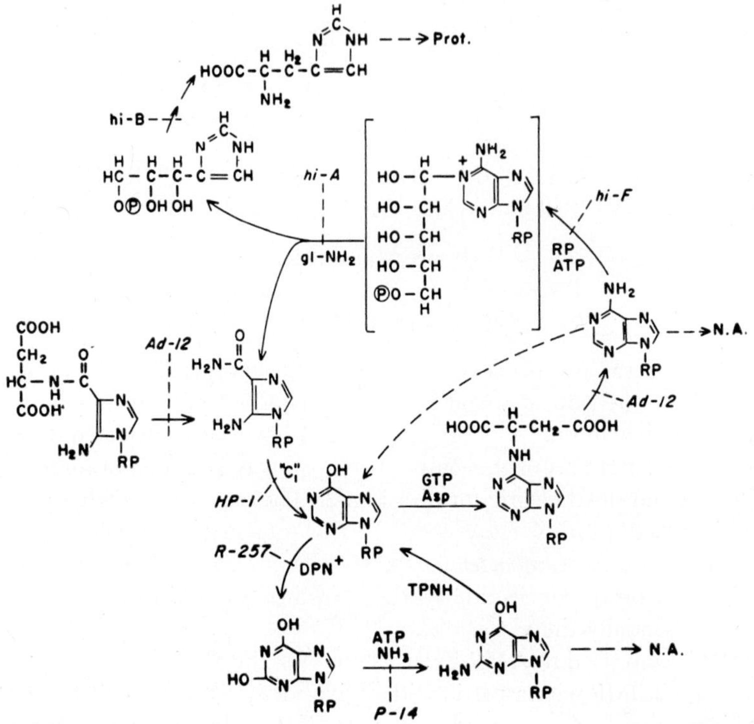

486   THE CHEMICAL BASIS OF DEVELOPMENT

Fig. 1. The cyclic interconversions of purine nucleotides and the biosynthesis of the histidine precursor imidazole glycerol phosphate. The dashed lines intersecting the arrows indicate genetic blocks; the symbols (R-257, P-14, etc.) at the dashed lines refer to mutants lacking the enzyme.

The AMP and GMP produced by these reactions may be converted to di- and tri-phosphates which are essential participants in many enzymatic reactions and also building blocks for nucleic acids.

Experiments with labelled guanine have shown that Enterobacteriaceae are capable of using this purine as a source of adenine nucleotides (3). This conversion cannot involve the same enzymes that are responsible for the formation of GMP from IMP, for it has been shown that the reactions catalyzed by these enzymes are irreversible, and that a mutant lacking XMP-aminase can nevertheless convert guanine to adenine nucleotides (2, 13). On the other hand, the conversion of guanine to adenine nucleotides must utilize the enzymes catalyzing the steps between IMP and AMP, for otherwise mutants lacking one of these enzymes should be able to grow not only on adenine, but also on guanine. The enzyme which enables the

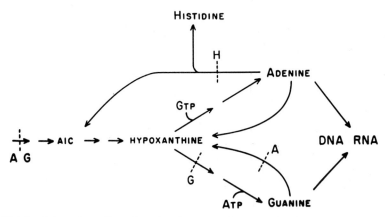

Fig. 2. Schematic outline of purine interconversions and histidine biosynthesis. For simplicity, the names of the bases are used in place of that of the corresponding nucleotides which are the actual substrates and products of the enzymatic reactions. The dashed lines intersecting the arrows indicate the site of negative feedback controls; the letters at the dashed lines refer to the compounds exerting the control by inhibiting the reaction. A, G, and H indicate derivatives of adenine, guanine, and histidine, respectively.

organism to use guanine for adenine biosynthesis must therefore catalyze a reaction in which a guanine derivative is converted to IMP. This enzyme has recently been identified as GMP-reductase; it catalyzes the irreversible reductive deamination of GMP to IMP with reduced triphosphopyridine nucleotide (TPNH) as the hydrogen donor (14).

The three enzymes, IMP-dehydrogenase, XMP-aminase, and GMP-reductase, together could theoretically catalyze the irreversible cyclic conversion of IMP via XMP and GMP to IMP, and would obtain the energy for the operation of this useless merry-go-round by the splitting of ATP. Actually, however, the cell possesses mechanisms which regulate the reactions of the cycle in such a way that AMP and GMP are produced only as they are needed for metabolic reactions.

The conversion of IMP to GMP is controlled by GMP or other guanine derivatives: GMP inhibits the action of IMP-dehydrogenase, and in addition the formation of this enzyme is suppressed by the presence of guanine in the culture medium (8). Consequently, a high intracellular level of GMP will prevent the formation of additional GMP from IMP, but will permit the conversion of GMP to IMP, and thus the formation of AMP at the cost of GMP. This conversion is in turn controlled by ATP, which inhibits the action of GMP-reductase (14); consequently, a high intracellular level of adenine nucleotides will prevent the formation of additional

AMP from GMP. In addition, adenine and guanine, or the corresponding nucleotides, control the synthesis of IMP de novo by inhibiting an early step of this biosynthetic pathway (5). Finally, the formation of AMP and of GMP may be coordinated by the reciprocal role of GTP and of ATP as energy sources for the conversion of IMP to AMP and GMP, respectively. These regulatory mechanisms are shown schematically in Fig. 2. The discovery of these controls explains why these organisms use exogenously supplied adenine or guanine preferentially to producing these compounds by synthesis de novo, and why when supplied with a mixture of adenine and guanine, they use each purine base preferentially for the production of the corresponding nucleotide component of the nucleic acids (2, 3).

In addition to its other functions ATP is the donor of a N-C fragment which becomes the $N_1$-$C_2$ portion of the imidazole ring of histidine (10, 13, 17, 18). The formation of the essential histidine precursor imidazole glycerol phosphate (1) from ATP, ribose-5-phosphate (RP), and glutamine is also shown in Fig. 1. Mutants which lack one of the enzymes catalyzing these reactions require histidine for growth (17). The first step in the formation of IGP is the attachment of RP to nitrogen 1 of the adenine ring; the complex is aminated by the amide group of glutamine and then cleaved to give IGP and the ribose nucleotide of 4-amino-5-imidazole carboxamide (AICAR) (17). AICAR, an intermediate in the synthesis of purine nucleotides de novo, can be converted to IMP by formylation with formyl tetrahydrofolic acid, followed by ring closure (4). This series of reactions constitutes a cycle from IMP via AMP and AICAR back to IMP. Every turn of the cycle results in the formation of one molecule of IGP from RP, a "formyl" group, and two amino groups, one supplied by aspartic acid, the other by glutamine.

The operation of this cycle is controlled by histidine. This amino acid inhibits the enzyme which catalyzes the condensation of ATP with PRPP, the first step on the pathway leading specifically to histidine (15); in addition, histidine suppresses the formation of several of the enzymes which catalyze reactions between ATP and IGP (15) (Fig. 2). The complete cycle will therefore operate just sufficiently to supply the cell with the histidine it requires for the synthesis of protein; however, by blocking the path leading from AMP to AICAR, histidine would also prevent the conversion of AMP to GMP, were it not for the existence of another enzyme(s), as yet not identified with certainty, which catalyzes the more direct formation of IMP from AMP.

The existence and operation of these two pathways leading from AMP to IMP can be demonstrated with a mutant of *S. typhimurium*, Ad-12, which has lost the ability to produce the single enzyme which catalyzes both the cleavage of AMP-S to AMP, and that of the ribose nucleotide of 5-amino-4-imidazole succinylcarboxamide to AICAR (6) (Fig. 1). This organism has therefore a specific requirement for adenine; it has to use this base as the source of AMP and GMP as well as of the $N_1$-$C_2$ portion of histidine. When this organism is grown in a medium containing adenine-2-$C^{14}$ (11), it is found that the molar radioactivities of the adenine of its nucleic acids and of the histidine of its protein are equal to that of the adenine fed, while the molar radioactivity of the guanine of its nucleic acid is only one-half that of the adenine fed. This result indicates that one-half of the guanine is unlabelled and has been formed from adenine which had given up its carbon 2 to form histidine, and had been converted to AICAR, which in turn had accepted an unlabelled "formyl" group to become IMP. But the amount of adenine which is converted to AICAR is apparently not sufficient to supply all of the guanine which the cell requires: the other half of the guanine must have come from adenine which had been converted to IMP without the loss of carbon 2. In another experiment histidine was added to the adenine-2-$C^{14}$-containing medium to suppress the formation of histidine from adenine. In this case, the histidine isolated from the cell protein was unlabelled, and the adenine and guanine isolated from the cell nucleic acid each had the same molar radioactivity as the adenine supplied in the medium. It appears that histidine has kept the AMP from entering the path leading to histidine and AICAR, and has forced it into the direct path to IMP.

It is evident that regulation by negative feedback provides a precise mechanism for the coordination of the complex metabolic mechanisms of the cell. It is not unlikely that the orderly progression of cellular differentiation may be controlled in a similar manner by small molecules elaborated by the cell.

## REFERENCES

1. Ames, B. N., *J. Biol. Chem.*, **228**, 131-143 (1957).
2. Balis, M. E., Brooke, M. S., Brown, G. B., and Magasanik, B., *J. Biol. Chem.*, **219**, 917-926 (1956).
3. Brown, G. B., and Roll, P. M., in *The Nucleic Acids* (Chargaff, E., and Davidson, J. N., eds.). Academic Press, New York, **2**, 341-392 (1955).
4. Flaks, J. G., Erwin, M. J., and Buchanan, J. M., *J. Biol. Chem.*, **229**, 603-612 (1957).
5. Gots, J. S., *J. Biol. Chem.*, **228**, 57-66 (1957).
6. ———, and Gollub, E. G., *Proc. Natl. Acad. Sci. U. S.*, **43**, 826-834 (1957).
7. Kornberg, A., Lieberman, I., and Simms, E. S., *J. Biol. Chem.*, **215**, 417-427 (1955).

8. Levin, A. P., and Magasanik, B., (unpub.).
9. Lieberman, I., *J. Biol. Chem.*, **223**, 327-339 (1956).
10. Magasanik, B., *J. Am. Chem. Soc.*, **78**, 5449 (1956).
11. ———, and Karibian, D., (unpub.).
12. ———, Moyed, H. S., and Gehring, L. B., *J. Biol. Chem.*, **226**, 339-350 (1957).
13. ———, ———, and Karibian, D., *J. Am. Chem. Soc.*, **78**, 1510 (1956).
14. Mager, J., and Magasanik, B., *Federation Proc.*, **17**, 267 (1958).
15. Moyed, H. S., *Federation Proc.*, **17**, 279 (1958).
16. ———, and Magasanik, B., *J. Biol. Chem.*, **226**, 351-363 (1957).
17. ———, and ———, *J. Am. Chem. Soc.*, **79**, 4812 (1957).
18. Neidle, A., and Waelsch, M., *Federation Proc.*, **16**, 225 (1957).

## DISCUSSION

Dr. POTTER: I should like to ask the preceding two speakers, as well as Dr. Umbarger, Dr. Magasanik, Dr. Vogel, Dr. Spiegelman, and any others who might care to comment, if any of them would care to make some suggestions as to the possible difference we find here between the bacterial organism and the cell of the highly differentiated adult mammalian organism. The point that I wish to make is that the operation of these feed-back mechanisms serves the purpose of economizing the efforts of the bacteria as far as producing certain substances which they need for growth. However, they do indeed grow, and that is the important point. They merely avoid making something they don't need. However, the problem in the adult differentiated tissue is how *not* to grow in the presence of both existing and inducible pathways with sufficient exogenous metabolites for de novo and preformed avenues of synthesis surrounding them. Presumably we do have feed-back mechanisms in this instance which are operative. I would like to hear any comments which anyone might have to make on this problem.

Dr. MAAS: In answer to Dr. Potter's question I think it is interesting that mammalian cells seem to behave somewhat differently in regard to feed-back inhibition of enzyme formation than do microorganisms. For instance, Dr. DeMars in Dr. Eagle's laboratory has studied the control of the formation of glutamine synthase in cultures of HeLa cells. When glutamine is supplied in the medium, the formation of glutamine synthase is depressed. If, after prolonged growth on glutamine, one removes glutamine and substitutes glutamic acid for it, the enzyme is not formed and the cells will not grow, except if one adds extremely large doses of glutamic acid. The point is, the HeLa cells do not seem to have the same high reserve capacity or "potential" for making enzyme that bacterial cells have. I think this is an interesting difference between mammalian cells and the bacterial systems I have described.

Dr. SPIEGELMAN: It may perhaps be useful to express an opinion on Dr. Cohn's suggestion that the solution found for the *E. coli* case is applicable to the "long-term adaptation" phenomenon we studied some 10 years ago in yeast. A comparison between the two may also serve to illustrate a somewhat philosophical issue which may be of more general interest to the purpose of the present symposium.

We found that certain "long-term adapting" stocks of yeast were characterized by heterogeneity with respect to the ability to adapt to galactose. That this heterogeneity was clonally distributed was easily demonstrated by the use of indicator-galactose agar plates. Two types of colonies developed; one composed of positive cells, the other of negative ones. We were able to show by essentially genetic experiments that positive cells possessed elements which could increase autocatalytically in the presence of galactose. Growth in the absence of inducer resulted in dilution of these elements. At each division they were distributed randomly between the mother and daughter cells.

Though these experiments had a certain air of ingenuity and decisiveness they failed to resolve the problem because we were unable to identify in biochemical terms the elements we were postulating and their function. At the time we were not in a position to identify and assay all of the enzymes involved and hence could not reduce the phenomenon to the precise enzymatic basis which has been recently achieved with the *E. coli* system. I am inclined to agree with Dr. Cohn that when the biochemical details of the yeast case are unraveled it will probably be similar to the mechanism found in *E. coli*.

In describing our results we were forced to use terms and constructs at a biological level that implied a more involved mechanism than probably exists. This brings me to the more general point I wanted to make; biological phenomena are simpler than they appear to be.

I have found it difficult to avoid the conclusion that many of the investigators concerned with morphogenesis are secretly convinced that the problem is insoluble. I get the feeling that many of the intricate phenomena described are greeted with a sort of glee as if to say, "My God, this is wonderful, it is so complicated we will never understand it."

It seems to me that perhaps the time has come to abandon this joyful pessimism and its attendant conviction of incomprehensible complexity. In particular, I should like to make a plea for a more optimistic view based on a belief in simplicity. The phenomena of morphogenesis can hardly be as complicated as implied by the welter of apparently unrelated observations constituting the literature of embryology. I suggest that the mechanisms underlying differentiation are in large part understandable and amenable to experimental illumination in terms of chemical components already known and identified. We are living in the era of the Watson-Crick model in which experiments have been performed, or are on the verge of execution, which will lead to the complete elucidation of the central problem of biology, the mechanism of duplication. This same era has seen the successful isolation in cell-free state of enzyme systems which can fabricate such complicated molecules as DNA, RNA, and protein. It is no longer relevant these days to phrase questions of cell physiology in terms of other than defined chemical entities. It seems to me that the same is true for morphogenetic events.

Dr. Umbarger: I would like to cite an example of feedback control which relates to Dr. Maas' presentation and which will illustrate the point Dr. Spiegelman has just made regarding the extremely simple mechanisms that are often found to underlie biological complexities. In the figure below I have shown diagrammatically on side A the essential features of the kind of negative feedback loop which Dr. Maas has described. The endproduct represented by $P$ prevents the formation of enzyme $A$, which converts intermediate $A$ to the next intermediate along the pathway to $P$. As Dr. Maas has emphasized, such a mechanism would make it possible for a microorganism to avoid synthesis of $P$ when $P$ is available in the environment. However, this kind of control mechanism is sluggish in its response to environmental change since any enzyme A previously present would continue to function.

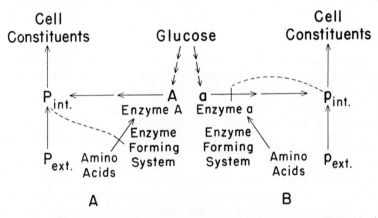

Studies on the kinetics of accumulation of metabolites preceding blocked reactions in bacterial mutants have indicated that biosynthetic pathways respond more quickly to changes in environment than mechanism A would permit. Analysis of several cases has now shown that it is the feedback mechanism shown in part B of the figure that is responsible for this rapid response. In this feedback loop, the activity of the initial enzyme, *enzyme a,* in the biosynthetic sequence is inhibited by the endproduct $p$. If an external source of the product ($p_{ex}$) is present, it will be concentrated by the cell and when the internal pool of product ($p_{int}$) is high enough the endogenous formation of $p$ from $a$ is prevented. The enzyme remains available for synthesis whenever the exogenous product is removed and the reversible inhibition is lifted.

The example with which I am most familiar is the mechanism whereby *E. coli* will stop synthesis of isoleucine whenever this amino acid has been added to the medium. The initial step leading specifically to isoleucine is the formation of α-ketobutyrate from threonine. The enzyme, L-threonine deaminase, is extremely sensitive to competitive inhibition by isoleucine. Thus, in the ab-

## DISCUSSION

sence of exogenous isoleucine, the level of isoleucine in the cell is maintained at a certain level and is continuously replenished from the threonine pool as isoleucine is removed for protein synthesis. In the presence of exogenous isoleucine, the deaminase is inhibited, threonine is spared and can be used exclusively for protein synthesis—essentially none being converted to $\alpha$-ketobutyrate.

It is of considerable biological significance, I think, that in this example, and in all examples, of feedback control that have been reported, the sensitive enzyme is the one that catalyzes the first step leading specifically to the endproduct. One can readily see that the selection of this step is the only one that nature could have selected if there was to be the nearly perfect adjustment between the cell and its environment based upon such a simple mechanism.

In the growing bacterial cell, of course, there are not only feedback loops preventing the oversynthesis of small molecules such as isoleucine but also mechanisms preventing overproduction of large molecules (enzymes) as well. And, in fact, we do find both mechanisms operating in the case of isoleucine biosynthesis that I have just referred to. Also, both mechanisms have been found in the closely related valine pathway. It will be quite surprising if both patterns will not also be found in animal cells.

DR. BESSMAN: I would like to discuss a clinical situation which might throw some light on the question raised by Dr. Potter earlier this afternoon. Some people, with disease of the liver, are unable to convert ammonia to urea at a normal rate, and become victims of ammonia poisoning. In these patients, the glutamine level of the blood is extremely high. The muscle removes a great deal of ammonia, and, when these individuals become emaciated and the muscle tissue is practically gone, they die of ammonia poisoning. We tried to find out what function the muscle might play in this disease. We assumed that the muscle might substitute for the liver by synthesizing glutamine from glutamic acid and ammonia. We tried to adapt rat muscle in the whole animal to form more glutamine synthetase by injecting salts of ammonia at frequent intervals for three days. Then we took out the muscle, made an extract and measured its ability to synthesize glutamine. Ordinary rat muscle tissue makes very little glutamine, but there is a several fold increase in its ability to make glutamine from ammonia and glutamate through the adaptive period of two to three days.

DR. MAGASANIK: I would like to reply to Dr. Potter's question by saying that the operation of control mechanisms in bacteria *can* restrict growth. Dr. Cohn has already mentioned the inhibitory effect of glucose on the synthesis of certain enzymes: glucose may prevent the synthesis of enzymes which destroy compounds which are essential for growth. We have studied a histidine-requiring mutant of *A. aerogenes* which grows well on a small supplement of histidine when glucose is the major carbon source, but not when other carbon compounds are substituted for the glucose. The reason for this is that carbon compounds other than glucose cannot prevent the formation of histidase, the cell makes histidase, and destroys the histidine it requires for growth. It is conceivable that mechanisms of this type are involved in the control of growth in the cells of higher animals: the cell cannot grow because an enzyme it produces destroys a compound which is essential for its growth; another compound may in turn prevent the formation of the enzyme, and thus stimulate growth.

## BIOSYNTHESIS OF UREA IN METAMORPHOSING TADPOLES [1]

### G. W. Brown, Jr. and P. P. Cohen

*Department of Physiological Chemistry*
*University of Wisconsin, Madison*

For some time it has been known that young tadpoles excrete mostly ammonia (35, 36). Sometime during metamorphosis, that dramatic event in which a fishlike larval tadpole is transformed into the adult, there results a shift in the excretory pattern from ammonia to urea. Prior studies have indicated that arginase activity of liver also increases (14, 35, 36). The arginase activity of the livers of young tadpoles is relatively high, and the increase in arginase activity does not pinpoint the locus responsible for the increased urea excretion. The fact that urea excretion increases while that of ammonia decreases suggests the induction of the Krebs-Henseleit ornithine-urea cycle (29) at metamorphosis.

The experimental demonstration that ammonia is indeed converted to urea in amphibia by the known enzymatic steps of the urea cycle has not to our knowledge been reported. Preliminary evidence for the operation of the cycle in the frog has been reported (14, 31, 35, 36). The nature of its induction at metamorphosis, however, must be sought in phenomena other than in the observed increase in arginase levels of the liver.

The urea cycle and closely allied ancillary reactions are depicted in Fig. 1. As illustrated, a highly integrated system consisting of nine enzymes is formed from three subcycles. The net reaction of this system under steady-state kinetics is the conversion of two moles of ammonia and one mole of bicarbonate to urea at the expense of three moles of ATP. Note that the oxidized DPN formed in the glutamic dehydrogenase reaction can be utilized in that catalyzed by malic dehydrogenase. Although most of the reactions are reversible, this has not been indicated in order to focus attention on directions obtaining in a steady state leading to a net synthesis of urea.

In order to explore the nature of the shift from ammonia to urea excre-

---

[1] Supported by grants from the National Science Foundation and the National Institutes of Health, U. S. Public Health Service.

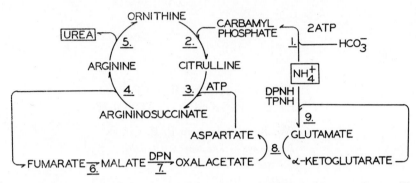

Fig. 1. Krebs ornithine-urea cycle. Enzymes: (1) carbamyl phosphate synthetase; (2) ornithine transcarbamylase; (3) argininosuccinate synthetase (condensing enzyme); (4) argininosuccinate cleavage enzyme; (5) arginase; (6) fumarase; (7) malic dehydrogenase; (8) aspartic-glutamic transaminase; (9) glutamic dehydrogenase. (Reversibility of reactions not indicated.)

tion taking place during metamorphosis of the tadpole, we have performed the following experiments:

(1) We have determined the ammonia and urea excretion of tadpoles in various stages of larval development;

(2) we have classified such tadpoles according to the indices of Taylor and Kollros (55) in order to describe accurately the stage of development in relation to external morphological features;

(3) we have developed techniques for the quantitative assay of individual steps of the urea cycle;

(4) we have assayed individual livers of metamorphosing tadpoles; and

(5) we have undertaken a study of the nature of induction of the urea cycle at metamorphosis.

### Experimental

*Animals*

In our initial studies, tadpoles of the giant bullfrog, *Rana catesbeiana* (var. Wisconsin, California, and North Carolina) were selected because of their ready availability and large size. The larger specimens weigh as much as 30 g. and often possess livers in excess of 1 g. in weight. Tadpoles of this species are available the year around because of the peculiarity that they may persist in the larval state for as long as three years in nature before undergoing metamorphosis (13).

## Classification of Stage of Development

Taylor and Kollros (55) have defined 25 larval stages of *Rana pipiens* on the basis of gross external morphology. Other classifications have been used for the pre-larval or embryonic stages. We have worked only with larval tadpoles. The dividing line between embryonic and larval stages in *Rana pipiens* is conveniently taken as the stage at which the tadpole acquires labial teeth. We have found the stages defined by the above investigators to apply, in the main, also to *Rana catesbeiana* tadpoles. Features of the various stages are briefly summarized in Table 1. The ratio of

TABLE 1

Stages in Larval Development of Rana catesbeiana
[After Taylor and Kollros (55)]

| General designation | Stage number | Synopsis of characteristics |
|---|---|---|
| Limb bud stages | I-V | Length of bud increases from slight elevation to 2 × diameter. |
| Paddle stages | VI-X | Flattening of limb bud to complete indentation between toes; 5th toe web directed to 3rd toe. |
| Foot stages (premetamorphic stages) | XI-XVII | 5th toe web reaches prehalix; nasolachrymal duct appears; proximal, middle, and distal toe pads appear. |
| Metamorphic stages | XVIII-XXV | Cloacal tail-piece disappears; front legs appear; larval mouth still present; labial fringes complete, angle of mouth tends toward posterior margin of eyeball, rapid decrease of tail length; tympanic cartilage ring perceptible; tail absent. |

hind limb length to tail length[2] is also a valuable index of the degree of metamorphosis, and in our early experiments only this ratio was employed as an index. The earlier experiments can be easily correlated with the later ones by reference to a curve in which Stage Number has been plotted vs. the hind limb to tail length ratio, as seen in Fig. 2. Tadpoles in various stages of development are shown in Fig. 3.

## Assay Procedures

The details of the assay procedures are to be reported elsewhere (7) and are modifications of existing techniques (8, 32, 43). The usual criteria of linearity of the reaction rate vs. enzyme concentration or time were ful-

[2] This is not the reciprocal of the allometric ratio suggested by Roth (51, cf. 14). Roth employs the ratio, body length/hind limb length, as a measure of the degree of metamorphosis.

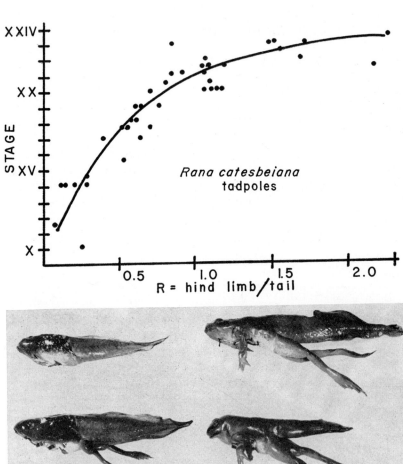

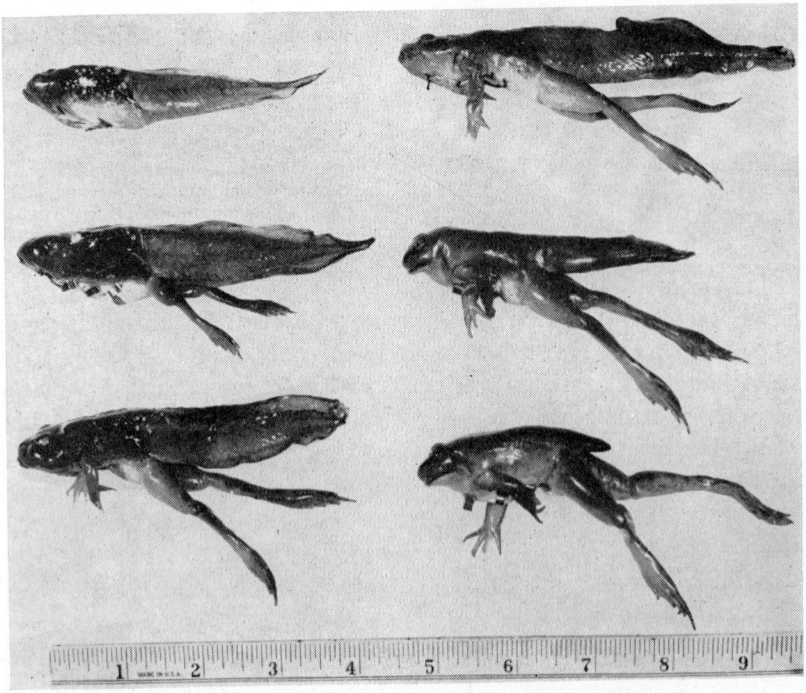

Fig. 2. Top, Correlation of stage of tadpole development with ratio of hind limb length to tail length in *Rana catesbeiana*.
Fig. 3. Bottom, Stages of development of *Rana catesbeiana* tadpoles. Top to bottom, left to right: Stages X, XVIII, XX, XX ½, XXIII, XXIV.

filled. Homogenates were prepared in the presence of 0.1 per cent cetyl trimethyl ammonium bromide (CTAB). CTAB appears to depress ATPase activity (cf. 59) and yields preparations of higher specific activity than are obtained in water extracts. Preparations of carbamyl phosphate synthetase of fairly high initial specific activity can be made by extracting the particulate fraction of a KCl homogenate with CTAB. Two such extractions are sufficient to remove over 90 per cent of the mitochondrial enzymes, carbamyl phosphate synthetase and ornithine transcarbamylase, and of arginase. Apparently all of the arginine synthetase activity (condensing and cleavage enzymes of Ratner et al., cf. 43) resides in the supernatant portion of a KCl or CTAB homogenate. CTAB has been found useful in our laboratory for the preparation of a highly purified carbamyl phosphate synthetase from frog liver (34).

## Ammonia and Urea Excretion by Tadpoles

Tadpoles at different stages of development were placed in beakers containing phosphate buffer (0.01 $M$, $pH$ 6.5), and 24 hours later the fluid was analyzed for ammonia and urea (36).

### NITROGEN EXCRETION BY TADPOLES

Fig. 4 shows the percentage of the ammonia + urea nitrogen excreted as urea nitrogen. Note that when the ratio of the hind limbs to tail length is approximately 1 there is a marked shift in the pattern of nitrogen excretion. Tadpoles prior to Stage XIX maintain a relatively low level of urea excretion, viz., about 10 per cent of the total (ammonia + urea). By the time Stage XXII is reached, the tadpole excretes over half of the total nitrogen as urea. According to Taylor and Kollros (55), the onset of true metamorphosis occurs at Stage XVIII, and Fig. 4 shows that the onset is followed almost immediately by a rapid shift in the nature of the excretion product. It is thus precisely at this time that one should expect to find induction of the urea cycle. It should be pointed out that the tadpoles (roughly of equivalent body weight) from which the data of Fig. 4 were obtained all were maintaining an excretion of between 1.4 and 2.8 mg. of nitrogen per day, and animals corresponding to points on the extreme ends of the graph excreted 1.94 and 1.95 mg. N/day. This points to an actual *shift* in the nature of the biochemical excretion mechanism (cf. 36) at metamorphosis and not to the mere development of an auxiliary mechanism which operates independently of the ammonia levels. The levels of excretion are comparable to those obtained by Munro (36) with *Rana*

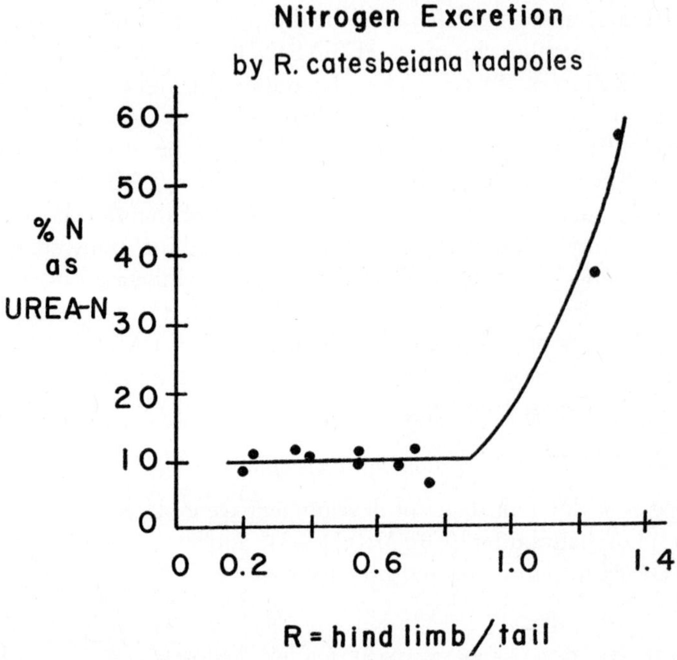

## Nitrogen Excretion
## by R. catesbeiana tadpoles

[Graph: %N as UREA-N vs R = hind limb/tail]

Fig. 4. Nitrogen excretion by *Rana catesbeiana* tadpoles. Percentage of total nitrogen (ammonia + urea) excreted as urea nitrogen. R = hind limb/tail length.

*temporaria* tadpoles undergoing precocious metamorphosis under the influence of thyroid extract.

### LEVELS OF UREA CYCLE ENZYMES DURING DEVELOPMENT OF *Rana catesbeiana* TADPOLES

*Definition of Units*

One unit corresponds to the production of 1 $\mu$mole of product per hour under assay conditions. The value, units/g., represents $\mu$moles of product produced per hour per g. wet weight of liver.

*Carbamyl Phosphate Synthetase*

This enzyme catalyzes the synthesis of carbamyl phosphate (25) from ammonia and bicarbonate (9, 10, 20-23, 32):

$$NH_4^+ + HCO_3^- + 2\,ATP \xrightarrow[Mg^{++}]{\text{Ac-glutamate}} H_2NCO \sim P + 2\,ADP + P_i \quad (1)$$

In a preliminary experiment on single CTAB extracts of KCl homogenates, the activity of carbamyl phosphate synthetase was found to increase markedly as the hind limb/tail length increased (Fig. 5). In this experiment quantitative recovery had not been achieved, and there was a question as to whether such values truly represented the relative amount of enzyme activity of the individual livers. A more refined study by the quantitative procedure, with a larger number of specimens classified ac-

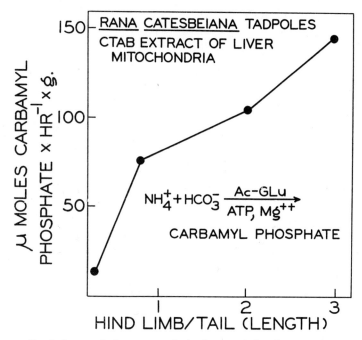

Fig. 5. Increase during metamorphosis of carbamyl phosphate synthetase in *Rana catesbeiana* tadpoles.

cording to Stage Number, fully confirmed the original observation and indicated that the activity increased markedly at the onset of metamorphosis. The results are shown in Fig. 6. This experiment has been repeated a number of times with practically the same results. Essentially the same type of curve obtains if the ratio, units/mg. protein (specific activity), is plotted on the ordinate. The activity of the enzyme increased from 17.4 units/g. at Stage X to over 500 units/g. at Stage XXIV, while the corresponding specific activities increased from 0.76 to 9.2. These data point to an actual increase in the quantity of this enzyme with metamorphosis.

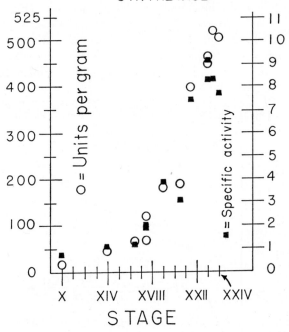

Fig. 6. Quantitative increase during metamorphosis of carbamyl phosphate synthetase in *Rana catesbeiana* tadpoles.

*Ornithine Transcarbamylase and Arginase*

These enzymes catalyze the following reactions (8-10, 20-23, 54; 11, 19, 40);

$$\text{carbamyl phosphate} + \text{ornithine} \xrightarrow{\text{O.T.C.}} \text{citrulline} + P_i \qquad (2)$$

$$\text{arginine} + H_2O \xrightarrow{\text{arginase}} \text{urea} + \text{ornithine} \qquad (3)$$

Plots of units/g. vs. Stage Number for these two enzymes show features similar to those obtained with carbamyl phosphate synthetase. There is initially a relatively low level of activity which rises sharply at about Stage XVIII, corresponding to the onset of metamorphosis. The activities of these enzymes are at all times much higher than those for carbamyl phosphate synthetase, ranging up to several thousand units/g. liver. Ornithine transcarbamylase in tadpoles prior to metamorphosis is considerable in amount (1000-2000 units/g.) and represents about one-fourth to one-third the activity realized in the young frog liver.

Data taken from a typical experiment, in which a series of tadpole livers were assayed for ornithine transcarbamylase, appear in Table 2. The ac-

TABLE 2

TADPOLE ORNITHINE TRANSCARBAMYLASE

(*Rana catesbeiana* var. Wisconsin, received July 1957. Liver preparation: double homogenization with 0.1 per cent CTAB. Conditions: ornithine, 20 μmoles; dilithium carbamyl phosphate (CP), 20 μmoles; glycylglycine, $pH$ 8, 90 μmoles. Final volume, 2.0 ml. T = 38° C, 15 min. incubation.)

| Stage | Body wt. g. | Liver wt. g. | Protein incub. μg. | System | Citrulline produced μmoles | Units per g. liver[1] |
|---|---|---|---|---|---|---|
| X | 11.7 | 0.330 | 40.4 | complete | 0.65 | 1260 |
| | | | | −ornithine | 0.23 | |
| | | | | −CP | 0 | |
| | | | | (boiled) | 0.30 | |
| XX | 23.0 | 0.735 | 40.8 | complete | 0.91 | 2770 |
| | | | | −ornithine | 0.07 | |
| | | | | −CP | 0 | |
| | | | | (boiled) | 0.14 | |
| XXIV | 19.2 | 0.900 | 66.0 | complete | 1.85 | 5660 |
| | | | | −ornithine | 0.14 | |
| | | | | −CP | 0 | |
| | | | | (boiled) | 0.28 | |

[1] Corrected for values obtained with boiled enzyme extracts.

tivity was found to increase from 1260 units/g. at Stage X to 5660 units/g. at Stage XXIV. Little or no activity could be demonstrated over that for boiled controls in the absence of either ornithine or carbamyl phosphate. Residual activity with the boiled enzyme in the otherwise complete system is due to nonenzymatic transcarbamylation. Our observation on an increase of arginase activity at metamorphosis is in agreement with the findings of Dolphin and Frieden (14) with *Bufo terrestris* and of Munro (35) with *Rana temporaria* in tadpoles undergoing precocious metamorphosis induced with thyroid preparations.

The levels of these two enzymes and those of carbamyl phosphate synthetase and arginine synthetase (see below) at various Stages are plotted in Fig. 7 as a percentage of that realized at Stage XXIV. The bar graph represents partial data averaged from several representative experiments. Some selection of the data was made, for a composite plot of experiments from groups of tadpoles employed at different times of the year and from various sources does not reflect the smoothness of curves obtained in indi-

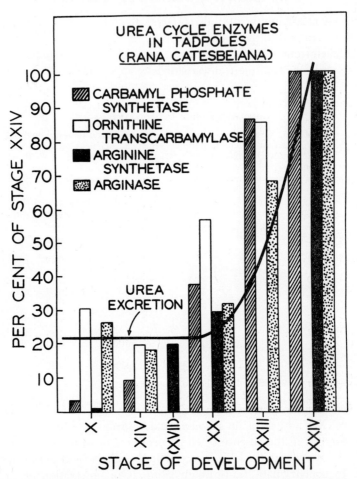

Fig. 7. Composite plot of activities of urea cycle enzymes. Urea excretion expressed in per cent of urea excreted at Stage XXIV. Assay for arginine synthetase (condensing plus cleavage enzymes) reflects activity of the rate-limiting component, probably the condensing enzyme.

vidual experiments. For comparison, the urea excretion is plotted as a percentage of the final value at Stage XXIV.

*Arginine Synthetase*

The term synthetase is used here to refer to the two enzymes (condensing and cleavage enzymes of Ratner et al. (42-49) which effect the conversion of citrulline and aspartic acid to arginine via argininosuccinate:

$$\text{ATP} + \text{citrulline} + \text{aspartate} \xrightarrow{\text{Mg}^{++}} \text{argininosuccinate} + \text{AMP} + \text{PP}_i \quad (4)$$

$$\text{argininosuccinate} \rightarrow \text{arginine} + \text{fumarate} \quad (5)$$

Little or no activity can be demonstrated for this overall system in tadpoles prior to Stage XX. With the maximum amount of extract permitted by our assay procedure and with an hour incubation, only from 0 to 2 units/g. can be demonstrated. However, beyond Stage XX the activity becomes appreciable ($> 5$ units).

About the middle of last July we assayed the arginine synthetase of a number of tadpoles which were rapidly developing toward the frog stage. We dutifully noted the external morphological characteristics and found these tadpoles to range from Stages XX to XXIV, inclusive. The data obtained are shown in Fig. 8. There was a hint in the scattered points that the enzyme system was increasing between Stages XX through XXIV, but we had so convinced ourselves that Stage Number, based upon morphologi-

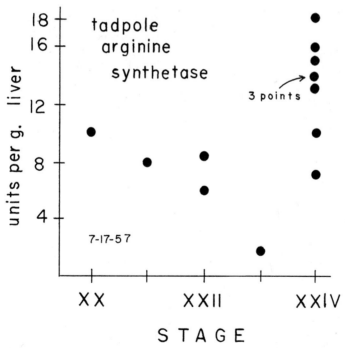

Fig. 8. Arginine synthetase (overall system = condensing enzyme and cleavage enzyme) vs. Stage Number, in metamorphosing tadpoles of *Rana catesbeiana*.

cal characteristics, was a better criterion of the degree of metamorphosis that we didn't bother to plot the activity vs. the ratio of hind limb length to tail length. In going over the data at a later time, we were convinced from the general smoothness of the plots of activities of other enzymes vs. Stage Number (as well as with the limb to tail ratio) that these values for arginine synthetase could not be as poor as they seemed. Upon plotting the activity vs. the hind limb/tail ratio (Fig. 9), we obtained a much better

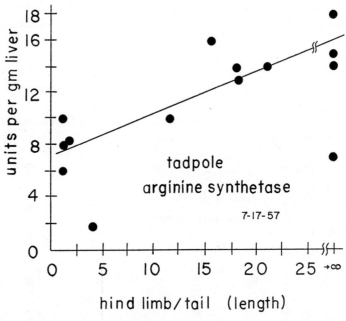

Fig. 9. Arginine synthetase (overall system = condensing enzyme and cleavage enzyme) vs. hind limb/tail length, in metamorphosing tadpoles of *Rana catesbeiana*.

representation of the phenomenon observed with the other enzymes. The two points lying low on the graph, while aberrant, nevertheless conform to the general trend.

The points illustrated by this study of arginine synthetase, other than the pertinence of the data regarding increase at metamorphosis, are two-fold. Firstly, when comparing tadpoles at the same Stage or at Stages close to one another such as these, use of the hind limb/tail ratio gives a better spread to the metamorphic variable. Secondly, metamorphosis as judged by a biochemical variable may not always be precisely in step with certain external morphological characteristics (cf. 14, 17). It is known, for example, that various tissues of the tadpole differ in their response to levels of thyroxine (1, 2, 6, 16, 39), and so one is hard-pressed to set up a rigid criterion of the metamorphic stage without knowing the level of thyroxine or other possible stimulators at the site of action in the tissues and what response to expect from a given level. Also, tadpoles of different species within the same genera apparently respond with different degrees to the induction of metamorphosis by thyroxine (2, 39). The enzyme levels in themselves afford a sort of *metabolic metamorphic index*, though this terminology leaves much to be desired.

*Rate-Limiting Enzyme of the Urea Cycle*

Failure to demonstrate unequivocally the presence of arginine synthetase in the premetamorphic tadpoles could be the result of low activity (or absence of activity) of either the condensing (Eq. 4) or cleavage (Eq. 5) enzymes. Dr. Ratner has kindly sent us several samples of argininosuccinate so that we could test for the activity of the cleavage enzyme, and we have found that this enzyme has considerable activity in premetamorphic tadpoles. An assay for the enzyme cleaving argininosuccinate carried out with livers pooled from tadpoles from Stages V to X yielded a value of 25.1 units/g. Clearly, then, the arginine synthetase as assayed in our hands contained the condensing enzyme as the rate-limiting component. This is in contrast to the finding of Ratner et al. (cf. 43) with mammalian tissue, in which the cleavage enzyme is usually rate-limiting in crude extracts.

We did not use an ATP-generating system in these assays of the overall reaction, and it is possible that the condensing enzyme is inhibited by substrate levels of ATP (cf. 43). Also, the rate may be dependent upon the levels of pyrophosphatase, and at present we do not know the level of activity of this enzyme in our crude extracts. In the event that the condensing enzyme of tadpoles is not found to be the rate-limiting component, the finding that the activity of this enzyme increases during metamorphosis will still be of significance because the same assay conditions were used on each specimen. Other evidence indicates that the rate-limiting step in mammals is concerned with the synthesis of arginine from citrulline (18, 41).

## Implications of the Findings

If the assays are a true reflection of the effective physiological levels of the urea cycle enzymes of the liver in vivo, some obvious implications exist. Before discussing these, it is worth emphasizing that the activity of an enzyme in vitro may differ markedly from that in vivo. Aside from the fact that in the cell the enzyme is not the freely suspended protein that it is in the test tube, concentrations of substrates and activators at the site of enzymatic action may be too low to saturate the enzyme, so that a maximum rate may not obtain. In general, assays of enzymes are carried out under conditions of pseudo zero-order kinetics because of the linearity afforded during successive time intervals; the concentrations of substrates are such as to maintain enzyme saturation during the course of the assay. As a first approximation, it would be expected that the observed rate would not differ by more than an order of magnitude from the corresponding activity in the cell.

Let us examine the urea excretion of a tadpole to see if the enzyme assays reflect the ability to synthesize a corresponding amount of urea. Prior to metamorphosis, *Rana catesbeiana* tadpoles weighing about 5 g. were found to excrete about 0.2 $\mu$moles of urea per hour. Tadpoles of this weight possess a liver which weighs approximately 300 mg. Thus approximately 0.7 $\mu$moles of urea were produced per hour per g. liver by these tadpoles. Hence if the urea cycle is operative in these premetamorphic tadpoles, one unit of the rate-limiting enzyme of the urea cycle should suffice to support this rate of urea synthesis via the urea cycle. This is just about the level of arginine synthetase (overall system) we have been able to measure without analytical difficulty, and it corresponds to about the *most* activity that could ever be demonstrated in premetamorphic tadpoles. By use of the same type of argument, we find that the levels of the urea cycle enzymes in the liver of tadpoles at any stage of development *beyond* metamorphosis can more than account for the overall rate of urea synthesis as measured by that excreted. Whether the urea cycle operates to a small extent prior to metamorphosis is at present equivocal. The excretion of small amounts of urea by premetamorphic tadpoles ($\sim$ 10 per cent of total N) may be due, in entirety or in part, to the action of liver arginase on dietary arginine or that synthesized by intestinal flora. It is of interest to note that arginase has always been found in larval tadpoles we have assayed, regardless of Stage Number. Some urea may also be formed during the degradation of purines.

The substrates employed in the enzyme assays carried out with tadpole livers are the same as those used to demonstrate reactions of the urea cycle with partially purified enzymes. Further, the properties of the enzymes from the tadpole, in the few instances in which they have been investigated, bear a close similarity to those of the mammalian enzymes. For example, acetyl glutamate is an obligate cofactor for tadpole carbamyl phosphate synthetase, as it is with the mammalian enzyme. In a single experiment the $K_m$ for argininosuccinate with the cleavage enzyme was $9.4 \times 10^{-4}$, a value not too different from that of $1.4 \times 10^{-3}$ reported by Ratner (43) with the mammalian enzyme. The enzymes also appear to be located in corresponding subcellular fractions of liver.

The following reaction for the overall synthesis of urea is consistent with the data presented here for amphibians, as well as with the observations on the reactions catalyzed by the urea cycle enzymes in mammalian systems:

$$NH_4^+ + HCO_3^- + \text{aspartate}^{\mp} + 3ATP + H_2O \xrightarrow[\text{ornithine}]{\text{Ac-glutamate}, Mg^{++}}$$

$$\text{urea} + \text{fumarate}^= + 2\,ADP + AMP + 2\,P_i + PP_i + H^+ \quad (6)$$

$$\Delta F^0\,(pH\,7) = -13.2\,\text{kcal}.$$

At present, any explanation for the observed increases in enzyme activities is a matter of speculation. It is our belief that we are indeed observing an enhanced synthesis de novo of these enzymes at the onset of metamorphosis. However, some alternative possibilities have yet to be explored, such as (1) do activators or inhibitors exist which are produced at different stages of larval development and which specifically influence the catalytic properties of the enzymes? and (2) are the actual catalytic and other physicochemical properties of the premetamorphic and postmetamorphic enzymes identical?

### EVOLUTIONARY SIGNIFICANCE OF THE UREA CYCLE

The evolutionary importance of the products of nitrogen excretion of various organisms has been dealt with at length by others (4, 5, 37, 38). We would like to expand somewhat upon these ideas with respect to the urea cycle in the light of observations reported here.

A simple form of excretion of metabolic nitrogen is that of ammonia elimination—a mechanism typical of most aquatic invertebrates and bony fishes. Ammonia is a relatively toxic compound for animal tissues and must be disposed of immediately. Aquatic forms possess a large "sink" in which to dispose of it. But ammonia excretion is not conducive to colonization of *dry* land, and if a group of organisms expects to spend considerable time away from the confines of water, it, or its progeny, must explore some new biochemical pathways to detoxify ammonia.

The main nitrogenous excretion product of the Devonian amphibian is not known. But there exist today three genera of Dipnoi, the lungfishes, which seem to be derived from the same or a similar stock as the first amphibians and which provide some information on this point. These primitive fishes are present only in South America, Australia, and Africa, and they live under conditions thought to be similar to those existing at about the time there appeared a form which could be recognized as an amphibian. They live in regions of seasonal drought and the streams they inhabit often dry up. When water is abundant the lungfish excretes ammonia; when faced with arid seasons it aestivates in the mud, breathes air, and stores up large amounts of urea which is excreted when it again returns to the water during the rainy season. An excellent account of this is given by Smith (53). The lungfish may have reached an evolutionary cul de sac, but some of its features are thought to be representative of those of near-cousins on the trunk of the evolutionary tree. Anatomically, lungfishes have many features in common with the modern salamanders (39), which we have also found to possess all of the enzymes of the urea cycle.

Some time in the Devonian, a period thought to have been plagued by extreme climatic changes (24), it is possible that the make-up of certain freshwater vertebrates was such that a survival value was afforded against periods of drought in the form of certain anatomical, physiological, and biochemical advantages. Ancestors of these vertebrates probably already had an inherent ability to effect at least some of the reactions of the urea cycle, but unfaced with such internal or external environmental stresses possibly continued to excrete mostly ammonia.

The urea cycle is generally considered today principally as a biochemical mechanism for the synthesis of urea. But the individual steps of the cycle probably did not arise simultaneously. More likely, through the gradual acquisition of the complement of enzymes, and under appropriate environmental conditions, a line of organisms incorporated these components into the cycle as we recognize it today. Attesting to an earlier and possibly more fundamental role of the cycle is the synthetic function (other than urea synthesis) of the urea cycle enzymes. Thus carbamyl phosphate, synthesized by enzymes found in bacteria and in animal liver, is a precursor (with aspartic acid) of ureidosuccinate (50) leading to pyrimidines in bacteria (30, 58, 60) as well as in animals (50, 58). There seems to be little question that pyrimidines are fundamental ingredients of living systems as we know them, and hence the nucleic acid constituents probably antedated the evolution of the rather elaborate cycle which synthesizes such a relatively simple compound as urea.

Other lines of evidence may be used to promote the thesis that before the evolvement of the urea cycle as a biochemical excretory mechanism, certain enzymes of the cycle were already being exploited in nature. For example, certain lactic acid bacteria possess an enzyme system capable of carrying out the non-oxidative generation of ATP from citrulline, e.g., *Streptococcus faecalis* (26). The component enzymes of this system catalyze reactions which are thought to be just the reverse of those catalyzed by the mammalian and amphibian liver enzymes, carbamyl phosphate synthetase (Eq. *1*) and ornithine transcarbamylase (Eq. 2). Thus the cycle (or part of it) in these bacteria serves as an energy-yielding device. In this connection it is to be noted that the steps requiring ATP, even in animal systems today, are catalyzed reversibly [the carbamyl phosphate synthetase in part, only (33)]. The mammalian counterpart of the bacterial ornithine transcarbamylase, while not reversible under physiological conditions, can be made to convert citrulline to ornithine by arsenolysis (27) but without providing any useful energy, as is the case with the bacterial enzymes. It is thus seen that these bacterial and mammalian enzymes bear a close resemblance to one another.

Note also that either or both of the enzymes which synthesize arginine from citrulline and aspartic acid also occur in yeast (42), jack beans (57), pea seeds (12), *Chlorella* (56, 57), *Escherichia coli* (42), and *Neurospora* (15). Arginase is fairly widely distributed (3, 19). Further, urea cycle enzymes have been demonstrated in the ciliated protozoan, *Tetrahymena pyriformis* S (52).[3] That the reactions of the urea cycle may be of more general significance than previously realized has been discussed by Ratner (42).

Enzymes of the urea cycle serve as a means of producing arginine for protein synthesis and for making available its amidine moiety for the synthesis of creatine and creatine phosphate. The relationship of the urea cycle to arginine and hence to the phospagens, arginine phosphate and creatine phosphate, of invertebrate and vertebrate muscle, respectively, may be more than a coincidence.

Considerable evidence can thus be marshalled attesting to the broad synthetic nature of reactions of the urea cycle, and even to an energy-yielding role in the case of bacteria. The urea cycle may occupy a position in nature similar to that of the tricarboxylic acid cycle, both having been considered by Krebs, Gurin, and Eggleston (28) to have evolved originally for the purpose of synthesizing metabolic intermediates. Both cycles, in part or in whole, now possess biosynthetic and energy-yielding features. The intimate relation of these cycles is revealed by the fact that the synthesis of urea in the intact, ureotelic organism requires the participation

---

[3] This observation has not been confirmed. See Dewey, V. C., Heinrich, M. R., and Kidder, G. W., *J. Protozool.*, **4**, 211-219 (1957).

of dicarboxylic acids of the tricarboxylic acid cycle, as indicated in Fig. 1. Reactions of the urea cycle currently contribute to the biochemical economy of diverse organisms, and evolution of the complete cycle has enabled amphibians to extend successfully their embryonic and larval confines.

## REFERENCES

1. Allen, B. M., *Anat. Record,* **54,** 45-64 (1932).
2. ———, *Biol. Revs.,* **13,** 1-19 (1938).
3. Baldwin, E., *Biol. Revs.,* **11,** 247-268 (1935).
4. ———, *An Introduction to Comparative Biochemistry,* The University Press, Cambridge, England (1937).
5. ———, *Dynamic Aspects of Biochemistry* (3rd ed.), The University Press, Cambridge, England (1957).
6. Blacher, L. J., *Trans. Lab. Exptl. Biol. Zoo-Park Moscow,* **14,** 172-173 (1928).
7. Brown, G. W., Jr., and Cohen, P. P., *Federation Proc.,* **17,** 197 (1958).
8. Burnett, G. H., and Cohen, P. P., *J. Biol. Chem.,* **229,** 337-344 (1957).
9. Cohen, P. P., and Grisolia, S., *J. Biol. Chem.,* **174,** 389-390 (1948).
10. ———, and ———, *J. Biol. Chem.,* **182,** 747-761 (1950).
11. Dakin, H. D., *J. Biol. Chem.,* **3,** 435-441 (1907).
12. Davison, D. C., and Elliott, W. H., *Nature,* **169,** 313-316 (1952).
13. Dickerson, M. C., *The Frog Book,* p. 234. Doubleday, Page and Co., New York (1906).
14. Dolphin, J. L., and Frieden, E., *J. Biol. Chem.,* **217,** 735-756 (1955).
15. Fincham, J. R. S., and Boylen, J. B., *Biochem. J.,* **61,** xxiii-xxiv (1955).
16. Fontès, G., and Aron, M., *Compt. rend. soc. biol. Paris,* **102,** 679-682 (1929).
17. Frieden, E., and Naile, B., *Science,* **121,** 37-39 (1955).
18. Gornall, A. G., and Hunter, A., *J. Biol. Chem.,* **147,** 593-615 (1949).
19. Greenberg, D. M., in *The Enzymes* (J. B. Sumner and K. Myrbäck, eds.), Vol. *I,* Part 2, p. 893-921, Academic Press, New York (1951).
20. Grisolia, S., in *Phosphorus Metabolism* (W. D. McElroy and B. Glass, eds.), **2,** 619-629, Johns Hopkins Press, Baltimore (1952).
21. ———, and Cohen, P. P., *J. Biol. Chem.,* **191,** 189-202 (1951).
22. ———, and ———, *J. Biol. Chem.,* **198,** 561-571 (1952).
23. ———, and ———, *J. Biol. Chem.,* **204,** 753-757 (1953).
24. Guyer, M. F., *Animal Biology,* p. 552. Harper & Bros., New York (1941).
25. Jones, M. E., Specter, L., and Lipmann, F., *J. Am. Chem. Soc.,* **77,** 819-820 (1955).
26. Knivett, V. A., *Biochem. J.,* **56,** 602-606 (1954).
27. Krebs, H. A., Eggleston, L. V., and Knivett, V. A., *Biochem. J.,* **59,** 185-193 (1955).
28. ———, Gurin, S., and Eggleston, L. V., *Biochem. J.,* **51,** 614-628 (1951).
29. ———, and Henseleit, K., *Hoppe-Seyler's Z. physiol. Chem.,* **210,** 33-66 (1932).
30. Lieberman, I., and Kornberg, A., *J. Biol. Chem.,* **207,** 911-923 (1954).
31. Mandersheid, H., *Biochem. Z.,* **263,** 245-249 (1933).
32. Metzenberg, R. L., Hall, L. M., Marshall, M. E., and Cohen, P. P., *J. Biol. Chem.,* **229,** 1019-1025 (1957).
33. ———, Marshall, M. E., and Cohen, P. P., (unpub.).
34. ———, ———, and ———, *J. Biol. Chem.* **233,** 102-105 (1958).
35. Munro, A. F., *Biochem. J.,* **33,** 1957-1965 (1939).
36. ———, *Biochem. J.,* **54,** 29-36 (1953).
37. Needham, J., *Chemical Embryology,* Vol. 1-3. The University Press, Cambridge, England (1931).

38. ———, *Biochemistry and Morphogenesis*. The University Press, Cambridge, England (1942).
39. Noble, G. K., *The Biology of the Amphibia*. McGraw-Hill Book Co., New York (1931).
40. Oppenheimer, C., *Die Fermente und ihre Wirkungen*, Vol. II, p. 793-795, Georg Thieme, Leipzig (1926).
41. Pearl, D. C., and McDermott, Jr., W. V., *Proc. Soc. Exptl. Biol. Med.,* **97,** 440-443 (1958).
42. Ratner, S., *Advances in Enzymol.,* **15,** p. 319-387 (1954).
43. ———, in *Methods in Enzymology* (S. P. Colewick and N. O. Kaplan, eds.), **2,** p. 356-367, Academic Press, New York (1955).
44. ———, Anslow, W. P., Jr., and Petrack, B., *J. Biol. Chem.,* **204,** 115-125 (1953).
45. ———, and Pappas, A., *J. Biol. Chem.,* **179,** 1183-1198, 1199-1212 (1949).
46. ———, and Petrack, B., *Arch. Biochem. and Biophys.,* **65,** 582-585 (1956).
47. ———, and ———, *J. Biol. Chem.,* **191,** 693-704 (1951).
48. ———, and ———, *J. Biol. Chem.,* **200,** 161-174 (1953).
49. ———, ———, and Rochovansky, O., *J. Biol. Chem.,* **204,** 95-113 (1953).
50. Reichard, P., and Lagerkwist, U., *Acta Chem. Scand.,* **7,** 1207-1217 (1953).
51. Roth, R., *Bull. museum natl. hist. nat. (Paris),* **11,** 99-109 (1939).
52. Seaman, G. R., *J. Protozool.,* **1,** 207-210 (1954).
53. Smith, H. W., *J. Biol. Chem.,* **88,** 97-130 (1930).
54. Smith, L. H., Jr., and Reichard, P., *Acta Chem. Scand.,* **10,** 1024-1034 (1956).
55. Taylor, C. A., and Kollros, J. J., *Anat. Record,* **94,** 7-24 (1946).
56. Walker, J. B., *Proc. Natl. Acad. Sci. U. S.,* **38,** 561-566 (1952).
57. ———, and Myers, J., *J. Biol. Chem.,* **203,** 143-152 (1953).
58. Weed, L. W., and Wilson, D. W., *J. Biol. Chem.,* **207,** 439-442 (1954).
59. Witter, R. F., and Mink, W., *J. Biophys. Biochem. Cytol.,* **4,** 73-82 (1958).
60. Wright, L. D., Valentik, K. A., Spicer, D. S., Huff, J. W., and Skeggs, H. R., *Proc. Soc. Exptl. Biol. Med.,* **75,** 293-297 (1950).

# PATTERNS OF NITROGEN EXCRETION IN DEVELOPING CHICK EMBRYOS

## Robert E. Eakin and James R. Fisher

*Department of Chemistry, The University of Texas, Austin*

BIOCHEMISTS INTERESTED in phenomena associated with development naturally would like to be able to present graphic illustrations of chemical differentiation which would appear as dramatic as pictures showing morphological changes. It is their desire thus to document and emphasize the concept that higher organisms in their embryonic development must recapitulate chemical patterns characteristic of their evolutionary ancestry. That such a concept should be true is a corollary to the accepted thesis that in living organisms chemical function cannot be divorced from biological structure.

Until just recently, relatively little has been known about changes in metabolic patterns occurring during development; hence there have been practically no examples suitable for illustrating this basic postulate. In early studies on metabolism in adult organisms most of the information obtained resulted from the investigation of excretion products under a variety of experimental conditions. In the case of embryonic development one finds an analogous situation, and of the few studies available for illustrating metabolic recapitulation during development the one usually chosen was the pattern of nitrogen excretion by chick embryos, based upon Needham's classical experiments reported in 1926 (3, 4).

One of the many graphs used by Needham in his treatise on *Chemical Embryology* (5) to present his data can give the impression that a chick embryo passes through three distinct stages during which ammonia, urea, and uric acid in turn are the dominant products of excretion. This one particular presentation of his experimental findings has been interpreted as showing that the developing embryo passes successively through an ammoniotelic type of metabolism, found in aquatic animals, then through a ureotelic phase characteristic of amphibia, and finally becomes uricotelic with a pattern of excretion characteristic of those terrestrial animals which develop inside a cleidoic egg. This presentation has been very widely publicized, at least among biochemists, because Ernest Baldwin has reprinted the graph in his extremely popular textbooks of biochemistry and has emphasized such an interpretation (1).

In our laboratories, Dr. Fisher was interested in studying possible mechanisms that might be involved when biochemical differentiation takes place during development. His first objective was to find a suitable system for studying the reasons for the apparent appearance or disappearance of different enzymes and the marked quantitative changes in enzymatic capacities. Because the classical nitrogen excretion pattern appeared to show such dramatic changes, he selected this system as a model one for further study.

As a byproduct of this study, certain data were obtained which raised some questions concerning the *interpretations* which have been placed upon earlier work (2). It is the purpose of this report to take up in turn these questions concerning the excretion of ammonia, urea, and uric acid by incubated chicken eggs during the first eleven days and to discuss what conclusions are warranted concerning the extent to which embryos recapitulate chemical mechanisms of more primitive vertebrates—that is, chemically, in their nitrogen excretion, do chick embryos pass through fish-like and amphibian-like stages?

## AMMONIA "EXCRETION"

Needham's plot indicates that there is an ammonia-excreting stage which reaches a peak on the fourth day. At this time the nitrogen excretion of the embryo might be comparable to that observed in aquatic animals, who, having an aqueous environment, need not be concerned with the toxicity of ammonia. The basic pathways of ammonia formation in these types of life may be summarized as follows:

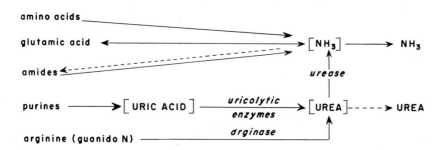

Data on the ammonia concentrations and contents of the allantoic fluid and the entire egg (Table 1) show:

TABLE 1

Ammonia in Allantoic Fluid and Entire Egg Contents during Development

| Incubation time (days) | Concentration of $NH_3$ ($\mu$g. N/ml.) | | Total $NH_3$ content ($\mu$g. N) | |
|---|---|---|---|---|
| | Allantois | Entire egg | Allantois | Entire egg |
| 0 | — | 8 | — | 350 |
| 2 | — | 10 | — | 450 |
| 4 | — | 11 | — | 500 |
| 5 | 8 | 10 | 3 | 450 |
| 6 | 8 | 8 | 10 | 350 |
| 7 | 8 | 8 | 19 | 350 |
| 8 | 6 | 9 | 21 | 400 |
| 9 | 6 | 8 | 28 | 350 |
| 10 | 10 | 7 | 55 | 300 |
| 11 | 10 | 9 | 64 | 400 |

(a) that the ammonia concentration of the entire egg remains relatively constant over the 11-day period studied;

(b) that the concentration of ammonia in the allantoic fluid is the same as in the entire egg;

(c) that the increase in total ammonia in the allantois merely reflects the increase in volume of the allantoic fluid;

(d) that the apparent abrupt drop in ammonia "excretion" (4) when the allantoic ammonia is plotted on the basis of dry weight of embryo is only a reflection of the weight increase of the developing embryo; and

(e) that any graphic indication of a peak in ammonia "excretion" is not warranted.

When exogenous ammonia is supplied by injecting ammonium sulfate into the yolk (Table 2) (a) the ammonia becomes uniformly distributed between the yolk and the allantois, and (b) the added ammonia disappears by the end of the second day (presumably because of an equilibrium between an "ammonia pool" and glutamic acid and other "ammonia carriers").

Urease, an enzyme needed for the complete degradation of purines and arginine, is commonly present among aquatic invertebrates; but no urease activity could be demonstrated in chick embryos even when measurements were made as early as the blastodisc stage of unincubated eggs.

Thus, it appears that there is *not* an active ammonia secretion mechanism during the early period, and that the small concentrations of ammonia ob-

TABLE 2

FATE OF AMMONIA INJECTED AS AMMONIUM SULFATE
(1.0 mg. N) INTO 6-DAY-OLD DEVELOPING EGGS

| Incubation time (days) | Concentration $NH_3$ ($\mu$g. N/ml.) | | | |
|---|---|---|---|---|
| | Allantois | | Yolk | |
| | $NH_3$ injected | $H_2O$ injected | $NH_3$ injected | $H_2O$ injected |
| 6 | — | 8 | — | 7 |
| 6.2 | 31 | 11 | 29 | — |
| 6.4 | 23 | 9 | 17 | — |
| 6.6 | 10 | 7 | 21 | 8 |
| 8 | 7 | 6 | 8 | 14 |
| 9 | 8 | 6 | 9 | 12 |
| 10 | 10 | 10 | 10 | — |
| 11 | 8 | 10 | 6 | — |

served in the yolk and allantois are in equilibrium with a metabolic "pool" of ammonia in the developing embryo.

## UREA EXCRETION

The characteristic pattern of nitrogen catabolism in ureotelic vertebrates (most adult amphibia and certain fish—the elasmobranchii) can be generalized as follows:

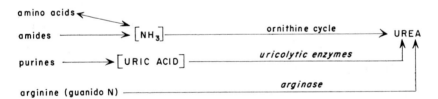

Typical urea values during development (Table 3) show that the urea picture differs from that of ammonia in two respects:

(a) the concentration of urea in the allantois does not remain constant but steadily increases; and

(b) the concentration of urea in the allantoic fluid is not the same as in the yolk but is several times greater.

The yolk does contain, however, about one-half of the total urea which has been formed. This yolk content might be due to a primary exchange between vessels of the yolk sac and the yolk or might arise secondarily by

## TABLE 3
### Urea in Allantoic Fluid and in Entire Egg Contents during Development

| Incubation time (days) | Concentration of urea ($\mu$g. N/ml.) | | Total urea content ($\mu$g. N) | |
|---|---|---|---|---|
| | Allantois | Entire egg | Allantois | Entire egg |
| 0 | — | 2 | — | 100 |
| 3 | — | 4 | — | 200 |
| 5 | 10 | 3 | 2 | 150 |
| 7 | 16 | 8 | 37 | 350 |
| 9 | 26 | 6 | 112 | 250 |
| 11 | 31 | 9 | 198 | 400 |

a "leakage" from the allantois. When the yolk urea content is built up by injection of an exogenous source of urea, it is observed that over a 5-day period half of the added urea has become concentrated in the allantoic fluid, a shift indicating some active excretory mechanism by the kidney (the mesonephros at this period of development) (Table 4).

## TABLE 4
### Fate of Urea (1.0 mg. N) Injected into 6-day-old Developing Eggs

| Incubation time (days) | Concentration of urea in the allantois ($\mu$g. N/ml.) | | Total amount of urea in the allantois ($\mu$g. N) | |
|---|---|---|---|---|
| | Urea injected | Water injected | Urea injected | Water injected |
| 5 | — | 10 | — | 2 |
| 6 | — | 11 | — | 14 |
| 6.2 | 50 | — | 70 | — |
| 7 | 84 | 16 | 190 | 37 |
| 8 | 77 | 21 | 250 | 69 |
| 9 | 94 | 26 | 400 | 112 |
| 10 | 93 | 29 | 490 | 154 |
| 11 | 92 | 31 | 580 | 198 |

However, is the mechanism of urea excretion here comparable to that observed in typical ureotelic organisms? Needham, Brachet, and Brown, in 1935, carried out the critical experiments to prove that the answer is "NO" (6). This they did by demonstrating in all stages from the second day of incubation onward that:

(a) the chick embryo did *not* produce urea from ammonia or amino acid nitrogen by the ornithine cycle;

(b) that the chick does not have the uric-acid-cleaving enzymes possessed by fish and amphibia and which are needed for conversion of purines to urea; but

(c) that all the urea produced arises from the action of a single enzyme, arginase, upon arginine.

At that time they also showed that the arginase activity *per unit weight* of chick embryo decreases dramatically from the second to the 12th day, then remains constant until hatching. (However, due to the rate of growth of the developing embryo, the total arginase activity of the embryo is increasing.)

A further difference between chick embryos and ureotelic animals (at least in the adult stage) is that in the latter the urea-forming enzymes are localized in the liver and kidney. Measurements of the distribution of arginase showed that no such localization occurs in developing chicks—an additional reason for questioning the existence during development of typical ureotelic mechanisms (5).

## Uric Acid Excretion

Terrestrial animals which develop in enclosed eggs—birds, many reptiles, as well as invertebrates—must have as their primary nitrogen waste product an osmotically inert substance. This they accomplish by excreting most of their nitrogen as uric acid (the increase in acidity of the allantoic fluid during the last half of the incubating period causing the precipitation of uric acid).

Uricotelic

amino acids ↘
$[NH_3]$ ↔ [glutamic acid, glutamine, aspartic acid, glycine] —purinogenic enzymes→ [nucleotides]
                                                                                  |
                                                                              nucleases
purines ——————————————————————————— ↓ xanthine oxidase → URIC ACID

arginine (guanido N) ——————————————— arginase → UREA

Quantitative studies on uric acid excretion have been carried out by a number of investigators, and widely varying values have been reported. Our data indicate that during the fifth day uric acid begins to accumulate in the allantois, with the *concentration* ($\mu$g./ml. of fluid) increasing each day at an arithmetical rate (Table 5).

## TABLE 5

FATE OF URIC ACID (0.5 mg. N) INJECTED INTO 6-DAY-OLD DEVELOPING EGGS

| Incubation time (days) | Concentration of uric acid in the allantois ($\mu$g. N/ml.) | | Total amount of uric acid in the allantois ($\mu$g. N) | |
|---|---|---|---|---|
| | Water injected | Uric acid injected | Water injected | Uric acid injected |
| 5   | 7   | —   | 2     | —     |
| 6   | 37  | —   | 48    | —     |
| 6.2 | 45  | 74  | 65    | 104   |
| 6.4 | 52  | 81  | 86    | 130   |
| 6.6 | 54  | 99  | 105   | 180   |
| 7   | 73  | 90  | 168   | 210   |
| 8   | 105 | 117 | 340   | 390   |
| 9   | 151 | 185 | 650   | 790   |
| 10  | 179 | 247 | 950   | 1,300 |
| 11  | 210 | 295 | 1,320 | 1,860 |

One can create a "peak of uric acid excretion" somewhere between the 9th and 11th days by plotting the uric acid content of the allantois vs. the dry weight of the embryo. Such a plot does little more than reflect the time when the rate of growth of the embryo starts to outpace the rate of uric acid production.

A comparison of control embryos with others into whose yolk exogenous uric acid was injected on the sixth day shows that this uric acid can be accounted for quantitatively by the increase observed in the allantoic sac. This is in contrast to the behavior of added ammonia (no concentration in allantoic fluid) or added urea (only partial recovery).

When does the embryo first exhibit a true uricotelic pattern of excretion? At the fifth day the ratio of urea:uric acid nitrogen in the very small volume of allantoic fluid parallels the ratio of these substances in the whole yolk. During the next twenty-four hours, however, a urea:uric acid pattern is established (Table 6) which continues until a day or two before hatching (5).

Circulation in the glomeruli of the mesonephric tubules does not commence until about the fifth day, and the developing kidney is not actively excreting until that time. Hence, it is likely that a uricotelic mechanism is operative in the embryo as soon as the kidney becomes a functional excretory organ. This conclusion (that from the beginning uric acid is the primary excretory product of the kidney) is in agreement with observations (a) that there is probably never any active secretion of ammonia, and

## TABLE 6
### Urea:Uric Acid Excretion Ratios in Developing Chicks

| Incubation time (days) | Allantoic fluid | |
|---|---|---|
| | Volume (ml.) | Urea N / Uric acid N |
| 0  | 0   | (1.3)[1] |
| 5  | 0.3 | 1.4 |
| 6  | 1.3 | .25 |
| 7  | 2.3 | .21 |
| 8  | 3.2 | .20 |
| 9  | 4.3 | .18 |
| 10 | 5.3 | .17 |
| 11 | 6.2 | .15 |

[1] Ratio of urea N:uric acid N in whole yolk.

(b) that all the urea, even in the early stages, originates entirely from arginine and not from other amino nitrogen.

On the basis of individual amino acids, arginine constitutes 8 per cent of yolk proteins; thus this one amino acid, because of its four nitrogen atoms, contributes one-quarter of the total nitrogen furnished by the yolk. If, as seems to be the case, practically all the guanido nitrogen is excreted as urea (very little creatinine is excreted), then one should expect this to be reflected in a rather constant urea:uric acid ratio of the magnitude observed, for the same egg proteins are supplying both the arginine from which urea is derived and the other amino groups which are excreted as uric acid. The fact that in the excreta of hens the urea to uric acid nitrogen ratio is approximately 1:16 rather than the 1:6 observed in embryos is most likely a consequence of the difference in arginine content of grain proteins (ca. 2 per cent) as compared with the much higher level in egg proteins. If this is the explanation, then one may say that the nitrogen excretion in the embryo at the time the kidney becomes functional is essentially the pattern observed in adult hens.

## Conclusions

It was disappointing to find that measurements of nitrogen excretion products of chick embryos do not present the clear-cut textbook example of chemical ontogeny which was originally postulated. However, that this one particular system does not provide us with a good illustration does not constitute a reason for altering the basic concepts of chemical recapitulation. There are several reasons why measuring nitrogenous end-products of chick embryos is a poor choice for studying metabolic development.

Although the concentrations are low, the *total amounts* of ammonia, urea, and uric acid in the yolk material are sufficiently large to make it impossible to measure any changes in content of these substances which could be produced by the small amounts of tissue in early embryos. Hence, no significant "excretion" values can be obtained until the fifth day, when nitrogenous products begin to accumulate in the allantoic fluid in concentrations high enough to stand out against the background levels of the yolk. But chick embryos develop at such a rapid rate that by the fifth day they have progressed through many of the critical early phases of development, and there is no practical method for slowing the rate of development, as can be done in amphibia.

Furthermore, the analysis of excretory products is a primitive method of studying metabolism compared to the many ingenious techniques available to modern biochemists, and though it served as a tool in the past, such analyses failed to give an adequate picture of metabolism. For example, early studies of this type failed to disclose the dynamic aspects of metabolism, and for a time established an unjustified distinction between anabolic and catabolic mechanisms.

The rate and complexity of development in chick embryos presents a drawback to their use as a model system for chemical studies which would not be encountered to the same degree in more primitive organisms; but there is, nevertheless, the challenge to have information concerning the most complex types of organisms. Modern techniques applicable to micro quantities of tissue can establish some of the more subtle changes occurring in basic metabolic patterns during development as the result of suppression or expression of chemical capacities. As such information accumulates, it should be possible to present more definitive pictures of chemical recapitulation in developing chick embryos.

## REFERENCES

1. Baldwin, E., *Dynamic Aspects of Biochemistry,* 3rd ed., Chapt. 12, Cambridge University Press, Cambridge, England (1957).
2. Fisher, J. R., and Eakin, R. E., *J. Embryol. Exptl. Morphol.,* **5,** 215-224 (1957).
3. Needham, J., *Brit. J. Exptl. Biol.,* **4,** 114-144 (1926).
4. ———, *Brit. J. Exptl. Biol.,* **4,** 145-154 (1926).
5. ———, *Chemical Embryology,* Vol. 2, Part III, Sec. 9, Cambridge University Press, Cambridge, England (1931).
6. ———, Brachet, Jean, and Brown, Robert K., *J. Exptl. Biol.,* **12,** 321-335 (1935).

## DISCUSSION

DR. NASON: I would like to raise a point with respect to Dr. Brown's paper. Are you considering trying to get effects with thyroxin in this system? As you know, thyroxin plays an intimate part in metamorphosis in amphibians and is probably a causal initiating factor in this process. It will be interesting to see if thyroxin would actually directly induce the formation of urea cycle enzymes in this particular system, and also to correlate this with the morphological stage of development in metamorphosis. This particular approach might also be extended to mammalian organisms. It would be also interesting to see if the initiation of the urea cycle in mammalian organisms is dependent upon thyroxin and if its continued functioning is in any way tied up with thyroxin production. Possibly the use of thyroidectomized animals would be of use in this kind of project.

DR. BROWN: Some of this work has already been started by Munro in England on *Rana temporaria*. He finds that the level of arginase increases in thyroid induced metamorphosis. The same thing has been found in this country by Dolphin and Frieden. We were interested primarily in setting up quantitative assay procedures for these enzymes, and we spent considerable time just trying to make sure that we could account for all the activity of these enzymes which is actually in the liver. We originally made several extractions of the liver in our early work to demonstrate that we were actually getting all the enzymes present in the liver out for our test system, and we decided at that time to use normal tadpoles in order to simplify the situation. I say this because in metamorphosis induced with thyroid-active agents there are certain complexities which arise. For example, it's been pointed out by several people, including Allen of the University of California, that each tissue has its particular threshold level for thyroxin and with a given concentration of thyroxin, very different results may occur in different tissues. Hence in thyroxin-induced metamorphosis certain anomalies may arise unless one takes particular care. In this connection, I might recall to you that the lower lip of the tadpole *Rana* contains three rows of labial teeth. One of these young tadpoles was placed in a triiodothyronine solution for two days and at the end of these two days it had lost all three rows of its labial teeth. This loss of the teeth corresponds to a stage change of X to XX; hence, by looking at the oral region of this animal you would expect to find it in stage XX, but if you looked at the hind legs, the animal had only progressed a stage or two by this criterion. Hence, considerable anomalies and gross deformities may be introduced. Consequently, when you try to use the criteria of morphological changes as an index for metamorphosis when you are inducing metamorphosis with these artificial agents, you run into this type of problem. However, by studying the levels of enzymes of the urea cycle in the liver, it seems that we have come up with a biochemical index of a degree of metamorphosis. However, we would like to carry out some experiments with thyroxin, triiodothyronine and desaminotetraiodothyronine, which, as shown by Kharasch, is 13,000 times as potent as thyroxin, but we really only recently began to employ these techniques and no results can be reported at this time. It will probably be rather complex trying to correlate our findings with external morphological features, but we hope some interesting developments will arise.

Dr. P. Cohen: I would like to speak about the synthesis of urea-cycle enzymes in the mammalian embryo. This work is being carried out by Dr. Kennan, who has done a number of assays for urea cycle enzymes in liver of rat embryos of various ages starting with the earliest possible age in which there was enough liver to work with. There was a progressive, almost linear, rise in ornithine transcarbamylase and arginase activity. However, there was a very dramatic increase in the carbamyl phosphate synthetase which occurred just a few hours after birth. The interesting thing is that the carbamyl phosphate synthetase takes off very dramatically in the postpartum period. We also tried to relate these findings to the kinds of things discussed by Dr. Melvin Cohn earlier this afternoon. We made a number of efforts, by the injection of arginine, ammonium chloride and similar compounds into the embryonic sac, to induce or alter the level of activity of the urea cycle enzymes. We found no consistent reproducible effect on the level of any of these enzymes by this particular type of treatment. However, we still feel that this system may respond to such influences and we are actively pursuing the problem at this time. It is interesting that the urea cycle in these animals is relatively late in appearing, and that the cycle really only becomes significantly operative in the postpartum period. We have obtained similar results with the pig embryo.

Dr. Weber: I would like to report similar experiments which we have done on the amphibian *Xenopus laevis*. We studied the proteolytic enzyme cathepsin in the tail where we find a decreasing specific activity during most of the growth. However, shortly before metamorphosis sets in there is a sudden change in that the specific activity of the enzyme increases progressively. And this continues throughout metamorphosis. Dr. Fruton from Yale University suggested to us the possibility that thyroxin might be uncoupling oxidative phosphorylation in this particular system. This uncoupling of oxidative phosphorylation results in a shift in the metabolism of this tissue and there is a consequent increase in the proteolytic activity. However, this point remains to be examined.

Dr. Brown: I think it is fairly clear that the action of thyroxin, although I will have to test this point, is not on the enzyme itself, but rather that thyroxin's effect is mediated in some manner which is not yet clear.

Dr. Popp: I would like to ask if some of these enzymes concerned in urea formation are found in the placenta? It seems that this would be a logical place to find these enzymes and that the organism might well find it expedient to concentrate its urea formation in this area.

Dr. P. Cohen: We have been pursuing this line of research, for the placenta probably does represent functionally a type of embryonic liver. It really provides the embryo with the same anatomical and physiological functions that a liver would in an adult animal. In the placenta Dr. Kennan has been able to dissect from the pig and rat, there is no evidence for the presence of any of the urea cycle enzymes. However, there appears to exist an inverse relationship between glutamine synthetase and glutamic dehydrogenase activity at certain stages of development of the placenta and liver. But there is no evidence of the existence of a urea cycle. However, we know, embryologically speaking, that the placenta is a rapidly growing and differentiating organ and by the time we can get at it, is is probably indeed highly differentiated to serve other functions. Dr. Kennan is pursuing this problem using pig embryos and placental tissues of known age.

Dr. Hadorn: I would like to comment on the question of the effect of thyroxin. Perhaps you might like to try administering thyroxin to some species of urodeles which cannot respond to thyroxin. Then you could have a very high thyroxin level without actually inducing metamorphosis. It would be interesting to see then if the metabolism of the liver—precisely, its urea cycle enzymes—are effected in this case.

Dr. Brown: We have obtained the species *Necturus maculosus* from our local area in Wisconsin and assayed its liver for the urea cycle enzyme. We found that all the enzymes were present in these livers even though they had not metamorphosed and in the adult stage remain as gilled forms. Carbamyl phosphate synthetase activity was low. You cannot make this species undergo metamorphosis even with the administration of thyroxin. We would like to extend this type of investigation however. I would like to refer now to the comment which Dr. Weber made on the work with *Xenopus*. This particular amphibian apparently has returned to the water for its adult life and, as Munro has pointed out, this particular amphibian excretes about 80% of its nitrogen in the form of ammonia. However, those Anura that spend a considerable part of their time on the land excrete almost 90% of their total nitrogen as urea. This raises the very interesting question as to what control the environment may have over the urea cycle in these livers. Certainly an organism which is in large part aquatic can afford to excrete large quantities of ammonia because they have a large reservoir in which to dilute out this toxic substance. Consequently, we would also like to take a look at the urea cycle enzymes of the aquatic species, *Xenopus*.

# IMMUNOCHEMICAL ANALYSIS OF DEVELOPMENT

## James D. Ebert

*Department of Embryology, Carnegie Institution of Washington*
*Baltimore, Maryland*

I TAKE as my point of departure the prescient article, "Immunological Studies in Embryology and Genetics," in which nearly a decade ago Irwin (37) looked forward to the time when the convergence of embryology, genetics, and immunology with biochemistry would enable a concerted attack to be made on the problems of development. The purpose here is to assess the extent to which a fruitful combination of these disciplines has been realized, to give some notion of current progress toward awareness of common conceptual problems. No effort will be made to be exhaustive in coverage; in the following remarks I shall wherever possible refer to my recently completed treatise on "The Acquisition of Biological Specificity" (25). The discussion will be based on the following premises, for which the pertinent evidence has been examined previously (25):

(1) The concept of tissue specificity extends to the molecular level. Although tissue specificity may, in part, reflect differences in associations and numbers of macromolecules which are alike in all tissues, we recognize by immunochemical and physiological techniques molecular types characteristic of each tissue. (2) The difference between two individuals is based on more than the difference in the sum of their tissue-specific proteins, or other antigenic molecules (or what Loeb (43) called a mosaic basis of individuality); it rests on the existence in each individual (or in all individuals of identical genotype, e.g. identical twins and members of a highly inbred strain) of specific molecules not restricted to certain tissues but common to all or almost all its parts. Loeb spoke of these characteristic molecules as part of the 'individuality differential'; we designate them as *isoantigens*. A third, but less secure, assumption states that similarly there are antigens common to all members of a species, the species-specific antigens.

Emphasis will be placed on the tissue-specific antigens and isoantigens. Among the questions worthy of consideration are the following: Are isoantigenicity and tissue specificity properties only of the adult, that is, do they represent end-points of differentiation, or are there tissue-specific molecules and isoantigens characteristic of the embryo, to be modified or replaced during development by related molecules of the adult? Do the isoantigens and tissue-specific molecules develop independently? Do they differ in site and time of formation? Are their synthetic mechanisms based on a single operative principle? When and how does the embryo or immature animal acquire the ability to recognize that a molecular type or tissue to which it is exposed is foreign, and just how foreign it is? Does every cell of the body have the ability to distinguish like and unlike, or is this ability the property of only a few cell types (25)? Aggregates of intermixed embryonic chick and mouse cells of the same type, e.g., chick and mouse chondrogenic cells, reconstruct a chimeric tissue consisting of interspersed cells of the two species. Aggregates of intermixed embryonic cells of different types become associated according to type, chick nephrogenic cells in one grouping, mouse chondrogenic cells in another (50). At one stage in development, then, a type of tissue-specificity dominates over species-specificity; cells aggregate histotypically regardless of generic origin.

During the past ten years there has been little confluence of the main streams of inquiry in the fields of immunoembryology and immunogenetics; questions like those just raised have been considered infrequently. Instead, research has been conducted at the level of chemical description: the enumeration of genes and antigens, and the establishment of a chemical inventory of tissue-specific proteins. It has been emphasized earlier (24, 25) that the application of immunological tools to embryological problems can be profitable. Among the more fruitful analyses have been those in which the synthesis of a specific protein has been studied under conditions in which the findings could be correlated with the results of experiments employing the methods of experimental morphology and information gained from studies of embryonic metabolism. In support of this point of view, I need cite only the recent studies of cardiac and skeletal myogenesis (18, 20, 27, 36) and of the formation of the lens proteins (42, 63, 65). Only in a few investigations, however, do we find points of contact with cell physiology and genetics. Several of these examples are off the beaten track; often the evidence is fragmentary, or at least deficient at one or more points. I have elected to emphasize them, however, because they may provoke more critical experimentation and thought and because they illustrate the convergence of several disciplines.

### Nature of the Antigenic Stimulus in Homograft Immunity

Although progress has been made in understanding the genetics of transplantable tumors (60), few laboratories have had the facilities, know-how and, it would appear, the desire to engage in the kind of long-term study that is urgently demanded if we are to understand the relationships of genes, isoantigens, and immune processes in the transplantation of normal tissues (25). Although the information available is incomplete, the inauguration of research directed at the chemical nature of transplantation antigens is encouraging, for, if the following arguments prove correct, even in part, biochemistry, genetics, and immunology converge at a common meeting ground, deoxyribose nucleoprotein.

From the evidence marshalled from a number of sources (4, 6, 25, 48) we may conclude that antigens issue from a homograft and reach the regional lymph nodes, presumably via the afferent lymphatics. Although whole cells may be transferred, it is generally believed that the antigenic message must be in the form of cellular fragments or ultramicroscopic particles. Medawar (48) has emphasized that antigens issue constantly from homografts, citing several lines of evidence, perhaps the most convincing being experiments in which successful homografts in tolerant animals are destroyed by the administration of lymph cells of normal mice. Clearly these experiments suggest that the homograft is constantly giving forth antigenic matter to which its tolerant host is unable to respond. Billingham, Brent, and Medawar (6; see also Brent, 11, and Medawar, 48) have advanced the hypothesis that the homograft antigens are deoxyribose (DNA) nucleoproteins. The evidence is far from complete; the findings cannot be evaluated fully. Yet the conclusions are so far-reaching that we are compelled to consider the evidence thus far advanced (25). The experiments consist in the injection of cells and tissue extracts from adult mice of one strain into adult mice of a second strain, after which the latter are challenged with homografts of the donor type. The investigators ask simply, is the inoculum capable of accelerating destruction of the grafts? In other words, what cells and fractions of cells contain the isoantigens? Whole blood is active, but its immunizing power rests in the leucocytes, being absent from red cells, platelets, and plasma. The antigens are stated to be present in all the nucleated cells of a single individual or the members of an inbred strain. Nucleated cells need be neither viable nor intact in order to function as antigens. Mouse spleen cells suspended in a variety of media can be disintegrated by ultrasonic irradiation without abolishing their capacity to immunize against skin grafts of the donor strain. Conventional fractionation studies lead Medawar and his group to conclude that cytoplasmic fractions are totally inactive, whereas nuclei are active. When cells are disintegrated in media in which nucleoproteins are insoluble, then the active fraction is found in the heavy "chromatin" fraction. The active fraction is destroyed or severely damaged by lyophilization, by repeated freezing and thawing, or by heating to 48.5°C for 20 minutes. It is insoluble in physiological salt solutions, but soluble in water. It is leached out of nuclei by extraction in 2 $M$ NaCl. All of these findings are consistent with the idea that the active principle is DNA nucleoprotein. The active fraction is destroyed by incubation with deoxyribonuclease (DNAase). Ribonuclease is ineffective in modifying its antigenicity. Since its vulnerability to DNAase is not as great as might be expected if DNA were the sole active factor, Medawar suggests as a possibility that it is DNA coated with protein and thus partly protected against the enzyme. The authors are careful to stress the preliminary nature of their findings. For example, in successive publications Medawar (48) and Brent (11) state that (1) although a protein fraction is removed by tryptic digestion, the antigenicity of the nuclear preparation is not destroyed; and (2) the antigens are susceptible to destruction by trypsin under certain conditions. Some uncertainty also exists concerning the possible involvement of lipoproteins, and polysaccharides or mucoproteins (71, 72). Although the authors appear to have shown that there are unstable transplantation antigens of nuclear origin, they have not proved that all isoantigens are nuclear. Inherent in most investigations on transplantable tumor homografts and skin homografts has been the assumption that the immune processes were essentially the same. For example, Billingham et al. (8) promptly applied to skin homografting the techniques and principles of adoptively acquired immunity established by Mitchison (49), using tumors. It is clear, however, that immunity to tumors may be evoked with a variety of nonnuclear agents including cell-free extracts (39, 48), red cell stromata (3), and microsome fractions (51). These conflicts are puzzling at the moment, but out of such contradictions may come clues to basic differences between resistance to tumors and resistance to normal tissues. One can only agree with Medawar that the ideas he has advanced are verifiable, by existing methods or by modifications of these methods (25).

Often the question is raised as to the bearing of the findings regarding the nature of the transplantation antigens on the concept of nuclear differentiation. For the moment, at least, the answer can be straightforward: the studies of Medawar and his associates neither affirm nor deny the possible significance of nuclear differentiation. The data do clearly suggest that nuclei contain some antigenic determinants distinct from those in the cytoplasm. Moreover, it is indicated that some of the nuclear antigens lack organ-specificity. Further comments do not appear to be justified at this time. Questions of the immunochemical specificity of DNA and other nuclear constituents are reviewed by Schechtman and Nishihara (56).

As I have pointed out recently, ideas from those studying the biochemistry of cell particulates, microbial genetics, and the use of clonal lines of tissue cells have yet to make their full impact on investigators working with immunochemical tools (25). To a considerable degree the converse is also true. For example, specific antibodies might be of unusual value in the study of nuclear, even of chromosomal, differentiation. As we shall see, they afford a precise, sensitive technique for following the synthesis of a specific protein in vitro by a nuclear or microsomal preparation. Are the enzymes or the structural constituents of the mitochondria (if, indeed, one can distinguish between enzymatic and structural proteins) tissue-specific? Are there clear antigenic differences in mitochondria and microsomes? Using specific antibodies, the embryologist should be able to perform "defect" experiments at the intracellular level. The direct immunochemical attack on the nucleus has been delayed not only by the poor antigenic properties of the nucleic acids and nucleoproteins, but also by difficulties in measuring their interactions with antibodies using conventional methods. Newer methods of preparing nucleic acids, and technical modifications, including the use of labeled antibodies, promise to overcome these obstacles (Blix, Iland, and Stacey, 9; Phillips, Brown, and Plescia, 54).

IMMUNOCHEMICAL STUDIES WITH CYTOPLASMIC PARTICULATES

In view of the complexity of microsomes and mitochondria, it may be necessary to isolate some of their components before specific antibodies can be developed (Albright, 1; Weiler, 67). However, carefully employed, even complex antisera may be valuable tools. For example, Weiler prepared antibodies against microsomes and mitochondria of normal rat liver. The antisera were powerful, but complete cross-reactivity in reciprocal tests with microsomes and mitochondria was the rule. Hepatomata induced by feeding dimethylaminoazobenzol failed to react with antibodies against the normal particulates. Weiler used the fluorescent antibody technique to great advantage: normal liver reacted with the labeled antibodies; hepatomata failed to react, indicating a loss of specific reactive groups. Even more striking was his observation that, when experimental animals were sacrificed at intervals shorter than necessary to produce the tumor, small groups of nonreactive cells were observed consistently among the normal cells, a clear instance of a change in immunological reactivity before overt malignancy.

*A Serum Albumin Precursor in Cytoplasmic Particles*

Focusing attention on the combination of biochemical and immunochemical techniques are the recent experiments which have led Peters (53) to conclude that he has discovered a precursor of serum albumin in a protein bound to cytoplasmic particles. Earlier Peters (52) had suggested the existence of an intermediate in the biosynthesis of serum albumin, basing his argument on studies of the kinetics of the appearance of $C^{14}$ in soluble serum albumin in liver slices incubated with labeled amino acids. In studying this "intermediate," Peters has found that the treatment of cytoplasmic particles of chicken liver with deoxycholate will release a nondialyzable substance which precipitates with antibodies against chicken serum albumin. Isotopic experiments indicate that the particulate (especially microsomal) serum albumin is the precursor of the soluble form. Peters has concluded, therefore, that the lag observed in the appearance of $C^{14}$ in serum albumin represents not the time required for synthesis via an intermediate stage, but rather the time required for release of the albumin from the bound form.

Although Peters' interpretation is consistent with the results presented, the experiments are not wholly convincing. Nevertheless, the experimental approach appears to hold considerable promise. Failure to find labeled carbon in the albumin from particulate fractions of kidney, spleen, heart, and intestinal mucosa after injection of labeled amino acids in vivo supports earlier conclusions that the liver is the sole source of serum albumin and lends confidence in the specificity of the technique used for isolation of this protein. Moreover, the treatment of all deoxycholate fractions with an unrelated antigen and its homologous antibody to remove complement and nonspecific coprecipitating proteins is a critical procedure designed to insure the specificity of the albumin recovered by precipitation with its antiserum. In contrast to the care exhibited in the immunochemical aspects of the study is an apparent lack of precision in defining the fractionation procedures and results. It is of more than casual interest to observe that yields of mitochondrial and microsomal fractions were not reproducible when the common homogenization and centrifugation methods were applied to liver slices which had been incubated in vitro for varying periods of time.

*Transfer by Nucleoprotein Fractions of Antibody-Forming Capacity*

With the exception of one recent study (32) in which the degeneration of Krebs mouse ascites tumor cells in the chick embryo in the latter quarter of the incubation period is attributed to the formation of a "natural antibody," there is no significant evidence of the ability of an embryo to form circulating antibody. Ebert (25) has concluded that the best available data indicate that the ability to produce an antibody can be detected first at about the time of birth. If one allows for species variation, and for the fact that only a few species have been studied and with few antigens, the tentative conclusion may be stated that the antibody-forming system reaches a state of functional development within a few days after birth, after which it matures rapidly. Although there is no direct correlation between the production of circulating antibody and transplantation immunity, there is a striking parallelism in the time course of their development. In one species, the sheep, homograft resistance has been demonstrated prior to birth (57); in several other forms, like the chicken and mouse, transplantation immunity is manifested within a few days after birth, whereas in still other animals, e.g., the rat, the immune mechanisms are not present for at least 14 days after birth (25). Sterzl and Hrubesova (61) have capitalized on the immunologic unresponsiveness of young rabbits in carrying out experiments of uncommon interest. These experiments indicate that it may be possible to transfer antibody-forming capacity into unimmunized recipients by injecting nucleoprotein fractions taken from animals capable of an immunologic response. The factual cornerstones on which the experiments are based are the following. (1) Neither complete nor incomplete antibodies can be demonstrated in cell suspensions of spleen or bone marrow prepared within 24 to 48 hours after the injection of an effective bacterial antigen into adult rabbits. Normally the first antibody response is detected at about 72 hours after injecting the antigen. (2) The five-day-old rabbit does not produce antibody. (3) When spleen or bone marrow cells of adult animals injected two days previously with bacterial antigen are transferred to unimmunized recipients, either adult or five-day-old rabbits, the recipients produce antibody within 24 hours. What elements in the spleen and bone marrow transfer antibody? Is antibody transferred in a bound form, bound to nucleoprotein, i.e., is there a stage in antibody formation when antibody is bound to nucleoprotein particles, as Peters believes to be the case for serum albumin? Sterzl and Hrubesova have been able to demonstrate that the capacity to form antibody can be transferred in partially purified nucleoprotein fractions. Heat-inactivated *S. paratyphi* were injected into a series of adult rabbits; after 24 and 48 hours, the spleens of these rabbits were removed and fractionated in one of two ways. From one group, nuclear and granular fractions were prepared; from a second group, DNA-protein and RNA-protein fractions were obtained, care being taken to specify methods of extraction and definition. These fractions were then injected into unimmunized animals, both adult and five-day-old rabbits, a total of 123 animals being used in this phase of the experiment. Among the controls were five-day-old rabbits injected with antigen alone. These did not produce antibody in the 27 days during which they were observed. In contrast, five-day-old rabbits injected with DNA-protein or RNA-protein from immunized animals produced antibody within two days. What is the significance of these findings? One thinks at once of the recent reports that enzyme-inducing capacity can be transferred by RNA. For example, Kramer and Straub (40) have stated that noninduced cells of *B. cereus* (pretreated with RNAase and 1 $M$ NaCl) will form penicillinase upon the addition of RNA from induced cells, amino acids, purines, and pyrimidines, in the absence of the inducer. RNAase destroys the ability of RNA to mediate the induction. The possibility certainly must be considered that Sterzl and Hrubesova have transferred an "activated" or "induced" nucleoprotein. On the other hand, perhaps a simpler explanation can be advanced. Is an intermediate, nucleic-acid–bound antibody transferred? Sterzl and Hrubesova advance tenuous evidence against this idea: in preliminary studies they have found that the amino acid composition of the protein in the nucleoprotein transferred differs from that of rabbit gamma globulin. The latter findings are too sketchy to be convincing. Meanwhile, there is much in these experiments and those of Peters that is relevant to the problems of development. It should be profitable to inquire whether in the formation of a tissue-specific protein, say myosin, it is possible to demonstrate a nucleo-myosin intermediate. Going one step further, it does not tax the imagination too greatly to inquire whether it might be possible to study inductive phenomena in experiments similar to those of Peters. Woerdeman (70) has presented a preliminary account of uncommon interest: a mixture in vitro of an extract of head ectoderm with an extract of young optic vesicle resulted in the formation of lens antigens not produced by either tissue in isolation. Woerdeman implies that precursors from the two sources may react in vitro to form an immunochemically characteristic adult protein. One might inquire, then, whether it would be worthwhile to examine appropriate mixtures of granular fractions from inducing and reacting components, in view of recent experiments (35, 55) which suggest that labeled proteins may be released from granular components in vitro.

### An Immune Reaction: "Graft-against-the-Host"

Incidental observations in studies of the destruction of kidney homografts led Dempster (15, 16) and Simonsen (58) to question whether a graft containing cellular elements normally active in immune processes might react against its host. Inasmuch as the "graft-against-the-host reaction" has become an important link in our understanding of graft-host relationships, I wish to consider briefly a series of recent independent findings in which cells and tissues involved in immune processes have been implanted into hosts which are incapable of an immune reaction. For a more extensive discussion of the phenomenon the reader is referred to several recent publications (5, 14, 23, 59).

#### In Bone Marrow Transplantation to Irradiated Animals

Jacobson (38) demonstrated that the lethal effects of whole body irradiation could be prevented by shielding the animal's spleen. He advanced several possible hypotheses to account for the success of spleen-shielding experiments, including humoral stimulation by a subcellular agent and colonization of damaged areas by cells of the shielded tissue. In extending these experiments, it was observed that several species of animals subjected to lethal doses of irradiation could be kept alive by intravenous injection of a normal bone marrow cell suspension after exposure (12, 44). Shortly thereafter, it was shown that blood-forming tissues are specific in causing recovery; hematopoietic cells are required. When the hematopoietic tissues of the host have been destroyed, "reseeding" by blood-forming cells is a requirement for recovery. A slight possibility remains that recovery can be obtained with cell-free extracts of hematopoietic tissues (28). The evidence strongly favors the involvement of cells, but the suggestion that cell-free extracts may be effective when the host's blood-forming tissues are not completely destroyed cannot be discounted altogether. The proof that bone marrow and blood cells in the animals receiving bone marrow injections are derived from donor cells was obtained by a variety of techniques, applying the basic principle that homologous or heterologous donor cells can be identified by virtue of differences between them and the corresponding host cells. The methods which have been employed are summarized by Congdon (12); perhaps the most elegant demonstration is that of Ford et al. (29, 30), who showed that dividing cells in bone marrow and lymph nodes of surviving host mice had chromosomes cytologically resembling the homologous donor cells which carried a chromosome marker.

When homologous or heterologous animals are employed as donors (rather than isologous animals), recovery and survival for 30 days is followed by the death of many of the hosts (13). The response is characteristic; the lymph nodes and to some extent the spleen undergo a granulomatous reaction, resulting in fibrosis and fibrinoid necrosis. This delayed effect is interpreted by most investigators as an immune response, in which the host tissues are affected adversely by reactions of the donor cells. Makinodan (47) believes this explanation to be an oversimplification, and, in fact, favors the view that death is due to a recovery of the radiosensitive antibody-producing cells of the host, which now react against the proliferating foreign bone marrow. Recognition is made also of the possibility that both reactions may occur. He bases his conclusions on several lines of evidence; in irradiated mice injected with rat bone marrow, he was unable to detect rat serum proteins, although he was employing a technique capable of detecting a concentration greater than 12 $\mu$g. of protein per milliliter of serum. He concludes that the graft is not producing antibodies against the host. Makinodan's argument holds only for circulating antibody, but ordinarily in heterologous combinations circulating antibodies can be detected. When normal isologous nucleated bone marrow cells were injected into irradiated mice along with rat red blood cells, circulating agglutinins against the rat cells were not detected. This finding, indicating inability of the bone marrow cells to produce antibody, is at variance with other recent reports. The time of the delayed bone marrow reaction and the severity of the reaction are dependent on the dosage of irradiation (faster recovery of immune reactions, following lower dosage, leading to a more severe reaction). Finally, he points out that if antibody-producing cells were derived from the transplant one should find an increase in death with an increase in bone marrow dose. This appears not to be the case. Makinodan has dealt largely with heterologous combinations in which the host reaction appears to predominate. Loutit (45) suggests that in homologous combinations the graft reaction takes precedence. Perhaps the ability of antibody-producing cells to recognize "foreignness" has quantitative implications.

## In Newborn Mice

Acquired tolerance may be induced by the intravenous injection of newborn mice with cell suspensions, the success of the operations being dependent on the strains and combinations of mice employed (5). Billingham and Brent found that a number of mice of strain A, made tolerant by injecting adult CBA spleen cells, died within 2 to 3 months after birth. Characteristically, the lymphoid tissues of these animals were aberrant:

the lymph nodes were lacking, the spleens fibrotic and deficient in Malpighian corpuscles; even Peyer's patches were missing. The findings are interpreted as a graft-against-the-host reaction.

*In the Chick Embryo and Larval Salamander*

When I prepared my notes for this Symposium several weeks ago, I planned to consider my own recent investigations briefly, and only as a dramatic illustration of the graft-against-the-host reaction. The direction of the discussion following my paper, and the orientation of the succeeding papers by Billingham and Harris, especially the latter's approving reference to my views on the uptake and transfer of macromolecules by embryonic cells, brought home to me that I had not made clear the relation between the latter hypothesis and the graft-versus-host reaction. Therefore, I propose to review the development of my current investigations; in doing so I shall draw heavily from previously published accounts (19, 21, 23, 25) and from research completed and only now being prepared for publication in detail (14, 26).

The transplantation of adult cells to an adult organism of different genetic makeup commonly results in an immune reaction, leading to the death of the implant. When adult cells are transplanted to the early embryo, in advance of the critical period for the induction of tolerance, another consequence is observed, a stimulation in the growth of the homologous tissue (68), exemplified by the increased growth of the embryonic chick spleen following transplantation of a fragment of adult chicken spleen to the chorio-allantoic membrane of the nine-day host embryo. This reaction is quantitatively organ-specific, i.e., of eleven donor tissues studied only three are capable of evoking the response; they are, in order of decreasing effectiveness: spleen > thymus > liver. The reaction is class-specific but not species-specific: adult mammalian spleen being ineffective, whereas fragments of spleen of other avian species do stimulate the growth of the chick spleen, although never to the same extent as does the homologous tissue. The age of the donor animal is a critical consideration, the effective factor or factors being absent from the embryonic spleen during the first two-thirds of the developmental period, and making their appearance first at about 14 days of incubation. Hence grafts of embryonic spleen from donors younger than 14 days are ineffective. The effective factors are accumulated slowly, reaching the "adult" level by about 6 weeks posthatching. The reaction proceeds without modification in the absence of the hypophysis (23, 26). Chorio-allantoic transplants made on the ninth day of incubation to embryos hypophysectomized by partial decapitation affect the homologous organ of the host equally as well as do grafts made to normal embryos.

Studies using radioactively labeled transplants showed that when the proteins of the donor tissue, e.g., kidney or spleen, were radioactively tagged by injecting into the intact animal $S^{35}$-methionine in quantities sufficient to insure maximal labeling of the tissue proteins without radiation damage, labeled materials from the graft were localized predominantly in the homologous organ of the host. The hypothesis was advanced that macromolecular components released by the graft were localized selectively in the homologous organ; it was held likely that, although these macromolecules were not incorporated directly into the host proteins, the amino acids and peptides into which they were fragmented before incorporation never entered the general amino acid pool. The results of this phase of the investigation, which embraces cytologic and autoradiographic studies and analyses of deoxyribonucleic acid in stimulated tissues, have shown that the stimulation does not result from a massive transfer of intact cells (14, 22, 25). On the other hand, they do not rule out the transfer of a seed population of viable cells, a possibility made attractive by the rapid development of other studies demonstrating the successful transplantation of bone marrow and other sources of hematoblasts (22). Several considerations argue against the importance of the transfer of viable cells in mediating the growth stimulation, for example: the observed difference in effectiveness between embryonic and adult grafts, whereas such a difference has not been observed in hematoblastic seeding; the absence of plasma cells, characteristic of the adult, in the affected host tissues; and the experiments of others, e.g., Andres (2), which indicate the stimulatory effectiveness of frozen, thawed triturated tissues. In fact, devitalization often has been shown to improve specificity and stimulatory effectiveness. Nevertheless, it cannot be stated that the requirement for intact cells was ruled out completely. Recent findings require a re-evaluation of the earlier conclusions (14, 25, 59).

In a small, but important, proportion of rapidly growing stimulated spleens, a distinct, characteristic change is noted beginning on or after the eighteenth day of incubation. This significant departure from the normal has been observed most frequently in cases in which the transplanted spleen was implanted somewhat earlier than usual, i.e., on the seventh rather than the ninth day of incubation. The cellular nature of this phenomenon and the conditions resulting in its expression lead DeLanney and

Ebert to postulate that shortly before hatching the transplanted adult spleen produces an immune response directed against its host. This hypothesis will be treated more fully in a later section.

### Localization of Injected Subcellular Fractions

It has been suggested that not only are devitalized tissues effective in promoting homologous organ growth, but also that in some instances such devitalized breis show an augmented specificity and stimulatory effect. In an effort to define more precisely the nature of the substances transported from graft to host tissue and the manner in which they act, the following experiments have been carried out: tissues of adult chickens were labeled, as in earlier experiments, by the injection of methionine labeled with sulfur[35]. These organs were stored in the frozen state, after which they were homogenized, and a small amount of the homogenate was injected directly into the chorio-allantoic vessels of the nine-day embryo. The results of these experiments, in which labeled kidney and spleen homogenates were injected on either the ninth or tenth day of incubation, with recovery after 24 hours, show that the exposure of the embryo to a very small quantity of tissue homogenate for only 24 hours results in a differential incorporation of the label in the homologous organ. The differential is not of the same order as that obtained with living transplants, acting over several days, but it is consistent and reproducible. The next step was the fractionation of the homogenate following standard procedures for the separation of nuclear, mitochondrial, microsomal, and supernatant fractions. The findings show clearly a selective localization only of the microsomal and supernatant fractions, and not of the nuclear and mitochondrial fractions. The properties of the microsomal and supernatant fractions which determine the specificity or which are essential to the localization are nondialyzable, and are destroyed by heating at 80°C for 5 minutes, and by ashing. The possible implications of the findings are apparent to all those who are familiar with the importance of the interaction of the enzymes and ribonucleic acid of the supernatant fraction and the ribonucleoprotein of the microsomes in the incorporation of amino acids and synthesis of protein. Yet, as the author has warned, the experiments have at least one major shortcoming. They deal only with one aspect of the problem, namely, *predominant localization*. They do not yet contribute to an understanding of why the homologous organ is stimulated. The small quantities of subcellular fractions tolerated by the embryo do not result in a significant growth stimulation. It appears that a continuing supply of injected materials may be necessary in order to bridge the gap between the transplantation and injection experiments. This objective has not yet been achieved.

*Serial Transfer of Growth-Promoting Activity*

The following experiments were initiated to determine the effect of grafts of adult chicken spleen made to the coelom of the host chick embryo before the appearance of the host spleen. The spleen of the host was stimulated greatly by the tenth day, and the question was raised as to whether such a stimulated 10-day embryonic spleen might not in itself be an effective donor. Normally the 10-day embryonic spleen is ineffective. This idea has been tested (23, 26). Grafts of spleen from embryonic, juvenile, and adult donors were made to the coelom of the four-day chick embryo, following a method described by Dossel (17). The findings confirmed earlier findings with respect to the importance of the age of the donor tissue. Utilizing the data obtained as a base line, grafts were made to a number of four-day embryos, which were permitted to develop until the tenth day, when the host spleens were harvested and used as the donor tissue for a second set of grafts, and so on, the experiment being terminated after the fourth set of grafts was recovered. The growth-promoting activity was not diluted by serial passage, but was maintained relatively constant, considering the qualitative nature of the experimental approach. Perhaps the simplest explanation of these observations is the colonization of the host spleen by whole cells from the donor, cells capable of reproduction. As in the earlier studies employing the chorio-allantoic technique, histological analysis of the host spleen denies the transfer of large populations of adult cells; in fact, the evidence is more clear, in view of the primitive architecture of the host spleen in the period covered by the experiment. Again, however, the transfer of a seed population of graft cells cannot be ruled out. Why should adult cells, which do not proliferate in the graft, proliferate extensively in the splenic primordium, whereas embryonic spleen cells proliferate when grafted but do not stimulate the host spleen? Simple "colonization" does not offer a complete explanation. The following possibilities have been considered (25): "(1) A predominant localization in the homologous host tissue of specific particulate or macromolecular constituents of the host; the evidence shows that such a predominant localization is possible under certain experimental conditions, but does not prove that the mechanism applies in transplantation. But what of the serial transfer of growth-promoting activity? The most economical explanation is a transfer of viable cells, but the transfer of subcellular particles is not ruled out. Moreover, the evidence does not permit the exclusion of a combination of colonization and cellular transfer of materials. (2) The growth of the homologous organ is the result of a colonization by living adult cells which, as a result of constituents transferred *from the host,* now acquire new properties which enable them to grow rapidly. (3) Circulating constituents of the host stimulate the graft in situ to activities directed toward the host. Thus, we are led to consider further the idea of continuous and reciprocal exchange between the tissues of the organism, both in the embryo and adult, and between the organism and an implanted tissue." This idea is not new; it is inherent in the theoretical discussions by Weiss (68, 69) and is stated explicitly in a stimulating paper by Medawar (48). Further, striking evidence of the impact on the graft of host "antigens" may be considered next.

Chorio-allantoic grafts of adult spleen result in the growth of the homologous organ; when grafts are allowed to continue beyond the seventeenth day, however, a small but consistent fraction of the host spleens undergoes a striking degenerative change characterized by the destruction or modification of the vascular bed, and leading to a stasis in flow of granulocytes, their accumulation in cystic masses in the spleen, and eventual death of the embryo (14, 23). The cystic condition of the spleen occurs with a frequency approximating that of the death of chicks at or shortly after hatching. I shall not attempt to present the detailed cytologic evidence for this conclusion; it will be found in a paper to be published by DeLanney and Ebert (14). Recently we have observed that, when intracoelomic grafts of adult chicken spleen are permitted to remain in position for at least 8 to 9 days, approximately 30 per cent of them exert a profound effect on the vascular system, leading to death (26).

Simonsen (59) has shown that a splenic enlargement typically follows the intravenous injection of leucocytes or spleen cells of adult chickens into the 18-day chick embryo. He concludes that splenic enlargement results from colonization by donor cells of the antibody-forming series. Simonsen has been able to propagate the cells through nine consecutive passages without significant loss of activity. These two findings seem to be of a conclusive nature. Simonsen, however, believes that there is no difference between the reaction of the spleen of the 9-day chick embryo to a graft of homologous spleen and the reaction of the spleen of the 18-day embryo to an intravenous injection of homologous cells. He argues that the recipients have to be of an age in which tolerance of homologous cells can be induced and that donors must be old enough to form antibodies. On another occasion I have pointed out the differences in the results of the two approaches (25). Simonsen remarks (59, p. 498), "Preliminary evidence in support of this hypothesis can be seen in experiments whereby 11-day or 18-day embryos of White Leghorn stock were inoculated intravenously with spleen cells of 18-day Brown Leghorn embryos. The group injected on the eleventh embryonic day comprised 5 animals. They were all killed at 3 days after hatching and showed no anatomical changes. . . . The group injected on the eighteenth embryonic day totalled 31 animals, 14 of which were likewise killed at 3 or 4 days after hatching and found them perfectly normal." And (p. 499), "Transplantation of embryonic chicken spleen cells give rise to no splenic enlargement in embryonic chicken hosts. . . ." His findings are not consistent with previous reports that implantation of splenic fragments from embryos 14 days and older produced a significant weight increase by the eighteenth embryonic day. Another inconsistency is found in Simonsen's observation that leucocytes of other avian species, pigeon and turkey, did not cause splenic enlargement in the chick. It is difficult to reconcile this observation with the graft-against-the-host hypothesis.

Finally, DeLanney's observations on the salamander will be mentioned only briefly, since DeLanney himself may wish to comment more extensively. Grafts of adult salamander spleen have been made either into the coelom or into a pocket in the dorsal fin of the larval salamander, *Taricha torosa*. Although there are differences in the behavior of the transplanted tissues in the two sites, grafts in both sites lead not to growth stimulation but to suppression of the host spleen. DeLanney has taken as his working hypothesis a "graft-against-the-host" immune mechanism.

Taken as a group, the graft-versus-host reactions present a further demonstration of graft-host interaction, and lend further support to the idea of a continuous circulation of antigens in the embryo.

From one point of view, lymphoid tissues provide a highly sensitive technique for the detection of possible organ-specific molecules or particles. However, in view of the ultimate destruction of the host tissues, I have questioned whether the use of organs like spleen for further tests of organ-specific regulation of differentiation and growth is advisable. Although organ-specific agents other than those involved in the immune response may be involved, the overwhelming power of the immune mechanism may render the description of other systems difficult. Clearly, the experiments of Ebert and of Simonsen should be repeated using a highly inbred line of chickens, in which incompatibility should not be manifested. Positive evidence should be sought of growth-stimulating effects of noncellular organ preparations. It is clear that such organ preparations exhibit the properties enabling them to localize predominantly in the homologous tissue (23, 64, 66). Evidence may be cited of the efficacy of frozen, thawed, and other nonviable preparations (2, 62); yet most of it falls short of being wholly convincing. Experiments employing filters which preclude cellular transfer from graft to host are indicated.

The graft-versus-host reaction has forced abandonment of plans for attempting to introduce lymphatic tissues into agammaglobulinemic patients in efforts to restore immunological capacity. In fact, Good (31) has described in such patients effects similar to those observed in experimental animals.

### Uptake and Transfer of Macromolecules by Embryonic Cells

In conclusion, I wish to return briefly to a question raised earlier, that of the ability of embryonic cells to recognize "foreignness," to distinguish unlike from like. At this Symposium, much attention has been paid to the nature of embryonic cell membranes and to their role in morphogenetic movements as well as in the uptake of macromolecules. Phenomena like the predominant localization of injected microsomal and supernatant fractions in the homologous organ of the recipient lead us to inquire whether other embryonic cells exhibit selectivity of the same order, and whether selective uptake must always imply a "selective activity" of the cell membrane. We are indebted to Brambell, Hemmings, and their associates for an unusually penetrating analysis of just such a situation (10, 33, 34). Maternal circulating antibodies enter the fetal blood in the rabbit by way of the uterine lumen and the vitelline circulation of the yolk-sac splanchnopleur. Uptake from the uterine lumen is selective in the sense that, when mixtures of homologous and various heterologous antitoxins are injected into it between the twentieth and twenty-eighth days of pregnancy, antitoxin can be detected in the fetal sera after twenty-four hours in concentrations which vary according to the species of the donor, in the order rabbit, man, guinea pig, dog, horse, and cow.

Where does the selectivity reside? Is entry selective? Does selection take place by exclusion of the heterologous molecules at the cell surface? Studies using iodinated rabbit and bovine gamma globulins have revealed that in the gravid uterus both rabbit and bovine globulins are almost entirely absorbed, a fact suggesting that these proteins are taken up by the fetal endoderm equally. Part of the protein, varying in magnitude with the species of origin, is passed on into the fetal circulation as protein, and the remainder is broken down. There is no evidence of differential treatment of homologous and heterologous protein after entry to the fetal circulation. Here, then, is an example in which selective "uptake" of macromolecular constituents results from a selection brought about *within* or *upon leaving* the cell, following a primary nonselective entry. I cite this example because it is often overlooked despite the clarity and precision of the analysis, in which substantial progress has been made over a period of several years. It conveys an important lesson: no doubt the "port of entry," the cell membrane, is important, but it is not all-important. Students of development interested in problems of cell and tissue affinities must be concerned with the entire cell, not just its boundaries.

## REFERENCES

1. Albright, J. F., *A Morphological and Serological Investigation of Heart Sarcosomes: An Adjunct to the Study of Differentiation*, Dissertation, Library, Indiana Univ., Bloomington, Ind. (1956).
2. Andres, G., *J. Exptl. Zool.*, **130,** 221 (1955).
3. Barrett, M. K., and Hansen, W. H., *Cancer Research*, **13,** 269 (1953).
4. Billingham, R. E., in *Cellular Mechanisms in Differentiation and Growth* (D. Rudnick, ed.), Princeton Univ. Press, Princeton, N. J. (1956).
5. ———, and Brent, L., *Transplant. Bull.*, **4,** 67 (1957).
6. ———, ———, and Medawar, P. B., *Phil. Trans. Roy. Soc. London, Ser. B*, **239,** 357 (1956).
7. ———, ———, and ———, *Nature*, **178,** 514 (1956).
8. ———, ———, ———, and Sparrow, E. M., *Proc. Roy. Soc. (London) B*, **143,** 43 (1954).
9. Blix, V., Iland, C. N., and Stacey, M., *Brit. J. Exptl. Pathol.*, **35,** 241 (1954).
10. Brambell, F. W. R., *Cold Spring Harbor Symposia Quant. Biol.*, **19,** 71 (1954).
11. Brent, L., *Ann. N. Y. Acad. Sci.*, **69,** 804 (1957).
12. Congdon, C. C., *Symposium on Antibodies: Their Production and Mechanism of Action*, *J. Cellular Comp. Physiol.*, **50** (*Suppl. 1*), 1 (1957).
13. ———, and Urso, I. S., *Am. J. Pathol.*, **33,** 749 (1957).
14. DeLanney, L. E., and Ebert, J. D., *Contrib. to Embryol., Carnegie Inst. Wash.* **37** (in preparation).
15. Dempster, W. J., *Brit. Med. J.*, **2,** 1041 (1951).
16. ———, *Brit. J. Surg.*, **40,** 447 (1953).
17. Dossel, W. E., *Science*, **120,** 262 (1954).
18. Duffey, L. M., and Ebert, J. D., *J. Embryol. Exptl. Morphol.*, **5,** 324 (1957).
19. Ebert, J. D., *Physiol. Zool.*, **24,** 20 (1951).
20. ———, *Proc. Natl. Acad. Sci. U. S.*, **39,** 333 (1953).
21. ———, *Proc. Natl. Acad. Sci. U. S.*, **40,** 337 (1954).
22. ———, in *Aspects of Synthesis and Order in Growth* (D. Rudnick, ed.), Princeton Univ. Press, Princeton, N. J. (1955).
23. ———, *Carneg. Inst. Wash. Yearbook*, **56,** 299 (1957).
24. ———, *Symposium on Embryonic Nutritional Requirements and Utilization* (D. Rudnick, ed.) Univ. of Chicago Press, Chicago, Ill. (in press).
25. ———, The Acquisition of Biological Specificity, in *The Cell* (J. Brachet and A. Mirsky, eds.), Academic Press, New York (in press).
26. ———, Mun, A. M., and Coffman, C., *Contrib. to Embryol., Carneg. Inst. Wash.* **37,** (in preparation).

27. ———, Tolman, R. A., Mun, A. M., and Albright, J. F., *Ann. N. Y. Acad. Sci.*, **60,** 968 (1955).
28. Ellinger, F., *Science,* **126,** 1179 (1957).
29. Ford, C. E., Hamerton, J. L., Barnes, D. W. H., and Loutit, J. F., *Nature,* **177,** 452 (1956).
30. ———, Ilbery, P. T., and Loutit, J. F., *Symposium on Antibodies: Their Production and Mechanism of Action, J. Cellular Comp. Physiol.,* **50** (*Suppl. 1*), 109 (1957).
31. Good, R. A., *Ann. N. Y. Acad. Sci.,* **69,** 767 (1957).
32. Green, H., and Lorincz, A. L., *J. Exptl. Med.,* **106,** 111 (1957).
32a. Haskova, V., and Hrubesova, M., *Nature,* **182,** 61 (1958).
33. Hemmings, W. A., *Proc. Roy. Soc. (London) B,* **148,** 76 (1957).
34. ———, and Oakley, C. L., *Proc. Roy. Soc. (London) B,* **146,** 573 (1957).
35. Hendler, R. W., *J. Biol. Chem.,* **229,** 553 (1957).
36. Holtzer, H., Marshall, J. M., and Finck, H., *J. Biophys. Biochem. Cytol.,* **3,** 705 (1957).
37. Irwin, M. R., *Quart. Rev. Biol.,* **24,** 109 (1949).
38. Jacobson, L. O., *Cancer Research,* **12,** 315 (1952).
39. Kidd, J. G., *Cold Spring Harbor Symposia Quant. Biol.,* **11,** 94 (1946).
40. Kramer, M., and Straub, F. B., *Biochim. Biophys. Acta,* **21,** 401 (1956).
41. Langman, J., *Acta Morphol. Neerland. Scand.,* **1,** 1 (1956).
42. ———, Schalekamp, M. A. D. H., Kuyken, M. P. A., and Veen, R., *Acta Morphol. Neerland. Scand.,* **1,** 142 (1956).
43. Loeb, L., *The Biological Basis of Individuality,* C. C. Thomas, Springfield, Ill. (1945).
44. Lorenz, E., Uphoff, D., Reid, T. R., and Shelton, E., *J. Natl. Cancer Inst.,* **12,** 197 (1951).
45. Loutit, J. F., *Symposium on Antibodies: Their Production and Mechanism of Action, J. Cellular Comp. Physiol.,* **50** (*Suppl. 1*), 343 (1957).
46. MacDowell, E. C., Potter, J. S., Taylor, M. J., Ward, E. N., and Laanes, T., *Carnegie Inst. Wash. Yearbook,* **40,** 245 (1941).
47. Makinodan, T., *Symposium on Antibodies: Their Production and Mechanism of Action, J. Cellular Comp. Physiol.,* **50** (*Suppl. 1*), 343 (1957).
48. Medawar, P. B., *Ann. N. Y. Acad. Sci.,* **68,** 255 (1957).
48a. ———, *Nature,* **182,** 62 (1958).
49. Mitchison, N. A., *Proc. Roy. Soc. (London) B,* **142,** 72 (1954).
50. Moscona, A., *Proc. Natl. Acad. Sci. U. S.,* **43,** 184 (1957).
51. Novikoff, A. B., *Cancer Research,* **17,** 1010 (1957).
52. Peters, T., Jr., *J. Biol. Chem.,* **200,** 461 (1953).
53. ———, *J. Biol. Chem.,* **229,** 659 (1957).
54. Phillips, J. H., Braun, W., and Plescia, O., *Records Genetics Soc. Am.,* **26,** 389 (1957).
55. Rabinovitz, M., and Olson, M. E., *Federation Proc.,* **16,** 235 (1957).
56. Schechtman, A. M., and Nishihara, T., *Ann. N. Y. Acad. Sci.,* **60,** 1079 (1955).
57. Schinckel, P. G., and Ferguson, K. A., *Australian J. Biol. Sci.,* **6,** 533 (1953).
58. Simonsen, M., *Acta Pathol. Microbiol. Scand.,* **32,** 36 (1953).
59. Simonsen, M., *Acta Pathol. Microbiol. Scand.,* **40,** 480 (1957).
60. Snell, G. D., *Ann. N. Y. Acad. Sci.,* **69,** 555 (1957).
61. Sterzl, J., and Hrubesova, M., *J. Microbiol., Epidemiol. Immunobiol. (U.S.S.R.),* **28,** 503 (1957). (Russian Journal published in English translation by Pergamon Press Inc., New York City.)
62. Teir, H., Larmo, A., Alho, A., and Blomquist, K., *Exptl. Cell Research,* **13,** 147 (1957).
63. Ten Cate, G., and Van Doorenmaalen, W. J., *Koninkl. Ned. Akad. Wetenschap., Proc.,* **53,** 3 (1950).
64. Tumanishvili, G. D., Djandere, K. M., and Svanedze, H. K., *Doklady Akad. Nauk. S.S.S.R.,* **106,** 1107 (1956).
65. Van Doorenmaalen, W. J., *Histo-serologisch Onderzoek betreffende de Localisatie van*

*Lensantigenen in de Embryonale Kippenlens,* Dissertation, Univ. of Amsterdam, Netherlands (1957).
66. Walter, H., Allmann, D. W., and Mahler, H. R., *Science,* **124,** 1251 (1956).
67. Weiler, E. Z., *Naturforsch.,* **11b,** 31 (1956).
68. Weiss, P., *Yale J. Biol. Med.,* **19,** 235 (1947).
69. ———, and Kavanau, J. L., *J. Gen. Physiol.,* **41,** 1 (1957).
70. Woerdeman, M. W., in *Biological Specificity and Growth* (E. G. Butler, ed.), Princeton Univ. Press, Princeton, N. J. (1955).

# IMMUNO-HISTOLOGICAL LOCALIZATION OF ONE OR MORE TRANSITORY ANTIGENS IN THE OPTIC CUP AND LENS OF *RANA PIPIENS*[1]

George W. Nace[2] and William M. Clarke[3]
*Department of Zoology, Duke University, Durham, N. C.*

The conviction that complex molecules possessing antigenic properties are intimately involved in the causative mechanisms of embryogeny has led to an extensive application of serological techniques in this field, as has been repeatedly noted in this symposium (for reviews see 14, 15, 27, 33, 37, 38, 40). These studies have centered on the in-vitro description of antigenic changes during embryogeny, the effect of antisera on developmental processes (e.g., among papers published in 1957, 20, 28, 29, 31, 38, 39, 41), and to a lesser extent, the histological localization of antigens (6, 7, 23) by the recently developed fluorescent antibody techniques of Coons (10).

An embryological desideratum would be to discover the precursors of the biochemical end-products of differentiation. This is a difficult problem, because our major store of biochemical knowledge is based on investigations of differentiated tissues. Applications of biochemical techniques to embryology have usually dealt with the adult systems as they are found during development, although numerous major efforts have been made to break through this limitation (cf. 8).

The first step in solving such a problem would be the identification of potentially significant entities characterizing specific stages of the differentiation process. Some tentative identifications of this type have been made (3, 12, 13, 16, 24, 30, 32, 34) but are still incomplete. No attempts have yet been successful in establishing the causal interrelationships of these materials.

This report is part of a study (27) utilizing antisera directed against

---

[1] This investigation was supported by a research grant (RG-3555) of the National Institutes of Health, United States Public Health Service, Bethesda, Maryland, and by the Duke University Research Council, Durham, North Carolina.

[2] Present address: Department of Zoology, University of Michigan, Ann Arbor, Michigan.

[3] Present address: School of Medicine, University of North Carolina, Chapel Hill, North Carolina.

embryonic antigens. Its purpose is: (1) to establish a descriptive base-line of the antigenic changes occurring during the embryogeny of *Rana pipiens;* (2) to use the same antisera as immuno-histological reagents for the localization of the antigens in embryonic tissues; and (3) to probe the functions of these antigens with the same antisera (28, 29).

In the course of adapting the immuno-histological techniques of Coons to studies of *Rana pipiens* development, fluorescein-conjugated anti-tailbud supernatant sera were found to localize in certain structures of the embryo. Some of these observations have been described by Clarke (4). A most suggestive localization occurred in the optic structures of stage 19 and 20 embryos but not in stage 17 or 21 embryos.

## Materials and Methods

### Antigen Preparation

Embryos were collected following standard pituitary injection of the frog, *Rana pipiens,* obtained from Vermont and maintained in hibernation. Complex antigenic mixtures were prepared from hatched tailbud embryos (Shumway, stages 19-20) which were washed with and homogenized in 0.05 $M$ phosphate-buffered saline at $pH$ 7.4. These standard homogenates were centrifuged at 1500 g with final clearing at 10,000 to 14,000 g, and the supernatants, which contained 1.6-2.0 mg. of total N per ml., were brought to 1/10,000 merthiolate and stored at 0-4°C until used as test or injection antigen.

Antigens from other stages of development were similarly prepared and treated for use as controls.

### Antiserum Production and Utilization

The tailbud supernatants were injected into four American Chinchilla rabbits using the adjuvant technique of Freund and Bonanto (18) for reasons discussed elsewhere (27). Each rabbit received a single inoculation of 400-600 $\gamma$ total N under each scapula. Antisera were collected at various times following injection and were characterized by the antigen dilution interface precipitin reaction and by a modification of the Ouchterlony agar diffusion method. Antisera to mature ovarian oocyte (large oocyte) supernatants were prepared in a similar manner.

On the basis of the observation (9) that adsorption with lyophilized liver powder reduces non-specific localization of fluorescent sera, 15-20 ml. aliquots of selected antisera were treated with 2 additions of 200 mg. each of such a mouse liver powder. They were shaken at 20°C for 1 hour, stored overnight at 4°C and cleared by centrifugation following each addition.

The globulins of these adsorbed sera were precipitated at 50% saturation with ammonium sulfate and were then redissolved and dialyzed using buffered saline at $pH$ 7.0. These globulins were compared with the crude antisera by the precipitin and agar diffusion tests.

The adsorbed globulins were brought to 1% protein concentration and conjugated with fluorescein isocyanate[4] according to the protocol of Coons and Kaplan (9). Uncombined fluorescein was removed by prolonged dialysis against buffered saline, and denatured proteins were removed by centrifugation.

The conjugates were brought to 1/10,000 merthiolate, stored at 0-4°C and characterized by the interface precipitin and agar diffusion techniques.

*Fluorescence Technique*

Both frozen and paraffin-embedded embryos fixed in a variety of reagents were tested with conjugates. Little variation in fluorescence localization could be attributed to the procedure used, although the intensity of autofluorescence seemed to depend upon the fixative (4, 19). Because of morphological distortion in the frozen tissues, the data presented in this report were obtained with embryos fixed in Smith's solution, embedded in paraffin, and sectioned at $10\mu$ according to the technique of Davidson (11). It has now been found that little or no autofluorescence of frog eggs is seen following use of Carnoy's fixative (19). Sections were passed through a hydration series ending with buffered saline prior to treatment with conjugate; several drops of the conjugate were placed on the section, which was then incubated at 37°C for 30 minutes in a humidified chamber. The sections were washed free of excess antibody with several changes of buffered saline and were mounted in 9 parts of glycerol to 1 part of buffered saline.

These preparations were examined with a Leitz Ortholux fluorescence microscope equipped with a Phillip's high pressure mercury vapor lamp CS 150 and with optics and filters as prescribed by the Leitz company (21). Photographs were taken with a Leica 35 mm. camera attachment using Eastman Plus-X film and uniform 5-minute exposures.

[4] Our preparations of fluorescein amine were made by Dr. Irwin Green and were compared spectrophotometrically before and after conjugation with a sample kindly provided by Dr. Albert H. Coons, Harvard Medical School.

## Controls and Tests

Difficulty of interpretation resulting from autofluorescence of the embryonic tissue as well as localized fluorescence produced by normal rabbit conjugate necessitated the following series of controls.

(1) Sections treated with buffered saline in place of the fluorescent reagents. These frequently showed a very low intensity blue-grey autofluorescence.

(2) Sections treated with fluorescent dialysate from the conjugation procedure to test the adequacy of washing. Only autofluorescence was seen in such sections.

(3) Sections treated with normal rabbit conjugate.

(4) Unconjugated normal rabbit globulin followed by antiserum conjugate.

(5) Unconjugated antiserum globulin followed by antiserum conjugate.

Following the preliminary analyses, a study was made of approximately 100 series of 3 adjacent sections per series cut from approximately 75 tailbud embryos (Shumway, stages 19 and 20). The middle section of each series was stained with conjugated antiserum alone. Each adjacent section was stained as a control, usually with normal rabbit conjugate, and the unconjugated antiserum globulin followed by the antiserum conjugate. After photographs had been taken the cover slip of the experimental section was removed in buffered saline; the section was washed to remove glycerol and then stained with hematoxylin and eosin following routine procedures. These served as histological controls.

As this investigation was primarily concerned with the development of techniques, an extensive comparison with embryos at various stages of development was not made. Where such comparisons were made, a minimum of five series was run from each of the other stages.

## RESULTS AND CONCLUSIONS

### Serological Specificity

All sera produced by the four rabbits injected with tailbud supernatant gave positive precipitin reactions at antigen dilutions of 0.80-0.16 $\gamma$ total N per ml. This was true for the crude sera, their adsorbed globulins, and the conjugates. Red, rather than white light, facilitated reading the titers of the colored conjugates. Tests with supernatant preparations from other developmental stages showed minor differences in end-point titer. Because of the complex composition of the antigenic mixtures and the resultant complexity of their antisera, the significance of the heterologous titer differences cannot be adequately evaluated solely on the basis of these interface reactions.

All normal rabbit serum reagents were negative in interface precipitin reactions.

Agar diffusion analyses were run on the crude antisera, their globulin fractions, and the conjugated globulin fractions. Considerable variation was found between rabbits and between the various bleedings of each rabbit. Since the data obtained from sera used in this study are included in a more extensive report on the changing pattern of antigens during embryogeny, in which the causes of variation in agar diffusion analyses are discussed (30), reasons for these differences will not be detailed here.

The anti-tailbud supernatant serum obtained from rabbit #30, 3 months following the injection, showed 8 major lines with tailbud supernatant in agar diffusion tests, whereas the other anti-tailbud supernatant sera showed 4 to 6 major lines. All globulin fractions produced an increased number of lines as compared with the crude antisera. Again, rabbit #30 produced the greatest number of lines. No differences in the line patterns could be referred to the adsorption with mouse liver powder.

Rabbit #30 conjugate showed only two lines in agar diffusion. As the unconjugated globulin diluted to the same extent, viz., 1% protein, also showed only two lines, conjugation *per se* probably did not destroy the antiserum specificity. The other anti-tailbud supernatant conjugates, which were prepared first, did not produce any lines in agar diffusion. Nitrogen determination showed them to be at less than 1 per cent concentration, which could be attributed to loss by denaturation at the time of conjugation.

The sera which did not produce lines in agar diffusion showed the same fluorescence localization patterns as did rabbit #30 conjugate. This would suggest that concentrations of conjugate inadequate for line production in agar diffusion are adequate for combination with antigen in the tissue sections. However, though the precise relationships between the antigenic systems responsible for line production in agar diffusion and those responsible for the specific localizations are not known, it seemed best to emphasize rabbit #30 conjugate in these studies because of the greater intensity of the fluorescence obtained with this reagent.

In agar diffusion tests with large oocyte supernatant, the crude anti-tailbud supernatant serum produced by rabbit #30 showed 2 or 3 lines less than with the homologous supernatant. The conjugate produced only 2 lines with the oocyte supernatant. These 2 were identical, on the basis of line continuity, to the 2 lines shown with tailbud supernatant. Since a similarly prepared anti-large-oocyte supernatant serum conjugate did not show the eye localizations reported below, it must be assumed that the antigenic systems involved in these localizations were not those which produced the agar diffusion patterns. Again, this result constitutes evidence that the systems too dilute to produce lines, could, however, produce fluorescence localizations.

*Fluorescence Specificity*

The effects of denaturing agents, such as fixatives, on the specificity of a given antigen cannot be predicted a priori. Thus, these immuno-histological studies were begun on frozen sections. However, due to histological distortion of frozen sections and the similarity of fluorescence localization and intensity in both the frozen and paraffin sections, the latter were used for most of this study. Because of this resistance to modification by fixatives, it is uncertain whether the specific antigens which were detected in these localizations are protein in nature or of some other composition. However, in support of a contention that they are not necessarily proteins, recent data (1) have demonstrated that many of the antigens in large oocyte supernatant and tailbud supernatant are not destroyed by a variety of precipitating reagents. When returned to solution the materials thus precipitated, as well as their unprecipitated supernatants, produced agar diffusion patterns containing lines which could be equated with lines in the patterns of the original antigenic mixtures.

The routine controls and tests were listed above. In all cases the embryonic tissue sections fluoresced at a very low intensity with mouse-liver-adsorbed normal rabbit conjugate. Occasionally this fluorescence was localized, but never in the regions of antibody localization described below. More frequently, it was fairly evenly distributed over the entire section. Controls of the washing procedure indicate that inadequate washing was not the cause of this "non-specific" fluorescence. In all cases, careful examination of these control sections in conjunction with the experimental sections was mandatory. The difference was usually quite apparent, especially when comparisons were made with the sections stained first with unconjugated normal rabbit globulin followed by the antiserum conjugate or by the unconjugated antiserum globulin followed by the antiserum conjugate. Thus, the necessity of a multiplicity of controls is evident.

The available anti-large-oocyte supernatant conjugates did not show localizations on sections of tailbud embryos, although there was some

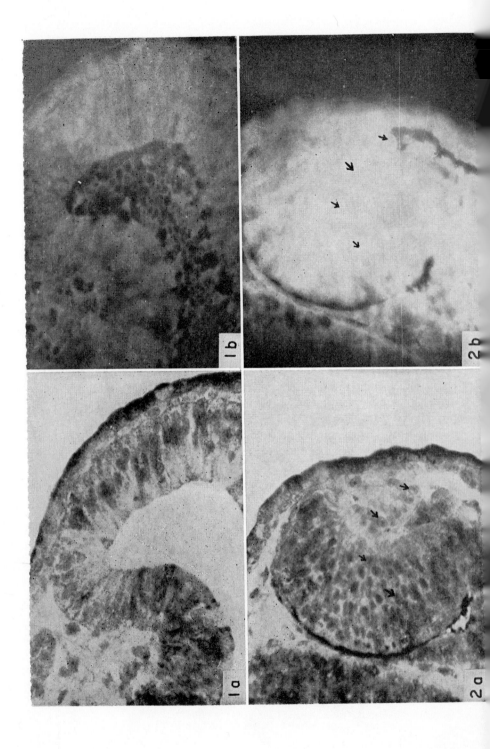

staining in the yolk mass which could not be clearly distinguished from that obtained with the "non-specific" normal rabbit conjugates. The anti-tailbud supernatant conjugates, hereafter designated simply as the "conjugate(s)," showed some staining of large oocytes but no consistent indication of localization.

## "Conjugate" Localization

Examination of tailbud sections revealed a differential distribution of the fluorescent staining with anti-tailbud supernatant conjugate. Specific localizations were observed in parts of the forebrain, certain portions of the notochord, the epidermis, and in the yolk mass (4). The details of these localizations will be presented in another publication.

An outstanding localization of the conjugate was seen in the eyes of stage 19 and 20 embryos. Figs. 2 and 3 are, respectively, transverse and parasagittal sections of these two stages. Figure "a" is, in each case, the hematoxylin and eosin control and "b" is the experimental. Frequently the optic regions, as well as the rest of the section stained with normal rabbit conjugate, showed a diffuse low intensity fluorescence, while the optic regions stained with the unconjugated antiserum followed by the conjugate showed a slightly more intense fluorescence in the region of the lens. This may represent a replacement of the unconjugated antibody molecules by conjugated molecules.

Comparison of Figs. 2a and 2b shows a localization of fluorescence in the sensory layer of the epidermis from which the lens placode developed, in the lens placode, in the optic cup, and in the optic stalk. Comparison of Figs. 3a and 3b indicates the occurrence of a phenomenon at stage 20 similar to that observed in stage 19 material. The lens vesicle fluoresced with a much higher intensity than did the margins of the cup, i.e., the upper portion of the figure, and the more central portion of the cup, i.e., the lower portion of the figure, fluoresced at an intermediate intensity.

Arrows on Figs. 2 and 3 indicate the position of nuclei. It is noteworthy that there is apparently a greater intensity of fluorescence in the nuclei of the cup than in the cytoplasm of the cup, whereas the reverse is true in the lens. The reality of this observation is somewhat obscured by the thickness of the sections ($10\mu$) and the photographic difficulty of demonstrating the contrast which was visible in the original material.

Fig. 1. Transverse section of *Rana pipiens* optic vesicle, stage 17, $\times$ 240. (a) Stained with hematoxylin and eosin. (b) Stained with fluorescent anti-tailbud supernatant serum.
Fig. 2. Transverse section of optic cup and lens, stage 19, $\times$ 240. (a) and (b) as for Fig. 1.

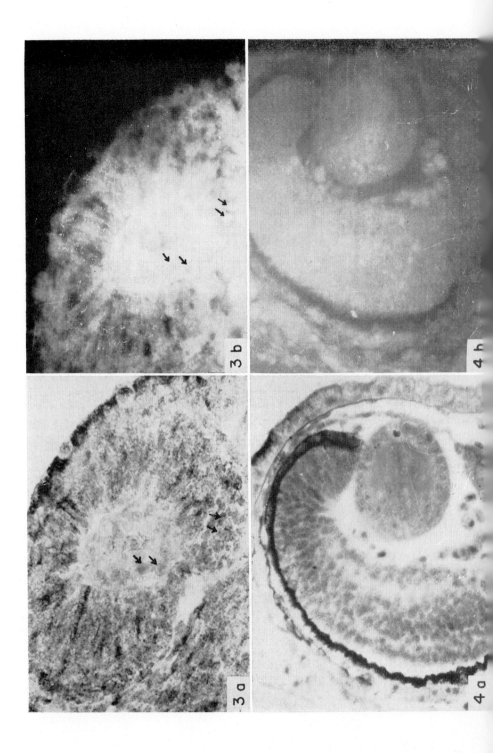

Because of the obvious relationship of the above observations to the body of literature on lens antigens in development, the optic region of a younger and an older stage were examined. Figs. 1a and 1b represent transverse sections of the optic vesicle when it has just come into contact with the presumptive lens region. Conjugate localization was not observed. The autofluorescence was of low intensity, as can be seen by the lack of contrast between the tissue and the background in Fig. 1b.

These data suggest the appearance of a lens antigen in the optic cup and lens concomitant with the morphological differentiation of the lens, and correspond with the in vitro observations of Ten Cate and Van Doorenmaalen (35), who demonstrated adult-type lens antigens in *Rana esculenta* at stages 19-20. However, Figs. 4a and 4b do not support the conclusion that adult-type lens antigen is being detected by this reagent. These figures show a transverse section of the cup and lens of a stage 21 embryo. It will be seen that the localization of fluorescence was no greater than in the stage 17 embryo. The autofluorescence was of very low intensity as evidenced by the difficulty in focusing the section and by the lack of contrast between the tissue and the background. The bright speck on the posterior margin of the cup was produced by a weakly fluorescent particle of denatured conjugate and suggests the intensity differences. The luminosity of the upper left quadrant of Fig. 4b represents incomplete absorption of the incident light and does not originate in the section itself.

These observations would suggest that the antigen(s) responsible for the elicitation of the localizing antibody or antibodies in these complex antisera is not of the adult lens type examined by Ten Cate and Van Doorenmaalen (35) and others, but is one characterizing this specific stage of development.

## Discussion

The extensive analytical background available on eye development (36) and the controversial reports by Guyer and Smith (22) have led to a particular interest in the lens antigens from both the descriptive (2, 3, 6, 7, 17, 35) and the functional viewpoints (17, 26, 40, 41). These in vitro and culture tests all confirm the appearance of adult lens antigen(s) concomitantly with the first cytological differentiation of this structure in the frog and chick.

---

Fig. 3. Parasagittal section of optic cup and lens, stage 20, × 240.
(a) and (b) as for Fig. 1.

Fig. 4. Transverse section of optic cup and lens, stage 21, × 240.
(a) and (b) as for Fig. 1.

Flickinger, Levi, and Smith (17) reported negative results in an attempt to produce abnormalities of the eye in frogs, rats, and chicks by using anti-lens sera in various types of in vivo situations. No other data pertinent to their culture observations on amphibians are available. Recently, however, similar tests have been made on the mammal by Wood (41). He successfully produced eye abnormalities by passive or active immunization of pregnant female rabbits and noted that careful and consistent preparation of the antigen is of utmost importance. More significantly, he observed that the time of exposure of the fetus to the presence of antibodies is critical. Lens anomalies will appear only if the antibody acts during the early part of the gestation period. This may be a possible explanation of the negative results obtained by Flickinger et al. (17). Their data are not explicit as to the antibody concentration or the time of exposure of the fetal rats. Also, variations between antigen preparations and variability of rabbit response to those antigens (30) may have been responsible for the discrepancies.

Further information comes from the success of Langman et al. (26) in obtaining highly specific and morphologically circumscribed eye defects by the use of anti-lens and anti-$\alpha$-crystallin sera in fluid culture of the optic structures of the chick. Their antibody concentrations were considerably higher than those used by Flickinger et al. Thus, it seems probable that Flickinger's sera were too dilute or were used in culture situations in which there was little opportunity for the antibodies actually to come into contact with the eye primordia.

Langman also noted that the appearance of eye anomalies was dependent on antiserum action upon the eye primordia at early developmental stages. At more advanced stages of differentiation (17-18 somites) the lens antigen is believed to be "withdrawn entirely from the effect of antibodies." Langman does not explain this expression, but apparently believes that at this time the antigen occupies a physiological space which is inaccessible to the antibody. He suggests further: "In early developmental stages a similar basic structure probably exists in the presumptive lens ectoderm and in the surrounding area. This basic structure is probably susceptible to the effects of anti-lens substances."

These reports, then, suggest at least a physiological difference between the early lens and later more well-differentiated lenses. Other evidence, found in the report of Burke et al. (3), suggests the existence of a transitory lens antigen in 120-130 hour chicks. Also, Beloff (2) has shown that "qualitative and quantitative differences in lens antigens can be demonstrated during the normal development of the chick's lens."

Thus, our immuno-histological demonstration of a transitory "lens" antigen is in harmony with previous work on the eye. Other examples of transitory antigens are to be found in reports from studies in vitro of various types, as cited in the introduction. One need not elaborate on the significance of such transitory materials. It has been repeatedly discussed by a number of authors (cf. 5, 12, 13, 26, 33, 35).

It remains now to speak of the nature of the transitory antigen(s) or combining group(s) (38) detected here, and to relate it to that of the adult lens antigens which also make their appearance at this time. Reference is made above to their possible chemical nature (1), but serological identification requires further characterization of the specificity of the antisera used and of the specificity of the localizations observed.

The frequent observation of staining by conjugated normal rabbit sera necessitated the adoption of rigid controls to eliminate the possibility that non-specific reagents in the antisera produced the specific localizations. It is believed the series of control reactions described above were adequate for this purpose.

It is difficult to define the specificity of the antisera used here on the basis of the tests in vitro. It has been amply demonstrated in precipitin and agar diffusion analyses that anti-tailbud supernatant sera will react with antigenic preparations from all stages of development and from the adult (16, 30). The antisera used in this particular study were shown to react with the large oocyte. In agar diffusion, the crude antisera produced a greater number of lines with homologous supernatants than with oocyte supernatants. However, following conjugation, only two lines were obtained and these were identical to two lines produced by anti-large-oocyte supernatant sera when tested against tailbud supernatant. Although the conjugated anti-large-oocyte supernatant sera did localize on tailbud sections, they did not localize on the optic structures. Thus, the antigen-antibody systems common to the large oocyte and tailbud preparations were not responsible for the localizations in the eye. As these were the only systems demonstrable in agar diffusion following conjugation, it must be concluded that the specific localizations were brought about by antibodies which had become too highly diluted to elicit the agar diffusion lines. For these reasons it is impossible to state with certainty whether or not any of the serological systems observed in the agar diffusion patterns, utilizing either crude sera or unconjugated globulins, were responsible for the localizations. The possibility exists that non-precipitating antibodies may be able to localize in the immuno-histological situation. One must depend, then, on other evidence to characterize more adequately the specificity of these fluorescent localization reactions.

Such evidence is found in an examination of the sections themselves. It is unfortunate that the antisera were not tested with adult lenses. However, the negative fluorescence localization in stage 21 embryos would suggest that such tests would have been negative. Similarly, the sequence of staining patterns observed between stages 17 and 21 would support a suggestion that antigenic changes of an unknown but specific nature were being detected, rather than "fortuitous" reactions of the type which produce fluorescence with a normal rabbit serum conjugate.

Relative to tissue specificity, it is uncertain whether or not other localizations observed in stage 19 and 20 embryos were produced by the antibodies which localized in the eye. Of these, the localizations in the epiphysis, ependyma, and cerebro-spinal fluid seem most likely to be related to those found in the optic structures; and attention is called to the heuristic interest of a possible serological relationship between the pineal body and the eye.

It is unfortunate that an intermediate stage taken at the time of thickening of the lens placode was not observed in order to indicate more clearly a possible relationship between the antigen(s) or combining group(s) detected here and the inductive relationships of the optic vesicle and presumptive lens ectoderm. The most pertinent observation is the weak fluorescence in the sensory layer of the ectoderm. Langman (25) reports pre-lens differentiation of this tissue is independent of optic vesicle contact.

It is of interest that this transitory antigen(s) of the optic structures is observed at the time the adult lens antigen(s) first becomes detectable by the most sensitive tests currently available. Its precise relationship to the adult lens antigen is obscure. But it is possible that an antiserum directed against a transitory substance might not react with the adult lens antigen though the adult antigen might produce antibodies capable of combining with the transitory substance(s). This would provide a basis for Langman's (26) suggestion of the existence of a "similar basic structure" in the young eye. The logic for this would depend upon the nature of the surface reactive groups of the antigens in question (38) and the variability of the rabbit response to those reactive groups (30). The transitory antigen(s) must have been a potent stimulus to antibody production, since it probably constituted only a limited portion of the injection antigen mixture. If so, on the basis of reactive group competition for antibody production sites, any adult-type antigenic groups in the mixture may not have elicited the development of corresponding antibodies, although these would be obtained by the injection of the more concentrated adult lens preparations.

It is necessary to note that Clayton's localization of lens antigen(s) in the eye of the developing mouse by utilizing fluorescent (6) and radioautographic (7) techniques seems to correspond to the localization observed here. She has not presented photographs of her fluorescent work and, since much later stages of development were analyzed, it does not seem appropriate at this time to make detailed comparisons of the two sets of results.

The suggestion of fluorescence localization in the nuclei and cytoplasm of the cup, but only in the cytoplasm of the lens, presents intriguing possibilities relative to problems of nuclear-cytoplasmic interaction. A number of possible interpretations concerning the control of transitory antigen production, the site of its synthesis, and the direction of its migration and thus its possible function come to mind, but the data are still too meager to attempt a resolution between them.

## Summary

1. The globulin fractions of antisera directed against supernatants of the tailbud (stages 19-20) of the frog *Rana pipiens* were conjugated with fluorescein isocyanate for use in adapting Coons' fluorescent antibody techniques to the analysis of development.

2. Such reagents were characterized by comparison with antisera against mature ovarian oocyte supernatants in a series of interface precipitin reactions and the Ouchterlony type of agar diffusion analysis.

3. Selected conjugates were applied to frozen and paraffin-embedded embryo sections, which were examined for specific localizations of the antibody.

4. Among others, localization was observed in the optic cup and lens of stage 19 and 20 embryos, but not in the optic vesicle or presumptive lens ectoderm of stage 17 or in any of the eye structures of stage 21.

5. Evidence is presented suggesting the localization of this reaction in the nuclei and cytoplasm of the optic cup, but only in the cytoplasm of the primordial lens.

6. The possible nature of this transitory antigen is discussed. Its relation to lens differentiation is viewed in the context of reports on the detection of adult lens antigens in similar and later stages of development.

## Addendum

After submission of this paper the interesting fluorescent antibody study by W. J. Van Doorenmaalen (*Acta Morphol. Néerl.-Scand.*, 2, 1 [1958]) using anti-adult chicken lens sera came to our attention. Instrumentation and techniques were very similar to those reported here. In contrast to earlier studies by other techniques, he could not demonstrate adult chicken lens antigens in embryos of 5 days of incubation or less. Although it is probable that technical difficulties were responsible for this failure, he also suggests the possibility of a transitory antigen.

## REFERENCES

1. Barber, M. L., M. A. thesis, Duke University (1958).
2. Beloff, R., Ph.D. thesis, Yale University (1957).
3. Burke, U., Sullivan, N. P., Peterson, H., and Weed, R., *J. Infectious Diseases,* **74,** 225 (1944).
4. Clarke, W. M., M.A. thesis, Duke University (1957).
5. Clayton, R. M., *J. Embryol. Exptl. Morphol.,* **1,** 25 (1953).
6. ———, *Nature,* **174,** 1059 (1954).
7. ———, and Feldman, M., *Experientia,* **11,** 29 (1955).
8. Cohn, M., in *Serological Approaches to Studies of Protein Structure and Metabolism* (W. H. Cole, ed.), p. 38, Rutgers Univ. Press, New Brunswick (1954).
9. Coons, A. H., and Kaplan, M. H., *J. Exptl. Med.,* **91,** 1 (1950).
10. ———, *Intern. Rev. Cytol.,* **5,** 1 (1956).
11. Davidson, M. H., *Turtox News,* **23,** 33 (1954).
12. Ebert, J. D., *Physiol. Zool.,* **24,** 20 (1951).
13. ———, *Ann. N. Y. Acad. Sci.,* **55,** 67 (1952).
14. ———, in *Aspects of Synthesis and Growth* (D. Rudnick, ed.), p. 69, Princeton Univ. Press, Princeton (1955).
15. ———, in *The Cell,* Vol. 1, (A. E. Mirsky and J. Brachet, eds.), *Academic Press,* New York (in press).
16. Flickinger, R. A., and Nace, G. W., *Exptl. Cell Research,* **3,** 393 (1952).
17. ———, Levi, E., and Smith, E. A., *Physiol. Zool.,* **28,** 79 (1955).
18. Freund, J., and Bonanto, M. V., *J. Immunol.,* **48,** 325 (1944).
19. Glass, L. E., Ph.D. thesis, Duke University (1958).
20. Gluecksohn-Waelsch, S., *J. Embryol. Exptl. Morphol.,* **5,** 83 (1957).
21. Gottschewski, G. H. M., *Zbl. mikroskop. Forsch. u. Meth.,* **9,** 147 (1954).
22. Guyer, M. F., and Smith, E. A., *J. Exptl. Zool.,* **38,** 79 (1923).
23. Holtzer, H., Marshall, Jr., J. M., and Finck, H., *J. Biophys. Biochem. Cytol.,* **3,** 705 (1957).
24. Inoue, K., *Japan. J. Exptl. Morphol.,* **8,** 45 (1954).
25. Langman, J., *Acta Morphol. Néerl.-Scand.,* **1,** 1 (1956).
26. ———, Schalekamp, M. A. D. H., Kuyken, M. P. A., and Veen, R., *Acta Morphol. Néerl.-Scand.,* **1,** 142 (1956).
27. Nace, G. W., *Ann. N. Y. Acad. Sci.,* **60,** 1038 (1955).
28. ———, *Twenty-third Ross Pediatrics Research Conf.,* p. 71, Ross Laboratories, Columbus, Ohio (1957).
29. ———, and Inoue, K., *Science,* **126,** 259 (1957).
30. ——— (unpub.).
31. Perlman, P., *Analysis of the Surface Structures of the Sea Urchin Egg by Means of Antibodies,* p. 5, Almqvist & Wiksells Boktryckeri Ab, Uppsala (1957).
32. Ranzi. S., in *The Beginnings of Embryonic Development* (A. Tyler, R. C. von Borstel, and C. B. Metz, eds.), p. 291, Am. Assoc. Advance. Sci., Washington, D. C. (1957).
33. Schechtman, A. M., in *Biological Specificity and Growth* (E. Butler, ed.), p. 3, Princeton Univ. Press, Princeton (1955).

34. Spar, I. L., *J. Exptl. Zool.*, **123**, 467 (1953).
35. Ten Cate, W. G., and Van Doorenmaalen, W. J., *Koninkl. Ned. Akad. Wetenschap.*, Proc., **53**, 894 (1950).
36. Twitty, U., in *Analysis of Development* (B. H. Willier, P. A. Weiss, and V. Hamburger, eds.), p. 102, W. B. Saunders Company, Philadelphia (1955).
37. Tyler, A., in *Analysis of Development* (B. H. Willier, P. A. Weiss, and V. Hamburger, eds.), p. 170, W. B. Saunders Company, Philadelphia (1955).
38. ———, in *The Beginnings of Embryonic Development* (A. Tyler, R. C. von Borstel, and C. B. Metz, eds.), p. 341, the *Am. Assoc. Advance. Sci.*, Washington, D. C. (1957).
39. Vainio, T., *Ann. Acad. Sci. Fennicae, Ser. A*, **35**, 1 (1957).
40. Woerdeman, M. W., in *Biological Specificity and Growth* (E. Butler, ed.), p. 33, Princeton Univ. Press, Princeton (1955).
41. Wood, D. C., *Twenty-third Ross Pediatrics Research Conf.*, p. 77, Ross Laboratories, Columbus, Ohio (1957).

# INFLUENCE OF ADULT AMPHIBIAN SPLEEN ON THE DEVELOPMENT OF EMBRYOS AND LARVAE: AN IMMUNE RESPONSE? [1]

Louis E. DeLanney

*Wabash College, Crawfordsville, Indiana*

Elsewhere in this symposium (see Billingham, Ebert) it was suggested that adult tissues capable of an immune reaction implanted into embryonic hosts (or into adult hosts experimentally deprived of their ability to form antibodies) may react against their hosts (1, 2, 4). The work to be reported here springs from studies of the influence on the homologous organ of avian chorioallantoic grafts of adult spleen (3). DeLanney and Ebert (2, 4) noted that extended residence of the graft on the extra-embryonic membrane often resulted in a deleterious effect on the host, including cystic infarcts in the host spleen and in the extra-embryonic membranes. They postulated that these effects were the product of action against the vascular bed of the host by antibodies produced by the graft, an argument consistent with the view of Pressman and Sherman (5, 6) that antibodies are produced against constituents of the vascular bed.

An attempt to duplicate the conditions noted in birds was made using the West Coast salamander *Taricha torosa* (*Triturus torosus*), since the larva is relatively pigment-free along its flank, the pellucid wall making it possible to observe the development of the spleen in situ. In brief, grafts of adult spleen made either to the coelom or to a pocket in the dorsal fin failed to demonstrate any convincing evidence of stimulation of the host spleen as seen in the chick. In contrast, a significant number of cases was observed in which the host's spleen could not be seen through the body wall. Histological examination showed the spleen to be fibrotic. It was postulated that this effect too is the result of antibody production by the graft against the host with the action preferentially localized in the larva's spleen. This work has been outlined briefly (DeLanney, in 4) and will be presented in detail elsewhere. However, these observations raised the question of the possible influence of grafts implanted at much earlier stages,

[1] Supported by a grant from the National Science Foundation.

before either coelom or fin are formed. It was believed that if the graft-against-host postulate had merit, it might be possible to observe similar aggressive action derived from the graft but acting earlier and expressing itself even before the host's spleen was developed; conversely, one might determine whether or not the reaction was strictly limited to the homologous organ. As it turns out, the first possibility appears to be the correct one and is discussed in more detail below.

## Methods and Results

Embryos of *Taricha torosa* (Twitty and Bodenstein stages 25 to 28) had implants of adult *T. torosa* spleen presented to them in either of two ways. (1) A trough was cut through the ectoderm, and tops of several somites and enough somite cells were cleaned away to allow a "rod" of adult spleen, trimmed to fit, to lie in the trough; since the graft spread the host tissues, it was necessary to cover the implanted spleen by a patch made of flank ectoderm taken from another embryo, a device more successful than forcing the wound closed with a glass bridge. Subsequently the following, and more satisfactory, method was devised. (2) A sharp needle was pushed into a somite at a flat angle so that it continued forward and penetrated through 3 to 4 somites in front of the point of entry. After the depth of penetration had been satisfactorily achieved, the needle was manipulated back and forth, always staying within the confines of the somites, so as to create a tunnel. The point of entry of the needle was slightly enlarged and then a "rod" of tissue cut from adult spleen was pushed into the tunnel. No bridge or other device was necessary to hold the graft in place. Efforts were made to keep as far back as possible from the anterior limb "field" so as to prevent limb reduplication and at the same time not to start so far to the rear that there was interference with the tail and posterior trunk elongation. This method of preventing reduplication was successful in all cases but one. The grafted spleen could easily be seen through the overlying tissues as a red or pink stripe. There is variation as to the length of time post-operatively that red color persists in the graft. All animals and tissues were maintained in Steinberg's solution (4), which was found to be eminently satisfactory for all of the work.

As a check on the observations of the writer, a senior student, Mr. C. D. McKeever, was assigned the same procedure as his objective, and his results were recorded separately. The student reported on 34 cases and their controls, the writer on 52 cases, their controls, and 13 sham-operated animals in which the tunnel was prepared in the somites but no material was inserted. Sham-operated animals without exception developed as normal

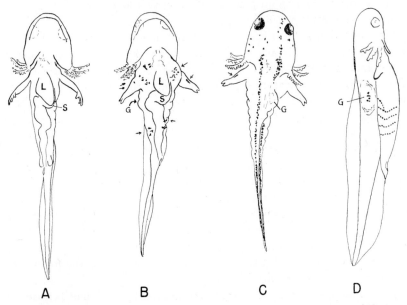

Fig. 1. Influence of grafts of adult *Taricha torosa* spleen implanted to somites of *T. torosa* embryos. A, B, and C (ca. stage 46), sham-operated, ventral view (A); grafted, ventral view (B) and dorsal view (C); drawn with camera lucida 23 days after implantation of adult spleen at stage 25. Sham-operated shows normal contours, lack of stasis; the animal which received the graft shows extreme stasis in the ventral pectoral area, in the limbs, and along the flanks, as shown by the arrows. No pigment cells were drawn for the ventral view. The dorsal view (C) shows the general contour, retarded limbs, and position of the graft; note the break in the dorsal pigment band at the site of the implant. (As-AdSo-58-33.) D. Free-hand sketch of a larva at ca. stage 39 (not to same scale as A-C), 8 days after receiving adult spleen along the right somites. No pigment cells have been drawn. Blood cells are indicated as being present in the graft and also populating parietal segmental vessels of the host; these first appeared on the flank 5 days post-implantation; their deeper red color and the known state of development of the host erythrocytes militates in favor of these being of donor origin. (As-AdSo-58-25.) G = graft area, L = liver, S = stomach.

larvae (Fig. 1A) and showed none of the results to be reported for the experimental cases.

The experimental animals showed two principal classes of results. First, as early as 4 days after the graft had been implanted, one sometimes saw bright red parietal vessels appear along the flanks (Fig. 1D). The blood islands of the hosts were not sufficiently advanced in their differentiation to account for this condition. These instances, which were not numerous, there being only four (LEDeL) and three (CMcK) recorded for the series, were often transient. In time it became impossible to see the vessels and their erythrocytes. Nevertheless, there can be no doubt but that donor cells can enter the host's vascular bed.

The second class of result was of greater pertinence, since it is believed that it is consistent with our previous results and with postulates made and based upon those results. In 22 instances of the 52 cases of the writer and in 15 cases of 34 of Mr. McKeever, there appeared to be a stasis of circulation in limited regions of the embryo. An extreme case of this is illustrated in Fig. 1B. These instances always occurred after some lapse of time, a single case being observed 7 days post-operatively but the rest ranging from 11 to 27 days before any sign of stasis was recorded. In several cases, such as is illustrated in Fig. 1B, C, there were other manifestations of host disorder, the most noteworthy being a broadening of the pectoral girdle area and a retardation of limb development; this was not seen in the controls or the sham-operated animals. The experimental series also showed a high mortality, 10 of the series dying by 35 days after the operation, although there was no visible evidence of contamination or infection, or of gross interference with the morphology of the animals as a result of the implantation procedure.

In addition to the two classes of result reported above, there is another observation that we make with less certainty but which is consistent with the observation of donor cell invasion of the early blood vessels. At least 8 of the writer's and 15 of McKeever's animals seemed more richly endowed with erythrocytes than the controls. When this was first observed, we turned our observation more critically to the gills and were led to believe that there was a mixture of yolk-laden, host-originating erythrocytes which were pale in color, interspersed with cells which seemed to be darker red. We tentatively believe that we are seeing a transient polycythemia with part of the erythrocyte population coming from the graft, part from the host. As the animals got older we were often forced to the opinion that there was no recognizable difference between the experimentals and the controls. Our caution in interpreting or recording this observation cannot be over-emphasized, since we have as yet found no certain way to insure that error of subjective judgment is not invloved.

## Discussion and Conclusions

### Colonization of the Host by Cells from the Graft

That we could clearly see erythrocytes in host blood vessels which would not have been colonized by their own erythrocytes until later in development seems obvious from the results such as those illustrated in Fig. 1D

This observation lends weight to the belief that at later stages we see a transient polycythemia with mixed intensities of erythrocyte color, since, if erythrocytes can enter early developing vascular channels fortuitously, later stages should have this probability increased if the donor cells survive, as they do.

It might also be postulated that donor cells could act as mechanical blocks in circulatory channels due to size differences, glutinosity, or for other reasons. This postulate seems unlikely, since we can draw absolutely no correlation between the incidence of stasis, suspected polycythemia, or early evidence of donor cell invasion. In fact, most of the 37 cases of stasis reported here show no sign of early invasion or concomitant polycythemia.

*Graft-Against-Host Reaction*

The results of stasis reported here are consistent with postulates previously made (2, 4) concerning the immune response of the graft against the host. Not only does one see visible suggestion of the reaction in the random appearance of sites of stasis but also in the general poor health of the animals affected by the grafts, with such evidences as anemia, edema, morphological abnormality like that illustrated in Fig. 1B and C, and the high incidence of deaths all reminiscent of the condition of serum sickness associated with an immune reaction.

The time of the appearance of the stasis is consistent with the known lapse in time necessary for antibody production. However, that the number of cases is not higher is possibly due to the relatively poor capacity of amphibians to produce antibodies.

The random site of the appearance of the stasis, sometimes in the limb, sometimes in the body wall, liver, gular region, or pericardium, leads one to question just what the mechanism of aggression is. If this is a circulating antibody, random appearance might be expected, but there is some question whether this is the nature of these antibodies. However, if not circulating antibodies, what is the nature of the transfer mechanism: host cells? graft cells? One could conjecture that a cell type of the host, comparable to the melanophage commonly seen in this species, has the capacity to transport antibody from the site of the transplant; however, one sees no localization of any specific cell type under the conditions of the experiments. Or, the transfer may be dependent upon some donor-orginating cell of the myelogenous strain which might emigrate from the graft, the freedom to emigrate possibly depending on the exact nature of the splenic implant, the exact environment around the implant, etc., which would condition whether or not the cell got into the larva's vascular channels.

Also whether the factor or factors at work in this relationship are acting intravascularly on the circulating components, to cause clumping, or on the endothelium of the channels, is unknown. The observations made upon *T. torosa* open up avenues for further exploration of the problem. For instance, it will probably be fruitful to utilize anuran donor material, taking advantage of the differences in nuclear sizes of the two orders. Steps are being taken at once to make this test, which is predicated on two assumptions: (a) that stasis will occur under these conditions of xenoplastic, heterochronic transplantation; and (b) that donor cells can be recognized clearly in histological preparations. With these two assumptions and their demonstration experimentally, it would be interesting to section sites of stasis to check for donor-originating cells. Also, through the courtesy of Dr. R. R. Humphrey, we have recently acquired the beginnings of a white axolotl stock which should greatly assist in extending the observations, since the larval white axolotl, even more beautifully than *T. torosa,* shows its spleen through its body wall. The much larger size of the animal should make an even greater quantitative challenge possible than can be done with *T. torosa*.

One observation stands out as distinct in the results reported here, a fact which has not previously been recorded in the work either with chicken or amphibian homologous organ transplantations. The early time of implantation and the timing of the results eliminates the host spleen as being involved at all, since the time of appearance of stasis was often before the host's spleen had begun to develop. This finding adds a new facet which must be considered in the interpretation of immune responses. There are many references in the literature relating to selective colonization of introduced cells (8).

We believe that the experimental approach reported may lead to the elucidation of the nature of the antigenic stimulus—if such it should turn out to be. Thus, in the spirit of this Symposium, it is believed that after clarifying some of the problems that can be approached by conventional methods of experimental embryology it will be possible thereafter to seek solutions through the techniques of analysis and synthesis of biochemistry.

### Summary

1. Embryos of *Taricha torosa* implanted with adult *T. torosa* spleen show gross abnormalities of the blood vascular system and secondary morphological variations.

2. Abnormalities occur under the influence of the grafted spleen before the host's spleen has made its appearance.

3. Postulation of an immune response, with the donor material producing antibodies against the host, is made as a working hypothesis for further investigation.

## REFERENCES

1. Billingham, R. E., and Brent, L., *Transpl. Bull.,* **4**, 67-71 (1957).
2. DeLanney, L. E., and Ebert, J. D., *Contrib. Embryol., Carnegie Inst. Wash.,* **37**, in prep. (1958).
3. Ebert, J. D., *Physiol. Zool.,* **24**, 20-41 (1951).
4. ———, *Carnegie Inst. Wash. Yearbook,* **56**, 324-325; 331-332 (1957).
5. Pressman, D., and Sherman, B., *Federation Proc.,* **10**, 416 (1951a).
6. ———, and ———, *Federation Proc.,* **10**, 568 (1951b).
7. Simonsen, M., *Acta Pathol. Microbiol. Scand.,* **40**, 480-500 (1957).
8. Weiss, P., and Andres, G., *J. Exptl. Zool.,* **121**, 449-487 (1952).

## DISCUSSION

DR. BILLINGHAM: To a layman it must seem rather peculiar that, despite the fact that the phenomenon of transplantation immunity has been studied for upwards of 50 years, we are still, as Dr. Ebert has just indicated, uncertain what the antigenic substances are that must be presumed to leave homografts of skin or other tissues, and be transported via the regional lymphatics to the regional lymph nodes and cause transplantation immunity. For several decades it was widely believed that only living homologous tissue cells could elicit transplantation immunity, a belief which gives some indication of the instability of the antigens. Although we now know how to disintegrate cells completely in a manner that does not destroy their antigenicity, and how to extract antigenically active material from cells and obtain it in true aqueous solution, the analysis of the nature of the antigens is proving a formidable task. Results have been slow in forthcoming and difficult to interpret. Although on the basis of our earlier studies my colleagues and I reached the provisional conclusion that the antigenic substances might be DNA-protein complexes, more recent evidence of Medawar and Brent suggests that they may be polysaccharides.

I fully sympathize with Dr. Ebert's reluctance to believe that these transplantation antigens are located solely in the nucleus of a cell. It is simply an empirical finding from various types of fractionation experiments that antigenic activity is associated with the nuclear derivatives. In our experience, particulate matter prepared from the cytoplasm of liver or kidney cells was devoid of demonstrable antigenicity. Obviously, however, antigens present in, or associated with, the nucleus must traverse the cytoplasm to leave the cell and elicit a response in a homologous host.

There is just one other comment I should like to make arising from the paper which we have just heard. Dr. Ebert stated that he cannot think of a

# DISCUSSION

single example of an embryo which can produce antibodies. However, we must not forget that the formation of humoral antibodies is not the only type of immunological response of which animals are capable. Foetal sheep can reject skin homografts quite promptly from adult donors transplanted to them as early as 40 days before birth, as Schinkel and Ferguson have shown. Here, then, is one perfectly good example of an embryo which does develop the adult mode of response towards at least one class of antigen long before birth.

DR. EBERT: If I may, Dr. Flexner, I would like to comment on Dr. Billingham's comments. First, I did not mean to imply that their work had gone slowly. It has proceeded much faster than my own. As a matter of fact, I thought I had made clear my opinion that, although the work is in a preliminary stage, it represents an exceptionally promising attack on a point at which we can bring together genetics, immunology and embryology. This approach may be more effective than the study of the synthesis of specific proteins such as alpha-crystallin, cholinesterase, myosin or any of the other proteins which I might cite. Second, with respect to the question of time of antibody formation, your response was the kind which I hoped I might evoke because it seems that the antibody-forming mechanism, or perhaps we should say the immune mechanism, provides an unusually favorable kind of system for experimental modification. It is important to study the synthesis of a specific protein in a descriptive way, but the immune mechanisms appear especially amenable to experimental analysis, to interference. We know, for example, that the work of Schinckel and Ferguson is at one end of the time scale with respect to the induction of immunity or tolerance; at the other extreme we have the rat in which immunity and tolerance mechanisms develop very slowly, providing a long interval [perhaps as much as two weeks after birth] in which experimental work can be done. Thus successful parabiosis and other operative procedures are the rule rather than the exception in the newborn rat. The duck also appears to fall at this end of the scale. In the chick and the mouse however these systems appear to be fixed, so to speak, shortly after birth. I also hoped that someone would rise to the occasion and discuss the possibility that the embryo may recognize different kinds of foreign substances at different times. The embryo, for example, may recognize heterografts far earlier than it would recognize homografts.

DR. BILLINGHAM: Obviously it would be difficult, if not impossible, to determine whether embryos or very young animals actually recognize different kinds or classes of antigen as being immunologically foreign at different stages or times during development. An organism's capacity to recognize a foreign substance can only be inferred on the basis of its carrying out some sort of reaction or response, e.g., by becoming tolerant or immune. There is, however, convincing evidence that young animals may develop the adult type of response towards some classes of antigen and yet still retain the embryonic, or tolerance-responsive, mode of response towards an antigen of a different class. For example, newborn calves react against skin homografts with a vigor which is not demonstrably inferior to that of adult cattle; yet they may still become tolerant of *Trichomonas foetus* antigen, as Kerr and Robertson's experiments show. Similarly, newborn rabbits have practically lost their tolerance-responsiveness of homologus tissue cells but may still become tolerant of bovine serum albumen injected into them as late as 12 days after birth.

DR. S. WRIGHT: I am interested in this fusion of genetics and immunology. There is one point you made at the beginning regarding individual specificity. You implied that that was not the result of a particular pattern of many antigens, but rather was really individual in that each individual has a specific antigen different from anybody else's. Now, in terms of genetics, would you contemplate this as due to an infinite number of alleles at certain loci? Of course, as far as pattern is concerned, we need only a few alleles, say 10 alleles at 100 loci to get 10 to the 100th power different patterns and get as much specificity as you want. You could also get individual antigens if the products of different loci converge on a single antigen in the fashion of Irwin's hybrid substance. I would be interested in hearing your comment.

DR. EBERT: First, this premise, although it is not fully proved represents to me the soundest approach to the problem. It is supported not only by the kind of work which Dr. Billingham and his associates have been doing, but by the research of others on the more or less ubiquitous distribution in the organism of the transplantation antigens. It stems from the observation that one can elicit homograft immunity by any kind of nucleated cells. Also pertinent is the work of Snell who has shown, with respect to certain tumor homografts, that compatibility or rejection of a tumor homograft is dependent upon the absence or presence of a single gene (the so-called histocompatibility loci). We have nothing comparable at all in the field of skin transplantation, simply because few, if any, laboratories are willing to undertake such long range, arduous and expensive kind of research. But I think it would be well worthwhile.

DR. NOVIKOFF: I am flattered by Dr. Ebert's reference to our work in this area. I want to emphasize the difficulty we are experiencing in reproducing the microsome effect. In only one experiment was a truly impressive inhibition of tumor growth obtained by previous injection of microsomes derived from the tumor. In several other experiments, there was a less dramatic, yet definite, inhibition of tumor growth; in others, there was little or no effect. It is of interest that Dr. Toolan finds, in a much more reproducible system, that the antibodies produced against a subcellular fraction sedimenting after the mitochondria have a cytotoxic effect on the tumor from which the fraction was derived. No such antibodies are produced with the nuclear fraction.

It should be remarked that in our fractions there is no DNA, as judged by the Dische diphenylamine reaction. I would expect this to be true of the fractions used by Dr. Toolan, and, in earlier work, by Syverton, Snell and Z'ilber. Thus, as Dr. Ebert has stressed, one ought not attempt generalization to transplantation antigens in tumors from Dr. Billingham's work with skin transplantation.

Electron microscopic examination of our tumor microsome fraction shows it to be a heterogeneous population of particles. There are many ribonucleoprotein granules, as expected, but there are also some intact and disrupted mitochondria, and numerous membranous vesicles of unknown origin. Among the latter are occasional ones not unlike viral particles, and, possibly more significant, structures which may be the microvillous portions of the cell membrane.

DR. M. COHN: I just want to point out that bacterial antigens, such as Salmonella, are not amenable to quantitative analysis since one doesn't know how many antibodies (or their combining properties) are involved in the agglutinin reaction.

Secondly, the experiments by Dixon seem to have been overlooked. He showed that if you take cells from adult rabbits which are immunizable, and transfer them into newborn rabbits, these cells will not produce antibodies. Dixon has concluded that the inability to produce antibodies is not the property of the cells but rather a property of the animal itself.

DR. EBERT: I am aware of Dixon's work; however, there are a number of studies which show clearly that if spleen or other cells capable of immune reactions are placed in newborn animals, they will produce effective immune reactions. For example, if fragments of spleen are inserted into a chick embryo at $3\frac{1}{2}$ to 4 days, they clearly produce an antagonistic response against the host. Here we come to your second question and that is, what should we call this response? Is it an antibody response or more generally an immune response? There is clearly a response in which the vascular bed of the embryo is destroyed by the so-called "graft versus host" reaction. Perhaps we tend to assign names to phenomena too readily, but these spleen cells do produce reactions within embryos. Thus I think Dixon may be wrong. I won't argue with your second point except to say that I believe you are sticking a little too closely to a narrow definition of antigen and antibody. I think we should adopt an operative terminology with respect to immune responses. You are thinking of an antigen only as something which produces a response that can be measured in terms of amount of gamma globulin produced, measurable in a precipitate, and I think this is a rather limited point of view. The hypersensitive reactions, presumably mediated by cells, are immune reactions; and one can talk about these substances which elicit them as antigens just as correctly as he uses the term for substances which stimulate the production of measurable quantities of gamma-globulin.

DR. WEISS: These last comments by Dr. Ebert and Dr. Billingham prompt me to bring in here a whole class of phenomena that have not been considered and I see no other place to discuss them. This is the subtle ability of cells to recognize each other, and the so-called affinities and disaffinities first pointed out by Holtfreter. A lot of information which shows actually what Dr. Billingham said, that there are several methods which can bring out the ability of cells to recognize each other's chemical constitution—obviously by surface contact since the nuclei do not come in contact in living cells—and that the immunological approach is only one. It should be valuable if at some later time we get a chance to discuss this problem, because it reveals a new and very subtle technique of finding out when and how these relationships develop which can be traced away back, of course, in the embryonic history, as Holtfreter, and myself with Andres, and a good many others have shown.

DR. LAUFER: Following his discussion of the work of Peters and Sterzl and Hrubesova, Dr. Ebert asked whether in the formation of a tissue-specific protein, say myosin, it might be possible to demonstrate a nucleo-myosin intermediate. In reply to his question, the following findings are pertinent (Laufer, H., *Anat. Rec.* **128**, 580-581, 1957; Laufer, H., An immunochemical analysis of limb regeneration in the adult newt, *Triturus:* Thesis, Library, Cornell University, Ithaca, New York, 1958): Antisera were prepared in rabbits against actomyosin and myosin isolated from the adult newt, *Triturus viridescens.* Analysis of these antisera by the agar diffusion method (Ouchterlony) yielded the following information:

(1) Antisera produced by the injection of actomyosin usually contained a minimum of three components (fig. 1A). The antigenic constituent which penetrated the agar at the slowest rate formed the major band which was believed to constitute the reaction of actomyosin with antiactomyosin. (2) The second and somewhat more rapidly migrating component was regarded to be myosin and its antiserum. (3) The most rapidly migrating antigen differed markedly from actomyosin and myosin; its properties indicate that it is nucleoprotein. Moreover, the serological evidence suggests heterogeneity in the nucleoprotein fraction. (4) The injection of myosin into rabbits elicited the formation of antibodies which when tested by agar diffusion proved to contain only one myosin band (fig. 1C). At times antimyosin sera were contaminated by antiactomyosin antibodies (fig. 1B).

In analyses of limb regenerates for muscle proteins by the agar diffusion method, both myosin and actomyosin were detected as early as the "palette" stage. Subsequently, in the digital stage, synthesis of these proteins was rapid. Histologically the appearance of muscle proteins occurred simultaneously with the time of myofibril formation. Cross striae characteristic of more mature muscle cells were not yet visible.

An immunochemical and spectrophotometric study of limbs revealed that mature tissues contained relatively little of the nucleoprotein fraction, whereas there was a striking increase in the fraction during the early phases of regeneration (fig. 2). The absorption curve of adult muscle lacks a defined peak in the range characteristic for nucleoprotein whereas sharp peaks exist when extracts of regenerates are measured in the ultraviolet range. The absorption was high-

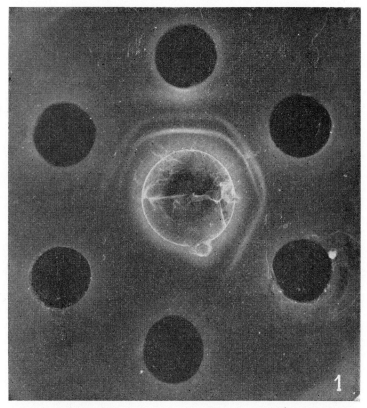

Reactions of a number of antimuscle sera tested against a mixture of muscle antigens in agar diffusion tests (Ouchterlony). The central reservoir contains a muscle extract as the antigen. The peripheral reservoirs are enumerated in clockwise order with "A" in the upper center of the figure.

| Reservoir designation | Reservoir contents |
|---|---|
| central | adult newt muscle extract |
| A | antiactomyosin serum |
| B | antimyosin serum |
| C | antimyosin serum |
| D | antiactin serum (low titer) |
| E | antiactin serum (low titer) |
| F | antiactomyosin serum |

Reactions are visible as white lines or bands of precipitate in the agar between the antigen and the antiserum-containing reservoirs.

est at the beginning of muscle differentiation, declining as the limb matured.

These findings, though not based upon as direct an approach as Dr. Ebert has suggested, do indicate that further studies would be fruitful in providing critical information on molecular interactions during growth and differentia-

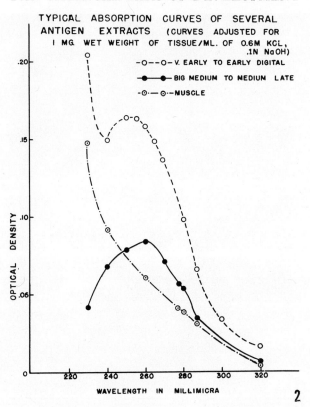

tion. Specific antisera can be produced by appropriate isolation, injection, and subsequent absorption procedures. Suitable labeling methods have also been described. Double or even multicolored fluorescent antiserum mixtures can be observed to localize specifically in tissue sections (Clayton, R. M., *Nature,* **174,** 1059, 1954). Autoradiograms of multilabeled radioactive antisera can be prepared (Buckaloo, G. W., and Cohn, D. V., *Science,* **123,** 333, 1956) so that a choice of procedures is now available for such studies.

# ACTIVELY ACQUIRED TOLERANCE AND ITS ROLE IN DEVELOPMENT

R. E. BILLINGHAM

*The Wistar Institute of Anatomy and Biology,
Philadelphia 4, Pa.*

## INTRODUCTION

The basic principle of immunology is the fact that when adult animals are confronted with substances that are foreign to them—i.e., antigenic—such as foreign proteins or polysaccharides, microorganisms, and living tissue cells of homologous or heterologous origin, certain highly specific reflex mechanisms are called into play. These include the production of conventional humoral antibodies, and various types of hypersensitivity reactions, which may have as their end result the elimination or inactivation of the particular agent responsible for their evocation. For example, adult birds and mammals will not permanently accept living tissue or cellular homografts from other genetically unrelated individuals of their own species. Although skin homografts heal in perfectly well at first, and may become functionally incorporated into the host's body, they are normally destroyed within a week or two as a consequence of the host's becoming sensitized to the foreign, genetically determined "transplantation" antigens (8) present in the grafts. Likewise, transfused red blood cells of homologous origin are eliminated, though here both the antigens concerned and the immunological reaction which they elicit differ profoundly from those involved in the reaction of animals against homografts of living tissues (30). Whereas red cell antigens are comparatively stable cytoplasmic substances, the isoantigens responsible for transplantation immunity are highly unstable and appear to be exclusively confined to nuclei (8).*

At the other end of the time scale, as immunologists have long been aware, when *embryos* or *very young* animals are exposed to antigens, they are incapable of reacting against them; their capacity to do so develops

---

* Notwithstanding their presence in nuclear fractions and in DNA-protein preparations, recent evidence suggests that the determinant groups of the "transplantation antigens" are related to the human blood group substances—amino acid polysaccharide complexes. (See Billingham, Brent and Medawar, *in press*, Transplantation Bulletin.)

very slowly. Until recently this state of affairs was ascribed, quite reasonably, to the fact that the immunological response or defense mechanism must undergo differentiation like other adult characteristics. The purpose of this paper is to show that this apparent inertness or unresponsiveness of very young animals to an antigenic stimulus is highly deceptive, for they may undergo a highly specific adaptive change of considerable, if not of permanent, duration. This expresses itself in later life as a specific inhibition or impairment of their capacity to respond to a later challenge with the antigen in question. It will also be argued that adaptive responses of this type may play an important role in development.

The important clue which has led to the development of this completely new field in immunology was Owen's remarkable discovery in 1945 of the phenomenon of red cell chimerism in twin cattle (37). In bovine twins, and indeed in triplets, quadruplets, and quintuplets too, all individuals usually share the same chorion, with the consequence that a continuous exchange of blood goes on from an early stage of fetal life until birth (28, 50). Not only are such animals born with mixtures of each other's red cells, but this state of chimerism may actually persist throughout their lives, indicating not only that the animals must have exchanged blood-forming cells in fetal life, but also that these cells must have survived contrary to the laws of transplantation immunity and continued to produce red cells of their own distinctive antigenic type. We now know that stable red cell chimerism is not restricted to cattle, for it is probably the rule in the dizygotic twin chicks which may sometimes be hatched successfully from doubly yolked and doubly fertile eggs (7), and it is also known to occur rarely in dizygotic twin sheep (48) and in man. In the latter species there are now three well-documented examples of persistent red cell chimerism in dissimilar twins (11, 19, 36). One of these is particularly interesting since it involved individuals of unlike sex (36). Among the leucocytes of the male some were cytologically definable as being of female type and must therefore have originated from cells received from the female. This has been cited as suggestive evidence of the interchange of primitive hemocytoblasts before differentiation of the different blood cell lineages had occurred.

Burnet and Fenner (12) were greatly influenced by Owen's discovery in the formulation of their theory of antibody formation. This theory led them to make the general prediction (12) that if embryos were to be exposed to antigenic substances, then their power to react against those substances when they grew up would be abolished or at least impaired—i.e., a state of specific immunological tolerance would result.

## The Experimental Induction of Tolerance

Following our demonstration that dizygotic cattle twins were mutually tolerant of homografts of each others' skin (1, 10)—a phenomenon which suggested, of course, that the tolerance conferred by the hemopoietic cells interchanged between them in embryonic life was not tissue specific—my colleagues and I carried out experiments to try to reproduce artificially a state of tolerance in various laboratory animals (6, 7). Embryos of chickens, mice, rats, and rabbits were injected with preparations of living homologous tissue cells. Then, after they had grown up and become immunologically mature, the animals were challenged with skin homografts from the original donor of the cells they had previously received, or, in the case of the mice, from another donor of similar genetic constitution. These experiments were successful: the exposure of embryos to the foreign tissue cells of a variety of different histological types, which included spleen, blood, and leucocytes of fetal or adult origin, induced tolerance as demonstrated by the acceptance of subsequent skin homografts. In some cases, when tolerance was complete, the grafts survived indefinitely. These findings have now been confirmed and extended by other workers.

Contrary to our earlier belief, it is now clear that at least in some species the inoculation of young animals with homologous cells soon *after* birth or hatching is not too late for homologous cells to confer tolerance, as Table 1 shows. From a purely technical viewpoint this has greatly simplified the problem of producing tolerant animals for experimental purposes, since it avoids the fairly high mortality associated with interference with embryos. In the mouse a high degree of tolerance may consistently be conferred if suspensions of living tissue cells—most conveniently prepared from the

TABLE 1

Upper Limit of Tolerance-Responsive Period with Respect to Living Homologous Tissue Cells[1]

| Species | Gestation or incubation period (days) | End of tolerance-responsive period |
|---|---|---|
| cattle | 283 | long before birth |
| man | 280 | ? time of birth |
| sheep | 150 | more than 50 days before birth |
| dog | 63 | 7 days after birth |
| rabbit | 32 | day of birth |
| rat | 22 | 14 days after birth |
| mouse | 21 | 2 days after birth |
| duck | 28 | 14 days after hatching |
| chicken | 21 | 2 days after hatching |

[1] For references, see text.

spleen, though of course this is not obligatory—are injected intravenously within 24 hours of birth, though beyond this time the proportion of animals in which tolerance is conferred falls rapidly (4). Rather surprisingly, the rat is a much more amenable subject, for, as Woodruff and Simpson (56) have shown, the subcutaneous inoculation of rats with spleen cells as late as 14 days after birth still confers tolerance on some. Indeed, merely grafting newborn rats with homologous skin confers at least some tolerance in a high proportion of them (33). The majority of newborn rabbits have already passed beyond the threshold of tolerance-responsiveness, for intravenous injection within a few hours of birth will confer feeble tolerance on about 25 per cent of the subjects (5).

A. Puza and A. Gomboš (41) have recently shown that following exsanguination of 3 to 4-day-old puppies and replacement of their blood by transfusion from adult donors, they will completely accept skin homografts transplanted from the same donors a few weeks later. Tolerance-responsiveness was still demonstrable in the 7-day-old puppy. At present we do not know whether the human infant is still tolerance-responsive at the time of birth; there is just a slight hint that it may be so to a feeble extent (55). In ungulates it appears that the tolerance response undergoes its metamorphosis into the immune response long before birth, since fetal sheep react vigorously against and destroy skin homografts transplanted to them as long as 50 days before birth (43), and newborn calves can destroy skin homografts with a vigor that is only slightly less dramatic than that of adult cattle (9).

In mammals, evidence has been sought of the possibility that tolerance may sometimes be induced naturally as a consequence of the accidental leakage or migration of maternal cells (such as leucocytes) into the fetus. Of the experimental animals so far investigated, which include mice, rabbits, cattle, and guinea pigs, only in the last species has evidence been obtained that tolerance of maternal homografts may rarely occur in this way (7). In man, clinical evidence suggests that this happens when melanomatous tumor cells of maternal origin invade the placenta and establish themselves in the embryo (53), ultimately being responsible for the death of the child at an early age. The finding of Lyndon Peer (40), that about 25 per cent of skin grafts exchanged between a fairly large series of mothers and their children have an unexpectedly long survival time may, at least in part, be a consequence of a maternally induced tolerance, though a satisfactory explanation is still required for the greatly weakened resistance of some of the mothers toward homografts from their offspring.

Lengerová's (27) recent demonstration that X-irradiation of the uterine horns of pregnant rats (the rest of their bodies were shielded) confers a high degree of tolerance on the offspring in respect to later homografts from their mother, but from no other donors, provides an excellent example of the consequence of what is almost certainly an experimentally reduced permeability of the placental barrier to cells. It may provide, moreover, means of conferring tolerance applicable to other mammals.

Birds are particularly suitable subjects for the induction of tolerance on account of the ease with which inocula comprising blood or tissue cell suspensions may be introduced directly into the blood stream of the embryos, or fragments of living tissues may be deposited on the chorio-allantoic membrane, where they become vascularized. Thanks to the ingenuity of Milan Hašek (21), a technique is available for synchorial parabiosis which facilitates the establishment of anastomoses between the blood circulations of the parabionts. Recently Cannon, Terasaki, and Longmire (15) have devised a method for carrying out exchange transfusions of blood between paired avian embryos. In chickens, a high degree of tolerance may be obtained by any one of these methods, though the technique of experimental synchorial parabiosis is probably the most effective for conferring complete tolerance of homografts. Moreover, tolerant animals produced in this way are normally red cell chimeras and remain so for as long as they are fully tolerant of skin homografts from their parabiont partners (7).

However, the newly hatched chick, like the newborn mouse, is still in the tolerance-responsive or adaptive phase of its life. Grafting with homologous skin soon after hatching still confers tolerance on a small proportion (14, 51). If newly hatched chicks are injected intravenously with homologous blood, the proportion which become tolerant of concomitantly or subsequently transplanted skin from the same donor is greatly increased (3, 7). The fact that intravenously injected blood cells are more effective in conferring tolerance upon a newly hatched chick than an orthotopic skin homograft is probably the outcome of the different times at which the two antigenic stimuli become effective. The stimulus provided by the injected cells presumably reaches the developing immunologically responsive tissues of the host—i.e., its lymphoid tissues—almost immediately, whereas that emanating from a skin graft can only become effective when vascular connections with the host have become established—a process which requires at least 48 hours. Consequently, grafting with skin alone at the time of birth is unlikely to induce tolerance in any animal which is within a day or so of the end of the adaptive period. The mouse is a good example, for skin homografts transplanted to the newborn do not confer tolerance, yet intravenous injections of cells usually do. In the newborn mouse subcutaneous injections, like skin grafting, are ineffective presumably because of the time taken for the antigenic stimulus to come into effect (4).

Unlike the chick, the duck remains tolerance-responsive up to two weeks after hatching (23). Skin homografts transplanted to 7-day-old ducks survive permanently.

The evidence presented in this brief survey of the tolerance-responsiveness of different mammalian and avian species toward the homologous cellular antigens responsible for transplantation immunity shows clearly that no rigid correlation or causal relationship exists between hatching or birth and the end of the tolerance-responsive period. The important fact appears to be the state of maturity of the animal at term. We do know that there is no sudden switch-over from the embryonic mode of response to homologous cells (i.e., acquired tolerance) to that of the adult (actively acquired immunity) which is expressed by an accelerated rejection of homografts. Careful experiments in which newborn mice of varying ages were injected with homologous spleen cells have shown that with the strain combination CBA → A, although injection within 24 hours confers tolerance on practically 100 per cent, by the fourth day only about 10 per cent are still just perceptibly tolerance-responsive (4). At this time the remainder of the mice were no longer tolerance-responsive and had entered the "neutral" period, since their response to subsequent skin homografts was indicative neither of tolerance nor of prior sensitization—they behaved as if they had never been injected at all. Not until the mice were 7 days old at injection did the inoculated cells elicit a feeble degree of sensitization. In the mouse, therefore, the neutral period is of about 5 days duration.

## The Fate of the Tolerance-Conferring Cells

The fact that naturally or artificially inoculated cells confer tolerance upon an embryo or newborn animal is most conveniently established on the basis of the fate of a skin homograft transplanted from the donor of the cellular inoculum to the recipient when it has become immunologically mature, i.e., has reached an age when it would normally be capable of rejecting a homograft. There is, however, a strong prima facie case for supposing that the cells which induced tolerance do in fact persist in their new host, provided that tolerance is complete. This rests upon the persistence of natural red cell chimerism in cattle, man, and chicken, and also of artificially produced red cell chimerism in the latter species. In all of these cases, it seems likely that the transposed homologous blood-forming cells settle out in the marrow of their host and continue to behave just as they would have done in their genetically proper host. Owen's studies (38) provide evidence that the foreign cells are at no selective disadvantage in cattle twins.

L. Brent and I (2, 4) have investigated the ultimate fate of the adult spleen cells, or their mitotic descendants, which had been the means of conferring complete tolerance upon newborn mice. The principle of the test employed (34) is very simple. Suppose we have an adult A strain mouse which is fully tolerant of skin from a CBA strain donor as a consequence of its injection at birth with CBA spleen cells. Cell suspensions are prepared from various organs of the tolerant A strain mouse and injected into *normal* adult A strain mice to see whether they will elicit transplantation immunity against subsequent CBA strain homografts; if they do, then the presence of CBA strain cells in the organs under test may be inferred. The persistence of foreign cells has been established in the spleens, lymph nodes, thymuses, bone marrow, and even in the leucocytes of the blood of fully tolerant mice, though no foreign cells persisted in incompletely tolerant mice after they had rejected their test grafts. Similar findings were obtained when the tissues of mice made tolerant by the injection of thymocytes were tested. These results show unequivocally that cells inoculated into young animals do persist and become incorporated in their hosts if complete tolerance is induced: fully tolerant animals produced by the methods described are true cellular chimeras, and test grafting with skin is simply a means of recognizing this fact.

It is of some interest that the tolerance-conferring cells appear to become established in anatomically appropriate tissues. It has yet to be shown whether this represents the outcome of a true homing instinct on the part of the cells, or simply that only those cells survive which land on biologically appropriate soil. A similar problem is posed by the demonstration by Weiss and Andres (52) that homologous melanoblasts injected into the blood stream of very young chick embryos appear to end up in anatomically appropriate places.

*The Specificity of Tolerance*

(a) *Tissue specificity*. It is now well established that animals made tolerant as a consequence of inoculation in early life with cells derived from one tissue are completely incapable of discriminating immunologically between the different tissues of the original donor of the tolerance-conferring stimulus, or of another donor of similar genetic constitution. For example, the tolerance induced in rats or mice by neonatal inoculation with spleen cells is completely effective with respect to homografts of skin, adrenal cortex (32, 57), ovarian tissue (26, 29), and thyroid (57). Moreover, as Simonsen (44) has shown, the mutual tolerance of dizygotic cattle twins toward each other's blood cells and skin also extends to kidney homografts exchanged between them. Such results, and others summarized in Table 2, show that there is a complete lack of tissue specificity, and lead to the im-

TABLE 2

Evidence that Acquired Tolerance of Homologous Cells is not Tissue-Specific[1]

| Species | Tissue which induced tolerance | Other tissues which animal also tolerates |
|---|---|---|
| cattle | fetal blood | skin (7), kidney (44) |
| dog | adult blood | skin (41) |
| mouse | adult blood | skin (7) |
|  | spleen cells | skin, adrenal cortex (32), ovary (26, 29) |
|  | mammary tumor | skin (7) |
|  | thymocytes | skin (2) |
|  | bone marrow | skin (2) |
| rat | spleen cells | skin (56), thyroid (57), adrenal cortex (57) |
| chicken | leucocytes | skin (7) |
|  | fetal blood | skin (7) |

[1] Numerals indicate references.

portant conclusion that all the living tissue cells of the body of an individual have exactly the same complement of transplantation antigens.

(b) *Individual specificity.* The skin grafting tests carried out on dizygotic twin cattle revealed that animals which were fully tolerant of grafts of each others' skin, as a consequence of their natural embryonic parabiosis, invariably rejected homografts from any other donors, including their mothers and full siblings of separate birth. The specificity of experimentally induced tolerance in mice and birds has likewise been shown to be complete if tolerance is complete; they will accept only grafts having the same genetic constitution as the cells which called the tolerant state into being. For example, if an A strain mouse which is fully tolerant of a CBA strain graft following its inoculation at birth with CBA spleen cells is grafted with skin from a completely unrelated strain AU, the latter graft is soon destroyed whereas the original tolerated graft remains unaffected. If, however, the second donor happened to share enough important antigens in common with the first donor, the residual antigens present in the second donor and not shared with the first might be expected to provoke a weaker response than normally. This state of affairs has now been realized experimentally. When A strain mice which were fully tolerant of C3H strain skin homografts, following their inoculation with C3H cells at birth, were challenged with CBA strain homografts, the latter survived for about 24 days instead of 11 days as on normal A strain hosts (5). Similar evidence has recently been forthcoming from a study of the responses of tetrazygotic quadruplet calves toward homografts from their dam (9).

*The Abolition of Tolerance*

From the evidence already presented it may appear self-evident that tolerance is the expression of an adaptation or transformation of the *host* and not of the inoculated cells which called the tolerant state into being, or of the skin homograft by means of which it was subsequently revealed. Indeed, if the fetally or neonatally inoculated cells had not retained their donor-specific properties, persistent red cell chimerism would never have been recognized after natural or artificial twinning, nor could the persistence of foreign spleen cells have been established in the various tissues of mice made tolerant by injection at birth with this type of cell. That skin homografts undergo no antigenic change or adaptation during the course of even long periods of residence on tolerant hosts follows from the facts that (a) such grafts are soon rejected if they are removed from a tolerant mouse and transplanted to a normal mouse of its *own* strain, and (b) they are accepted if transplanted to a normal mouse of their strain of origin.

The most decisive evidence that the reactivity of a tolerated graft remains unchanged derives from the following simple experiment. Consider an A strain mouse which has been made tolerant of CBA tissue by injecting it at birth with CBA spleen cells, and which now bears a healthy CBA skin homograft of long standing. If this mouse is injected with lymph node or spleen cells from a normal A strain mouse which has itself been actively immunized against CBA tissue, the hitherto tolerated graft will begin to deteriorate within a few days, and breakdown is soon complete. This indicates not only that the reactivity of the tolerated graft has remained unchanged, but also that a tolerant animal is perfectly capable of implementing an immunity of adoptive origin.

If instead of implanting the tolerant A strain mouse with *sensitized* lymphoid tissue cells from an actively immunized mouse of its own strain, it is inoculated with *normal* lymphoid tissue from an untreated A strain donor, breakdown of the tolerated graft is still brought about, though the process is much slower. This experiment shows that the foreign cells present in the tolerant mouse must have been producing and liberating antigens all the time, but that the host was unable to recognize them as such. If normal immunologically competent cells are introduced into a tolerant mouse they respond to these antigens with the result that the graft and presumably all the established cells of the neonatal inoculum as well are destroyed.

These experiments show that tolerance represents a specific failure of the host's immunological response mechanism. It is at the level of the lymphoid tissue cells that we must seek the mechanism of tolerance.

### Tolerance with Respect to other Types of Immunological System

So far I have considered the concept of tolerance only as it applies to the isoantigens which are contained within living tissue cells and which are responsible for transplantation immunity—i.e., the destruction of tissue or cellular homografts—since it is from the analysis of this system that the basic principles of the phenomenon have been worked out most fully.

Tolerance can also be conferred with respect to both living tissue cells and red cells of *heterologous* origin, though at present it appears improbable that complete tolerance is attainable with respect to either of these cellular antigens. Egdahl and Varco (20) have recently succeeded in procuring significant prolongations of survival of rabbit skin heterografts on rats by inoculating the hosts in utero with spleen cells from the future skin donors. Likewise Willys Silvers and I have induced a sufficient degree of tolerance in Syrian hamsters, by injecting them at birth with mouse spleen cells, to permit some of them to accept murine skin heterografts in a state of complete normality for upward of 20 days. Some of the skin heterografts regenerated crops of new hairs.

Birds appear to be more favorable subjects for the induction of tolerance with heterologous tissue cells or red cells. Following their parabiotic union in embryonic life with chick embryos, ducks have been known to accept skin grafts from their parabionts for as long as 45 days (7). Furthermore, the injection of newly hatched ducks with goose spleen cells completely suppresses agglutinin formation and also enables the ducks to accept skin grafts from their heterologous donors for about 25 days (22). Highly significant prolongations of survival of skin heterografts have also been obtained in turkeys following their injection with chicken's blood and grafting with skin from the same donor on the day of hatching, for the grafts survived for 3 to 4 weeks (3). Hraba (24) has reported transient red cell chimerism of 8 weeks duration in a chicken which had been placed in embryonic parabiosis with a turkey. In experiments involving heterologous cellular antigens the taxonomic relationship between the donor and recipient is clearly of great importance.

One particularly important study carried out with heterologous erythrocytes is that of Simonsen (45), who investigated the influence of injections of human blood into chick embryos on their subsequent power to respond to active immunization by the formation of heteroagglutinins. He found that whereas injection of blood into 17 to 19-day-old embryos induced some measure of tolerance, the injection of 15-day-old embryos was less effective, and the injection of 10-day embryos was completely ineffective. As Medawar (31) has commented, this evidence strongly suggests that in deliberating at what age to inject an embryo to confer tolerance, the "earlier the better" is not a valid rule. It may be inferred from Simonsen's experiment that to confer tolerance the antigen must be present during the transition period from the embryonic to the adult mode of response. The time of inception of this period and its duration will clearly depend upon the species concerned, and, as will be shown later, upon the nature of the antigen. Too early a presentation of the antigen may result in its being completely eliminated or broken down *before* the embryo has entered the adaptive phase. Obviously this argument should be inapplicable to the tolerance induced by living homologous tissue cells, and possibly to that conferred by living cells of *heterologous* origin, for however early these are injected into an embryo they should survive and proliferate unharmed until they can exert their immunological influence when the embryo has entered the adaptive phase.

Evidence that tolerance may be induced by more familiar and more definable heterologous antigens of various classes has now been presented by numerous workers, and much of it has been brought together or reviewed elsewhere (18, 38). To cite only one example: it is now well established that baby rabbits injected with serum protein antigens of heterologous origin may not acquire the ability to make antibodies against these antigens when challenged after they have grown up. Moreover, it has been convincingly demonstrated that tolerance with respect to these antigens may be just as specific as with the transplantation immunity class of antigens (16).

It has already been hinted that the adaptive period may vary considerably according to the antigen concerned. For even *within* the transplantation antigen system there is evidence of variation: although newborn A strain mice are still uniformly tolerance-responsive toward living cells from strain CBA, C57 mice at birth have already passed the period during which exposure to A strain cells will confer tolerance. The differences in timing of the adaptive period in ontogeny may become much greater when antigenic substances of *different* classes are compared. For example, at birth the calf has passed beyond the adaptive period and acquired the adult mode of response toward transplantation antigens, yet it may still become tolerant of *Trichomonas foetus* antigen if it is injected as late as 4 weeks after birth (30). Similarly, although newborn rabbits are at the extreme limit of their tolerance-responsiveness so far as homologous cells are concerned, even the 17-day-old rabbit is still tolerance-responsive to bovine serum albumen (47).

### The Possible Role of Tolerance in Normal Development

So far the phenomenon of tolerance has been considered as it applies to antigens of extraneous origin which accidentally gain access to, or are experimentally introduced into, very young animals. However, as Burnet and Fenner were well aware, some explanation is required as to why the cells which mediate an animal's immunological responses do not normally react against some of the differentiated constituents of its *own* body, i.e., why autoantibodies are not formed. Paul Ehrlich appreciated this philosophical problem more than half a century ago when he formulated his law of "Horror autotoxicus," which implies that autoantibodies are incompatible with normal life.

As all cells of the body, including those of the lymphoid tissue system, possess in common from a very early stage those antigens responsible for transplantation immunity, this problem does not concern them. However, many of the cells and tissues of the body do develop specialized substances as they differentiate which they obviously do *not* share in common with the antibody-producing cells and which might therefore appear foreign or antigenic to the latter. The most plausible reason why an individual's reticulo-endothelial system does not react against these substances is that it becomes exposed to them at a fairly early stage of development, and so becomes tolerant of them. It follows from this argument that if any specialized tissue constituents either are not produced until after the functional maturation of an animal's immunological response mechanism, or develop in what is effectively a physiological quarantine, so that the host's reticulo-endothelial system fails to be exposed to, and therefore does not become tolerant of, such ingredients, then such substances should be at least potentially auto-antigenic and therefore iso-antigenic (7, 30). Examples of both categories of substance are known. Spermatozoa develop relatively late in life, and if appropriately administered, will elicit the formation of autoantibodies against some of their distinctive constituents: of course they never get an opportunity to do this naturally. The lens of the eye, on the other hand, develops at a fairly early stage, but it is encapsulated and physiologically isolated from the reticulo-endothelial system—it has no vascular supply. It is scarcely surprising, therefore, that it possesses protein ingredients which are autoantigenic if administered appropriately. The protein, thyroglobulin, appears to belong to this category, too. It develops in what appears to be an effective physiological quarantine—the follicles of the thyroid gland—from which it does not normally escape into the circulation. Consequently this substance remains potentially autoantigenic. Witebsky (54) and his associates established this fact when they demonstrated the autoantigenicity of thyroglobulin when it was administered to experimental animals under appropriate conditions. Just occasionally in man, presumably as a consequence of leakage or a lesion, it appears that thyroglobulin does escape, whereupon the regional or draining lymph nodes are stimulated to elaborate antibodies which circulate in the blood stream, and there is an infiltration of the thyroid gland itself with cells believed to be capable of forming antibodies. The result is a progressive thyroiditis and impairment of glandular function (42). If the reasoning presented in this section is correct, it follows that a specialized ingredient of the body which for some cause, e.g., genetic, becomes differentiated later in development than usual (so that the host fails to become tolerant of it), might then be autoantigenic and elicit an auto-immune response.

One recent example of experimentally induced tolerance which seems appropriate for consideration in the present section concerns distinctive ingredients of nervous tissue. It has long been known that various species of laboratory mammals develop an allergic encephalomyelitis, with characteristic lesions which are confined to the central nervous system, a few weeks after injection with an appropriate preparation of mammalian nervous tissue. Patterson (39) has recently discovered that injection of *newborn* rats with a homogenate of homologous spinal cord results in a marked decrease in their capacity to develop allergic encephalomyelitis if they are injected with this antigen 8 to 10 weeks later. The protective effect of the neonatal inoculation is ascribed to its ability to confer tolerance with respect to the distinctive antigens in nervous tissue which are responsible for the paralytogenic response. This experiment suggests very strongly that brain tissue, like that of the lens, contains potentially auto-antigenic ingredients, since the organism never gets an opportunity to become tolerant of them. This is understandable, for the brain has no lymphatic drainage nor is it served by lymph nodes. On the other hand, susceptibility of the brain to a reaction provoked by injection of the animal with the appropriate antigens was to be expected on account of its rich vascular supply, which provides the efferent pathway for the immunological response.

## The Reaction of Injected Homologous Lymphoid Tissue Cells against the Host

To confer tolerance, the most obvious and convenient sources of living cell suspensions suitable for introduction into the blood stream of very young animals comprise lymphoid or closely related tissues—nodes, thymus, spleen, bone marrow, and leucocyte concentrates of whole blood. With some donor/recipient strain combinations of mice it has been observed that injection of the newborn animals with suspensions of any one of these types of tissue cells causes varying degrees of involution of the lymphoid tissue of some of the hosts, grossly impairs the development of others so that they remain chronic runts, and is responsible for the premature death within 2 to 3 weeks of others (4). Careful investigation has shown that the severity of this disease, which we have referred to as "runt disease," is mainly dependent upon the genetic (or antigenic) relationship between the inoculated, immunologically competent, cells and their host.

Likewise in the chicken, as Simonsen (46) has shown, the inoculation of adult spleen cells or leucocytes into embryos results in a marked splenomegaly accompanied by pathological lesions in the spleen and other organs of the body, and usually causes death within a week or two after hatching. A wealth of evidence has been amassed indicating that these deaths and abnormalities in both the mice and the chickens are the consequence of immunological reactions on the part of the inoculated, immunologically competent, adult homologous cells against the "foreign" transplantation antigens of their young hosts. The tolerance which these cells confer in their hosts secures them from rejection as the host grows up. The tissues which appear to be most severely damaged are those belonging to the reticuloendothelial system, i.e., are those in which inoculated homologous lymphoid cells are known to establish themselves.

The possible existence of such graft-versus-host reactions was predicted a few years ago on the basis of rather slender histological evidence. The inoculation of immunologically competent cells into very young animals has provided the means for verifying this prediction and for a detailed analysis of the phenomenon.

There is overwhelming evidence that tolerance is not causally related to the involution of the host's lymphoid tissue which follows the inoculation of newborn mice with homologous spleen cells (2, 7); for example, animals made tolerant by inoculation with *embryonic* cells show no abnormalities, as one would expect, for in such cases not only does the host become tolerant of the antigens of the inoculated cells, but the latter in their turn presumably become tolerant of their host's antigens.

By way of concluding this section, I shall mention only two of many important aspects or implications of the concept that grafted cells may be able to react against their hosts. Firstly, the demonstration that leucocyte concentrates prepared from the circulating blood of an adult mouse will cause "runt disease," if inoculated into newborn mice of certain strains, has an important embryological bearing. In the relationship between the mammalian mother and her unborn young, not only must the placental barrier prevent the passage of fetal cells into the maternal blood to avoid the danger of the fetus immunizing the mother, a fact which is familiar to all; but it now seems important that the placenta should prevent the passage of maternal leucocytes into the fetal circulation. If these maternal cells gained access to the fetus, besides conferring tolerance, they would give the immunologically competent ingredients among such cells an opportunity to become established in the lymphoid tissues of the new host and to proceed to react against it.

Secondly, Cannon and Longmire, and their associates (13, 14, 51), who have made detailed and important investigations of the induction of tolerance in newly hatched chicks, especially by grafting the chicks with homologous skin at or soon after birth, have obtained evidence which has led them to postulate that the tolerance obtained may be, to a greater or less extent, the outcome of a change in the tissue specificity of the *graft*—i.e., adaptation—rather than an induced change on the part of the host (13). Among several of their experimental findings which appear to support their thesis are the following: (1) chickens made tolerant by grafting at birth with skin from very young donors nearly always reject "repeat" homografts from their original (now adult) donors made in later life, yet the original graft continues to thrive; and (2) although 32 per cent of skin homografts 7 sq. cm. in area from *2-day-old* donors transplanted to 2-day-old recipients were still alive after 14 weeks, when homografts of comparable size and thickness were transplanted from *adult* donors to 2-day-old recipients, none survived for as long as 14 weeks, and the majority broke down very much earlier (51).

Recently Silvers and I have obtained evidence that even the skin of adult chickens, although not that of newly hatched chicks, contains immunologically competent cells. We find that if thin grafts of adult skin are transplanted to the chorioallantoic membrane of 10-day chick embryos, the grafts heal in rapidly and become vascularized. Careful examination of the hosts at about the time of hatching reveals pathological lesions in their spleens and other organs resembling those observed by Simonsen following the injection of chick embryos with adult spleen cells or leucocytes. In addition, we have evidence of a local proliferative invasion and reaction of cells deriving from the skin graft, in the underlying host tissue. On the basis of these findings, which are now being extended, we feel that the short survival time of some of Cannon and Longmire's adult skin homografts may not be the consequence of an immunological reaction on the part of the host, but of the *grafts'* reacting against their host, and rejecting themselves (it is tempting to say, rejecting their hosts) as a consequence of the predominantly local nature of their response. The most direct experimental procedure for confirming or refuting the hypothesis of graft adaptation in these chicken experiments would require inbred strains. The revealing test of adoptive immunization could then be carried out on tolerant birds, as on tolerant mice.

The intervention of unsuspected graft-versus-host reactions in experiments involving the transplantation of homologous tissues can be very misleading. For example, it now appears that the significance of some of Danchakoff's (17) and Murphy's (35) classical experiments, in which they transplanted adult splenic tissue to chick embryos, must now be reappraised.

### Epilogue: The Mechanism of Tolerance

So far I have cited a great deal of evidence which indicates—indeed, virtually proves—that tolerance represents a specific central failure of the immunological response mechanism, i.e., is the outcome of some kind of induced functional change in the immunologically competent cells. Unfortunately, at present there is no evidence or hint as to how tolerance comes about—it is simply an empirical fact of immunology which any theory of immunological response or antibody formation must take into account. The various theories of antibody production have recently been critically reviewed (49) and need not be considered here. According to Burnet and Fenner's theory, which takes tolerance into consideration, there are functional units located in the "scavenger" or reticulo-endothelial cells of the body (which are also responsible for antibody production) concerned with the breakdown of substances which form part of the expendable cells of an animal's own body. In embryonic life these units become functionally adapted to the substances concerned so that they can "recognize" self-marker components of expendable body cells and eliminate them without the formation of autoantibodies.

Humphrey (25) has suggested an explanation of tolerance in terms of an acquired ability to break down the particular foreign antigenic substance fast enough to prevent its acting as an antigen. According to this hypothesis, ability to recognize "self" would be the ability to destroy completely and possibly reutilize self-made products. It is presumed that these processes would take place in the cells normally involved in mediating immune responses. An essentially similar hypothesis has also been put forward by Haldane. According to him, embryos may have the capacity to metabolize and so dispose of substances which the adult animal cannot metabolize, but against which it makes antibodies instead. Tolerance, therefore, would be the consequence of retention of the embryonic type of response, possibly because of the persistence of certain embryonic enzymic mechanisms.

Finally, it must be emphasized that although the concept of tolerance seems to apply to immunological reactions provoked by antigens of widely different classes, the mechanisms underlying the tolerance-responses to these various kinds of antigen may differ considerably. This is to be expected from our knowledge of the different types of immunological response which antigens of different types exact from adult subjects.

### Acknowledgment

Some of the work described in this paper was supported by a grant (C-3577) from the National Institute of Health, U. S. Public Health Service.

## REFERENCES

1. Anderson, D., Billingham, R. E., Lampkin, G. H., and Medawar, P. B., *Heredity,* **5,** 379-397 (1951).
2. Billingham, R. E., *Ann. N. Y. Acad. Sci.* (in press).
3. ———, and Brent, L., *Proc. Roy. Soc. (London)* B, **146,** 78-90 (1956).
4. ———, and ———, *Transpl. Bull.,* **4,** 67-71 (1957).
5. ———, and ———, *Proc. Roy. Soc. (London)* B (in press).
6. ———, ———, and Medawar, P. B., *Nature,* **172,** 603-606 (1953).
7. ———, ———, and ———, *Phil. Trans. Roy. Soc. London, Ser. B,* **239,** 357-414 (1956).
8. ———, ———, and ———, *Nature,* **178,** 514-519 (1956).
9. ———, and Lampkin, G. H., *J. Embryol. Exptl. Morphol.,* **5,** 351-367 (1957).
10. ———, ———, Medawar, P. B., and Williams, H. Ll., *Heredity,* **6,** 201-212 (1952).
11. Booth, P. B., Plaut, G., James, J. D., Ikin, E. W., Moores, P., Sanger, R., and Race, R. R., *Brit. Med. J.,* **i,** 1456-1460 (1957).
12. Burnet, F. M., and Fenner, F., *The Production of Antibodies,* MacMillan, Melbourne (1949).
13. Cannon, J. A., *Transpl. Bull.,* **4,** 22-24 (1957).
14. ———, and Longmire, W. P., *Ann. Surg.,* **135,** 60-68 (1952).
15. ———, Terasaki, P., and Longmire, W. P., *Ann. Surg.,* **146,** 278-284 (1957).

16. Cinader, B., and Dubert, J. M., *Proc. Roy. Soc. London*, B, **141**, 18-33 (1956).
17. Danchakoff, V., *Am. J. Anat.*, **20**, 255-328 (1917).
18. Discussion on Immunological Tolerance, *Proc. Roy. Soc. London*, B, **141**, 1-92 (1956).
19. Dunsford, I., Bowley, C. C., Hutchison, A. M., Thompson, J. S., Sanger, R., and Race, R. R., *Brit. Med. J.*, ii, 81 (1953).
20. Egdahl, R. H., and Varco, R. L., *Ann. N. Y. Acad. Sci.* (in press).
21. Hašek, M., *Československ. Biol.*, **2**, 25-27 (1953).
22. ———, and Hašková, V., *Folia Biol.* (in press).
23. Hašková, V., *Folia Biol.*, **3**, 129 (1957).
24. Hraba, T., *Folia Biol.*, **2**, 165 (1956).
25. Humphrey, J. H., *Proc. Roy. Soc. London*, B, **146**, 34-36 (1956).
26. Krohn, P. L., *Nature*, **181**, 1671 (1958).
27. Lengerová, A., *Folia Biol.*, **3**, 333-335 (1957).
28. Lillie, F. R., *J. Exptl. Zool.*, **23**, 371-452 (1917).
29. Martinez, C., Aust, J. B., and Good, R. A., *Transpl. Bull.*, **3**, 128-129 (1956).
30. Medawar, P. B., *Proc. Roy. Soc. London*, B, **146**, 1-8 (1956).
31. ———, *Transpl. Bull.*, **4**, 72-74 (1957).
32. ———, and Russell, P. S., *Immunology* **1**, 1-12 (1958).
33. ———, and Woodruff, M. F. A., *Immunology* **1**, 27-35 (1958).
34. Mitchison, N. A., *Brit. J. Exptl. Pathol.*, **47**, 239-247 (1956).
35. Murphy, J. B., *J. Exptl. Med.*, **24**, 1-6 (1916).
36. Nicholas, J. W., Jenkins, W. J., and Marsh, W. L., *Brit. Med. J.*, i, 1458-1460 (1957).
37. Owen, R. D., *Science*, **102**, 400-401 (1945).
38. ———, *Federation Proc.*, **16**, 581-591 (1957).
39. Patterson, P. Y., *Ann. N. Y. Acad. Sci.* (in press).
40. Peer, Lyndon A., *Ann. N. Y. Acad. Sci.* (in press).
41. Puza, A., and Gomboš, A., *Transpl. Bull.* **5**, 30-32 (1958).
42. Roitt, I. M., and Doniach, D., *Proc. Roy. Soc. Med.*, **50**, 958-961 (1957).
43. Schinkel, P. G., and Ferguson, K. A., *Austr. J. Biol. Sci.*, **6**, 533-546 (1953).
44. Simonsen, M., *Ann. N. Y. Acad. Sci.*, **59**, 277-466 (1955).
45. ———, *Acta Pathol. Microbiol. Scand.*, **39**, 21-34 (1956).
46. ———, *Acta Pathol. Microbiol. Scand.*, **40**, 480-500 (1957).
47. Smith, R. T., and Bridges, R. A., *J. Exptl. Med.* **108**, 227-250 (1958).
48. Stormont, C., Weir, W. C., and Lane, L. L., *Science*, **118**, 695-696 (1953).
49. Talmage, D., *Ann. Rev. Med.*, **8**, 239-256 (1957).
50. Tandler, J., and Keller, K., *Deut. tierärztl. Wochschr.*, **19**, 148-149 (1911).
51. Terasaki, P. I., Longmire, W. P., and Cannon, J. A., *Transpl. Bull.*, **4**, 112-113 (1957).
52. Weiss, P., and Andres, G., *J. Exptl. Zool.*, **121**, 449-487 (1952).
53. Wells, H. G., *Arch. Pathol.*, **30**, 535-601 (1940).
54. Witebsky, E., *Proc. Roy. Soc. Med.*, **50**, 955-958 (1957).
55. Woodruff, M. F. A., *Transpl. Bull.*, **4**, 26-28 (1957).
56. ———, and Simpson, L. O., *Brit. J. Exptl. Pathol.*, **36**, 494-499 (1955).
57. ———, and Sparrow, M., *Quart. J. Exptl. Physiol.*, **43**, 91-105 (1958).

## DISCUSSION

Dr. Harris: Dr. Billingham, will mature red blood cells alone induce tolerance in embryos?

Dr. Billingham: I am quite sure that although red blood cells alone will confer tolerance with respect to the distinctive isoantigens of red cells themselves, they are unable to induce tolerance of living tissue cells because the important "transplantation" antigens present in the latter are not present in the red cells. Conversely, however, there are good grounds for believing that normal tissue cells, for example spleen cells, are able to confer tolerance in respect of subsequently inoculated red cells since antigens of the red cell type—those responsible for the formation of iso-agglutinins—are also present in most living tissue cells of the body.

DR. SAUNDERS: I would like to hear, Dr. Billingham, your ideas on the behavior of homografts between chick embryos. As I look around the room, I see that there are many of us who have made numerous grafts of this kind. I think we have in these grafts an opportunity for satisfying many of the conditions which are supposed to lead to tolerance: we have living cells, they are put on very early. Yet, whereas these grafts are often retained, a very high percentage of them are rejected. I think this is worth noting because, although laws of tolerance are so beautifully worked out, in the chick we do not see them operating in the manner we might expect.

DR. BILLINGHAM: I'm very glad that you have drawn attention to the fact that embryologists have been repeatedly producing evidence of tolerance of homologous, or even heterologous, tissue grafts for many years. Some of the best examples of tolerant birds I have ever seen were in Dr. Mary Rawles' laboratory. They were melanocyte chimeras. However this evidence, invariably produced *indirectly* during the course of other investigations, suffered from one serious shortcoming; it did not reveal whether the tolerance of the foreign tissue was the result of an adaption on the part of the graft or on the part of the host.

I think that the high proportion of homografts made between chick embryos which, as you say, are ultimately rejected is the result of two factors. Firstly, solid tissue grafts are probably less effective in conferring tolerance than cell suspensions, since wide dissemination of the antigenic stimulus appears to be necessary to induce a high degree of tolerance. Secondly, many of the environments into which embryologists implant their grafts may be unsuitable for inducing tolerance. For example, although a high degree of tolerance may be induced if homologous spleen cells are injected intravenously or intraperitoneally into newborn mice, injection of the cells subcutaneously is completely ineffective. Moreover, with avian embryos, genetic variability is probably another source of variation in the degree of tolerance obtained.

DR. HUGHES: I want to ask if it is possible to destroy tolerance by the host's system by the active transfer of antibodies or is this tolerance permanent?

DR. BILLINGHAM: Although humoral antibodies are formed when adult hosts are confronted with homografts there is no convincing evidence that these antibodies play any role in the rejection of homografts of normal tissue such as skin. All the available evidence suggests that if antibodies are responsible for homograft-rejection they are cell-bound to cells of lymphocytic type. Consequently the transfer of serum antibodies to a mouse tolerant of a homograft does not destroy the tolerant state; the foreign graft continues to thrive.

However, transplantation immunity can be transferred by means of cells obtained from the regional nodes of specifically sensitized hosts. A state of tolerance can be abrogated completely if a tolerant animal is inoculated with specifically sensitized lymphocytic cells of appropriate genetic constitution.

DR. ALLFREY: In some of the experiments you have mentioned there is a suggestion that there is a nuclear differentiation which is irreversible. In the case of the red cell, I believe you pointed out that nucleated red cells are not able to confer tolerance to other cells. This suggests that somehow the nucleoprotein of the nucleated red cell differs from that of other cells in the same organism. If the immune response is DNA-nucleoprotein dependent, as your experiments have suggested, then one could deduce that the DNA-protein of the erythrocyte is differentiated and altered beyond the extent of differentiation in other cells. Will an earlier stage in the development of the erythrocyte, e.g. the nucleated reticulocyte, confer tolerance?

DR. BILLINGHAM: So far as I'm aware this has not been investigated.

DR. WILLIER: I would like to comment on a very significant question which John Saunders has raised in connection with acquired tolerance in chickens. For instance if we take the limb bud from a White Leghorn embryo and transplant it into the coelom of the Barred Rock embryo, it undergoes differentiation into a normal-appearing limb which is covered with white down feathers. If the skin of this limb is then transplanted to the back of a newly hatched White Leghorn chick, in many cases it sloughs off, but in a number of cases it does not. In the most famous one, the bird lived for six years and is probably the one you saw, Dr. Billingham, when you were here. As to how to explain these results, we haven't had any satisfactory explanation, but I would like to call attention to the fact that Danforth long ago, before we began working on this question, transplanted skin from one newly hatched chick to another of different breed and got many cases of successful "takes." Are we justified in explaining these successful skin grafts solely on the basis of acquired tolerance? How are we to explain the manifold cases in which the grafts are rejected? We had once considered the possibility that the embryos and chicks used in most of our experiments came from a flock of White Leghorn chickens which were closely related genetically owing to much interbreeding and hence likely to give a high frequency of successful grafts.

DR. BILLINGHAM: I can't comment on that most interesting finding in the bird which I did have the privilege of seeing, Dr. Willier. However, I think in my table I indicated that the newborn chick was tolerance responsive for at least two days after hatching and it is true that Danforth, I think as long ago as 1929, and perhaps before, did discover that a reasonable proportion of skin homografts made to newly hatched chicks did survive indefinitely. That, I think, fits in perfectly well with the concept of tolerance. The newborn chick is still tolerance responsive and is rather like to newborn mouse in this respect. However, the tolerance responsiveness of the newly hatched chick is of a little longer duration than the mouse. I have grafted many newborn mice with skin homografts and they have always been rejected, but recently Woodruff and Medawar took the much more favorable newborn rat and they found that a proportion accepted their homografts, the rat will accept it for quite a long time—I think 70 or 80 days. However the degree of tolerance fell short of being complete. In those various experiments my interpretation is that the bird or newborn rat is still tolerance responsive and as long as the graft gets hooked up with a good vascular and lymphatic drainage soon enough, it will provide the proper antigenic stimulus necessary to touch off the tolerance response. I think Dr. Danforth was one of the earlier discoverers of tolerance and since his work, Cannon and Longmire and their associates also working in California have been using what they call "Danforth preparations."

# SELECTIVE UPTAKE AND RELEASE OF SUBSTANCES BY CELLS [1]

## Morgan Harris

*Department of Zoology, University of California, Berkeley*

### Introduction

The transfer of materials into and out of cells directly conditions a wide array of physiological processes and is especially crucial for the phenomena of growth and development to which this symposium is devoted. At the population level, cell transfers may determine the extent to which growing systems are dependent on a feedback equilibrium between cells and internal environment. The dynamic nature of these adjustments is particularly conspicuous for cell populations maintained in tissue culture, or in the evolution of neoplastic growths in vivo. Conceptual schemes have been proposed for the regulation of growth and differentiation, based on a reciprocal balance between intracellular synthesis and the concentration of inhibitors in the extracellular fluid (Weiss, 140-143; Tyler, 130; Rose, 108). A theoretical model is available for the expression of these relations in mathematical terms (Weiss and Kavanau, 144).

The transition from theory to specific mechanisms is limited in part by scarcity of information on selective uptake and release of substances by cells. The present discussion deals with two aspects of this broad field: (1) uptake or passage of proteins into cells, especially as related to growth mechanisms; and (2) the release of lactic acid as a measure of glycolysis and indicator of the energetics of growing cell systems. In each case the principal focus of interest lies in the behavior of cell populations maintained in tissue culture. As working models in cellular ecology, these systems offer an experimental approach to population dynamics.

### Sources of Intracellular Protein

The conventional view of protein synthesis assumes that amino acids are the immediate precursors for larger structural units, although the

---

[1] The original investigations described from the author's laboratory were aided by grants from the Atomic Energy Commission, Division of Biology and Medicine, and by Cancer Research Funds of the University of California.

mechanisms of assembly are not yet clear (Askonas, Simkin, and Work, 1; Loftfield, 85). Experimental proof for this assumption is available for microorganisms from studies on adaptive enzymes. In *E. coli* Hogness, Cohn, and Monod (68) have demonstrated that $\beta$-galactosidase is not derived from preformed proteins and is not in dynamic equilibrium with other proteins in the cell. In yeast the formation of adaptive enzymes is prevented by administration of amino acid analogues (Halvorson and Spiegelman, 61; Spiegelman and Halvorson, 118), a finding which suggests that adaptation involves protein synthesis de novo from the amino acid level.

In animals the synthesis of tissue proteins from amino acids follows as a natural consequence of protein breakdown during digestion and the distribution of free amino acids through the blood. Moreover, a number of specific studies of protein synthesis also indicate direct assembly from the amino acid level. From a study of relative isotope concentration with $C^{14}$-labeled glycine, Green and Anker (59) concluded that serum globulin is not a direct precursor of antibody and that antibody is synthesized from amino acids. Similarly, Simpson and Velick (117) studied the formation of aldolase and glyceraldehyde-3-phosphate dehydrogenase in the rabbit, using a mixture of five labeled amino acids. The results indicated synthesis of both enzymes from a common pool of amino acids, although the rates were different. Christensen (16) and Loftfield (85) in more comprehensive discussions express the general view that protein synthesis in tissue cells, as in microorganisms, occurs by direct combination of amino acids as structural units.

The work of Whipple and his collaborators (for summaries see Madden and Whipple, 87; Whipple, 148) on the dynamic equilibrium of body proteins provides a different outlook. A long series of reports (Whipple, 148) has suggested that plasma proteins are in equilibrium with tissue proteins and if necessary can serve as a sole nitrogen source. Terry, Sandrock, Nye, and Whipple (128) demonstrated that dogs could be maintained in good health and weight equilibrium for one to three months by intravenous or intraperitoneal administration of dog plasma, combined with a protein-free diet of fat, carbohydrates, and minerals. Yuile, Lamson, Miller, and Whipple (151) and Yuile, O'Dea, Lucas, and Whipple (152) extended this study by using plasma proteins labeled with $C^{14}$-lysine. If administered orally, the labeled lysine was utilized as if fed in the free amino acid form, with rapid loss of the label as respiratory $CO_2$. In contrast, if the labeled proteins were injected intravenously the respiratory loss was only one-tenth as great and about 60 per cent remained after a week, two-thirds of this in the body tissues. In these experiments there was no attempt to determine the degree of conversion of plasma proteins to tissue proteins. Humphrey and McFarlane (74) subsequently found with the transfusion of labeled globulins that injected antibodies entered the body tissues and rapidly underwent transformation to some other kind of intracellular protein, losing in this process the original specific immunologic configuration.

The dynamic equilibrium of proteins, as described by Whipple, may be a complementary phenomenon to the more common origin of proteins from amino acids. Little is known as yet of metabolic alternatives in protein synthesis. The use of precursors other than the free amino acid pool needs further examination under varied conditions, especially in open systems, i.e., embryonic organs, tissue cultures, and tumors. Some available data on this point are outlined in the following sections.

### Evidences of Protein Uptake

According to conventional views of permeability, proteins and other macromolecules do not enter or leave cells except under special circumstances (e.g., the secretion of digestive enzymes). This concept requires modification in the light of experimental studies which indicate penetration of cells by proteins under a variety of conditions, without preliminary breakdown. Schechtman (115) has reviewed this field broadly; additional aspects are covered by Ebert (39), Danielli (23), and Brambell and Hemmings (5). The principal lines of information can be summarized under the following headings.

#### Uptake of Proteins by the Avian Ovum

Knight and Schechtman (76) injected crystalline bovine serum albumin intravenously into hens and examined the protein content of freshly laid eggs during the following days. Large amounts of bovine protein accumulated in the egg yolk and could be estimated by serologic methods. Up to 15-20 mg. bovine protein per yolk were found in eggs laid 5 to 8 days after the original intravenous injection. Additional studies showed that the bovine proteins from egg yolk retained their normal electrophoretic mobilities and appeared to be transferred into the ovum with little or no structural alteration.

#### Immunochemical Studies

Coons and his collaborators (Coons and Caplan, 20; Coons, 19) have developed a method for visual detection of antigens through the use of fluorescent antibodies. Antisera are conjugated with fluorescein isocyanate and added to tissue sections containing the corresponding antigens. Fluorescence of the antigen-antibody precipitate indicates visually the specific localization of antigens within cells. Following injection of heterologous proteins (crystalline egg albumen, human gamma globulin, crystalline bovine albumen), Coons (19) found that antigen localizes within the parenchymal cells of the liver, in lymphocytes, fibroblasts, and renal epithelial cells, as well as in the connective tissue and the reticulo-endothelial system generally.

Gitlin, Landing, and Whipple (56) used the fluorescent antibody technique to demonstrate the presence of $\beta$-lipoprotein and other human plasma proteins in human tissue cells. Both nuclear and cytoplasmic localization were observed in some cases. A dynamic shift in blood proteins was also found in studies on patients with specific plasma protein deficiencies (Gitlin, 55). In patients lacking blood gamma globulin, fluorescence was essentially absent in biopsy tissues treated with the corresponding antibody. After a transfusion with normal human serum, however, strong fluorescence was observed in extravascular sites. Similarly, tissues from patients with congenital deficiency of fibrinogen showed only a faint response to specific antibody. After infusion of fibrinogen strong fluorescence was observed in biopsies of skin or muscle. These results are consistent with findings from earlier immunological studies using dye-labeled proteins, but indicate that passage of proteins from blood to tissues is not confined to uptake by the reticulo-endothelial system.

## Use of Proteins Labeled With Radioisotopes

Several investigators have studied the distribution of $I^{131}$ in animals injected with proteins carrying this label. Crampton and Haurowitz (22) injected $I^{131}$-labeled ovalbumin and beef serum pseudoglobulin into rabbits. Cell fractions were obtained from liver by differential centrifugation. The results indicated that the greater part of the activity was associated with the mitochondria. In similar experiments with human gamma globulin and serum albumin in mice, Fields and Libby (44) also observed high activity in the mitochondria. More recently, Haurowitz and associates (127) have injected homologous proteins into rabbits, using serum albumin and serum globulin doubly labeled with $S^{35}$ and $I^{131}$. The $S^{35}:I^{131}$ ratio was essentially unchanged for several days in the plasma proteins but rose rapidly in the cell proteins of liver, lung, spleen, and other tissues. The authors conclude that the labeled proteins enter tissue cells as intact molecules but are rapidly degraded to smaller units before incorporation into cellular proteins.

Busch, Greene, and their collaborators (9, 10) have presented evidence for the selective uptake of plasma proteins by tumor cells of the Walker carcinosarcoma in rats. Homologous serum albumin and serum globulin were labeled with $C^{14}$-lysine and injected intravenously into tumor-bearing rats. Normal and tumor tissues were subsequently fractionated by differential centrifugation and the distribution of the label determined. In animals injected with labeled albumin the specific activity of proteins in the whole homogenate was three times those of the liver, kidney, spleen, and testis, but not significantly greater than that of the lung. The specific activity of the microsomal proteins of the tumor was four to six times that of other tissues, including the lung. A similar contrast was evident in the whole homogenates of tumor and normal tissues from animals injected with labeled globulins, although the differential was less.

*Transfers From Organ Transplants*

The enlargement of the spleens of chick embryos as a result of chorioallantoic grafts of adult chicken spleen was first reported by Murphy (95) and has been studied by a number of workers, especially Ebert (38, 39). Ebert labeled the donor spleen with $S^{35}$-methionine and observed a rapid transfer of label from graft to the proteins of the host. This effect was not obtained when free $S^{35}$-methionine was injected into the yolk sac, and administration of unlabeled methionine did not affect the transfer of the protein-bound label. Grafts of organs other than spleen were less effective in stimulating the host spleen, and transplants of mouse spleen did not cause enlargement of the host organ. Specific transfer of protein-bound label to the corresponding host structure was also observed with kidney transplants. These results suggest the transfer of tissue-specific molecules from donor to host tissues. Over a three-day period following transplantation, as much as 15 per cent of the proteins in the host spleen may be derived at least in part from the proteins of the donor (Ebert, 39). These experiments dealing with a balance between tissues of differing physiologic age illustrate particularly well the possibilities of metabolic regulation in growth and protein synthesis.

*Protein Uptake by Kidney Cells*

Fractionation of kidney cells by differential centrifugation of homogenates yields a "droplet" component, rich in enzymes, at the bottom of the mitochondrial fraction (Strauss, 120, 121). A number of investigators have proposed that these droplets from kidney cells are associated with absorption and metabolism of foreign or native proteins (Gerard and Cordier, 49; Oliver, MacDowell, and Lee, 97; Rather, 106). In experiments with rats, Strauss and Oliver (124) showed that kidney droplets isolated after the injection of egg white contained several times as much serologically reacting egg white as the other fractions of the kidney homogenate. This accumulation was associated with an increase in size of droplets and a release of enzymes from the droplets into the supernatant fluid (Strauss, 122). Subsequently, Strauss (123) provided further proof of the ability of kidney droplets to accumulate extracellular protein by injecting rats with a plant enzyme, horse-radish peroxidase, which could be determined colorimetrically in cell fractions. The results indicated specific activities of peroxidase which were five to eight times higher in droplet fractions than in other fractions of the kidney homogenates.

## Mechanisms of Entry of Macromolecules

The well-known phenomenon of phagocytosis has offered a provisional explanation to immunologists and others for the selective uptake of antigens or other proteins. This view is expressed typically by Haurowitz (67), who contemplates the entry of large antigen molecules into cells as "a process akin to phagocytosis, an engulfing of the antigen molecules by pseudopodia or protuberances of the cytoplasm."

In the classical sense, phagocytosis refers to the entrapment of particles and appears to be restricted to certain cell types, such as blood elements, macrophages, and sinusoidal cells of the reticulo-endothelial system. The closely related phenomenon of pinocytosis, however, seems to occur much more generally among cells. This process, discovered by Lewis (83, 84), deals with the uptake of fluid globules by the cell surface. The globules move toward the interior of the cell and eventually shrink and disappear, leaving the contents in the cytoplasm. Comprehensive studies have revealed that pinocytosis can occur widely among cells in tissue culture. Gey (52) states that in most living animal cells which can be cultured, lapsed-time photography reveals "rapid extrusion and retraction of microfibrils at the cell surface, and formation of membranous pseudopodia or ruffles, with some aspects of these structures accomplishing occasional and sometimes abundant pinocytosis." These observations stress the dynamic nature of the cell surface and provide a useful counterbalance to interpretations of permeability based solely on physicochemical characteristics of the plasma membrane.

There appears to be no lower limit in the size of globules taken in by pinocytosis, and in amoebae Holter and Marshall (69) found globules varying from 10μ down to the limits of detection with the light microscope. That pinocytosis or a related phenomenon also occurs at the submicroscopic level is suggested by studies of the cytoplasm and cell membranes of tissue cells with the electron microscope (Howatson and Ham, 71). In electron micrographs the free surfaces of cells often appear convoluted. It has been suggested by several investigators (Howatson and Ham, 70; Palade, 98), that the invaginations commonly seen at the cell surface may pinch off and move toward the center of the cell as part of the endoplasmic reticulum, a system of cytoplasmic membranes and vesicles (Porter and Kallman, 104).

It thus appears that an array of mechanisms may exist for the penetration of tissue cells by macromolecules under appropriate conditions, and without conflict with the conventional picture of the cell membrane as a lipid-protein matrix of limited permeability. The participation of other processes, e.g., facilitated diffusion and active transport mechanisms (Danielli, 23), in the entry of proteins remains to be explored.

### Protein Growth Factors in Tissue Culture

The role of extracellular proteins for cell growth has received little systematic attention in the past, although in the context of the preceding discussion this problem deserves careful study. A large number of pioneer investigations have been published dealing with the effects of tissue extracts on cells in vitro (see list of references in Murray and Kopech, 96), but specific information is limited since these materials as well as the assay systems have been poorly defined for the most part.

*Embryonic Extracts*

The early studies of Carrel (11) and other workers showed that extracts from various tissues, and especially from embryos, stimulated proliferation and cell migration in vitro. These experiments were done primarily with chick fibroblasts cultivated in a plasma clot and indicated activity in both dialyzable and nondialyzable components of embryo extract (Fischer, 45; Harris, 62). The macromolecular component appeared to be associated with a nucleoprotein fraction (Fischer and Astrup, 46; Davidson and Waymouth, 24). Details of these early investigations may be found in reviews by Fischer (45), Willmer (149), Stewart and Kirk (119), and Biggers, Rinaldini, and Webb (4). More recently, Kutsky (78) has isolated a ribonucleoprotein fraction of high activity from chick embryos by selective precipitation with streptomycin from saline extracts. This fraction stimulates marked proliferation of chick fibroblasts in plasma cultures and shows a synergism with dialysate growth factors (Harris and Kutsky, 65).

Mechanisms for the utilization of embryonic extract by explanted tissues have been studied by Winnick and his collaborators in a series of investigations. Gerarde, Jones, and Winnick (50) demonstrated extensive net synthesis of protein in roller tube cultures of chick lung tissues maintained in embryonic extract. Incorporation of $C^{14}$-labeled amino acids into cellular proteins was observed but occurred more rapidly under conditions of tissue autolysis than under conditions favorable to growth. No net increase in protein could be demonstrated for tissues cultivated in synthetic media lacking protein. In 1953 Francis and Winnick (47), using a balance sheet analysis, showed that protein increases in chick heart cultures maintained in embryonic extract could be accounted for by corresponding decreases in protein and amino acids in the medium. During the initial stages of growth, tissue protein was synthesized chiefly from the amino acids, but after a few days over half of the tissue protein was derived from protein of the embryo extract. In other experiments, the proteins of embryonic extract were labeled by injecting $C^{14}$-labeled amino acids into the eggs at an early stage of development. From experiments in which relatively large concentrations of nonradioactive amino acids were added as metabolic traps, it appeared that most of the $C^{14}$ of the labeled extract proteins could be transferred from the nutrient medium to the heart tissue protein without the release of free amino acids. Structural analogues depressed the uptake of free amino acids from the medium, but the most potent of these, fluorophenylalanine, did not inhibit the transfer of $C^{14}$ from phenylalanine-labeled protein of the medium to tissue protein. These results have been extended in further investigations (Winnick and Winnick, 150; Lu and Winnick, 86) and show in general that cultures of chick embryonic tissues can utilize protein of the nutrient medium extensively without complete hydrolysis to the amino acid level.

The recent studies of Walter and Mahler (133) suggest a similar pattern for the growth of chick embryonic cells in vivo. Radioactive homologous and heterologous proteins, peptides, and amino acids were injected into the yolk of hens' eggs, and the embryos were harvested after five to nine days incubation. Dry powders were prepared from embryonic proteins. From a study of incorporation in these materials the authors conclude that proteins are the preferred precursors for chick embryonic proteins, in accord with the view of Francis and Winnick.

## Nucleoproteins From Adult Tissues

Growth stimulation by adult as well as embryonic tissue extracts was noted by Trowell and Willmer (129) and has been extensively studied by Doljanski and his collaborators (25, 26, 89). Saline extracts or acetone precipitates were prepared from a variety of adult chicken organs and induced marked proliferation in plasma assay cultures. Using more selective separations, Kutsky and Harris (79) were able to obtain active nucleoprotein fractions (NPF) from adult chicken brain, liver, heart, and spleen. These fractions caused marked increases in the size and glycolytic activity in plasma cultures of fibroblasts from chick heart, skeletal muscle, and frontal bone. No indications of organ specificity were observed and the specific growth-promoting activity of nucleoprotein fractions remained unchanged from embryo to adult.

To study the effects of these nucleoproteins on growth rates, appropriate procedures were adapted for routine culture of freshly isolated chick skeletal muscle fibroblasts as monolayers on glass (Harris, 63). Although growth stimulation could be obtained with nucleoprotein fractions from all adult organs tested (Harris and Kutsky, 66), the effect of adult spleen NPF was particularly marked (Fig. 1). The addition of spleen NPF to cultures containing chicken serum and a synthetic nutrient, Medium 199 (Morgan, Morton, and Parker, 94) regularly produced logarithmic growth, and the cultures could be maintained for a number of serial passages in this medium with no diminution in growth rate or change in appearance of the cultures. In the experiments mentioned, growth did not occur in chicken serum + 199 alone. In this medium, the cell populations rapidly became heteromorphic and could not be maintained in serial subculture. Likewise it was not possible to cultivate these cells in spleen NPF and Medium 199 alone, without the presence of chicken serum. Even if the total protein concentration was adjusted over a considerable range, the cell populations in NPF-199 remained in a contracted state on the glass surface, without proliferation, and glycolysis was completely inhibited. Omission of Medium 199 from the complete medium did not prevent normal proliferation, but growth took place at a reduced rate. These observations suggest that serum proteins and cellular proteins are not equivalent in growth-promoting properties, and indicate an interaction of protein and crystalloidal growth factors.

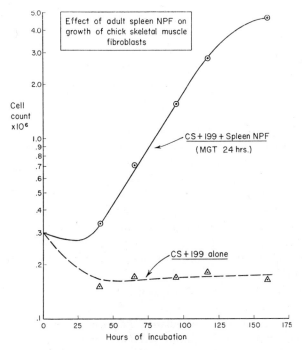

Fig. 1. The effect of a nucleoprotein fraction (NPF) from adult chicken spleen on the growth of chick skeletal muscle fibroblasts. Medium includes 10% chicken serum, 50% Medium 199, and 0.4 mg./ml. spleen NPF.

## Utilization of Serum Proteins

With the introduction of methods for culturing cells on a glass substrate (Evans, Earle, Sanford, Shannon, and Waltz, 43; Dulbecco, 27; Younger, 153) or in free suspension (Earle, Schilling, Bryant, and Evans, 37; Kuchler and Merchant, 77; Siminovitch, Graham, Lesley, and Nevill, 116) the classic plasma coagulum could be eliminated for certain types of nutritional studies. For a number of established cell strains the macromolecular components of tissue extracts also proved dispensable (Sanford, Waltz, Shannon, Earle, and Evans, 112; Eagle, 31; Siminovitch et al., 116), the cell populations growing reproducibly in dilute serum supplemented only by crystalloidal nutrient factors.

In these simplified media the contribution of serum proteins to cell growth can be more readily determined. Westfall, Peppers, and Earle (145) maintained a strain of human malignant epithelial cells (Strain

HeLa) in mass shaker flasks and determined the sources of intracellular protein. The medium contained 30 per cent human placental serum and a protein-free ultrafiltrate from fresh hens' eggs. Calculations based on the decrease in amino acids in the medium indicated that only one-third of the cellular protein was derived from free amino acids, the remainder apparently coming from proteins of the serum. Similarly, Kent and Gey (75) have presented evidence for direct utilization of serum protein by rat tumor cells maintained in roller tubes in 50 per cent human placental serum plus balanced saline. Protein analyses of the supernatant fluid indicated a progressive drop in serum protein during cell growth, and Kjeldahl determinations showed that this decrease was paralleled by a corresponding drop in total nitrogen in the medium.

Using the Cohn low temperature fractionation procedure, Sanford, Westfall, Fioramonti, McQuilkin, Bryant, Peppers, Evans, and Earle (113) studied the effect of a variety of fractions from horse serum on the proliferation of Strain L mouse cells in glass substrate cultures. All protein fractions isolated were found to promote the growth of Strain L cells when tested in a protein-free basal medium. Growth stimulation by individual serum protein fractions was inversely proportional to the amount of chemical and physical manipulation necessary to separate and process the fractions. The gross globulin fraction could be substituted for the large molecular portion of horse serum to yield comparable rates of increase in cell numbers. Protein uptake was not investigated in these experiments. Graff and McCarty (58), however, have maintained L cells in continuous fluid suspension cultures, and report that protein nitrogen determinations indicate a utilization of serum protein.

The investigations of Eagle and coworkers are oriented in a different direction than the foregoing discussion and have stressed the contribution of crystalloidal rather than protein components of the culture medium for cell growth. Eagle (28) found that L cells would proliferate in a mixture of amino acids, vitamins, salts, carbohydrates, and cofactors if the medium was supplemented with a small amount of dialyzed protein. Omission of a single essential component resulted in the early death of the culture. By this means it has proved possible to determine amino acid requirements as well as a number of essential vitamins and other nutritional factors for several cell strains (Eagle, 28-32; Eagle, Oyama, and Levy, 33; Eagle, Oyama, Levy, and Freeman, 34). The role of the dialyzed serum protein in the system has not yet been clarified. It seems apparent (Eagle, 29) that the dialyzed protein is not providing free amino acids in significant amounts, in view of the clear-cut inhibition of growth if a single essential amino acid is omitted in the synthetic supplement. The possibility that the protein contributes trace elements or vitamins to the medium has been suggested (Eagle, 30), an explanation also favored by Sato, Fisher, and Puck (114) in a related study with HeLa cells. The possible physical role of extracellular protein in preserving the integrity of cell surfaces has also been stressed (Biggers, Rinaldini, and Webb, 4).

There is another possible explanation of Eagle's findings which may reconcile his data with other investigations indicating utilization of serum proteins. In cultures of chick tissues, utilization of proteins from embryo extract is substantial in the presence of simultaneous amino acid uptake and incorporation (Francis and Winnick, 47). Growth of similar cultures in dialyzed embryo extract, however, was completely suppressed (Lu and Winnick, 86). Glucose was available in the saline fraction, and it is unlikely that the direct inhibition of growth was caused by vitamin deficiencies, which develop more gradually as intracellular stores are depleted (Eagle, 30). It seems possible that a minimal synthesis of protein from amino acid precursors may be an essential condition for the utilization of media proteins by direct incorporation. Thus a deficiency of lysine or other essential amino acid, as in Eagle's experiments, might block both amino acid and protein utilization, but would not constitute conclusive proof that serum protein could not be utilized in the presence of a complete medium. It should be emphasized that the minimal levels of dialyzed serum found necessary by Eagle for optimal growth are appreciable, particularly at the 5 per cent level required for HeLa cells. Inasmuch as Westfall et al. (145) have presented evidence that HeLa cells can utilize large quantities of serum proteins under some conditions, a further study of the role of dialyzed serum in Eagle's system seems indicated. Nitrogen balance studies as well as labeled proteins and metabolic traps may be useful in resolving this problem.

*Evolution of Cell Strains*

It is apparent from the literature that cell populations differ considerably in their requirements for growth and maintenance in vitro. In part this may reflect species differences as well as the innate characteristics of various cell types. Parshley and Simms (100, 101) presented evidence that human adult epithelium and fibroblasts have different ionic requirements. Chick fibroblasts require embryo extract for proliferation, but monocytes grow progressively in dilute serum (Carrel and Ebeling, 12; Baker, 2). Puck,

Cieciura, and Fisher (105) found that freshly isolated epithelioid cells from human tissues will proliferate in serum alone plus a synthetic nutrient, while fibroblastic cells grow best if embryonic extract is added.

A more important factor, however, is the possibility of a progressive shift or evolution in the nutritional requirements of a cell strain during extended serial cultivation in vitro. This process is illustrated particularly well by Earle's well-known Strain L mouse sarcoma cells (Earle, 35), which originated in 1940 from normal subcutaneous connective tissue of an adult C3H strain mouse. The initial cultures were established in a chicken plasma coagulum, with a supernatant fluid containing horse serum and chick embryo extract. In early stages this strain was exposed to 20-methylcholanthrene, and the cell populations underwent permanent alterations in morphology, eventually producing sarcomas on injection into C3H mice (Earle, 36). Similar morphological changes and neoplastic transformation, however, took place in mouse fibroblast cultures not exposed to carcinogens (Sanford, Earle, Shelton, Schilling, Duchesne, Likely, and Becker, 110). In 1947, Strain L cells were first cultivated on a glass substrate (Evans and Earle, 42) and a clone culture was subsequently obtained of single-cell origin (Sanford, Earle, and Likely, 109). At this stage, experiments showed that the clone strain no longer required the macromolecular component of embryo extract, although serum protein was essential for progressive growth, in combination with ultrafiltrate from embryo extract (Sanford, Waltz, Shannon, Earle, and Evans, 112). Eventually, it proved possible to dispense entirely with serum proteins for certain substrains of the L line, the cells multiplying progressively in completely synthetic media for indefinite periods (Evans, Bryant, Fioramonti, McQuilkin, Sanford, and Earle, 40; Evans, Bryant, McQuilkin, Fioramonti, Sanford, Westfall, and Earle, 41). In a recent study McQuilkin, Evans, and Earle (91) examined the transitional changes in adaptation of Strain L cells to a protein-free medium. Fifty-six sublines from clone 929 were started in the defined nutrient and ten were successfully adapted. In practically all cases the cultures underwent a gradual or abrupt change in cellular morphology, rate of population growth, and level of population density. If a critical period of variable onset and duration was passed, the cultures became stable and grew progressively in the synthetic nutrient, although at a lower rate than in the presence of horse serum. Cultures adapted to the synthetic nutrient could also be cultivated in a serum–embryo-extract medium, but if then returned again to the protein-free nutrient the cell populations grew directly without a period of adaptation.

These results suggest strongly that the L strain has undergone progressive simplification in cultural requirements during the long period of culture in vitro, specifically with respect to extracellular protein. Progressive growth in protein-free nutrients has not been reported for any freshly isolated cell strain, and suspensions of chick fibroblasts are lysed if directly inoculated into synthetic nutrients (Biggers, Rinaldini, and Webb, 4). The changes seen in the evolution of the L strain are not typical of phenotypic adaptation since they are apparently irreversible. Although the data suggest successive mutation and selection, specific evidence on this point is still lacking.

As other recent evidence has accumulated, it appears that the evolution of cell lines in vitro is a general phenomenon. Such lines can be initiated from fresh explants or monolayers maintained in serum with supplemental nutrients, commonly embryo extract (Parker, Castor, and McCullock, 99; Swim and Parker, 126; Syverton and McLaren, 127; Chang, 13; Zitcer and Dunnebacke, 154; Westwood, MacPherson, and Titmuss, 146; Westwood and Titmuss, 147; Berman, Stulberg, and Ruddle, 3). After a number of passages in which growth may proceed vigorously there is a diminution in growth rate, the cultures go into a decline, and the majority of cell strains are lost. If the cell strain survives this phase, however, a new cell type may emerge with altered morphology, often from a localized area of a single culture. These altered cells do not resemble cell types obtained from normal tissues but may proliferate indefinitely in vitro as established strains. Recent studies indicate that in addition to morphological changes, these stable cell lines are also atypical in other respects. All established strains thus far examined have proved to have markedly abnormal chromosome numbers, usually hypotetraploid (Hsu and Moorhead, 72; Berman, Stulberg, and Ruddle, 3; Westwood and Titmuss, 147; Chu and Giles, 17). The significance of the polyploid state is not yet apparent, although it has been suggested that genotypic diversity may favor the emergence of a stable type adapted for growth under conditions of culture (Hsu and Moorhead, 72). Neoplastic transformations in cell strains have been demonstrated by Gey (51), by Gey, Gey, Firor, and Self (53), and by Goldblatt and Cameron (57), as well as in the studies of Earle described above. Moore, Southam, and Sternberg (93) and Moore (92) have presented additional evidence of tumorous characteristics in supposedly "normal" cell lines. That variation in established cell strains is not restricted to the initial conversion to a stable line is indicated by the studies of Sanford, Likely, and Earle (111). Starting with a clone line of L cells, sublines were eventually obtained which differed markedly in morphology and in ability to produce tumors. Similarly, Chang (14) has been able to obtain variant

HeLa cell populations which can proliferate continuously on D(+)xylose, D-ribose, or sodium lactate as a carbohydrate source, although these substrates did not support growth initially in stock cultures.

The picture which emerges suggests that so-called stable strains cannot be considered as normal cells but rather are variant populations stabilized for a particular set of cultural conditions, and capable of additional variation in still other directions if subjected to appropriate selection pressure (e.g., protein-free media or alternative carbohydrate sources). The nutritional characteristics and other properties of these specialized strains are of considerable interest but are not necessarily representative of cell populations in general, especially those freshly isolated from the intact organism.

## Glycolysis and Growth

### Energy Sources for Growth in Vitro

The available evidence indicates that glucose or other carbohydrate is the principal energy source for isolated tissue cell populations. A respiratory quotient of 1.0 has been observed in a number of different tissue cultures (Phillips and Feldhaus, 102), and high rates of glucose utilization are characteristic for HeLa cells (Westfall, Peppers, and Earle, 145), Walker rat sarcoma cells (Gemmill, Gey, and Austrian, 48), and L cells (Graff and McCarty, 58), as well as a variety of other normal and malignant cell strains (Leslie, Fulton and Sinclair, 82). Cells in vitro exhibit selectivity in the utilization of substrates other than glucose. Chick heart fibroblasts can utilize D-fructose or D-mannose as well as D-glucose. L-glucose, pentoses, sucrose, and lactose are inert for these cells (Harris and Kutsky, 64). Essentially similar results have been reported for stock cultures of HeLa cells, although these cells can utilize D-galactose as well (Chang and Geyer, 15). The production of variant HeLa strains able to grow on D(+)xylose, D-ribose, or sodium lactate as a carbohydrate source (Chang, 14) has already been described.

Both respiration and glycolysis occur in isolated cell populations, although the relative contributions of these two processes for growth have seldom been evaluated. Oxygen consumption has not generally been regarded as an index of cell growth (Willmer, 149), but Gifford, Robertson, and Syverton (54) found in HeLa cultures that the rate of change of oxygen uptake was correlated with cell multiplication. Leslie, Fulton, and Sinclair (82) analyzed the total glucose utilization of normal and malignant cell strains. A high correlation was observed between increases in nucleic acid phosphorus and the portion of glucose which did not appear as lactic or keto acids (glucose "oxidized"). Other evidence shows that some cell strains, especially from the early embryo, will proliferate at least temporarily under anaerobic conditions if supplied with glucose (Burrows, 8; Laser, 80; Harris and Kutsky, 64). Cells in tissue culture are also for the most part less sensitive to cytochrome inhibitors such as carbon monoxide or cyanide (Swann, 125) and are unaffected by malonate or fluoroacetate at moderate concentrations (Hughes, 73). On the other hand, iodoacetate and fluoride, which inhibit glycolysis, selectively block mitotic activity in cultures (Pomerat and Willmer, 103; Hughes, 73). Swann (125) expresses the view that glycolysis may be essential for cell division, and that where glycolytic powers are poor, respiration through the Krebs cycle is also essential.

*Production of Lactic Acid in Tissue Cultures*

The interrelations of glycolysis and growth remain quite obscure, in part because conventional manometric studies of glycolysis have not revealed the nature of this association. It is difficult within the limitations of the Warburg technique to follow the glycolysis of intact, growing populations over a period of time. Moreover, there is no yardstick for evaluating the objectivity of glycolytic rates determined manometrically, particularly since the composition of suspending medium may affect the values obtained (Warburg, 135), and even the ratio of tissue to fluid in the Warburg flasks can greatly influence glycolysis in some cases (McIlwain and Rodnight, 90; Rodnight and McIlwain, 107).

As an alternative, direct measurements of glycolysis can be made on intact growing populations, using a chemical rather than manometric method for determination of lactic acid production. For such studies, skeletal muscle fibroblasts obtained from chick embryos are particularly appropriate. As freshly isolated strains, these cells grow logarithmically, forming a stable monolayer on a glass substrate with little or no cell loss into the supernatant fluids (Harris, 63; Harris and Kutsky, 66). Lactic acid is conveniently determined from samples obtained by a daily 50 per cent change of the supernatant fluid. From these data and a value for the blank medium a cumulative curve for total lactic acid can be constructed over the corresponding time interval (Fig. 2). The cumulative curve for lactic acid may taper off abruptly if the population stops increasing, a relation which suggests that the production of lactic acid noted is essentially an aerobic glycolysis. A Pasteur effect, moreover, can be observed in such cultures by the conspicuous resumption of glycolysis if the air phase over maximal populations is replaced by nitrogen. It should be noted, however, that

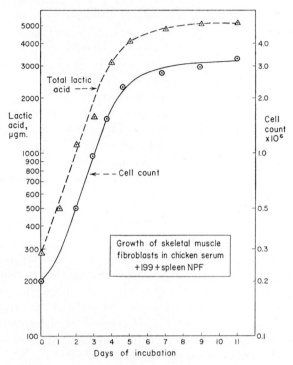

Fig. 2. Cumulative curves for lactic acid and cell number in cultures of chick skeletal muscle fibroblasts. Medium as described for Fig. 1.

some cell strains which reach large limiting cell numbers may continue to glycolyze at a reduced rate in the maximum stationary phase, presumably owing to a partial anaerobiosis in the crowded cell sheets.

A low aerobic glycolysis in these limiting populations might be expected from the wide array of data on manometric measurements of tissue slices or other preparations in the Warburg apparatus (Burk, 6; Greenstein, 60; Warburg, 135). These data indicate low values for the aerobic glycolysis of most nongrowing normal tissues. On the other hand, the data shown in Fig. 2 and from many similar experiments stress the constant association of a high aerobic glycolysis with the logarithmic growth phase of isolated cell populations. In experiments with chick skeletal muscle fibroblasts, we have found it convenient to express glycolytic rates during logarithmic growth in terms of an index, $\beta$, representing the release of lactic acid/cell/ unit time. Since the total cell volume is a negligible fraction of the medium, the values obtained approximate the production of lactic acid by cell populations. The derivation of $\beta$ for cultures in logarithmic growth can be indicated as follows.

Let $K$ = per cent growth rate in terms of cell multiplication,
and $\beta$ = lactic acid produced/cell/unit time,
and $N_0$ = an arbitrary cell number obtained by extrapolating the log phase curve to zero time.

If $A_1$ and $A_2$ represent the total lactic acid at intervals $T_1$ and $T_2$ during logarithmic growth,

$$A_2 = A_1 + \beta \int_{T_1}^{T_2} N_0 e^{KT} \, dt \tag{1}$$

By further evaluation,

$$A_2 - A_1 = \frac{\beta}{K} [N_0 e^{KT_2} - N_0 e^{KT_1}]$$

$$A_2 - A_1 = \frac{\beta}{K} [N_2 - N_1]$$

$$\beta = K \left[ \frac{A_2 - A_1}{N_2 - N_1} \right] \tag{2}$$

The relative growth rate, $K$, can be expressed as

$$K = \frac{\ln N_2 - \ln N_1}{T_2 - T_1}$$

or in terms of common logarithms,

$$K = \frac{2.303 \, (\log N_2 - \log N_1)}{T_2 - T_1}. \tag{3}$$

Both $K$ and $\beta$ can be evaluated from equations (2) and (3), using data obtained during logarithmic growth phase on cell counts and total lactic acid. The validity of this approach can be illustrated with measurements from cultures maintained in homologous serum supplemented with adult spleen nucleoprotein. In the experiment shown in Fig. 3 the production of lactic acid remained constant at about $5.5 \times 10^{-5}$ µg./cell/hr. until late log phase, after which glycolysis declined abruptly. This strain of cells grew vigorously for a number of serial passages in vitro without diminution of growth rate or alteration in the cell population.

In the case of other cell strains maintained in chicken serum spleen nucleoprotein, the glycolytic rate often declined earlier in log phase, and for cultures grown in serum and embryo extract glycolysis decreased from

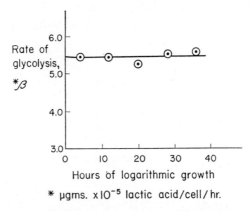

Fig. 3. Aerobic glycolysis during logarithmic growth of chick skeletal muscle fibroblasts. Medium as described for Fig. 1.

the outset. This is illustrated by the trend of the curves for total lactic acid and cell number shown in Fig. 4. In these cultures the decline in glycolytic rate could be expressed by a series of $\beta$ values, estimated successively in log phase by assuming $\beta$ as constant for short intervals. Using these values in a test of validity, total lactic acid curves could be reconstructed as accurately as for cultures with a constant rate of glycolysis (Fig. 5).

On the basis just described, we have calculated glycolytic rates for chick skeletal muscle fibroblasts during logarithmic growth in a variety of media. In early log phase these values range from approximately 1.0 to $7.0 \times 10^{-5}$ $\mu$g. lactic acid/cell/hr. As seen in Table 1, cultures in chicken-serum–spleen-NPF showed rapid growth as well as high glycolysis, while for cultures in serum–embryo-extract the rates of these two processes appeared to have an inverse relationship. Further experiments confirmed this interpretation (Figs. 6, 7). In either homologous or heterologous serum at 10 per cent embryo extract, the initial level of glycolysis was proportional to the serum concentration, although the rate of decline was of the same order of magnitude in each case. The growth rates of these cultures, as in previous studies (Harris, 63) decreased as the concentration of serum was raised from 20 per cent to 40 per cent, although the difference in rate of proliferation was not great at 10 per cent and 20 per cent serum, respectively. The data in Fig. 7 show further that glycolysis during log phase is inversely

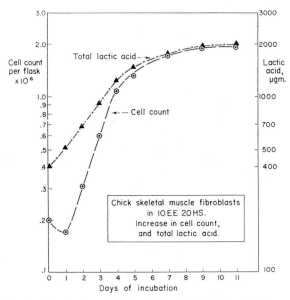

Fig. 4. Lactic acid and cell number for cultures maintained in 20% horse serum, 10% chick embryo extract, 70% balanced saline (Earle's solution).

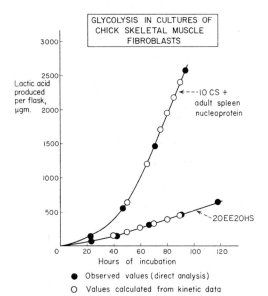

Fig. 5. Comparison of observed lactic acid levels with values calculated from kinetic data (see text).

TABLE 1

Glycolysis and Growth Rates in Cultures
of Chick Skeletal Muscle Fibroblasts

| Medium[1] | Glycolytic rate, $\beta$,[2] early log phase | Logarithmic growth rate,[3] K |
|---|---|---|
| 20 EE 20 HS | 1.7 | 3.4 |
| 10 EE 40 HS | 6.7 | 1.3 |
| 10 EE 20 CS | 3.4 | 2.8 |
| 10 CS + adult spleen nucleoprotein | 5.5 | 3.2 |

[1] EE = embryo extract; HS = horse serum; CS = chicken serum.
[2] $\mu$g. $\times 10^{-5}$ lactic acid/cell/hr.
[3] Per cent per hour.

related to the concentration of embryo extract in the medium. A rough correlation was observed between the duration of logarithmic growth and the estimated interval for the decline of glycolysis to low levels. Cultures in 10 EE 40 HS (Fig. 7) grew logarithmically for a protracted period, in comparison to the brief log phase observed for cultures in 20 EE 20 HS. These findings suggest that the decline in glycolysis may be a function of population density, although the nature of this relation is not yet clear. In any event, the eventual cessation of glycolysis or diminution to a low

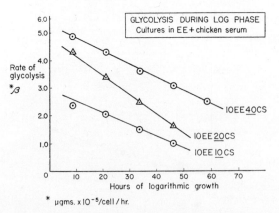

Fig. 6. Glycolytic rates of chick skeletal muscle fibroblasts maintained in 10% chick embryo extract (EE) with varying percentages of chicken serum (CS). Balance of medium, physiological saline (Earle's solution).

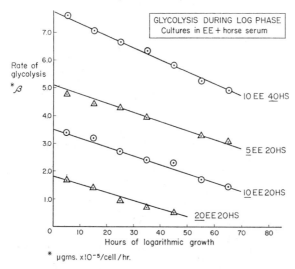

Fig. 7. Glycolytic rates of chick skeletal muscle fibroblasts in cultures with varying percentages of embryo extract (EE) and horse serum (HS).

level in maximal populations is consistent with Swann's suggestion (125) that glycolysis may be essential for cell division.

*Glycolysis in Normal and Malignant Cells*

In 1930 Warburg (134) published extensive studies by manometric methods on the metabolism of normal and neoplastic tissues. These investigations revealed a high level of glycolysis in tumors generally, as contrasted to normal tissues, and this differential was especially marked for measurements made in the presence of oxygen, i.e., aerobic glycolysis. Warburg maintained that the high level of glycolysis represented a special tumor metabolism, resulting from some damage to the oxidative metabolism of these cells during the neoplastic transformation. Subsequent investigations have not provided concrete evidence for deficient oxidation in tumors, although quantitative depletion of enzymes and reserve capacity has been noted (Greenstein, 60; Weinhouse, 138). On the other hand, with some exceptions (e.g., the high aerobic glycolysis of retina), the general contrast in glycolytic activity of normal and malignant tissues has been repeatedly confirmed by manometric methods (Burk, 6; Greenstein, 60; Warburg, 135, 136; Weinhouse, 139; Burk and Schade, 7).

Warburg has consistently maintained that the production of lactic acid under aerobic conditions by normal tissues does not occur except in damaged cells (Warburg, 134, 135). The data presented in the preceding section, however, demonstrate a high aerobic glycolysis during the growth of freshly isolated chick fibroblasts, and Leslie, Fulton, and Sinclair (82) have reported similar findings on cell strains obtained from normal human tissues.

In order to compare our results with conventional manometric data, dry weights were determined for chick skeletal muscle fibroblasts and representative $\beta$ values converted to the equivalent Q values used in manometric studies. The figures obtained (Table 2) indicate that the aerobic glycolysis

TABLE 2

GLYCOLYTIC RATES OF TUMORS AND NORMAL CELL STRAINS

| Cells and medium[1] | Type of measurement | Glycolytic rate per cell, $\beta^2$, early log phase | Aerobic glycolysis, $Q_A^{O_2}$ [3] |
|---|---|---|---|
| Ascites tumor cells, mouse | Manometric (Warburg) | — | 30 |
| L sarcoma cells, mouse<br>Line 1742<br>Line 2049 | Manometric (Burk et al.) | —<br>— | 30<br>10 |
| Skeletal muscle<br>fibroblasts, chick<br>20 EE 20 HS<br>10 EE 40 HS<br>10 CS + adult<br>spleen NPF | Chemical analysis<br>of culture medium | 1.7<br>6.7<br>5.5 | 18<br>69<br>57 |

[1] EE = embryo extract; HS = horse serum; CS = chicken serum; NPF = adult spleen nucleoprotein.
[2] $\mu$g. $\times 10^{-5}$ lactic acid/cell/hr.
[3] $\mu$l. $CO_2$ equivalent/mg. dry wt./hr.
Dry weight for chick skeletal muscle fibroblasts = 240 $\mu$g./$10^6$ cells.

of intact actively growing populations may be as high as or higher than the maximal values for tumors in the Warburg apparatus (Warburg, 135, 136). Similarly, Leslie, Fulton, and Sinclair (82), from a different method of calculating glycolytic rates, found that the aerobic glycolysis of normal human cells in tissue culture substantially exceeded that of several malignant strains maintained in parallel. Moreover, these results parallel the work of LePage (81), who demonstrated with homogenates that the potential glycolytic rates of normal tissues are as great or greater than those of malignant tissues. While differences in methods and material as well as species differences (Burk and Schade, 7) make the significance of quantitative comparisons with manometric data uncertain, the main point is that both in our experiments and in those of Leslie, Fulton, and Sinclair aerobic glycolysis has been associated with active growth of healthy cells rather than with damaged tissues, as postulated by Warburg.

It seems possible that apparent discrepancies in the aerobic glycolysis of normal tissues under different conditions may relate merely to the state of dissociation of the corresponding cell populations. In our experiments glycolysis was highest in the dispersed cells of early log phase and decreased either continuously, or abruptly at the end of log phase, when maximal population densities are approached. It also seems significant that the highest values for aerobic glycolysis of tumors have been obtained with cell suspensions rather than solid tumors, using ascites cells or harvested populations of L cells from tissue culture (Warburg, 135, 136). Greenstein (60) states that one of the most puzzling aspects of glycolysis in tumors has been the accumulation of lactic acid in these tissues, even in vivo (Voegtlin, Fitch, Kahler, Johnson, and Thompson, 131; Cori and Cori, 21). From our special point of view, this suggests that tumor cells, although they may be associated physically in a common mass, behave physiologically as if dissociated (i.e., with an active aerobic glycolysis). Correlated studies on the metabolism and population dynamics of tumor cell populations will be required to assess this possibility more fully.

### Summary

In this discussion we have developed two examples of metabolic regulation, a concept which seems fully as important in cell growth as its classical counterpart in embryonic differentiation. The possibility of shifting or alternating patterns in protein synthesis may include participation by extracellular protein precursors as well as free amino acids. As Humphrey and McFarlane (74) point out, this need not imply direct incorporation without change, but merely that the entire array of substrates for protein synthesis is not necessarily in complete equilibrium with the extracellular amino acid pool. The clearest indications of direct utilization of protein are seen in the "open" systems discussed in the present paper, i.e., isolated cell populations, transplants, embryonic tissues, and tumors. The adjustment of synthetic mechanisms to available substrates is illustrated by HeLa cell populations, which in protein-rich media utilize serum proteins extensively (Westfall), although the same cells in media of minimal protein content grow primarily, at least, from free amino acids (Eagle).

In a somewhat similar sense, the glycolytic patterns of growing cells may shift in accordance with the state of dispersal of the population. Within the embryo or adult organism the extent of regulation may be small, but the adjustments observed in vitro reveal new degrees of freedom in cellular metabolism. As an information factor, the production of lactic acid is a useful index of population dynamics in growing systems.

## REFERENCES

1. Askonas, B. A., Simkin, J. L., and Work, T. S., in *The Structure of Nucleic Acids and Their Role in Protein Synthesis* (E. M. Crook, ed.), p. 32, Cambridge University Press, New York (1957).
2. Baker, L. E., *J. Exptl. Med.*, **58,** 575 (1933).
3. Berman, L., Stulberg, C. S., and Ruddle, F. H., *Cancer Research,* **17,** 668 (1957).
4. Biggers, J. K., Rinaldini, L. M., and Webb, M., *Symposia Soc. Exptl. Biol.,* **11,** 264 (1957).
5. Brambell, F. W. R., and Hemmings, W. A., *Symposia Soc. Exptl. Biol.,* **8,** 476 (1954).
6. Burk, D., *Cold Spring Harbor Symposia Quant. Biol.,* **7,** 420 (1939).
7. ———, and Schade, A. L., *Science,* **124,** 270 (1956).
8. Burrows, M. T., *Am. J. Physiol.,* **68,** 110 (1924).
9. Busch, H., and Greene, H. S. N., *Yale J. Biol. Med.,* **27,** 339 (1955).
10. ———, Simbonis, S., Anderson, D., and Greene, H. S. N., *Yale J. Biol. Med.,* **29,** 105 (1956).
11. Carrel, A., *J. Exptl. Med.,* **17,** 14 (1913).
12. ———, and Ebeling, A. H., *J. Exptl. Med.,* **38,** 513 (1923).
13. Chang, R. S., *Proc. Soc. Exptl. Biol. Med.,* **87,** 440 (1954).
14. ———, *Proc. Soc. Exptl. Biol. Med.,* **96,** 818 (1957).
15. ———, and Geyer, R. P., *Proc. Soc. Exptl. Biol. Med.,* **96,** 336 (1957).
16. Christensen, H. N., *Ann. Rev. Biochem.,* **22,** 233 (1953).
17. Chu, E. H. Y., and Giles, N. H., *J. Natl. Cancer Inst.,* **20,** 383 (1958).
18. Coons, A. H., in *The Nature and Significance of the Antibody Response* (Pappenheimer, A. M., ed.), p. 200, Columbia University Press, New York (1953).
19. ———, *Intern. Rev. Cytol.,* **5,** 1 (1956).
20. ———, and Kaplan, M. H., *J. Exptl. Med.,* **91,** 1 (1950).
21. Cori, C. F., and Cori, G. T., *J. Biol. Chem.,* **64,** 11 (1925).
22. Crampton, C. F., and Haurowitz, F., *Science,* **112,** 300 (1950).
23. Danielli, J. F., *Symposia Soc. Exptl. Biol.,* **8,** 502 (1954).
24. Davidson, J. N., and Waymouth, C., *Biochem. J. (London),* **39,** 188 (1945).
25. Doljanski, L., Hoffman, R. S., and Tenenbaum, E., *Compt. Rend. Soc. Biol. Paris,* **131,** 423 (1939).
26. ———, and Hoffman, R. S., *Growth,* **7,** 67 (1943).
27. Dulbecco, R., *Proc. Natl. Acad. Sci. U. S.,* **38,** 747 (1952).
28. Eagle, H., *J. Biol. Chem.,* **214,** 839 (1955).
29. ———, *J. Exptl. Med.,* **102,** 37 (1955).
30. ———, *J. Exptl. Med.,* **102,** 595 (1955).
31. ———, *Science,* **122,** 501 (1955).
32. ———, *Arch. Biochem. and Biophys.,* **61,** 356 (1956).
33. ———, Oyama, V. I., and Levy, M., *Arch. Biochem. and Biophys.,* **67,** 432 (1957).
34. ———, ———, ———, and Freeman, A. E., *J. Biol. Chem.,* **226,** 191 (1957).
35. Earle, W. R., *J. Natl. Cancer Inst.,* **4,** 165 (1943).
36. ———, in *Am. Assoc. Advance. Sci. Research Conference on Cancer,* (F. R. Moulton, ed.), p. 139. Am. Assoc. Advance. Sci., Washington, D. C. (1945).

37. ———, Schilling, E. L., Bryant, J. C., and Evans, V. J., *J. Natl. Cancer Inst.*, **14**, 1159 (1954).
38. Ebert, J. D., *Proc. Natl. Acad. Sci. U. S.*, **40**, 337 (1954).
39. ———, in *Aspects of Synthesis and Order in Growth* (D. Rudnick, ed.), p. 69, Princeton University Press, Princeton (1954).
40. Evans, V. J., Bryant, J. C., Fioramonti, M. C., McQuilkin, W. T., Sanford, K. K., and Earle, W. R., *Cancer Research*, **16**, 77 (1956).
41. ———, ———, McQuilkin, W. T., Fioramonti, M. C., Sanford, K. K., Westfall, B. B., and Earle, W. R., *Cancer Research*, **16**, 87 (1956).
42. ———, and Earle, W. R., *J. Natl. Cancer Inst.*, **8**, 103 (1947).
43. ———, ———, Sanford, K. K., Shannon, J. E., and Waltz, H. K., *J. Natl. Cancer Inst.*, **11**, 907 (1951).
44. Fields, M., and Libby, R. L., *J. Immunol.*, **69**, 581 (1952).
45. Fischer, A., *Biology of Tissue Cells*, Hafner Publishing Co., New York (1946).
46. ———, and Astrup, T., *Pflüger's Arch. ges. Physiol.*, **247**, 34 (1943).
47. Francis, M. D., and Winnick, T., *J. Biol. Chem.*, **202**, 273 (1953).
48. Gemmill, C. L., Gey, G. O., and Austrian, R., *Bull. Johns Hopkins Hosp.*, **66**, 167 (1940).
49. Gerard, P., and Cordier, R., *Biol. Revs.*, **9**, 110 (1934).
50. Gerarde, H. W., Jones, M., and Winnick, T., *J. Biol. Chem.*, **196**, 51 (1952).
51. Gey, G. O., *Cancer Research*, **1**, 737 (1941).
52. ———, *Harvey Lectures*, **50**, 154 (1954-1955).
53. ———, Gey, M. K., Firor, W. M., and Self, W. O., *Acta Unio Intern. contra Cancrum*, **6**, 706 (1949).
54. Gifford, G. E., Robertson, H. E., and Syverton, J. T., *J. Cellular Comp. Physiol.*, **49**, 367 (1957).
55. Gitlin, D., in *Serological Approaches to Studies of Protein Structure and Metabolism* (W. C. Cole, ed.), p. 23, Rutgers University Press, New Brunswick (1954).
56. ———, Landing, B. H., and Whipple, A., *J. Exptl. Med.*, **97**, 163 (1953).
57. Goldblatt, H., and Cameron, G., *J. Exptl. Med.*, **97**, 525 (1953).
58. Graff, S., and McCarty, K. S., *Exptl. Cell Research*, **13**, 348 (1957).
59. Green, H., and Anker, H. S., *Biochim. et Biophys. Acta*, **13**, 365 (1954).
60. Greenstein, J. P., *Biochemistry of Cancer*, Academic Press, New York (1954).
61. Halvorson, H. O., and Spiegelman, S., *J. Bacteriol.*, **64**, 207 (1952).
62. Harris, M., *J. Cellular Comp. Physiol.*, **40**, 279 (1952).
63. ———, *Growth*, **21**, 149 (1957).
64. ———, and Kutsky, P. B., *J. Cellular Comp. Physiol.*, **42**, 449 (1953).
65. ———, and Kutsky, R. J., *Exptl. Cell Research*, **6**, 327 (1954).
66. ———, and ———, *Cancer Research*, **18**, 585 (1958).
67. Haurowitz, F., in *Serological Approaches to Studies of Protein Structures and Metabolism* (W. H. Cole, ed.), p. 2, Rutgers University Press, New Brunswick (1954).
68. Hogness, D. S., Cohn, M., and Monod, J., *Biochim. et Biophys. Acta*, **16**, 99 (1955).
69. Holter, H., and Marshall, J. M., Jr., *Compt. rend. trav. lab. Carlsberg, Sér. chim.*, **29**, 7 (1954).
70. Howatson, A. F., and Ham, A. W., *Anat. Record*, **121**, 436 (1955).
71. ———, and ———, *Can. J. Biochem. Physiol.*, **35**, 549 (1957).
72. Hsu, T. C., and Moorhead, P. S., *J. Natl. Cancer Inst.*, **18**, 463 (1957).
73. Hughes, A. F. W., *Quart. J. Microscop. Sci.*, **91**, 252 (1950).
74. Humphrey, T. H., and McFarlane, *Biochem. J. (London)*, **57**, 195 (1954).
75. Kent, H. N., and Gey, G. O., *Proc. Soc. Exptl. Biol. Med.*, **94**, 205 (1957).
76. Knight, P. F., and Schechtman, A. M., *J. Exptl. Zool.*, **127**, 271 (1954).
77. Kuchler, R. J., and Merchant, D. J., *Proc. Soc. Exptl. Biol. Med.*, **92**, 803 (1956).
78. Kutsky, R. J., *Proc. Soc. Exptl. Biol. Med.*, **83**, 390 (1953).

79. ———, and Harris, M., *Growth,* **21,** 53 (1957).
80. Laser, H., *Biochem. Z.,* **264,** 72 (1933).
81. LePage, G. A., *Cancer Research,* **10,** 77 (1950).
82. Leslie, I., Fulton, W. C., and Sinclair, R., *Biochim. et Biophys. Acta,* **24,** 365 (1957).
83. Lewis, W. H., *Bull. Johns Hopkins Hosp.,* **49,** 17 (1931).
84. ———, *Amer. J. Cancer,* **29,** 666 (1937).
85. Loftfield, R. B., *Progress in Biophysics and Biophys. Chem.,* **8,** 348 (1957).
86. Lu, K. H., and Winnick, T., *Exptl. Cell Research,* **9,** 502 (1955).
87. Madden, S. C., and Whipple, G. H., *Physiol. Revs.,* **20,** 194 (1940).
88. Marcus, P. I., Cieciura, S. J., and Puck, T. T., *J. Exptl. Med.,* **104,** 615 (1956).
89. Margoliash, E., and Doljanski, L., *Growth,* **14,** 7 (1950).
90. McIlwain, H., and Rodnight, R., *Biochem. J., (London),* **56,** xii (1954).
91. McQuilkin, W. T., Evans, V. J., and Earle, W. R., *J. Natl. Cancer Inst.,* **19,** 885 (1957).
92. Moore, A. E., in *Cellular Biology: Nucleic Acids and Viruses* (T. M. Rivers, consult. ed.), *N. Y. Acad. Sci. Special Publ.,* **5,** 321 (1957).
93. ———, Southam, C. M., and Sternberg, S. S., *Science,* **124,** 127 (1956).
94. Morgan, J. F., Morton, H. J., and Parker, R. C., *Proc. Soc. Exptl. Biol. Med.,* **73,** 1 (1950).
95. Murphy, J. B., *J. Exptl. Med.,* **24,** 1 (1916).
96. Murray, M. R., and Kopech, G., *A Bibliography of the Research in Tissue Culture (1884-1950),* Academic Press, New York (1953).
97. Oliver, J., MacDowell, M., and Lee, Y. C., *J. Exptl. Med.,* **99,** 589 (1954).
98. Palade, G. E., *J. Biophys. Biochem. Cytol.,* **1,** 59 (1955).
99. Parker, R. C., Castor, L. N., and McCulloch, E. A., in *Cellular Biology: Nucleic Acids and Viruses* (T. M. Rivers, consult. ed.), *N. Y. Acad. Sci. Special Publ.,* **5,** 303, (1957).
100. Parshley, M. S., and Simms, H. S., *Anat. Record,* **94,** 486 (1946).
101. ———, and ———, *Am. J. Anat.,* **86,** 163 (1950).
102. Phillips, H. J., and Feldhaus, R. J., *Proc. Soc. Exptl. Biol. Med.,* **92,** 478 (1956).
103. Pomerat, C. M., and Willmer, E. N., *J. Exptl. Biol.,* **16,** 232 (1939).
104. Porter, K. R., and Kallman, F. L., *Ann. N. Y. Acad. Sci.,* **54,** 882 (1952).
105. Puck, T. T., Cieciura, S. J., and Fisher, H. W., *J. Exptl. Med.,* **106,** 145 (1957).
106. Rather, L. J., *Medicine,* **31,** 357 (1952).
107. Rodnight, R., and McIlwain, H., *Biochem. J. (London),* **57,** 649 (1954).
108. Rose, S. M., *Am. Naturalist,* **86,** 337 (1952).
109. Sanford, K. K., Earle, W. R., and Likely, G. D., *J. Natl. Cancer Inst.,* **9,** 229 (1948).
110. ———, ———, Shelton, E., Schilling, E. L., Duchesne, E. M., Likely, G. D., and Becker, M. M., *J. Natl. Cancer Inst.,* **11,** 351 (1950).
111. ———, Likely, G. D., and Earle, W. R., *J. Natl. Cancer Inst.,* **15,** 215 (1954).
112. ———, Waltz, H. K., Shannon, J. E., Jr., Earle, W. R., and Evans, V. J., *J. Natl. Cancer Inst.,* **13,** 121 (1952).
113. ———, Westfall, B. B., Fioramonti, M. C., McQuilkin, W. T., Bryant, J. C., Peppers, E. V., Evans, V. J., and Earle, W. R., *J. Natl. Cancer Inst.,* **16,** 789 (1955).
114. Sato, G., Fisher, H. W., and Puck, T. T., *Science,* **126,** 961 (1957).
115. Schechtman, A. M., *Intern. Rev. Cytol.,* **5,** 303 (1956).
116. Siminovitch, L., Graham, A. F., Lesley, S. M., and Nevill, A., *Exptl. Cell Research,* **12,** 299 (1957).
117. Simpson, M., and Velick, S., *J. Biol. Chem.,* **208,** 61 (1954).
118. Spiegelman, S., and Halvorson, H. O., in *Adaptation in Micro-organisms.* (E. F. Gale and R. Davies, eds.), p. 98, Cambridge University Press, New York (1953).
119. Stewart, D. C., and Kirk, P. O., *Biol. Revs.,* **29,** 119 (1954).
120. Strauss, W., *J. Biol. Chem.,* **207,** 745 (1954).

121. ———, *J. Biophys. Biochem. Cytol.*, **2**, 513 (1956).
122. ———, *J. Biophys. Biochem. Cytol.*, **3**, 933 (1957).
123. ———, *J. Biophys. Biochem Cytol.*, **3**, 1037 (1957).
124. ———, and Oliver, J., *J. Exptl. Med.*, **102**, 1 (1955).
125. Swann, M. M., *Cancer Research,* **17**, 727 (1957).
126. Swim, H. E., and Parker, R. F., in *Cellular Biology: Nucleic Acids and Viruses,* N. Y. Acad. Sci. Special Publ., **5**, 351 (1957).
127. Syverton, J. T., and McLaren, L. C., *Texas Reports Biol. Med.*, **15**, 577 (1957).
128. Terry, R., Sandrock, W., Nye, R., and Whipple, G. H., *J. Exptl. Med.*, **87**, 547 (1948).
129. Trowell, O. A., and Willmer, E. N., *J. Exptl. Biol.*, **16**, 60 (1939).
130. Tyler, A., *Growth* **10** (Suppl. 6th Sympos.), 7 (1946).
131. Voegtlin, C., Fitch, R. H., Kahler, H., Johnson, J. M., and Thompson, J. W., *U. S. Public Health Service, Public Health Bull.*, **164**, 15 (1935).
132. Walter, H. F., Haurowitz, F., Fleischer, S., Lietze, A., Cheng, H. F., Turner, J. E., and Friedberg, W., *J. Biol. Chem.*, **224**, 107 (1957).
133. Walter, H. W., and Mahler, H. R., *J. Biol. Chem.*, **230**, 241 (1958).
134. Warburg, O., *The Metabolism of Tumors,* Constable & Co., London (1930).
135. ———, *Science,* **123**, 309 (1956).
136. ———, *Science,* **124**, 269 (1956).
137. Waymouth, C., *Texas Reports Biol. Med.*, **13**, 522 (1955).
138. Weinhouse, S., *Advances in Cancer Research,* **3**, 269 (1955).
139. ———, *Science,* **124**, 267 (1956).
140. Weiss, P., *Yale J. Biol. Med.*, **19**, 235 (1947).
141. ———, in *The Chemistry and Physiology of Growth* (A. K. Parpart, ed.), p. 135, Princeton University Press, Princeton (1949).
142. ———, *J. Embryol. Exptl. Morphol.*, **181**, 1 (1953).
143. ———, in *Biological Specificity and Growth* (E. G. Butler, ed.), p. 195, Princeton University Press, Princeton (1955).
144. ———, and Kavanau, J. L., *J. Gen. Physiol.*, **41**, 1 (1957).
145. Westfall, B. B., Peppers, E. V., and Earle, W. R., *Am. J. Hyg.*, **61**, 326 (1955).
146. Westwood, J. C. N., MacPherson, I. A., and Titmuss, D. H. J., *Brit. J. Exptl. Pathol.*, **38**, 138 (1957).
147. ———, and Titmuss, D. H. J., *Brit. J. Exptl. Pathol.*, **38**, 587 (1957).
148. Whipple, G. H., *The Dynamic Equilibrium of Body Proteins,* Charles C Thomas, Springfield (1956).
149. Willmer, E. N., *Tissue Culture,* Methuen & Co., London (1954).
150. Winnick, R. E., and Winnick, T., *J. Natl. Cancer Inst.*, **14**, 519 (1953).
151. Yuile, C. L., Lamson, B. G., Miller, L. L., and Whipple, G. H., *J. Exptl. Med.*, **83**, 539 (1951).
152. ———, O'Dea, A. E., Lucas, F. V., and Whipple, C. H., *J. Exptl. Med.*, **96**, 274 (1952).
153. Younger, J. S., *Proc. Soc. Exptl. Biol. Med.*, **85**, 202 (1954).
154. Zitcer, E. M., and Dunnebacke, T. H., *Cancer Research,* **17**, 1047 (1957).

## DISCUSSION

DR. MARKERT: In connection with the difference in the rate of glycolysis, is this a reflection of the growth potential or just the state of aggregation? That is, a state of aggregation in your tissue culture, when it is high, tends to diminish the rate of multiplication. Perhaps it is primarily the rate of multiplication rather than the state of aggregation that determines the rate of glycolysis.

DR. HARRIS: There appears to be no direct relation between glycolytic and growth rates during log phase. In our assay system by appropriate selection of medium we can obtain rapid growth with high glycolysis, low growth rate and high glycolytic rate, or rapid growth with low glycolysis.

DR. MARKERT: Doesn't that in turn then affect the rate of multiplication? That is, as the population becomes dense, does the rate of cell multiplication decrease?

DR. HARRIS: Yes. Aggregation appears to be the significant factor, however, rather than cessation of growth alone. This is in harmony with Warburg data on tissue slices which show, for example, that fetal rat liver which grows rapidly has a low aerobic glycolysis.

DR. EAGLE: As Dr. Harris has indicated, we have been working entirely with mammalian cell systems; and any differences in our results may perhaps be referable to the cell substrate. First of all, with respect to protein utilization, we have seen very little indication that the cells with which we work do indeed use protein for growth. When cells are grown in a medium with labeled essential amino acids, the specific activity of the amino acid residues in the cell protein is precisely that of the amino acids in the medium. We are in the process now of determining to what degree, if any, the cell does use prelabeled protein from the medium. I was greatly interested in the fact that one can get from the adult chick tissues a nucleoprotein factor which has a growth-promoting activity not possessed by chick serum. It would be of great interest to find out just what this is.

I would like to say a few words with respect to glycolysis. In cultures which are growing at essentially the same rate, one may have as much as a 20-fold difference in glycolysis by appropriate modification of either the carbohydrate substrate or its concentration. In this respect there has been no difference whatever between cells deriving from normal tissues and cells deriving from malignant tissues. The discrepancy between these data in cell cultures, in which the normal and malignant cells are growing at the same rate, and those obtained with normal and malignant tissues freshly isolated from the animal, is interesting and is under present study.

DR. HARRIS: I agree with Dr. Eagle that certain differences in our results may in part reflect the assay systems used. While he has used established cell strains we have worked with stabilized populations of freshly isolated cells. The established cell strains in general use differ from freshly isolated populations in morphology, and have abnormal chromosome complements, usually in the hypotetraploid region. It would not be surprising if these altered cell strains also differed nutritionally from freshly isolated cells in the utilization of protein.

DR. LIPMANN: With respect to glycolysis, I favor to argue that a high rate of glycolysis does not really relate with malignancy as such but rather with one aspect of malignancy, namely adaptation to survival with a poor supply of oxygen. There are instances of non-malignant embryonic and adult tissues with high anaerobic and aerobic glycolysis which are non- or poorly vascularized. One might, therefore, classify the tumor with poorly vascularized tissues in need of anaerobic energy supply. One would assume, then, that a metabolic structure adapted to anaerobic survival may be part, although probably not an indispensable part, of the malignancy syndrome, the cause of which we do not understand.

Dr. Harris: I think there is no question that the level of glycolysis is very significantly higher in the absence of oxygen. In our cultures glycolysis definitely increases if they are made anaerobic by replacing the air phase with nitrogen.

Dr. Niu: I would like to furnish Dr. Harris some preliminary observation regarding the role of protein in tissue culture. When we grow presumptive ectoderm in hanging drops in the presence of ribonucleoprotein on the one hand, and in the presence of ribonucleic acid on the other hand, we find that the outgrowth in the two series is different. In cultures containing nucleoprotein, we usually get big outgrowths. In cultures containing ribonucleic acid the outgrowths are smaller and compact, yet the type of differentiation in both cases is the same. From this experiment we tentatively came to the conclusion that protein plays an important role in tissue culture both by promoting growth and by facilitating migration.

Dr. Harris: This is certainly consistent with the results which we obtain when we separate the nucleic acid and protein fractions in our own assay system.

Dr. Brachet: I would just like to say a word about the experiments which we have been making on the action of ribonuclease on living cells. The RNAase penetrates and there is absolutely no doubt that this protein gets into the cell. However, this may be the special case of a protein which can get into a large variety of cells in a very short time. In the amoebae, anyway, we have shown, quantitatively of course, that if you treat them with ribonuclease in a minimal medium, within ten minutes the ribonuclease content of the amoebae is increased four or five times. You can do the same sort of experiment, of course, with labeled ribonuclease. The kinetics of the penetration of labeled ribonuclease have been studied by Dr. Schumaker who studied in my lab last year. The conclusions seem to be that there is first the formation of a complex between the protein and the surface of the cell, and then you have the pinocytosis mechanism. But it can be shown that the enzyme is still bound to the surface of the pinocytosis vacuole. You do get some protein in solution in the vacuole but most of it is attached to the membrane. Now there is another thing which seems to come out of the work being done presently, in our laboratory, on a number of different types of cells regarding this ribonuclease uptake: that is that there seems to be a correlation between the amount of ribonuclease which will get into a cell and the initial amount of ribonuclease. I think that this may be an important factor.

Dr. Eagle: I think we have to distinguish very clearly between the uptake of protein from the medium into the cell, and its utilization for cell protein synthesis. The two do not necessarily go hand in hand. We have no data on the uptake of proteins in cell cultures; but I am very much intrigued by the fact that ribonuclease is concentrated four fold. For all the essential amino acids, and for all of the cell cultures with which we have worked with, the concentration factor is four to ten fold.

Dr. Harris: One thing that you might emphasize, Dr. Eagle, is that the growth rates which you have reported in the literature are not maximal for the cell strains used, and that your data as well as that of Earle's group and Graf and McCarty show marked stimulation of growth rates as the concentration of serum protein is increased from minimal levels. One can certainly conclude from your results that in the absence of essential amino acids protein utilization does not occur either.

Dr. Eagle: Well, we are getting cell doubling in less than 24 hours; and we can't improve on that rate of growth by adding more protein.

Dr. Brown: Is it clear that there is no secretion of an extracellular protease in these instances of apparent or actual protein utilization by the cells?

Dr. Harris: We ourselves have not studied this directly but in experiments of Francis and Winnick, the use of metabolic traps and amino acid analogues seems to rule against a breakdown to the amino acid level. In our cultures, growth rates and glycolysis are not affected by the addition of crystalline soybean trypsin inhibitor at the level of one mg. per ml.

Dr. Herrmann: In connection with Dr. Harris' interesting suggestion of a modification of the interaction of cells I would like to recall the observation of Dr. Dale Coman who found that malignant cells, in those cases where a test could be made, show less adhesiveness, less calcium in the surfaces, and a different cell surface structure in the electron microscope.

Dr. Harris: I think this is consistent with many aspects which we know of tumor cells which indicate their autonomous qualities.

Dr. Magasanik: I just wanted to make one point. Is there any evidence that the protein which gets into the cells is utilized for protein synthesis without first being broken down into amino acids?

Dr. Harris: This is difficult to say. The results of Haurowitz and his collaborators for the uptake of labeled protein from the plasma suggest to them a breakdown within the cell. The point I would like to make is merely that if such intermediates do occur, they need not be in equilibrium with the free amino acid pool outside.

# Part IV

*CONTROL MECHANISMS IN DEVELOPMENT*

# CHEMICAL CONTROL OF DEVELOPMENT

Nelson T. Spratt, Jr.
*Department of Zoology,*
*University of Minnesota,*
*Minneapolis, Minnesota*

It seems rather probable to the writer that our comprehension of the phenomenon of development will ultimately depend upon our knowledge and understanding of the properties (organization) of the mature egg or other reproductive unit and subsequently upon our understanding of: (1) how these properties become parcelled out to the increasing cellular population (segregation); (2) how they are maintained through many cell generations (cellular heredity and stability); (3) how new properties arise through (a) interaction of parts and cells (cellular interaction) and (b) the influence of the internal and external environment (position effects); and (4) how all of these mechanisms remain properly geared to one another so as to maintain the integrity of the whole organism (gradient-field or other control mechanism). Indeed, it is the orderliness of the sequence of developmental events which characterizes the phenomenon and presents us with the core problem of the mechanism of its control.

Actually, little is known about the control mechanisms which operate in normal development, but some information about the external, artificial control of the developmental system via manipulation of its chemical environment is gradually accumulating (8-12, 21). Although successful artificial chemical control of the developing organism does not imply that it is this kind of control which actually operates in the normal situation, it does show that the developing organism and its parts and processes are amenable to this type of control. All we can do at present is to cite examples of the chemical control of cellular behavior and surmise that some of the effective procedures *may be* used by the developmental system.

It is the purpose of the present contribution to the symposium to describe briefly some of the effects of various changes in the external environment (essentially the internal environment in many respects) upon the development of the young (1 to 2-day-old) chick embryo. Space does not permit a review of similar studies done on fish (21) and mammal (3) embryos. Nor would it be feasible to review the numerous examples of chemical control of the development of invertebrate embryos and non-embryonic developmental systems. Fortunately, the early chick embryo is very sensitive to manipulation of its chemical environment, with the result that rapid and often dramatic responses may be observed. By way of conclusion, a chemical mechanism which may contribute to the control of normal development will be suggested.

## Necessity for Exogenous Nutrients

The nutrient requirements of an embryo may be built into its cells in the form of yolk or yolk-like substances (e.g., as in amphibian embryos) or largely separated from it as extracellular yolk (e.g., as in the highly telolecithal eggs of birds and many fishes). In the latter case it is possible to study directly the nutrient requirements of the embryo by separating it from its natural, exogenous supply and explanting it upon various media in vitro, the chemical composition of which may be controlled at will. The general, as well as any differential, nutrient or chemical requirements for particular developmental processes may thus be discovered.

If one explants an early (e.g., 18-30 hours' incubation) chick blastoderm to a non-nutrient, buffered Ringer (saline) or Ringer-agar medium, almost all developmental activity rapidly ceases. When relatively large (ca. 5 ml.) volumes of medium are used this is always accompanied by a type of degeneration of the explant: a breakdown of visible organization at the *tissue* and *organ levels* and a rounding up and dispersal of the individual cells (8; and Figs. 1, 2, 3). In some explants this may begin within 30 minutes at room temperature (ca. 20°C), particularly when the $pH$ of the medium is in the range 8.5-9.0. Subculture of the dispersed cells to adequate (e.g., blood-plasma–embryonic-extract) nutrient media up to about 20-24 hours of starvation reveals that these cells are not dead, since they respond by dividing and moving about and may even reaggregate to form tissues (8, p. 352 and 14). On the other hand, the explanted blastoderm, if starved for no more than about 4-6 hours or when explanted to a relatively small (ca. 0.1-0.2 ml.) volume of medium, is able to recover and continue its organ- and tissue-forming activity when subcultured to a nutrient medium (14, 17), *provided* there has been no appreciable loss of the dispersed cells. It is also interesting to note that developmental processes already under way at the time of explantation to a non-nutrient medium (e.g., notochord or somite formation in a 1 to 2-somite blastoderm) may continue for a short period of time (2-3 hours) during which endogenous nutrients are used up, but that processes not under way at the time of explantation (e.g., notochord and somite formation in the younger, primitive streak blastoderm) are not initiated in the absence of exogenous nutrients.

From the above evidence it may be concluded that maintenance of the organ and tissue levels of complexity requires a greater and more continuous supply of exogenous nutrients than does maintenance of the viability of isolated, dispersed cells of the same tissues and organs. The effects of severe starvation upon the histogenetic capacities of the cells, whether dispersed or not, are currently under investigation, and there are strong indications that some kinds of cells possess a remarkably large "safety-factor." In general, it is clear that a certain measure of artificial control leading to differential effects in the developing system may be achieved by simply reducing or removing the source exogenous nutrients.

### Control of the Quality of Exogenous Nutrients

*Complex and Synthetic Media*

A number of complex nutrient media will support continued development of explanted blastoderms. In approximate order of their comparative effectiveness the media tested are as follows: yolk-albumen > yolk > albumen > blood plasma + embryonic extract = White's synthetic medium (Fig. 1). The exact composition of these media is given in references (8, 9). Development may continue on the more optimal media without subculture for about three days, after which shrinkage, differential degeneration of organs or structures, vacuolation, and eventual necrosis of the explants begin for reasons not entirely understood. The complex media differ mainly in respect to their adequacy for supporting over-all growth, modeling and histogenesis of the central nervous system, somite formation, and development of an embryonic-extraembryonic circulation. For example, the latter develops frequently in whole or anterior half blastoderms on a yolk-albumen medium but has never been observed in explants to blood-plasma or synthetic media (18). Development of definitive streak–15-somite blastoderms explanted on yolk-albumen or albumen media is remarkably normal during the first 24 hours' cultivation and does not differ significantly, with respect to either *rate* of formation or *size* of many of the organ primordia, from that of blastoderms developing within the egg shell.

A synthetic medium (e.g., such as White's) made up without glucose will not support further development of explants to it (9). The embryos rapidly exhibit the degenerative changes characteristic of those on non-

632    THE CHEMICAL BASIS OF DEVELOPMENT

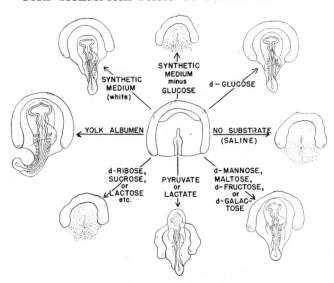

Fig. 1. Summary of the results obtained when anterior halves of early chick blastoderms are explanted to non-nutrient and various nutrient media. Media containing yolk-albumen extracts provide the most nearly optimal nutrient conditions, yet a glucose-saline medium is remarkably effective in supporting many of the early developmental changes. See text for further details.

nutrient Ringer media. On the other hand, a simple Ringer-glucose medium will support some of the early developmental processes in embryos explanted to it (Fig. 1). Although these results attest to the great importance of glucose, they obviously do not show that glucose alone is entirely adequate. It is remarkable that glucose will do so much for the explant, yet it is to be noted that such explants (deprived of an exogenous source of nitrogen) often become smaller during cultivation and have fewer and more poorly formed somites compared with those on either "complete" synthetic or yolk-albumen media (Figs. 1 and 2). Nevertheless, it seems probable (in view of some preliminary evidence) that one of the most important nutrient functions of exogenous glucose lies in its utilization by the cells of the embryo to prevent the leaching out of protein and amino acids into the medium (which does occur in explants on non-nutrient media). This problem is under study by one of my students.

### Carbohydrate Requirements

Glucose is not the only carbohydrate that is effective in supporting some of the early developmental activities of the blastoderm (10). Out of 20-odd carbohydrates (some indicated in Fig. 1) and metabolic intermediates used as the sole carbon source, only the naturally occurring hexoses: D-glucose, D-mannose, D-fructose, D-galactose; one disaccharide, D-maltose; sodium pyruvate and sodium lactate showed any signs of being utilized for supporting the developmental changes (Fig. 1). In terms of their relative effectiveness at optimal concentrations the following order was found: glucose > mannose > maltose > fructose > galactose > pyruvate = lactate. Note the relatively small size, poorly developed brain, heart, and somites in the pyruvate and lactate explant. In contrast to explants on the various sugar media, these show a cessation of all progressive changes after about 15 hours of cultivation (12).

## Heart and Brain

In general, the brain is both quantitatively and qualitatively more exacting in its carbohydrate requirements than is the heart. Thus, only glucose (possibly mannose) among the several substrates tested approaches complete adequacy as a carbohydrate source for supporting continued histogenesis (for 1-2 days) of the brain, whereas all of the sugars in this group are almost equally effective in supporting development and pulsation of the heart. Also, much greater molar concentrations of fructose, galactose, and maltose compared with glucose are necessary to prevent degeneration of the brain in contrast to the heart, which requires only slightly higher concentrations of these sugars (11).

## Isolated Node Center

Some of the results obtained by explanting the isolated (excised) node area of the early (primitive streak–10-somite stage) blastoderm are summarized in Fig. 2. Note the distinct superiority of albumen compared with glucose media in supporting elongation, growth, and somite formation, particularly in the smaller node explants. Another interesting feature of the results is the capacity of an explant starved for about 12 hours on a nonnutrient (saline) medium to recover when subcultured to an albumen medium (also to some degree on a glucose medium). A third feature of note in the results is the response (recovery) of the explant to an albumen medium after 12 hours on a glucose medium. One possible meaning of some of these results will be suggested below.

### SENSITIVITY TO ADVERSE ENVIRONMENTAL CONDITIONS

Some of the effects of various adverse environmental conditions are summarized in Fig. 3. In general, it is to be noted that cells of the node center,

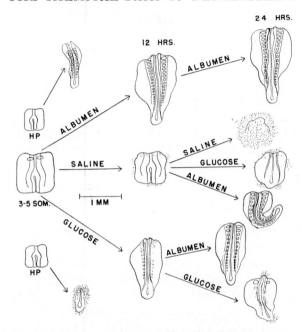

Fig. 2. Results obtained by explanting the isolated node area (organizer center) of the early (1-5 somite) blastoderm on nutrient and non-nutrient media. See text for further explanation.

in contrast to other cells, exhibit greater sensitivity to calcium removal, low sodium medium ($2 \times 10^{-2}\,M$), starvation, presence of an injurious substrate (succinate), carbon dioxide deficiency, and oxygen deficiency (Fig. 4). All axial (spinal cord, notochord, somite) organ-forming activities of the node center cease, and this is usually preceded by a dispersal of the node center cells (19). In contrast, other regions of the embryo may continue to undergo the progressive changes characteristic of their development. Recovery (continued axial organ formation) is possible in all these cases provided the period of exposure to the adverse condition is brief (about 2 to 3 hours). It is interesting that the deleterious effect of a $10^{-2}\,M$ succinate medium can be prevented by the simultaneous presence of $10^{-3}\,M$ glucose (Fig. 3). Among further interesting differential effects, that obtained in explants to a low ($6 \times 10^{-3}\,M$) potassium medium is worth mentioning. A brain may develop in these explants, but the heart is either very small or absent in contrast to the large and well-formed pulsating heart of control explants on a $6 \times 10^{-2}\,M$ potassium medium (see below for a similar effect obtained by use of a metabolic inhibitor).

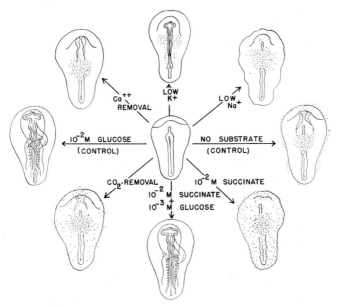

Fig. 3. Summary of results obtained when whole early embryos (exclusive of the opaque area) are explanted under various unfavorable environmental conditions. Note that it is the node center which is most sensitive to the adverse conditions in almost all instances.

## Minimal Environmental Requirements

On the positive side it can be stated that the minimal requirement for some degree of continued development of the early embryo is the presence of optimal levels of glucose, sodium, potassium, calcium, carbon dioxide, and oxygen in its environment. An approach to normal development would, of course, require the presence of additional, presumably more complex nutrients (amino acids, proteins, etc.). In any case, it seems clear that manipulation of any one or more of the parameters of its chemical environment enables us to control differentially some of the processes of development going on in the embryo. As a specific example one may cite the increasing evidence for the importance of carbon dioxide, a relatively simple environmental component.

Many of the beneficial developmental effects of crowding several embryos or fragments of embryos into a small culture space (as opposed to cultivating a single embryo or fragment in a relatively large volume of medium) are apparently the result of an increased carbon dioxide tension surrounding the cells (17). A similar situation has been described as ex-

636   *THE CHEMICAL BASIS OF DEVELOPMENT*

isting in the case of explanted fish (*Fundulus*) embryos (21). It is also probable that the better histogenesis which occurs in large as opposed to small, isolated populations of embryonic cells (also in a small vs. a large culture space) is partially (but not entirely) the result of a closer approach to an optimal carbon dioxide tension (17). Other examples of the profound importance of carbon dioxide tension in the control of developmental processes in non-embryonic systems are known but cannot appropriately be described here.

### DIFFERENTIAL EFFECTS OF METABOLIC INHIBITORS

Results obtained with metabolic inhibitors added to glucose-Ringer media are summarized in Fig. 4. Note that iodoacetate, malonate, cyanide,

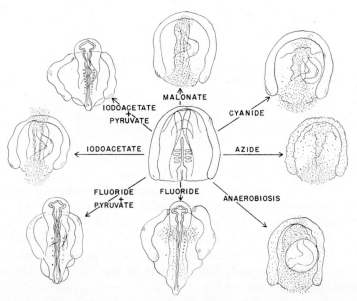

Fig. 4. Results obtained with metabolic inhibitors added to glucose-Ringer media. See text for details.

and anaerobiosis may completely suppress development of the brain and spinal cord without having any appreciable effect upon formation and pulsation of the heart (12). Azide is the only inhibitor which as yet has shown no clear-cut differential effects. In contrast, fluoride, unlike any of the other inhibitors used, may completely suppress heart formation without appreciably affecting further development of the brain. This differential effect of sodium fluoride on the heart and some other mesodermal organs has been studied more recently by others (1). As noted in Fig. 4, the effects of iodoacetate and of fluoride can be prevented by the substitution of sodium pyruvate for glucose in the culture medium. In addition to these differential effects upon brain and heart formation, all inhibitors used (except azide) give rise at certain concentrations to a characteristic pattern of differential degeneration (cell dispersal) in the blastoderm, a pattern which coincides with its pattern of developmental activity, the node center being the most sensitive region and the most developmentally active during the first two days of development.

We have here another example of the chemical control of development, achieved in this case by interfering with the metabolic machinery of the cells. Presumably, the blocking of certain pathways permits some processes to proceed but inhibits or blocks others. If one concedes that the observed differential effects of the metabolic inhibitors are indicative of the existence of different quantitative (possibly qualitative) metabolic activities underlying different developmental activities, the possibility of the control of development via interference with a mechanism controlling cellular metabolism becomes somewhat likely. We shall return to a consideration of this below.

### Inhibition by Organs and Organ Homogenates

It seems unlikely that changes in the simple chemical constituents (nutrients, $CO_2$, $O_2$ etc.) of the embryonic environment (although admittedly important for cell survival) are the primary means by which chemical control of normal development is achieved. Features such as the orderliness, exact timing, and sharp localization of the developmental events suggest the operation of a more precise and specific chemical communication among the parts of the system. Such a mechanism may operate in the many instances of cellular interaction (e.g., induction of neural plate by underlying chorda-mesoderm, etc.). Pending an approach in which interactions between cells of the same developmental age are to be studied, an indirect approach to this problem was attempted in the experiments summarized in Fig. 5. When whole blastoderms or isolated node areas were explanted to the cut surface of *living* 14-day embryo heart or liver slices (partially embedded in an albumen-agar medium), all axial organ-forming activities of the node area ceased. No specific (homologous) tissue or organ inhibition was found, but the effect was regionally specific in that only the node center activity in whole blastoderms (in contrast to continued progressive changes in other regions) was inhibited (19). A similar result was

638  THE CHEMICAL BASIS OF DEVELOPMENT

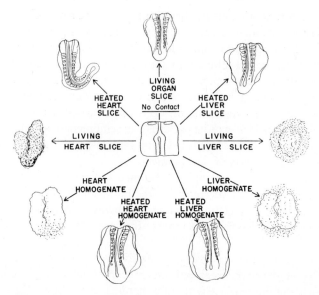

Fig. 5. Diagrams illustrating the results obtained when node areas are explanted onto a living or heat-killed slice of 14-day chick liver or heart and onto egg albumen media containing a heated or unheated liver or heart homogenate from a 14-day chick embryo. See text for explanation.

obtained with homogenates of 14-day hearts and livers added to an albumen-agar medium. Note that heating (60°C for 15 minutes) of the slices or homogenates completely destroys their inhibitory effect. Of special interest is the fact that inhibition by the living tissue slices was *only* manifest when there was very close contact between the slice and the explanted blastoderm or node area. This contact consisted, in all cases examined, in the presence of *adhesions* or *cellular bridges* connecting the older and younger cells. In the homogenate experiments chemical communication presumably does not depend upon the formation of such connections. Further study of the exact nature and distribution of these connections, as well as of the chemical nature of the inhibitor agent(s), will be necessary. Whether or not exchange occurs, and if so what may be exchanged between the reacting cells, is a question of great general importance in experimental embryology. Experiments designed to give us an answer are under way.

QUANTITATIVE DIFFERENCES IN CARBOHYDRATE REQUIREMENTS

By controlling the exogenous *quantity* of carbohydrate available to the explanted whole blastoderm or isolated node area, it is possible to demonstrate that the basic physiological activities (maintenance, cell movement, differentiation, and growth) as well as the special activities leading to the formation of specific tissue types or organs (e.g., heart, brain, notochord, etc.) have *quantitatively* different nutrient requirements. It is thus possible to achieve to some degree an artificial separation of these general physiological processes such that cell movement, for example, may be supported in the absence of any significant differentiation (histogenesis) or growth. Fig. 6 summarizes in diagrammatic form the results of explanting the node

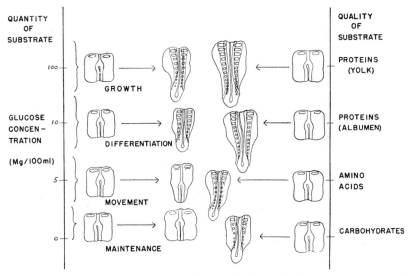

Fig. 6. Diagrammatic summary of results of explanting isolated node areas of 1 to 5-somite blastoderms on media containing (a) different quantities of glucose and (b) different general types of nutrients. Note that some degree of separation of the basic developmental processes is achieved by changing the quantity and/or quality of the substrate.

area to media containing: (1) *different amounts* of exogenous glucose (glucose-Ringer-agar medium) and (2) *different kinds* of nutrients. Artificial chemical control of these general, developmental processes can thus be achieved by quantitatively or qualitatively changing the composition of the explantation medium.

### Localization of the Energy-Substrate Supply System

It does not seem unreasonable to suspect that the nutrient and other chemical requirements of cells must somehow be closely related to the developmental work in which the cells are engaged. Indeed, there is a fairly large body of evidence which seems to justify such a suspicion. The results of several studies (11, 12) have demonstrated both qualitative and quantitative differences in the nutrient requirements for formation and continued development of different organs and tissues. But perhaps the clearest demonstration of the correlation between developmental activity and exogenous nutrient requirements lies in the coincidence between the effects of changes in concentration of inhibitors, oxygen tension, substrates, etc., available to the explanted embryo, on the one hand, and the spatial pattern of differential, developmental activity in the blastoderm, on the other. Fig. 7 schematically summarizes this correlation.

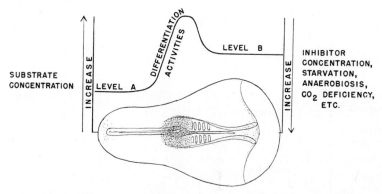

Fig. 7. Diagram illustrating the correlation of developmental activity with exogenous nutrient requirements, sensitivity to adverse environmental conditions, and metabolic activity. The density of stippling indicates the degree of reducing enzyme and other metabolic activities *per cell*. See text for further description.

It need only be pointed out here that in regions of the embryo where the cells are actively engaged in the work of constructing the axial organs, that is, in notochord, neural tube, somites, etc. (the node area), the cells are more sensitive to adverse environmental conditions and require higher concentrations of substrates (nutrients) than do cells in relatively more quiescent regions (where axial organs have already been formed, viz., level B; or where the cells are concerned with simpler construction tasks, e.g., extraembryonic membrane formation, viz., level A).

## Biochemical Properties of the Node Center

As one might expect, differential nutrient requirements of the node area cells compared with other cells of the embryo seem to be associated with differential enzymatic activities. Use of various enzyme indicator dyes (e.g., neotetrazolium chloride, methylene blue, etc.) in media containing a single kind of substrate as the sole nutrient source has shown that the node cells exhibit greater activities in more different kinds of dehydrogenase enzyme systems than cells of any other region of the blastoderm (13, 15, 16, 18, 20).

Node center cells also exhibit: (a) greater cytochrome oxidase (indophenol blue) activity (6, 16); (b) greater cytoplasmic basophilia (ribonucleic acid concentration?) when stained with toluidine blue (2, 16); and (c) greater staining with and reduction of Janus green, presumably a fairly specific indicator of mitochondrial activity (5, 7, 20). Since the major locus of tetrazole reduction within cells seems to be the mitochondria (4, 16, 17, 19, 20), the various studies cited above strongly suggest that the quantitatively greater enzyme activities peculiar to node center cells may be the result of their possession of greater numbers or of more active mitochondria, or both.

*An Endogenous Chemical Control Mechanism*

Numerous observations, primarily made on developing eggs of marine invertebrates, suggest that mitochondria, as important structural components of the energy-substrate supply system of the cell, may, through their differential distribution among cells of the embryo, function as important components of an endogenous chemical control mechanism. Thus, the mere possession of a larger or enzymatically more active population of mitochondria (provided exogenously required chemicals are not limiting) might enable one cell in contrast to another with fewer and/or less active mitochondria to engage in profoundly greater synthetic activities. It does not seem inconceivable that quantitative differences between cells or regions of an embryo in the level of operation of the energy-substrate supply mechanism could lead to recognizable differences between the otherwise similar (visibly indistinguishable) cells or regions.

It may well be that at least part of the precise chemical control mechanism required by the orderly phenomenon of development may ultimately have to be sought in the still unknown mechanism of differential segregation and/or accumulation (synthesis?) of mitochondria or their precursors. These suggestions as to the nature of part of the control mechanism which, undoubtedly, operates in the normal development of many organisms are presented here simply as possible avenues of attack upon this most intriguing problem. Obviously, no case is being made for the exclusive role of mitochondria, nor would any be attempted for any other single group of cellular particulates.

## REFERENCES

1. Duffey, L. M., and Ebert, J. D., *J. Embryol. Exptl. Morph.*, **5**, 324-339 (1957).
2. Gallera, J., and Oprecht, E., *Rev. suisse zool.*, **55**, 243-250 (1948).
3. Grobstein, C., in *Aspects of Synthesis and Order in Growth* (D. Rudnick, ed.), p. 233-256, Princeton Univ. Press, Princeton (1955).
4. Kun, E., *Proc. Soc. Exptl. Biol. Med.*, **78**, 195-197 (1951).
5. Miller, J., *Anat. Record,* **79,** Suppl., 47 (1941).
6. Moog, F., *J. Cellular Comp. Physiol.*, **22**, 223-231 (1943).
7. Rulon, O., *Protoplasma,* **24,** 346-364 (1935).
8. Spratt, N. T., Jr., *J. Exptl. Zool.*, **106**, 345-365 (1947).
9. ———, *J. Exptl. Zool.*, **107**, 39-64 (1948).
10. ———, *J. Exptl. Zool.*, **110**, 273-298 (1949).
11. ———, *J. Exptl. Zool.*, **114**, 375-402 (1950).
12. ———, *Biol. Bull.*, **99**, 120-135 (1950).
13. ———, *Anat. Record,* **109,** 384-385 (1951).
14. ———, *Anat. Record,* **111,** 553-554 (1951).
15. ———, *Anat. Record,* **113,** 602 (1952).
16. ———, *Ann. N. Y. Acad. Sci.*, **55**, 40-50 (1952).
17. ———, unpub.
18. ———, *J. Exptl. Zool.*, **128**, 121-164 (1955).
19. ———, in *Aspects of Synthesis and Order in Growth* (D. Rudnick, ed.), p. 209-231, Princeton Univ. Press, Princeton (1955).
20. ———, *J. Exptl. Zool.* (in press).
21. Trinkaus, J. P., and Drake, J. W., *J. Exptl. Zool.*, **132**, 311-347 (1956).

## DISCUSSION

Dr. Spiegelman: There is one thing that puzzles me. You said that in order to get the inhibitory effect you have to have extremely close contact and you implied that this might involve actual cytoplasmic bridges. Yet you obtained the effect with an homogenate.

Dr. Spratt: Those are the results. You can get the inhibitory effect by using the slice, and in all cases intimate contact is prerequisite to the effect, and there are cell bridges. This is not an implication: this is an observation. Now you ask me why the homogenate has this effect. I presume that in the homogenate there is very intimate contact between the cells and the active substance. This is assuming that the homogenate really is a homogenate and not just a mash or brei. When an explant is cultured on a slice of liver, if the explant is separated from the slice by a thin film of fluid, no bridges are formed and there is no effect. The explant develops normally. But if the explant is drawn down very tightly against the slice, then bridges are formed. When I used the homogenate, I presume that whatever is being transferred into the explant (assuming that there is a transfer of something) is already in intimate contact with the cell surfaces. Whatever is working as the active agent is immediately in contact with the explant system in the case of the homogenate and need not be in direct contact when you are using the slice—unless bridges are formed.

DR. SPIEGELMAN: You haven't done a membrane experiment have you?

DR. SPRATT: No, that would be an obvious thing to do. If one were really interested in what was transpiring here, one would have to introduce "millipore" filters or the like, to determine the nature of the transfer. All I can say is that these bridges are present and you only get the inhibitory effect with organ slices when they are present. I have never seen the effect otherwise. Also, the effect is restricted to just the region of contact. For instance, if a slice of liver which is smaller than the area of an entire blastoderm placed on it, dispersal occurs only in the region of the blastoderm in contact with the liver. The rest of the blastoderm undergoes normal differentiation.

DR. DEHAAN: I would like to comment on your point about intercellular fibers. After you visited our lab last year and we talked about these fibers, as soon as you left I ran to the microscope and tried your technique of squashing down the chicks. It is very obvious that if you look for these bridges you can see them, especially between cells or masses of cells that are being drawn apart. Thus I would certainly confirm the existence of fibrous intercellular contacts. However, whether these represent filopodial cell processes, with or without protoplasmic continuity, or possibly just drawn-out extracellular mucous material, I cannot say.

The other point that I wanted to worry about a little is your dispersal of cells. In the initial experiments where you simply take chicks and put them on a nutritionless medium, that is, just agar without glucose, I presume you are using a phosphate buffered system. I am not surprised that they disaggregate in the sense I was talking about yesterday. Now in those systems where you spoke of the use of Versene, of amino acid, or of succinate, was this also without glucose or was this in the face of glucose?

DR. SPRATT: With succinate, this was the sole nutrient source and phosphate buffer was present. With versene now, understand that this can be done in a glucose medium or it can be done in a simple saline solution before the embryo is explanted. I have done it both ways.

DR. DEHAAN: Have you ever found a protection against versene with glucose?

DR. SPRATT: I don't know about glucose effects on versene activity, but I do know that glucose has a protective effect against simple starvation and against succinate dispersal and that resulting from a number of other agents. We also know that glucose prevents the leaching out of protein from the cells into the medium. This protein can be measured, and it is apparently nucleoprotein as well as other protein.

DR. DEHAAN: Then, the last question concerns those slides in which you showed for example a chick with only a beating heart, and the rest of the embryo drawn dotted. Did you mean to imply by this that the rest of the embryo is dispersed in the same sense that the nutritionless embryo is and that you simply have a heart amidst a mass of disaggregate cells?

DR. SPRATT: Yes, this is the general picture: cell dispersal everywhere except for a little vesicle in which the heart continues to develop. That is the typical picture. I might mention by the way that "dispersal of the cells" might be a misleading phrase. If one uses agents like trypsin, versene, or high phosphate, etc., one obtains the first stages of cell dispersal only. I think all people who have worked with cell dispersal recognize this fact. What one gets is a loosening up of the cell system, but actually there are connections between the cells which hold them together. The only way one can get complete dispersal from this stage on is to mechanically disrupt the system. As a matter of fact a very easy way of observing these intercellular connections is to use a calcium chelating agent, which loosens up the system. I presume this may be taking away the intercellular matrix if it is present but is not attacking these connections. Trypsin, versene and other agents do not digest these. You have to break them mechanically.

DR. GRANT: The presence of these protoplasmic bridges in the intact blastoderm suggests a very intriguing mechanism for the explanation of "field" phenomena which we observe in embryology. I am just wondering if you find any quantitative distribution of protoplasmic bridges in various areas of the blastoderm.

DR. SPRATT: No, I have not. Unfortunately, as I indicated, there are almost too many of these connections. Very few cells seem not to be connected in some way or another with other cells. There are a few free cells, I am sure, that wander around, aside from the obvious free blood cells. However, the balance of the cells are connected, and I have seen no great difference in the number of connections in different regions; and in fact I have not studied this in detail at all.

DR. WEISS: Dr. Spratt probably knows what I am going to say. I wanted to say that this discussion has side-tracked from the very beautiful demonstration of specialized metabolic effects and particularly of the need for intimate contact. The particular hypothesis we heard of the manner of contact—protoplasmic continuity—is based on what I consider an inconclusive technique, namely, the limited resolution of the microscope. I am sure you agree that all of these bridges exist as far as one can see under the microscope. We have seen all of these fibers here. They mark intercellular contact points. However, so far, every single one of those that I have been able to investigate under the electron microscope has proved not to be a protoplasmic continuum but rather had a little disc intercalated, usually of the order of magnitude of 100-150 Ångstroms, which would entirely remove it, of course, from the realm of microscopic observation. Before we enter into speculations about transfer from cell to cell, I think we ought to wait until Dr. Spratt has actually carried out his observations on the transfer of microscopic granules from one cell to another, which at this time I would doubt very much occurs.

DR. HERRMANN: I would like to emphasize one point in commenting on the type of explantation technique used by Dr. Spratt. It seems to me that in this system we have the simplest experimental condition in which at least a part of normal morphogenesis of the chick embryo will occur in vitro. While Dr. Spratt and other investigators have concentrated on the exploration of the qualitative aspects of this system I should like to point out that these explants lend themselves to a far-reaching quantitative analysis. For example, we have found it possible to obtain quantitative data on protein accumulation in the whole embryos and in individual primordia, to vary protein accumulation depending upon explantation conditions from a very low rate to a rate which is close to that found for the embryo in the egg, and to measure the rate of uptake of labelled amino acids into the embryonic proteins and into the free amino acid fraction in the absence and presence of amino acid analogs. It may be mentioned that the different organ primordia differ in their sensitivity to these amino acid analogs. For example, leucine analogs lead to a fusion of somites in the explants with hardly any disturbance in other parts of the embryo (Herrmann, Konigsberg and Curry, *J. Exptl. Zool.*, **128,** pp. 359, 1955). At the same time quantitative measurements show that in the affected somites incorporation of labelled amino acids in the somite cells is not impeded but there is a net loss of protein indicating an increased rate of protein or cell breakdown in these cells. Other tissues which are not affected morphogenetically do not show this biochemical defect. (Schultz and Herrmann, *J. Embryol. Exptl. Morph.*, in press). Together with similar measurements on isolated organ primordia and single cell cultures there seem to be possibilities in these systems to arrive at least at elementary kinetics of protein formation which may contribute to bridging the gap between the microbiological approach to protein formation and the analysis of analogous processes during morphogenesis of these explants.

DR. SUSSMAN: I would just like to mention that some time ago we began a collection of morphogenetically deficient mutants of slime mold which could not complete the morphogenetic sequence. When we put them together they completed the sequence, but when we separated them by an agar membrane two cell diameters thick, they could not synergize. In accordance with Dr. Cohn's terminology, we have termed this nongratuitous non-induction.

# CHEMICAL STIMULATION OF NERVE GROWTH [1][2]

### Rita Levi-Montalcini
*Department of Zoology,
Washington University,
St. Louis, Missouri*

GROWTH IN THE NERVOUS SYSTEM is the result of an interplay between two forces. One is internal: the intrinsic capacity of the nerve cells to develop and to grow according to the constitutional growth potentialities of each cell type. The other is external: the physico-chemical environment of the nerve cell. It is the aim of the present paper to discuss the effects of the extrinsic agents on the differentiation and growth of nerve cells. Such effects have been intensively investigated in recent years in the embryo and, to a lesser extent, also in the mature organism. These studies have made it apparent that the growth rate and the size of the nerve cells are under the continuous regulation of peripheral factors which may exert their influence through the afferent and the efferent ends of the nerve cells and directly also on the nerve cell bodies. The influence of peripheral factors extends throughout the life of the nerve cell. While this role of the periphery has been acknowledged by all students of the nervous system, its importance in development is not agreed upon.

It was recently stated that such influences "do not determine activity, growth or size of a neuron in any absolute sense, but simply enhance or depress its inherent activities, and within relatively narrow limits at that" (40, p. 365). It is implied in this concept that the inherent growth capacities of nerve cells set qualitative and quantitative limits on the influence of the extrinsic factors, limits not to be transcended under any conditions. Experimental neuroembryology corroborated this concept, which finds general acceptance among students of developmental neurology. In recent years, however, it was discovered that chemical factors isolated from a number of biological materials not only can "force" some nerve cells to produce nerve fibers far in excess of the normal range, but that they can

---

[1] This work has been supported by grants from The National Institutes of Health, Public Health Service, The National Science Foundation, and The American Cancer Society. Our recent work, outlined here, will be reported more extensively elsewhere.

[2] Dedicated to Prof. Giuseppe Levi on the occasion of his 86th birthday.

also divert the outgrowth of such nerve fibers and endow them with the capacity to transcend barriers normally impermeable to their penetration. The way these agents operate, their biological effects, and their structural characteristics will be discussed in the following pages. Before presenting these data, we propose to review the evidence in favor of chemical stimulation of nerve growth as gathered from other sectors of experimental neurology.

If we wish to evaluate growth processes in the nervous system, the first problem we face is the choice of the material and of the developmental stage of the nerve structures to be examined. There are various levels of organization at which growth processes may be investigated in any living organism. No growth process, however, presents the student with so many variables and different levels as the nervous system. To mention only some of these variables, we should keep in mind that the growth potentialities of the nervous system of lower vertebrates are much broader than those of higher forms. In the same way, growth capacities decrease dramatically from embryonic to adult life in the same organism; hence different criteria should be applied in the evaluation of growth in lower and higher forms as well as in the embryonic and mature nervous system. An additional problem faced by the student of growth in the nervous system is to be found in the heterogeneity of nerve structures.

"The concept of a nerve cell," writes V. Hamburger, "is a rather pale abstraction. There is a greater variety of cell types in the nervous system than in all other tissues taken together, and the adaptive features of neurons with respect to the size, number and branching of their processes, and the modes of their synaptic connections with each other and with end organs are countless" (14, p. 263).

The structural heterogeneity is paralleled by a similar heterogeneity in function and capacity to react to environmental conditions. It is an everyday experience of the neuropathologist that nerve cells react in a specific and differential way to adverse agents, and it is equally well known to the neuroembryologist that different nerve centers behave in a different way when exposed to the same extrinsic conditions. It follows that the same agent which may enhance growth in one particular type of cell may be ineffectual or even detrimental to another type of nerve cell. This aspect of growth will be elaborated more in detail in a following section.

Still another perplexing factor in dealing with growth processes in the nervous system is provided by the very structure of the nerve cells. Nerve cells are undoubtedly the most highly specialized cells in the animal kingdom. Not only do they differ from each other in their structural and functional complexity, but significant structural differences are to be found in different parts of the same cell. In some cells, like the Mauthner cells, the elaboration of structural complexity has reached such a refinement as to make one wonder whether such cells are more akin to other cells or to miniature nerve centers. Growth processes in such cells may be very hard to evaluate unless all the constituent parts of the cell are carefully inspected. A similar problem in defining growth is also faced in other situations familiar to all neurologists. We have in mind the regenerative processes which follow the transection of a peripheral sensory or motor nerve fiber. The gradual increase in length of the transected nerve fiber is attained through synthesis of new axoplasm, and it is therefore by definition a growth process. However, while the nerve fiber is actively increasing in length, chemical and morphological processes which could hardly be defined as "growth processes" take place in the nerve cell body. It is therefore a rather difficult task to attempt to evaluate growth processes in the nervous system. The difficulty is not only semantic; it stems also from our lack of information and from the inadequacy of our methods of analysis. As we shall see, the problem of defining growth is particularly acute in the mature nervous system. In the embryo, growth occurs in "telescopic" spatial and temporal dimensions. Furthermore, the response of nerve cells to adverse or favorable conditions is much more rapid than in the adult nerve cell. We shall therefore consider at first the evidence for stimulation of nerve growth in the embryo and thereafter in the full-grown organism.

## Factors Stimulating Nerve Growth in the Embryo

### Effects of the Periphery on the Embryonic Nervous System

Growth in any sector of the early embryonic nervous system is characterized by increase in number of cells, increase in size of individual cells, and elongation of their neurites. Mitotic activity declines rapidly and then ceases in the primary motor and sensory nerve centers; in most of the higher brain centers it continues throughout embryonic life. Increase in size of nerve cells continues after fetal life and comes to an end only with the attainment of the full size of the organism; the same may be said of the length of nerve fibers.

Neuroembryologists focused their analysis mainly on primary motor and sensory nerve centers in amphibian larvae and bird embryos. Few experiments were performed on secondary sensory centers and on that unique type of nerve cell known as the Mauthner cell. The techniques used were the classical techniques of experimental embryology, involving extirpation, grafting, and interchange of peripheral end organs. It was the primary goal of these experiments to attempt to define the role played by intrinsic and extrinsic structures in the development of the nervous system. Such experiments revealed the complexity of the apparently simple response of nerve centers to changes in the peripheral field of innervation (16).

Two aspects of this phenomenon are worth mentioning since they are pertinent to the topic to be discussed. (1) Different nerve centers react in a different and unpredictable way to peripheral changes; sensory primary centers exhibit more plasticity than primary motor centers in adapting themselves to an enlarged peripheral area. Increase in mitotic activity and in size of individual neurons was actually observed in overloaded ganglia (15), whereas the changes in the primary motor neurons in a similar situation were not clearly established. (2) The effects of peripheral overloading are restricted to the ganglia supplying nerve fibers to the enlarged periphery; it was therefore concluded that the observed effects (hyperplasia and hypertrophy) are mediated by nerve fibers. However, in spite of the large amount of work invested in the analysis of this problem the mechanism of action of the periphery and the kind of "information" conveyed by nerve fibers to their centers have not been determined. The only conclusion one could draw from these experiments was that proliferation, differentiation, and growth in the nervous system are in some unspecified way under the control of the peripheral field of innervation.

## Hormonal Effects on the Embryonic Nervous System

While the experiments mentioned above provided evidence for the role played by nerve fibers in mediating the effects of the periphery on the development of the associated nerve centers, other experiments indicated that the embryonic nerve centers are also directly affected by changes in the surrounding medium. The effects of thyroid hormone on the development of brain centers in amphibian larvae were explored by P. Weiss and F. Rossetti (41). It is well known that amphibian metamorphosis is under the control of the thyroid hormone. It was shown that such effects are in some instances restricted to tissues immediately adjacent to the grafted gland or to implants of a piece of agar soaked with the hormone (29). This same technique, consisting of grafting a piece of gland or an inert carrier of the thyroid hormone, was then used by Weiss and Rossetti to test the effects of the thyroid hormone on the development of the hindbrain centers. The authors reported a decrease in size of the Mauthner cells and an increase of other hindbrain cells. Similar effects are observed in these cells during metamorphosis. The size decrease of the Mauthner cells during metamorphosis could be ascribed to shrinkage of their peripheral field of innervation (tail resorption) or to direct hormonal effects on the cells. The experiments by Weiss and Rossetti provided evidence in favor of the second alternative. Results to be reported in the second part of this paper corroborate their statement that nerve cells are sensitive to factors present in the surrounding media.

### Factors Stimulating Nerve Growth in the Adult Organism

In mammals the capacity of mature nerve fibers to regenerate is well established for motor and sensory nerves but is still controversial in the case of fibers in the central nervous system (12). The clinical aspect of this process prompted an extensive search for nerve growth-stimulating agents. We shall briefly present some of the experimental work performed along this line.

Different agents were tested and found effective in promoting nerve regeneration. One such agent is the "NR factor" isolated by von Muralt and his associates from the white matter of the brain of rabbit and calf (27). The speed of regenerative processes was compared in control and experimental animals (rabbit) receiving the NR factor through daily intraperitoneal injections. The Jent test (25) consisted of comparing the speed of regeneration of the sensory fibers innervating the cornea after a circular cut of the cornea. Restoration of the reflex on touching the cornea provided the criterion of the achieved regeneration. It was reported that the regeneration time in animals receiving the NR factor was reduced from an average of 20 days to 6 to 8 days (28). Histological examination gave evidence of a more extensive branching of the transected nerves in treated animals. The NR factor was identified as a low molecular weight substance, which is water- and alcohol-soluble, dialyzable and heat-stable (27).

Other agents with specific effects on nerve regeneration were described from the same laboratory. They are dinitriles and piromen. Both, when assayed with the cornea test, were found to stimulate nerve regeneration in a manner similar to the NR factor. The investigation of the stimulating activity of dinitriles (malonitrile and succinitrile) was performed by M. P. Konig (30) and by V. Martini and J. Pattay (38). It was prompted by a report by Hydén and Hartelius (24), who tested the effect of malonitrile on the PNA content of nerve cells. As reported by Hydén: "An increase in the PNA content of nerve cells has been shown to occur within one hour of injection into animals of small amount of malonitrile" (23, p. 228). Martini and Pattay correlate their own findings with the results of Hydén and Hartelius and favor the hypothesis of a direct action of dinitriles on the nucleoprotein metabolism of the nerve cells. Such an hypothesis seems at the present time highly speculative, since the two phenomena were observed in different experimental material and under different experimental conditions.

The effect of piromen (a bacterial polysaccharide complex) in enhancing nerve regeneration was first investigated in the central nervous system and then in the peripheral nerves. Its activity on the central nervous system was systematically studied by a team of workers under the leadership of Windle (42). Evidence was presented for its indirect effect on the nervous system: the inhibition of scar formation at the lesion site. This would in turn favor nerve growth, thus speeding regenerative processes in transected nerve fibers. The same mechanism is claimed by Hoffman, who investigated the effect of piromen in peripheral regeneration (20). Other authors suggest instead a direct action of this drug on nerve cells to explain the enhanced speed of regeneration in peripheral nerve fibers (17).

## Collateral Regeneration

The reinnervation of a partially denervated muscle by the adjacent intact nerve fibers was first described at the end of the past century by Exner (11). It was reinvestigated in recent years by a number of authors. We shall mention here only the extensive work performed by Edds and by Hoffman, who independently studied the histological and the chemical aspects of this phenomenon and attempted an explanation of the mechanism involved in the regenerative processes. Since the sprouting of additional branches from the intact nerve fibers occurs not only at the tip but also from the distal preterminal portion of the nerve, this process is known as "collateral regeneration." We refer for a detailed analysis of the phenomenon to the lucid review by Edds (9), and to the subsequent contributions by him (10) and by Hoffman (21). Briefly, the experiment consists of partial denervation of a muscle in rats or other mammals. It was noticed that the partially denervated muscle still retains an unexpectedly large fraction of its normal weight and strength. The hypothesis that "the phenomenon may represent a peripheral extension of the processes of viable axones to neighboring muscle fibers" (18, p. 52) was confirmed by neurohistological studies by Edds (7, 8) and by Hoffman (19). Extensive branching of collaterals from intact nerves which grow into vacant motor end-plates of adjacent muscle fibers was actually observed by these authors. In an attempt to identify the active agent responsible for collateral regeneration, Hoffman injected extracts of different tissues into intact muscles.

He reported the occurrence of collateral sprouting from intact nerves when an ether extract of ox spinal cord was injected, thus resulting in hyperneurotization. The active substance appeared to be a lipid. The author suggested that "such substances act by disorganizing the cell membrane and liquefying the axogen, resulting in outflow of axoplasm" (20, p. 424). Among other agents tested by the same author, piromen (see above) and pironin, a xanthine dye, proved to possess also the capacity of stimulating excessive nerve branching from intact nerves.

Agents of different chemical structure were therefore credited with such growth-stimulating activity, and different mechanisms were suggested by the different authors. In all instances only the peripheral branches of the nerves seemed to be affected by the assayed agents. It is doubtful, as pointed out by Edds (9), whether growth increase did actually occur in the proximal part of the nerve fibers and in the cell bodies. None of the experiments reported presented evidence to this effect. It seems to us that more extensive information on these stimulating agents and on their effects on nerve cells should be gathered before attempting an evaluation of these results.

### Biological Aspects of Nerve Growth-Promoting Proteins Isolated from Sarcomas, Snake Venom, and Salivary Glands

Experimental neuroembryology has given evidence of the remarkable plasticity of the embryonic nervous system, which is ready to adopt a periphery quantitatively and qualitatively different from its own. It also showed the tolerance of embryonic tissues in admitting nerves issuing from different regions of the spinal cord, or even nerves supplied by the nervous system of a different species. The neuroembryologists took advantage of this situation by planning and performing a vast array of heteroplastic transplantations and exploring the innervation of the chimaeric organisms.

Observations in the adult organisms proved at the same time that other tissues, like tumors, which do not fit into the economy of the organism, are normally not innervated, although in some instances a limited number of nerves may find their way among the neoplastic cells (26, 6). At variance with such previous observations, Bueker (1) reported in 1948 that a mouse tumor implanted in the body wall of chick embryos had been extensively invaded by sensory nerve fibers from the host. Such a report should have therefore sounded very surprising.

Embryologists, however, are so much used to the extraordinary performances of embryos that this communication apparently did not stir up too much curiosity. The discoverer seemed to consider it more as an interesting documentation of the potentialities of the embryonic nervous system, than as an infraction of the rules which regulate the relations between nerve fibers and peripheral structures. In the same way, when we repeated the experiments in 1951 (34), and found new facets of this phenomenon, we were bewildered but still not prepared to consider it an exceptional event. It was not until the following year, 1952, that the phenomenon revealed itself in all its singularity and defied all efforts to fit it into our preconceived scheme of interrelations between nervous and non-nervous structures (31). From that moment to the present day, many new and more perplexing aspects of this phenomenon were discovered. We are however now prepared, or at least we believe we are, to take up the challenge, and in a better position to direct questions to the nervous system, even though we may still not understand its answers.

The discovery by Bueker and the interpretation given by him, as well as our own interpretation based on a more extensive analysis, were the objects of many reports in past years. We shall therefore give only a short account of the first part of this work and refer the reader to the previous publications and review articles (1 2, 31, 34, 35, 13, 32).

The original experiment was conceived in the general frame of experimental neuroembryology; it was aimed at the investigation of the effects of the peripheral field of innervation on the associated nerve centers. It was known from previous experiments that mammalian tumors grow on the chorio-allantoic membrane of the chick embryo. By transplanting mouse tumors intra-embryonically, Bueker wanted to test the capacity of nerve fibers to invade tissue of a different class and of high growth potentialities. He reported that a fragment of mouse sarcoma 180 implanted in the body wall of a 3-day chick embryo became invaded by sensory nerve fibers emerging from the adjacent spinal ganglia. These ganglia appeared considerably enlarged 4 to 5 days later. The over-all increase was the result of an increase in cell number and an increase in size of individual neurons. No motor fibers were traced into the tumor, and no enlargement was observed in the motor column of the spinal cord. The author concluded that histochemical properties of the tumor favored the branching of sensory nerve fibers, and thus called forth the observed increase in size and number of the sensory neurons (1). Our investigation (34) corroborated these observations and added the following new facts. (a) The sympathetic system contributes nerve fibers to the tumor and does so even to a larger extent than the sensory system. (b) The increase in size of the sensory and sympathetic ganglia is far more extensive than that observed in limb transplantation experiments. (c) The over-all enlargement of ganglia is

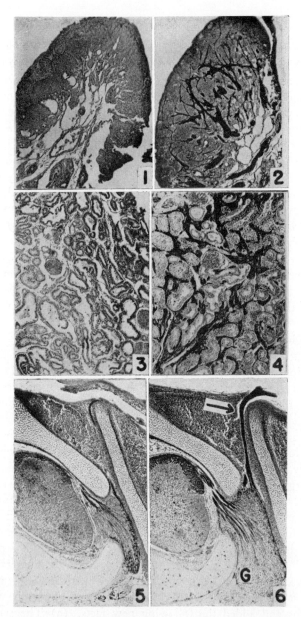

Figs. 1-6. Microphotographs of silver-impregnated control and experimental embryos.
Figs. 1, 3. Ovary and mesonephros of control embryos of 16 days.
Figs. 2, 4. Same organs in 16-day embryos bearing a transplant of mouse sarcoma 180.
Figs. 5, 6. Lumbar spinal ganglion in a 9-day control embryo and in a 9-day embryo injected with salivary gland extract. Note the size increase in the experimental ganglion (G). Arrow points to the dorsal ramus.

[Figs. 1-4 reproduced from *Ann. N. Y. Acad. Sci.*, **55**, 330 (1952)].

the result not only of an increase in cell size and cell number, but also of an acceleration of differentiative processes in sensory and sympathetic cells. (d) Sensory and sympathetic nerve fibers branch in all directions into the tumor but *do not* establish synaptic contact with the neoplastic cells. (e) The two types of cells that are present in the spinal ganglia (15, 36) are affected differentially. One class of neurons seemed to be either slightly affected or not affected at all, whereas the other was highly hypertrophic and hyperplastic. (f) The effect of the tumor is a cumulative one. Embryos observed in later developmental stages showed a much more impressive response than embryos fixed in earlier stages. The effect was not apparent before the sixth day of incubation. (g) In most instances the tumor called forth not only an increase in size of the sympathetic ganglia, but also the formation of additional ganglionic complexes on the fringe of the tumor. (h) Two mouse sarcomas were found to possess such activity: sarcoma 180 and sarcoma 37. The effect of these tumors varied considerably from one experiment to another and was not correlated with the size attained by the grafted tumor in the embryo. Other tumors, like neuroblastoma and mammary carcinoma, which exhibit equal or even more vigorous growth potentialities than mouse sarcomas 180 and 37, did not elicit any growth effects in the nervous system of the host. Hence the observed phenomena could not be attributed only to a general non-specific growth-promoting effect of rapidly expanding tissues. Subsequently, a systematic screening by Bueker of different mouse and other mammalian tumors revealed that other mouse sarcomas, besides the two mentioned, possess the capacity of stimulating nerve growth when implanted in the embryo. A mild growth-promoting activity was also found in three rat sarcomas (Wc, S6, Fs), whereas the epithelial mouse tumors and all other mammalian tumors investigated were consistently negative (2).

We shared with Bueker the belief that the effect of the tumor was exclusively mediated by nerve fibers. This belief was shaken a few months later (1951) by the following observation; it was found that sympathetic ganglia which are too distant from the tumor to participate in its neurotization were also enlarged, and that fibers emerging from them had invaded the adjacent viscera. These viscera are not innervated in normal embryos of the same stages (Figs. 1, 2, 3, 4). The observation suggested that the tumor releases in the blood stream of the host an agent which directly affects the embryonic nerve cells. This hypothesis received full confirmation from the experiment which followed: a fragment of tumor was transplanted onto the allantoic membrane of 4-day chick embryos, in such a position as to prevent sensory and sympathetic nerve fibers from invading it. The results were in all respects similar to the previous results: the sympathetic ganglia were hyperplastic and hypertrophic and the viscera were hyperneurotized. In both groups of experiments of intra- and extra-embryonic transplantation of the tumor, another most unusual phenomenon was observed: nerve fibers emerging from the enlarged sympathetic ganglia perforated the walls of large veins and bulged into the lumina of the vessels (see Fig. 20). In some instances many dozens of such neuromas were observed in the same embryo. Such findings gave decisive evidence of the singularity of the effects elicited by the tumor. They differed indeed in so many respects from all effects observed in previous experiments of homoplastic or heteroplastic transplantations of embryonic tissues as to make comparison hardly possible (31, 35).

The next problem we were confronted with was to identify the immediate target of the agent released by the tumor. Did this agent hit the nerve cells selectively, speeding their differentiative processes and "forcing" them to produce more fibers than normally, or did the agent permeate the entire organism and in this way made it abnormally receptive to the entrance of nerve fibers? In order to decide between these alternatives, we tested the effect of the tumor in vitro, using the standard hanging drop technique. The medium consisted of chicken plasma and embryonic extract. A fragment of mouse sarcoma 180 or 37 was explanted in such medium in proximity to but not in contact with a sensory or a sympathetic ganglion explanted from 7- to 9-day chick embryos. In control experiments we used, instead of a tumor, a fragment of heart isolated from a chick or a mouse embryo. The ganglia explanted together with the mouse tumor produced an extraordinary number of nerve fibers within the first 24 hours, whereas control ganglia cultured with fragments of chick heart grew out a very limited number of nerve fibers in the same time. The growth pattern of nerves branching out from the ganglia exposed to the effect of the tumor in vitro differed qualitatively as well as quantitatively from the growth pattern of controls (Figs. 7, 9). In the experimental ganglia the fibers grow in a radial pattern and are extremely dense on the side facing the tumor. On this side they are shorter and stop abruptly in front of the neoplastic tissue as though hampered by some invisible barrier (Fig. 9). In the control series, where the ganglia were explanted with a fragment of heart of mouse embryo, an intermediate result was obtained. The nerve fibers were more numerous than in the previous controls with embryonic chicken heart, but much less numerous than in ganglia cultured with tumor. A consistent increase in the number of nerve fibers was observed on the side of the ganglion facing the mouse tissue. Such results suggested that "the

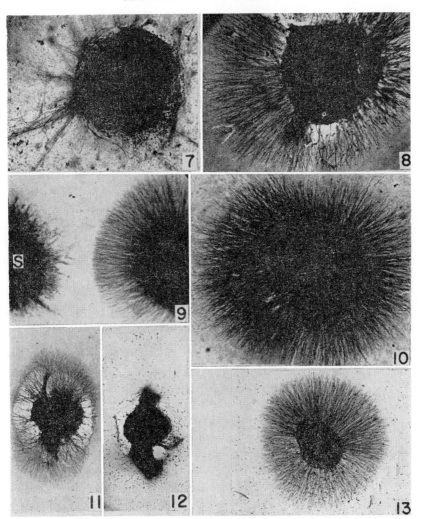

Figs. 7-13. Microphotographs of silver-impregnated sensory ganglia of a 7-day chick embryo in vitro, comparing the effects of a fragment of mouse tumor (fig. 9), extract of mouse tumor (fig. 8), snake venom (fig. 10), and salivary gland extract (fig. 13), with control (fig. 7). Effect of salivary gland on ganglion of rat embryo (fig. 11) compared with control (fig. 12).

[Figs. 7-10 reproduced from *Proc. Natl. Acad. Sci. U. S.*, **42**, 695 (1956).]

sarcoma agent may be present in low concentration in normal mouse tissue" (37, p. 56).

The significance of this observation regarding the effect of mouse tissue was not clear at that time. Four years had to elapse before another discovery by Dr. Stanley Cohen, to be reported below, called for a reconsideration of this "mouse effect."

Although the effects of mouse sarcomas in vivo and in vitro differed from each other in many respects, they had one outstanding aspect in common: the capacity to enhance nerve growth from the same types of nerve cells (sensory and sympathetic), and to be ineffective on other nerve cells, namely, the motor somatic cells. Furthermore, the tumors which did not possess such activity in the embryo also lacked it in vitro; neuroblastoma and mammary carcinoma did not elicit fiber outgrowth from ganglia either in the embryo or in vitro. Such similarity in the effects suggested that the same agent released by the tumor was responsible for the in-vivo and in-vitro effects. We shall point out the two most significant results of these experiments: (1) they gave conclusive evidence in favor of the hypothesis that the growth agent present in mouse tumors affects primarily the nerve cells, since such effects were observed also outside of the embryo; and (2) they provided a simple and rapid bioassay to investigate the biological properties of this agent.

A protein fraction which was isolated from these tumors by Dr. Stanley Cohen showed the same growth-promoting activity in ganglia explanted in vitro as did the growing tumor (4). In order to exclude the possibility that the activity was due to traces of nucleic acids still present in the "protein fraction," this active fraction was incubated in the presence of a mixture of nuclease and phosphodiesterase. Crude snake venom was used as the source of phosphodiesterase. We were prepared to see a sharp decrease in nerve growth in vitro, if the degradation of the nucleic acids had destroyed the activity of the agent present in the tumor extract, or to see no changes in the effect, if the activity was not associated with nucleoproteins. We were not prepared to see what we did see on the following day: in control cultures in which only the snake venom but not the tumor factor had been added to the medium, the ganglia had produced overnight a tremendous number of nerve fibers (Fig. 10) (4). The chemical and biological properties of the snake venom agent, as tested in vitro, will be reported in this Symposium by Dr. Cohen (3). We shall instead mention here the results of the in-vivo experiments with the purified snake venom. Injections of microgram quantities into the yolk of 6- to 9-day chick embryos replicated the effects observed in embryos bearing intra- or extra-embryonic transplants of mouse sarcomas (33). Not only were the sympathetic ganglia considerably enlarged and the viscera filled with nerves (Figs. 14-19), but also the unusual phenomenon of sympathetic nerve fibers trespassing upon the endothelial barriers and bulging into the lumina of the veins of the host, was again observed (Fig. 21). Such experiments therefore gave conclusive evidence in favor of our hypothesis that the in-vivo and the in-vitro effects are elicited by the same agent.

A new facet of the phenomenon, and perhaps a clue to the understanding of its very nature, was recently discovered. We shall devote the last part of this paper to its analysis.

Possibly one of the most mysterious and even disturbing aspects of the phenomenon was the striking similarity between the chemical and the biological properties of the two agents isolated from two such unrelated biological objects: the mouse sarcomas and the snake venom. Since the venom is produced by glands which are modified salivary glands, it occurred to Dr. Cohen to assay the salivary glands of the mouse. The biologist is generally not used to seeing his bold hypotheses confirmed by nature, and therefore we were very much surprised at the outcome of this experiment. Sensory ganglia explanted in vitro together with small fragments of mouse salivary glands produced the same exuberant outgrowth of nerve fibers as the ganglia exposed to the effect of snake venom (Fig. 13). Recently, Dr. Cohen found that the extract of salivary glands, as well as the snake venom, promotes exuberant nerve growth in vitro not only from chicken but also from mouse and rat ganglia (Figs. 11, 12).

Some preliminary data on the chemical characteristics of this newly isolated factor will be dealt with in the following communication (3). Here we shall report briefly on the effects of the extract of mouse salivary glands on the chick embryo.

Microgram quantities of the purified extract of mouse submaxillary glands were injected into the yolk of 6- to 8-day chick embryos, with the same technique used in the experiments with snake venom. All the embryos injected showed effects similar to those described in embryos bearing mouse sarcomas or injected with snake venom. A distinct enlargement of the sensory and of the sympathetic ganglia was already apparent in 7-day embryos which had received two injections of the salivary gland extract during the sixth day of incubation. Some unusual features deserve mentioning.

In previous experiments with mouse sarcomas, the sensory and sympathetic ganglia adjacent to the tumor were very much enlarged. Sympathetic ganglia not contributing nerves to the tumor were also hyperplastic

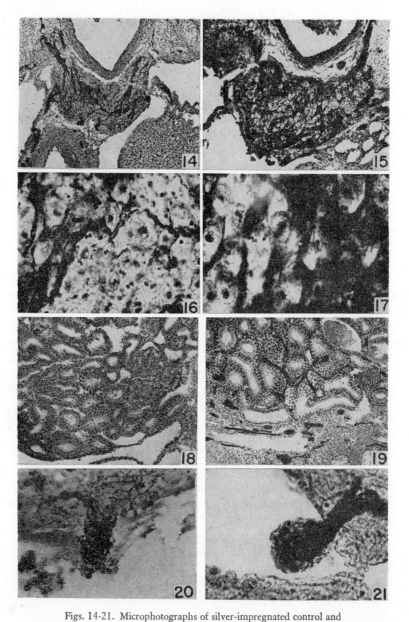

Figs. 14-21. Microphotographs of silver-impregnated control and experimental chick embryos.
Fig. 14. Prevertebral sympathetic ganglion of a normal 10-day embryo.
Fig. 15. Prevertebral sympathetic ganglion of a 10-day embryo injected with snake venom.
Figs. 16, 17. Enlarged sectors of Figs. 14 and 15, respectively, to show increase of cell size and precocious fibrillar differentiation.

and hypertrophic. A similar response had been elicited later with allantoic transplants of the tumor and with injections of snake venom. A remote effect on the sensory system had not been clearly established in any of these experiments. On the contrary, in the present experiments, the injections of the salivary gland extract into the yolk sac of the embryos resulted in a remarkable increase of both the sensory and the sympathetic ganglia. Planimeter measurements showed that sensory ganglia in 8-day embryos were about three times larger than controls. The nerves emerging from these hyperplastic and hypertrophic ganglia collected in thick nerve bundles directed to exteroceptive fields. Particularly impressive was the size increase in the dorsal ramus (Figs. 5, 6) which branches in the mesenchyme beneath the skin of the dorsal aspect of the embryo. The sensory innervation of the muscles did not seem to differ from controls. A differential response was observed in the two types of nerve cells described in the sensory ganglia, as it had been found in experiments of intra-embryonic transplantations of mouse sarcomas (see p. 230). The increase in size of the sympathetic ganglia was very much similar to the increase observed in experiments with sarcomas and snake venom. The peripheral distribution of nerves emerging from the enlarged ganglia differed however in some embryos from the one described in previous experiments. In embryos receiving a large dose of the salivary extract, we could not trace any nerve in the viscera, although the ganglia were very much increased in size. The thick nerve bundles emerging from these ganglia branched instead in the mesenchyme surrounding the vertebrae, or gave origin to a dense fibrillar mat of nerve fibers encasing the mesonephroi. Sympathetic nerves of considerable size were also traced to the skin of the dorsal aspect of the embryo. Different results were obtained with injections of a more dilute salivary extract. In these experiments, sympathetic nerve fibers were found in the viscera in large numbers.

## Conclusions

It seems premature to evaluate the significance of these results, since the investigation is still in progress and from past experience we may anticipate that future experiments may uncover new facets of this perplexing

Figs. 18, 19. Mesonephros of a control (left) of an injected (right) 10-day embryo. Note the sympathetic fibers between the mesonephric tubules in the injected embryo.
Fig. 20. Neuroma projecting into the lumen of a vein of a 9-day embryo carrying an extra-embryonic sarcoma 37.
Fig. 21. Same in a 10-day embryo injected with snake venom.
[All figures reproduced from *Proc. Natl. Acad. Sci. U. S.*, **42**, 695 (1956).]

phenomenon. We shall therefore make only a few comments on some of the multiform aspects of this research.

Perhaps the most remarkable feature of this investigation was to prove the possibility of a peaceful collaboration between an embryologist and a biochemist. Such a possibility seems to have been questioned by some participants in the Symposium.

Other aspects are worth mentioning. The substance endowed with the property of enhancing nerve growth appears to be organ-specific and restricted to the salivary glands of different vertebrates. Its occurrence (in a much more dilute concentration) in mouse sarcoma does not find an explanation at present. Many instances are known in the literature, however, of tumors manufacturing vitamins or other substances.

The specificity of the phenomenon extends also to the reacting system: only two types of nerve cells were found to be receptive to the growth-promoting factors: the sensory and the sympathetic ganglionic nerve cells.

Another aspect should be emphasized: these experiments revealed in nerve cells unsuspected growth potentialities which do not materialize under normal conditions, but do so under the impact of these agents. The influence of extrinsic factors on the differentiation and growth of nerve cells seems therefore (at least in our material) to extend well beyond a limited range. Both the quantitative and the qualitative nerve-growth dimensions were dramatically changed by three agents: mouse sarcomas, snake venom, and a salivary gland extract.

Finally, we should like to say that if two of the outstanding aspects of this phenomenon are its specificity and the magnitude of the effects, its most mysterious and challenging aspects are its significance and the mechanism of action of this proteinic agent.

What is the function—if any—of this agent present in the salivary glands of some vertebrates on the differentiation and growth of embryonic nerve cells under normal conditions? What is its function—if any—on mature nerve cells? It is conceivable that the salivary glands play a role neither in the differentiation of embryonic nerve cells nor in the maintenance and growth of differentiated nerve cells. If so, does this protein with such a potent effect have any function?

The lack of information about the functional significance of such an agent need not detract from its biological interest, which resides in the discovery of a clear-cut response of given nerve cells to well-defined chemical agents.

## REFERENCES

1. Bueker, E. D., *Anat. Record*, **102**, 369-390 (1948).
2. ———, *Cancer Research*, **17**, 190-199 (1957).
3. Cohen, S., this symposium.
4. ———, and Levi-Montalcini, R., *Proc. Natl. Acad. Sci. U. S.*, **42**, 571-574 (1956).
5. ———, ———, and Hamburger, V., *Proc. Natl. Acad. Sci. U. S.*, **40**, 1014-1018 (1954).
6. Duncan, D., and Bellegie, N. J., *Texas Reports Biol. Med.*, **6**, 461-469 (1948).
7. Edds, M. V., *J. Exptl. Zool.*, **112**, 29-48 (1949).
8. ———, *J. Exptl. Zool.*, **113**, 517-552 (1950).
9. ———, *Quart. Rev. Biol.*, **28**, 260-276 (1953).
10. ———, *J. Exptl. Zool.*, **129**, 225-248 (1955).
11. Exner, S., *Arch. ges. Physiol.*, **36**, 572-576 (1885).
12. Gerard, R. W., in *Regeneration in the Central Nervous System* (W. F. Windle, ed.), Charles C Thomas, Springfield, Ill. (1955).
13. Hamburger, V., in *Cellular Mechanisms in Differentiation and Growth* (D. Rudnick, ed.), pp. 191-212, Princeton Univ. Press, Princeton (1956).
14. ———, *Am. Scientist*, **45**, 263-277 (1957).
15. ———, and Levi-Montalcini, R., *J. Exptl. Zool.*, **111**, 457-502 (1949).
16. ———, and ———, in *Genetic Neurology* (P. Weiss, ed.), U. of Chicago Press, Chicago (1950).
17. Hammer, H., and Martini, V., *Pflügers Arch. ges. Physiol.*, **257**, 308-317 (1953).
18. Hines, H. M., Wehrmacher, W. H., and Thompson, J. D., *Am. J. Physiol.*, **145**, 48-53 (1945).
19. Hoffman, H., *Australian J. Exptl. Biol. Med. Sci.*, **28**, 383-397 (1950).
20. ———, *J. Comp. Neurol.*, **100**, 441-460 (1954).
21. ———, in *Regeneration in the Central Nervous System* (W. F. Windle, ed.), Charles C Thomas, Springfield, Ill. (1955).
22. ———, and Springell, P. H., *Australian J. Exptl. Biol. Med. Sci.*, **29**, 417-424 (1951).
23. Hydén, H., in *Neurochemistry* (K. A. C. Elliott, I. H. Page, and J. H. Quastel, eds.), Charles C Thomas, Springfield, Ill. (1955).
24. ———, and Hartelius, H., *Acta Psychiat. et Neurol. Scand.*, **48**, (Suppl.), 1-117 (1948).
25. Jent, M., *Helv. Physiol. et Pharmacol. Acta*, **3**, 65-69 (1945).
26. Julius, H. W., *Brit. J. Exptl. Pathol.*, **10**, 185 (1929).
27. Koechlin, B. A., in *Regeneration in the Central Nervous System* (W. F. Windle, ed.), Charles C Thomas, Springfield, Ill. (1955).
28. Koechlin, B. A., and v. Muralt, A., *Helv. Physiol. et Pharmacol. Acta* **3**, 38-39 (1945).
29. Kollros, J. J., *Physiol. Zool.*, **16**, 269-279 (1943).
30. Konig, M. P., *Helv. Physiol. et Pharmacol. Acta*, **11**, 1-19 (1953).
31. Levi-Montalcini, R., *Ann. N. Y. Acad. Sci.*, **55**, 330-343 (1952).
32. ———, in *Decennial Review Conference on Tissue Culture* (P. R. White, ed.), Woodstock, Vermont, *J. Natl. Cancer Inst.*, **14**, (1957).
33. ———, and Cohen, S., *Proc. Natl. Acad. Sci. U. S.*, **42**, 695-699 (1956).
34. ———, and Hamburger, V., *J. exptl. Zool.*, **116**, 321-362 (1951).
35. ———, and ———, *J. exptl. Zool.*, **123**, 233-288 (1953).
36. ———, and Levi, G., *Pontif. Acad. Sci. Acta*, **8**, 528-568 (1944).
37. ———, Meyer, H., and Hamburger, V., *Cancer Research*, **14**, 49-57 (1954).
38. Martini, V., and Pattay, J., *Arch. intern. Pharmacodynamie*, **99**, 314-327 (1954).
39. Stuart, E. G., in *Regeneration in the Central Nervous System* (W. F. Windle, ed.), Charles C Thomas, Springfield, Ill. (1955).

40. Weiss, P., in *Analysis of Development* (B. Willier, P. Weiss, V. Hamburger, eds.), W. B. Saunders, Philadelphia & London (1955).
41. ———, and Rossetti, F., *Proc. Natl. Acad. Sci. U. S.*, **37**, 540-556 (1951).
42. Windle, W. F., Clemente, C. D., Scott, D., and Chambers, W. W., *Arch. Neurol. Psychiat.*, **67**, 553 (1952).

# A NERVE GROWTH-PROMOTING PROTEIN [1]

STANLEY COHEN

*Department of Zoology, Washington University*
*St. Louis, Missouri*

WE HAVE REPORTED the presence, in certain mouse sarcomas (2, 3) and in snake venom (1, 5) of a factor with specific growth-promoting effects on sympathetic and spinal ganglia of the chick embryo. It has been found recently that the salivary glands of species other than the snake contain a similar growth factor. It is present in the submaxillary glands of the mouse and rat, the saliva of the mouse, and the venom of the Gila monster. The biological effects of these factors are reviewed in this Symposium by Dr. Levi-Montalcini. This paper will be concerned with our studies on the nature and mode of action of the growth factor. Most of the biochemical work has been performed using the venom of the moccasin snake, *Agkistrodon piscivorus*.

## ASSAY AND DISTRIBUTION OF THE NERVE GROWTH-PROMOTING FACTORS

The tissue culture assay (3) for the detection of nerve growth-stimulating activity consisted of explanting sensory ganglia from 8 to 9-day chick embryos into hanging drop tissue cultures containing plasma, synthetic medium 1066 (with thrombin), and the material to be assayed in a final volume of 0.075 ml. The effects were observed after 18 hours of incubation at 37°C, and the amount of fiber outgrowth was recorded on a semi-quantitative scale as 1+ to 4+ (see Figs. 1-4). The assay was sensitive to twofold changes in concentration of the active material; smaller changes were not detected by gross observation of the ganglia.

The specific activities of the crude venom and of unfractionated homogenates of the submaxillary gland of the mouse were very much greater than that shown by homogenates of the mouse tumor, where the factor was initially studied. Equivalent growth stimulation effects were obtained by 15,000 $\mu$g. of a sarcoma 180 homogenate, 6 $\mu$g. of snake venom, and 1.5 $\mu$g.

---

[1] This work has been supported by grants from The National Institutes of Health, Public Health Service, The National Science Foundation, and The American Cancer Society. Our recent work, outlined here, will be reported more extensively elsewhere.

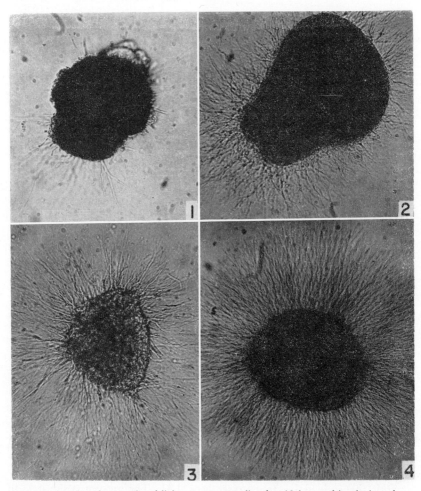

Figs. 1-4. Microphotographs of living sensory ganglia after 18 hours of incubation, showing the nerve growth-promoting effect of increasing concentrations of the venom protein. Fig. 1, a control culture. The cultures shown in Figs. 2, 3, and 4 were recorded as 1 plus, 2 plus, and 4 plus, respectively. The relative concentrations of the venom protein were 1:2:8.

of submaxillary gland of the mouse per ml. of tissue culture medium. The potency of the mouse submaxillary gland is illustrated by the fact that its biological effect can easily be detected in a saline extract of one pair of the salivary glands diluted to 50 liters. The rat submaxillary gland was much less active and approximated the sarcoma in specific activity. Homogenates of other mouse tissues (liver, kidney, brain, pancreas, intestine, stomach, and mammary gland) and of whole chick embryos were inactive. In one of seven experiments, an extract of mouse muscle showed some activity (2).

The growth-stimulating effects of the venoms from species of the three families of poisonous snakes are shown in Table 1. All of the venoms

TABLE 1

Distribution of the Nerve Growth-Promoting Factor in Venoms of Various Species

| Family | Species | Venom required to show a 3+ response |
|---|---|---|
| | | µg./ml. tissue culture medium |
| Elapidae | *Naja naja* | 3 |
| | *Sepedon haemachates* | 3 |
| Viperidae | *Vipera russellii* | 1.5 |
| | *Bitis gabonica* | 3 |
| | *Vipera aspis* (yellow or white) | 1.5 |
| | *Vipera ammodytes* | 3 |
| Crotalidae | *Agkistrodon piscivorus* | 6 |
| | *Crotalus horridus* | 6 |
| | *Crotalus adamanteus* | 6 |
| | *Bothrops atrox* | 6 |
| | *Bothrops jararaca* | 6 |

examined were effective. The crude venoms of the Elapidae and Viperidae were from two to four times as potent as those of the Crotalidae.

### Properties of the Growth Factor in the Venom of the Moccasin Snake

The growth factor has been purified approximately 40-fold (on the basis of protein content; 6) by means of standard procedures of protein fractionation. The purification steps involved ammonium sulfate fractionation (the active material precipitated between 45 and 59 per cent saturation) and adsorption and elution from the anionic and cationic cellulose resins, diethylaminoethyl and carboxymethyl cellulose. Approximately 25 per cent of the nerve growth-promoting activity present in 1 g. of the crude venom was recovered in 5.2 mg. of protein.

The growth factor was non-dialyzable, heat-labile (5 minutes at 90°C in isotonic saline, $pH$ 7.4), destroyed by acid (0.1 $N$ HCl for 1 hour at 26°C), stable to alkali (0.1 $N$ NaOH for 1 hour at 26°C) and stable to 6 $N$ urea (1 hour at 0°C) (1).

668  THE CHEMICAL BASIS OF DEVELOPMENT

The behavior of the material was examined in a Spinco analytical ultracentrifuge. The results are shown in Fig. 5. Only a single component was detectable, having an $S_{20}$ of 2.22; the molecular weight was estimated to be on the order of 20,000. The centrifugation was continued until the boundary had completely sedimented. The supernatant fluid was removed, and the residue redissolved in an isotonic saline. Tissue culture assays of these fractions showed that the supernatant fluid did not contain the growth factor: all of the activity was present in the sedimented fraction.

The ultraviolet absorption spectrum of the material is shown in Fig. 6.

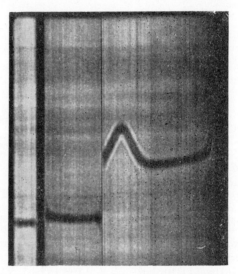

Fig. 5. Ultracentrifugation pattern of the purified protein after 64 minutes at 59,780 r.p.m. The protein concentration was 6 mg./ml. in 0.1 $N$ NaCl.

The 280/260 absorption ratio was found to be 1.3, the low value suggesting that some non-protein material might still be in the fraction. A solution containing 1.0 mg./ml. in distilled water showed an optical density of 1.03 at 280 m$\mu$ in a cell with a path length of 1 cm.

Upon acid hydrolysis (6 $N$ HCl for 10 hours in an autoclave) and two-dimensional paper chromatography of 150 $\mu$g. aliquots, the amino acid pattern was qualitatively identical to a similar chromatogram prepared with crystalline bovine albumin. No reducing sugars could be detected on the chromatograms, although the original material showed the presence of 1.6 per cent hexose as determined by the orcinol procedure, using galactose as a standard at 540 m$\mu$ (9).

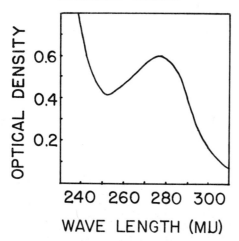

Fig. 6. Absorption spectrum of the purified protein at a concentration of 0.58 mg./ml. in distilled water.

Two additional lines of evidence support the view that the biological activity is associated with a protein: (a) the destruction of the biological activity upon incubation with proteolytic enzymes; and (b) the loss of biological activity upon incubation with antiserum to snake venom.

The biological activity of the purified growth factor was completely destroyed by incubation with each of the crystalline proteolytic enzymes, trypsin, chymotrypsin, and papain. In these experiments, 80 $\mu$g. of the purified protein were incubated with 20 $\mu$g. of each of the proteases in 0.2 ml. of 0.1 $M$ phosphate buffer, $p$H 7.0 for 4 hours at 28°C. (The papain digestion mixtures also contained 60 $\mu$g. of cysteine.) Controls were run in which the enzyme was omitted. After the incubation was completed, aliquots were assayed. Under these conditions there was no diminution of the activity in the control tubes, whereas only traces of activity could be detected after proteolytic digestion. Incubation with ribonuclease or deoxyribonuclease did not affect the biological activity.

The possibility that the proteolytic enzymes were acting by inhibiting the response of the ganglia was excluded, since unincubated mixtures of the growth factor and the proteases were as active as the control. The failure of the proteases to act on the factor in the tissue cultures is understandable, in view of the dilution of the enzyme mixture in the cultures (1:2000) and the presence in the culture of plasma proteins. The possibility that enzymes were liberating an inhibitor of some sort during the 4-hour incubation period was eliminated, at least in the case of trypsin, by adding soybean trypsin inhibitor and fresh growth factor after the enzymatic incubation had been completed. No inhibition of the added growth factor was detected.

The biological activity of the crude venom was abolished by antiserum to snake venom (1). These results were confirmed, using the purified growth factor. 40 µg. of the protein were pre-incubated for 1 hour at 26°C with varying amounts of commercial antiserum in 10 ml. of isotonic saline, and aliquots were then assayed. The biological activity was completely inhibited by 5 mg. of the antiserum preparation. In control experiments the further addition to the mixture of an equivalent amount of the growth factor restored approximately 50 per cent of the original activity, indicating that the antiserum was not inhibiting the response of the ganglia in some nonspecific fashion. In addition, normal horse serum had no effect.

All of these data indicated that the biological activity resides in an intact protein molecule. Whether the growth factor consists exclusively of amino acids residues remains to be demonstrated.

### Enzymatic Activities of the Purified Growth Factor from Venom

The possibility that the growth-promoting properties of venom were due to the activity of one of the enzymes known to be present in snake venom was then investigated. The enzymatic composition of animal toxins has been reviewed by Zeller (10).

The results of the enzymatic assays on crude venom (*Agkistrodon piscivorus*) and the 40-fold purified growth factor are shown in Table 2. It

TABLE 2

Enzymatic Activities of Crude Snake Venom and Purified Nerve Growth Factor

The nerve growth-promoting activity of the purified fraction was 40 times greater than that of the crude venom.

| Enzyme | Crude venom | Nerve growth factor |
|---|---|---|
| | units/mg. | units/mg. |
| Phospholipase A | 6.2 | <0.4 |
| Phosphodiesterase | 11. | <0.2 |
| 5-Nucleotidase | 19. | <0.1 |
| L-Amino acid oxidase | 14. | <0.1 |
| Deoxyribonuclease | 4.2 | <0.2 |
| Diphosphopyridine nucleotidase | 13.8 | <0.2 |
| Adenosine triphosphatase | 19.5 | <0.1 |
| Hyaluronidase | 15.6 | <0.5 |
| Ribonuclease | 6.0 | 1.5 |
| Protease | 16.7 | 12.6 |

was expected that if the growth activity was due to a particular enzyme, its specific activity (units per mg.) should also be approximately 40 times higher in the purified fraction than in the crude venom. The results show that the specific activities of all of the enzymes examined were lower in the purified fraction. Indeed, only two of the enzymes, protease and ribonuclease, showed any appreciable activity. The possibility that these enzymatic activities might still be associated with the growth factor was further reduced by the fact that whereas the growth factor is stable in 0.1 $M$ NaOH (26°C, 1 hour), the protease and ribonuclease activities are completely destroyed by this treatment.

Crystalline crotoxin (phospholipase A) showed no growth-promoting activity. The fact that both the white and yellow venom of *Vipera aspis* showed similar biological activity confirms the elimination of L-amino acid oxidase, since the former venom is free of this enzyme. Cholinesterase, while present in the Elapidae, is not found in the Viperidae or Crotalidae. Similarly, catalase, while found in some venoms, is not present in the venom of *Vipera russellii* (7). The addition of beef liver catalase to the cultures had no effect. The protease activity of *A. piscivorus* venom has been reported to be 30 times greater than that of *C. adamanteus* (4), while both had identical growth-promoting activity.

These data, of course, do not exclude the possibility that the biological effect is due to a still unidentified enzyme present in the venom.

### Site of Action of the Growth Factor

The outgrowth of nerve fibers under the influence of the venom protein might be due (a) to a direct effect on the ganglion itself, or (b) to an indirect effect as a result of some interaction of the growth factor with the medium. Conceivably, the latter might involve the removal of an inhibitor or the liberation of some stimulating substance. The following experiments were performed to decide between these alternatives.

3 ml. of plasma and 3 ml. of synthetic medium (containing thrombin) were incubated separately with 2 $\mu$g. of the growth factor in 1.5 ml. of isotonic saline for 8 hours at 37°C. (This period of time was sufficient to allow for the initiation of nerve growth in control cultures.) Equal volumes of the two solutions were mixed to assay for biological activity. In other aliquots, the antiserum to the venom was added before mixing for the assay. The presence of the antiserum completely abolished the nerve-stimulating effect of the mixture. These results indicate that the venom does not remove an inhibitor from the medium and that if the venom produces some other stimulating substance it must have a transient existence.

The possibility that the venom was inducing the cells of the ganglion to produce (and secrete into the medium) a growth stimulant was examined by incubating 40 ganglia for 8 hours at 37°C in 0.1 ml. of the synthetic medium containing the venom protein. The medium was then assayed for biological activity in the presence and absence of the antiserum. Since the addition of the antiserum completely abolished the activity, this hypothesis was ruled out.

We have reported (2) that the growth factor must be present continuously in the cultures to produce its effect. These experiments (performed with the nerve growth factor partially purified from sarcoma 180) showed that exposure of the ganglia for periods up to 20 hours to the growth factor did not induce nerve growth when the ganglia were subsequently cultured in the standard medium; the presence of additional amounts of the factor was required.

### Nutritional Requirements for Nerve Fiber Outgrowth

We have reported (2) that the outgrowth of nerve fibers in tissue culture resulting from the presence of the tumor factor, although completely inhibited by iodoacetate ($10^{-4}$ $M$), was not prevented by fluoride ($10^{-2}$ $M$), cyanide ($10^{-3}$ $M$) or dinitrophenol ($10^{-4}$ $M$). These results have been confirmed using the growth factor obtained from venom.

The possibility was considered that the outgrowth of fibers was due simply to a redistribution of existing cytoplasm with no associated metabolic effects. Therefore, we tried to see (a) whether there were any nutritional requirements for nerve outgrowth, and (b) whether the growth factor had any effect on the oxidation of glucose and/or synthesis of protein and nucleic acids in the cells of the ganglion.

To examine the carbohydrate requirements for nerve growth, cultures were prepared with plasma which had been dialyzed for 24 hours against isotonic saline containing 200 units of penicillin and 20 µg. of streptomycin per ml. Eagle's synthetic medium (containing thrombin) was prepared with the substitution of sodium chloride or a variety of possible "energy sources" for glucose. The carbohydrates and other added metabolites were present in the cultures at a final concentration of $10^{-2}$ $M$. The cultures were prepared as described, with 0.025 µl. of plasma, 0.025 µl. of synthetic medium, and 0.025 µl. of the growth factor (0.01 µg.) in isotonic saline.

The presence of glucose or mannose is required for the outgrowth of nerve fibers. In the absence of any added "energy source" the outgrowth of fibers was initiated but then ceased abruptly. The glucose could not be replaced by D-fructose, L- or D-arabinose, D-ribose, D-galactose, glucuronic acid, gluconic acid, malic acid, α-ketoglutaric acid, succinic acid, or fumaric acid. Lactate and pyruvate could partially replace the glucose requirement. It is not known to what extent the failure of some of the substrates to support growth was due simply to permeability.

To examine the amino acid requirements for nerve growth, a basal culture medium was prepared containing the dialyzed plasma and Earle's salt solution (with glucose and thrombin). However, even under these conditions, in the absence of any added amino acids, the growth factor elicited the usual halo of outgrowth of nerve fibers after 18 hours of culture. We were thus presented with two alternatives: (a) amino acids were not required for the outgrowth of the nerve fibers; or (b) amino acids were required, but could be obtained from the protein of the medium, from degenerating cells of the explant, or from internal reserves.

The following experiment, involving the use of the amino acid analogue p-fluoro-phenylalanine, indicated that at least one amino acid, phenylalanine, was required for the nerve fiber outgrowth. Under the above basal conditions the addition of 0.2 m$M$ DL-p-fluorophenylalanine almost completely inhibited nerve growth. This inhibition was specifically reversed by the addition of L-phenylalanine (0.5 m$M$). The addition of a mixture of amino acids (arginine, cysteine, histidine, isoleucine, leucine, lysine, methionine, threonine, tryptophan, valine, and glutamine) each at a concentration of 0.5 m$M$, did not reverse the inhibition.

Although other amino acid analogues have not been tested, it seems probable that there is a more general requirement for amino acids during nerve fiber growth.

### Effect of the Venom Factor on the Oxidation of Glucose and the Synthesis of Protein and Nucleic Acids

Isotopically labelled substrates were used to obtain, more directly, information concerning the synthetic and metabolic events occurring in the nerve cells under the influence of the growth factor. For these experiments the cultures were prepared in large depression slides in pairs, each culture containing five identical contralateral sensory ganglia isolated from the lumbosacral plexus of a 9-day chick embryo. The total protein content of the five ganglia averaged 30 μg., and the protein content of the two sets of ganglia agreed to within 10 per cent.

The effect of the growth factor on the incorporation of $C^{14}$-lysine into the protein of the ganglion was then examined. Each set of five ganglia was treated in an identical manner except for the omission of the growth factor in the control cultures. (The medium contained dialyzed plasma,

Eagle's synthetic medium prepared with the omission of lysine, thrombin, and $C^{14}$-DL-lysine in a final volume of 0.15 ml.) The sealed cultures were allowed to incubate for 20 hours at 37°C, and the total protein fraction (the fraction precipitable by trichloroacetic acid) was separated and counted.

The effect of the growth factor on the incorporation of 8-$C^{14}$-adenine into the RNA and DNA of the ganglia was examined in a similar manner. The RNA and DNA fractions were separated by a modified Schmidt-Thannhauser-Schneider procedure (8).

The oxidation of 1-$C^{14}$ and 6-$C^{14}$-glucose to $C^{14}O_2$ was also examined. The results of these experiments are shown in Table 3. Under the influ-

TABLE 3

Effect of the Growth Factor on the Incorporation of Lysine into Protein, Adenine into Nucleic Acid, and Oxidation of Glucose to Carbon Dioxide

The quantities and counts/minute of the added labelled substrates per culture were: DL-lysine, 5.8 μg., 10,670 c./m.; adenine, 4.05 μg., 25,420 c./m.; glucose (C-1 and C-6-$C^{14}$) 150 μg., 160,000 c./m. These counts were made in a manner identical to that used for counting the respective samples to eliminate the variations due to self-absorption.

| Labelled substrate | Culture | Fraction examined[1] | Exp. 1 | Exp. 2 | Exp. 3 |
|---|---|---|---|---|---|
| | | | c./m. | c./m. | c./m. |
| 1-$C^{14}$-DL-lysine | Control | Protein | 801 | 823 | 644 |
| | Venom | Protein | 1264 | 1351 | 1110 |
| 1-$C^{14}$-DL-lysine | Control + NaF | Protein | 475 | 458 | |
| | Venom + NaF | Protein | 922 | 841 | |
| 8-$C^{14}$-adenine | Control | RNA | 238 | 216 | 204 |
| | Venom | RNA | 334 | 306 | 345 |
| 8-$C^{14}$-adenine | Control | DNA | 1 | 2 | 4 |
| | Venom | DNA | 4 | 3 | 3 |
| 1-$C^{14}$-glucose | Control | $CO_2$ | 3752 | 3577 | 3990 |
| | Venom | $CO_2$ | 5777 | 5040 | 6015 |
| 6-$C^{14}$-glucose | Control | $CO_2$ | 2262 | 2623 | 2890 |
| | Venom | $CO_2$ | 2840 | 2943 | 3352 |

[1] In order to check the washing procedures, control experiments were run in an identical manner except that the ganglia were killed by boiling before culturing. In duplicate runs the protein, RNA, and DNA fractions each contained less than 1 c./m.

ence of the growth factor the incorporation of lysine into protein was increased 58 to 72 per cent. In view of the fact that sodium fluoride does not inhibit the outgrowth of the nerve fibers, but does inhibit the secondary migration of cells from the explant, the effect of 0.01 $M$ NaF on the incorporation of lysine into the protein fraction was also examined. It can be seen from Table 3 that although the control level of lysine incorporation was reduced, the growth factor increased the incorporation by 84-94 per cent.

The incorporation of adenine into RNA was increased by 40 to 69 per cent. Only traces of radioactivity could be detected in the DNA fractions of both the control and venom-treated ganglia. Whether or not there is any effect of the venom on the synthesis of DNA cannot be seen from these data.

Similarly, the factor increased the oxidation of 1-$C^{14}$-glucose by 41 to 54 per cent. Somewhat lower amounts of 6-$C^{14}$-glucose were oxidized by the ganglia, and the increase due to the presence of the venom ranged from 12 to 26 per cent.

The data indicate that the outgrowth of nerve fibers in tissue culture is associated with an increased oxidation of glucose, incorporation of lysine into protein, and adenine into RNA. Together with the observations on the nutritional requirements for nerve growth, these results indicate that the growth factor stimulates the synthesis of cytoplasm, and not merely its redistribution.

No data are available to distinguish between net synthesis and turnover. The interpretation is further complicated by the facts that the ganglia are not pure cultures of nerve cells, and that a variable amount of necrosis occurred in the central parts of the ganglia during the incubation period.

### The Nerve Growth Factor of the Submaxillary Gland of the Mouse

The nerve growth factor present in extracts of the submaxillary gland of the mouse appears to be similar to that of the venom and tumors in that all are heat-labile, nondialyzable, stable in 0.1 $N$ NaOH and unstable in 0.1 $N$ HCl. However, neither the tumor factor nor the salivary gland factor is inhibited by the antiserum to the snake venom, thus indicating that the active molecules are not antigenically identical.

The salivary gland factor has been purified (on a protein basis) approximately 10 to 15-fold by means of alcohol fractionation, ammonium sulfate fractionation, and adsorption and elution from cellulose ion-exchange resins previously mentioned. This material has a specific activity between 1 and 2 times that of our most purified venom protein. It has a typical protein absorption spectrum with 280/260 absorption ratio of 1.7. The salivary gland factor has been found to stimulate nerve growth not only of the ganglia of the chick embryo, but also the sensory ganglia isolated from both mouse and rat embryos.

SUMMARY

A protein, with nerve growth-stimulating properties, has been purified from snake venom. The protein appears to act directly on the nerve cell. The presence of an "energy source" (glucose or mannose) is required for the continuation of nerve growth. Indirect evidence indicated the necessity of at least one amino acid, phenylalanine. Experiments with isotopically labelled substrates show that, concomitant with the outgrowth of nerve fibers, there is an increased incorporation of lysine into protein, of adenine into RNA, and an increased oxidation of glucose to carbon dioxide. The initial step involved in the mechanism by which this factor stimulates the growth of nerve fibers, both in tissue culture and in the chick embryo, is not known.

Factors with similar growth-promoting properties have been found in the salivary glands of the mouse and rat, in certain mouse sarcomas, and in Gila monster venom. Although it is tempting to think of them as hormones, the physiological role (if any) played by these growth factors in the normal economy of the organism awaits further study.

## REFERENCES

1. Cohen, S., and Levi-Montalcini, R., *Proc. Natl. Acad. Sci. U. S.*, **42**, 571 (1956).
2. ———, and ———, *Cancer Research*, **17**, 15 (1957).
3. ———, ———, and Hamburger, V., *Proc. Natl. Acad. Sci. U. S.*, **40**, 1014 (1954).
4. Deutsch, H. F., and Diniz, C. R., *J. Biol. Chem.*, **216**, 17 (1955).
5. Levi-Montalcini, R., and Cohen, S., *Proc. Natl. Acad. Sci. U. S.*, **42**, 695 (1956).
6. Lowry, O., Rosebrogh, N. J., Farr, A. L., and Randall, R. J., *J. Biol. Chem.*, **193**, 265 (1951).
7. Slotta, K., in *Progress in the Chemistry of Organic Natural Products*, **XII**, 406, Springer, Wien (1955).
8. Volkin, E., and Cohn, W., in *Methods of Biochemical Analysis* (D. Glick, ed.) p. 287, Interscience Publishers, New York (1954).
9. Winzler, R. J., in *Methods of Biochemical Analysis* (D. Glick, ed.) **II**, 279, Interscience Publishers, New York (1955).
10. Zeller, E. A., *Advances in Enzymol.* **8**, 459 (1948).

## DISCUSSION

DR. MARKERT: I should like to point out an apparent similarity between the growth responses just discussed and the developmental responses discussed earlier by Dr. Yamada. Both of these responses are elicited by apparently protein materials from extremely bizarre origins (guinea pig kidney and snake venom). It seems unlikely that these proteins could play any normal, specific role in embryonic development. More probably, these foreign proteins intrude upon the metabolism of the cell and upset the normal metabolism by provoking or inhibiting reactions. The changed metabolic pattern of the cell would then lead through abnormal pathways to the type of development observed. I do not wish to detract from the developmental significance of these dramatic experiments but only to caution against interpreting the results in terms of normal development. In analyzing the effects of these exotic proteins it would, of course, be most useful to identify them in their effective positions in the responding cells. In this connection I wonder if it would be possible to measure the persistence of these proteins in the cell?

DR. S. COHEN: As far as the bizarre origins of the growth factor, it seems to us, that except for the small amounts in mouse sarcomas, it is striking that all of the other sources have been salivary glands or their secretions. With respect to the specificity, only certain nerve cells appear to be affected and only at a specific stage in their differentiation. We cannot say, as yet, whether or not these factors play a role in normal development. Our experiments indicate that the factor is not stored in the nerve cells in sufficient amounts to enable them to grow autonomously. I might also say that we have tried to get at this problem by coupling the protein with fluorescein isocyanate, but this completely inactivates the material.

DR. MARKERT: How about isotopes?

DR. S. COHEN: That's a possibility to be considered, but we have not done it yet.

DR. MIRSKY: How about labeling the antibody?

DR. S. COHEN: Yes, that's certainly the next thing to be done. I believe that by using fluorescent dye-labeled antibody we could probably detect the presence of this substance in the cell. That remains to be done however.

DR. PAPPENHEIMER: Does the sodium hydroxide treatment change any of the properties of the nerve growth-stimulating factor?

DR. S. COHEN: I don't know, but it is still heat labile and non-dialyzable after sodium hydroxide treatment.

DR. WEISS: What effects do these types of substances have on other kinds of cells?

DR. S. COHEN: Well, we have tried to find some kind of effect on other cells. We didn't see any stimulation of the growth of explants of other tissues in vitro. However, we may be hiding effects by using this kind of system. We did not see any difference in the migration of these cells when the growth factor is added. I think we'll have to pursue this further, though, before we can rule out this possibility.

DR. LEVI-MONTALCINI: I really have nothing to add to that. We were not able to detect in the embryo any other growth effect besides the one described in the sensory and sympathetic ganglia. I should, however, like to mention that in some embryos which received a high dose of the salivary gland extract, we observed a loosening of the ectoderm from the sub-ectodermal tissues.

Dr. Weiss: I was only going to suggest that this be done with the cell suspension technique. That is where surface changes could probably most easily be detected. I wonder to what extent this growth phenomenon is based on a surface change in the neuroblast.

Dr. S. Cohen: Well, again, we don't know what the initial effects of this material are. We do observe effects of the snake venom on single cells. However, most of the cells die under the conditions which we have used, and I don't know why. Those that survive produce longer and more branched fibers than control cells.

Dr. Markert: Have you tried to get induction in amphibian gastrulae with this material?

Dr. S. Cohen: No.

Dr. Sussman: Did you attempt to get activity out of chick homogenates?

Dr. S. Cohen: Oh, I am sorry I forgot to mention many of the controls which we used. There was no activity in chick embryo homogenates, mouse liver, brain, pancreas, mammary gland, pig liver, pigeon heart, etc.

Dr. M. Cohn: I wonder if you would care to comment, Dr. Markert, on precisely what you mean by the lack of specificity in this reaction?

Dr. Markert: Well, of course, I can't discuss specificity in chemical terms since we know nothing as yet of the chemical effects of this substance. However, the origin of this growth-stimulating factor from snake venom makes it seem unreasonable, from an embryological point of view, that this substance is specifically responsible for growth responses in nerves during normal development. Of course, it may be that the effective substance is a molecule of very wide biological distribution which plays a variety of roles in many different kinds of cells.

Dr. S. Cohen: You have to remember that extracts of mouse salivary gland are effective on mouse ganglia. So it is not quite as far-fetched as one might at first suppose.

Dr. Brachet: Does the agent have an effect on mitosis, and what is the situation with regard to the formation of polyploid cells?

Dr. Levi-Montalcini: Yes, the agent affects the mitotic activity in the sensory ganglia. We did not investigate its effect on the mitotic activity of the sympathetic ganglia. In the sensory ganglia of embryos bearing a transplant of mouse tumor we observed increase in the mitotic activity between six and seven days. At seven days the mitotic activity ceases both in control and in experimental ganglia. In subsequent stages the growth agents appear therefore to affect the cell size and the nerve outgrowth but not the number of nerve cells. The occurrence of polyploidy in nerve cells is very difficult to ascertain and we have not investigated this point.

## DISCUSSION

Dr. Brachet: Yes, but your nuclei are much larger in these tissues.

Dr. Levi-Montalcini: I should like to mention some investigations which are now in progress, in association with Dr. A. DeLorenzo, and have a bearing on this problem. We studied the sensory and sympathetic ganglia of experimental and control embryos with the electron microscope. We observed a remarkable increase in the size of the nucleoli of both sensory and sympathetic neurons. We also observed in both types of cells changes in the structure of the ergastoplasm and of the mitochondria.

Dr. Brachet: Do you know what the DNA content per nucleus is?

Dr. Levi-Montalcini: No, I can't tell you that.

Dr. Maas: Does antibody against the snake venom react with and neutralize the growth-promoting activity of the other factors you have isolated?

Dr. S. Cohen: Antibody to the snake venom does not inhibit either the tumor factor or the factor extracted from mouse salivary gland.

Dr. Hartman: If you think you have an enzyme here but do not know the substrate, you might try a uniformly or randomly $C^{14}$-labeled cell residue which has been dialyzed, and upon incubation of this with your factor see if you can detect any materials being released into the acid-soluble or dialyzable fractions. You might have a hydrolytic enzyme, considering the sources of your active preparations.

Dr. S. Cohen: Well, I really don't believe that we have an enzyme here. I don't know of any enzyme which will resist $0.1\ N$ alkali. If anyone else does, I would like to hear about it.

Dr. McElroy: DPNase will resist this sort of treatment.

Dr. S. Cohen: As far as the snake venom factor is concerned it is not DPNase.

Dr. Coulombre: Because it might have some bearing on the type of mechanism with which you are dealing, I would like to ask whether the effect of the agent is a trigger effect or whether you need the continued presence of the agent.

Dr. S. Cohen: We have tried such experiments in which we expose the ganglia to the venom for varying lengths of time up to 20 hours, and then we remove the ganglia to a standard medium without growth factor. No especially luxuriant outgrowth was observed in any of these experiments. If you add the growth factor, however, the ganglion immediately responds. We interpret this to mean the continued presence of the factor is necessary. We could check this, but we have not done it yet, by putting in the venom, letting the fibers grow, and then adding antiserum to the factor to the culture to see if their outgrowth then ceases.

# NEOPLASTIC DEVELOPMENT

LESLIE FOULDS

*Chester Beatty Research Institute, Institute of Cancer Research,
Royal Cancer Hospital, London*

## INTRODUCTION

The stubborn persistence amongst pathologists of the discredited idea that cancer entails a "reversion to embryonic type" is not accidental or merely perverse; it stems mainly from a misinterpretation of phenomena that show the operation of similar biological laws at all stages of normal and neoplastic development. Nicholson (42) interprets neoplasia as *continued development*. Whilst accepting Nicholson's concept of "development" and his unequivocal rejection of "reversion," I dissent from the implication that neoplasia is a continuation of a normal process. The view to be presented here differs fundamentally from Nicholson's in attributing neoplasia to local diversions of development into new, abnormal pathways. Normal development begins with fertilization of the ovum, advances through embryonic and post-embryonic life to maturity and onwards to senescence and death unless it is distorted by injury, in the widest sense, or unless some cells are diverted into the pathological "neoplastic" path. The diversion depends upon an irreversible, heritable, qualitative change in cells; it can occur during embryonic life or at any time thereafter and in any cells capable of proliferation, but it is impossible to induce neoplasia in cells that cannot multiply under normal or pathological conditions. The point of diversion, marking the onset or initiation of neoplasia, is not histologically recognizable; it precedes manifest neoplastic growth by long periods extending sometimes up to 40 or 50 years in man. The clinical stage of lumps and "growths" covers only a fraction of neoplastic development, which extends far backwards to inconspicuous beginnings and which can be extended forwards, without set limit, by serial transplantation in animals or by culture in vitro.

Even under the most favorable conditions of laboratory experiment, a comprehensive study of neoplasia from beginning to end has proved extremely difficult. The main contributions to a better understanding of neoplasia as a process of development have come from two dissimilar but complementary branches of research, namely, the study of chemically induced neoplasia of the skin in rabbits and mice and of the abundantly occurring spontaneous mammary neoplasia in inbred strains of mice. This review summarizes these investigations and the ideas they have generated, applies the latter to a few selected examples of neoplasia in man, and discusses the mechanisms and general principles of neoplastic development.

### EXPERIMENTAL INVESTIGATIONS

#### 1. Chemical Carcinogenesis in the Skin of Rabbits and Mice

Studies of the induction of neoplasia in the skin of rabbits and mice have led to a two-stage theory, according to which carcinogenesis comprises two distinct processes, known as *initiation* and *promotion*. Initiation, it is believed, is a rapid action and brings about a *subthreshold neoplastic state* characterized by the presence of *latent* or *dormant* tumor cells which proliferate only when stimulated during a subsequent and more prolonged phase of promotion (7, 8, 18). Some substances, of which croton oil is the most used, are strong promoting agents but weak or ineffective as initiating agents. Conversely, urethane and some other substances are effective initiating agents but have little or no promoting action. These substances are called *incomplete carcinogens* (50, 51). Tar and the carcinogenic hydrocarbons are complete carcinogens that exert both initiating and promoting actions in varied proportions. It is remarkable that complete and incomplete carcinogens effect initiation when administered internally, so that tumors develop where croton oil is applied to the skin (9, 31, 43, 44).

The separation of initiating and promoting phases opens up new approaches to the histological analysis of carcinogenesis. The most notable finding so far is that urethane, even after prolonged application, produces no histological changes in the skin (50, 51). The decisive change in reactivity effected by the initiating agent is manifest histologically only after application of a promoting agent. The whole area of skin subjected to the initiating action of a carcinogen reacts to croton oil with hyperplasia differing in some details from that evoked by non-specific irritants (49). The essential actions of promoting agents, in Berenblum's opinion, are to induce cell proliferation and to delay cell maturation. The effects of the promoting agents are, for a considerable time, reversible, but if the action is maintained an irreversible change takes place at one or more foci within the field of hyperplasia, and at those sites neoplastic growth becomes evident (7, 8, 36, 48).

The first, most specific, and most decisive step in neoplasia involves no proliferation of cells and is not recognizable histologically, but it effects a change in reactivity that apparently is permanent and irreversible; it brings about a change in competence and resembles *cryptodifferentiation, premorphological differentiation,* or *determination* in embryonic development.

There is some evidence to show that although initiation ordinarily may be a rapid process it is not so of necessity; under some circumstances cells seem to acquire "latent neoplastic potentialities" and advance automatically but slowly and gradually to the neoplastic state, so that tumors emerge after long periods, extending to two or three years in rabbits, after withdrawal of carcinogenic stimuli (23). The promoting phase always extends over a substantial period of time accounting, probably, for most of the "latent period" in carcinogenesis. During this period another step in neoplastic development is accomplished at one or more foci in the initiated tissue, but not diffusely; at some stage a second irreversible heritable change occurs.

The neoplastic lesions that develop from initiated skin after an adequate period of promotion are varied in structure, behavior, and prospective fates. They are classified as benign tumors, most of them being warts or papillomas of several kinds. Some of the benign growths regress and disappear, some persist for a long time without much growth, some grow but retain their benign qualities, some change into other kinds of benign growth, and some give origin to carcinoma. One particular type of papilloma in mice is subject to at least eight different fates (54). In rabbits, some warts regress during the period of application of the carcinogen, and others regress when carcinogen is withdrawn. These *conditional* or *dependent* tumors, which grow only so long as the carcinogen or another promoting agent stimulates them, have all the other characteristics of neoplasms and some of them, called *carcinomatoids,* are invasive and histologically like carcinomas (45, 46, 60). Carcinoma, not dependent on extrinsic stimulation, develops in several different ways. Most often carcinoma develops from a benign lesion, usually a wart, but sometimes it emerges from apparently normal skin with all the qualities of carcinoma present at the outset. There are several *indirect* paths to carcinoma passing through benign precursors, and there is a *direct* path which traverses no intermediate stages. The intermediate lesions on the indirect paths are *potential* precursors of carcinoma; they do not inevitably develop into carcinoma and as a rule the majority do not. When the carcinogenic stimulus is optimal, carcinoma develops in almost every mouse, but not from every benign lesion. The change from wart to carcinoma is not a mere extension in space or exaggeration of properties formerly present but an irreversible heritable and qualitative change leading to a new kind of tumor (4, 22, 25, 46).

It is evident that, contrary to wide belief, the course of neoplasia is not one of relentlessly progressive growth with qualities fixed irrevocably from the outset. The first step of initiation involves no growth at all. The second step leads often to conditional tumors that do not grow without extrinsic stimulation. Neoplasia may halt permanently at the first step; when it advances it does so along either a *direct* path or one of a choice of *indirect* paths. At intermediate stages along the indirect paths there are choices between several courses leading to regression, to inert persistence, to growth without qualitative change, to qualitative change without loss of benign habits, or to malignant carcinoma. Malignant neoplasia thus is often the culmination of a discontinuous development through a sequence of heritable qualitative changes in neoplastic cells, but it may spring directly from initiated but seemingly normal cells.

## 2. Mammary Neoplasia in Mice

Mammary tumors in mice develop in several different ways. Some spring from apparently normal breast tissue and some originate within ducts, but most of them develop from precursor lesions, namely *hyperplastic nodules* or *plaques*. The nodules, which are of minute size and not detectable in the living animal, are the chief sources of mammary tumors in most inbred strains of mice with a high natural incidence of mammary cancer, and usually they are very numerous. The nodules are of several types, acinar hyperplasia being the most common. Many of the nodules regress, some persist without notable growth or qualitative change, and a minority give origin to carcinomas. In a few strains of mice the majority of mammary tumors originate from plaques, which are about ten times as big as nodules and are palpable during life. The plaques are often multiple but not numerous. They are *conditional, dependent,* or *responsive* tumors that grow only in response to some unidentified stimulus, presumably hormonal, operative during pregnancy; they grow during pregnancy and regress after parturition. Some of the plaques continue in this way without net increase in size over a long period of time, some increase in size but remain responsive to pregnancy, and some give origin to carcinomas that grow progressively without the stimulus of pregnancy. Carcinoma develops most often by focal change within a plaque (17, 19).

Plaques and nodules are comparable stages along two indirect paths of mammary neoplasia. The whole breast tissue is liable to neoplasia but, as in skin carcinogenesis, the first lesions that are recognizably neoplastic are focal. Of these focal changes some regress, some persist inertly, and some give origin to carcinoma. Only one or a few of the numerous nodules and as a rule only a portion of a plaque develops into carcinoma. If the earliest tumors are excised, new ones almost always develop, an outcome showing that more tissue is capable of developing into carcinoma than ordinarily does so. Gross tumors are liable to further qualitative change evidenced by an enhanced growth rate or by structural alterations; and histological examination shows that this change, too, is usually focal (Foulds, unpub.). At each step in the neoplastic sequence there is a restriction of the field of further advance. Only a portion and usually only a small fraction of the available "competent" tissue actually advances to the next stage in the sequence.

## 3. Progression

Development along indirect pathways in the skin and in the breast advances discontinuously by irreversible heritable changes in one or more of the characters of neoplastic cells. The term *progression* is used for these changes to avoid premature identification with one or other of the known mechanisms of heritable cell change. Mammary neoplasia, as manifested in plaques, is especially useful for the analysis of progression because two dissimilar characters, namely, growth rate and responsiveness to hormones, are available for observation and measurement in multiple sites in the intact living animal. Progression in both characters is demonstrable. Some general principles of progression derived empirically from these observations are widely applicable to neoplasia in animals and in man (17, 18, 21). The principles are here summarized.

I. *Independent progression of tumors.* It is the general rule that only one of multiple tumors, or a small percentage if they are numerous, undergoes progression at one time. Progression depends upon a qualitative change within the tumor cells and is not a reaction to an altered environment, although the possibility remains that progression is "triggered" by extrinsic stimuli. This principle can be extended to include progression at one or a few foci within a wide expanse of diffusely "prepared" or "competent" tissue.

II. *Independent progression of characters.* Different identifiable characters of a tumor undergo progression independently of one another. Responsiveness and growth rate in mammary tumors in mice are liable to independent progression (17), and so are histological type, growth rate, and invasiveness in chemically-induced skin tumors (54). This leads to a more general proposition that the structure and behavior of tumors are determined by numerous *unit characters* that within wide limits are independently variable, capable of highly varied combination, and liable to independent progression. These correspond, perhaps, with the theoretical "reactons" which Selye (53) defines as the smallest targets capable of selective biological reactivity or with relatively independent biochemical systems within cells. "Malignancy" is an abstract concept, applying to a lethal mode of clinical behavior, which depends on many characters, including rate of growth, type and degree of cellular differentiation, invasiveness, powers of metastasis, and responsiveness to hormones. These characters, within wide limits, are independently variable, combined together in a great variety of ways in different tumors, and liable to independent progression. Discussion of "malignancy" as though it were an indivisible entity or quality of tumor cells is meaningless.

III. *Progression is independent of growth.* There is substantial evidence that progression occurs in stationary, dormant, or regressed tumors, but the reservation must be made that in some forms of neoplasia continued proliferation seems to favor progression and may be essential for it (18, 24, 34). In accordance with the first principle, progression usually occurs focally in the diffusely proliferated tissue (3, 18).

IV. *Progression is continuous or discontinuous.* It may advance slowly and gradually, or, more usually, by abrupt steps.

V. *Progression follows one of alternative paths of development.* A choice between qualitatively different paths of development is an important feature of early stages of neoplasia. In many forms of neoplasia there is a choice between a *direct* and an *indirect* path and, often, between several indirect paths. Moreover, on the indirect paths there are repeated choices between regression, persistence without qualitative change, or progression, which almost always leads to more aggressive behavior. One type of lesion can develop along several different paths to diverse end-points; several different paths can lead from diverse starting points to tumors of a single kind. Progression may result in step-like advance along one path or it may switch neoplasia into a new pathway.

VI. *Progression does not always reach an end-point within the lifetime of the host.* Progression occurs in advanced stages of neoplasia and continues, often conspicuously, during serial transplantation in animals. It is dubiously correct to presume that there is an end-point. A capacity for progression seems to be a universal characteristic of neoplastic cells at every period of their history (18, 33, 34).

## Neoplasia in Man

The several types of neoplasia discussed in this section are chosen to illustrate the application of general principles of tumor development to special cases about which a useful amount of information is available. This section is based on a more detailed analysis, with references, published elsewhere (21).

### 1. Epidermal Neoplasia

Neoplasia of the skin is most accessible to direct observation, and the occupational forms are closely comparable with chemically-induced neoplasia of the skin in animals. Dermatologists describe various affections of the skin as "precancerous lesions" or "precanceroses" with varied degrees of justification.

Despite differences in detail, the pattern of neoplastic development is similar in most of the "environmental" neoplasias, including the occupational neoplasias induced by oils, tars, and pitches, the solar and senile neoplasias, and the neoplasias attributable to the medicinal use of arsenic. One common feature is the long induction period averaging, for example, about 20 years in workers exposed to coal-tar products and 50 years in those exposed to mineral oils. Moreover, neoplasia may appear for the first time up to 40 years after the worker has left the industry and had no further exposure. Various inflammatory and hyperplastic lesions develop on exposed skin. Many of the lesions ultimately regress, others persist for a long time, and some warts, papules, and ulcers eventually undergo progression to carcinoma. Most of the cancers develop in this way from "precanceroses," but some emerge, with their definitive properties established, from apparently normal skin and may do so long after all precanceroses have disappeared and the skin has regained a normal appearance. New cancers may develop successively from this normal-looking skin for the rest of the patient's life. The initiation of neoplasia is evidently far more extensive than the early "precancerous" lesions indicate. Only a fraction of the precancerous lesions undergo progression to carcinoma, and they are not obligatory precursors of cancer.

The rare disease xeroderma pigmentosum almost invariably terminates in cancer. The victims have an inherited hypersensitivity to sunlight, and the disease usually begins during childhood after exposure to the sun and consists of various pigmentary, inflammatory, and hyperplastic changes in the skin. Carcinoma develops most frequently by progression in keratoses but also may emerge from apparently normal skin; more rarely malignant melanoma develops from pigmented lesions. Carcinoma develops, sometimes before puberty and often during adolescence or early adult life. The disease is incurable; if cancers are removed, new ones develop successively elsewhere, but progression to carcinoma takes place in only one place, or a few places at a time, and not in all the potential precursor lesions.

Bowen's disease and some related conditions, formerly classed as precanceroses, are now described more usually as carcinoma in situ. This popular term, synonymous with preinvasive carcinoma, noninvasive carcinoma, and intraepithelial carcinoma, needs more precise definition. It is best restricted to a persistent overgrowth of epithelial cells with the cytological characteristics of carcinoma cells but limited to the epithelial surface without invasion of neighboring tissues. Carcinoma in situ lacks invasive power, which it acquires only by qualitative progression. Bowen's disease fulfills these criteria. Progression to carcinoma takes place with a frequency variously estimated but probably in 20 per cent or more of the patients. Bowen's disease has more of the stigmata of carcinoma than the precanceroses previously mentioned, but like them it does not become invasive cancer save by progression.

Malignant melanoma develops most frequently by progression in a junctional or compound naevus and much less often from apparently normal skin or from pigmented areas in senile or solar dermatitis or xeroderma pigmentosum. Several other fates are open to junctional naevi, and although most malignant melanomas originate from them the vast majority of these common lesions do *not* undergo progression to melanoma.

## 2. Neoplasia in the Urinary Bladder

Occupational neoplasia of the bladder in dye workers develops in conformity with the usual pattern of industrial cancer after long periods of exposure averaging about 19 years, and at varied times up to 12 years after the final exposure. Some tumors are malignant when first detected; others make their appearance as benign tumors and either persist for a long time in the benign state or undergo progression to carcinoma. In some patients, although not in all, there is a strong liability to develop new tumors after destruction of the original ones, and new growths develop consecutively in various parts of the bladder throughout the patient's life. The clinical observations show that the carcinogen induces a widespread and permanent alteration in vesical epithelium; the alteration persists, without visible signs, after withdrawal of the carcinogen as the basis upon which benign and malignant neoplasia develops focally at unpredictable times and places.

### 3. Familial Intestinal Polyposis

In this familial disease, multiple tumors develop in the intestine during childhood or early adult life. At first all the tumors seem benign in structure and behavior but eventually one or a few of them undergo progression to carcinoma. The disease resembles xeroderma pigmentosum in having an hereditary basis, in the early age-incidence, and in the progression to carcinoma in most patients but in only a small proportion of the multiple benign lesions.

### 4. Neoplasia of the Uterine Cervix

Neoplasia of the uterine cervix is of outstanding current interest on account of the efforts being made to detect and cure it before the stage of invasive carcinoma. Many details of the neoplastic sequence are still obscure. It seems that the usual course is from normal cervical epithelium through irregular hyperplasia, to which *dysplasia* and many other terms are applied, to carcinoma in situ and thence to invasive carcinoma. Most frequently dysplasia regresses; alternatively it persists for a long time or changes to carcinoma in situ or, rarely, directly to invasive carcinoma. Carcinoma in situ, similarly, may regress, persist unchanged for a long time, or undergo progression to carcinoma. Alternative paths, circumventing dysplasia or carcinoma in situ or both, are probably available. There are substantial grounds for the belief that most invasive carcinomas are preceded by dysplasia or carcinoma in situ, or both, but the evidence that these lesions are consecutive steps in one process leading to invasive carcinoma is mainly circumstantial. The various lesions are often present simultaneously, but it is at least suggestive that dysplasia is found on the average about 6 years earlier than carcinoma in situ, which, in turn precedes invasive carcinoma by about 10 years.

### 5. Neoplasia of the Breast

The course of neoplasia in the human breast is highly complex and controversial. Two main paths of neoplasia are distinguishable, firstly, a *centrifugal* path characterized by differentiation towards the myoepithelial and connective tissue cell types and, secondly, a *centripetal* path with differentiation of predominantly epithelial type. The centrifugal path leads to fibroadenoma and thence sometimes to sarcoma or mixed tumor; the more dangerous centripetal path leads to carcinoma.

Centripetal development pursues several different courses. There is evidence for a *direct* path from morphologically unaltered mammary epithelium to carcinoma and for several *indirect* paths traversing diverse precursor lesions. Controversy has centered especially on the ill-defined disease, comprising a variety of lesions, known as "chronic mastitis." The interpretation of available information is still in dispute. It seems that a disorder called "cystic hyperplasia" in young women can regress, persist, develop into chronic cystic disease, or undergo progression to benign neoplasia manifested usually as papilloma. Many papillomas regress, some persist without qualitative change, and some undergo progression to carcinoma. According to this interpretation, chronic cystic disease, as manifested clinically, is the product of a side-chain of development from cystic hyperplasia and has lost the capacity for progression to papilloma or carcinoma. Chronic cystic disease and papilloma are alternative fates of cystic hyperplasia and not consecutive steps in one process. Most probably development can circumvent the stages of cystic hyperplasia or papilloma and can advance, also, along other indirect paths through intermediate lesions of a different kind. The relative frequency of the different paths of neoplasia is unknown.

## 6. *Characteristics of Neoplastic Development in Man*

As shown by the examples cited and by others discussed recently by Hamperl (30) and by Wakeley and Graves (57), there are many different varieties of neoplastic development and a single type of tumor may result from development along several different paths, but a general pattern is beginning to emerge. In many forms of neoplasia, in animals and in man, there is a wide field of "prepared" or "predisposed" tissue irreversibly altered in reactivity and potentialities, presumably by initiating stimuli, but not necessarily neoplastic or at all abnormal by morphological standards. Tumors develop focally within this field. Some of them, including carcinomata, emerge with all their definitive properties established from the outset, by *direct* development from the prepared tissue. Others result from *indirect* development along paths traversing varied "precancerous" or "precursor" lesions before the definitive neoplasm is established. At each step along indirect paths there is a choice between regression, inert persistence, growth without qualitative change, and progression toward malignant neoplasia. Indirect development is notably unpredictable. At all intermediate stages the relentless advance to a predetermined outcome often attributed to neoplasia is conspicuously absent. The lesions on the indirect paths are potential precursors of cancer but not obligatory precursors. They do not advance to cancer automatically, but only as a result of irreversible progression in one or more of their characters. Even when the fate of the patient is predictable, the fate of individual lesions is not, and the majority of "precancerous" lesions do *not* undergo progression to cancer. The probability of progression differs widely in different kinds of neoplasia, but in most individual lesions, with the possible exception of carcinoma in situ, the probability is low. Progression may halt at any stage along indirect paths; and often there are alternative indirect paths to a common end-point, and almost always, it seems, the possibility of *direct* development, which, in some forms of neoplasia, is the most usual course.

## Mechanisms of Neoplastic Development

The concept of progression is an empirical one embodying no presumptions about its mechanism. The most precise evidence about mechanisms of progression comes from studies of transplantable tumors, especially those in the ascites form. In a valuable review, Klein and Klein (34) remark on the extreme plasticity and changeability of malignant tumor cells and their ability to become independent of almost any known kind of stimulation or inhibition. It is very improbable, in their opinion, that any single master cellular change is responsible for the various kinds of progression even in a single type of neoplasm. A comprehensive analysis of neoplastic development must take into account the earliest stages of initiation and promotion and the clinical course of tumors in their original hosts as well as the behavior of transplanted tumors. It must take into account also two remarkable phenomena, namely, the progression in normal and neoplastic cells cultivated in vitro, and the reproduction by subcellular particles of the filterable chicken tumors first described by Rous.

### 1. Progression in Transplanted Tumors

Klein and Klein (34) present evidence for two mechanisms basically different in principle, namely, a spontaneous random change occurring during continued cell proliferation and a change induced by environmental stimuli.

The progressions involved in the transformation of tumors from the solid to the ascites form, in drug resistance, and in the acquirement of the capacity to grow in hosts of foreign genotype seem to depend on the selection of specific variant cell-types differing from the original majority of cells in having a higher survival value in the unusual environmental conditions to which the population has been exposed. ". . . . The cellular change may be called a mutation, provided that this term is used in its broadest sense, simply meaning stable, heritable changes without regard to their intra- or extrachromosomal nature. . . ." (34).

Klein and Klein confirm the original observation of Barrett and Deringer (5, 6) that a heritable cell change, of the nature of an adaptation, is induced in a tumor of an inbred strain by short residence in hybrid mice, one of whose parents belongs to the tumor's strain of origin; and they show that the change is effected by diffusible humoral factors of the hybrid host. No other example of this mechanism is yet known but it possibly operates, alone or in combination with the "mutation" mechanism, in other forms of progression.

Other experiments by Klein and Klein (34) show that cells of seemingly independent tumors may still prove dependent when transplanted in small numbers. A tumor may thus become independent at the population level but not at the cell level, a fact suggesting that progression to states of independence may sometimes depend on the cell's acquiring, by way of adaptation or by selection of new variants, the capacity to produce and utilize endogenous substances capable of substituting for the exogenous stimuli previously needed. The resulting continued proliferation may, in turn, increase the chances of formation of new variants and so lead the tumor onwards to increased autonomy. The same experiments suggest that progression may sometimes depend on continued cell proliferation per se, although it is not implied that this is a general or usual mechanism of progression. Progression in hormone-induced neoplasia possibly depends on maintained cell proliferation (3, 18, 24), but there is strong evidence that progression is often independent of growth.

## 2. Mechanisms of Initiation and Promotion

Reasoning from observations on experimental skin carcinogenesis, Berenblum (8) advances a hypothesis which comprises two main propositions: (1) initiating action brings about a sudden irreversible change in the potentialities of a few isolated epidermal cells, hidden as a rule in hyperplastic epidermis, and converts them into dormant tumor cells; (2) promoting action encourages the proliferation and delays the maturation of the dormant cells and, by so doing, allows their undifferentiated daughter cells to accumulate to a critical colony size, whereafter they continue to proliferate without further promoting stimuli.

Berenblum rightly emphasizes the need to account for the focal emergence of tumors from a wide area of skin exposed to carcinogenic action, but his hypothesis is not a necessary or convincing alternative to the "field" theory, which he rejects. The hypothesis does not account for the localization of carcinogenic action to a few scattered cells without the presumption that these cells are predetermined to react differentially to the carcinogenic stimulus. The changed reactivity manifested by a specific histological response to promoting agents is diffuse (51). There is no conclusive evidence that initiation is completed suddenly, and some strong indications that under certain circumstances it is not (23, 50). The adequacy of scattered dormant tumor cells to account for the observed phenomena has been questioned on several grounds (36, 40, 49). The present review provides several indications in varied types of neoplasia that initiation is far more extensive than the earliest emergent neoplastic foci reveal. Far more cells are permanently altered and able to develop into visible tumors than ordinarily do so without experimental interference in animals or surgical intervention in man.

The term *promotion* seems to cover more than one kind of phenomenon (18). In rabbits neither papillomas nor invasive carcinomas become independent of promoting agents by virtue of their size. Promotion in mice resulting in tumors independent of promoting action effects a heritable qualitative change, which is not convincingly attributable to critical colony size.

A hypothesis more comprehensively applicable to many forms of neoplasia in animals and in man is here proposed as an alternative to Berenblum's, from which it differs basically in presuming a second heritable qualitative change or progression during the stage of promotion.

The suggested sequence is as follows:

(1) Initiation effects at varied rates an irreversible change in reactivity and potentialities of all or most of the cells exposed to the initiating stimulus.

(2) During promotion the altered cells proliferate and a heritable qualitative change or progression occurs focally in the hyperplastic tissue. The focal progression in a small percentage of the available or "competent" cells corresponds with the "mutational" form of progression described by Klein and Klein in ascites tumors, and it is consistent with the first general principle of tumor progression that all competent tissue does not undergo progression simultaneously.

(3) Newly established neoplastic foci possibly inhibit progression in neighboring cells, but these cells retain their competence and, as many examples quoted in this review indicate, they often undergo progression at a later date.

(4) The change effected during promotion may establish the definitive properties of the tumor (*direct* development), but often this stage is reached only after a sequence of progressions each occurring focally within the competent tissue.

This hypothesis interprets the whole of neoplastic development from the stage of promotion onwards to a series of progressions but leaves open the nature of the primary change at initiation. It is beyond the scope of this review to discuss the conclusions suggested by chemical and immunological investigations of carcinogenesis, for an account of which reference may be made to a recent article by Griffin (27). Briefly, the current trend is to interpret carcinogenesis in terms of material loss by deletion of gene proteins and interference with gene synthesis (28, 29); by deletion, by combination with carcinogens, of key proteins essential for the control of growth (41); or by loss of critical proteins as a result of immune reactions (26). It is convenient to describe the results of later progression in terms of a gain of characters, as for example "invasiveness"; these characters are phenotypic ones and justify no inferences about genetic gain or loss. In previous publications I have described progression in hormone reactivity in terms of loss of "responsiveness" as being more generally applicable than the more usual descriptions of acquirement or gain of "autonomy." Many other characters of tumors can be similarly interpreted in terms of either loss or gain.

## 3. Transmission of Chicken Tumors by Filterable Agents

The transmission of sarcomas and leukemias of chickens by replicating subcellular particles (filterable agents; viruses) differs from all other inductions of neoplasia in several important ways. (1) The transformation of normal cells into fully-characterized tumor cells is rapid or almost immediate (39); there is no more or less protracted "latent period" such as occurs in chemical or physical induction. (2) The transformation is *direct*, without intermediate progressions. (3) The transformation results in exact reproduction of the particular type of tumor from which the agent was derived. (4) The agents are recoverable in vastly increased quantity from the tumors they produce.

The filterable sarcomas and leukemias are diverse in structure and behavior; each transmissible tumor line has distinctive properties and each has its specific agent, which exactly reproduces the particular tumor type (12, 13). The cellular change effected by the agents is permanent and heritable, being transmitted to daughter cells under circumstances which preclude an extracellular action of the agent or the transformation of normal cells into tumor cells (14). It is not possible to account for the individualities of all the diverse tumors by presuming for each a different normal cell of origin; there are not enough different kinds of normal cell to go round. The cell of origin imposes limits on what kind of tumor *can* develop, but within those limits the filterable agent, and not the cell of origin, determines what particular kind of tumor *does* develop. The agent impresses a qualitative specificity of structure and behavior on the cells it renders neoplastic and, by imposing a particular degree and type of differentiation, the agents may override the natural proclivities of the cells and direct them into a path of differentiation, which, aside from neoplasia, they follow only under exceptional circumstances (16).

The main characteristics of tumor transmission have been known for a long time (14), but their interpretation remains obscure. Insufficient attention has been given to the qualitative specificity of the agents. It was suggested many years ago that in order to exert its specific effect the agent, or part of it, must form an intimate union with the cell constituents concerned with differentiation as well as with division, and that the "filterability" of a tumor depends upon the break-down of the cell-agent complex in a particular way to yield an agent capable of independent survival for a sufficient time to allow transmission to new hosts. A circumscribed particle of the size attributed to the agent of the Rous sarcoma might stimulate cell division or even induce neoplasia, but it is hardly credible that it could *re*-produce a preexisting tumor with high precision and impose specific heritable characters upon cells. To do this the agent must lose its separate identity within the cell and become, in effect, part of the cell (15). Particles corresponding in size with that of extractable agent are demonstrable by electron microscopy in some Rous sarcoma cells but not in most of them. It is impracticable to examine every portion of a tumor cell by electron microscopy, so that it is arguable that all the particles are in the portions of the cell that are not examined. Nevertheless, there are some indications that the absence of particles from most cells is real. Particles are visible more frequently in irradiated, damaged or dying tumor cells than in "healthy" ones and commonly they are visible in large numbers or not at all. Epstein (11) has shown that the number of visible particles is proportional to the quantity of extractable agent. The most reasonable interpretation on the basis of the information now available is that the moiety of the extracellular agent that determines and directs the neoplastic behavior and specific activities of the tumor cells is not present in the form of discrete particles demonstrable by electron microscopy or transmission experiments. Claude and Murphy (10) described the filterable agents as "transmissible mutagens" on account of their ability to transmit specific heritable cell characters. More recently the analogy between transmission by tumor agents and transduction in bacteria has been emphasized (47). The extractable agents are composed of ribose nucleic acid, protein, and lipoid

(32). In conformity with what is now known about transduction, in its broadest sense, and about viruses, it is justifiable to suspect that the active and specific moiety of extractable agents is the ribose nucleic acid, which is incorporated in heritable mechanisms of the host cell (37, 38, 55). For the maintenance of neoplasia, it is essential only that replication of specific material shall keep pace with cell division. When replication exceeds this minimum the excess is combined with protein and is thereby segregated from the metabolism of the cell and rendered capable of survival outside it. The extractable "virus" is merely waste.

The outstanding importance of the chicken tumors is not their bearing on a generalized virus theory of cancer but the opportunity they give for investigating the intracellular mechanisms of neoplasia and the evidence they provide that all the characteristics of neoplasia may be imposed on a normal cell by the incorporation in it of a specific pentose nucleic acid derived from another neoplastic cell.

*4. Progression in Vitro*

Conversion of normal connective tissue cells into sarcoma cells and further divergent differentiations into distinctive sublines are becoming commonplace events during the prolonged cultivation of cells in vitro (52). The transformations occur unpredictably after varied periods of cultivation and without identifiable carcinogenic stimulus; they provide the extreme contrast to the rapid, direct, and highly specific transformations effected by the filterable agents of chicken tumors.

### Summary and Discussion

The description and analysis of neoplastic development is hampered by the lack of an appropriate terminology. Concentration on late stages of neoplasia is natural on account of their dire significance in man and of their experimental convenience in animals, but the concepts and nomenclature based on clinical cancer alone are inadequate for the analysis of neoplasia as a dynamic process of development, a space-time continuum in which the classifiable lesions and lumps are conspicuous incidents. As Weiss (58) writes of normal cells, tissues, and organs, a "precancerous lesion," a benign tumor, or a cancer is to be studied, ". . . not as a static feature but as a phase, transitory or terminal, in a continuous chain of transformations. . . ." Neoplastic growth is not necessarily "uncontrolled"; normal hormones control the growth of some tumors that have all the cardinal features of malignant neoplasia. Neoplastic growth is not "chaotic"; in most tumors there is an organized pattern of structure approximating sometimes to the perfection of a normal tissue (20). Neoplastic tissue can carry out normal functions and produce, for example, physiologically effective hormones. The rate of neoplastic growth is not extraordinary; at its maximum, in some transplanted tumors, it approaches but does not exceed that of embryonic tissues (35). The first steps in neoplasia do not inevitably entail any growth at all. Neoplasia is not sustained hyperplasia, from which it differs as differentiation differs from modulation. Sustained hyperplasia is not an essential, or even usual, precursor of neoplasia, which often develops long after hyperplasia has subsided and in atrophic tissue. The essence of neoplasia is not growth *per se* but an irreversible and heritable cell change determining that growth, under those circumstances that allow the neoplastic cells to proliferate at all, is not controlled or coordinated in harmony with the needs and safety of the whole organism. The menace of neoplasia results from a failure of integration, and on qualitative rather than on quantitative characteristics of growth. The unsolved problem of supracellular integration is of equal concern to embryology and to cancer research, and both disciplines may contribute to its solution.

The onset of neoplasia is marked by an irreversible determination, or change in competence, not necessarily manifested quickly or at all by histological change or by growth without additional extrinsic stimulation. At least three different modes of origin are distinguishable. (1) The filterable agents of chicken tumors effect a transformation of normal cells into tumor cells that is rapid or immediate, direct, complete, and specific. So far, no exact counterpart of this phenomenon, akin to transduction, has been demonstrated in neoplasia in mammals. (2) Chemical carcinogenesis, according to present views, is a two-stage process. The first stage of *initiation* seems to consist of an irreversible change in normal cells effected by a direct chemical action on heritable materials or mechanisms in the cells. The change may be interpreted as an injury of a distinctive kind resulting from specific chemical action. The second stage of *promotion* is more prolonged, and accounts probably for most of the latent period or induction time in experimental carcinogenesis. A similar mechanism seems to operate in the occupational neoplasias of man. There is statistical evidence consistent with a two- or multi-stage progress in other forms of neoplasia in man but other interpretations are possible (1, 2). (3) In many kinds of natural or induced neoplasia in animals and in man, there is no evidence of direct and injurious chemical action on cell constituents. Hormonal induction of neoplasia is not, apparently, a two-stage process like chemical carcinogenesis (3). The induction is attributable to imbalance of physiological stimuli and the preponderating hormone is dubiously comparable with a chemical carcinogen. The high incidence of varied types of neoplasia in inbred strains of mice is due to physiological rather than "carcinogenic" stimuli whose effects are conditioned by the genotype. No specific carcinogen can be implicated in the transformation of normal into neoplastic cells in vitro, and no stage of initiation is apparent. Tumor induction, in these examples, is a slow process. The long delay may correspond with the time needed either for the occurrence and selection of random cell variants or for the "build-up" of some self-reinforcing process to the stage of irreversibility. The primary change is not an alteration of cell structures by direct chemical injury but a reversible modulation which, only after a long time, becomes established as an irreversible, heritable progression. Irreversible differentiations in normal development are similarly interpreted as produced by way of initially reversible modulations (58).

This analysis implies that the heritable cell change upon which neoplasia depends may be effected in several different ways, including transduction, direct chemical or physical alteration of cell constituents, and the build-up of an initially reversible modulation to an irreversible heritable state. This change being established, neoplasia advances by progression. The early emergent lesions are not always neoplastic as judged by the usual histological criteria; some are conditional or dependent tumors with all the cardinal features of benign or malignant neoplasia save independence of extrinsic stimulation, and some are independent benign or malignant tumors. Some tumors emerge, by *direct* development, with their definitive characters established from the outset; other acquire their characters by *indirect* development through a series of progressions. The steps along indirect paths are variously, and often inappropriately, described as precanceroses, preneoplastic or precancerous lesions, benign neoplasms, or carcinomas in situ. They are all stages in a process of neoplasia which advances only as a consequence of progression in one or more of the characters of the neoplastic cells. Progression is not inevitable or predictable; usually it seems to occur at random in only a small proportion of the available competent tissue. At intermediate stages along indirect paths there is a choice between diverging pathways leading to various kinds of tumors, to persistence or growth without qualitative change or to regression. Progression continues in gross tumors and, indefinitely, in transplanted tumors, even after several decades of serial passage. The capacity for progression is an important characteristic of neoplastic cells, and it enables them to adapt themselves to almost all unfavorable circumstances to which they can be exposed.

The course of neoplasia can be described in terms applicable to normal development. Like normal differentiation as described by Waddington (56), neoplasia is step-wise progressive, it is usually canalized into one of a number of possible channels, it is self-reinforcing, and it is more or less irreversible. (The irreversibility of neoplasia is probably complete; it can halt at any stage or regress to extinction, but there is no convincing evidence of true reversion.) The three basic principles of cyto-differentiation enunciated by Weiss (59), namely, "discreteness," "exclusivity," and "genetic limitation," are equally applicable to neoplasia. Neoplasia usually entails stricter canalization, discreteness, and exclusivity than obtains in normal embryonic or adult tissue (16). Genetic limitation is conspicuous and often, as in inbred mice, the decisive factor in determining the frequency and type of natural and induced neoplasia. The frequent limitation of progression to small portions of the competent tissue has its counterpart in normal organogenesis. Wigglesworth (61), for example, emphasizes the inhibition exercised by a dominant center, once initiated, on the formation of other centers with the same potential activities. The onset of neoplasia can be interpreted as the diversion or deflection (59) of development into a new, abnormal pathway. The diversion depends upon a heritable cell change which, it seems, can be effected in more than one way. Thereafter neoplasia advances, by further heritable cell changes or progressions, to varied extents along one or another of a considerable number of possible branches of the neoplastic path. Progression, in a wide sense, probably includes several kinds of heritable cell changes, of which only those occurring in transplanted tumors have yet been analysed. It is not yet clear whether similar mechanisms operate in neoplasia advancing naturally in the original host. The close analogies between neoplastic progression and normal differentiation suggest that similar basic mechanisms are concerned in both.

## REFERENCES

1. Armitage, P., and Doll, R., *Brit. J. Cancer*, **8**, 1-12 (1954).
2. ———, and ———, *Brit. J. Cancer*, **11**, 161-169 (1957).
3. Axelrad, A., and Leblond, C. P., *Cancer*, **8**, 339-369 (1955).
4. Bang, F., *Acta Unio Intern. contra Cancrum*, **7**, 52-58 (1951).
5. Barrett, M. K., and Deringer, M. K., *J. Natl. Cancer Inst.*, **11**, 51-59 (1950).
6. ———, and ———, *J. Natl. Cancer Inst.*, **12**, 1011-1017 (1952).
7. Berenblum, I., *Advances in Cancer Research*, **2**, 129-175 (1954).
8. ———, *Cancer Research*, **14**, 471-477 (1954).
9. ———, and Haran-Guera, N., *Brit. J. Cancer*, **11**, 77-84 (1957).

10. Claude, A., and Murphy, J. B., *Physiol. Revs.*, **13**, 246-275 (1933).
11. Epstein, M. A., *Brit. J. Cancer*, **10**, 33-48 (1956).
12. Foulds, L., *11th Sci. Rep. Imp. Cancer Research Fund, London*, pp. 1-13, (1934).
13. ———, *11th Sci. Rep. Imp. Cancer Research Fund, London*, pp. 15-25 (1934).
14. ———, *11th Sci. Rep. Imp. Cancer Research Fund, London, Suppl.*, 1-14. (1934).
15. ———, *Am. J. Cancer*, **31**, 404-413 (1937).
16. ———, *Am. J. Cancer*, **39**, 1-24 (1940).
17. ———, *Brit. J. Cancer*, **3**, 345-375 (1949).
18. ———, *Cancer Research*, **14**, 327-339 (1954).
19. ———, *J. Natl. Cancer Inst.*, **17**, 713-753 (1956).
20. ———, *J. Natl. Cancer Inst.*, **17**, 755-781 (1956).
21. ———, *J. Chronic Diseases*, **8**, 2-37 (1958).
22. Friedewald, W. F., and Rous, P., *J. Exptl. Med.*, **80**, 101-126 (1944).
23. ———, and ———, *J. Exptl. Med.*, **91**, 459-484 (1950).
24. Furth, J., *Cancer Research*, **13**, 477-492 (1953).
25. Glücksman, A., *Cancer*, **5**, 385-400 (1945).
26. Green, H. N., *Brit. Med. J.*, ii, 1374-1380 (1954).
27. Griffin, A. C., in *Cancer* (R. W. Raven, ed.), Vol. I, 123-160, Butterworth, London, (1957).
28. Haddow, A., in *Physiopathology of Cancer:* (F. Homburger and W. H. Fishman, eds.), pp. 441-551, Academic Press, New York (1953).
29. ———, *Ann. Rev. Biochem.*, **24**, 689-742 (1955).
30. Hamperl, H. F., *Problems Oncol.*, **3**, 131-139 (1957).
31. Haran, N., and Berenblum, I., *Brit. J. Cancer*, **10**, 57-60 (1956).
32. Harris, R. J. C., *Advances in Cancer Research*, **1**, 233-271 (1953).
33. Horava, A., and Skoryna, S. C., *Can. Med. Assoc. J.*, **73**, 630-638 (1955).
34. Klein, G., and Klein, E., *Symposia Soc. Exptl. Biol.*, **11**, 305-328 (1957).
35. ———, and Revesz, L., *J. Natl. Cancer Inst.*, **14**, 229-277 (1953).
36. Klein, M., *J. Natl. Cancer Inst.*, **14**, 83-89 (1953).
37. Lederberg, J., *J. Cellular Comp. Physiol.*, **45**, Suppl. 2, 75-107 (1955).
38. ———, *Texas Rep. Biol. Med.*, **15**, 811-826 (1957).
39. Loomis, L. N., and Pratt, A. W., *J. Natl. Cancer Inst.*, **17**, 101-108 (1956).
40. Marchant, J., and Orr, J. W., *Brit. J. Cancer*, **7**, 329-341 (1953).
41. Miller, J. A., and Miller, E. C., *Advances in Cancer Research*, **1**, 339-396 (1953).
42. Nicholson, G. W. de P., *Studies on Tumour Formation*, Butterworth, London; Mosby, St. Louis (1950).
43. Ritchie, A. C., *Brit. J. Cancer*, **11**, 206-211 (1957).
44. ———, and Saffiotti, U., *Cancer Research*, **15**, 84-88 (1955).
45. Rous, P., and Kidd, J. G., *J. Exptl. Med.*, **69**, 399-424 (1939).
46. ———, and ———, *J. Exptl. Med.*, **73**, 365-389 (1941).
47. Rusch, H. P., *Cancer Research*, **14**, 407-417 (1954).
48. Salaman, M. H., *Brit. J. Cancer*, **6**, 155-159 (1952).
49. ———, and Gwynn, R. H., *Brit. J. Cancer*, **5**, 252-263 (1951).
50. ———, and Roe, F. J. C., *Brit. J. Cancer*, **7**, 472-481 (1953).
51. ———, and ———, *Brit. J. Cancer*, **10**, 363-378 (1956).
52. Sanford, K. K., Likely, G. D., and Earle, W. R., *J. Natl. Cancer Inst.*, **15**, 215-237 (1954).
53. Selye, H., *Texas Rep. Biol. Med.*, **12**, 390-422 (1954).
54. Shubik, P., Baserga, R., and Ritchie, A. C., *Brit. J. Cancer*, **7**, 342-351 (1953).
55. Stanley, W. M., *Texas Rep. Biol. Med.*, **15**, 796-810 (1957).
56. Waddington, C. H., *Symposia Soc. Exptl. Biol.*, **2**, 145-154 (1948).
57. Wakeley, C., and Graves, F. T., *Lancet*, i, 329-333 (1958).
58. Weiss, P., *Quart. Rev. Biol.*, **25**, 177-198 (1950).
59. ———, *J. Embryol. Exptl. Morphol.*, **1**, 181-211 (1953).

60. Whiteley, H. J., *Brit. J. Cancer*, **11**, 196-205 (1957).
61. Wigglesworth, V. B., *Symposia Soc. Exptl. Biol.*, **2**, 1-16 (1948).

## DISCUSSION

DR. POTTER: As one of the originators of the deletion hypothesis, I should like to make a few comments on the current status of this idea from the biochemical standpoint. I agree with Dr. Foulds' interpretation of the cancer problem. The idea of the heterogeneity of tumors and the idea of progression or of stages of tumor development are almost universally accepted ideas for many types of tumor at this time. I think what is not really understood by a great number of people is that the enzyme patterns of tissues are not fixed by the framework of the constant genome. They are subject to tremendous variations within this constant genic background, i.e. without gene mutation. This was the kind of thing which was discussed by Dr. Maas yesterday. The presence of a particular enzyme could be completely suppressed by a feedback mechanism even in the absence of gene mutation. The DNA of the gene transmits its activity to the RNA which in turn is the enzyme-forming system, and this gives us a capacity for tremendous variation in the enzyme pattern of a tissue. It has recently been shown by Rosen et al. (*Science,* 1958) that cortisone in rats can act through an enzyme-forming system in liver giving as much as a six-fold increase in a specific glutamic-pyruvic transaminase. Also, the administration of thyroxin will greatly increase the TPN-cytochrome $c$ reductase in rat liver (Phillips and Langdon, *Biochim. Biophys. Acta,* 1956). So there is a tremendous plasticity in the enzyme patterns because of the potential for variability at the level of the enzyme-forming system. Nevertheless, the constant potentiality resides back in the original DNA pattern. The question then becomes, if there is a deletion, or if there is a progression of cancer from one stage to another, what is the deletion and what is the progression based upon? Here I wish to make the point that, in the case of a building block for growth such as thymidylic acid, which can go through several steps to form DNA, we have alternative pathways of synthesis. The thymidylic acid may be formed from uridylic acid and the uridylic acid in turn may be formed from either aspartic acid or from uracil. Thymidylic acid may also be formed from thymidine. The point I wish to raise is an extension of the question which I pointed out yesterday afternoon. A normal cell in adult tissue has the problem of how not to duplicate itself in the presence of an abundance of these substances needed for cell duplication. It is my contention that normal growth is limited by a variety of catabolic enzymes some of which are localized in liver and which maintain such low concentrations of critical intermediates that growth is impossible. Superimposed on this basic mechanism we must have specific feedback mechanisms which act on critical enzyme-forming systems directly. I think in many cases these feedback functions may be carried out by regulatory proteins for specific tissues acting on enzyme-forming systems for particular biosynthetic pathways, and thus we get a regulation of the quantity of compounds which are necessary for growth. Now it is unfortunate that there are so many different kinds of cancer tissue, because if this idea of a succession of deletions is correct, then it seems very likely that an identical succession of deletions is not required. There would be many deletions which are not strategic. But there would be some deletions which are strategic, and I think these must include the control mechanisms for the formation of DNA, and must also include the synthesis of the factors which are concerned in the recognition of the tissues and cells for one another. Here we have a situation where DNA is perhaps directly involved in the specification of surface proteins concerned with the recognition of one cell by another. We can set up certain minimal requirements, then, for the conversion of a normal cell to a cancer cell. There ought to be the deletion of a DNA concerned in the formation of proteins by which cells recognize one another, and in addition we should have deletions in pathways concerned with the catabolism of substances concerned in normal growth. I think that the kind of tumors that Dr. Foulds was talking about are the less malignant tumors because certain synthetic reactions have been maintained, but in the case of the more malignant tumors there is also an additional deletion in the formation of surface proteins so that the cells can no longer recognize one another. I think that the important fact here is the succession of events.

In order to avoid the idea that this is completely speculation, I would like to cite the work which has recently been done on the Novikoff hepatoma which is believed to be derived from parenchymal liver cells (cf. Potter, *Fed. Proc.* 1958). I might say here that in any biochemical studies of the mechanism of carcinogenesis, you must have a comparison between the cancerous tissue and the normal tissue from which it is derived; so it is essential to know the derivation of various tumor cell lines. A number of catabolic enzymes have been studied in the Novikoff hepatoma. The TPN-cytochrome $c$ reductase and transhydrogenase are found to be missing in this tumor, and we are extending our work on the study of this pathway. Others have studied the pathway of the formation of carbon dioxide from uracil and certain hepatomas are deficient in this pathway. Dr. Auerbach has demonstrated that a whole series of amino acid-catabolizing enzymes are missing. Moreover, he has demonstrated that the enzyme-forming systems for some of these enzymes are missing, because in the tumor the enzymes will not respond to the addition of substrate to the medium but these enzymes are inducible in normal liver tissue. The group working in Montreal has also demonstrated an absence of glucose-6-phosphatase, xanthine oxidase, and uricase in this tissue; so there are important deletions in the pathways for purines, pyrimidines, amino acids, and hydrogen transport. So I merely wish to say that I feel Dr. Foulds has advanced some very sound ideas here this afternoon and that the biochemical and immunological bases of tumor progression are beginning to become apparent.

DR. WEISS: I think that this has been the most beautiful biologically confirmable theory and exposition of cancer I have ever heard. I should like to add to this simply that the cell recognition is the principle which ordinarily gives trouble because its loss gives rise to metastases. I believe firmly that cancers arise by the superposition of small single steps of deviations in a lineage of cells, and we should turn from the idea of looking at cancer as a single trigger reaction to the probability calculus of the occurrence of these individual steps in generations of cells, and the probability of finding accumulations in the same cell strains, without which the result which you have described would not occur.

Dr. Foulds: I am dealing primarily with phenotypic characters, and I don't mind whether it's addition or deletion, but nearly everything can be interpreted either one way or the other, and I have no antagonism whatever to the deletion idea. I was using the term in a rather different sense. I was using it to describe the rather violent knocking out of bits of chromosomes, or something of that sort, which is quite a different matter. All I would suggest is that this deletion, if it is a deletion, can occur in other ways rather than by this violent knocking out by drastic chemical treatment. It may occur also simply by random change. I am glad that reference was made to the surface proteins. One thing I think we must get away from is this preoccupation with mere growth. There is really nothing extraordinary about the growth of tumors. They don't grow as fast as embryonic tissues as a rule, and it is not uncontrolled growth. Many of them are controlled by hormones. Also, it is not unorganized growth; it can be most elaborately organized and malignant tumors can be physiologically active and produce hormones. One example, and I admit it is only one, is that of a patient who suffered from hormone deficiency when the tumor was removed. So long as his tumor was in, he was happy and well. As soon as the surgeon took his cancer out he suffered from lack of the hormone which that tumor was producing. There are two points as I see it. There is an irreversible heritable cell change, and this entails a failure of integration. The growth is not integrated in harmony with the needs and safety of the whole organism. That seems to be the crux of the matter, not the mere capacity for growth, because other tissues can grow. It has been pointed out many times that you cannot make a tumor out of any tissue which is incapable of growing. Only tissues will become cancerous which have the capacity to grow inherent in them from the beginning.

Dr. P. Cohen: I should like to report an example of what I would like to think of as an example of biochemical dedifferentiation which applies in the case of hepatomas, and to certain other tissues as well. A reaction which carbamyl phosphate plus aspartate undergoes to form the precursor substance for orotic acid, is mediated by an enzyme aspartic transcarbamylase, which has already been referred to. In all the tumors which we have studied, the level of this enzyme is relatively high. If one takes the Novikoff transplantable tumor and compares it with normal liver, its activity is 2 to 8 times higher. Dr. Auerbach, in experiments referred to by Dr. Potter, did this experiment recently in my laboratory using techniques which we recently developed for assaying this enzyme. We have also done this independently in a more systematic study than that done with Dr. Auerbach. Now the interesting thing is that if one measures the level of this enzyme, at about 24 to 36 hours after partial hepatectomy, this particular enzyme goes up about three fold, reaches a peak, and then slowly decreases so that when regeneration is complete the enzyme actively has returned to the normal level. However, at the same time, the ornithine transcarbamylase activity which is competing with the aspartic transcarbamylase for carbamyl phosphate and is important in the urea cycle goes down. The urea cycle represents a unique type of a highly specialized biosynthetic mechanism which is characteristic of the fully matured and developed hepatic cell. In this particular case, the cell has the problem of what to do with the small amount of carbamyl phosphate it can form. In the face of regeneration or more rapid growth it has to forego the highly specialized function of making urea in order to make pyrimidines. Thus the urea-synthetic activity decreases rapidly from tumors at the onset of neoplasia. On the other hand, the aspartic transcarbamylase activity increases so that while you may have deletion of many enzymes during neoplasia or regeneration, you have here a case in which an enzyme is actually increased. In both cases, one is dealing with a process of dedifferentiation both at the cytological as well as the enzymatic level.

DR. POTTER: This is a beautiful example, and I had hoped that Dr. Cohen would bring it up because it is the counterpart of the feedback mechanism which was demonstrated in microbiology, and we think the type of thing which was discussed formerly in terms of alternative pathways is just one level of sophistication, and the next level is the identification of the feedback mechanism which regulates the level of the synthetic enzyme. It seems apparent that in the case mentioned by Dr. Cohen the enzyme-forming system has had some of the brakes removed from it and that you actually build up an increased amount of the aspartate transcarbamylase. It is going to be interesting to see if the mechanism is identical to the mechanism which has been demonstrated in *E. coli* by Yates and Pardee. I am sure that there are a number of adaptive enzymes which increase in cancer tissue provided that they are on the synthetic pathway. This will come up again tomorrow in the case of the regenerating liver.

DR. UMBARGER: I would wonder if this enzyme, aspartic transcarbamylase, which Dr. Cohen has just mentioned, is inhibited by any pyrimidine groups.

DR. P. COHEN: This has not been tested as yet. We also plan to test carbamyl aspartate. Mr. C. Schmidt, a medical student working with me one summer, fed some animals high levels of arginine and ornithine while they were being fed a carcinogenic diet. The purpose was to try to divert the available carbamyl phosphate from carbamyl aspartic synthesis to citrulline synthesis and thus possibly prevent pyrimidine synthesis and possibly neoplasia. The results of this preliminary experiment, however, revealed no significant effect on either incidence or time of hepatoma formation.

# ORGAN DIFFERENTIATION IN CULTURE

E. BORGHESE

*Institute of Histology and General Embryology,
University of Cagliari, Cagliari, Italy*

## INTRODUCTION

The culture in vitro of embryonic organs, which for a long time was only a collateral application of the tissue culture method, seems now to have reached its maturity and thus to have realized an old aspiration of embryologists working on warm-blooded animals. It is now providing a very effective tool to attack the problem of the behavior of isolated developing organs and of the mutual influences acting during embryonic life to assure a harmonious development.

The effectiveness of this method is proved by the steadily increasing amount of publication in this field and of the research work still going on, so that it becomes more and more difficult to collect and organize all the literature thus far available.

The main advantages of the method are as follows: (1) embryonic organs can be subtracted from the influence of contiguity with other organs and of the nervous system, and to a greater or smaller extent, depending on the method, also to the action of blood components; (2) the development can be followed under direct microscopical observation; (3) growth can occur in all directions without any limitation but those depending on the exhaustion of the intrinsic capabilities of the organ; (4) the culture medium may be changed ad libitum so as to analyse nutritive needs, and the action of hormones and vitamins; and (5) the parts of an organ, or different organs may be combined so as to study many kinds of reciprocal influences.

The first observation that chick embryonic organs, cultivated in vitro, go on growing and differentiating is usually ascribed to Thomson (234, 235) who, in order to distinguish this type of growth from that known at that time of the purely peripheral migration of cells, called it "somatic growth." In contradistinction to tissue culture proper or "histiotypic cultures (Maximow, 170), cultures of embryonic organs were called "artificial organisms" (Fischer, 64) or "organotypic cultures" (Maximow, 170). The latter expression is at present the most frequently in use.

The organotypic culture does not require in principle methods fundamentally different from those applied in the histiotypic cultures, but rather the choice of the most suitable developmental stage and some attention in the preparation of the explants.

The usual medium of chick plasma and embryonic extract was largely used as a basic medium, and appeared to be very good, as a rule, also for mammalian organs. If there is any reason to avoid embryonic material, an extract of the heart of adult animals can be used (Moscona and Moscona, 181). In research on human organs substances of homologous origin were preferred (Gaillard, 72; Zaaijer, 303).

The oldest techniques of tissue culture, such as hanging drop or Carrel flasks, may give excellent results in many cases. However, organ culturists have developed new techniques, which proved frequently more appropriate for organotypic development. The first attempt, made by Strangeways and Fell (230) to introduce for large explants a new technique of culture in tubes was soon improved by the more convenient one of the watch glass (Fell and Robison, 62), based on the principle of laying the explant on the surface of a large clot of plasma and embryo extract in a humid atmosphere. This method is still widely used. However, certain changes were suggested, consisting in maintaining around the explant some amount of fluid by means of a piece of albuginea (Martinovitch, 163) or by laying the explant on glass rods (Martinovitch, 165), or even in using a completely liquid medium where the explant is kept floating on a piece of tissue paper (Trowell, 242; Chen, 31). The use of a cellulose-acetate fabric instead of paper, suggested by Shaffer (211), together with the return to the plasma clot, is the latest step and allows, like Chen's method, the transfer of explants onto a new, fresh medium without disturbing the peripheral cells of the organ.

If transferring to completely fresh medium is not desired, a supply of fresh nutritive medium may be obtained by frequently changing a layer of liquid medium at the surface of the culture. This may be done if the culture is kept in Carrel flasks (Grobstein, 99) or with a system of continuous perfusion, such as that suggested by De Haan (37, 38), which is now little used because of being exceedingly complicated. The same advantage is offered by the roller tube culture method, devised by Gey (84) for histiotypic cultures and only very recently applied to organotypic culture (Gonzales, 93; Ito and Endo, 124; Gliozzi, 88) with even better results, as proved by Tateiwa (233) for bone by means of growth measurements.

A temperature lower than that of the animal supplying the explant was found by Martinovitch (161, 164) to favor a longer survival of some mouse organs, which were cultured at 33-34°C for periods as long as many months.

The method suggested by Leighton (143) of cultivating organs on commercial cellulose sponge in roller tubes, is supposed to offer conditions more similar to normal ones, as a tridimensional frame is supplied to the tissues instead of a flat clot, and the enormous increase of plasma surface would seemingly offer mechanical pathways to metabolic exchange. As a matter of fact, however, this method, which gave very good results in cultures of liver and other organs of 9 to 10-day chick embryos, was in subsequent research limited only to the study of problems concerning tumoral growth (Leighton, 144, 145; Leighton and Kline, 146; Leighton, Kline, Belkin, and Tetenbaum, 149; Leighton, Kline, Belkin, Legallais, and Orr, 147; Leighton, Kline, Belkin, and Orr, 148).

A method was especially devised for organ culture by Wolff and Haffen (275, 276) consisting in a non-nutritive basic medium of agar, supplemented by mineral salts, glucose, and embryonic extract. The first idea (Wolff, 263) was to avoid the excessive richness in nutritive materials contained in plasma, supposed to encourage the peripheral migration of cells at the expense of the preservation of organ structure. In time the method proved to offer further advantages, one of which is to permit the study of the nutritive needs of embryonic organs. If instead of embryonic extract mixtures of known amino acids are added, a medium can be obtained which, although giving in any case quantitative results inferior to those of natural media, has revealed itself extremely suitable to study the influence of given substances on the growth and differentiation of particular organs.

As long as only plasma media were used, it was considered very important, in order to get organotypic development, to respect the integrity of the organs, taking them as wholes and without damaging their surfaces, so as to prevent the peripheral migration of cells, which tends to disorganize the organ and to injure the original structure. For skeletal rudiments the suggestion was made (Fell, 46; Jacobson and Fell, 125; Chen, 29) to remove the ectoderm covering the piece explanted, which otherwise would tend to cover the explant and to keratinize, inducing the cultivated cartilage to degenerate rapidly. An exception was found by Zaaijer (303), who found it advantageous to preserve a layer of epidermis around phalanges of 8 to 15-week human embryos: such an epidermis encourages in turn the development of thick lamina of connective tissue, very favorable to the development of the cartilages.

At present Wolff and Haffen's method (276) has proved so suitable in preventing any migration of cells that not only entire organs, but even fragments of a rather large organ, such as liver, give organotypic growth without disruption of their typical structure. In such a medium many tissues exhibit a strong tendency to cover the surface with an epithelium-like layer and thus to preserve the original structure (Wolff, 270).

The survival of the culture, in most of the research here considered concerning the embryonic development of organs, has not gone beyond two or three weeks; however, it should be remembered that Martinovitch (161) has succeeded in obtaining much longer survival, even of some months. In most cases, anyway, a period of culture of 8 to 9 days or even less is sufficient for many purposes; in fact, for those species generally employed, such as chick and mouse, this time corresponds to about half of the developmental period prior to birth.

The present review is strictly limited to the development of embryonic organs in vitro, and in relation to embryological problems sensu strictu. For this reason many related subjects are excluded, such as the following.

(1) Research on the development of embryonic organs isolated by means of techniques different from culture in vitro, such as chorioallantoic grafts, implantation in the chick intraembryonic coelom, or grafts in the anterior eye chamber. Many results obtained by these methods are included in a previous review by the present author (Borghese, 19).

(2) Research done by means of the culture method and on embryos, but for different purposes from those here treated. These concern: (a) early developmental stages, studied for different purposes, as for example problems of determination (Abercrombie, 1; Abercrombie and Bellairs, 2; Bellairs, 5, 6; Grobstein, 95-98; Grobstein and Zwilling, 111; Simon, 214-217; Wolff and Simon, 295; New, 184; Spratt, 220, 221, 226, 227; Nicholas and Rudnick, 186; Waterman, 248, 249; Waddington and Waterman, 247; Waddington, 246) or the nutritional needs of whole embryos (Spratt, 222-225) or of parts of them (Harrison, 119, 120); and (b) problems related to histogenesis, for which a review by Fell (48) may be consulted.

(3) Research done by means of the culture method on postnatal organs, whether juvenile or adult, whatever the problems studied may be: postnatal differentiation of testes (Gaillard and Varossieau, 79) or of eye (Tansley, 232); secretory activity of endocrine glands such as the thyroid (Seamen and Stahl, 206), hypophysis (Gaillard, 66; Martinovitch, 162, 166; Martinovitch and Vidovic, 168; Martinovitch and Radivojevitch, 167; Gaillard and Varossieau, 80), adrenal glands (Martinovitch, 167; Schaberg, 201), the action of hormones and vitamins on explanted vagina (Kahn, 127; Hardy, Biggers, and Claringbold, 118), the metabolism of lymph

nodes (Trowell, 241, 243), maintenance of structure in vitro of juvenile organs such as the epiphysis (Martinovitch, 163; Milkovic-Zulj, 171), and inductive action of the posterior hypophysis explanted after birth (Gaillard, 67).

(4) Research done by means of the culture method and related to cancer research, such as the action of carcinogenic substances on adult prostatic gland (Lasnitzki, 137-140, Lasnitzki and Pelc, 142) or on lung (Lasnitzki, 141); or the invasion of embryonic organs by malignant tissues (Wolff, 266-268; Wolff and Schneider, 293, 294).

Other reviews, collecting some previous literature of this kind, have been published by Fell (47, 49), by Borghese (19), by Gaillard (77), and by Wolff (269).

### Morphogenesis of Isolated Organs

Limb buds explanted from chick embryos at 72 to 80 hours of incubation, when they are formed of ectoderm and mesenchyme only (Strangeways and Fell, 230) produce in two days skeletal cartilaginous rudiments, which continue to develop for 7 days but do not ossify. In limb buds explanted from 3-day chick embryos and cultivated as long as 28 days (Fell, 44) cartilages differentiate. If the explantation takes place at the 8th day, when cartilages are already present, these grow so as to reach 3 to 4 times their original size, epiphyses are formed, and histological differentiation continues as far as the stage of chondroblastic hypertrophy. Endochondral bone and marrow do not appear, but periosteal bone is formed. The appearance of epiphyses only in cartilages explanted at the 8th day suggests that epiphysis determination may occur between the 3rd and the 8th day, possibly under the influence of a factor extrinsic to the cartilage.

The average length increase of the femur rudiment (Fell, 45) in 27 days of culture is 226 per cent when they are explanted at $5\frac{1}{2}$ days, 124 per cent when explanted at 6 days. In femur rudiments cultivated in vitro (Fell and Robison, 62), an alkaline phosphatase reaction can be shown, in association with the appearance of cellular hypertrophy.

Hindlimb buds of 6-day chick embryos (Momigliano Levi, 175) containing tibial cartilage and the still undifferentiated shank after three days of culture develop tarsal and metatarsal rudiments, and tibial epiphyses are delineated.

In limb buds of 6 to 12-day chick embryos, ossification appears in vitro (Gliozzi, 88) if not yet present, and goes on if already begun. The bone appears first as a thin lamina adhering to the cartilaginous shaft, then bony arches are formed, and extend more and more into the surrounding mesenchyme. Deformations were observed owing to disproportion between the epiphyses, where interstitial growth goes on, and the shafts, in which growth is delayed. Fusion may occur between parallel bone rudiments.

Epiphyses of 10 to 14-day chick embryos, explanted separately from shafts and cultivated as long as 50 days (Sevastikoglu, 210) are surrounded by a fibrous tissue similar to perichondrium. After 20-40 days of culture small cells derive from the bordering chondrocytes and from the perichondral layer; later metachromasia, deposition of preosseous tissue, and calcification were observed. It is however doubtful that this osteogenesis can be called, as the author claims, endochondral in the generally accepted sense.

Presumptive tissue of chick patella, explanted as undifferentiated mesenchyme at the 9th day of incubation (Niven, 189) produces cartilage in three days in vitro. After three further days this cartilage acquires its characteristic triangular shape. However, it does not ossify in 20 days of culture, although, it is true, this would not occur in vivo in the same length of time.

The comparison between the development in vitro of the palatus quadratus and Meckel's cartilage, explanted from the 5 to 6-day chick embryo (Fell and Robison, 63; Fell, 45) shows that the ossifying of the former, and lack of it in the latter, are at this stage intrinsic: in fact, the former develops hypertrophic cells in vitro and becomes surrounded by periosteal bone, whereas the latter does not show either of these changes.

The development of the chick mandible in vitro was studied by Jacobson and Fell (125) from its presumptive tissue explanted between the 3rd and the 7th day. The osteogenic zones of the mandible, already determined at the 4th day, are independent from Meckel's cartilage, but this has a morphogenetic action inasmuch as it distends the branchial arch and gives to the bone its characteristic elongated shape.

The femur and tibia of the mouse embryo near full term, with advanced ossification (Niven, 187) grow in length no more than 33 per cent in 22 days of culture. In the epiphyses, where only small cells are present at explantation time, after 16 days of culture cellular hypertrophy occurs, followed by perichondral ossification, as in vivo. On the other hand, many bony lamellae of the shaft and hemopoietic tissue, present at explantation time, disappear and are replaced by fibrous tissue.

The metacarpals, metatarsals, and phalanges of hand and foot, explanted as cartilages from human embryos between the 6th and 7th weeks and cultivated in vitro for 28 days (Zaaijer, 303) continue development: the epiphyses increase in volume and cellular hypertrophy appears at the center of the diaphysis.

710 THE CHEMICAL BASIS OF DEVELOPMENT

The development of the avian sternum in vitro was studied by Fell (46). Research on the prospective potencies of different parts of the lateral body wall of the budgerigar showed that sternum does not originate from ribs, nor from the coracoid bone, but from two paired rudiments formed independently in the body wall (Fig. 1, a) In vitro they move together, form

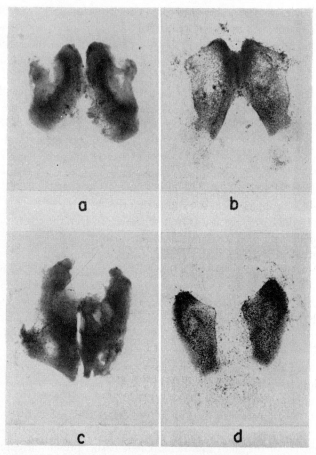

Fig. 1. Analysis of the convergence of the sternal plates under the active movement of the intermediate mesenchymal cells. a. Paired sternal rudiments of a budgerigar embryo, explanted at the 8th day of incubation, separated along the midline, when put in culture in normal position with the medial margins facing together. b. The same after 48 hours of cultivation; the two fragments have moved together and fused. c. Sternal rudiments, as above, but put in culture in reverse position so that the lateral margins face together. d. The same after 48 hours of culture; the two rudiments moved apart. Living cultures. × 16. From Fell (46) by courtesy of the Royal Society.

the keel, develop posterolateral processes and the characteristic sternal curvature, so as to show that all such phenomena are intrinsic to the sternum itself, and do not depend on the action of pectoral muscles, nor upon visceral pressure, nor upon rib action. However, extrinsic factors seemingly play a part in vivo in perfecting the shape of the sternum, as is shown by the fact that in vitro its curvature becomes successively deformed and the angle between keel and sternal body tends to flatten and to disappear.

The two lateral plates, studied in vitro by means of reference marks, move together toward the midline, not under the effect of their own expansion nor of the elongation of the ribs, but following an active migration of the intermediate mesenchymal cells (Fig. 1, b). In fact, if the two lateral margins face together, they are drawn by mesenchymal cells of the intermediate zone, now lateral, and instead of moving together they separate (Fig. 1, c, d).

The behavior of the mouse sternum is much the same. When the lateral body wall of mouse embryo is explanted at 10 to 11 days (Chen, 29), in 6 to 7 days of culture a pair of mesenchymal condensations appear, which move towards each other, come together at the cranial end in 4 days and fuse in 6 to 8 days. This occurs in spite of the absence of ribs, a fact which shows that their elongation has no influence in the coalescence of the two sternal rudiments. If the explantation takes place at the 13th day, when the plates are already moving together, the two halves fuse in vitro, and cellular hypertrophy and periosteal ossification occur.

Also in the mouse (Chen, 30) the movement toward the midline of the two sternal plates is due to the active movement of intermediate cells, and likewise, when the sternal rudiments are inverted, so that the ventral margins face outwards, they move apart instead of approaching one another. Sternal segmentation occurs in mammals only, and first appears as a segmental hypertrophy in cartilaginous cells in the intervals between ribs. It may be proved that segmentation is due to ribs, inasmuch as they inhibit the cartilaginous cellular hypertrophy wherever they contact the sternum. In fact, if the sternum is explanted at the small-celled stage, after removing ribs, cellular hypertrophy occurs throughout its length, and so does the formation of periosteal bone. If the ribs are removed on one side only, segmentation is limited to the opposite longitudinal half of the sternum (Fig. 2, a, b); extra ribs implanted on the sternum increase the number of segments, as they form new zones of inhibition of hypertrophy. The mechanism of rib action is not known; in any case it is a specific one, as it may be obtained with extra ribs, but not with other skeletal pieces, such as

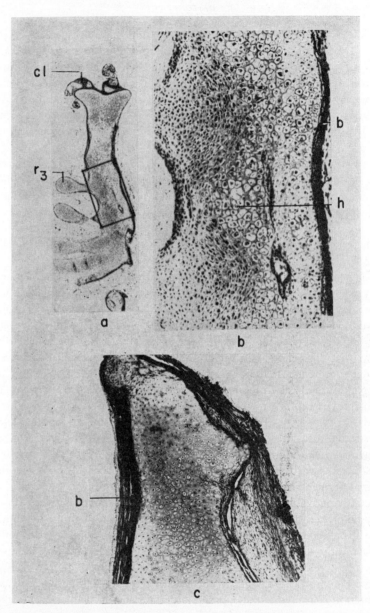

Fig. 2. a, b. Appearance of segmentation in mouse sternum under the influence of the ribs. a. Sternal rudiments of a mouse embryo, explanted at the 14th day of gestation, after removing all ribs on the right hand side, while some are left on the left, and cultivated for 7 days ($\times$ 18). (cl: clavicle; $r_3$: third rib). b: the zone in the rectangle seen at higher magnification ($\times$ 100). In the left half, similarly to normal development, cartilage is hyper-

clavicle or a fragment of another sternum, a boiled rib, or a glass rod.

When a hindlimb bud of the chick embryo is explanted at the 4th day, at the stage of condensed mesenchyme, and is cultivated for 17 days (Fell and Canti, 52, 53), cartilaginous rudiments of femur, tibia, and fibula appear, and become more and more defined, leaving an intermediate space, which represents the anlage of the knee joint. After 4 days of culture the articular line is quite clear; articular surfaces are covered with a condensed layer of cells which turns to cartilage, whereas in the interval loose connective tissue only is present. However, after 6 days of culture, the articular cartilage disappears, the skeletal rudiments fuse, and no articular cavity is produced. This proves that the first anlage of the joint may appear in spite of the absence of any muscular action, blood circulation, and nerves, but under such conditions it cannot complete its development as far as the formation of the cavity. Such inability to complete its differentiation may be due to the lack of muscular action.

While the first appearance of cartilaginous rudiments, their initial differentiation, and the first stages of perichondral ossification largely depend on intrinsic factors, further changes appear to be rather bound to extrinsic factors. In fact, when ossifying cartilage is exposed in vitro to pressures and tensions, changes of normal structure occur.

If the growth in length of a chick embryo cartilage is prevented in vitro, so as to be forced to bend (Glucksmann, 89), the cells on its concave, compressed side orient with their major axis perpendicular to the surface; on its convex side they preserve the normal orientation and flatten out. The homogeneous matrix is reduced or disappears on the concave side and is replaced by divergent fibers, whereas some cells may come out from the cartilage and turn to fibroblasts.

The laying down of perichondral bone on such artificially bent cartilage shows a different behavior according to the developmental stage in which the deforming action is applied. If ossification first appears when the cartilage is already bent, bone is laid down more on the concave part, where under the action of epiphysis growth, perichondrium tends to become de-

---

trophic ($h$) only in the interval between the ribs, whereas hypertrophy is inhibited in their contact; in the right half, where ribs were removed, hypertrophy is continuous as well as the perichondral bone ($b$). Frontal section. From Chen (30), by courtesy of *J. Anat., Lond*.
c. Action of epiphysial traction on the formation of the perichondral bone. Phalanx of an 11-day chick embryo, from which the right half of the epiphysis was removed; cultivated in vitro for 10 days. On the left, where the perichondrium is still subject to the normal tension exerted by the epiphysis, bone ($b$) has developed regularly; it is lacking on the right, where tension is missing. Longitudinal section. $\times$ 90. From Glucksmann (89) by courtesy of *J. Anat., Lond*.

tached from cartilage. Bony trabeculae are produced along the tension lines, perpendicular to the cartilage axis. If ossification has already begun when the mechanical action is applied, bone is formed more abundantly at the convex side, which is subject to a greater tension parallel to the surface.

The growth of the epiphyses exerts on the periosteal envelopment of the shaft a tension which acts not only along the length but also in the perpendicular direction. Glucksmann (89) showed that such a tension stimulates ossification under normal conditions. In fact, if half of the epiphysis is removed from an ossifying cartilage, which is then cultured, bone is laid down on the intact side only (Fig. 2, c). It may be concluded that tension determines not only the direction of bony trabeculae, but also their very formation.

## Digestive and Respiratory Apparatus

Tooth rudiments of rat embryos of 18 to 21 days of gestation, consisting of the papilla covered by the enamel organ (Glasstone, 85) differentiate odontoblasts, when not yet present at explantation time. Odontoblasts in turn produce dentine. Odontoblasts are formed in the presence of and in contiguity with the enamel organ, and reproduce the shape of the cusps if these are already delineated. However, if odontoblasts were already present in explants, they can produce dentine even if the enamel organ is removed, but in such a case they invaginate into the pulp, and dentine forms in irregular strips. Mesenchymal cells of the papilla, completely deprived of odontoblasts and of contact with ameloblasts, are not able to differentiate into odontoblasts.

Tooth rudiments from 17-day rat and 20-day rabbit fetuses (Glasstone, 86), explanted a day before the appearance of cusps, show the latter forming through invagination of the internal enamel epithelium into the papilla. In rat cultures the cusps have the same number and position as in vivo; in rabbit cultures, on the other hand, a supernumerary cusp may appear, which is interpreted as one which may be transitory under normal conditions and is maintained in vitro because the explant is more free to expand in all directions. Calcification of dental rudiments of the rat and rabbit takes place in vitro (Hughes, 123) and their phosphatase activity is similar to that of bone.

The submandibular and sublingual glands of mice, explanted between the 13th and 16th days of gestation (Borghese, 12, 13) begin in vitro the branching process and continue it if already begun (Fig. 3). Any initial bud divides into two, which in turn divide dichotomously many times.

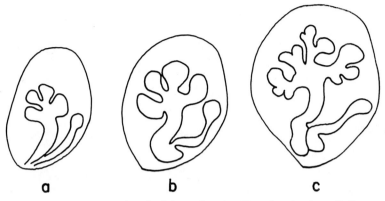

Fig. 3. Branching process of a glandular rudiment, taking place in vitro. Rudiments of submandibular and sublingual glands a, at explantation time; b, after one day; c, after two days. Camera lucida drawings. × 40. From Borghese (15), by courtesy of *Arch. ital. Anat. Embriol.*

Each of them differentiates into a swollen end, which is the anlage of the acinus, and a narrow part destined to form a secondary duct. The chief duct rudiment is recognizable even from the earliest single-bud stage as a solid epithelial cord. A glandular tree develops very similar to that formed in vivo (Fig. 4). Acini and ducts, at first solid, acquire a cavity and show a secretion, which is mucous not only in the sublingual but also in the submandibular gland, although the latter is exclusively serous in vivo.

With a slightly different technique Grobstein (99) obtained development of the gland even when explantation is carried out at the earlier stage of 12 days, when it is a primary bud. However, in this case no secretion appeared, unless the explants were grafted, after some time in culture, in the anterior eye chamber.

Formation of villi was observed in intestine cultures from rabbit embryos of 13 to 17 mm. (Chlopin, 33). Fragments of chick intestine explanted before hatching and reversed so that the epithelium is turned outside, become transparent spherules covered by villi (Fischer, 64). These disappear later in culture unless in a liquid medium.

Peristaltic activity is seen in cultures of fragments of chick embryo intestine, explanted earlier, for instance at the 11th day (Bisceglie, 9). Even earlier stages were utilized by Gozzi (94), who noted that contractions start after 12 to 24 hours when the explantation has been made at the 4th day, but start at once if the explant comes from a 5-day embryo. Contractions begin before the appearance of myofibrils.

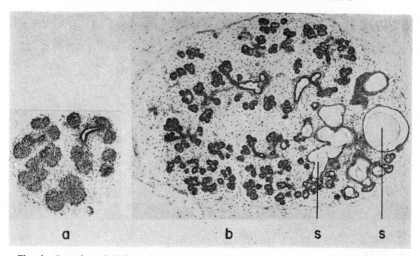

Fig. 4. Growth and differentiation in vitro of a glandular rudiment. a, Submandibular gland of a 16-day mouse embryo, consisting of some solid acini and a hollow excretory duct. b, Submandibular gland of an embryo of the same litter, after 5 days of cultivation. The gland has enlarged, the acini are very numerous and some have a cavity; secreted material (s) is present in a secondary duct as well as in the main duct. Sections. × 60. From Borghese (13), by courtesy of *J. Anat., Lond.*

According to De Jong and De Haan (39), intestinal contractions begin when the organ, explanted at 4 to 5 days of incubation, has reached a total age of 7 days. They seem to be neurogenic, as they appear only when autonomic nervous tissue has differentiated.

Keuning (129), using a less traumatizing technique, saw the first contractions appear much earlier, after 24 to 36 hours, in cultures of 4-day intestine. He would rather admit a myogenic nature of the first contractions, which appear while muscles and nervous cells are being formed from a common mesenchymal blastema.

In cultures of esophagus-stomach-trachea complexes explanted from chick embryos at the 4th day, Keuning (128, 129) studied the origin of autonomic nerve cells. During culture, some cells still belonging to a mesenchymal syncytium begin to emit processes. They are first isolated, then aggregate with others arising in the same way; silver impregnation reveals neurofibrils in their cytoplasm. From this moment on they should be considered as neuroblasts; later they separate completely from the mesenchyme (Fig. 5, a, b, c).

No difference in formation of such nerve cells was seen, whether the culture contains a vagus ganglion or not. Later on, in any case, the neuroblasts form a plexus which after 4 days of culture has moved toward the muscular cells and become connected with them through the interposition of a mesenchymal cellular syncytium. Among the various theories concerning the origin of peripheral autonomic nerve cells, viz., migration from the vagus ganglion, origin from the neural crest, or formation in loco from mesenchymal cells, Keuning (129) believes that his findings exclude the first and he seems to incline toward the last. However, the second is not altogether excluded, and in any case is not incompatible with the last, as it is known also that mesenchymal cells may derive from the neural crest. It remains open whether some of these are from the beginning determined to form nerve cells.

Liver fragments from 13 to 15-day embryos (Chen, 31) continue to differentiate for 6 to 15 days after explantation. Trabeculae of epithelial cells become more and more regular. Hemopoietic tissue tends to disappear, as in vivo. In liver explants of 12-day mouse embryos (Borghese, 20) cultivated for 4 days, epithelial differentiation goes on in vitro, but the hemopoietic tissue appears to undergo massive degeneration rather than to follow a normal developmental pattern.

In the pancreas of the chick embryo, explanted between the 6th and 10th days (Black and Comolli, 10), acini already formed go on branching repeatedly. The undifferentiated part of the gland is formed by cellular masses without cavities, some similar to normal islets, others appearing intermediate in character between endocrine and exocrine tissue. If cultures are made from more advanced stages, at the 11th to 14th days, when islets are already present, more and more islets are formed from the undifferentiated masses. In pancreas from 13 to 15 day rat embryos (Chen 31) new tubules and acini are formed, in which zymogen granules appear: in addition, a number of islets differentiate from lateral buds of the ducts.

In explants of the primitive intestinal tube of chick embryos taken at the 4th-5th day of incubation and consisting only of epithelium and condensed mesenchyme (De Jong and De Haan, 39), trachea and bronchi develop and form discontinuous rings of cartilage.

In cultures of material explanted from more advanced rudiments, from the 7th to the 11th day of incubation (Loffredo Sampaolo and Sampaolo, 153, 154), cartilages of the trachea and syrinx are formed, or continue their development if already present; and a normal pattern of syrinx cartilages becomes recognizable. The chick lung, which at the 7th day of embryonic life looks like a tubular gland with the tubules covered by a monostratified columnar epithelium, does not change its structure after cultivation for a maximum of 15 days except for the formation of folds of epithelium and

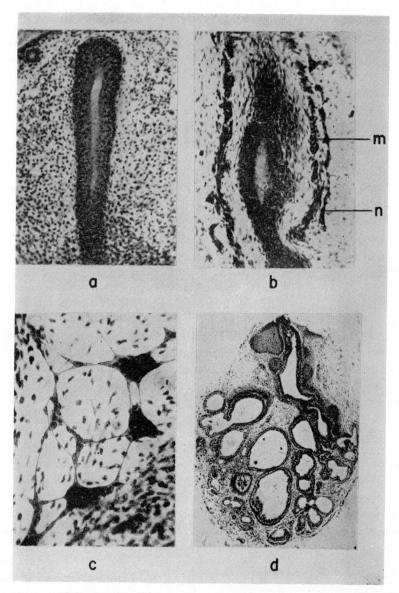

Fig. 5. Differentiation in vitro of parts of the digestive and respiratory apparatus. a, Esophagus anlage of a 4-day chick embryo, cultivated for 12 hours. The epithelial tube is still contained in undifferentiated mesenchyme. Transverse section. $\times$ 115. b, Esophagus anlage as in a, cultivated for 4½ days. In the mesenchyme a nearly continuous muscular layer ($m$) has been formed and groups of neuroblasts ($n$) have appeared. Transverse section. $\times$ 140. c, Detail of the mesenchymal field of a culture as in b, showing neuroblast groups, stained with Bodian technique. Tangential section. $\times$ 325. From Keuning (129), by

epithelial tufts. Some cells show in sections a club-like shape which is interpreted as a morphological sign of movement.

Lung rudiments isolated from the 10-day rat embryo (Chen, 31), when they consist of trachea and primary bronchi with two or three buds, go on branching dichotomously in vitro and form in 6 days a complex tree-like structure (Fig. 5, d).

## Heart

Heart was explanted as a whole from chick embryos at 7 to 11 days of incubation (Loffredo Sampaolo, 151, 152) and cultivated for as long as 15 days. Muscular fibers were seen to continue their differentiation, becoming longer and thinner, whereas myofibrils acquired a nearly complete striation.

## Urogenital Apparatus

Metanephric rudiments of the chick embryo explanted at the 6th to 8th days continue their development in vitro (Rienhoff, 196). Collecting tubules grow longer and branch, whereas metanephrogenic mesenchyme divides into masses, each of them giving origin to coiled tubules. Their formation can be followed under the microscope. The compact cellular mass is at the beginning comma-shaped, where the tail bends to form two loops. In the loop nearer to the head of the comma the lumen appears as the cells move apart, the lumen extends further in both directions, and becomes at the proximal end the cavity of Bowman's capsule. In its contact, in spite of lack of any blood circulation, glomerular vessels and islands of hemopoietic cells appear. The connection between the convoluted tubule and the collecting one may be seen to take place in vitro.

Chick embryo ovaries, explanted at 10 to 12 days, when all germ cells are at the oogonial stage (Galgano, 81) contain, after 4 days of culture, ovocytes at the leptotene and synaptene stages. This seems to show that the hypophysial hormone has no influence in early ovocyte differentiation, and this conclusion is confirmed by other cultures in which, in order to eliminate the small amount of hormone contained in plasma, this was diluted (Galgano, 82), or the cultures were made without plasma in an extract of decapitated, 7-day chick embryos.

---

courtesy of *Acta neerl. Morph.* d, Lung rudiment explanted from a 10-day rat embryo and cultivated for 6 days. A bronchial tree has formed and cartilage rings have differentiated around the trachea. Longitudinal section. × 36. From Chen (31), by courtesy of *Exptl. Cell Research.*

When cultures are explanted at the 17th day, the ovocytes in 4 days (Galgano, 83) reach the pachytene and even the strepsinema stages.

Duck gonads explanted before sexual differentiation (Wolff and Haffen, 273, 277) at 8 to 9 days of incubation and cultivated for 4 to 6 days, differentiate according to their genetic sex. If this is male, the columnar germinal epithelium turns flat, and medullary cords acquire a testicular appearance, developing Sertoli cells and spermatogonia. If female, left gonads become ovaries: the epithelium changes to a cortex with high cells and develops ovogonia, whereas the medullary cords become a lacunar tissue with necrotic remnants. Right female gonads, on the other hand, undergo regression and turn sterile. However, sometimes it may occur that male gonads and the left female ones differentiate well so far as their sex is concerned, but remain sterile, i.e., without gonocytes.

Duck gonads explanted at the 10th day, when sexual differentiation has already taken place (Wolff and Haffen, 273) continue their development in vitro. Testicular cords grow, a Sertoli epithelium is formed, spermatogonia multiply. Left female gonads produce a typical cortex with ovogonia: medullary cells are often necrotic and lacunae are bigger than in vivo. Right female gonads undergo the normal regression.

Chick gonads (Wolff and Haffen, 281) behave differently from those of the duck. When explanted before sexual differentiation, between the 6th and 8th days, they develop into testes if they are genetically male. If they are female, the left ones form a normal cortex, but the medulla produces testicular cords, with lumen and Sertoli epithelium; in this way an ovotestis is produced, although with scarce gonocytes. The right female gonad turns into a testis. This abnormal behavior may be explained as due to an alteration *in vitro* of the normal hormonal balance of the gonads resulting in a lack of female hormone, which normally inhibits the testicular development of the medulla in the female.

In fragments of mouse testes explanted from the 13th to the 18th day of gestation (Borghese and Venini, 21) the tubules develop their convoluted shape and spermatogonia differentiate in vitro. If explanted in later stages and cultivated longer (Martinovitch, 158, 160) spermatogonia may give origin to spermatocytes which commence meiosis and reach the diplotene stage.

Rat or mouse ovaries, explanted at 15 to 16 days of gestation (Martinovitch, 159) when ovogonia and ovocytes are present, may be cultivated as long as 30 days. Meiosis begins and goes on as far as the diplotene stage (Fig. 6, a). Then, similarly to what occurs in vivo, a long period begins during which the chromatin acquires a "resting" appearance and some ovocytes commence the growth process which lasts till the 18th day after explantation, i.e., to an age corresponding to the 13th day of postnatal life. Then the ovocytes degenerate and are replaced by a new wave. Such growth, in successive waves, is known to occur also in vivo. If cultures are started from later stages, between the 17th day and birth (Martinovitch, 158, 159), the differentiation of the ovary proceeds further, and after two weeks primary follicles are found.

If the cultures are incubated at 34°C (Martinovitch, 161) the growth of the explant is slowed down but the rate of differentiation of germ cells is not altered. The ovary may be cultivated longer, as long as 70 to 80 days or more, and a greater number of growth waves are observed; and even the meiotic metaphase may be reached. In rat cultures egg cells survived as long as 103 days, indirectly favoring the hypothesis that also in vivo germ cells may be accumulated and kept in reserve, waiting to grow when required.

Mouse ovaries explanted at the early stages of 12 to 13 days (Wolff, 262) show the progress of differentiation. Ovogenesis follows its normal course and after 5 days of culture the maturation of ovocytes begins. When explanted between the 13th and 18th days (Borghese and Venini, 21) the differentiation goes further and may reach the stage of primary follicles.

If mice gonads are explanted at the 10 to 11-day stage, when sex is not yet recognizable (Wolff, 262), they differentiate into testes and ovaries, with normal structure of the somatic cells; but the germ cells degenerate. This would indicate that factors of sexual differentiation are already present, but the culture medium appears to be in some respect unfavorable for the development of germ cells. On the other hand, such degeneration does not occur if they are explanted at the 12th day (Borghese, 20).

A different behavior was described for human ovary. Fragments of ovarial cortex explanted from fetuses of 16 to 36 weeks become covered (Gaillard, 71), when in vitro, by a cuboidal or columnar epithelium; the central parenchyma degenerates first, so that only mesenchymal cells are left. Succesively from the epithelium, cellular cords proliferate toward the inside of the organ. In material of the early stages, after 4 to 6 days a great many ovocytes are recognizable in these cords (Fig. 6, b), presumably derived from the apparently undifferentiated cells of the epithelium. All stages of meiotic prophase develop, and even metaphases and primary follicles with a layer of covering flat or cuboidal cells. If the explants come from fetuses of the more advanced stages, central degeneration may affect only

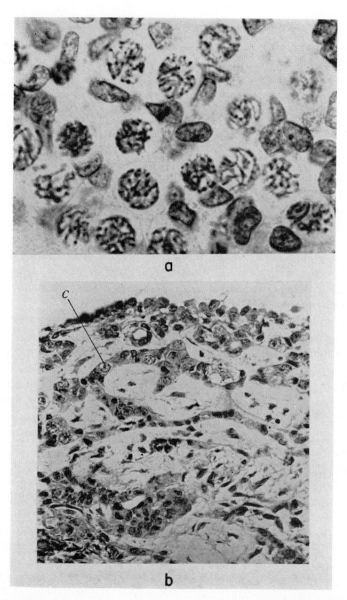

Fig. 6. Differentiation in vitro of mammalian ovary. a, Rat ovary explanted at the 16th day of gestation, when no meioses are yet present, and cultivated for 6 days. Most ovocytes are in pachytene stage. Section. × 1200. From Martinovitch (159), by courtesy of *Proc. Roy. Soc.* b, Portion of caudal pole of an ovary explanted from a 12-week human fetus and cultivated for 9 days. The cellular cords containing ovocytes (c) have been regenerated in culture from the superficial epithelium. Section. From Gaillard (73), by courtesy of *Intern. Rev. Cytol.*

germ cells, whereas follicular cells are preserved. From these new ovocytes may derive, which go through meiotic prophase and reach anaphase. Also in this case, however, no follicles go beyond the one-layered stage.

*Endocrine Glands*

The *thyroid gland* of the chick embryo, explanted at the 8th day of incubation when it is formed of solid cellular cords (Carpenter, 25, 26), after two days in vitro show colloid granules. At the third day the granules come together and form primitive follicles, similar to those present at the 11th day of development in vivo. In the following days follicles increase in number and size, through multiplication of cells and fusion of adjacent follicles. Similar events were seen in thyroids of 20-day rat embryos cultivated in vitro (Chen, 31).

The physiological activity of cultivated chick thyroid, such as capacity to concentrate iodine selectively, has been repeatedly studied with the aid of radioactive $^{131}$I, added to the medium. In such conditions, as Gaillard (70, 73) showed by means of autoradiography, and Carpenter, Bettie, and Chambers (28) both in the same way and also with the counting technique, absorption of iodine was proved to increase in thyroid glands explanted at the 8th day and grown for 8 days in vitro, at the same rate as it would have occurred in vivo.

Gonzales (92) supplied $^{131}$I to the thyroid of 7-day chick embryos, 24 hours after the beginning of the culture, then extracted the cultures after 1 to 5 days, and made chromatograms. The result was that monoiodothyronine and inorganic iodine are recognizable after one day of culture, diiodothyronine and thyroxine after 2 days. As far as this stage the behavior is as in vivo; afterwards such substances diminish. However, if the roller tube technique is used (Gonzales, 93) the synthetic activity is more intense and lasting. Chromatography shows after three days of culture monoiodothyronine, diiodothyronine, and thyroxine. The production rate grows in time and reaches a peak at 6 days of culture. Some production, although reduced, is still demonstrable at the 14th day.

*Hypophysis* rudiments explanted from 6-day chick embryos, at the vesicular stage with a wall of columnar epithelium (Moscona and Moscona, 181) form epithelial cords in vitro; after 8 days of culture the first granules in acidophilic and basophilic cells may be seen, similar in shape and distribution to those appearing in vivo at the same age. The Hotchkiss positive reaction in the basophilic cells seems to prove a secretory activity.

The hypophysis of 20-day rat embryos (Chen, 31), already containing acidophilic and basophilic cells, remains healthy and shows little change in 13 days of cultivation.

The *pineal body* explanted from the 4-day chick embryo, at the stage of epithelial evagination (Vidmar, 245), gives origin in culture to epithelial buds and to follicles. The differentiation goes on normally and after 7 to 10 days in vitro two types of cells are recognizable, high columnar ones and small ones, situated around the bases of the former. Also if explanted at 9 to 12 days, a regular differentiation occurs.

*Skin and Adnexa*

Skin of the chick embryo, explanted at the 6th day, when the epidermis is one- or two-layered (Miszurski, 172), gives origin in a month of culture to a multistratified epidermis, where tonofibrillae, eleidin, and keratin are produced.

Feather buds of the chick embryo already emerged from the skin explanted from the 7th to the 17th day of culture grow in length through proliferation of apical cells (Rossi and Borghese, 197, 198). Feather papillae appear in skin cultivated in vitro (Lepori, 150) explanted from chick and duck embryos of 5 to 6 days, when their rudiments have not yet appeared. In explants of 7 to 14-day embryos, where they are already present, they go on differentiating. However, according to Sengel (208, 209), feather buds develop in vitro from chick embryo skin only when the first dermal buds are already present at explantation time, i.e., after $6\frac{1}{2}$ days of incubation.

The uropygial gland of the duck develops in vitro (Gomot, 90) from 8 to 11-day explants. Epithelial invagination appears in culture, and alkaline phosphatase may be shown histochemically in the mesenchyme.

In fragments of guinea pig embryo lips, cultivated in vitro, hairs continue their development (Strangeways, 229). Hairs of rat embryos (Murray, 183) develop in vitro from the first stages as far as their emergence through the canal, and form pigment.

The skin of mouse embryos, including ectoderm and mesenchyme explanted between the 10th and 18th days of gestation (Hardy, 115), produces in vitro hairs which go through complete histogenesis, from the appearance of the first epidermal buds as far as the emergence of hairs at the surface. In the hair cortex and cuticular scales are formed, and a complete root with outer sheath. Henle's and Huxley's layers and the cuticle of the inner sheath are recognizable. If a pigmented breed of mouse is used, pigment is formed in melanophores and as granules in epithelial cells. Rudiments of sebaceous glands appear, but do not reach a complete development. In mesenchyme smooth muscular fibers are found, but they do not adhere to the follicle and do not form a true *arrector pili*.

Tactile hairs or vibrissae (Hardy, 117; Davidson and Hardy, 36) develop in culture from explants of 11- to 15-day embryos and follow a substantially similar behavior as trunk hairs, from epithelial buds as far as the outgrowth from the skin surface of keratinized hairs (Fig. 7, a). Some features which distinguish them from the trunk hair appear also in vitro: greater size from the beginning, an hour-glass shape of the external root sheath (Fig. 7, a), a greater density of the dermal sheath. This result proves them to be intrinsic characters, not depending on the special blood and nerve supply of the zone. On the other hand, features which do not appear in culture may be due to extrinsic factors: presence of blood sinuses, of special nervous endings, of *arrectores pilorum*.

Primary buds of the mammary glands develop in body ventral wall of mice embryos explanted at the 10th day (Balinsky, 3). The buds appear in one to three days of culture, as in vivo or with a slight retardation. If explanted from the 11th to the 14th day, when the rudiments are already present, they go on branching. After the cavity in the primary sprouts is formed (Hardy, 116) secondary and later buds appear, which in turn acquire a lumen, so that a complete canalization occurs as far as the terminal sprouts. The growth is slow in the first 6 days of culture (Balinsky, 4) and then becomes quicker; the same occurs in vivo.

The behavior in culture of the mammary gland shows that such early stages develop without requiring sexual or hypophysial hormones.

Brown fat body develops in culture of the dorsal interscapular area explanted from 15 to 17-day rat embryos (Sidman, 212). In the explants of 15 days, containing loose mesenchyme only, many mesenchymal cells retract their ameboid processes and aggregate in small islands clearly demarcated. If the explantation takes place at $16\frac{1}{2}$ days, when morphologically differentiated fat cells are already present, they grow larger and deposit lipids, but no glycogen. In 18-day explants glycogen is formed, but only when serum or glucose is contained in the medium.

The droplets of neutral fat, present in the cytoplasm at explantation time, become larger and after 2 days in vitro coalesce to form typical multilocular adipose cells.

The fact that brown adipose tissue differentiates in vitro from interscapular mesenchyme in 15-day embryos, suggests that such tissue is already determined in the mesenchyme of a specific zone before appearing morphologically and should therefore be considered a specific tissue.

In synthetic media a progressive differentiation of typical brown adipose

726  THE CHEMICAL BASIS OF DEVELOPMENT

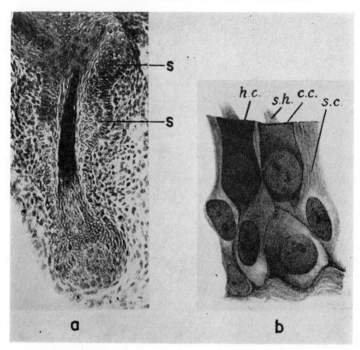

Fig. 7. Differentiation in vitro of skin adnexa and sense organs. a, Vibrissa follicle developed in lip skin of mouse embryo, explanted at the 15th day and cultivated for 6 days. In culture the outer root sheath developed two swellings (*s*) so as to acquire the hour-glass shape which is characteristic of the vibrissae in contradistinction to ordinary hairs. Longitudinal section. × 200. From Davidson and Hardy (36), by courtesy of *J. Anat., Lond.*
b, Detail of an otocyst of a chick embryo, explanted at the 3rd day of incubation as a simple vesicle with uniform epithelium and cultivated for 12 days. Two areas of sensory epithelium have differentiated, one of which, here reproduced, is similar to that of the organ of Corti, as it contains supporting cells (*s.c.*) hair cells (*h.c.*) with cellular cuticle (*c.c.*) and sensory hairs (*s.h.*) Section. From Fell (43), by courtesy of *Arch. exp. Zellforsch.*

tissue takes place, whereas in serum the tissue becomes more similar to white adipose tissue. This would suggest that no basic difference exists between brown and white adipose tissue.

*Sense Organs*

The otocyst of the chick embryo, explanted at the 3rd day, when it is an oval sac, lined by uniform epithelium and provided with a small appendix representing the rudiment of the endolymphatic duct, continues its histological differentiation in vitro (Fell, 43). The epithelium of endolymphatic duct grows and becomes folded; that of the otocyst contains, after 5 days, sensory cells. After 8 days all sensory components of the labyrinth can be seen: elongated cells with long cilia, similar to those of a normal *crista* or *macula acustica,* and cells similar to those of the organ of Corti, with short hairs covered by a membrane with fibrillar structure, resembling the tectorial membrane (Fig. 7, b). In any group of sensory cells, supporting cells are also present. After 11 to 12 days histological differentiation is nearly complete, and sometimes an amorphous mass enclosing the sensory hairs may be seen as it normally occurs in the otolithic membrane or ampullar cupula. After 14 days in vitro the histological structure of the parts of labyrinth is very similar to that of a normal embryo of 17 days. However, no corresponding morphological development occurs.

Eye rudiments from 64 to 72-hour chick embryos were cultivated by Strangeways and Fell (231). At explantation time the optic cup shows its external layer, sometimes already pigmented, and an internal, partially stratified layer; the lens vesicle is nearly or completely closed. In vitro a great amount of pigment is formed. After two days of culture the optical portion of the retina becomes recognizable from the *pars coeca* and differentiates in 8 to 9 days into the *membrana limitans interna,* nervous cells, layer of fibers, internal plexiform layer, and internal granular layer. After 13 days also cones and rods and the external plexiform layer are formed. The lens frequently degenerates. In contradistinction to such nearly normal histogenesis, the external shape is grossly altered and soon becomes unrecognizable.

A similar very regular stratification of the retina, accompanied by an alteration of form, was found in the cultures of Dorris (41), developing from whole eye rudiments or parts of them explanted at 1 to 7 days.

Good retinal differentiations were obtained also in cultures of chick eye rudiments made by Harrison (119), starting from the 14 to 22-somite stage.

Reinbold (195) cultivated the eyes of chick embryos explanted at the 3rd day, when a non-pigmented optic cup is present, or at the 4th day when some pigment has developed. In the first 5 days in vitro the eye grows, pigmentation appears or progresses, but only in contact with the air. After 5 days the eye does not grow any more, but the differentiation continues for 10 to 13 days while a regular stratification of the retina occurs and an iris rudiment is formed.

Prescleral mesenchyme of the chick, explanted from the 4th day onward, produces in vitro (Weiss and Amprino, 252) scleral cartilage which proves capable of self-differentiation inasmuch as it develops its characteristic features. On the other hand, if it is explanted at the 3rd day, no cartilage is formed. The architecture of scleral cartilage may be modified under the action of tensions applied in vitro. If cartilage has not yet appeared in the explants, a strong tension causes a thin lamina to be formed, one in which no cartilage is produced, but connective tissue only. If tension is moderate, cartilage is formed, but with a small number of cells and a reduced matrix. Even internal tensions of scleral cartilage influence its architecture, and produce at the convex surface a thick perichondrium, with tangentially directed fibers and cells, and at the concave surface a loose perichondrium with fibers radially oriented.

## Regulation in Explants

In most organs so far studied, development in vitro has occurred according to the prospective potencies of the organ explanted and of its parts, and no regulative tendency has appeared. In other words, their "mosaic" behavior has proved that their determination had become irreversible at explantation time.

A few exceptions, however, were found. Some degree of regulation, more of a histological than of a morphological nature, was found in very early cartilage stages (Niven, 187, 188). Femur and tibia cartilages of chick embryos, cut at the center of the shaft and cultivated with the cut surfaces in contact, fuse altogether and no trace of lesion is left, provided they are explanted not later than the 5th day. Such regenerative capacity lasts for epiphyses until the 6th day, perhaps because they are less differentiated. Meckel's cartilage also preserves in vitro until the 6th day the ability to regenerate a lesion completely. But if long bone rudiments of the chick embryos are sectioned at $5\frac{1}{2}$ days or later, when the central cells are already hypertrophied, cartilaginous cells close to the lesion degenerate and the gap is filled by osteoblasts which produce osteoid tissue. Similarly, if Meckel's cartilage is explanted at the 8th day, the repair is no longer complete and the gap is filled with fibroblasts.

When the ossification is already advanced, a partial process of repair by a connective, noncartilaginous callus may be observed in long bones of chick and mouse embryos (Weil, 251; Bucher, 23). The repair may be accelerated if an extract of bones of young animals is added to the medium (Weil, 251; Bucher and Weil, 24).

The determination of the submandibular gland of the mouse, when differentiating in vitro, has been repeatedly studied. Its early rudiment is formed by an epithelial cord and a bud, destined respectively to form the chief duct and the remaining glandular tree. If the rudiment, explanted at this age, is divided so as to separate the two epithelial parts, without diminishing the total amount of connective tissue, no regulative capacity appears (Borghese, 11). The cord becomes hollow, but does not produce acini; the bud gives origin to other buds and secondary cords but not to the chief duct (Fig. 8, a-d). However, in a later research (Grobstein, 30) it was found that, although such behavior is most frequent, some formation of buds from the cord may be observed provided that its distal end is put in contact with a sufficient amount of condensed mesenchyme of a submandibular gland. In no case was a chief duct regenerated from the bud.

An exceptional case of nearly complete regulation, in spite of a greatly advanced developmental stage, has been found in cultures of tooth rudiments. The first and second temporary molar of the rabbit (Glasstone, 87), explanted at the 20th or 21st day of gestation, namely, on the day before or on the same day in which cusps should appear, is divided into two halves, which are cultivated. Each of the two parts gives origin to a complete set of cusps, although smaller than those which are formed in a cultivated whole tooth rudiment. At the 22nd day, when the first intercuspal groove has become deeper, the regulation does not take place any more and from a half rudiment only half a tooth is formed.

The lens, as in vivo, also regenerates very easily from tissues different from those which normally produce it, as shown by Dorris (41), and van Deth (40). It should be noted, however, that not all results obtained in this connection are equivalent, as confusion is possible between true lenses and "lentoid bodies" (Danchakoff, 35) whose resemblance to lenses is only apparent.

### Rate of Growth and Differentiation

The development of explanted embryonic organs, even when morphogenesis and histogenesis are sufficiently regular, very often proceeds slower than in vivo. However, a distinction should be made (Hoadley, 122) between growth, due to cellular multiplication, which is more frequently slowed down, and differentiation, which in many organs occurs at a normal rate. Some examples are given here.

Limb-bone rudiments grow and differentiate in vitro slower than in vivo, but growth is much more retarded than is differentiation (Strangeways and Fell, 230; Momigliano Levi, 175; Fell, 44; Gliozzi, 88). Avian and mouse sternum (Fell, 46; Chen, 29, 30) always remains smaller in vitro than that of the same age normally developing in vivo; and growth lags strongly behind chondrification, so as to produce also an alteration of the normal shape.

A curious exception concerning bone rudiments was found by Zaaijer (303) in hand and foot cartilages explanted from human embryos. If they

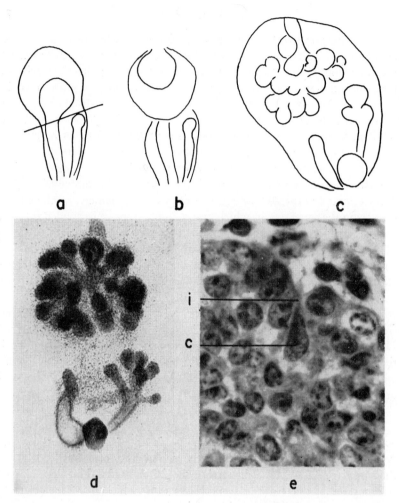

Fig. 8. a, b, c, d, "Mosaic" structure of a glandular rudiment developing in vitro. a: Rudiment of a submandibular and sublingual gland of a 13-day mouse embryo, at explantation time, when it is sectioned (a) so as to separate the bud from the cord of the submandibular gland; the former is then rotated (b) by 180° in order for the epithelial parts to remain separated, whereas the amount of connective tissue is not changed. After 2 (c) and 3 days (d) of culture, the submandibular gland has branched extensively, whereas the cord has yielded only a canal and has not regenerated any secretory part. Camera lucida drawings (a, b, c) × 39. Section. (d). × 48. From Borghese (19), by courtesy of *Monit. zool. ital.* e: Culture of a submandibular gland of a mouse explained at the 13th day of gestation and cultivated for 4 days, showing a club-like cell ($c$), presumably fixed in the act of moving inwards and producing an inflection of the surface ($i$) corresponding to the dichotomous division of a bud. Section. × 1200. From Borghese, (15) by courtesy of *Arch. ital. Anat. Embriol.*

come from the 9 to 15-week stage, they follow the general rule, growing and differentiating slower than in vivo; but if they are explanted at 6 weeks, they grow in vitro much quicker than in vivo. A similar exception, but concerning differentiation only—which takes place at an abnormally high rate—was observed in rabbit intestine by Maximow, (170) and in chick skin by Miszurski (172).

In the rat tooth a strong retardation was found not only in growth but also in differentiation, as dentine formation takes place at an exceedingly low rate. After 20 days in vitro (Glasstone, 85) no more dentine is found in rat incisors and molars than would be formed in vivo 4 days after the appearance of odontoblasts. In rat molars explanted at 17 days (Glasstone, 86) odontoblasts are not formed in less than 6 days in vitro, whereas in vivo at the same total age odontoblasts and dentine are abundant. On the other hand morphogenesis, as indicated by the number and size of the cusps, is not delayed more than 24 hours. The behavior of rabbit teeth of a corresponding age is different; odontoblasts and dentine are formed at about the same rate as in vivo; however, growth is slower and the cusps remain small.

The differentiation rate of germ cells in chick ovary (Galgano, 81; Borghese and Venini, 21) is not different in vitro from in vivo, but the size of the organ remains smaller. Growth of the egg cells in the rat ovary in vitro (Martinovitch, 159) is slightly delayed, but the total size of the ovary after one month of culture has become only twice as great as when explanted.

The rate of differentiation is slowed down in the chick heart (Loffredo Sampaolo, 151, 152).

In cultivated thyroid (Carpenter, 26) the formation of colloid occurs at the same rate or even quicker than in vivo; but a certain delay affects the appearance of follicles. As to the size of the explants, not only does it not increase in culture, but it even diminishes, a fact which is explained by the loss of connective tissue at each renewal of the medium.

The differentiation of hypophysis (Moscona and Moscona, 181) is slightly delayed in vitro: its size also decreases as in the case of the thyroid. The pineal body (Vidmar, 245) differentiates at the normal rate, but its growth is progressively delayed.

Hairs (Hardy, 115) have in vitro a differentiation rate sometimes lower and sometimes equal to that in vivo; their size remains normal until the age corresponding to birth; subsequently their growth stops, while in vivo the size is doubled or trebled in the first 15 days of postnatal life. Vibrissae (Davidson and Hardy, 36) differentiate at about the same rate as in vivo:

and their size also increases in vitro as in vivo, and much quicker than for trunk hairs. This shows that the greater growth rate which in normal conditions characterizes vibrissae in comparison with trunk hairs is intrinsic and does not depend on the milieu.

Histological differentiation of the otocyst (Fell, 43) proceeds normally in vitro. Instead, growth is strongly delayed, even if allowance is made for the fact that the real increase is greater than it externally appears, a large part of the volume being employed in abnormal folding.

Growth is retarded also in eye cultures (Strangeways and Fell, 231; Dorris, 41) although it may be normal in the first 5 days of culture (Reinbold, 195).

The different behavior of growth and differentiation in vitro shows that in development a distinction should be made between those factors acting on cellular multiplication and those which affect differentiation.

The development of the ability to assimilate amino acids of the medium has been measured by means of a quantitative analysis of the nitrogen contained in the cultivated organs (Wolff, Haffen, Kieny, and Wolff, 282, 283). Also in this respect a discrepancy between differentiation and growth was found. In some cases an alteration of the two factors was established. For example, gonads explanted before sexual differentiation differentiate regularly but grow very little; on the other hand, if explanted after such a stage has been passed through, they show a better growth. It would seem that the beginning of differentiation uses up the whole available energy of the cultured organ; when such an obstacle is left behind, growth becomes easier.

The explanations which could be given of the slowness of growth are purely a matter of speculation. A general factor could be the reduced supply of nutritive substances; for some cases, with a particularly slow rate of differentiation, such as occurs in bone or teeth, a tendency of the $pH$ of the medium to turn acid may be invoked, as it would inactivate alkaline phosphatase, the optimum $pH$ of which is 8.4-9.4 (Fell, 44).

In some cases the slowness of development in vitro has proved to be useful in interpreting some of the processes of the normal embryonic development. For instance, in bones of chicks affected by the Creeper anomaly (Fell and Landauer, 55) the delay in growth and differentiation, further artificially increased by means of a reduction of nutritive supply, sheds some light on the relationship between the hypertrophy of cartilaginous cells and the beginning of perichondral ossification. If the differentiation rate of cartilage is reduced to such an extent that hypertrophy is limited to central cells and does not reach the periphery, no perichondral bone is formed. This would favor the idea that hypertrophic cells play an important role in the first appearance of perichondral bone.

Advantage was taken of the general slowing down in the development of mouse sublingual and submandibular glands in vitro (Borghese, 15, 16) in order to investigate the elementary morphogenetic processes. Their slowness allows one to take note of some details which are likely to escape or to remain obscure when studied in normal embryo sections.

In this way two phenomena were, for example, particularly well illustrated: the appearance of club-shaped cells where a glandular bud is about to inflect at the beginning of a dichotomous division (Fig. 8, e); and the mode of lumen formation in excretory ducts through the moving apart of the cells.

MESENCHYMO-EPITHELIAL INDUCTION

The development in vitro of organs consisting of different tissues offers an experimental device to study the reciprocal influence exerted by the two tissues in collaborating in the harmonious development of an organ. Such an action may be named "inductive" in analogy to the well-known phenomena of the earlier developmental stage.

In tooth development, ameloblasts are necessary for the formation of odontoblasts (Glasstone, 85); and hair does not commence its development (Hardy, 115) if the association of ectoderm with mesenchyme is missing.

A wider investigation of the action of mesenchymal tissue on epithelium was carried out in cultures of mouse submandibular glands. As said above, if the whole rudiment of the gland, including the epithelial bud and the mesenchymal capsule, is explanted at the 13th day of gestation, the epithelial bud starts branching and gives a glandular tree. On the other hand, if the greatest part of the mesenchymal capsule is taken away (Borghese, 14, 19) the epithelial component does not branch regularly, remains undivided, and eventually is reduced to a small epithelial mass or cyst (Fig. 9). This shows that the mesenchymal capsule is necessary in order to obtain a normal morphogenetic process. However, if after most of the mesenchymal capsule is removed, and a fragment of limb bud or of spinal cord is added, the development becomes normal again. The first impression one gets is of the mesenchymal action not being specific, since different formations may be substituted for it. But it should be noted that in such a type of experiment the subtraction of mesenchyme is not total, and the tissue added may only act to stimulate the regeneration of the missing mesenchyme from its remnants.

Further progress was possible when a way was found by means of tryp-

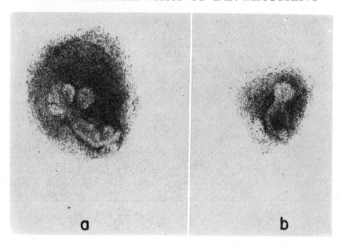

Fig. 9. Action of the mesenchymal capsule on the normal branching of the epithelial component of a glandular rudiment. a, Submandibular rudiment explanted from a 13-day mouse embryo when at the one-bud stage and cultivated for one day with intact mesenchymal capsule. The branching has gone on regularly and four adenomera have been formed. b, A similar rudiment in which most mesenchymal tissue was removed at explantation time. After one day of cultivation the epithelium has not branched. Living cultures. × 48. From Borghese (19), by courtesy of *Monit. zool. ital.*

sin to separate completely the epithelial and mesenchymal components, so as to recombine them in different ways (Moscona, 176). Epithelial rudiments of mouse submandibular glands, explanted at the 13th day of gestation and completely separated from their capsular mesenchyme (Grobstein, 100) lose all morphogenetic capacity and spread out as thin sheets; but if after being isolated they are combined with "autogenous" mesenchyme, i.e., mesenchyme derived from the submandibular capsule, they go through a regular morphogenesis. No "heterogenous" mesenchyme taken from lung, metanephros, or undifferentiated stages of younger embryos has the same action; the epithelium associated with it turns to a small, round mass or a cyst. A comparison between the autogenous and the heterogenous mesenchyme may be made in the same culture if the epithelial rudiment is placed between a piece of each. After two days the epithelium has branched only where it is in contact with the autogenous element. The latter preserves its inductive capacity even if cultivated for 10 to 20 days by itself and then associated with a fresh epithelial rudiment; however, its action in this case is weaker, and the branching takes place in a slower, less regular way.

A similar inductive interaction may be shown between epithelium and mesenchyme of the metanephros. In this organ two inductive actions occur in opposite directions. The mesenchyme represented by the caudal end of the nephrogenic cord acts on the epithelium of the ureteral bud, which forms the collecting tubules; on the other hand, the epithelium induces in the mesenchyme the formation of coiled tubules. If the epithelial rudiments of the metanephros of the 11-day mouse embryo are separated from its mesenchyme by means of trypsin and are combined with it in culture (Grobstein, 101), morphogenesis occurs regularly and both collecting and coiled tubules arise. On the other hand, epithelial morphogenesis does not occur if epithelium is associated with heterogenous, submandibular mesenchyme. This result shows that the effect of metanephric mesenchyme is specific, as it cannot be replaced by any other mesenchyme. On the contrary, nephrogenic cord associated with submandibular epithelium continues its development and produces coiled tubules. This shows that whereas "autogenous" mesenchyme is necessary for normal morphogenesis of the two epithelia, the coiled tubules may be induced also by a non-specific tissue, such as mandibular epithelium and even, as further research has proved, by spinal cord (Grobstein, 102, 104).

Spinal cord was discovered to have also another inductive action (Grobstein and Parker, 110). Somite mesoderm explanted from mouse embryos at the 12 to 24-somite stage, and cultivated alone, remains homogeneous; but if it is combined with spinal cord, it produces in culture, cartilage nodules (Fig. 10, a), which may successively turn to bone if implanted in the anterior eye chamber. Differently from the two inductive systems reported above, in this case induction occurs at some distance from the inductor.

Grobstein and Holtzer (109) showed that the cartilage-promoting activity belongs also to cerebral parts of the nervous system, but not to other tissues, whether alive or killed. The chondrogenic activity of spinal cord is strong from the 9th to the 11th day, declines remarkably in the following days, and disappears altogether at the 15th to 16th days. It is limited to the ventral part, corresponding to the original basal plate, and is lacking in the dorsal part, derived from the alar plate. On the contrary the latter, which is from the beginning associated with the lateral axial mesoderm, is the most active in inducing renal tubules.

A reciprocal alternation between mesenchymal and epithelial inducing activity was found in feather buds of chick embryos at the 6th and 7th days of incubation. Sengel (208, 209), by means of trypsin, separated the epithelial and the dermal rudiments and recombined them so as to associate epidermis and dermis of different ages. It resulted that as early as the

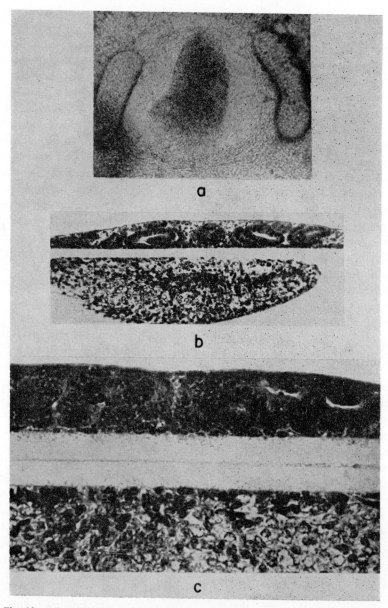

Fig. 10. Inductive action of the spinal cord. a, Ventral half of spinal cord of a 9-day mouse embryo, associated in culture with somites of an embryo of the same age and cultivated for 6 days. Two cartilaginous nodules have been formed in the somite tissue, separated from the spinal cord by rarefied areas, through which the inductive influence seems to have spread. Living culture. × 40. From Grobstein and Holtzer (109), by courtesy of *J. exp. Zool.* b, Spinal cord (below) has induced coiled tubules in metanephrogenic mesenchyme

6th day no determination is yet present in the dermis, although dermal buds may appear under the action of a non-specific inductor, such as neural tube. When the dermal bud is formed, at the beginning of the 7th day, it becomes an induction center, and acts on the overlying epidermis, which thickens so that a complete feather bud is formed. The inducing activity of the dermis is soon lost, but it is acquired by the epidermis, and acts in the reverse sense. Actually, if epidermis taken at the end of the 7th day is combined with a not yet differentiated 6th day dermis, a feather may be formed.

Epithelial or mesenchymal rudiments of duck uropygial gland, explanted from embryos of 9 to 11 days (Gomot, 91; Lutz and Gomot, 155) and separated from each other, do not undergo any morphogenesis in vitro. But caudal mesenchyme, or its subendodermal portion only, may induct glandular invagination in ectoderm taken from any other part of the body. On the other hand, caudal ectoderm, transferred onto dorsal mesenchyme, loses its ability to form a gland and contributes instead to feather buds. If a mesenchymal tissue from another part of the body, and not yet containing feather buds, is placed between caudal mesenchyme and ectoderm, the inducing activity crosses the interposed tissue, and epithelial invagination takes place.

Something more about the inducing agent was learned through researches done by means of a porous membrane filter, which, while preventing direct contact between two tissues, does not under certain conditions stop the passage of the agent. A very favorable system for this research proved to be the mouse spinal cord associated with metanephrogenic mesenchyme. In the latter, kidney tubules are induced and show a pattern quite comparable to what develops in direct contact with the inductor.

If membrane filters about 20-30 $\mu$ thick and of two different porosities are used, the inducing activity from spinal cord can be shown to cross regularly the filter with the higher porosity, of about 0.8 $\mu$ (Fig. 10, b), whereas some reduction of the effect, as indicated by a lower number of the tubules induced, occurs through a filter of a porosity of about 0.45 $\mu$ (Grobstein, 106). If the filters are sectioned and studied by means of phase contrast microscopy, it is revealed that no cell has crossed the filter, although cyto-

---

(above) through a membrane filter (white space in the middle). Section. $\times$ 160. From Grobstein (106), by courtesy of *Exptl. Cell Research*. c, A similar experiment where two filters of porosity 0.8$\mu$ were superimposed so as to reach a total thickness of 40$\mu$. Cellular processes are seen to protrude from both tissues to approximately one-third of the thickness of the filter. 3 days of cultivation. $\times$ 275. From Grobstein (107), by courtesy of *Exptl. Cell Research*.

plasmic processes and granular flocculent material may be seen to penetrate it to some extent.

When the thickness of the membrane filter is increased (Fig. 10, c), the transmission of activity undergoes a gradual reduction between 30 and 80 $\mu$ of thickness, and is eliminated altogether at the latter distance (Grobstein, 107). This result favors the idea that the inductive effect of spinal cord on metanephrogenic mesenchyme does not depend on cytoplasmic contact, since induction still occurs through a total thickness of 60 $\mu$ of a filter, of a porosity around 0.45 $\mu$, and at a distance at which the sections show that cytoplasmic contact is surely excluded.

If a third type of membrane filter of the lowest porosity of 0.1 $\mu$ is used, in comparison with the two previously employed (Grobstein and Dalton, 108), the activity is shown to cross regularly filters also of this porosity, provided that they are no more than 20 $\mu$ thick, but only occasionally and weakly when the thickness is 30-40 $\mu$. The filters used, studied by the electron microscope, show that whereas all possibility of a cytoplasmic contact is not excluded for the highest porosity filters (0.8 $\mu$) the penetration of cellular processes does not go beyond a third of the thickness of the filter of medium porosity (0.45 $\mu$), so that intermingling and contact of cytoplasm does not occur regularly. With the filter of lowest porosity (0.1 $\mu$), cytoplasmic penetration is exceedingly shallow and the possibility of cytoplasmic contact is practically excluded. It may be concluded that induction of tubules by action of spinal cord may take place in the absence of cytoplasmic contact between the interacting layers, as shown by the fact that the results are practically alike with the filters of highest and medium porosity in spite of the strong reduction of cytoplasmic penetration in the latter; and this is further confirmed by the results of the filters of lowest porosity, where induction, although reduced, may occur although no cytoplasmic penetration is present in any case.

It is clear that the inductive influence is at least potentially extracytoplasmic. As the pores prevent the passage of cells but should not hinder dissociated molecules from going through, Grobstein (100) suggests as an intermediary mechanism the transfer of big molecules containing glycoprotein, as a cohesive strictly associated with the cells; in other words, it may provisionally be supposed that the inductive agent is contained in a sort of intercellular matrix or ground substance.

### Interaction between Different Organs

Two different organs of embryos of the same species, of equal or different ages, and even organs belonging to distinct species, can be associated in vitro. The organs unite, and do not reveal any incompatibility, even when they belong to distinct zoological groups, such as birds and mammals. Many problems can be faced with the aid of this type of culture. So far, data on morphological reciprocal behavior, and on hormonal action, are available.

*Morphological Behavior*

Bresch (22), Wolff and Bresch (272), and Wolff (265) have combined in different ways, two by two, mesonephros, liver, lung, spleen, and thyroid, explanted from chick embryos between the 6th and the 10th day of incubation. In such "heterogeneous" combinations connective tissue and epithelium of the two organs tend to intermingle, sometimes one organ overgrows the other, and liver tissue penetrates very actively into the mesenchyme of the other organ. Tissues of the two organs may contribute to form common structures: bronchial canals may be completed by mesonephric or thyroid cells of the hepatic parenchyma, which acquire thereby an epithelial appearance. In addition, in the vicinity of hollow organs, such as a mesonephric or bronchial canal, organs still consisting of solid cords, e.g., liver and thyroid, occasionally develop a lumen. This feature can be interpreted as an induction effect exerted by the more differentiated hollow organ.

In the "anachronic" associations of lung and liver, or lung and thyroid, of chick embryos of different ages, performed by Vakaet (244), bronchial lumina appear to remain continuous or to repair the lesions with their own cells, which can in contact with solid cords of the other organ flatten out and acquire an endothelioid aspect. No cells of the other organs were seen to take part in the reorganization of bronchial walls, although once a continuation was found between a cavity excavated in liver tissue and a bronchial lumen. Hepatic digitations became insinuated under the basement membrane of bronchial canaliculi. Solid hepatic or thyroid cords occasionally acquired a cavity in the vicinity of bronchial canals, perhaps following a sort of assimilative induction.

"Xenoplastic" associations, where organs of chick or duck were united with mouse organs, did not reveal any incompatibility and provided some information about the affinities of tissues of similar types. The cells belonging to the two species remain usually recognizable in such "chimeras" because of their different cytological properties, such as nuclear size or stainability. In associations of gonads of 8 to 11-day chick embryos and 13 to 16-day mouse embryos, performed by Wolff and Weniger (296, 297), connective tissue cells of mouse testes migrated inside the duck testis, especially following the basement membranes of the epithelia; seminiferous tubules of the two species communicated and formed a Sertoli epithelium where duck and mouse cells took part together and spermatogonia were intermixed. If ovaries of both species are combined, somatic cells of the mouse penetrate among nests of duck ovogonia and ovocytes; a lesser intermingling occurs between the gonocytes of the two species. When a mouse ovary is cultivated with a duck testis, mouse connective tissue cells invade the ovary, particularly insinuating themselves under the cortical epithelium of the duck organ; mouse ovocytes may be found in testicular tubules of the duck. In reverse association mouse testicular tubules may contain duck ovogonia intermingled with mouse spermatogonia (Fig. 11, b).

In the xenoplastic associations carried out by Wolff (265) and by Vakaet (244), where lungs of 8 to 10-day chick embryos were cultivated with lungs of 12 to 16-day mouse embryos, mouse bronchi branched and developed very readily in the stranger connective tissue of the chick embryo, and a collaboration of epithelia of both origins was observed in the formation of bronchial walls (Fig. 11, a).

Among the features of the tissues growing together in such associations, attention should be particularly called to certain facts (Wolff, 265, 270): the attraction of cells of different types as proved for instance by the tendency of connective tissue to spread under epithelial structures; the readiness of epithelia to grow in a foreign mesenchyme which even seems to exert, in this case also, an inductive action; the collaboration of epithelial cells even of systematically very distant species so as to form combined structures, where the mesenchyme is on the other hand excluded. This fact points out an affinity of cells of the same type, although of different species, which may correspond to a similarity also in molecular protein structure.

### Hormonal Action

Hormones, produced by embryonic organs from very early stages, act on other organs and regulate their morphological and functional differentiation. The hormone-producing organ may be combined in vitro with the receptor organ which is responsive to its action. In this way any secondary action arising from other endocrine glands or the general organism is eliminated. A further investigation may be carried out when, instead of associating the two organs, the isolated hormone is added to the medium where the receptor organ is cultivated.

In this way some hypophysial, parathyroid, and sex hormones have been studied. The study of the effect of thyroid hormones and of insulin on developing bone, although done only with hormone preparations, also belongs here.

Explants of the anterior *hypophysis* of the adult chicken, or preparations of thyreotropic hormone, added in culture to thyroid gland of 8 ½-day chick embryos (Gaillard, 73), stimulate intracytoplasmic formation of colloid, but do not influence the differentiation of follicles.

The hypophysis of 15 to 17-day chick embryos, associated in culture with 8-day thyroid (Tixier-Vidal, 236), stimulates the production of colloid. Association in culture of a 5, 6, or 7-day thyroid with a 15-day hypophysis (Tixier-Vidal, 237, 238) shows that the thyroid becomes sensitive to the thyreotropic hormone already at the 6th day, i.e., 4 days before the normal appearance of colloid. If hypophysis and thyroid of the same age, (Fig. 11, c, d) are combined, thyreotropic activity is shown to develop in culture and is demonstrated at the 7th day. The fact that in vivo colloid is not produced before the 10th day may be due to distance and resulting dilution. In the functional differentiation of the thyroid two factors are to be distinguished: an intrinsic one, the sensitivity to the thyreotropic hormone, appearing at the 6th day, and a hypophysial factor, appearing at the 7th day and gradually increasing in intensity.

The response of the thyroid is favored (Tixier-Vidal, 239) by contact with muscle and liver, but is inhibited by spleen or hypothalamus from a more advanced embryo or a newborn chick. A research on combinations of thyroid and hypophysis of the same age, from 5 to 8 days, and with and without hypothalamus (Tixier-Vidal, 240), shows that the hypophysis differentiates its thyreotropic activity in vitro and that the hypothalamus of the same age does not exert any influence on its formation.

The gonadotropic hormone has been proved to be secreted in vitro by the embryonic hypophysis. In fact, testes of 16 to 17-day chick embryos, associated in culture with the cephalic lobe of the anterior hypophysis of the same age (Moskowska, 182) show in three days cavitation of sexual cords, as well as an increase in their size and in the number of gonial mitoses.

The secretion of embryonic and new-born parathyroid glands may be demonstrated in culture conditions. These glands, isolated from human new-borns, cultivated for a week in human plasma and embryonic extract, and then transferred for 10 to 21 days in human plasma and serum, continue to produce their hormone (Gaillard, 69, 73), as is proved by their

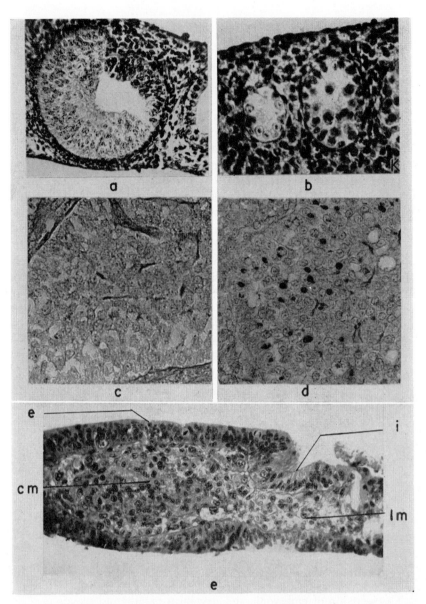

Fig. 11. Xenoplastic associations of embryonic organs of different species. a, Mouse and chick lung. A bronchiole whose epithelium consists of a part of chick cells, with light nuclei, and a part of mouse cells, with dark nuclei. Section. × 245. From Wolff (265) by courtesy of *Bull. Soc. Zool., France*. b, Duck ovary with mouse testis. In testicular tissue of mouse, two tubes are seen: right, a normal mouse testicular tube; left, a mouse testicular tube, containing some oogonia of duck. Section. × 340. From Wolff and Weniger (297), by courtesy of *J. Embryol. exptl. Morphol.*

ability to cure postoperative tetany when grafted, at this culture stage (Kooreman and Gaillard, 136), into the vascular lymphatic spaces of the axillary arteries and veins of patients suffering from this affliction.

Resorption of bone tissue, accompanied by the appearance of osteoclasts, was seen in cultures of parietal bone of mice embryos, when cultivated with parathyroid glands of human newborns or young mice (Gaillard, 74) or with parathyroid gland tissue derived from 9 1/2 to 12 1/2-day chick embryos or even with fluid expressed from the medium where such embryonic parathyroids had been cultivated (Gaillard, 75, 76). The alteration of bone structure in such conditions, the action of osteoclasts, and also the formation of new bone matrix, occurring in some sites, were investigated in the cultures by means of microcinematography (Gaillard, 78). The addition of commercial Parathormone to the medium produced bone resorption of essentially the same type (Gaillard, 75, 76).

If embryonic gonads of opposite sexes are associated in culture, it can be shown that the gonads secrete, from their earliest developmental stages, a hormone which influences their differentiation.

When left gonads of two different duck embryos, explanted at 7 to 8 days before sexual differentiation, are cultivated together for 4 to 6 days (Wolff and Haffen, 274, 278, 280; Wolff, 263), the female gonad affects the male one. The ovary develops in such cases normally, whether the combination is homosexual or heterosexual. The male gonad differentiates normally in the former case, but if united with a female one, loses its purely testicular character and becomes an *ovotestis,* with luxuriant development of the germinal epithelium, which invades the medullary part of the organ with cords containing ovogonia. On the other hand, testicular cords belonging to the medulla differentiate very weakly (Fig. 11, e). Such change is interpreted as due to the passage of a hormone from the female gonad into the male one, stimulating epithelial proliferation and partially inhibiting the testicular differentiation of the medulla. If earlier gonads of

---

Hormonic action in vitro. c, Thyreotropic hormone. Thyroid of a 6-day chick embryo, cultivated for 3 days. c, the control thyroid, associated with cardiac muscle, did not develop any colloid; d, the thyroid associated with hypophysis of the same age developed a great many colloid drops, stained by the McManus method. Section. × 425. From Tixier-Vidal (238), by courtesy of *Arch. Anat. Morphol. exptl.*

e, Sexual hormones. Association of two gonads of different sex, explanted from a 7-day duck embryo and cultivated for 6 days. The testis (left) has changed to an *ovotestis,* consisting of high epithelium (*e*) and a compact medulla (*cm*), without any differentiation into testicular cords. An inflection at the surface (*i*) indicates the transition to ovary, which is normal, with a loose medulla (*lm*) containing lacunae. Section. × 300. From Wolff and Haffen, first published in Borghese (19). By courtesy of *Monit. zool. ital.*

6 to 7 days are used, the feminizing action on the testis is even stronger, so that it becomes nearly altogether similar to an ovary. That a hormonal action is involved is proved by the fact that if 6 to 8-day prospective male gonads are cultivated in a medium where estradiol benzoate or a synthetic estrogenic hormone is added, a similar alteration of the testis in the direction of an ovary occurs. This does not take place, however, in a uniform way, but is limited to the parts of the cultured testis which come in contact with the hormone drops.

Combination in culture of the left gonad of the duck with the right one of another embryo, explanted between the 7th and the 10th day of incubation, demonstrates a hormonal activity of the female right gonad (Wolff and Haffen, 279). When, as revealed by subsequent behavior, the left gonad proved to be male and the right one female, the latter in spite of its imperfect development as an ovary, turned out to exert a feminizing action on the testis. In fact, if the explantation was made at an early stage, preceding sexual differentiation, the testis changes to an *ovotestis;* however, if at explantation time male differentiation had already begun, the resulting intersexuality is weaker. The right female gonad is not influenced by the testis and preserves its typical, sterile structure, characterized by an atrophic cortex and a loose, vacuolated medullary structure.

Sexual hormones also influence in vitro the development of the duck syrinx. This organ has in such a bird a sexual dimorphism which becomes recognizable at the 10th day of normal development. It differentiates into a symmetrical organ in female embryos, whereas in male ones its left side becomes more expanded than the right. It was already known (Wolff and Wolff, 298) that castrated embryos, whatever their genetic sex may be, acquire a male type of syrinx. A similar effect occurs in vitro. At the 7th to 8th day of embryonic life (Wolff, Wolff, and Haffen, 302; Wolff and Wolff, 299-301) the syrinx is represented by a slight swelling of the trachea, formed only by epithelium and undifferentiated mesenchyme. Explanted at this age and cultivated in the usual medium, the cartilages differentiate; and whatever the genetic sex of the donor embryo, it acquires the male asymmetrical form, with a left chamber much wider than the right one (Fig. 12). It is therefore clear that the asymmetrical syrinx, which is produced in the absence of sex hormones, either male or female, is a neutral or asexual syrinx: the differentiation in the female direction, which occurs during the normal development, is due to the ovarian hormone. In fact, if to the medium of a culture of syrinx, explanted at the same age, estradiol benzoate is added (Wolff and Wolff, 299, 300) the organ tends to acquire the symmetrical, female form.

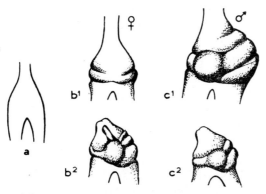

Fig. 12. Action of female hormone in the development of the duck syrinx, proved by the comparison between the behavior in vivo and that in vitro in non-hormonal medium. a, Sexually undifferentiated syrinx at the 7th day of incubation. $b^1$, $c^1$, Syringes differentiated it vivo, as they appear at the 13th day of incubation; the female one ($b^1$) is symmetrical, the male one ($c^1$) is asymmetrical. $b^2$, $c^2$, Syringes explanted at the 7th day and cultivated for 6 days. In spite of the different genetic sex, female for $b^2$ and male for $c^2$, both have developed the male asymmetrical type. Schematic drawings. From Wolff, Wolff, and Haffen (302), by courtesy of *Compt. rend. Acad. Sci.*

Presumably such an action takes place through an inhibition of development, as is shown by the poor differentiation of cartilages and the relative smallness of the organ. Even the paradoxical effect of very strong doses of male hormone, which produce a feminizing effect (Wolff and Wolff, 300) can be interpreted as a non-specific action, inasmuch as the organ would reduce its ability to assimilate nutritive substances.

If the syrinx is explanted at 9 to 12 days of incubation (Wolff, Wolff, and Haffen, 302), when it has already acquired the asymmetrical male type or the symmetrical female one, it continues its predetermined development according to its genetic sex.

The genital tubercle of duck behaves very similarly to the syrinx, both in conditions of embryonic castration (Wolff and Wolff, 298) and in vitro. (Wolff and Wolff, 261, 300). This organ, between the 7th and the 10th day of incubation, is identical in both sexes and consists of a simple swelling. Explanted between the 7th and 9th days, it differentiates in culture in any case according to the male type, and at the age corresponding to the 15th to 16th day of normal development has become a helicoid, asymmetrical organ with a spiral groove. When explanted at the 10th day, it develops according to the genetic sex of the embryo to which it belongs: if male, it becomes asymmetrical; if female, it atrophies and turns to a small symmetrical body with a straight groove. This shows that also for the genital tubercle, the male type is the neutral or asexual form, whereas the hormone emitted by the female gonad causes its atrophy. However, if it has already undergone the action of the female gonad, the development continues in this direction in spite of absence of the hormone.

Also in the differentiation of Müllerian ducts in vitro, the action of gonadal hormones is involved. Müllerian ducts, explanted from male or female rat embryos at the 15th to 16th day of gestation and incubated in non-nutritive, fluid medium (Jost and Bergerard 126), persist for 48 to 72 hours, whereas when developing in vivo in a male embryo they strongly regress in an equal period of time.

A more extensive investigation was made on the Müllerian ducts of the chick. In vivo the epithelium of the Müllerian ducts of male embryos (Wolff and Lutz-Ostertag, 287) start to atrophy at the 9th day, the age at which gonad differentiation begins; the connective tissue survives until the 13th day, after which the ducts disappear altogether. If they are explanted at about the same age and incubated in non-nutritive medium (Wolff, Ostertag, and Pfleger, 291) they do not change whether they come from male or female embryos of 8 days or from females of 9 or $9\frac{1}{2}$ days. They regress, however, if they are taken from male embryos of 9 or $9\frac{1}{2}$ days. The histological study (Wolff, Ostertag, and Pfleger, 292) shows that the regression begins at the caudal end as connective tissue and epithelial degeneration, accompanied by loss of the lumen.

Also in true culture (Wolff and Lutz-Ostertag, 287, 288); Wolff, Lutz-Ostertag, and Haffen, 289; Wolff, 263), Müllerian ducts explanted between the 7th and 9th days develop equally, regardless of the sex of the embryo, and sometimes produce at their caudal ends a dilatation similar to the shell gland. If explanted at $9\frac{1}{2}$ days, when sex is already recognizable, they go on differentiating in the same way, if they come from female embryos; whereas if they are male they degenerate beginning at the caudal end. All these facts suggest that the regression is caused by the male hormone, produced by the gonad at the beginning of sexual differentiation. Its effect, once it has occurred, persists in any case, even when ducts are cultured in vitro, and thereby abstracted from the gonadal influence.

If testosterone propionate or androsterone is added to the medium of cultures of Müllerian ducts explanted at the 7th day, when sex is not yet recognizable (Wolff and Lutz-Ostertag, 288; Wolff, Lutz-Ostertag, and Haffen, 290), they undergo a necrosis like that above described, beginning at the caudal end (Fig. 13 a, b). If the same hormone is added to Müllerian ducts explanted at 9 days and already differentiated in the female direction, necrosis occurs, but it proceeds in cephalo-caudal direction. Wolffian ducts used as controls suffer no alteration; this fact indicates that the action of the hormone is specific.

Regression of Müllerian ducts can be obtained by association in vitro with testes (Wolff, Lutz-Ostertag, and Haffen, 290). The necrosis then occurs only in immediate contact with the testis, as a local effect of the testis hormone. If on the other hand male Müllerian ducts explanted at $9\frac{1}{2}$ days are treated in culture with estradiol propionate (Wolff and Lutz-Ostertag, 288), regression does not occur regularly, as the action already exerted by the male hormone before explantation is partially counteracted by the female hormone. The duct is changed into a series of vesicles, connected by strands of mesenchyme only.

Treatment with nitrogen, carried out by Pfleger (193) on Müllerian ducts of $9\frac{1}{2}$ days, incubated in physiological glucose solution, and by Scheib-Pfleger (202, 205) on true cultures of different ages, seems to indicate the existence of a proteolytic enzyme as immediate agent of their regression.

This enzyme seems to be inactivated by ultrasonic waves. In fact, if Müllerian ducts are explanted from male embryos at $9\frac{1}{2}$ days, when they have already undergone the hormonal action, and are treated with ultrasonic waves and then cultivated (Lutz and Lutz-Ostertag, 156) their regression begins but stops very soon. Intact parts of the ducts persist, thus imitating the action of the female hormone.

Ultrasonic waves, however, do not affect the action of the male hormone. If undifferentiated Müllerian ducts of $6\frac{1}{2}$ to $7\frac{1}{2}$ days, or sexually differentiated ducts of 7 to $7\frac{1}{2}$ days, are associated with testes (Lutz and Lutz-Ostertag, 157) previously treated with ultrasonic waves, the regression produced by the male hormone does not undergo any change.

Under normal developmental conditions, in female embryos the right Müllerian duct begins to regress at the 13th day in a cephalo-caudal direction. If the ducts are explanted from a female embryo at such an age (Scheib-Pfleger, 203), the right one becomes opaque and loses in 4 days a third of its initial amount of total nitrogen. This suggests that the regression, evidently hormonally conditioned, has already begun, and isolation from the general organism does not prevent its progress further.

However, it should be noted that the left duct, although surviving for 24 to 48 hours, does not grow; neither does it go on differentiating; and it even undergoes a nitrogen loss although much less than the right one. It may therefore be supposed that in contradistinction to other organs, Müllerian female ducts cannot develop regularly when excluded, as occurs

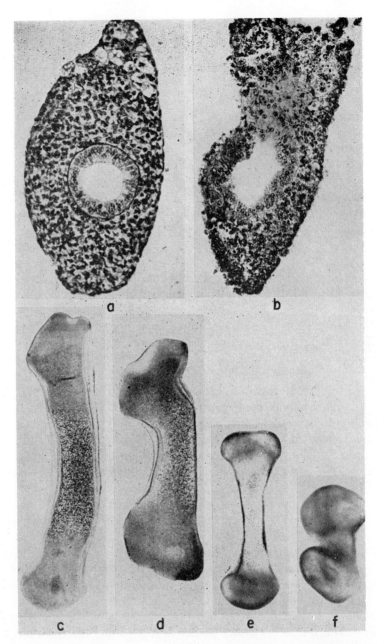

Fig. 13. Hormonal action in vitro. a, b, Action of male hormone on chick Müllerian duct, explanted at the 7th day and cultivated for 5 days in non-hormonal medium (a), where connective tissue and epithelial tissue develop normally; or in medium containing testosterone

in vitro, from a continuous supply of female hormone. In fact, the right female Müllerian duct explanted at the 13th day and cultivated in a medium without hormones (Scheib-Pfleger, 204) continues to regress in vitro and loses about one-third of its total nitrogen. If a male hormone is added, the regression and nitrogen loss are intensified and reach about half. On the other hand, an addition of female hormone to the medium exerts a protective action on the tissues and favors the survival of the organ with a gain of its total nitrogen of 7 to 17 per cent. The left female Müllerian duct, explanted at 13 days and cultivated in a medium without hormones, survives and does not show any degeneration; but it does not differentiate and it loses in 4 days 15 per cent of its nitrogen. The addition of male hormone results in its degeneration and a nitrogen loss of about 40 per cent in 4 days; the female hormone, on the contrary, favors its growth and differentiation and increases its nitrogen content in the same time by 18 to 25 per cent. It should therefore be concluded that the female hormone, although not necessary to the survival of female Müllerian ducts, prevents in vitro the regression of the right one and promotes the development of the left one. The male hormone, on the other hand retains its necrotizing action also at this relatively late stage.

The effect of *thyroxine* on bone development was tested by Fell and Mellanby (58), using limb buds of chick embryos at 4 to 7 days of incubation, i.e., at stages ranging from a purely mesenchymal condensation to that of bone rudiments with chondroblastic hypertrophy. The results differed according to the stage, for different bones. In a general way, it may be stated that thyroxine has a stimulating effect on the differentiation of diaphysial hypertrophying chondroblasts so as to produce first an increase of length; but a toxic action follows, the stronger as the developmental stage of the cartilage is more advanced, so that a serious retardation of growth finally occurs. (Fig. 13, c, d).

---

propionate (b), in which the connective tissue has completely degenerated and the epithelium has already lost its normal structure. Transverse sections. × 250. From Wolff and Lutz-Ostertag (288), by courtesy of *Compt. rend Assoc. Anat.*

c, d, Toxic effect following the stimulatory one, of thyroxine on bone. A tibia explanted from a chick embryo at the stage in which chondroblastic hypertrophy had not yet begun and cultivated for 6 days in non-hormonal medium shows (c) a normal growth and appearance of hypertrophy in shaft cells. The opposite tibia of the same embryo, cultivated for an equal time in medium in which L-thyroxine was added, shows (d) a short shaft, although cellular hypertrophy is present. Longitudinal sections. × 17. From Fell and Mellanby (58), by courtesy of *J. Physiol.*

e, f, Action of insulin on bone. Tibia of 7-day chick embryos cultivated for 8 days in control medium (e) and in medium in which insulin has been added (f). In the latter case, the shaft is shortened and deformed, the epiphyses enlarged. Living cultures. × 8. From Chen (32), by courtesy of *J. Physiol.*

*Triiodothyronine,* supplied to the culture medium of bone rudiments of the same stages (Fell and Mellanby, 59) has, at the concentration used, a qualitatively similar effect on the growth rate of the diaphysis as that of thyroxine, but is four times as effective. Epiphyses, on the other hand, become larger, following an increase of matrix formation and cell size.

Chen (32) added *insulin* to the medium where long-bone rudiments of 6 to 7-day chick embryos, explanted at the stage of chondroblastic hypertrophy, were growing. A retardation in their increase in length was observed, probably through specific disturbances of diaphysial development and of differentiation of hypertrophic and flattened cells. However, no general inhibition of growth occurred as epiphyses became larger than normal (Fig. 13, e, f). Additional cartilage, as a secondary shaft, was sometimes formed. The cartilage became soft and flexible.

Fragments of interscapular brown fat body, explanted from 17 to 21-day rat embryos or newborn, and cultivated in medium containing serum with addition of *insulin* (Sidman, 213) show that the hormone stimulates in vitro the synthesis of glycogen. The activity is no less if insulin is put in contact with the explant for only 15 minutes and then removed. If the culture is carried out in a synthetic medium without nitrogen, where the only sources of carbon are nitrogen and bicarbonate, insulin, in addition to favoring glycogen synthesis, prolongs the life of the tissue. The capacity to stimulate glycogen synthesis does not fail even if the tissues are previously cultivated for 4 days in synthetic media, a result which would suggest the independence of insulin action from other endocrine factors.

## REAGGREGATION OF DISSOCIATED CELLS

If an embryonic tissue is submitted to the action of trypsin so thoroughly as to reduce it to a suspension of discrete cells (Moscona, 176) and these are cultivated in vitro in an appropriate medium, they may reaggregate and in a surprising manner reconstitute the normal pattern. This new line of research, which has already given a few very promising results, permits an approach to the problem of the degree of determination and stability reached, at a given stage, by an organ on the cellular scale. It is evident (Grobstein, 105) that, when the cells are completely freed from their original associations and then reaggregate, the reappearance of the original type would prove their cytodifferentiation to have become definitely fixed at explantation time. If, on the other hand, the determination depends on factors extrinsic to the cells, a reconstitution of the original pattern would not be possible.

Cells of 7 to 8-day chick embryos, reduced to a cellular suspension, when reassembling not only reform the epidermis, but even typical feather buds which develop clearly identifiable rows of barb rudiments (Weiss and James, 253).

From dissociated mesoblastic cells of limb buds, cartilage is produced in culture, and from mesonephric cells the characteristic nephric tubules (Moscona and Moscona, 180).

If after dissociation, cells of different organs are recombined in culture, they appear to keep their original determination. Mesonephric and limb-bud cells from chick embryos of 3 to 5 days, reduced to cellular suspensions, reaggregate in culture (Moscona, 177) and produce cartilage, nephric tubules, and groups of epithelial cells showing precocious keratinization. If different types of cells are mixed in varying ratios, each follows its own pattern of differentiation, provided that the amount of each type is not under a certain liminal concentration. If the cells to reaggregate are "heterochronic," that is to say, of different age, as when mesonephric and limb-bud cells of embryos respectively of 3 and 5 days, are intermingled, cartilage and nephric tissue are formed as in the other cases.

Doubt may arise, in considering these experiments, whether the tissues formed in culture really derive from the cells destined to produce them in vivo, as no means is available to follow their individual behavior during the 4 to 6 days of culture. No objection of this kind can be made when isolated cells of mouse and chick are associated, as they are always recognizable by means of their difference in size and the staining characteristics of their nuclei. Chondrogenic, mesonephric, and hepatic cells of 3 to 5-day chick embryos and of 11 to 13-day mouse embryos (Moscona, 178), reduced to suspensions and recombined in culture, maintained their original character. Cells of the same type ("isotypic") produced after reaggregation cartilage (Fig. 14, a), kidney tubules, or liver cords, in which cells of the two species are interspersed with each other, in a quite comparable way to what was observed in the chimeras obtained by Bresch (22) and Vakaet (244), cited above. If, on the contrary, cells of two different types ("heterotypic") are combined, as for instance chondrogenic cells of the chick and hepatic cells of the mouse, they segregate according to their origin, so that in the same culture cartilaginous pieces composed of chick cells and hepatic tissue composed of mouse cells are found. A mixture of chick mesonephric cells and mouse chondrogenic cells produced mouse cartilage and chick nephric tubules (Fig. 14, b). The type specificity prevailed over the species specificity.

In cultures of isolated retinal cells the determination was found not to be stable, and the original course of differentiation may be altered. Retinal

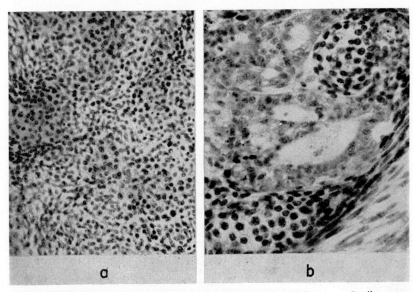

Fig. 14. Reaggregation of dispersed cells of chick and mouse embryos. a, Cartilage mass formed in vitro by reaggregation of interspersed limb-bud cells of a 4-day chick embryo (light nuclei) and limb-bud cells of a 12-day mouse embryo (dark nuclei), after 4 days of cultivation. × 120. From Moscona, (178) by courtesy of *Proc. Natl. Acad. Sci. U. S.* b, Aggregate formed in 3 days of cultivation by a suspension of mouse chondrogenic and chick mesonephric cells. The mouse cells, with dark nuclei, formed two cartilaginous pieces, whereas the chick cells, with light nuclei, formed mesonephric tubules. × 220. From Moscona in Gaillard (77), by courtesy of *J. Natl. Cancer Inst.*

tissue from a 6 to 7-day chick embryo, reduced to a suspension of discrete cells, when reaggregating (Moscona, 179) produces not only rosettes which develop further as sensory tissue, but also, when in conditions of overcrowding, gives rise to lentoids, consisting of pear-shaped cells surrounded by small ones. Perhaps this exception to the general rule is related to the pattern of reassociation in rosettes, one not existing in the original condition, and to the presence of cytolysed epithelial cells.

## Action of Vitamin A

It is known that vitamin A (Wolbach, 254) administered in high dosage to growing animals, causes alterations in bone development. If vitamin A is added in excess to the plasma where precartilaginous bone rudiments of 5 to 6-day chick embryos are cultured (Fell and Mellanby, 56; Fell, 50) growth is prevented, while differentiation is slightly slowed down. The cartilaginous matrix retracts, turns soft, and loses its basophilia and metachromasia. If bones of the mouse embryo, explanted at 17 to 20 days of gestation, are similarly treated, the bone matrix becomes rarefied and disappears completely after 10 days of culture, whereas the cartilaginous matrix behaves as above (Fig. 15, a, b).

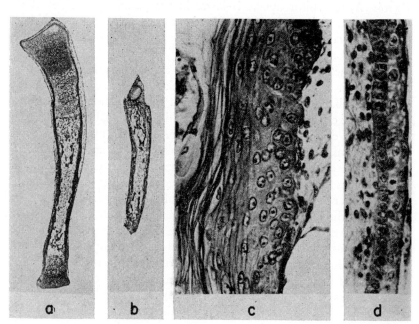

Fig. 15. Effect of vitamin A on bone and skin. a, Radius of a mouse fetus at a late stage of gestation cultivated for 9 days in normal medium. b, Opposite radius of the same embryo cultivated for 9 days in a medium containing about 1,000 i. u. of added vitamin A per 100 ml. Note the nearly complete disappearance of the cartilaginous ends and the resorption of diaphysial bone. $\times$ 25. From Fell (50), by courtesy of *Brit. Med. Bull.* c, Skin of a 13-day chick embryo cultivated for 6 days in normal medium. A thick stratum corneum has developed. d, Skin of the same age cultivated for 6 days in medium containing 1500 i. u. of added vitamin A per 100 ml. The superficial cells have degenerated and sloughed; the rest of the epithelium has reorganized into a secretory epithelium. $\times$ 550. From Fell (51), by courtesy of *Proc. Roy. Soc.*

If, after a period of culture in a medium enriched with vitamin A, the bone rudiments are transferred to a normal medium (Herbertson, 121) the damage is arrested, and epiphyses start growing again through an abnormally high mitotic rate; but if the shaft had stopped growing, no further elongation takes place. Its width may increase, probably through new formation of periosteal bone, but cartilage matrix is not reformed, nor is the lost basophilia recovered.

If labeled sulfate $Na_2{}^{35}SO_4$ is added to the cultures of bone rudiments (Fell, Mellanby, and Pelc, 61) and studied by means of autoradiography, the sulfate is found after 2 to 4 hours mainly in the cells; later, considerable amounts also appear in the matrix. In medium containing an excess of vitamin A, the sulfur uptake is lower in the peripheral regions of the epiphyses and of the shaft. If transferred to an unlabeled medium, a loss in the periphery of the shaft was observed, prior to the disappearance of metachromasia from the matrix.

Strong concentrations of vitamin A have also a characteristic action on skin cultures. Skin of trunk or limb of chick embryos, explanted at 6 to 7 days of incubation (Miszurski, 172), when the epithelium consists of a deep layer of cylindrical or cuboidal cells and a superficial one of flat cells, continues its development and keratinizes normally or even precociously. If high doses of vitamin A are added to the plasma (Fell and Mellanby, 57; Fell, 51), the epithelial cells do not keratinize; they acquire the ability to secrete mucus as intracellular drops or in large intercellular cavities looking like glandular lumina. After 14 days of culture, superficial cells are detached and the basal layer turns into epithelium of secreting cells, some ciliated and very similar to those of the nasal mucosa. Such transmutation is reversible, as, when epidermis is transferred to a medium without vitamin A, the secreting cells are dropped and keratinization starts again. It may be concluded that also in normal development different concentrations of vitamin A in various epithelia may be instrumental in inhibiting keratinization. Different thresholds of sensitivity to the same concentrations might explain the different behavior of the various tissues of the same organism.

Skin of more advanced developmental stages, such as shank and foot of 13, 14, and 18-day chick embryos (Fell, 51) behaves in the same way (Fig. 15, c, d). Keratinization is inhibited in all explants; cells that had already begun to cornify produced a substance with staining properties intermediate between keratin and mucin. The secretory epithelium was formed by the deepest and least differentiated layers of the epidermis. Dermis, too, showed some alteration and formed less fibrous tissue.

The same effect was obtained when skin cells of 7 to 8-day chick embryos, after being dispersed by means of trypsin, were treated for a short time, 15, 30, or 60 minutes, with vitamin A and then cultivated (Weiss and James, 253). The cells reaggregated, an epithelial cyst developed, in which no keratinization occurred, and the cells tended to preserve their nuclei. Such metaplastic behavior lasted at least 5 days. In order to prolong it, "booster" treatments were made, similar to the first but without redispersing the fragments, such as 30-minute baths in vitamin A at 2-day intervals. In such conditions, after 20 days of cultivation, goblet-shaped cells were found in the cysts. It was concluded that vitamin A does not act as a growth factor altering the cell metabolism, but rather as an inductive agent which modifies the cellular pattern of differentiation. In fact, it acts not only on the cells directly exposed to the vitamin, but also on their descendants for at least 5 days, and therefore not only on the terminal stages—keratin and mucus—but also on the initial stages, so as to "deflect" the determination of the cells. This interpretation is further supported by the fact that the return to the normal state, seen by Fell and Mellanby (57), does not affect the already modified cells, but only the newly proliferated cell generations.

In order to study the sulfur metabolism of the skin thus changed, ectoderm and underlying mesenchyme of 7-day chick embryos were cultivated by Fell, Mellanby, and Pelc (60) in a normal medium and in another to which vitamin A was added. After 7 or 10 days of culture, labeled $^{35}SO_4$ was added and left for 24 hours, after which autoradiographs were taken. In controls sulfur was found in connective tissue and basement membrane, but very little in the epithelium, and none in keratin; in the experimental culture, enriched with vitamin A, basal cells have a strong sulfate content, and secretory cells are particularly charged with it. This shows that sulfur metabolism is profoundly changed by vitamin action and resembles closely that of a normal mucous membrane.

## Action of Antimitotic Substances

Some antimitotic substances have a specific action either on the cortex or on the medulla of gonads differentiating in vitro.

Tripaflavin, supplied to cultures of left gonads of 8 to 10-day chick embryos (Salzberger, 199) inhibits the development of the male part, represented by the medulla. If the gonad is an ovary which, as seen above, tends to turn into an ovotestis when cultivated in a normal medium, tripaflavin induces the formation of lacunae in the medulla and a reduction of testicular cords, whereas the cortex is not modified and preserves its normal ovarian type. The result is an organ which looks very much like a normal ovary. If, on the other hand, the gonad is a testis, only a partial suppression of its characteristics takes place: Sertoli cells remain normal, the germ cells are partially preserved, and lacunae appear under the epithelium.

Colchicine, hydroquinone, and sodium cacodilate show an opposite action, inasmuch as they totally or partially destroy the cortex whereas they affect less severely the medulla. Left gonads of 10-day chick embryos

(Salzberger, 200) cultivated in a medium containing one of the said substances, show in 24 hours a nearly total degeneration of the cortex, whereas the testicular cords of the medulla are hardly differentiated and show some pycnoses and necrotic zones. Whether such gonads are genetically ovaries or testes, the result is a testis-like gonad.

### GENETICAL PROBLEMS

The disturbances caused in developmental processes by deleterious mutations have been in some cases submitted to investigation by means of culture in vitro of the organs concerned.

The Creeper anomaly in the development of bones in the chick embryo has been studied in two different ways. First, an attempt was made to reproduce the same defects in hindlimb buds of normal 3-day chick embryos by a strong reduction of the amount of the embryonic extract in the medium (Fell and Landauer, 55). After 21 to 22 days, reduced growth, irregular morphogenesis of cartilage, delay in ossification, and fusion of fibula and tibia were observed, very similar to those appearing spontaneously in Creeper embryos. The conclusion was drawn that the mechanism of the anomaly is a general retardation in growth, since it may be reproduced by a lack of nutritive substances.

A more direct approach was made by Wolff and Kieny (286) by explanting tibias of 6 to 7-day normal and Creeper embryos and cultivating them in media containing extracts made from normal and Creeper embryos. The linear growth of Creeper tibias is minimal in Creeper extract. It is remarkably improved in the extract from normal embryos, although they never reach the length of normal tibias. On the other hand, normal tibias grow less if cultured in Creeper extract, but are never so short as the Creeper tibias. This proves that in the Creeper embryos, used for the extract, there is a factor which prevents the growth of both normal and Creeper tibias. This factor may be an inhibiting substance or merely the lack of something promoting growth.

A lethal mutation of rats, discovered by Grüneberg (112), leads to death as a consequence of hyperplasia of the ribs and of the tracheal cartilages. If cartilages of 4 to 13-day post-natal rats are explanted and cultivated for 4 to 17 days (Fell and Grüneberg, 54) the abnormalities, represented by thick capsules around the chondroblasts, and active proliferation and chondrogenesis in perichondrium are maintained in vitro, although progressing very little during the cultivation period.

The mice affected by the homozygous combination of the gene $W$ (white spotting) are anemic and also sterile (Preti, 194; Coulombre and

Russell, 34; Borghese, 17). An attempt was made to decide, by means of organ culture, whether the two abnormalities are one the consequence of the other, or whether they are produced independently. The answer was in favor of the second alternative. The gonads of all embryos of each litter resulting from a mating of two heterozygous mice, and therefore containing 25 per cent of anemic-sterile individuals were explanted at the 12th day (Borghese, 18, 20), before the anemia and sterility are recognizable. After 4 days of culture the gonads differentiated sexually and some became fertile and some sterile (Fig. 16), in a ratio not significantly different from the 3:1 which would have occurred had they been allowed to continue their development in vivo. It was concluded that sterility is already definitely determined before the appearance of the anemia, since it develops in the gonads differentiating outside the general organism. In other words, the gonads become sterile not because of some influence coming from anemic blood but under the direct action of the gene.

## Nutritive Medium

In the nutritive medium supplied to the organs cultivated extrinsic developmental factors are contained, the nature of which can be subjected to some extent to experimental analysis.

The action of embryonic extract may be analyzed in different ways. First, the value of its various constituents has been investigated in relation to the age of the embryos employed for its preparation. Both in normal and in experimental conditions, phases in which growth prevails alternate with phases of predominant differentiation. This may be explained by an incompatibility of the two phenomena, but there are some indications of the existence of different substances contained in embryonic fluids of various ages, which favor the one or the other.

If tibias of 7-day chick embryos are cultivated in embryo extracts of different ages, distinct factors can be recognized (Miszurski, 173): in 7-day embryos a growth-stimulating factor, in 13-day ones another favoring differentiation, represented by ossification. At 19 days a third factor appears, inhibiting both growth and differentiation. The last factor may be eliminated from the extract (Miszurski, 174) if this is kept at 0-2°C for three days and then centrifuged again. The supernatant, tested on cultures of tibias, appears to have lost the inhibiting factor, whereas it still contains those of growth and differentiation.

The importance of the age of the embryos from which extracts are made was studied by Gaillard (65), first on histiotypic cultures of frontal bones of chick embryos and later on organotypic cultures (Gaillard, 73). The

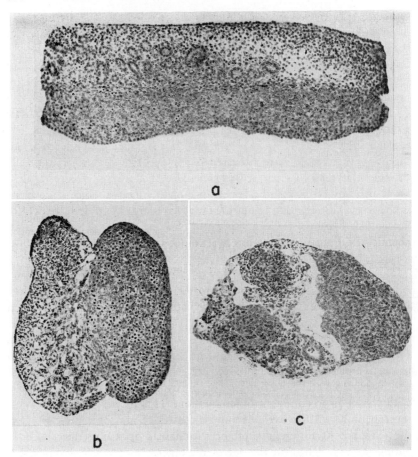

Fig. 16. Development in vitro of gonads explanted from mice embryos resulting from a $Ww \times Ww$ mating. a, Sexually undifferentiated gonad (bottom) with mesonephros (top), at explantation time at the 12th day of gestation. After 4 days of cultivation the gonads have differentiated into testes or ovaries, some fertile and some sterile, in the same ratio as would have occurred in vivo. Here two ovaries are reproduced. b, A fertile ovary, which has developed a normal shape and contains numerous ovocytes in meiosis. c, A sterile ovary, which is strongly reduced in volume, is irregular in shape, and does not contain ovocytes, but somatic cells only. Section. $\times$ 110. From Borghese (20), by courtesy of *Sympos. genet., Pavia*.

conclusion was drawn that the best way to favor development of tissues and organs in culture, and to prevent their dedifferentiation is to use in successive stages of culture an "ascending range" of nutritive materials, corresponding as well as possible to the actual age of the explants. This would seem to be confirmed by the fact that embryonic extract does not prove favorable to the culture of post-natal organs, as shown by the example of the hypophysis of the 3-months-old rabbit, cultivated in a medium containing embryo extract of different ages (Gaillard, 68, 69). If the basic medium is pure fibrinogen, instead of plasma, to which an extract of 2 to 6-day rabbit embryos is added, cultures immediately degenerate. If the extract derives from 13-day embryos, degeneration occurs after 2 days. The normal structure of explants is better preserved with an extract of 19-day embryos, but the best results are obtained when embryonic extract is altogether excluded and is replaced by plasma of an adult rabbit. Under such conditions the structure of the organ does not change for a long time. Chromophilic cells are maintained for a few days, chromophobic ones for 3 to 5 months.

The femur of the chick embryo, explanted at 6 days (Carpenter, 27) may indicate the effectiveness of dried culture media in promoting growth. In fresh plasma the average growth in 6 days is 124 per cent of the original length, the maximum 246 per cent. In bovine fibrinogen the maximum values are 48 per cent, 134 per cent in dried bovine plasma, and 224 per cent in dried chick plasma. Only in the last case does the growth correspond to that in fresh plasma; however, no artificial medium reaches the maximum effectiveness of the latter.

A systematic investigation of the nutritive needs of developing organs was initiated by Wolff, and his coworkers (282, 283), and is still going on very actively (Wolff, 271). The principle is to use a synthetic culture medium, made out of agar as a non-nutritive substrate, to which embryonic extract, serum of adult animals, or known mixtures of amino acids are added. The results obtained under such different conditions are then compared. Three criteria are used to evaluate the effect: survival, differentiation, and growth. The duration of survival can easily be measured; the degree of differentiation is estimated from morphological and histological characters; growth is calculated from mitosis number, weight increase, and content of total nitrogen. Skeletal cartilages, gonads and the syrinx were studied.

The tibia of chick embryos explanted at 7 days and cultivated in a medium containing agar and embryonic extract (Kieny, 130) continues to differentiate, developing articular faces and progressing in cellular hypertrophy. In addition, it increases in length and weight through protein synthesis, as the content of total nitrogen proves. If embryonic extract is replaced by a mixture of the 9 amino acids isolated from fibrin by Bergmann and Niemann (7) with addition of novocain, morphogenesis goes on, articular faces appear, but the increase in length and weight is less. Total nitrogen does not increase and is even reduced. If novocain is not added, the loss is greater.

If a richer mixture, consisting of 21 amino acids, is supplied, the results are no better. The increase in volume and length of the tibia in synthetic media is inferior to that obtained with embryo extract and is due presumably only to water and mineral salt absorption. Protein assimilation is in any case negligible and does not even compensate for the disassimilation, as total nitrogen content shows.

The effectiveness of the synthetic medium is mostly due to novocain; and further research (Kieny, 131) showed that the active factor contained in novocain is para-aminobenzoic acid, which can also improve the increase in weight when added to the natural nutritive medium containing embryonic extract. On the other hand, alone it is not able to produce growth, so that it presumably acts as a catalyzer which promotes protein synthesis and makes the explant capable of utilizing the amino acids contained in the medium.

A mixture of a limited number of amino acids as a minimum basic medium sufficient for the successful culture of tibia as well as of gonads and syrinx, was defined by Kieny, Stenger-Haffen, and Wolff (135). A further simplification was attempted, for the tibia, by Kieny (132); and a minimum medium was defined, consisting of arginine, methionine, glutamic acid, histidine or lysine, and para-aminobenzoic acid. On the other hand, the association lysine-histidine has an inhibitory effect upon linear growth. Such a basic medium can be improved by addition of threonine, asparagine, hydroxyproline, serine, and taurine. Cysteine (Kieny, 133) may be useful if alone, but inhibits growth in the presence of other amino acids.

Para-aminobenzoic acid can be replaced (Kieny, 134) by carnitine or dicarnitine, which have a similar or even greater effect on linear and weight increase. However, the effects of such compounds do not summate.

The replacement of embryonic extract by an equal amount of horse serum or chick plasma (Wolff and Kieny, 285) does not favor the growth of the tibia in vitro, differently from what occurs, as will be seen later, for the syrinx. The linear growth and increase of total nitrogen are less than in embryonic extract and are similar to that in synthetic medium. A partial replacement of embryonic extract with horse serum or chick plasma, however, furthers length increase and morphogenesis of the tibia of 5 to 6 days, the latter only of the tibia of 7 days.

A different approach to the study of synthetic media for the femur and tibia of chick embryos of 6 to 7 days was undertaken by Biggers, Webb,

Parker, and Healy (8). In their media, consisting of known mixtures of amino acids, the bone rudiments elongate about at the same rate as in a natural medium containing serum, but they are distorted, translucent, and lighter. The utilization of amino acids was studied in the medium by recovery after 4 days of culture and analysis for amino nitrogen, which was shown to be considerably diminished. Bone rudiments cultivated in a medium containing glycine labelled with $C^{14}$ showed incorporation of glycine in the explants and some synthesis of serine from glycine. The amino acids "essential" for growth were sought in 20 media, each deficient in a single amino acid, and the response, estimated in wet weight, was compared with that in a medium containing 20 amino acids. A list of 11 essential amino acids was made, and some non-essential ones were identified by means of labeled glucose as being synthesized from glucose. The results do not agree completely with the research of Kieny but the discrepancy can be explained by the greatly different technique.

Duck gonads (Wolff, Haffen, and Wolff, 284) cultivated in a synthetic medium containing glucose as the only nutritive material, do not survive beyond 3 to 4 days. Sexual differentiation continues if it has already begun, but is not initiated. In Bergmann and Niemann's mixture (7) survival goes beyond 5 days, growth continues, and sexual differentiation goes on if already begun. If the explantation occurs before sexual differentiation, female gonads differentiate regularly; genetically male ones, on the contrary, do not become testes but develop a cortex of female type, while the medullary cords remain of indefinite structure. The deficiency of nutritive substances acts much like an insufficiency of male hormone, as was reported earlier.

Chick gonads (Haffen, 113) explanted at the 8th to 9th day of incubation, before sexual differentiation, and cultivated in a medium containing Bergmann and Niemann's mixture instead of embryonic extract, follow an incomplete development: testicular tubes, although containing gonocytes and Sertoli cells, do not grow; the left female gonad becomes an ovotestis, whereas the right one resembles a testis with incompletely formed cords.

The mixture of 21 amino acids gives results more similar to those obtained in the natural medium containing embryo extract. In the testis cords grow, and germ cells are more numerous and mitoses abundant. The left female gonad becomes an ovotestis, with a more complete differentiation and numerous mitoses both in somatic cells and in ovogonia; the right female gonad produces undifferentiated tubules and others with a testicular aspect. Growth, estimated through mitosis number and nitrogen amount, is negative or nil in the Bergmann mixture, reaches 13 per cent of the original size with 21 amino acids, and 20 per cent if vitamins are added.

The chick gonad is therefore able, if explanted before differentiation, and in contradistinction to the tibia, to utilize partially the amino acids of the culture medium. However, growth is in any case less than in cultures in natural medium, where it reaches 60 per cent of the original size.

The basic medium cited above, which was found by Kieny, Stenger-Haffen, and Wolff (135) to be sufficient for development of the tibia, gonads, and syrinx, gives improved results (Stenger-Haffen, 228) when the gonads are cultivated with addition of further amino acids. Ornithine, threonine, hydroxyproline, and glycine improve survival and differentiation. On the other hand, glutamic acid, tryptophan, and phenylalanine are specifically favorable to differentiation.

The nutritive needs of the chick syrinx, explanted between $7\frac{1}{2}$ and 9 days of incubation and cultivated in an entirely synthetic medium, were studied by Wolff (255). To the basic medium, representing the minimum requirement for tibia and syrinx and containing as the only nitrogen sources cysteine, arginine, methionine, lysine, and histidine, 18 other amino acids and glutathione were added one by one (Wolff, 257). Some amino acids were found to act favorably only on one or two of the characters considered, survival, differentiation, and growth; but asparagine, serine, phenylalanine, glycine, and glutathione improve all three characters; the favorable amino acids are not the same as for the tibia. The difference in behavior can be explained by the fact that the tibia is already cartilaginous at explantation time, whereas the syrinx must synthesize cartilage.

Whereas the embryonic extract of 7 to 9-day embryos is sufficient for the development of the tibia and gonads, the syrinx requires additional substances. The syrinx of the 7 to 8-day chick embryo or of the 8 to 9-day duck embryo survives, grows, and differentiates only in a medium where embryonic extract is partially or even totally replaced by a heterologous adult material such as horse or bovine serum (Wolff, 256). Further research aimed at identifying the active substances in the serum showed (Wolff, 258) that it cannot be replaced by its ultrafiltrate. Evidently the amino acids contained in it are not in an appropriate proportion. Embryonic extract, on the other hand, may be improved by the addition of cysteine, arginine, or glutathione separately. Methionine alone is even more favorable. The best mixture is one of methionine, arginine, histidine, glutathione or cysteine, and glutamic acid or lysine.

The sexually dimorphic duck syrinx differentiates (Wolff, 260, 261) according to the asymmetrical male, or neutral, type only when cultivated in serum or in serum with embryonic extract, or in embryonic extract supplemented by an optimum mixture of amino acids. If the medium contains embryonic extract and a few acids corresponding to the minimum requirements, it develops a nearly symmetrical form, very similar to the female type (Fig. 17). This convergence of the effects of the female hormone with that of insufficient nutrition may explain the normal mechanism

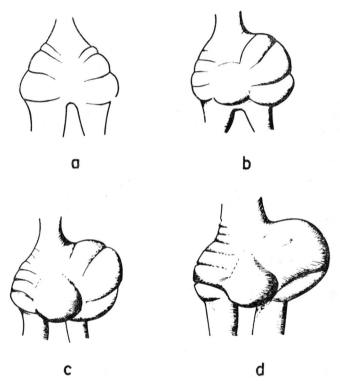

Fig. 17. Action of the composition of the nutritive medium on the sexual differentiation of the duck syrinx cultivated in a non-hormonal medium. The syrinx was explanted before sexual differentiation and has developed in vitro. a, In a deficient medium containing embryonic extract only, where it acquires the symmetrical female aspect. In media containing increasing amounts of nutritive substances the asymmetry typical of the male aspect appears and gradually increases, as in following figures. b, With embryonic extract and methionine. c, With embryonic extract, cysteine, arginine, lysine, and histidine. d, With embryonic extract and serum, or with serum alone. Schematic drawings. From Wolff (260), by courtesy of *Arch. Anat. microscop. Morph. exptl.*

of the female hormone, which presumably acts to select nutritive substances or to inhibit metabolism.

In only one case did the growth in synthetic medium appear to be superior to that in the natural one, although the shape was altered. In cultures of chick feathers (Lepori, 150) where the Bergmann and Niemann mixture of 9 amino acids causes the papillae to be more voluminous than in natural medium, the addition of 21 amino acids makes them grow longer and thinner (Fig. 18). On the other hand pigmentation of the

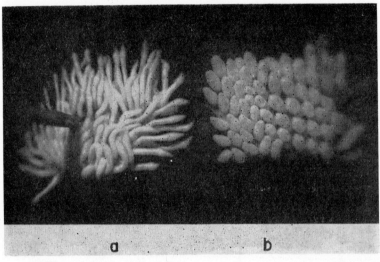

Fig. 18. Behavior of chick feather papillae in synthetic medium. Skin of chick embryo explanted at 10 days of incubation and cultivated for 7 days. a, Papillae grown in synthetic medium containing 21 amino acids. They are thinner and longer than those grown in b, natural medium containing embryonic extract. × 15. From Lepori (150), by courtesy of *Arch. Anat. microscop. Morph. exptl.*

same feather buds (Sengel, 207) is favored in culture by arginine and is inhibited by the mixtures most favorable to growth.

## Conclusions

The cultivation in vitro of embryonic organs has made it possible to apply on a vast scale to the more delicate embryos of birds and mammals a method of experimental research that had already given very important results in cold-blooded animals, especially the amphibia. The problem of keeping alive the organs of warm-blooded animals without disturbing their development had already been tackled by using methods of partial isolation, such as grafts on the chorioallantoic membrane, in the coelomic cavity or in the anterior chamber of the eye. Culture in vitro has supplied a straightforward method of isolating them altogether. In this way organs not only survive but also keep on differentiating in a medium which, although still containing some substances from the general organism, may be more and more simplified and even made synthetic so as to have an exactly known composition.

The first stage of this research, to obtain "isolated" differentiation, at least in a medium which still contains some substances from the general organism, gave satisfactory results for many organs. It is what in classical experimental embryology is called an "isolation" experiment. The list of organs which, when explanted at an appropriate stage, are capable of going on growing and differentiating, and also to a certain extent of functioning, is increasing day by day. It even seems that interest in this direction is waning, since the differentiation in vitro of any new organ tested is almost to be expected as a matter of course. However, it should be remembered that generalizations are always dangerous, and there may still be certain exceptions.

On the other hand, differences in behavior of many organs, in comparison with their development in vivo, are found, and these promise to offer clues to the interpretation of normal development. Some of these discrepancies between development in vitro and in vivo may be rather general and therefore of minor interest, as for example the slowing down of growth and sometimes also of differentiation; but many others concern individual details of morphogenesis and are more instructive. For instance, the deformations of skeletal rudiments in vitro can be related to the lack of the morphogenetic effect of contiguous organs upon them, or to disturbance of the normal sequence of developmental stages. Some examples have been mentioned, and others may be found in the papers cited. In any case, if the researcher would prefer more exciting fields, some are already open and are far from being exhausted.

One of these concerns the problem of the origin of the harmonious structure of any organ through the regular, balanced development of its components. The first attempt in this connection was made with a technique corresponding to some extent to what in classical experimental embryology is called a "recombination" experiment; that is to say, one of the type which, when carried out on amphibian embryos, led to the discovery of the organizer (Spemann and Mangold, 218). Embryonic compound organs were separated into their epithelial and mesenchymal rudiments, first by rather crude mechanical means, then with the more refined usage of trypsin, which may even be pushed to the point of a complete disintegration of the tissue. The parts or cells were then recombined so as to put in contact the components of the same organ, or of different ones. The existence of an inductive action from connective tissue on epithelium and vice versa was proved, a process which in vivo may be necessary in shaping the normal structure of an organ. It seems to be established that extracellular substances are often involved, but a more precise chemical definition of the substances is still to be attained.

Not only do organs differentiate in vitro, but very often they start functioning, as is proved most easily in the case of endocrine glands. In such cases the target organs are rather well known; but doubt may persist whether the hormone directly stimulates the organ concerned or only initiates a chain of reactions which are realized through activities of the general organism. The direct association in culture of the two organs has proved in the cases so far tested that the hormonal action is immediate or direct.

If research on such associations is especially focussed on the receptor organ rather than on the effector, the experiment may be modified and the latter replaced by a purified, natural or synthetic, hormone. Sometimes under these conditions only further confirmation was obtained of results obtained with the effector organ itself, but an altogether independent line of research was in this way developed for embryonic organs. This approach has been extended to the actions of many types of substances, of hormones as well as vitamins and antimitotic substances, and has also been applied to adult organs.

Finally, if between the two components of a culture, the organ and the medium, chief attention is directed to the latter, the attempt can be made to define as precisely as possible what substances are necessary to the development of embryonic organs in general, or for any particular organ or tissue according to its structure and function. In other words, the problems of the nutrition and metabolism of cells in "histiotypic" culture, recently reviewed by Waymouth (250) and by Evans (42), may be transferred to the study of whole embryonic organs while maintaining a normal structure and reciprocal relationship of their components. It is a time-consuming work, as it requires the testing of a larger number of synthetic chemically defined media, and the difficulty is increased by the fact that the degree of growth and differentiation in such media has so far turned out to be inferior to that which takes place in plasma or embryonic extract.

However, some progress has been made, and lists of amino acids or other components of the medium are now available, which indicate the substances that by themselves or in definite associations favor or inhibit the differentiation and growth of many organs.

These are some of the lines of research which at present seem to be the most fruitful, among those to which the culture in vitro of embryonic organs has become a very valuable and effective instrument.

## REFERENCES

1. Abercrombie, M., *Nature,* **144,** 1091-1092 (1939).
2. ———, and Bellairs, R., *J. Embryol. exptl. Morphol.,* **2,** 55-72 (1954).
3. Balinsky, B. J., *Trans. Roy. Soc. Edinburgh,* **62,** 1, 1-31 (1949-50).
4. ———, *J. Anat.,* **84,** 227-235 (1950).
5. Bellairs, R., *J. Embryol. exptl. Morphol.,* **1,** 115-124 (1953).
6. ———, *J. Embryol. exptl. Morphol.,* **1,** 369-385 (1953).
7. Bergmann, M., and Niemann, C., *J. Biol. Chem.,* **115,** 77-85 (1936).
8. Biggers, J. D., Webb, W., Parker, R. C., and Healy, G. M., *Nature,* **10,** 825-828 (1957).
9. Bisceglie, V., *Arch. exptl. Zellforsch.,* **12,** 86-101 (1932).
10. Black, L., and Comolli, R., *Arch. anat. microscop. morphol. exptl.,* **43,** 276-281 (1954).
11. Borghese, E., *Boll. soc. ital. biol. sper.,* **25,** 86-101 (1949).
12. ———, *Monit. zool. ital.,* **57,** Suppl., 229-232 (1949).
13. ———, *J. Anat.,* **84,** 287-302 (1950).
14. ———, *J. Anat.,* **84,** 303-318 (1950).
15. ———, *Arch. ital. Anat. Embriol.,* **55,** 182-214 (1950).
16. ———, *Monit. zool. ital.,* **58,** Suppl., 71-73 (1950).
17. ———, *Gaz. méd. portug.,* **7,** 261-269 (1954).
18. ———, *Compt. Rend. Assoc. Anat.,* **52,** 339-346 (1955).
19. ———, *Monit. zool. ital.,* **63,** Suppl., 50-138 (1955).
20. ———, *Sympos. genet., Pavia,* **5,** 84-130 (1956).
21. ———, and Venini, M. A., *Sympos. genet., Pavia,* **5,** 69-83 (1956).
22. Bresch, D., *Bull. biol. France et Belg.,* **139,** 179-188 (1955).
23. Bucher, O., *Acta Anat.,* **14,** 98-107 (1952).
24. ———, and Weil, J. T., *Experientia,* **7,** 38-45 (1951).
25. Carpenter, E., *Anat. Record,* **75,** Abstracts, 101 (1939).
26. ———, *J. exptl. Zool.,* **89,** 407-431 (1942).
27. ———, *J. exptl. Zool.,* **113,** 301-315 (1950).
28. ———, Beattie, J., and Chambers, R. D., *J. exptl. Zool.,* **127,** 249-269 (1954).
29. Chen, J. M., *J. Anat.,* **86,** 387-401 (1952).
30. ———, *J. Anat.,* **87,** 130-149 (1953).
31. ———, *Exptl. Cell Research,* **7,** 518-529 (1954).
32. ———, *J. Physiol.,* **125,** 148-162 (1954).
33. Chlopin, N., *Arch. mikroskop. Anat. u. Entwicklungsmech.,* **96,** 435-493 (1922).
34. Coulombre, J. L. and Russell, E. S., *J. exptl. Zool.,* **126,** 277-296 (1954).
35. Danchakoff, V., *Contrib. Embryol. Carneg. Inst.,* **18,** 63-78 (1926).
36. Davidson, P., and Hardy, M. H., *J. Anat.,* **86,** 342-356 (1952).
37. De Haan, J., *Bull. histol. appl. et tech. microscop.,* **4,** 293-317 (1927).
38. ———, *Acta Neerl. Morphol.,* **1,** 12-23 (1937).
39. De Jong, B. J., and De Haan, J., *Acta Neerl. Morphol.,* **5,** 26-51 (1943).
40. Deth, J. H. M. G. van, *Acta Neerl. Morphol.,* **3,** 151-169 (1940).

41. Dorris, F., *J. exptl. Zool.*, **78**, 385-415 (1938).
42. Evans, V. J., *J. Natl. Cancer Inst.*, **19**, 539-545 (1957).
43. Fell, H. B., *Arch. exptl. Zellforsch.*, **7**, 69-81 (1928-29).
44. ———, *Arch. exptl. Zellforsch.*, **7**, 390-412 (1928-29).
45. ———, *Arch. exptl. Zellforsch.*, **11**, 245-252 (1931).
46. ———, *Phil. Trans. Roy. Soc. London, Ser. B.*, **229**, 407-463 (1939).
47. ———, *J. Roy. Microscop. Soc.*, **60**, 95-112 (1940).
48. ———, in *Cytology and Cell Physiology* (G. H. Bourne, ed.), 2nd. ed., p. 419-433, Clarendon Press, Oxford (1951).
49. ———, *Science Progr.*, **162**, 212-231 (1953).
50. ———, *Brit. Med. Bull.*, **12**, 35-37 (1956).
51. ———, *Proc. Roy. Soc. (London)*, B., **146**, 242-256 (1957).
52. ———, and Canti, R. G., *Arch. exptl. Zellforsch.*, **15**, 311 (1934).
53. ———, and ———, *Proc. Roy. Soc. (London)*, B., **116**, 316-351 (1935).
54. ———, and Grüneberg, H., *Proc. Roy. Soc. (London)*, B, **127**, 257-277 (1939).
55. ———, and Landauer, W., *Proc. Roy. Soc. (London)*, B, **118**, 133-154 (1935).
56. ———, and Mellanby, E., *J. Physiol.*, **116**, 320-349 (1952).
57. ———, and ———, *J. Physiol.*, **119**, 470-488 (1953).
58. ———, and ———, *J. Physiol.*, **127**, 427-447 (1955).
59. ———, and ———, *J. Physiol.*, **133**, 89-100 (1956).
60. ———, ———, and Pelc, S. R., *Brit. Med. J.*, II, 611-612 (1954).
61. ———, ———, and ———, *J. Physiol.*, **134**, 179-188 (1956).
62. ———, and Robison, R., *Biochem. J. (London)*, **23**, 767-784 (1929).
63. ———, and ———, *Biochem. J. (London)*, **24**, 1905-1921 (1930).
64. Fischer, A., *J. exptl. Med.*, **36**, 393-397 (1922).
65. Gaillard, P. J., *Protoplasma*, **23**, 145-174 (1935).
66. ———, *Protoplasma*, **28**, 1-17 (1937).
67. ———, *Acta Neerl. Morphol.*, **1**, 3-11 (1938).
68. ———, *Arch. exptl. Zellforsch.*, **22**, 65-67 (1939).
69. ———, *Symposia Soc. exptl. Biol.*, **2**, 139-144 (1948).
70. ———, *Ned. Tijdschr. Geneesk.*, **93**, 385-388 (1949).
71. ———, *Proc. Acad. Sci. Amsterdam*, **53**, 1300-1316, 1337-1347 (1950).
72. ———, *Methods in Med. Research*, **4**, 241-246 (1951).
73. ———, *Intern. Rev. Cytol.*, **2**, 331-401 (1953).
74. ———, *Exptl. Cell Research*, 3 (Suppl.), 154-169 (1955).
75. ———, *Koninkl. Ned. Akad. Wetenschap, Proc.*, Ser. C, **58**, 279-293 (1955).
76. ———, *Acta Physiol. Pharmacol. Neerl.*, **4**, 108 (1955).
77. ———, *J. Natl. Canc. Inst.*, **19**, 591-607 (1957).
78. ———, *Schweiz. med. Wochschr.*, **87**, 217-228 (1957).
79. ———, and Varossieau, W. W., *Acta Neerl. Morphol.*, **1**, 313-327 (1938).
80. ———, and ———, *Arch. exptl. Zellforsch.*, **24**, 141-168 (1942).
81. Galgano, M., *Monit. zool. ital.*, **49**, 235-243 (1938).
82. ———, *Monit. zool. ital.*, **49**, Suppl., 60-63 (1938).
83. ———, *Monit. zool. ital.*, **50**, 94-100 (1939).
84. Gey, G. O., *Am. J. Cancer*, **17**, 752-756 (1933).
85. Glasstone, S., *J. Anat.*, **70**, 260-266 (1936).
86. ———, *Proc. Roy. Soc. (London)*, **B**, **126**, 315-330 (1939).
87. ———, *J. Anat.*, **86**, 12-15 (1952).
88. Gliozzi, M. A., *Monit. zool. ital.*, **6**, Suppl., 364-371 (1958).
89. Glucksmann, A., *J. Anat.*, **76**, 231-239 (1942).
90. Gomot, L., *Compt. rend. soc. biol.*, **150**, 910-913 (1956).
91. ———, *Compt. rend. Acad. Sci.*, **243**, 2142-2144 (1956).
92. Gonzales, F., *Texas Rep. Biol. Med.*, **12**, 828-832 (1954).

93. ———, *Exptl. Cell. Research,* **10,** 181-187 (1956).
94. Gozzi, R., *Arch. ital. Anat. Embriol.,* **43,** 229-254 (1940).
95. Grobstein, C., *J. exptl. Zool.,* **115,** 297-313 (1950).
96. ———, *J. exptl. Zool.,* **116,** 501-525 (1951).
97. ———, *J. exptl. Zool.,* **119,** 355-379 (1952).
98. ———, *J. exptl. Zool.,* **120,** 437-456 (1952).
99. ———, *J. Morphol.,* **93,** 19-43 (1953).
100. ———, *J. exptl. Zool.,* **124,** 383-414 (1953).
101. ———, *Science,* **118,** 52-54 (1953).
102. ———, *Nature,* **172,** 869-871 (1953).
103. ———, in *Aspects of synthesis and order in growth* (D. Rudnick, ed.), p. 233-256, Princeton Univ. Press, Princeton (1955).
104. ———, *J. exptl. Zool.,* **130,** 319-339 (1955).
105. ———, *Ann. N. Y. Acad. Sci.,* **60,** 1095-1106 (1955).
106. ———, *Exptl. Cell Research,* **10,** 424-440 (1956).
107. ———, *Exptl. Cell Research,* **13,** 575-587 (1957).
108. ———, and Dalton, A. J., *J. exptl. Zool.,* **135,** 57-73 (1957).
109. ———, and Holtzer, H., *J. exptl. Zool.,* **128,** 333-357 (1955).
110. ———, and Parker, G., *Proc. Soc. exptl. Biol. Med.,* **85,** 477-481 (1954).
111. ———, and Zwilling, E., *J. exptl. Zool.,* **122,** 259-284 (1953).
112. Grüneberg, H., *Proc. Roy. Soc. (London),* B, **125,** 123-144 (1938).
113. Haffen, K., *Compt. rend. soc. biol.,* **147,** 861-864 (1953).
114. ———, and Wolff, Em., *Compt. rend. Acad. Sci.,* **237,** 754-756 (1953).
115. Hardy, M. H., *J. Anat.,* **83,** 364-384 (1949).
116. ———, *J. Anat.,* **84,** 388-393 (1950).
117. ———, *Ann. N. Y. Acad. Sci.,* **53,** 546-561 (1951).
118. ———, Biggers, J. D., and Claringbold, P. J., *Nature,* **172,** 1196-1197 (1953).
119. Harrison, J. R., *J. exptl. Zool.,* **118,** 209-241 (1951).
120. ———, *J. exptl. Zool.,* **127,** 493-510 (1954).
121. Herbertson, M. A., *J. Embryol. exptl. Morphol.,* **3,** 355-365 (1955).
122. Hoadley, L., *Biol. Bull.,* **46,** 281-315 (1924).
123. Hughes, S., *Arch. exptl. Zellforsch.,* **19,** 329 (1937).
124. Ito, Y., and Endo, H., *Endocrinol. Japonica,* **3,** 106-115 (1956).
125. Jacobson, W., and Fell, H. B., *Quart. J. Microscop. Sci.,* **82,** 563-586 (1941).
126. Jost, A., and Bergerard, Y., *Compt. rend. soc. biol.,* **143,** 608-609 (1949).
127. Kahn, R. H., *Nature,* **174,** 317 (1954).
128. Keuning, F. J., *Acta Neerl. Morphol.,* **5,** 237-247 (1944).
129. ———, *Acta Neerl. Morphol.,* **6,** 1-35 (1948).
130. Kieny, M., *Compt. rend. soc. biol.,* **147,** 868-872 (1953).
131. ———, *Compt. rend. Acad. Sci.,* **236,** 1920-1922 (1953).
132. ———, *Compt. rend. soc. biol.,* **149,** 418-421 (1955).
133. ———, *Compt. rend. soc. biol.,* **149,** 1482-1485 (1955).
134. ———, *Compt. rend. Acad. Sci.,* **243,** 320-321 (1956).
135. ———, Stenger-Haffen, K., and Wolff, Em., *Compt. rend. Acad. Sci.,* **239,** 1426-1427 (1954).
136. Kooreman, P. J., and Gaillard, P. J., *Arch. Chir. Neerl.,* **2,** 326-355 (1950).
137. Lasnitzki, I., *Brit. J. Cancer,* **5,** 345-352 (1951).
138. ———, *Cancer Research,* **14,** 632-639 (1954).
139. ———, *J. Endocrinol.,* **12,** 236-240 (1955).
140. ———, *Brit. J. Cancer,* **9,** 434-441 (1955).
141. ———, *Brit. J. Cancer,* **10,** 510-516 (1956).
142. ———, and Pelc, S. R., *Exptl. Cell Research,* **13,** 140-146 (1957).
143. Leighton, J., *J. Natl. Cancer Inst.,* **12,** 545-561 (1951).

144. ——, *J. Natl. Cancer Inst.*, **15**, 275-293 (1954).
145. ——, *Texas Rep. Biol. Med.*, **12**, 847-864 (1954).
146. ——, and Kline, I., *Texas Rep. Biol. Med.*, **12**, 865-873 (1954).
147. ——, ——, Belkin, M., Legallais, F., and Orr, H. C., *Cancer Research*, **17**, 359-363 (1957).
148. ——, ——, ——, and Orr, H. C., *Cancer Research*, **17**, 336-344 (1957).
149. ——, ——, ——, and Tetenbaum, Z., *J. Natl. Cancer Inst.*, **16**, 1353-1373 (1956).
150. Lepori, N. G., *Arch. Anat. microscop. morphol. exptl.*, **42**, 194-208 (1953).
151. Loffredo Sampaolo, C., *Boll. soc. ital. biol. sper.*, **32**, 1580-1581 (1956).
152. ——, *Quaderni anat. pratica*, **12**, 415-434 (1957).
153. ——, and Sampaolo, G., *Boll. soc. ital. biol. sper.*, **32**, 797-801 (1956).
154. ——, and ——, *Boll. soc. ital. biol. sper.*, **33**, 173-176 (1957).
155. Lutz, H., and Gomot, L., *Compt. rend. soc. biol.*, **150**, 2055-2056 (1956).
156. ——, and Lutz-Ostertag, Y., *Compt. rend. Acad. Sci.*, **242**, 557-560 (1956).
157. ——, and ——, *Arch. anat. microscop. morphol. exptl.*, **45**, 218-235 (1956).
158. Martinovitch, P. N., *Nature*, **139**, 413-414 (1937).
159. ——, *Proc. Roy. Soc. (London)*, B, **125**, 232-249 (1938).
160. ——, *Arch. exptl. Zellforsch.*, **22**, 74-76 (1939).
161. ——, *Proc. Roy. Soc. (London)*, B, **128**, 138-143 (1939).
162. ——, *Nature*, **165**, 33-35 (1950).
163. ——, *Methods in Med. Research*, **4**, 237-240 (1951).
164. ——, *Methods in Med. Research*, **4**, 240-241 (1951).
165. ——, *Exptl. Cell Research*, **4**, 490-493 (1953).
166. ——, *J. Embryol. exptl. Morphol.*, **2**, 14-25 (1954).
167. ——, *J. exptl. Zool.*, **129**, 99-127 (1955).
168. ——, and Radivojevitch, D. V., *Nature*, **175**, 308-309 (1955).
169. ——, and Vidovic, L., *Bull. Inst. Nuclear Sci. "Boris Kidrich,"* **48**, 131-137 (1953).
170. Maximow, A., *Contrib. Embryol. Carneg. Inst.*, **16**, 47-113 (1925).
171. Milkovic-Zulj, K., *Acta Med. Yugoslavica*, **7**, 255-260 (1953).
172. Miszurski, B., *Arch. exptl. Zellforsch.*, **20**, 122-139 (1937).
173. ——, *Arch. exptl. Zellforsch.*, **22**, 130-131 (1939).
174. ——, *Arch. exptl. Zellforsch.*, **23**, 80-83 (1939).
175. Momigliano Levi, G., *Boll. soc. ital. biol. sper.*, **5**, 134-136 (1930).
176. Moscona, A., *Exptl. Cell Research*, **3**, 535-539 (1952).
177. ——, *Proc. Soc. exptl. Biol. Med.*, **92**, 410-416 (1956).
178. ——, *Proc. Natl. Acad. Sci. U. S.*, **43**, 184-194 (1957).
179. ——, *Science*, **125**, 598-599 (1957).
180. ——, and Moscona, H., *J. Anat.*, 287-301 (1952).
181. Moscona, H., and Moscona, A., *J. Anat.*, **86**, 278-286 (1952).
182. Moskowska, A., *Arch. anat. microscop. morphol. exptl.*, **45**, 65-76 (1956).
183. Murray, M. R., *Anat. Record*, **57**, Abstracts, 74 (1933).
184. New, D. A. J., *J. Embryol. exptl. Morphol.*, **3**, 326-331 (1955).
185. Nicholas, J. S., *Quart. Rev. Biol.*, **22**, 179-195 (1947).
186. ——, and Rudnick, D., *J. exptl. Zool.*, **78**, 205-232 (1938).
187. Niven, J. S. F., *J. Pathol. Bacteriol.*, **34**, 307-324 (1931).
188. ——, *Arch. exptl. Zellforsch.*, **11**, 253-258 (1931).
189. ——, *Roux' Arch. Entwicklungsmech. Organ.*, **128**, 480-501 (1933).
190. Olivo, O., *Arch. exptl. Zellforsch.*, **1**, 427-500 (1925).
191. ——, *Anat. Anz.*, **66** (Erg.), 108-118 (1928).
192. ——, *Compt. rend. Assoc. Anat.*, **23**, 357-374 (1928).
193. Pfleger, D., *Compt. rend. soc. biol.*, **145**, 425-429 (1951).
194. Preti, A., *Sympos. genet., Pavia*, **3**, 127-139 (1952).

195. Reinbold, R., *Compt. rend. soc. biol.*, **148,** 1493-1495 (1954).
196. Rienhoff, W. F., *Johns Hopkins Hosp. Bull.*, **33,** 392-406 (1922).
197. Rossi, F., and Borghese, E., *Arch. Anat. microscop.*, **31,** 459-478 (1935).
198. ———, and ———, *Compt. rend. Assoc. Anat.*, **30,** 440-443 (1935).
199. Salzberger, B., *Compt. rend. Acad. Sci.*, **236,** 1306-1308 (1953).
200. ———, *Compt. rend. soc. biol.*, **149,** 190-192 (1955).
201. Schaberg, A., *Transplantation Bull.*, **2,** 145-146 (1955).
202. Scheib-Pfleger, D., *Compt. rend soc. biol.*, **147,** 872-875 (1953).
203. ———, *Compt. rend. soc. biol.*, **147,** 875-878 (1953).
204. ———, *Compt. rend Acad. Sci.*, **236,** 1446-1448 (1953).
205. ———, *Bull. biol. France et Belg.*, **89,** 404-490 (1955).
206. Seaman, A. R., and Stahl, S., *Exptl. Cell Research*, **11,** 220-221 (1956).
207. Sengel, P., *Compt. rend. soc. biol.*, **149,** 1032-1035 (1955).
208. ———, *Compt. rend. soc. biol.*, **150,** 2057-2059 (1956).
209. ———, *Experientia*, **13,** 177-182 (1957).
210. Sevastikoglu, J., *Exptl. Cell Research*, **12,** 80-91 (1957).
211. Shaffer, B. M., *Exptl. Cell Research*, **11,** 244-248 (1956).
212. Sidman, R. L., *Anat. Record*, **124,** 581-601 (1956).
213. ———, *Anat. Record*, **124,** 723-739 (1956).
214. Simon, D., *Arch. anat. microscop. morphol. exptl.*, **45,** 290-301 (1956).
215. ———, *Compt. rend. soc. biol.*, **150,** 11-15 (1956).
216. ———, *Compt. rend. soc. biol.*, **150,** 415-417 (1956).
217. ———, *Compt. rend. Acad. Sci.*, **244,** 1541-1543 (1957).
218. Spemann, H., and Mangold, H., *Arch. mikroskop. Anat. u. Entwicklungsmech.*, **100,** 599-638 (1924).
219. Spratt, N. T., *J. exptl. Zool.*, **85,** 171-209 (1940).
220. ———, *J. exptl. Zool.*, **89,** 69-101 (1942).
221. ———, *J. exptl. Zool.*, **106,** 345-365 (1947).
222. ———, *J. exptl. Zool.*, **107,** 39-64 (1948).
223. ———, *J. exptl. Zool.*, **110,** 273-298 (1949).
224. ———, *J. exptl. Zool.*, **114,** 375-402 (1950).
225. ———, *Biol. Bull.*, **99,** 120-135 (1950).
226. ———, *J. exptl. Zool.*, **120,** 109-130 (1952).
227. ———, *J. exptl. Zool.*, **128,** 121-163 (1955).
228. Stenger-Haffen, K., *Compt. rend. Acad. Sci.*, **240,** 1018-1019 (1955).
229. Strangeways, D. H., *Arch. exptl. Zellforsch.*, **11,** 344-345 (1931).
230. Strangeways, T. S. P., and Fell, H. B., *Proc. Roy. Soc. (London)*, B, **99,** 340-366 (1926).
231. ———, and ———, *Proc. Roy. Soc. (London)*, B, **100,** 273-283 (1926).
232. Tansley, K., *Proc. Roy. Soc. (London)*, B, **114,** 79-103 (1934).
233. Tateiwa, K. *J. Japan. Orth. Surg. Soc.*, **30,** 1-22 (1956).
234. Thomson, D., *Proc. Roy. Soc. Med.*, **7,** Marcus Beck Lab. Reps., 21 (1914).
235. ———, *Proc. Roy. Soc. Med.*, **7,** Marcus Beck Lab. Reps., 71 (1914).
236. Tixier-Vidal, A., *Compt. rend. soc. biol.*, **149,** 1377-1379 (1955).
237. ———, *Compt. rend. soc. biol.*, **150,** 1371-1375 (1956).
238. ———, *Arch. anat. microscop. morphol. exptl.*, **45,** 236-253 (1956).
239. ———, *Compt. rend. soc. biol.*, **151,** 5-8 (1957).
240. ———, *Compt. rend. soc. biol.*, **151,** 40-43 (1957).
241. Trowell, O. A., *Exptl. Cell Research*, **3,** 79-107 (1952).
242. ———, *Exptl. Cell Research*, **6,** 246-248 (1954).
243. ———, *Exptl. Cell Research*, **9,** 258-276 (1955).
244. Vakaet, L., *Arch. anat. microscop. morphol. exptl.*, **45,** 48-64 (1956).
245. Vidmar, B., *J. Embryol. exptl. Morphol.*, **1,** 417-423 (1953).
246. Waddington, C. H., *The Epigenetics of Birds*, Cambridge, at the University Press (1952).

247. ———, and Waterman, A. J., *J. Anat.*, **67,** 356-370 (1933).
248. Waterman, A. J., *Am. J. Anat.*, **53,** 317-345 (1933).
249. ———, *Arch. Biol. (Liége)*, **48,** 273-290 (1937).
250. Waymouth, C., *J. natl. Cancer Inst.*, **19,** 495-504 (1957).
251. Weil, J. T., *Schweiz. Z. allgem. Pathol. u. Bakteriol.*, **14,** 205-224 (1951).
252. Weiss, P., and Amprino, R., *Growth,* **4,** 245-258 (1940).
253. ———, and James, R., *Exptl. Cell Research,* **3,** 381-394 (1955).
254. Wolbach, S. B., *J. Bone and Joint Surg.*, **29,** 171-192 (1947).
255. Wolff, Em., *Compt. rend. soc. biol.*, **147,** 864-868 (1953).
256. ———, *Compt. rend. soc. biol.*, **148,** 2078-2080 (1954).
257. ———, *Compt. rend. Acad. Sci.*, **240,** 1016-1017 (1955).
258. ———, *Compt. rend. Acad. Sci.*, **243,** 2154-2156 (1956).
259. ———, *Bull. biol. France et Belg.*, **91,** 271-283 (1957).
260. ———, *Arch. anat. microscop. morphol. exptl.*, **46,** 1-38 (1957).
261. ———, and Wolff, Et., *Compt. rend. soc. biol.*, **146,** 492-493 (1952).
262. Wolff, Et., *Compt. rend. Acad. Sci.*, **234,** 1712-1714 (1952).
263. ———, *Rev. Sci., Paris,* **90,** 189-198 (1952).
264. ———, *Schweiz. med. Wschr.*, **83,** 171-175 (1953).
265. ———, *Bull. Soc. zool. Fr.*, **79,** 357-368 (1954).
266. ———, *Compt. rend. Acad. Sci.*, **242,** 1537-1538 (1956).
267. ———, *Brux. méd.*, **36,** 2235-2243 (1956).
268. ———, *Experientia,* **12,** 321-322 (1956).
269. ———, in *Les facteurs de la croissance cellulaire. Activation et inhibition* (J. A. Thomas, ed.), p. 157-188. Masson & Cie., Paris (1956).
270. ———, *Acta anat.*, **30,** 952-960 (1957).
271. ———, *Symposia Soc. exptl. Biol.*, **11,** 298-304 (1957).
272. ———, and Bresch, D., *Compt. rend. Acad. Sci.*, **240,** 1014-1016 (1955).
273. ———, and Haffen, K., *Compt. rend. soc. biol.*, **145,** 1388-1391 (1951).
274. ———, and ———, *Compt. rend. Acad. Sci.*, **233,** 439-441 (1951).
275. ———, and ———, *Texas Rep. Biol. Med.*, **10,** 463-472 (1952).
276. ———, and ———, *Compt. rend. Acad. Sci.*, **234,** 1396-1398 (1952).
277. ———, and ———, *J. exptl. Zool.*, **119,** 381-403 (1952).
278. ———, and ———, *Compt. rend. soc. biol.*, **146,** 109-111 (1952).
279. ———, and ———, *Compt. rend. soc. biol.*, **146,** 1772-1774 (1952).
280. ———, and ———, *Arch. anat. microscop. morphol. exptl.*, **41,** 184-207 (1952).
281. ———, and ———, *Ann. Endocrinol.*, **13,** 724-731 (1952).
282. ———, ———, Kieny, M., and Wolff, Em., *Compt. rend. Acad. Sci.*, **236,** 137-139 (1953).
283. ———, ———, ———, and ———, *J. embriol. exptl. morphol.*, **1,** 55-84 (1953).
284. ———, ———, and Wolff, Em., *Ann. nutrition et aliment.*, **7,** 5-21 (1953).
285. ———, and Kieny, M., *Compt. rend. Acad. Sci.*, **243,** 2152-2154 (1956).
286. ———, and ———, *Compt. rend. Acad. Sci.*, **244,** 1089-1091 (1957).
287. ———, and Lutz-Ostertag, Y., *Arch. Anat., Strasbourg,* **34,** 459-465 (1951-52).
288. ———, and ———, *Compt. rend. Assoc. Anat.*, **39,** 214-228 (1952).
289. ———, ———, and Haffen, K., *Compt. rend. soc. biol.*, **146,** 1791-1793 (1952).
290. ———, ———, and ———, *Compt. rend. soc. biol.*, **146,** 1793-1795 (1952).
291. ———, Lutz-Ostertag, Y., and Pfleger, D., *Compt. rend. Acad. Sci.*, **229,** 1263-1265 (1949).
292. ———, ———, and ———, *Compt. rend. soc. biol.*, **144,** 280-282 (1950).
293. ———, and Schneider, N., *Compt. rend. soc. biol.*, **150,** 845-846 (1956).
294. ———, and ———, *Arch. anat. microscop. Morphol. exptl.*, **46,** 173-197 (1957).
295. ———, and Simon, D., *Compt. rend. Acad. Sci.*, **241,** 1994-1996 (1955).
296. ———, and Weniger, J. P., *Compt. rend. Acad. Sci.*, **237,** 936-938 (1953).

297. ———, and ———, *J. Embryol. exptl. Morphol.*, **2**, 161-171 (1954).
298. ———, and Wolff, Em., *J. exptl. Zool.*, **116**, 59-97 (1951).
299. ———, and ———, *Compt. rend. soc. biol.*, **146**, 111-113 (1951).
300. ———, and ———, *Bull. biol. France et Belg.*, **86**, 325-349 (1952).
301. ———, and ———, *Poultry Sci.*, **32**, 348-351 (1953).
302. ———, and ———, and Haffen, K., *Compt. rend. Acad. Sci.*, **233**, 500-502 (1951).
303. Zaaijer, J. J. P., *Embryonale pijpbenties van de mens in vitro.* Proefschrift, Leiden, 1953.

# A STUDY OF SEX DIFFERENTIATION IN THE FETAL RAT [1]

DOROTHY PRICE AND RICHARD PANNABECKER[2]

*Zoology Department, The University of Chicago, Chicago, Illinois*

IN THIS RESEARCH, the method of organ culture has been used to investigate the mechanisms which may control normal differentiation of the Wolffian ducts, Müllerian ducts, and accessory reproductive glands in the rat. The specific problem is: are sex hormones from fetal gonads decisive agents in turning the indifferent or ambisexual stage of the fetal reproductive tract in the direction of the male or the female type? Organ culture is especially advantageous as a method for such a study in the fetus of the placental mammal. Reproductive tracts can be isolated at early stages, cultured on a controlled medium and observed during the culture period. The gonads can be removed at explantation with a minimum of traumatic injury so that this important phase of the study can be fairly easily explored.

Three hundred reproductive tracts ranging in age from 13.5 days to 18.5 days post coitum were explanted and cultured by the watch-glass method for periods of one to six days. The standard medium was fowl plasma and chick embryo extract, to which micropellets of testosterone or estradiol were added in some experiments. The results have clarified certain aspects of the problem in the rat (3-6). Some of the most significant results can be summarized with reference to one stage—tracts explanted at 17.5 days of fetal age and cultured for 4 days on standard medium (Fig. 1).

## RESULTS IN THE MALE

At this stage, the male (Fig. 1, A) still possesses both sets of ducts. Dilatations of the Wolffian ducts near the urogenital sinus are well marked. From these, the seminal vesicle primordia normally develop at about 18 days. The Müllerian ducts have begun to retrogress at the anterior ends.

---

[1] This investigation was aided by a grant from the Dr. Wallace C. and Clara A. Abbott Fund of the University of Chicago and by Research Grant 2912 from the National Institute of Health, Public Health Service.

[2] Present address, Biology Department, Ohio Northern University, Ada, Ohio.

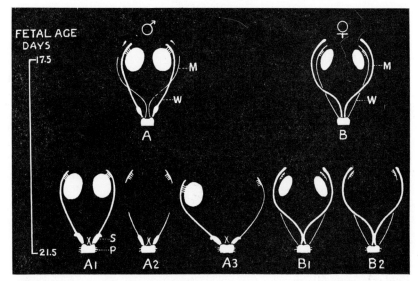

Fig. 1. Diagrammatic representation of the results of culturing fetal rat reproductive tracts; M, Müllerian ducts; P, prostate gland; S, seminal vesicles; W, Wolffian ducts. From *Endocrines in Development* (R. Watterson, ed.), in press.

## Explants with Both Testes

When 17.5-day-old tracts were dissected from the fetus and cultured for 4 days, an essentially normal male differentiation occurred, although the tract did not enlarge in size (Fig. 1, A1). Wolffian ducts persisted, seminal vesicles formed, and prostatic buds grew out from the urogenital sinus. The Müllerian ducts retrogressed antero-posteriorly but continued their posterior development and joined to form the prostatic utricle near the sinus, as in vivo.

## Explants without Testes

If the testes were removed at explantation by severing the efferent ducts, the Wolffian ducts retrogressed antero-posteriorly in vitro, no seminal vesicles developed, and there were fewer prostatic buds (Fig. 1, A2). The retrogression of the Wolffian ducts is not attributable to surgical damage, for if one or both testes were removed and immediately replaced in position, the Wolffian ducts persisted. In the absence of the testes, the addition of testosterone micropellets to the medium resulted in maintenance of the Wolffian ducts and the development of seminal vesicles and prostatic buds, as in A1. The Müllerian ducts in the explants (Fig. 1, A2) underwent anterior retrogression and posterior development as in A1.

### Explants with One Testis

The removal of one testis produced a unilateral effect if the two sides of the tract were spread apart on the clot (Fig. 1, A3). The Wolffian duct was less well maintained at a distance from the testis, and the seminal vesicle was smaller or failed to develop. However, in experiments in which the sides of the tract were close together, the Wolffian ducts and seminal vesicles were equally well developed on both sides. These two types of experiments illustrate the presence of a gradient of diffusion of the effective agent from the testis.

## RESULTS IN THE FEMALE

The double set of ducts is present at 17.5 days and the Wolffian ducts are still complete (Fig. 1, B).

### Explants with Ovaries

In culture a normal type of female differentiation took place. The Müllerian ducts continued to develop and to form the utero-vaginal canal. The Wolffian ducts retrogressed antero-posteriorly (Fig. 1, B1).

### Explants without Ovaries

Removal of the ovaries apparently did not affect the course of differentiation (Fig. 1, B2).

## DISCUSSION AND CONCLUSIONS

The results of culturing indicate that in the male, Wolffian duct maintenance and the development of the primordia of seminal vesicles and prostate glands depend upon the presence of fetal testes or an exogenous male hormone such as testosterone. However, the hormone which diffuses from fetal testes in culture has not been identified. The fact that fetal rat testes secrete hormone late in gestation and that the testis-secreted hormone stimulates the growth of accessory reproductive glands was established by the method of fetal castration (7, 8). In female explants, the Müllerian ducts appeared to be independent of the ovary. There is no indication from culture experiments that the fetal rat ovary secretes hormone.

Although certain aspects of the study are not yet clearly understood, it is probable that the general pattern of sex differentiation in the rat may be similar to that of the rabbit (1, 2).

In conclusion, in the male, the presence of testis-secreted hormone may determine the maintenance of Wolffian ducts, the development of accessory reproductive glands, and the degeneration of Müllerian ducts. In the female, the absence of male hormone may result in degeneration of the Wolffian ducts and persistence of the Müllerian ducts.

## REFERENCES

1. Jost, A., *Arch. Anat. micr. Morph. exptl.*, **39**, 577-607 (1950).
2. ———, *Josiah Macy, Jr. Foundation Trans. Third Conf. Gestation*, N. Y., 129-171 (1957).
3. Pannabecker, R., *Ph. D. Thesis*, Univ. Chicago (1957).
4. Price, D., *Josiah Macy, Jr. Foundation Trans. Third Conf. Gestation*, N. Y., 173-186 (1957).
5. ———, in *Endocrines in Development*, (R. Watterson, ed.), Univ. Chicago Press, Chicago (in press).
6. ———, and Pannabecker, R., *Ciba Foundation Colloquia on Ageing*, **2**, 3-13 (1956).
7. Wells, L. J., *Josiah Macy, Jr. Foundation Trans. Third Conf. Gestation*, N. Y., 187-227 (1957).
8. ———, and Fralick, R. L., *Am. J. Anat.*, **89**, 63-107 (1951).

## DISCUSSION

Dr. McElroy: Would someone clarify for me the Grobstein experiment? Did I understand correctly that contact across the membrane is essential?

Dr. Holtfreter: If I remember correctly in Grobstein's experiments, in which the cell layers were separated by films of varying pore size, no direct cell contact through the pores was required to activate the tissue transformation observed. In a corresponding way, in the case of the transformation of explanted chick somites into vertebral cartilage, Holtzer found that the inducing spinal cord affects the prospective somite cells through an intervening zone of non-cellular material. It appeared that this material (intercellular ground substance?) mediates or possibly further specifies the chondrogenic agent. Interestingly enough, the vertebrae in the normal embryo develop at a slight distance and not in direct contact with the spinal cord.

Much emphasis has been placed on the observation that in other instances of embryonic induction, such as in neural and lens induction, intimate contact between inducing and reacting tissues is required. To be sure, this occurs under normal conditions, and many experiments supported this view. It implied that neurogenic agents, unlike hormones, do not diffuse freely. However, in this instance too, several other experiments have shown that the conversion of gastrula ectoderm into neural, muscle and other tissues can occur through the application of non-cellular tissue extracts. Just a while ago, Dr. Yamada gave us the most impressive demonstration of it. So did Dr. Niu and myself.

This would mean that probably in all cases of embryonic induction, it is not so much a matter of stereometric fittingness of the molecular populations of the adjacent cell surfaces—the inducing cells calling forth and re-organizing the molecular structuration at the surface of the competent reacting cells (P. Weiss) —but rather a matter of chemico-physical reactions occurring inside the cytoplasm. This notion does not minimize the essential role which the cell membrane plays in transmitting or blocking the entrance of activating or inducing agents. It appears that many of the effectively neuralizing agents, such as alkali, alcohol or distilled water, simply increase the permeability of the cell membrane, allowing for a flooding of the cell with water. But the essential metabolic switch-over created by these "unspecific" external agents and leading to neuralization of the affected ectoderm, seem to occur intracellularly rather than at the cell surface. The latter follows suit and secondarily provides the mechanisms to shape a cell into a neuroblast, myoblast or other cell types.

Dr. Hamburger: In conclusion I should say that we are very grateful to Dr. Borghese for his comprehensive and lucid review of the field of organ culture. In this particular technique, we have a system which enables us to study tissue interactions under somewhat simpler conditions than we find in the organism. The recent advances in this field are very promising, indeed. I wish to thank Dr. Borghese and the other speakers of this afternoon for their participation.

# ROLE OF GENES IN DEVELOPMENTAL PROCESSES

### Ernst Hadorn

*Zoologisch-vergl. anatomisches Institut der Universität Zürich,
Zürich, Switzerland*

Among the immense problems which the student of developmental genetics tries to solve only a few can be treated in this report. We shall confine ourselves to a discussion of the following questions. When do genes exert their specific action during development? How are the character-forming processes related to the primary action of genes? On what mechanism is cell-specific and organ-specific genic action based? How are the different characters (phenes) of a pleiotropic pattern of manifestation interrelated? Though our symposium deals with the chemical basis of development, adequate biochemical information can be given in only a few cases.

## Genic Action along the Time Axis of Development

All chromosomal genes are present from the moment of fertilization on, throughout the whole development. Do all of them act at all times, or can we assign to the different loci restricted periods of activity? Certainly in one respect all genes are active simultaneously and during a much extended developmental range: they reproduce at each mitotic cycle. But we do not know whether this fundamental process of forming identical replicas has anything to do with the formation of those primary gene products which are at the base of the character-forming processes.

The interesting observation made on salivary gland chromosomes of dipterous insects, that definite sections show cyclic developmental changes, might demonstrate that there are limited periods of genic action (Beermann, 1952, 1956; Pavan and Breuer, 1955). The swelling and regression of such "puffs" could in fact stand for the dynamics of genic loci. One has, however, to be cautious in interpreting any cytomorphological observations as signs of primary activities. Some details of "nuclear differentiation" might be nothing more than the consequence of mere reactivity towards extragenic cellular conditions.

There are, however, facts available which help us to answer a subquestion of our general problem. It is definitely proved that all the genes of an organism are not needed for the successful achievement of definite developmental processes. Poulson (1945) showed that genotypes of *Drosophila melanogaster* deficient for chromosomal blocks, including besides the *Notch* locus up to 45 further bands of the X-chromosome, are able to cleave, to form a blastoderm, and to initiate organogenesis and histogenesis. After this performance, it is true, development comes to a standstill and the embryos die. Deficiencies affecting other chromosomal segments allow still further development. Thus the *white* deficiency, which can remove up to 10 bands in addition to the *white* locus, does not interfere with normal differentiation of the ectoderm; and embryos carrying the *scute* deficiency (up to 11 bands) develop into fully differentiated larvae which die only at hatching time. We conclude from these cases that loci missing in the respective genotypes have no essential functions with respect to those developmental steps preceding the lethal effect.

On the other hand, the deficiencies do reveal the existence of genes which are absolutely necessary for definite developmental achievements. These mutants show us further at what time in ontogeny the action of individual genes becomes essential (Hadorn, 1948b, 1955). Thus all of the *Notch* mutants, among which deficiencies of quite different extent, from one to 45 bands, are represented, manifest the same pattern of damage; and all come to a standstill after the 6th hour of embryonic life. The common factor in all *Notch* deficiencies is the absence of the locus $3c7$, the action of which must be essential for development beyond the stage designated. Apparently the *white* and *scute* deficiency mutants reach a higher developmental level because in them no locus of early action, such as $3c7$, is missing. But here too the later breakdown of development must be due to the absence of some definite genic locus. Generalizing, we conclude that the damaging effect of a deficiency is caused by the absence of whatever genic locus involved has the earliest essential role in normal development.

This interpretation is in accordance with the "principle of stepwise insertion of the products of different genes in ontogeny" (Hadorn, 1948b, 1955). At the beginning of development only a limited number of genic loci would be needed. The further ontogenesis proceeds, the more genes would have to set in with a specific and indispensible function. Taking this principle for granted, we can infer the following facts:

1. An extended deficiency or loss of whole chromosomes causes in *Drosophila* an early embryonic death, because the more extended a deficiency, the higher the probability that some locus of early essential action has been lost.

2. Nevertheless, groups of overlapping deficiencies of quite differ-

ing lengths can cause, as shown before, identical developmental effects. This uniformity would be due to the loss of that locus, common to all members of the group, which is first to exert its essential action.

3. There exist along the ontogenetic time axis of an organism limited sensitive phases during which there is a characteristically enhanced probability that any mutant will lead to a standstill of development. Therefore, we find that the developmental crises of the various lethal factors known in *Drosophila* cluster mainly at the border between embryonic and larval life, at the onset of metamorphosis, and at the hatching time of the fly from the pupa (Hadorn, 1948a; Hadorn and Chen, 1952). Such phases especially sensitive to gene mutation can easily be demonstrated in other organisms as well (Hadorn, 1955). We assume that these phases indicate periods of development where many new systems are first differentiated, or where such systems must stand the test of normal function. The observed clustering of effects of mutations would indicate that in each of these sensitive phases groups of many genes for the first time furnish their essential and locus-specific contribution to development.

4. *Merogonic hybrids* are systems in which the enucleated egg cytoplasm of one species is combined with the sperm nucleus of another species. Experiments on amphibia and echinoderms have shown that such chimeric cellular systems develop up to a definite embryonic or larval stage before a lethal crisis sets in (cf. Baltzer, 1940). The phase which can be reached is correlated with the taxonomic relationship of the combined species. Thus the cytoplasm of the newt *Triton palmatus* develops with the nucleus of *Triton taeniatus* up to a late larval stage, with the nucleus of *Triton alpestris* only up to a much younger larval stage; when the *T. palmatus* cytoplasm is combined with the nucleus of *Triton cristatus,* development stops in the embryo shortly after neurulation (Hadorn, 1934).

On the basis of our principle of stepwise insertion of genic function we can explain the differences between various merogonic hybrids in the following way. The more remote two species are taxonomically, the lower is the probability that all genic loci of early essential action required by the cytoplasm are present in the foreign nucleus. A harmonious collaboration of the specific factors present in cytoplasm and nucleus is in such cases possible only during a highly restricted ontogenetic phase. Between more closely related species the nuclei can be mutually interchanged over a longer developmental period. But here, too, the merogonic system will eventually meet a task for which one or more genes or mutated states of genes are needed which are not available in the foreign nucleus.

To explain the differently timed breakdown of different merogonic hybrids in terms of individual genes is by no means the only possible or reasonable way. But our interpretation has the advantage of being in some accord with experience gained from the behavior of the deficiency mutants. There is furthermore some biochemical evidence which is reasonably compatible with our interpretation. Zeller (1956), working on the early lethal merogon *T. palmatus*-cytoplasm × *T. cristatus*-nucleus, found this system able to synthesize normal amounts of RNA. According to Brachet (1952, 1957) this RNA would be produced by the nucleus. The genic makeup of *T. cristatus* might release into the cytoplasm of *T. palmatus* a molecular population of RNA different in one or more respects from that which a *T. palmatus* nucleus would produce. Even single gene differences between the two species might be decisive. Zeller's findings led moreover to the conclusion that the cytoplasm of the merogonic hybrid cannot utilize the foreign RNA for metabolic purposes. In fact, in this merogon the RNA accumulates to an abnormally high level with the onset of the lethal crisis.

### Early Embryonic Action of Genes

In interpreting deficiencies in *Drosophila* we have concluded that although all genes are not needed for embryonic development, early ontogenetic processes are nevertheless dependent on the locus-specific action of individual genes. Since the validity of this concept is not generally accepted, we have to look for further and more direct corroborating evidence.

Among the numerous genes of *Drosophila* two are known, the action of which takes place almost immediately after fertilization of the egg. The Mendelian factor *Suppressor of erupt* (*Su-er*) prevents the malforming action of another gene *erupt* (*er*) on the imaginal eyes. X-irradiation of eggs as early as 8 minutes after fertilization inactivates or destroys an agent, the formation of which depends on the presence of the *Su-er* gene in the zygote (Glass and Plaine, 1950). This demonstration that the suppressor gene functions immediately subsequent to its introduction by fertilization is especially important since it indicates activity of the gene long before the imaginal discs are formed.

The second factor to be mentioned here is *deep orange* (*dor*); its developmental action was investigated by Counce (1956). In eggs from homozygous *dor* mothers, fertilized by *dor* sperms, development ceases during completion of the maturation divisions or in an early cleavage stage. If such eggs are fertilized by a sperm introducing the normal allele of *dor* (*dor*$^+$), normal development ensues. The *dor* case demonstrates a paternal gene becoming active in the egg system even prior to the formation of the zygotic nucleus.

In mammals, several factors have been found which cause early embryonic death in a strict phase-specific manner (cf. Hadorn, 1955). Well known is the classical case of the *Yellow* factor of the house mouse. Here the homozygotes degenerate between the 5th and the 6th day of embryonic life, that is, shortly after the blastocyst comes into contact with the epithelium of the uterus (Robertson, 1942). Within the complex *Tailless* locus (Dunn, 1956) different sites of mutation have been demonstrated, each of which causes, when homozygous, a characteristic phase-specific lethality. The $T/T$ fetus dies between the 10th and the 11th day, whereas $t^0/t^0$ embryos come to a standstill at the 6th day (Dunn and Glueksohn-Waelsch, 1952). A still earlier breakdown was recently described by Smith (1956) for the $t^{12}$ factor. The homozygous carriers of this mutation become abnormal during the fourth day after fertilization. They die as morulae, the few cells disintegrating before a blastocyst can be formed. From this observation we can conclude that within the chromosomal segment where the $t^{12}$ mutation took place, there must be a genic constitution essential for the developmental steps leading from the cleavage stage to the early blastocyst. In general, there is now ample evidence to indicate that even the earliest and most fundamental ontogenetic performances depend on the specific action of Mendelian genes.

### Primary Activity versus Manifestation

The fact that definite genes become functionally indispensable only during specific sensitive phases, does not imply that such genes may not have been active long before. The concept of the stepwise insertion of genic function in ontogeny is not necessarily at variance with the assumption that all genes produce their specific agents all the time. As long as we know so little about the nature of the primary gene products, it is in most cases impossible to determine the onset of activity of any gene. It is also a possibility that a genic locus, after passing through a functional period, might later become inactive. There is however a line of investigation which promises to supply some information. If we speculate that enzymes or antigens are close to being primary gene products, or may even be the primary products themselves, the detection of such a specific substance and its localization during ontogeny would be extremely useful. As an example of such an approach, we may discuss the action of a gene in *Drosophila*.

The mutant *rosy* (*ry*) is externally characterized by brownish eyes, hardly differing in respect to this phene, from several other eye-color mutants. By chromatographing an *ry* allele, which newly appeared in our cultures ($ry^2$), we found the *ry* gentoype unique with regard to an unexpected biochemical phene. *Ry* does not contain isoxanthopterine (Hadorn and Schwinck, 1956a, b). Instead, the immediate precursor of this compound, 2-amino-4-hydroxypteridine, accumulates in *rosy* flies to an abnormally high level. There are other pteridines the quantities of which are much affected by the *ry* mutant. In collaboration with Dr. H. K. Mitchell, of Pasadena, it was further found that *ry* does not form any uric acid but instead forms hypoxanthine as an excretion product (cf. Hadorn, 1956b). At least parts of this pattern of biochemical pleiotropy are now explained by the fact that the *ry* genotype lacks the enzyme xanthine oxidase or xanthine dehydrogenase (Glassman and Mitchell, 1958).

It is known that this polyvalent enzyme catalyzes the oxidation of hypoxanthine through xanthine to uric acid and the formation of isoxanthopterine from 2-amino-4-hydroxypteridine (Forrest, Glassman, and Mitchell, 1956). The xanthine dehydrogenase present in a normal *Drosophila* can be considered as a more or less direct product of the activity of the normal allele of rosy ($ry^+$). We might call it an $ry^+$ substance. We can now ask when, in development, does the $ry^+$ locus come into action. A student taking notice only of the visible eye color will find the gene active not earlier than at the end of metamorphosis. If we consider other pteridine phenes, we come to the conclusion that the locus-specific action takes place at the onset of metamorphosis. In fact, however, the $ry^+$ gene is active much earlier, since the presence (in $ry^+$) or the absence (in *ry*) of the gene-conditioned xanthine dehydrogenase can be demonstrated throughout larval life (Glassman and Mitchell, 1958). The enzyme may be present even in embryos: this remains to be investigated.

We learn from the *rosy* mutant that a gene-conditioned product can be present long before any easily detectable phenes become apparent. Thus, as far as the action on pteridines is concerned, the xanthine dehydrogenase has to wait until adequate amounts of the substrate 2-amino-4-hydroxypteridine are available. This compound is found only after pupation (Hadorn and Ziegler-Günder, 1958). The same reasoning holds true for the above-mentioned action of the *Suppressor of erupt*. Here the phenotypic manifestation cannot take place before the formation of the eye anlagen. We have to wait for further combined biochemical and developmental studies before definite statements as to the timing of gene activity in ontogeny can be made. One fact is clear however: the stepwise insertion of genic products in the character-forming processes by no means corresponds with a time sequence of primary genic activities.

## The Cell-Specific and Organ-Specific Pattern of Action

A given gene becomes manifest not only in a phase-specific manner, it also exerts different effects in different cellular and developmental systems. Thereby a complicated pattern of manifestation will arise which usually includes many different cell- and organ-specific phenes. Numerous mutants showing such pleiotropic syndromes may be cited (Hadorn, 1955). The *wingless* (*wg*) factor of the chick, for instance, leads to the following phenes: the wings do not develop; the legs grow out, but their toes are much malformed. Feathers are poorly differentiated. Lungs and air sacs are missing. There is no ureter, and as a consequence of its absence no metanephros differentiates (Waters and Bywaters, 1943; Zwilling, 1949). Other embryonic organs develop apparently more or less normally. We have no idea what sort of biochemical or physiological mechanism might be involved in producing such a strange pleiotropic pattern of damage.

In studying other lethal mutants we meet similar difficulties. The factor *lethal giant larvae* (*lgl*) of *Drosophila melanogaster* causes the breakdown of development in homozygous carriers at the beginning of metamorphosis. By transplanting the different imaginal blastemas from lethal donors into normal hosts it was found that some cellular systems, e.g., the somatic cells of the gonads, the primordium of the hindgut, and the corpora allata and corpora cardiaca, are able to metamorphose. On the other hand, the imaginal rings of the proventriculus and the salivary glands, the male germ cells, and the big imaginal discs of the thorax lack the potencies for imaginal differentiation (cf. Hadorn, 1955). Why does a genic constitution present in all these cells determine the developmental fate and viability in different blastemas so differently? Perhaps the biochemical requirements of the cells of the hindgut are different from those of the foregut. One system might therefore be sensitive to the dysfunction or the nonfunctioning of the *lgl* locus, whereas the others would not be affected.

Apparently we need information on the extent to which mutations can alter differentially the biochemical makeup in various cell systems of an organism. In the meal moth *Ephestia kühniella* by using chromatographic methods we found a mutation *biochemica* (*bch*) in which imaginal eyes contain only two pteridines: isoxanthopterine and 2-amino-4-hydroxy-pteridine; whereas in normal genotypes many other pteridines are present in the eyes (Hadorn and Kühn, 1953; Hadorn and Egelhaaf,

1956). In the *biochemica* abdomen, however, the whole inventory of substances characteristic for the wild type is found. On the other hand, in the red-eyed mutant (*a*) the pteridines of the eyes differ from the wild type only quantitatively, whereas in the abdomen qualitative differences in pteridines are found between *a* and $a^+$.

Using these mutants as models for cell-specific patterns of damage we can argue as follows: if one or the other of the pteridines, affected by *bch* or by *a*, were essential for cell life (which is not in fact the case) we should find in *biochemica* a localized breakdown of development in the eyes but a normal functioning of abdominal organs; whereas in the *a* genotype the abnormal phenes should be restricted mainly to the abdomen. Chromatographic studies on other mutants of *Drosophila* have confirmed the finding that genes influence the inventory of substances differently in different cell systems. For instance, in the *rosy* mutant biopterine is present in the eyes and testes but hardly any is found in Malpighian tubes, whereas in the Malpighian tubes of the wild type this compound is plentiful. Furthermore, *rosy* females have a normal amount of kynurenine in the ovaries, whereas *rosy* males, in contrast to the wild type, lack this derivative of tryptophan in the Malpighian tubes (Handschin, unpub.).

From the phenomenon of cell specificity of genic manifestation problems arise similar to those discussed already with regard to phase specificity. Are the revealed biochemical patterns due to corresponding differences in primary genic activities, or are they merely the consequences of an unequal reactivity towards an over-all and identical genic function (Hadorn, 1955)? If we consider enzymes as being close to primary gene products, we might gain some information with respect to our question by determining whether or not a gene-conditioned enzyme is present in different cellular systems of an organism.

## Extrinsic Supply of an Enzyme?

The fact that the mutant *rosy* lacks xanthine dehydrogenase and consequently forms no isoxanthopterine has already been discussed. What will happen in an *ry* testis if we transplant it into a normal host in which xanthine dehydrogenase is present? And how will a normal testis react within an *ry* abdomen? We transplanted larval testes at a time when they still contained no isoxanthopterine and we measured the content of this compound, which is synthesized only after pupation, in the implanted and in the host testes in one-day-old flies. The reciprocal transplantations yielded clear-cut results (Hadorn, Graf, and Ursprung, 1958).

An *ry* testis, implanted in a normal female or male larva, contained after metamorphosis the same quantity of isoxanthopterine as a control implant of $ry^+$ constitution. On the other hand, an $ry^+$ testis was not able to form any isoxanthopterine when it had to develop within an $ry$ host. Thus we are dealing here with one of those rare cases of the non-autonomous manifestation of an hereditary character. How are we to interpret these results? We know first of all that 2-amino-4-hydroxypteridine, which is the direct precursor of isoxanthopterine, is present in $ry$ as well as in $ry^+$ testes, and we have found in measurements that in those testes which did not contain isoxanthopterine the precursor accumulated to abnormally high levels. Certainly the biochemical machinery of a testis is not self-sufficient in producing isoxanthopterine. Some agent which is only available in $ry^+$ has to be furnished from other cell systems.

In character-forming processes which often are based on chain reactions, we have to distinguish between intrinsic and extrinsic links. The *intrinsic links* are derived from the genic makeup of the character-forming cells, whereas the *extrinsic links* are provided from outside. They are products of other cellular systems within the same organism (Hadorn, 1950, 1955). The $ry^+$ substance is an example of such an extrinsic link. What is its nature? The most reasonable assumption so far, the one least at variance with the available evidence, is to consider the xanthine dehydrogenase as being the active $ry^+$ substance.

Besides the fact that the *rosy* mutant is the first case in *Drosophila* where an enzyme can be identified as a locus-specific product (Glassman and Mitchell, 1958), we here come across a developmental system in which such an enzyme seems to be formed in only certain cellular systems. From further experiments we know that the Malpighian tubules and the fat bodies are potent sources of the $ry^+$ agent (Hadorn and Schwinck, 1956b; Hadorn, 1956a; Hadorn and Graf, 1958). Finally we have to ask why the $ry^+$ locus, though present in the testes, does not provide intrinsically the enzyme xanthine dehydrogenase. Either this locus-specific enzyme is not a primary genic product, or else the rosy case is at variance with a concept postulating an overall function of genic loci in all cellular systems.

## Autophenes and Allophenes

In many pleiotropic patterns of genic action so far analyzed, characters (phenes) are met which behave autonomously in transplants whereas other phenes can be shown to be determined by the genic specificity of extrinsic links (cf. Hadorn, 1955). The locus-specific characteristics of the first category, the *autophenes,* are intrinsically conditioned. In contrast, *allophenes,* of which the described presence or absence of isoxanthopterine in *Drosoph-*

*ila* testes is an example, are due to gene-specific influences of foreign cellular systems. The relevance of these concepts has been experimentally tested in many cases of pleiotropic systems in *Drosophila* by the author and his collaborators. From the work on higher vertebrates only one example will be presented.

The exhaustively studied *Creeper* factor (*Cp*) of the chick causes a pleiotropic pattern of damage in those homozygotes which die at the end of incubation. The extremities remain stunted and malformed in a phocomelic fashion. Microphthalmia and coloboma are the *Cp* phenes of the eyes (Landauer, 1933). In transplants the characters of the *Cp* limbs behave as irreparable autophenes (Hamburger, 1941; Rudnick, 1945), whereas the abnormalities of the eyes, being allophenes, are not manifest in normal hosts (Gayer and Hamburger, 1943).

## The Biochemical Approach

Compared with the immense amount of information gained from the study of microorganisms, we know indeed very little with regard to the biochemical action of genes in the development of higher organisms. On account of the fact that in highly organized plants and animals a great variety of different cell systems are laid down and differentiated, the manifestation of genes becomes extremely complicated. Furthermore, the methods suitable for detecting the biochemical bases of mutants are far less favorable in higher organisms than in microbiological objects.

As a rule the student of genic actions working on multicellular systems has to start his analysis from some easily detectable and superficial "indicator-phenes." They are usually of morphological nature. He will then try to relate these hereditary traits to more fundamental genic actions which might be found on the physiological level. Being aware of the fact that each gene-conditioned process must ensue from one or more primary biochemical products, his final goal will be to find these basic or "initial-phenes."

Such an approach, which succeeded however only in exploring half of the way which leads from the gene to the pleiotropically conditioned phenes, was undertaken in analysing several lethal mutants of *Drosophila* (Hadorn and Mitchell, 1951; Hadorn and Stumm-Zollinger, 1953; Chen and Hadorn, 1955; Hadorn, 1956a; Chen, 1956). The mutant *letal-translucida* (*ltr*) is characterized by its bloated body, which is due to an accumulation of a tremendous amount of hemolymph. Starting from this indicator-phene we were able to show that the *ltr* genotype becomes unable to synthesize proteins before metamorphosis. Correspondingly, the free amino acids reach an abnormally high level in the hemolymph. Another mutant, *letal-meander* (*lme*), in which growth comes to an end in the middle of the third instar and which never pupates, starves to death because, beginning at a definite sensitive phase, it cannot digest the proteins of the food. In both cases some enzyme systems might be affected. But certainly we have not discovered the initial phenes.

By starting an analysis on the outermost twigs of the pleiotropic tree of phenes one might succeed in reaching a few bigger branches and finally the stem. Is there really only one stem? Here we have to face the much discussed question (cf. Grüneberg, 1948; Wagner and Mitchell, 1955; Hadorn, 1955) of whether all pleiotropy is of a secondary nature, or whether one single locus might produce in one or in different cell systems several different primary products. Certainly the "one gene, one enzyme" concept of Beadle (1945) was and is still of heuristic value. Unfortunately, the science of our day does not know the individual structure of even one concrete locus, and nobody has so far demonstrated a really primary genic product. But if the template concept, according to which primary products are formed on the "surface" of the gene molecules, is in fact correct, one might consider it possible that a gene could condition several primary products. A certain gene-specific sequence of amino acids determined by the template might then be incorporated as building-blocks in different peptides or proteins. From such a mechanism a genuine primary pleiotropy could be initiated.

## REFERENCES

1. Baltzer, F., Ueber erbliche letale Entwicklung und Austauschbarkeit artverschiedener Kerne bei Bastarden. *Naturwiss.*, **28**, 177-187; 196-206 (1940).
2. Beadle, G. W., Biochemical genetics. *Chem. Revs.*, **37**, 15-96 (1945).
3. Beermann, W., Chromomerenkonstanz und spezifische Modifikationen der Chromosomenstruktur in der Entwicklung und Organdifferenzierung von *Chironomus tentans*. *Chromosoma*, **5**, 139-198 (1952).
4. ———, Nuclear differentiation and functional morphology of chromosomes. *Cold Spring Harbor Symposia Quant. Biol.*, **21**, 217-232 (1956).
5. Brachet, J., The role of the nucleus and the cytoplasm in synthesis and morphogenesis. *Symposia Soc. Exptl. Biol.*, **6**, 173-200 (1952).
6. ———, *Biochemical Cytology*, Academic Press, New York (1957).
7. Chen, P. S., Elektrophoretische Bestimmung des Proteingehaltes im Blut normaler und letaler (*ltr*) Larven von *Drosophila melanogaster*. *Rev. suisse Zool.*, **63**, 216-229 (1956).
8. ———, and Hadorn, E., Zur Stoffwechselphysiologie der Mutante letal-meander (*lme*) von *Drosophila melanogaster*. *Rev. suisse Zool.*, **62**, 338-347 (1955).
9. Counce, S. J., Studies on female-sterility genes in *Drosophila melanogaster*. 1. The effects of the gene deep orange on embryonic development. *Z. indukt. Abstamm.-u. Vererbungslehre*, **87**, 443-461 (1956).

10. Dunn, L. C., Analysis of a complex gene in the house mouse. *Cold Spring Harbor Symposia Quant. Biol.,* **21,** 187-195 (1956).
11. ———, and Gluecksohn-Waelsch, S., The genetic behavior of seven new mutant alleles which arose in one balanced lethal line in the house mouse. *Genetics,* **37,** 577-578 (1952).
12. Forrest, H. S., Glassman, E., and Mitchell, H. K., Conversion of 2-amino-4-hydroxypteridine to isoxanthopterin in *Drosophila melanogaster. Science,* **124,** 725-726 (1956).
13. Glass, B., and Plaine, H. L., The immediate dependence of the action of a specific gene in *Drosophila melanogaster* upon fertilization. *Proc. Natl. Acad. Sci. U. S.* **36,** 627-634 (1950).
14. Glassman, E., and Mitchell, H. K., Mutants of *Drosophila melanogaster* deficient in xanthine dehydrogenase. Manuscript (1958).
15. Gayer, K., and Hamburger, V., The developmental potencies of eye primordia of homozygous Creeper chick embryos tested by orthoptic transplantation. *J. exptl. Zool.,* **93,** 147-180 (1943).
16. Grüneberg, H., Genes and pathological development in mammals. *Symposia Soc. Exptl. Biol.,* **2,** 155-176 (1948).
17. Hadorn, E., Ueber die Entwickungsleistungen bastardmerogonischer Gewebe von *Triton palmatus* ♀ x *Triton cristatus* ♂ im Ganzkeim und als Explantat in vitro. *Wilhelm Roux' Arch. Entwicklungsmech. Organ.,* **131,** 238-284 (1934).
18. ———, Gene action in growth and differentiation of lethal mutants of *Drosophila. Symposia Soc. Exptl. Biol.,* **2,** 177-195 (1948a).
19. ———, Genetische und entwicklungsphysiologische Probleme der Insektenontogenese. *Folia biotheor.,* **3,** 109-126 (1948b).
20. ———, Physiogenetische Ergebnisse der Untersuchungen an *Drosophila-Blastemen* aus letalen Genotypen. *Rev. suisse Zool.,* **57,** 115-128 (1950).
21. ———, Developmental action of lethal factors in Drosophila. *Advances in Genet.,* **4,** 53-85 (1951).
22. ———, *Letalfaktoren in ihrer Bedeutung für Erbpathologie und Genphysiologie der Entwicklung,* Thieme Verlag, Stuttgart (1955).
23. ———, Patterns of biochemical and developmental pleiotropy. *Cold Spring Harbor Symposia Quant. Biol.,* **21,** 363-373 (1956a).
24. ———, Erbkonstitution und Merkmalsbildung. *Verhandl. schweiz. naturforsch. Ges.,* **1956** (Basel), 52-66 (1956b).
25. ———, and Chen, P. S., Untersuchungen zur Phasenspezifität der Wirkung von Letalfaktoren bei *Drosophila melanogaster. Arch. Julius Klaus-Stift. Verebungsforsch. Sozialanthropol. u. Rassenhyg.,* **27,** 147-163 (1952).
26. ———, and Egelhaaf, A., Biochemische Polyphänie und Stoffverteilung im Körper verschiedener Augenfarb-Genotypen von *Ephestia kühniella. Z. Naturforsch.,* **11b,** 21-25 (1956).
27. ———, and Graf, G. E., Weitere Untersuchungen über den nicht-autonomen Pterinstoffwechsel der Mutante rosy von *Drosophila melanogaster. Zool. Anz.* **160,** 231-243 (1958).
28. ———, ———, and Ursprung, H., Der Isoxanthopterin-Gehalt transplantierter Hoden von *Drosophila melanogaster* als nicht-autonomes Merkmal. *Rev. suisse Zool.,* **65,** 335-342 (1958).
28a. ———, and Kühn, A. Chromatographische und fluorometrische Untersuchungen zur biochemischen Polyphänie von Augenfarb-Genen bei *Ephestia kühniella. Z. Naturforsch.,* **8b,** 582-589 (1953).
29. ———, and Mitchell, H. K., Properties of mutants of *Drosophila melanogaster* and changes during development as revealed by paper chromatography. *Proc. Natl. Acad. Sci. U. S.,* **37,** 650-665 (1951).

30. ———, and Schwinck, I., A mutant of *Drosophila* without isoxanthopterine which is non-autonomous for the red eye pigments. *Nature*, **177**, 940-941 (1956a).
31. ———, and ———, Fehlen von Isoxanthopterin und Nicht-Autonomie in der Bildung der roten Augenpigmente bei einer Mutante (rosy²) von *Drosophila melanogaster*. *Z. indukt. Abstamm.-u. Vererbungslehre*, **87**, 528-553 (1956b).
32. ———, and Stumm-Zollinger, E., Untersuchungen zur biochemischen Auswirkung der Mutation "letal-translucida" (*ltr*) von *Drosophila melanogaster*. *Rev. suisse Zool.*, **60**, 506-516 (1953).
33. ———, and Ziegler-Guender, I., Untersuchungen zur Entwicklung, Geschlechtsspezifität und phänogenetischen Autonomie der Augen-Pterine verschiedener Genotypen von *Drosophila melanogaster*. *Z. Verebungslehre* **89**, 221-234 (1958).
34. Hamburger, V., Transplantation of limb primordia of homozygous and heterozygous chondrodystrophic ("Creeper") chick embryos. *Physiol. Zool.*, **14**, 355-365 (1941).
35. Landauer, W., Untersuchungen über das Krüperhuhn. IV. Die Missbildungen homozygoter Krüperembryonen auf späteren Entwicklungsstadien. (Phokomelie und Chondrodystrophie.) *Z. Mikroskop. anat. Forsch.*, **32**, 359-412 (1933).
36. Pavan, C., and Breuer, M. E., Differences in nucleic acid content of the loci in polytene chromosomes of "*Rhynchosciara angelae*" according to tissues and larval stages. *Symposium on Cell Secretion*, Belo Horizonte, 90-99 (1955).
37. Poulson, D. F., Chromosomal control of embryogenesis in *Drosophila*. *Amer. Nat.*, **79**, 340-363 (1945).
38. Robertson, G. G., An analysis of the development of homozygous yellow mouse embryos. *J. Exptl. Zool.*, **89**, 197-231 (1942).
39. Rudnick, D., Differentiation of prospective limb material from Creeper chick embryos in coelomic grafts. *J. Exptl. Zool.*, **100**, 1-17 (1945).
40. Smith, L. J., A morphological and histochemical investigation of a preimplantation lethal ($t^{12}$) in the house mouse. *J. Exptl. Zool.*, **132**, 1-83 (1956).
41. Wagner, R. P., and Mitchell, H. K., *Genetics and Metabolism*, John Wiley & Sons, New York (1955).
42. Waters, N. F., and Bywaters, J. H., A lethal embryonic wing mutation in the domestic fowl. *J. Heredity*, **34**, 213-217 (1943).
43. Zeller, Ch., Ueber den Ribonukleinsäure-Stoffwechsel des Bastardmerogons *Triton palmatus* ♀ × *Triton cristatus* ♂. *Wilhelm Roux' Arch. Entwicklungsmech. Organ.*, **148**, 311-335 (1956).
44. Zwilling, E., The role of epithelial components in the developmental origin of the "wingless" syndrome of chick embryos. *J. Exptl. Zool.*, **111**, 175-187 (1949).

## DISCUSSION

Dr. Kalckar: What happens when a wild-type egg of *Drosophila* is fertilized by a mutant deep orange sperm?

Dr. Hadorn: This gives rise to a normal embryo.

Dr. Weber: I would like to ask Dr. Hadorn if he has also studied the enzymes concerned with protein metabolism in the mutant letal-translucida? This would be very interesting because of the deficit in proteins in this mutant.

Dr. Hadorn: No, we haven't started on this particular line of research yet. However, we have begun work on the lethal mutant letal-meander.

Dr. Hartman: I wonder, Dr. Hadorn, if your rosy mutant might not have an enzyme inhibitor or possibly another enzyme which more rapidly utilizes an essential precursor of the eye pigments.

Dr. Hadorn: I believe that this possibility is almost completely ruled out by some work we have done in vitro. Thus I don't think the rosy mutant can be explained by an inhibitor. Dr. H. K. Mitchell and coworkers came to the same conclusions.

Dr. Kalckar: I would like to comment on the beautiful specimen, the rosy mutant. About 10-15 years ago the late Prof. Heinrich Wieland showed that xanthine oxidase can also act as a pterine oxidase. Indeed it is the same enzyme. In this particular case, realizing that the biosynthesis of pterine comes later in development, I don't know if this is a valid explanation, but I do feel that one could venture to formulate the situation as follows: The mere lack of specificity of xanthine oxidase constitutes the molecular basis for the pleiotropic effect of the gene.

Dr. Hadorn: Yes, I agree, we try to explain the rosy phenes by the lack of the polyvalent xanthine oxidase. However, we do have difficulty in interpreting certain morphological facts. This is almost always the case in pleiotropic patterns. Usually it is rather easy to explain some of the results in biochemical terms but then one runs into difficulties.

Dr. Maas: I have a rather vested interest in eye pigments for I worked on them many years ago. I would like to ask about the sepia mutant. In sepia I feel you just have one pigment, as far as the red component is concerned.

Dr. Hadorn: In the sepia eye we find besides the visible pigments which are ommochromes a yellow pteridine and several other pteridines but no drosopterine (red eye pigments).

Dr. Maas: In the case of sepia-vermilion or sepia-cinnabar, several components of the red pigment are not formed and instead a different component appears. I wonder if this has been analyzed in biochemical terms.

Dr. Hadorn: Well, it's a long story and it's extremely difficult to know exactly what the sequence of pteridine biosynthesis is. There is an extremely complex interrelationship of pteridines and in addition to this the pathways for the syntheses of ommochromes and the drosopterins can become interlinked at the very beginning of their biosynthesis.

Dr. Suskind: I think, if I am not mistaken, Glassman and Mitchell reported in some xanthine dehydrogenase mutants of *Drosophila* that an antigen immunologically related to this enzyme, but enzymatically inactive, appeared. Is this your rosy mutant?

Dr. Hadorn: I think it is maroon—like the biochemical phenotype which is very similar to rosy.

Dr. Strehler: Dr. Hadorn, are there any clear-cut cases in which you feel that one is compelled to assume that there is a primary pleiotropic effect as you mentioned at the end of your talk, or does this just represent a distinct possibility?

## DISCUSSION

Dr. Hadorn: Of course, I don't think it is more than an attractive hypothesis, which is, however, very difficult to prove; but we should regard this as a possibility.

Dr. Gluecksohn-Waelsch: It might be of interest in connection with Dr. Hadorn's discussion of the role of genes in development, to mention here briefly some results we have obtained recently with a mutation in mice which affects kidney development. Using methods of organ tissue culture, we have studied the behavior in vitro of kidney rudiments from mouse embryos homozygous for the mutation *Sd* which in homozygous condition causes agenesis of kidneys. Descriptive studies of the development of such kidneyless homozygotes had indicated that a failure of inductive interaction between ureter and metanephrogenic blastema might be responsible for the abnormality. Experiments in organ tissue cultures where normal ureteric primordia were combined with mutant metanephrogenic tissue, and mutant ureteric primordia with normal metanephrogenic tissue, have shown that mutant metanephrogenic tissue is able to form kidney tubules in combination with normal ureter; at the same time, mutant ureter is able to induce normal metanephrogenic tissue to form kidney tubules when the two are combined in culture. Thus, agenesis of kidneys in embryos homozygous for *Sd* is neither due to absence of inductive power on the part of the mutant ureter, nor to inability to react on the part of the mutant metanephrogenic tissue. There are quantitative differences between normal and mutant kidney rudiments in the time of onset of differentiation in tissue culture and the number of kidney elements formed. The organ culture experiments have shown that the potencies of kidney rudiments of *Sd* homozygotes when isolated are greater than realized when left in the intact embryo.

# HORMONAL REGULATION OF INSECT METAMORPHOSIS

CARROLL M. WILLIAMS

*The Biological Laboratories, Harvard University,
Cambridge, Massachusetts*

"My mind is bent to tell of bodies changed into new forms."

Ovid, *Metamorphoses*

As I UNDERSTAND IT, my task is to consider how insects achieve "better things for better living through chemistry." The story begins about 250 million years ago, when some of the more enterprising insects began to experiment with what has come to be known as metamorphosis. They found it increasingly prudent to subdivide their lives into two parts—to live, as it were, two successive lives. The first life could then be devoted to nutrition and the future of the individual; the second, to sex and the future of the species.

From the outset the trend in the evolution of insects was to separate these preoccupations ever more sharply within the life-span. The earth-bound early stages built enormous digestive tracts and hauled them around on caterpillar treads. Later in the life-history these assets could be liquidated and reinvested in the construction of an essentially new organism—a flying-machine devoted to sex.

This reorganization put a heavy load on the developmental machinery. Nature met this challenge by carrying over into the post-embryonic life of the insect the morphogenetic mechanisms of the embryo. To students of development this is the particular fascination of insect metamorphosis. In the metamorphosing insect embryological mechanisms become accessible in a conveniently post-embryonic setting.

### PROGRAMMING OF CELLS

During the early development of the insect egg the coded genetic information begins to impress itself on the embryonic system. In some unknown manner the cells and tissues are "programmed" for a concatenated series of future happenings. A cell is first in stage A; it then moves on to stage B, and subsequently, to stage C. It may then proceed to stage D, which, in the concrete case, may even correspond to "biological death" (31, 53).

Nowhere is the acting out of a program of determination seen to better advantage than in insects that undergo advanced degrees of metamorphosis. Consider the skin of an insect. It is one cell thick—a preformed histological section one cell thick. Each epidermal cell secretes on its outer surface a little island of cuticle. In so doing it leaves its "finger-prints" on the cuticle it secretes. By a simple inspection of the intact insect, one can see what is going on in the underlying cells.

If the skin is, say, that of the abdomen of a silkworm, the cells first secrete a larval cuticle which is smooth, green, and shiny. Later on, at the time of metamorphosis, the old cuticle is detached. The very same cells now secrete a pupal cuticle that is thick, brown, and rugose. Still later, the process repeats itself and the cells now secrete an adult cuticle equipped with scales and hairs. Substantially the same picture is seen in almost incredible detail in the development of the flattened pouch of skin that corresponds to the butterfly wing (10, 28, 38, 56).

It was Wigglesworth (57, 60, 61, 63) who introduced us to the beauty of the insect skin as an experimental object; in his recent monograph (64) he probes deeply into the programming of insect cells and tissues. Like Snodgrass (49), he emphasizes that metamorphosis is not a progressive differentiation of a relatively undifferentiated organism. At each stage the insect is no less than a highly differentiated and specialized creature. What we see in metamorphosis is, not progressive differentiation, but the unfolding of a sequence of highly differentiated states.

The programming of insect cells shows a close kinship to events encountered in vertebrate embryology. As Bodenstein (4, 5) has emphasized, it is a step-by-step affair. The polarity of the body axes is already determined in the unfertilized ovum. Shortly thereafter, the early embryo is mapped out in terms of future body regions and primary embryonic fields. These are then divided and subdivided. And so it continues from hour to hour and into the post-embryonic period. The "epigenetic landscape" (56), even in an insect, is not impressed in one swift stroke: it is a progressive affair which is first roughed out and then refined. For any particular cell the outlook and potentialities are ever-narrowing. The end-result is that the cells and tissues are ultimately committed to a program of future states and functions—a kind of flow-sheet of things to come. Consequently, in the insect as in the vertebrate, the programming of cells is something that begins at the supercellular level and proceeds, so to say, centripetally.

What is going on here in terms of physics and chemistry is as baffling in insects as in other forms. Professor Hadorn has just shown us that the genes within the individual nuclei possess powerful leverage on these cellular and supercellular happenings. This conclusion is further affirmed in the work of Poulson (39), Waddington (56), and Ede (12). Perhaps, as Medawar (30, 31) and Burnet (7) suggest, the day is not far distant when we must learn to think with greater precision about extracellular gene products, "self-markers" and "recognition units."

## Coordination of Development

According to this analysis, the central feature of metamorphosis is the acting out of a sequence of latent commitments which, days, weeks, or months earlier, has somehow been impressed on cells and tissues. Manifestly, this is a matter that must be subject to overall control and coordination. The cells must act together. Therefore, the developing insect must have at its disposal an effective mechanism for pacing the growth and metamorphosis of the cellular community.

Insects have solved this problem with customary ingenuity. It turns out that there is a certain molecule which the cells require in order to grow and to proceed with their developmental programs. During the period of embryonic development the ability to synthesize this molecule centers more and more exclusively in a certain tissue. This particular tissue therefore becomes an endocrine organ; the substance, itself, a hormone. In this manner the insect equips itself with an agency of overall control—a pacemaker and coordinating force in post-embryonic development.

## Developmental Standstill

If the cells of the insect become less than self-sufficient developmental units, what, then, if an interruption occurs in the supply of the factor which has become a growth hormone? Nature perennially conducts this experiment on a grand scale in the course of the life histories of thousands of species of insects.

An excellent example is a favorite experimental animal, the Cecropia silkworm. In early summer the Cecropia caterpillar, liberally supplied with growth hormone, develops rapidly, increasing its weight 5,000-fold. In mid-summer it spins a cocoon and pupates. At this point further secretion of growth hormone comes to an abrupt end. The pupal tissues, now deprived of hormone, stop right there; the animal as a whole enters a prolonged period of developmental arrest termed "diapause" (29). Months later, the hormone is again secreted and metamorphosis continues where it had left off. Consequently, in the pupal diapause one sees a clear-cut dissociation of development (which requires hormone) from "maintenance" (which does not require hormone).

## THE PROTHORACIC GLANDS

The endocrine source of the growth hormone was a matter of considerable debate and confusion for about twenty years. Finally, in a series of decisive experiments described in 1940, the Japanese investigator Fukuda (15, 16), demonstrated that the hormone comes from a pair of glands in the thorax—the so-called "prothoracic glands." This finding was soon confirmed for numerous other species (37, 43, 62, 67, 69).

Since the prothoracic glands are in the anterior of the insect, they are carried away with the anterior end when the insect is transected by ligation or surgery. Under this circumstance the anterior end is often able to continue its development. However, the hind end, now deprived of further resources of hormone, lapses into a state of developmental arrest equivalent in every way to diapause. Abdomens isolated in this manner may live for one or more years without any further trace of development. However, at any time one can cause them to develop by implanting living, active prothoracic glands (71).

## THE PROTHORACIC GLAND HORMONE

Prior to the discovery of the endocrine role of the prothoracic glands, studies were under way the outcome of which would be the isolation of the hormone itself. This trend had its beginning in the work of Gottfried Fraenkel (14), who, in 1935, made an apparently trivial discovery. He found that when he ligated mature fly larvae, only the front end was able to form a puparium; that is, only the front end showed the hardening and darkening of the larval skin peculiar to the higher Diptera just prior to pupation. Fraenkel found a way to induce this change in the isolated abdomen, namely, by transfusing the abdomen with blood obtained from the anterior end. He concluded that puparium formation was dependent on a hormone secreted somewhere in the anterior end.

Shortly thereafter, Becker and Plagge (2, 3) in Germany initiated chemical studies of the hormone inducing puparium formation. They prepared active extracts, not only from fly larvae, but also, curiously enough, from the pupae of Lepidoptera. This work was resumed in 1943 by Butenandt, Karlson, and Hanser at Tübingen. Large masses of silkworm pupae (*Bombyx mori*) were extracted and fractions obtained which were highly active in causing puparium formation (24, 25).

Early in 1953 Professor Butenandt shipped a small mass of this dark gummy material to Harvard for testing on diapausing Cecropia pupae. To my surprise, the implantation of a bit of this substance caused the termination of diapause and initiation of adult development. For the first time it was evident that the factor that induces puparium formation in flies is also a hormone in other insects.

Butenandt and Karlson redoubled their efforts and were now able to assemble and extract a half-ton of silkworm pupae. The fractions that were most active in inducing puparium formation were shipped to Harvard for assay on diapausing silkworm pupae and isolated pupal abdomens. These tests made it increasingly certain that the hormonal activity corresponded to the prothoracic gland hormone (72).

Finally, in 1954, Karlson obtained a crystalline fraction which showed maximal activity both in Germany and the United States. A total of 25 mg. of crystals was obtained from the 500 kg. of pupae—a 20-million-fold purification (8, 20).

Karlson named the hormone "ecdysone" for the reason that it is a ketone whose earliest effect is to cause a shedding or "ecdysis" of the cuticle (21). Chemical studies resolved two closely related molecules, $\alpha$-ecdysone and $\beta$-ecdysone. Elementary analyses showed that the hormone contains only carbon, hydrogen, and oxygen, and suggested the empirical formula $C_{18}H_{30}O_4$. Its UV and IR spectra suggest the presence of alpha and beta ketone groups (21, 22). Up to the present time the crystalline material has not been available in sufficient amounts to permit even Butenandt to resolve its structural formula. As I understand it, they are working up another half-ton of silkworms.

Ecdysone is the first invertebrate hormone to be isolated in pure form. It turns out to be a new and distinctive molecule—a result consistent with the finding that mammalian hormones (with the exception of the neurohumors) are inert when tested on invertebrates (70). Thus far, ecdysone has apparently been tested only on insects where, in each case, it mimics the function of living, active prothoracic glands. However, Karlson (22, 23) has been able to detect ecdysone in extracts prepared from the shrimp, *Crangon,* and it seems altogether likely that ecdysone is the principal growth hormone, not only of insects, but also of other arthropods.

So then, we see that the insect is equipped with central control of developmental events at the cellular level. When the prothoracic glands withhold the secretion of ecdysone, the result is developmental standstill. When the glands secrete ecdysone, the cells proceed to grow and to act out their programs of determination.

### Formation of Muscles

Some years ago we tripped over a phenomenon which is worth considering at this point. Howard Schneiderman and I were attempting to measure the extremely low metabolism that is peculiar to diapausing silkworm pupae. From time to time the pupae performed callisthenics within the respirometers, twirling their abdomens and frustrating our measurements. As a rather desperate countermeasure we decided to make the animals sit still by removing their central nervous systems.

This proved to be a relatively simple procedure. One goes in through an opening in the anterior end and removes the brain, subesophageal ganglion, and three thoracic ganglia; and then through the tip of the abdomen, and cuts out the chain of eight abdominal ganglia and connectives. Curiously enough, when resealed with wax and plastic slips, the pupa survives this extensive surgery and lives for months thereafter. Needless to say, one is no longer troubled by spontaneous motion of the abdomen; however, it may be noted that the intersegmental muscles do not break down even after months of denervation.

It happened by chance that some of these nervous-systemless pupae were caused to develop into adult moths. Development proceeded according to the usual time-table to produce moths which were flaccid, but wholly normal to external examination, even to the minute details of segmentation and pigmentation.

However, a surprise was in store when the moths were examined internally. There were no muscles! The thorax, which is normally crammed with flight muscles, contained tracheal sacs and yellowish fatbody, but no muscles. So, also, in the abdomen the old pupal muscles had broken down in the usual manner, but no adult muscles had formed (78).

As is so frequently the case, we subsequently learned that the same sort of phenomenon had been observed years earlier by Kopeč (27) and Suster (51). They found that after excision of individual ganglia, everything developed normally except the denervated muscle anlage. Evidently the development of new muscles requires, not only ecdysone, but also some further factor which the myoblastic tissues receive from motor nerves.

To test this possibility several "loose" nervous systems were reimplanted into nervous-systemless pupae. Then, several weeks later, the preparations were caused to metamorphose. Most generally, the result was a moth without muscles. Consequently, it seems unlikely that the nervous system factor can be transported in the blood. However, an occasional preparation showed the full and complete development of a single unilateral muscle mass. Presumably, the anlage of this muscle had been reinnervated by outgrowth from one of the implanted nervous systems.

Dr. Hans Nüesch, of Basel, was at this point introduced to the problem. Subsequently, in a wonderfully detailed series of investigations, Nüesch (32-36) was able to show that the anlage of each muscle is innervated and that this nerve supply is prerequisite for the development of the muscle itself. Moreover, the nerve factor proves to be necessary during only the first few days of adult development, i.e., during the precise period when ecdysone is released from the prothoracic gland and circulates in the blood.

My present purpose is to emphasize the extraordinary behavior of somatic muscle—a matter which has been further documented by Finlayson (13) and Wigglesworth (65). What the nervous system factor may be is a complete mystery. Evidently, it is delivered directly from motor nerve to myoblastic tissue. In several respects the phenomenon is reminiscent of the critical effects of nerve on the regeneration of the amphibian blastema (48).

## The Brain Hormone

The analysis to this point has focused attention on ecdysone—the secretory product of the prothoracic glands. The somatic muscles, as we have just seen, develop only when supplied with both ecdysone and a further factor delivered to them by motor nerve. But this is quite exceptional: the non-muscular tissues require only ecdysone. In this sense the prothoracic glands are charged with impressive authority over the insect as a whole.

Now the principles of endocrinology are such that one may immediately predict that the prothoracic glands must, in turn, be subject to control. This prediction is confirmed in all insects that have been carefully studied. The prothoracic glands are under the control of yet another hormone whose source is the brain itself.

The endocrine function of the insect brain was first suggested by the Polish biologist, Kopeč (26), nearly forty years ago. It turns out that insect brains are equipped with certain specialized neurones whose function is the synthesis and secretion of a brain hormone (46). Most generally, there are four clusters of these neurosecretory cells in the brain—a pair of medial groups and a pair of lateral groups (68). In many insects the cell bodies of the neurosecretory cells can be seen in the living brain. In the Cecropia silkworm there is a total of twenty-six cells—eight in each of two medial groups and five in each of the lateral groups. Especially in the medial groups the cell bodies are filled with a whitish colloid material which gives them a semi-opaque character.

The insect brain, like brains in general, is surrounded by a substantial blood-brain barrier. There is a sheath which envelops the brain and is reflected along the nerves (18, 19, 41, 42, 54). When tested for its permeability, one finds this sheath to be permeable to water but highly impermeable to most other molecules, including both electrolytes and non-electrolytes (77). It is worth asking how the neurosecretory cells, deep in the brain, can get their secretion into the blood.

The answer to this question now seems clear. The secretory products escape from the brain by passing down the axons of the individual neurosecretory cells (1, 44, 52). The course of these axons is therefore of special interest and importance. The axons of the medial cells decussate across the mid-line and then emerge from the rear of the brain as a pair of tiny nerves. The axons of the lateral cells do not decussate; they join together on each side and emerge from the rear of the brain as a second pair of tiny nerves (11, 17). So, then, there are these four little nerves—a pair of medial nerves connecting with the medial neurosecretory cells and a pair of lateral nerves connecting with the lateral neurosecretory cells.

What about the endings of these axons? The answer to this question has now been resolved. The medial and lateral nerves on each side join together to make a definite organ called the corpus cardiacum (11, 17). The corpus cardiacum is a plexus of endings of the neurosecretory axons; in many insects the endings can be seen to be distended with secretory material. The corpus cardiacum is a good example of a "neurohaemal organ" (9)—a place where neurosecretory products can be accumulated and dumped into the blood. Evidently, at this point the blood-brain barrier must be permeable or even absent.

In the Cecropia silkworm there is clear-cut evidence that both the medial and lateral neurosecretory cells are necessary for the formation of the brain hormone (68, 70). When the brain is cut into pieces and tested, activity is retained only if one uses a fragment containing both medial and lateral neurosecretory cells. In such a fragment the axons of medial and lateral cells seem to come together and regenerate a plexus of endings analogous to that in the normal corpus cardiacum (50). Consequently, the medial and lateral neurosecretory cells appear to be qualitatively different. It is our belief that medial and lateral cells secrete two precursors of the brain hormone and that the hormone itself is first formed by the conjunction of these precursors in the corpora cardiaca.

### Role of the Brain Hormone

The brain hormone seems to have a limited role within the insect. Its only clear-cut function is to control the prothoracic glands (71). It is a tropic hormone for the prothoracic glands in just the same way as, in the mammal, ACTH is a tropic factor for the adrenal cortex. The prothoracic glands are brought under control of the brain by virtue of their dependency on the brain hormone. Consequently, the removal of the brain has the same net result as the excision of the prothoracic glands. In both cases, the secretion of ecdysone comes to an end and the insect lapses into a state of developmental standstill or diapause.

In point of fact, this seems to be a device which has been exploited by insects for millions of years. In diapausing insects the brain is found to be "turned-off" (66). Diapause is terminated only after the brain recovers its endocrine function. As a rule, this recovery is vastly hastened when the brain is subjected to low temperatures, as during winter (73).

### Control of the Brain

In our progress to this point we have assembled a miniature endocrine system: brain hormone driving prothoracic glands, and prothoracic glands driving the tissues. In terms of control engineering the matter resolves itself into the fresh problem of controlling the brain.

On the basis of present information it is clear that the neurosecretory cells are not aloof to nerve impulses conveyed to them by afferent and internuncial neurons. The neurosecretory cells are tuned to inflowing nerve impulses generated by sensory stimuli. This fact is amply documented in the outstanding recent monograph by Lees (29).

The early studies of Wigglesworth take on new meaning in this connection. Thus, in *Rhodnius,* the effective stimulus for molting is the distension of the abdomen following a blood meal. Yet, as Wigglesworth (58) showed in 1934, the molt does not take place if the nerve cord is transected in the thorax. This implies that the neurosecretory cells are triggered by nerve impulses arising from abdominal proprioceptors and conveyed by nerve cord to brain (64).

Additional insight is derived from studies of diapausing pupae of the Cecropia silkworm. Here the neurosecretory cells, after a suitable period of exposure to low temperature, are able to function even when all connections between brain and nervous system have been severed. However, it is still possible that the neurosecretory cells are modulated by nerve impulses arising within the brain itself. The recent studies of Van der Kloot (55) have an important bearing on this point.

Van der Kloot finds that the diapausing Cecropia brain is not only endocrinologically inactive, but also electrically "silent." Microelectrodes fail to record any nerve impulses in the diapausing brain, even when the nerve cord is stimulated electrically. In effect, therefore, the neurosecretory cells of the diapausing brain are imbedded in an inexcitable matrix.

The brain at this time shows no detectable cholinesterase and only a trace of a cholinergic substrate presumed to be acetylcholine. What chilling does to the brain is to promote the progressive accumulation of this substrate. Then, when the brain is exposed to higher temperatures, a prompt and apparently inductive synthesis of cholinesterase takes place. At this point the brain recovers its nervous activity and endocrine function.

In summary, we begin to see an intimate coupling between the neurosecretory cells and the rest of the nervous system. Nervous system and endocrine system get together in the neurosecretory cell, which is both (45). In this sense the neurosecretory cell is a device for translating nerve impulses into hormonal action.

## The Juvenile State

The picture that takes shape reveals a system of control engineering which presides over the secretion of ecdysone by the prothoracic glands. When ecdysone is released into the blood, it promotes not only growth, but also the progress in development which we recognize as metamorphosis.

At this point we would be well advised to have a look at the actual life history of an insect. This history has plenty of change in it—from embryo, to larva, to pupa, to adult. However, for a considerable period the immature insect resists metamorphosis. It feeds, grows, sheds its cuticle a number of times, but fails to advance to the next stage.

Indeed, it is during this juvenile state that an insect does most of its growing and attains maximal size. It is a distinguishing feature of insects that the major growth period precedes metamorphosis; in other invertebrates, metamorphosis is a very early event which precedes growth.

It is easy to show that ecdysone is the prime-mover for the growth of the juvenile insect. The growth and molting of the larva are propelled by ecdysone. But somehow there is at work a conservative force that opposes change without interfering with growth itself.

## The Juvenile Hormone

Twenty-four years ago, V. B. Wigglesworth (58) suggested that this conservative force was a "juvenile hormone." Moreover, on the basis of experiments performed on *Rhodnius,* he (59) concluded that the juvenile hormone was secreted by a pair of tiny cephalic glands we have not had to mention up to this point—the corpora allata.

Confirmation of Wigglesworth's conclusions was soon forthcoming, in France, Germany, England, Japan, and the United States. By going through the underside of the "neck" of caterpillars, Bounhiol (6) in France, and Fukuda (16) in Japan were able to remove the corpora allata from immature silkworms. Instead of continuing the juvenile period of growth, the insects underwent precocious metamorphosis and developed into midget pupae and adults weighing as little as one-fortieth of normal.

So, in the immature insect we find a total of three hormones at work—brain hormone, ecdysone, and juvenile hormone. Since metamorphosis is blocked by the juvenile hormone, it follows that in the normal course of events the corpora allata must decline in activity late in larval life. However, under experimental conditions the titer of hormone can be raised by the implantation of one or more active corpora allata obtained from younger larvae. The insect is then forced into extra larval instars and develops into a giant (64). As many as six extra larval instars have been induced in certain species (40), and insects of sensational size have been produced by this maneuver.

Several years ago (74) the first active extracts of juvenile hormone were prepared from the Cecropia silkworm. Subsequently, in collaboration with Howard Schneiderman and Lawrence Gilbert, the crude extract has been highly purified. When assayed on previously chilled pupae of the Polyphemus silkworm, our best preparations are active at the level of one part of hormone per million parts of insect.

The transformation of a pupa into an adult moth requires the presence of ecdysone and the absence of juvenile hormone. If one injects a high dose of juvenile hormone, ecdysone is no longer able to cause metamorphosis. Instead we get only a growth response, as signalled by the molting of the pupa to a second pupal stage. Lower doses of juvenile hormone result in a bizarre mixture of pupal and adult characters in which some parts move forward and others do not (76).

The hormone, as extracted from Cecropia, is active when tested on other families and orders of insects. The hormone itself belongs to the general class of lipids. It is a very stable molecule, as testified by its recovery from dried museum specimens which have been dead for eight years (47, 75). Our present problem is to accumulate enough of the purified material to determine its chemical characteristics.

Here, then, is an agent of unusual biological action—a juvenile hormone which somehow imposes a biochemical restraint on the concatenated gene action which surely must underlie the morphogenetic program of the insect. There can be little doubt that the juvenile hormone has much to tell us about the chemical engineering of insect development—and I hope I have been able to persuade you that insects have much to tell us about development in general.

## REFERENCES

1. Arvy, L., and Gabe, M., *Z. Zellforsch.*, **38**, 591-610 (1953).
2. Becker, E., *Biol. Zentr.*, **61**, 360-388 (1941).
3. ———, and Plagge. E., *Biol. Zentr.*, **59**, 326-341 (1939).
4. Bodenstein, D., in *Insect Physiology* (K. D. Roeder, ed.), p. 879-931, John Wiley & Sons, New York (1953).
5. ———, in *Analysis of Development* (Willier, Weiss, and Hamburger, eds.), pp. 337-345, W. B. Saunders, Philadelphia (1955).
6. Bounhiol, J. J., *Bull. biol. France et Belg.*, Suppl. **24**, 1-199 (1938).
7. Burnet, F. M., *Enzyme, Antigen, and Virus*, Cambridge, at the Univ. Press (1956).
8. Butenandt, A., and Karlson, P., *Z. Naturforsch.*, **9b**, 389-391 (1954).
9. Carlisle, D. B., and Knowles, F. G. W., *Nature*, **172**, 404 (1953).
10. Caspari, E., *Quart. Rev. Biol.*, **16**, 249-273 (1941).
11. Cazal, P., *Bull. biol. France et Belg.*, Suppl. **32**, 1-227 (1948).
12. Ede, D. A., *Wilhelm Roux' Arch. Entwicklungsmech. Organ.*, **148**, 416-451 (1956).
13. Finlayson, L. H., *Quart. J. Microscop. Sci.*, **97**, 215-233 (1956).
14. Fraenkel, G., *Procs. Roy. Soc. (London)*, B, **118**, 1-12 (1935).
15. Fukuda, S., *Proc. Imp. Acad. (Tokyo)*, **16**, 414-416; 417-420 (1940).
16. ———, *J. Fac. Sci. Univ. Tokyo, Sect. IV*, **6**, 477-532 (1944).
17. Hanström, B., *Biol. Generalis*, **15**, 485-531 (1942).
18. Hoyle, G., *Nature*, **169**, 281-282 (1952).
19. ———, *J. Exptl. Biol.*, **30**, 121-135 (1953).
20. Karlson, P., *Verhandl. deut. zool. Ges.*, **1954** (Tübingen), 68-85 (1954).
21. ———, *Ann. sci. nat., zool. et biol. animale*, **18**, 125-137 (1956).
22. ———, *Vitamins and Hormones* **14**, 227-266 (1956).
23. ———, *Boll. zool. agrar. e bachicolt., Univ. studi Milano*, **22**, 65-80 (1957).
24. ———, and Hanser, G., *Z. Naturforsch.*, **7b**, 80-83 (1952).
25. ———, and ———, *Z. Naturforsch.* **8b**, 91-96 (1953).
26. Kopeč, S., *Biol. Bull.*, **42**, 322-342 (1922).
27. ———, *J. Exptl. Zool.*, **37**, 15-25 (1923).
28. Kühn, A., *Vorlesungen über Entwicklungsphysiologie*. Springer, Berlin (1955).
29. Lees, A. D., *The Physiology of Diapause in Arthropods*. Cambridge, at the Univ. Press (1955).
30. Medawar, P. B., *Biol. Revs. Cambridge Phil. Soc.*, **22**, 360-389 (1947).
31. ———, *The Uniqueness of the Individual*. Methuen, London (1957).
32. Nüesch, H., *Rev. suisse zool.*, **59**, 294-301 (1952).
33. ———, *J. Morphol.*, **93**, 589-608 (1953).

34. ———, *Rev. suisse zool.*, **61**, 420-428 (1954).
35. ———, *Rev. suisse zool.*, **62**, 211-218 (1955).
36. ———, *Zool. Jahrb. Abt. 2. Anat. u. Ontog. Tiere,* **75**, 615-642 (1957).
37. Pflugfelder, O., *Verh. deut, zool. Ges.*, 169-173 (1949).
38. ———, *Entwicklungsphysiologie der Insekten*, Akademische Verlag, Leipzig (1952).
39. Poulson, D. F., *Amer. Nat.*, **79**, 340-363 (1945).
40. Radtke, A., *Naturwiss.*, **30**, 451-452 (1942).
41. Roeder, K. D., *Ann. Rev. Entomol.*, **3**, 1-18 (1958).
42. Scharrer, B., *J. Comp. Neurol.*, **70**, 77-88 (1939).
43. ———, *Biol. Bull.*, **95**, 186-198 (1948).
44. ———, *Biol. Bull.*, **102**, 261-272 (1952).
45. Scharrer, E., *Scientia*, **6**, 177-183 (1952).
46. ———, and Scharrer, B., in *Handbuch microscop. Anat. des Menschen* (W. Bargmann, ed.), Springer, Berlin **6**, No. 5, 953-1066 (1954).
47. Schneiderman, H. A., and Gilbert, L. I., *Anat. Record*, **128**, 618 (1957).
48. Singer, M., *Quart. Rev. Biol.*, **27**, 169-200 (1952).
49. Snodgrass, R. D., *Insect Metamorphosis. Smithsonian Inst. Publs., Misc. Collections*, **122**, No. 9 (1954).
50. Stumm-Zollinger, E., *J. Exptl. Zool.*, **134**, 315-326 (1957).
51. Suster, P. M., *Zool. Jahrb. Abt. 3. Allgem. Zool. u. Physiol.*, **53**, 49-66 (1933).
52. Thomsen, E., *J. Exptl. Biol.*, **31**, 322-330 (1954).
53. Tiegs, O. W., *Trans. Roy. Soc. S. Australia*, **46**, 319-527 (1922).
54. Twarog, B. M., and Roeder, K. D., *Biol. Bull.*, **111**, 278-286 (1956).
55. Van der Kloot, W. G., *Biol. Bull.*, **109**, 276-294 (1955).
56. Waddington, C. H., *Principles of Embryology*, Macmillan, New York (1956).
57. Wigglesworth, V. B., *Quart. J. Microscop. Sci.*, **76**, 269-318 (1933).
58. ———, *Quart. J. Microscop. Sci.*, **77**, 191-222 (1934).
59. ———, *Quart. J. Microscop. Sci.*, **79**, 91-121 (1936).
60. ———, *J. Exptl. Biol.*, **14**, 364-381 (1937).
61. ———, *Symposia Soc. Exptl. Biol.*, **2**, 1-16 (1948).
62. ———, *J. Exptl. Biol.*, **29**, 561-570 (1952).
63. ———, *J. Embryol. Exptl. Morphol.*, **1**, 269-277 (1953).
64. ———, *The Physiology of Insect Metamorphosis*, Cambridge, at the Univ. Press (1954).
65. ———, *Quart. J. Microscop. Sci.*, **97**, 465-480 (1956).
66. Williams, C. M., *Biol. Bull.*, **90**, 234-243 (1946).
67. ———, *Biol. Bull.*, **93**, 89-98 (1947).
68. ———, *Growth*, **12**, (Suppl., 8th Sympos.) 61-74 (1948).
69. ———, *Biol. Bull.*, **97**, 111-114 (1949).
70. ———, *Harvey Lectures*, **47**, 126-155 (1953).
71. ———, *Biol. Bull.*, **103**, 120-138 (1952).
72. ———, *Anat. Record*, **120**, 743 (1954).
73. ———, *Biol. Bull.*, **110**, 201-218 (1956).
74. ———, *Nature*, **178**, 212-213 (1956).
75. ———, *Anat. Record*, **128**, 640-641 (1957).
76. ———, *Sci. American*, **198**, 67-75 (1958).
77. ———, *unpub.*
78. ———, and Schneiderman, H. A., *Anat. Record*, **113**, 54-55 (1952)

# THE METABOLIC CONTROL OF INSECT DEVELOPMENT

## G. R. WYATT

*Department of Biochemistry, Yale University,
New Haven, Connecticut*

As THE WORK of Williams and others has shown, the developing insect may provide a tractable subject for study of the biochemical control of processes of diapause, growth, and metamorphosis. We have begun some experiments with the aim of elucidating the nature of the action of the insect growth hormone, and I should like to mention some preliminary results from these. Although the effect of the juvenile hormone is even more intriguing in its implications, we have directed our attention first to the breaking of diapause in the Cecropia pupa, which apparently results from the action of ecdysone. We feel that this step may be the simpler to analyse, and that an understanding of it may be prerequisite to analysing the action of the juvenile hormone.

It seemed possible that the breakdown and resynthesis of the cytochromes, which Williams and his group have demonstrated in Cecropia, might not be unique. It might be merely an example of changes undergone by many enzymes, which could result from changes in the activity of protein-degrading and protein-synthesizing systems in the animal. With this in mind, and considering current views on the probable relation between ribonucleic acid and protein synthesis, we have made some measurements on changes in RNA in Cecropia at the initiation of adult development from the diapausing pupa. For this, we have used the wing epithelium, a tissue which starts extensive proliferation at this time.

The second day of morphological development was recognized by the retraction of the leg epidermis (5). At this stage, the ratio of RNA to DNA in the wing epithelium was 7.4. It had increased from the diapausing level of 5.6, so as to indicate an early synthesis of RNA. This ratio subsequently declines somewhat, as cell division and synthesis of DNA commence. The incorporation of injected inorganic $P^{32}$ into wing tissue RNA was measured, and at the second day of development it was found to have increased 5.1-fold (average, during the interval 4-24 hours after injection) over the rate in diapause (7).

This did not of course prove a primary effect of the growth hormone on RNA synthesis, since greater availability of RNA precursors could easily result from some other change, such as raised cytochrome levels or even perhaps increased transport of phosphate into the cell.

With the hope of getting some information on this last possibility, Mr. Frank Carey in my laboratory has been measuring the rate of disappearance of injected inorganic $P^{32}$ from Cecropia blood. Some results are shown in Fig. 1. Each curve represents the specific activity of inorganic phosphate

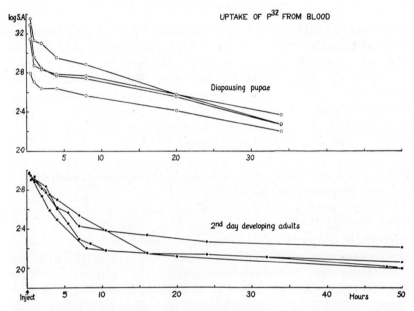

Fig. 1. The disappearance of inorganic $P^{32}$ from Cecropia blood. Each curve represents one animal. Diapausing animals had been debrained and fitted with facial windows; developing animals were intact, and development was recognized by the retraction of the leg epidermis. After injection of isotope, animals were incubated at 25°C, and blood samples (about 0.05 ml.) were taken. These were deproteinized, and orthophosphate was extracted for counting (4).

in successive blood samples from a single animal. In the diapausing animals, there is an initial drop in activity due to the time taken for distribution of the isotope through the blood space, followed by a prolonged phase in which the decline is roughly linear on the logarithmic scale. We assume that the slope of this phase of the curve represents the rate of uptake into the tissue. In the developing animals, blood circulation is much more rapid (as shown by experiments with $C^{14}$-inulin), and the approximately linear phase starts almost at once, but with much steeper slope and shorter duration than in diapause. In this experiment, the average rate of phosphate uptake in the second-day developing animals is 5.5 times that in diapause.

The rates of respiration of these same animals were measured in the Warburg apparatus. Using the values of $Q_{O_2}$ obtained before injection of the isotope, the developing animals had on the average 7.1 times the rate of respiration of the diapausing ones. However, this value may be too high for comparison with the change in rate of phosphate uptake, since the respiration of the diapausing animals increased during the experiment owing to the effect of injury. We do not yet know to what extent the metabolism due to injury may affect phosphate uptake.

Thus, so far as we can interpret these preliminary results, it looks as though several phases of metabolism increase in rate more or less synchronously during early natural development, and we cannot infer from these experiments whether respiratory metabolism, phosphate uptake, or RNA synthesis is nearest to the primary action of the growth hormone. This is perhaps not surprising, since the initiation of biochemical development does not coincide with that of morphological change. Rather, there is gradual biochemical change during late diapause, preceding overt development by some weeks, as Shappirio and Williams' studies on the cytochrome system have shown (6).

We plan next to examine the changes which follow the injection of partially purified ecdysone into diapausing animals, and we are more hopeful of detecting specific effects in this way.

From quite another line of work, we have come upon some evidence which may have a bearing on the nature of the control of metabolism during diapause. This is the accumulation of glycerol in Cecropia blood, found by W. L. Meyer and myself (8). There is evidence that many insects are almost lacking in lactic dehydrogenase, but contain high levels of $\alpha$-glycerophosphate dehydrogenase (9), and that some insect tissues under anaerobic conditions produce $\alpha$-glycerophosphate rather than lactate (2). We have found $\alpha$-glycerophosphate in Cecropia blood (8). Presumably, the reduction of triose phosphate to glycerophosphate serves to regenerate diphosphopyridine nucleotide from its reduced form, as is necessary for continuation of glycolysis. This is reminiscent of the fermentation by yeast in the presence of sulfite, which led to production of glycerol by dephosphorylation of glycerophosphate (3). If this analogy is correct, accumulation of glycerol should signify anaerobic conditions or lack of electron acceptor owing to enzyme deficiency. In Cecropia, we find by analysis that glycerol is virtually absent from the larva, then it builds up steadily during the earlier months of pupal diapause, and then, coincident with restored cytochrome levels in the developing adult, it rapidly disappears again (Fig. 2). This seems to me to lend some support to the idea expressed by

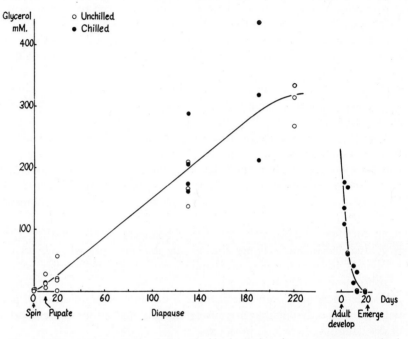

Fig. 2. The glycerol content of Cecropia blood. Each point represents one animal. Unchilled and developing animals were kept at 25°C; chilling was at 6°C for 30 days or more. Glycerol was isolated by paper chromatography and determined with periodate.

Williams that the cytochrome system, or certain members of it, may limit metabolism during diapause.

It is interesting in this connection that glycerol, as well as sorbitol, is also found in high levels in the diapausing egg of the silkworm, *Bombyx mori* (1). However, to prevent our too facile generalization on this subject, we have the fact that certain saturniids related to Cecropia (for example, *Samia walkeri*) contain no detectable glycerol during the pupal diapause. Plainly, we are only beginning the study of these matters.

## REFERENCES

1. Chino, H., *Nature,* **180,** 606-607 (1957).
2. Kubista, V., *Nature,* **180,** 549 (1957).

3. Neuberg, C., and Reinfurth, E., *Biochem. Z.*, **92**, 234-266 (1918).
4. Nielsen, S. O., and Lehninger, A. L., *J. Biol. Chem.*, **215**, 555-570 (1955).
5. Schneiderman, H. A., and Williams, C. M., *Biol. Bull.*, **106**, 238-252 (1954).
6. Shappirio, D. G., and Williams, C. M., *Proc. Roy. Soc. (Lond.)*, B, **147**, 233-246 (1957).
7. Wyatt, G. R. (unpub.)
8. ———, Meyer, W. L., and Kropf, R., *Federation Proc.*, **17**, 340 (1958).
9. Zebe, E., *J. Gen. Physiol.*, **40**, 779-790 (1957).

## DISCUSSION

Dr. Sussman: How many silkworms did you use to prepare this extract?

Dr. Williams: The male Cecropia moth builds up an enormous concentration of hormone in its abdomen. You see, nature does most of the work and from a single abdomen one gets 0.2 to 0.3 ml. of active golden oil. There are perhaps 50 animals represented here in this vial of extract. How much do you get from one animal, Dr. Schneiderman?

Dr. Schneiderman: From one animal we get about 0.2 ml. of crude extract which contains enough hormone to affect 200 other animals in the assay we use.

Dr. Borghese: What you say about the influence of the central nervous system upon the insect muscle reminds me of some research I did many years ago on the development of silkworm (*Boll. Soc. ital. Biol. sper.* **21,** 212-219, 1946). I took out from silkworm larvae the fifth abdominal ganglion. When such larvae continued their development and went through the metamorphosis, not only the resulting moths lacked the third abdominal ganglion—derived from the fifth of the larva—but also most somatic muscles of the 5th and 6th abdominal segment were missing. This shows that the somatic abdominal muscles of the imago do not develop when the part of the nervous chain which supplies their nerves is not there.

Dr. Williams: Yes, I think that that is an extremely interesting observation. I wonder how generic it is among animals. I am reminded, of course, of Singer's work on the effect of nerves upon the regeneration of the amphibian leg blastema. I think that this is part of the same package.

Dr. Schneiderman: Dr. Williams' story has been very interesting and I almost hesitate to append a postscript to it, but I would like to tell you a little bit about some recent findings which we have turned up at Cornell University. Mr. Lawrence Gilbert and myself have been able to extract juvenile hormone activity from animals other than the silk worm. We have examined 40 species scattered over 12 phyla, and we were astounded to find that from vertebrates we can extract materials which have juvenile hormone activity. In particular, beef adrenal cortex had a really remarkable effect when we injected it into silkworm pupae. The injected pupae precociously molted into strange creatures which were intermediate between pupa and adult, just as when we injected juvenile hormone extracted from insects themselves. The adrenal extract was water-soluble and it seemed to duplicate in all details the action of the juvenile hormone itself. I think the thing that Gilbert and I find most exciting is that, so far as we are aware, this is the first instance of a substance extracted from a vertebrate affecting the growth of an invertebrate. It seems very likely that the active principal of the adrenal extracts and the juvenile hormone itself are steroids. This would agree with all that we presently know about the chemistry of the juvenile hormone. The active principal is not, however, any one of perhaps 50 pure crystalline steroids which we have tested so far. Also, we cannot duplicate in vertebrates any effect of a known vertebrate hormone by injecting the juvenile hormone. I have just learned from Dr. Williams that he and Miss Lynn Moorehead at Harvard have been successful in extracting material from newborn rats which also has juvenile hormone activity. It seems possible then that we have a molecule which is not restricted to insects, and perhaps is quite widespread.

Dr. Williams: Though mammals have evolved their own hormones, they still may be mindful of their biochemical sensitivities prior to backbones.

Dr. Yamada: Is ecdysone the only hormone that the prothoracic gland gives out or are there any more? Are the alpha and beta types different in their effects?

Dr. Williams: I don't know about this alpha and beta subdivision. The alpha is far more active than the beta and it is even possible that the beta is produced as a kind of artifact in the isolation procedure. I think the alpha form is the growth hormone of arthropods.

Dr. Sussman: Have you examined for any alternative sources of ecdysone?

Dr. Williams: Do you mean microorganisms, or something like that?

Dr. Sussman: Yes.

Dr. Williams: I would like to get some microbiologist interested in the matter.

Dr. Potter: Do you know the ratio of DPN-cytochrome c reductase to TPN-cytochrome c reductase and to transhydrogenase at the various stages through here?

Dr. Williams: I wish we did! The recent work of Talalay and Williams-Ashman (*Proc. Nat. Acad.*, **44**, 15-26, 1958) is surely one of the most exciting things in a long time, wouldn't you say? It is interesting that the transhydrogenase and its steroid prosthetic group are concerned with electron transfer—a thing which we have been trying to pinpoint in relation to hormonal action in the insect.

# THE MECHANISM OF LIVER GROWTH AND REGENERATION

ANDRÉ D. GLINOS

*Growth Physiology Laboratory,
Walter Reed Army Institute of Research,
Washington, D. C.*

INFORMATION ON LIVER GROWTH has greatly increased in late years. The largest part of this information stems from studies on liver regeneration after partial hepatectomy, mostly on the rat; observations on the growth of the liver during development are far less numerous. While formerly attention was centered on description at the morphological level, more recent investigations have dealt predominantly with the chemical features of the regenerative process. The vast amount of diversified information thus collected has recently been integrated in the excellent review by Harkness (41).

Direct attempts to gain insight at the chemical level into the mechanism controlling the growth of the liver have been made by using two distinct and in some ways opposite methodological approaches. The first approach consists of the study of the effects on the growth of the liver of more or less clearly defined chemical agents thought to be related to growth processes on the basis of other available evidence. Nutritional factors, hormones, and a variety of tissue extracts are among the agents thus investigated. The net outcome of studies of this type has been essentially negative, the most important conclusion to be drawn being that neither nutritional nor endocrine factors seem to form critical links in the chain of reactions which leads to liver regeneration (see review, 41).

In the second approach no a priori assumptions are being made concerning the identity of the factors involved in the control of liver growth and regeneration. On the contrary, a step-by-step analysis is undertaken aiming to obtain information on the existence of systemic chemical factors controlling liver growth and regeneration, their mode of action at the physiological level in terms of stimulation and inhibition, their chemical identity and, finally, their mode of action at the chemical level in terms of molecular kinetics.

The results from studies of this type will be reviewed and certain interpretations and hypotheses will be formulated. Next, some consequences of these concepts will be derived and examined as to their conformity with the available evidence. This will provide the means of evaluating our interpretations and will furnish guide-lines for further conceptual formulation and experimental design.

### The Problem of the Existence and Physiological Mode of Action of Systemic Chemical Factors

Evidence for the existence of systemic chemical factors controlling the growth of the liver following partial hepatectomy was obtained by the application of two independent and widely different experimental techniques, tissue culture and parabiosis.

*Evidence from Tissue Culture*

The first attempt to explore the possibility that liver regeneration is controlled by chemical factors present in the blood was made in 1923 by Akamatsu (2), who studied the growth of adult rabbit liver in cultures in vitro containing blood plasma taken before and after partial hepatectomy. The cultures with normal plasma, after a latent period of 8 days, showed growth of connective tissue elements followed, in subsequent days, by a moderate epithelial outgrowth in 4 out of 15 explants. In the cultures with plasma taken 24 hours after partial hepatectomy the latent period was reduced to 6 days, and was followed by a very extensive outgrowth of connective tissue and epithelium in 12 out of 15 explants.

These results were fully corroborated thirty years later in our own comparable tissue culture experiments with rat liver (35), conducted without awareness of Akamatsu's work.

*Evidence from Parabiosis*

The parabiotic technique was applied to the study of chemical factors controlling liver regeneration independently and almost simultaneously by two research teams (16, 24). Virtually identical results were obtained which were further confirmed by others (99). In the very thorough study of Bucher, Scott, and Aub (16) partial hepatectomies were performed on one partner of each of 14 parabiotic pairs and on two partners of each of 3 sets of parabiotic triplets. The mitotic activity in the intact livers of the non-hepatectomized partners at 48 or 72 hours post-operatively was compared with the activity of the control livers removed at the time of the hepatectomy. The mitotic rate of the latter was well within the range characterizing normal adult resting liver, thus demonstrating that the parabiosis per se was without effect. The mitotic rate of the intact livers of the non-hepatectomized partners showed an increase over the normal by a factor of 6 in the parabiotic twins and a factor of 50 in the triplets.

These experiments in tissue culture and parabiosis provide compelling evidence for the presence in the blood of partially hepatectomized animals of chemical factors involved in the control of liver regeneration.

## The Problem of the Physiological Mode of Action

The next question to be considered concerns the action of the controlling factors in the physiological terms of stimulation and inhibition. In the tissue culture work of Akamatsu, as well as in the studies in parabiosis, an answer to this question was provided by postulating the presence or increase of growth-stimulatory chemical substances in the blood of the partially hepatectomized animals. This answer is far from being unequivocal, as the alternative interpretation of the absence or decrease of growth-inhibitory substances would fit the data of these experiments equally well.

A more direct approach to this problem was made by Friedrich-Freksa and Zaki (31), who injected into normal rats small amounts of serum taken from other rats 24 to 72 hours after partial hepatectomy. They reported that the mitotic activity of the intact livers of the injected rats was increased over the normal control range by a factor of 40. This finding was interpreted as definitive evidence for the presence of growth-stimulatory factors in the serum, unless a more complex control system is envisaged by postulating that the factors in the serum act by neutralizing inhibitory substances in the liver.

Given these important implications, an attempt was made in our own laboratory to reproduce the findings of Friedrich-Freksa and Zaki by following a procedure identical with theirs in all respects except the strain of rats used. The results of these experiments are shown in Table 1. No increase in the mitotic activity of the liver of normal rats was obtained by injections of serum from partially hepatectomized rats, and unless Friedrich-Freksa and Zaki's results are confirmed, no advantage will be gained by including them in further considerations.

In our tissue culture studies already mentioned (35), a comparable outgrowth was observed in a high concentration of serum from partially hepatectomized rats and in a low concentration of normal serum. A high concentration of normal serum showed inhibitory effects. These findings

TABLE 1

The Effect on Liver Mitosis of Injections of Sera from Normal and Partially Hepatectomized Rats

| | Serum donors | | Serum recipients |
|---|---|---|---|
| No. | Young normal rats | No. | Young normal rats |
| 1 | 0.52 | 6 | 0.03 |
| 2 | 0.17 | 7 | 0.02 |
| 3 | 0.61 | 8 | 0.11 |
| 4 | 0.01 | 9 | 0.03 |
| 5 | 0.01 | 10 | 0.27 |
| No. | Young partially hepatect. (48h) rats | No. | Young normal rats |
| 11 | 3.05 | 16 | 0.04 |
| 12 | 2.54 | 17 | 0.02 |
| 13 | 1.38 | 18 | 0.04 |
| 14 | 0.97 | 19 | 0.03 |
| 15 | 1.42 | 20 | 0.13 |

Male Wistar rats weighing from 150 to 200 g. were used. Nos. 1-5 were bled, sacrificed, and the liver fixed. The serum obtained was injected intraperitoneally into rats Nos. 6-10, 2 ml. per rat. Forty-eight hours later, these rats were sacrificed and the liver fixed. The same procedure was followed with rats Nos. 11-15 and 16-20 with the difference that Nos. 11-15 were partially hepatectomized 48 hours prior to the bleeding. Mitotic counts were performed on all livers, and the figures express the results in per cent mitotic indices.

were considered to indicate an inverse relationship between the concentration of certain serum constituents and liver growth. In order to verify this hypothesis two obvious consequences were put to test. If the postulated relationship between the concentration of certain blood constituents and liver growth has a factual basis, it should be possible to induce cell division in the resting liver of an intact animal by lowering the concentration of serum constituents. Conversely, by increasing the concentration of serum constituents it should be possible to inhibit cell division in the regenerating liver of a partially hepatectomized animal. Plasmapheresis and fluid intake restriction were used respectively in order to test these possibilities.

In the plasmapheresis experiments blood was withdrawn every 12 hours corresponding to 31-38 per cent of the initial total blood volume of the animal in a first group and to 39-46 per cent in a second. The bleedings were followed by reinjection of the blood cells suspended in an equal volume of saline. Under these conditions cell division was induced in the resting liver of adult rats and was intensified with increasing rate and duration of plasmapheresis (Table 2, Figs. 1, 2).

In the fluid intake restriction experiments two groups of rats were

## TABLE 2

### Induction of Cell Division in the Resting Liver by Plasmapheresis

| Rate | Time in hours | | | |
|---|---|---|---|---|
| | 36-48 | 60 | 72-96 | Mean |
| 0% | <.002<br>.003 | .002<br><.002 | .004<br>.002 | .002 |
| 31-38% | <.002<br>.022 | .002<br>.048 | .009<br>.002 | .014 |
| 39-46% | .168<br>.002 | .019<br>.054 | .441<br>.197 | .147 |
| Mean | .048 | .031 | .163 | |

A total of eighteen adult male rats was used. The rate of plasmapheresis is expressed as the percentage of the initial total blood volume of the animal replaced by saline per 12 hours. In the control group 0 rate refers to the fact that blood was merely withdrawn and reinjected, with the animals submitted to the same stressful conditions of restriction, anesthesia, venipuncture, etc. as the experimental groups. The mitotic activity was obtained by counting 50,000 cells and expressed as the per cent mitotic index. When no mitosis was found, the mitotic index was recorded as <0.002.

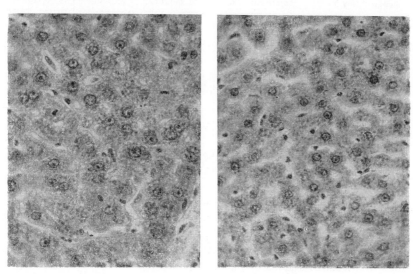

Fig. 1. Left, Intact liver of an adult rat plasmapheresed for 48 hours at the rate of 45% of total blood volume/12 hours. Increased nuclear and nucleolar size, increased basophilia and variability of its distribution among the different cells, and mitotic figures (two prophases, one metaphase), render this liver indistinguishable from regenerating liver (from ref. 35).

Fig. 2. Right, Plasmapheresis control liver. Same magnification as Fig. 1. Typical picture of normal resting liver (cf. Fig. 1).

used, differing with regard to the extent of the partial hepatectomy. All animals were partially hepatectomized and tube-fed an identical isocaloric diet containing 3 per cent water. The controls were given drinking water ad libitum, but the experimental animals were deprived of drinking water for the duration of the experiment, which was 64 hours, starting 16 hours prior to the operation and continuing for 48 hours post-operatively, at which time the animals were sacrificed. A measure of total body water loss is given by the difference in weight change between experimental and control animals in each group. This difference did not exceed 10 per cent of the body weight and is in agreement with the finding that no appreciable metabolic disturbances occur in the rat after 3 days of water intake restriction (101). In both the experimental groups an effective inhibition of liver cell division was obtained (Table 3). Control

TABLE 3

INHIBITION OF CELL DIVISION IN THE REGENERATING LIVER BY FLUID RESTRICTION

| Degree of hepatectomy | Treatment | No. of rats and weight | Weight % change | Mitoses/ 50,000 cells |
|---|---|---|---|---|
| 30% (Median lobe) | Controls | 10 457 | −7.9 | 81.4 |
| | Fluid restriction | 9 458 | −12.9 | 18.0 |
| 10% (Caudate lobe) | Controls | 8 331 | −9.1 | 16.5 |
| | Fluid restriction | 8 313 | −17.6 | 0.1 |

The experimental variables are defined in the text. The per cent changes refer to differences in weight between the values obtained before treatment and those obtained at sacrifice.

observations on the intestinal epithelium revealed no inhibitory effects on the mitotic activity of this tissue.

The evidence provided by these experiments indicates that the action of the factors controlling liver growth is inhibitory. The simplest way, then, to visualize a system of automatic self-regulation would be to postulate that the site of synthesis of these factors is the liver itself. Partial removal of the liver would result in a lowering of the extracellular concentration of the inhibitors and this in turn would initiate growth. As the size of the liver would increase the level of the inhibitors would also increase, and this would lead to a progressive decrease of the rate of regeneration and eventual cessation of growth. The same progressive decrease of the growth rate should also be manifested during the growth of the intact liver in prenatal and postnatal development.

Data from relevant observations are consistent with this postulate. In the case of the regenerating liver, the increase in the number of liver cells is very rapid during the first three days after partial hepatectomy, becomes progressively slower in subsequent times, and stops two to three weeks later (Fig. 3, ref. 14). In the case of the growth of the intact liver

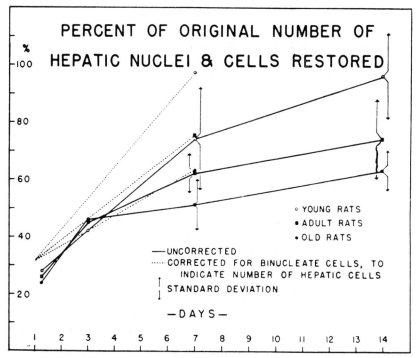

Fig. 3. Increase of total number of liver cells after partial hepatectomy. Irrespective of age, the rate of increase is maximum between 24 and 72 hours and declines rapidly at later times (from ref. 14).

during development the ratio of the number of liver cells to body weight decreases progressively in the embryo (27), and this decrease continues after birth until in the adult animal all growth ceases (Table 4, ref. 14). This indicates that the mechanism controlling the growth of the liver

### TABLE 4

THE DECLINE OF THE RATE OF LIVER CELL PROLIFERATION DURING DEVELOPMENT
(Data from refs. 14, 27)

Relative hepatic cell population = $\dfrac{\text{Total hepatic cell population}}{\text{Body-weight in grams}} \times 100$

| Prenatal, sheep (from Dick 1957) | | Postnatal, rat (from Bucher and Glinos 1950) | |
|---|---|---|---|
| Age | R H C P $\times 10^6$ | Age | R H C P $\times 10^6$ |
| 35 days | 1600 | 35 days | 725 |
| 150 days term | 600 | 180 days | 280 |

cannot be equated with the factors which limit the growth of the organism as a whole.

## THE PROBLEM OF THE CHEMICAL IDENTITY OF THE FACTORS

Before discussing the problem of the chemical identity of the factors controlling the growth of the liver, it is necessary to consider in chemical terms the essential features of the growth process as it occurs after partial hepatectomy. Special emphasis will be placed on the proteins and nucleic acids; storage products, such as glycogen and neutral fat, will not be considered.

### Some Chemical Features of Liver Regeneration

Following partial hepatectomy there is an initial period of approximately 24 hours during which no cell division occurs. The principal morphological change in the cells during this period consists of a 50 to 100 per cent increase in volume of the cytoplasm, nucleus, and nucleolus (41). During this initial period synthesis of the various cellular constituents is activated. This is manifested by an increase of the rate of incorporation of labelled precursors. As a result of increased synthesis the net amounts of the various cellular constituents increase. Because of variations in the time reactions of the different components, their relative concentrations tend to deviate from the normal values, and the composition of the liver appears altered (41). Cell division starts rather abruptly at approximately 24 hours post-operatively and reaches its maximum frequency between 24 and 48 hours (12). Subsequently, the rates of the increase of the cellular constituents as well as the rate of mitosis decline until the stationary phase is reached.

The time sequence of the activation of synthesis and net increase of the cellular proteins and nucleic acids as it emerges from the extensive work of Potter and his collaborators (6, 45, 46, 88, 92) and others (4, 5, 39, 40, 43, 75) is shown in Table 5. The estimates in this table are only

TABLE 5

The Time Sequence of the Activation of Synthesis and Net Increase of Liver Cell Constituents after Partial Hepatectomy

| Component | Regeneration | | | |
|---|---|---|---|---|
| | Activation of synthesis | | Net Increase | |
| | Initiation | Maximum | Initiation | Maximum |
| DNA | 12-18 | 20-24 | 12-18 | 24-30 |
| RNA nuclear | 12-14 | 26 | 24 | 20-30 |
| RNA cytoplasmic | 6-8 | 14-18 | 6 | 20-30 |
| Protein | 6-10 | 24-40 | 12-16 | 12-24 |

Figures are hours after partial hepatectomy. Terms defined in text (data from ref. 4-6, 39, 40, 43, 45, 46, 75, 88, 92).

approximate, as the results of various authors are not without some inconsistencies. Differences in the nutritional conditions of the animals, e.g., starvation (44), ad lib. feeding (43), or force-feeding (40), and differences in the method of administration of a variety of precursors probably account for these inconsistencies. Nevertheless, the data clearly indicate that the response to partial hepatectomy seems to follow a certain order, with cytoplasmic RNA and protein preceding nuclear RNA and DNA. Most noteworthy is the prompt and sustained response of cytoplasmic RNA, which can explain the marked increase of RNA concentration found at 2 days after partial hepatectomy (5, 74).

The inherent inability of the chemical methods to provide information relative to the response of individual cells renders highly desirable the addition of the histochemical approach to these studies. The most interesting autoradiographic study of Moyson (72) was limited to only one time interval, 24 hours after partial hepatectomy, and was concerned only with differences between parenchymal and littoral cells, without reference to the topographical localization of the hepatic cells in the lobule.

Because of the promptness of its response, its role in protein synthesis (10), and certain features of its organization in the cytoplasm, RNA seems the logical choice for a histochemical study of the activation of individual liver cells following partial hepatectomy. Electron microscopy and ultracentrifugation studies have revealed that most of the RNA is found in the cytoplasm in combination with protein, forming complexes of varying sizes (76, 78). In the normal resting liver cells the largest of these ribonucleoproteins are found to be closely associated with the membranes of the endoplasmic reticulum. Aggregation of such ribonucleoprotein-associated membranes in certain areas of the cytoplasm is responsible for the discrete basophilic bodies seen with the light microscope in normal resting liver cells (50, 76, 84). Fragmentation of these membranes during ultracentrifugation gives rise to the so-called microsome fraction (77).

In agreement with now generally accepted concepts concerning the role of nucleic acids in protein synthesis (10), it has been shown that these different RNA's mediate in the stepwise transfer of the activated amino acids to the peptide linkage, low molecular weight RNA being involved in the early, and high molecular weight microsomal RNA in the late phases of the process (49, 102). The metabolic activity of the different RNA's seems to be inversely related to their size (Fig. 4, ref. 53). During growth the number of the larger RNA complexes seems to

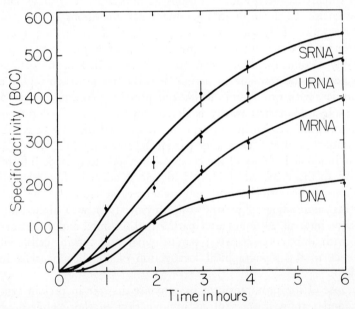

Fig. 4. Specific activity of supernatant RNA, ultramicrosomal RNA, microsomal RNA, and DNA of regenerating mouse liver at various times after $P^{32}$ injection, 58 hours after partial hepatectomy (from ref. 53).

decrease and the number of the smaller ones to increase (78, 79). Instead of the large aggregates of RNA-associated membranes found in the resting cells, in growing cells the ribonucleoprotein particles are freely dispersed in the cytoplasm. At the level of the light microscope, this change is manifested by the disappearance of the discrete basophilic bodies, the cytoplasm staining evenly with basic dyes (50, 76, 84). Changes in the organization of cytoplasmic ribonucleoproteins detectable with basic dyes may then be used as a sensitive indicator for the activation of the synthetic mechanisms of individual cells. The one condition sine qua non under which this can be done is that the histochemical findings must be considered in conjunction with other morphological and chemical data supplying direct information on the activation process, since changes in cytoplasmic basophilia are known to occur in situations other than growth (11, 55, 58-60).

In our studies rats were sacrificed at frequent intervals after partial hepatectomy, and the livers were fixed and stained with gallocyanin chromalum. Within 30 minutes after the operation the ribonucleoprotein-associated basophilic bodies began to disappear from the cells in the periportal area of the lobule, their cytoplasm staining evenly with the dye. This change proceeded gradually toward the center of the lobule, so that at 8 hours after partial hepatectomy even the cells adjacent to the central vein were affected, although to a lesser degree than the peripheral ones. Thus a topographical gradient with regard to both the order of appearance and the extent of the changes among individual liver cells was present at all times. This gradient was maintained during the reappearance of the basophilic bodies which took place at subsequent times. At 16 hours and even more at 24 hours such bodies could be discerned in the cells of the centrolobular area, whereas cells in the middle and the periphery of the lobule remained altered (Figs. 5-10). At subsequent times increasing numbers of cells in these areas showed reappearance of the basophilic bodies until the entire lobule returned to normal. The intimate connection of these changes with cell growth and division was strikingly demonstrated in our material because, confirming an earlier observation by Harkness (42), we found that mitosis, which began at 24 hours after the operation, started also in the periportal area and gradually expanded toward the center of the lobule.

Consideration of these results in conjunction with the chemical features of the regenerative process previously discussed, with studies in vitro of the metabolic activity of RNA complexes isolated from regenerating liver (20, 52, 94), and with analytical electron microscopy (8) and

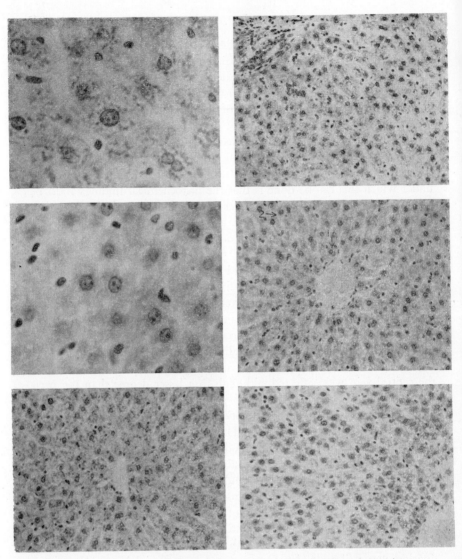

Fig. 5. Top Left, Normal liver cells. The cytoplasm contains ribonucleoprotein-associated bodies reacting with basic dyes.

Fig. 6. Top Right, Regenerating liver 30 minutes after partial hepatectomy. The basophilic bodies begin to disappear from the cytoplasm of the cells in the periportal area (arrow).

Fig. 7. Middle Left, High magnification of cells from the periportal area (cf. Fig. 5).

Fig. 8. Middle Right, Regenerating liver 8 hours after partial hepatectomy. Centrolobular area. Note that the extent of cytoplasmic changes follows a gradient with cells adjacent to the central vein exhibiting rudimentary basophilic bodies (arrow 1), whereas no such bodies can be seen in the cytoplasm of the great majority of the peripheral cells (arrow 2).

ultracentrifugation (78) findings concerning RNA organization in regenerating liver, leaves no doubt that we are here dealing with the activation of the synthetic mechanisms of the liver cells. The evidence conclusively shows that this activation starts immediately after partial hepatectomy and that the liver cells are not activated simultaneously or randomly but according to a precise gradient determined by their position in the lobule.

Based on the preceding considerations and since the physiological action of the factors controlling the growth of the liver was shown to be inhibitory, we can now define a minimum number of physiological conditions which should be satisfied for any chemical constituents of the blood likely to be considered as involved in the control of liver growth. Thus, the site of production of these constituents should be the liver, their concentration in the blood should decrease after partial hepatectomy, and their decrease in the immediate environment of the liver cells, i.e., the interstitial fluid of the liver, should be such that the time and space parameters of the activation of the liver cells could be accounted for. Systematic examination with reference to these conditions of the available evidence concerning known blood plasma constituents would by a process of progressive elimination indicate if any of these constituents are likely to be involved. Thus treated, the evidence seems to implicate the plasma protein, and the degree of satisfaction of the conditions stated will now be discussed.

*Plasma Proteins and Partial Hepatectomy*

The classical plasmapheresis studies of Madden and Whipple (63) establishing the liver as the main site of plasma protein synthesis, and the confirmation of this fact by numerous studies using a variety of different methods (see review, 65) are too well known to necessitate a detailed discussion here. It is sufficient to mention that, using organ perfusion in vitro, Miller and his collaborators have conclusively shown that all the plasma albumins and almost all the alpha and beta globulins are, under normal conditions, synthesized in the liver, whereas gamma globulin synthesis is extrahepatic (67-69). In-vivo isotope studies as well

---

Fig. 9. Bottom Left, Liver from a sham hepatectomized control rat 8 hours after the operation. Centrolobular area (cf. Fig. 8).

Fig. 10. Bottom Right, Regenerating liver 24 hours after partial hepatectomy. Central vein at lower right corner. Note gradient, with cells in the centrolobular area exhibiting numerous rather well-formed basophilic bodies, whereas cells in the middle and the periphery of the lobule remain altered. Two mitotic figures in the middle zone.

as in-vitro organ perfusion work indicate that the replacement rate of plasma proteins in the rat is very rapid, between 15 and 30 per cent daily (see review, 65). Within 24 hours after partial hepatectomy involving 70 per cent of the total liver, the level of plasma protein decreases by 20 per cent (22, 86) with the globulins declining first and, albumin being

TABLE 6

Percentage Composition of Plasma Proteins in the Rat after Partial Hepatectomy as Determined by Electrophoresis
(Data from ref. 26)

|  | Days after hepatectomy | | | | | |
|---|---|---|---|---|---|---|
|  | 0 | 1 | 2 | 3 | 4 | 5 |
| No. animals | 4 | 4 | 4 | 4 | 4 | 4 |
| Albumin | 53 | 39 | 37 | 37 | 37 | 39 |
| $\alpha$-Globulin | 8 | 12 | 15 | 18 | 16 | 16 |
| $\beta$-Globulin | 8 | 13 | 14 | 14 | 13 | 15 |
| Fibrinogen | 22 | 16 | 15 | 14 | 18 | 13 |
| $\gamma$-Globulin | 9 | 20 | 19 | 17 | 16 | 17 |

the fraction longer affected (Table 6, ref. 26, 28). This decrease is not due to an increase of the volume of circulating plasma (61).

## Plasma Proteins and Interstitial Fluid of the Liver

Recent investigations in the rat and the rabbit have demonstrated that the protein concentration of the hepatic lymph, like that of the lymph from other parts of the body, depends upon the rate of lymphatic flow. Furthermore, it was shown that the protein concentration at the average rate of lymph flow is approximately 30 per cent lower than the concentration of protein in the plasma and that the albumin–globulin ratio tends to be higher in the lymph than in the plasma. On the basis of this and other evidence Benson, Kim, and Bollman (7) and Friedman, Byers, and Omoto (30) have questioned the widely held view that the capillaries of the liver are almost freely permeable to plasma proteins. The significance of these findings, however, is that, even if it is accepted that the liver capillaries are relatively more permeable, transfer of water and protein across the capillary wall proceeds in the liver in a manner which is essentially the same as in other parts of the body. Without any particular reference to liver capillaries, the basic features of this process as it occurs generally in the organism have recently been discussed by Gitlin (32).

Since from a practical viewpoint it may be considered that water is transferred from the capillary to the interstitial fluid at the proximal end and from the interstitial fluid to the capillary at the distal end, a gradient of increasing protein concentration of the interstitial fluid would tend to develop along the direction of the capillary. In tissues where the direction of the capillaries is random, the actual development of this concentration gradient will be prevented, since capillaries with opposite directions would cancel each other's gradients. In tissues, however, with uni-directional capillaries the gradient would be expected to be actually present. This is precisely the situation in the liver lobule, where the direction of the intralobular capillary flow creates the conditions necessary for the development of a gradient of increasing protein concentration of the interstitial fluid from the periphery to the center of the lobule.

Decrease of the concentration of the interstitial fluid protein can be brought about by a greater net transfer of water from the capillary. Such greater transfer of water occurs in cases where the plasma protein concentration is decreased or the hydrostatic pressure is moderately increased (32). With marked increases of hydrostatic pressure the capillary permeability to proteins increases and therefore the concentration of protein in the interstitial fluid will tend to increase also. That these general hemodynamic principles apply also to the liver is indicated by the results of Morris (71), who found that, whereas decrease in the oncotic pressure and moderate increase in portal pressure resulted in lower protein concentration in the hepatic lymph, a greatly increased portal pressure produced a marked increase of lymph protein concentration.

After partial hepatectomy plasma protein decreases (22, 86) and portal pressure increases (38). As we have seen, the decrease of plasma protein reaches its maximum 24 hours after partial hepatectomy. Portal pressure increase, however, would occur immediately after the operation because of the reduction of the size of the vascular bed of the liver. This would be expected to result in an immediate decrease of the protein concentration of the interstitial fluid along the gradient already described.

Thus partial hepatectomy would cause a decrease of the protein concentration of the interstitial fluid of the liver in a manner that would account for the time and space parameters of the activation of liver cells were the plasma proteins to be involved in the control of the growth of the liver.

*Plasma Proteins and Control of Liver Growth*

In the preceding section we have seen how certain physiological considerations suggest as a reasonable postulate that the inhibitory plasma constituents controlling the growth of the liver are the plasma proteins. A number of consequences will now be derived from this postulate and will be examined as to their conformity with the evidence available. Where appropriate observations are lacking we shall indicate the type of experiments which should be performed in order to reach a final decision regarding the acceptance or rejection of the proposition that the plasma proteins regulate the growth of the liver.

Among the more obvious consequences of this proposition are the following. (a) Decrease of the protein concentration in the environment of the liver cells induces activation of the synthetic mechanisms of the cells, growth, and eventual cell division. (b) The activation and growth of the liver cells which follows partial hepatectomy is due to the decrease of the protein concentration of the interstitial fluid of the liver. (c) The decrease of the protein concentration of the interstitial fluid of the liver is initiated by increased pressure in the portal system. Portal pressure, therefore, would be expected to be of paramount importance in the control of liver regeneration. (d) In the later stages of regeneration, as well as during the growth of the intact liver in normal development, local hemodynamics are of no importance since the vascular bed of the liver is of a normal size. In these cases, therefore, interstitial fluid protein concentration need not be considered apart from the concentration of protein in the circulating plasma, which would be expected to rise as the stationary phase of growth of the liver is approached. (e) Artificial increase of the concentration of plasma proteins by intravenous administration of isolated plasma protein fractions would be expected to inhibit growth in the regenerating liver as well as in the intact growing liver during development. The relative efficiency of the various plasma protein fractions in inhibiting the growth of the liver could thus be ascertained.

*Consequence a.* As is well known, a decrease of the protein concentration of the circulating plasma leads to a decrease in the protein concentration of the interstitial fluid. In the first section of this paper direct evidence was presented that plasmapheresis, the classical method for depleting serum proteins, induced active growth in the intact liver of adult rats. Further experimental verification was obtained by showing that changes in the cytoplasmic ribonucleoprotein of the cells in the periportal area identical with the changes observed after partial hepatectomy, appear rapidly after a sudden decrease of the concentration of serum proteins (Table 7). Under the conditions of this experiment the means used to lower the serum protein level, dextran injection, or replacement of

TABLE 7

INDUCTION OF CYTOPLASMIC RIBONUCLEOPROTEIN CHANGES
IN THE LIVER BY PLASMA DILUTION

| Treatment | Fluid | Serum protein change | Liver ribonucleoprotein change |
|---|---|---|---|
| Addition | Saline | −11.8 | 0 |
|  | Dextran | −31.2 | + |
|  | Serum | +7.9 | 0 |
| Replacement | Saline | −19.2 | + |
|  | Dextran | −37.8 | + |
|  | Serum | −8.7 | 0 |

A total of 6 male adult rats was used. Addition refers to a single intravenous injection of 5.5 ml. of fluid. Replacement refers to a 5.5 ml. single plasmapheresis treatment. All animals were sacrificed 2 hours after treatment. Serum protein change refers to the per cent difference between the values obtained before treatment and those obtained at sacrifice. Liver ribonucleoprotein change refers to the disappearance of the basophilic bodies from the cytoplasm of the cells in the periportal area.

plasma by saline or dextran, appear to be without significance. It appears that the liver cells become activated when the level of serum proteins is lowered by more than 12 per cent. In another experiment, gavage with a mixture of isotonic saline and glucose was used in order to induce and maintain a low intra- and extra-vascular fluid protein concentration over a period of two days (Table 8). In this case the total serum protein level was reduced by over 30 per cent. In agreement with the findings of Friedman et al. (30) and of Bollman et al. (7, 25), total protein concentration of the hepatic lymph was lower than the concentration of the serum by 25 per cent, and whereas the serum was low in albumin and high in globulins the reverse was true for the lymph. Under these conditions the liver cells exhibited an active process of growth and division.

Additional evidence that decrease of the protein concentration in the environment of the liver cells activates the synthetic mechanisms of the cells is provided by the experiments of Ziegler and Melchior (103), who found that incorporation of $S^{35}$-labeled methionine into the proteins of liver slices was considerably higher when the slices were preincubated in Ringer solution than when they were preincubated in serum (Table 9).

*Consequence b.* There is no direct evidence available bearing on the

TABLE 8

The Effect of Lowered Plasma Protein Concentration on the Protein Content and Composition of the Extravascular Compartment of the Liver and on Liver Cell Growth

| Conditions | Compartment | Total protein | Albumin | Globulin | | | | Liver Mitoses/ 50,000 cells |
|---|---|---|---|---|---|---|---|---|
| | | | | $\alpha 1$ | $\alpha 2$ | $\beta$ | $\gamma$ | |
| Normal 24 h. starvation | Intravascular (Blood serum) | 6.59 | 3.38 | 1.17 | 0.69 | 0.84 | 0.49 | — |
| Experimental 48 h. starvation Gavage 45 ml. | Intravascular (Blood serum) | 4.52 | 1.57 | 0.96 | 0.91 | 0.90 | 0.15 | 90 |
| | Liver Extravascular (Liver lymph) | 3.45 | 1.76 | 0.52 | 0.49 | 0.55 | 0.10 | |

See text. Figures refer to grams of protein/100 ml. serum.

problem of the concentration of the interstitial fluid protein in the liver after partial hepatectomy. Investigations now in progress on the protein composition of the hepatic lymph during liver regeneration, as well as autoradiographic studies, are expected to provide the necessary information.

*Consequence c.* Attention to the role of the portal vein in the growth of the liver was drawn by Rous and Larimore (87), who showed that tying a branch of the portal vein in rabbits produced atrophy of the corresponding lobe and hypertrophy of the other lobes. Subsequent work by Mann and his collaborators (64) and others have demonstrated the

TABLE 9

The Effects of Ringer's Solution and Serum on the Incorporation of $S^{35}$-Methionine into Liver Proteins

| Rat no. | Conditions | Methionine $S^{35}$ per gm. nitrogen, $\mu$moles |
|---|---|---|
| 3 | Control Ringer's<br>Preincubated Ringer's<br>Preincubated Ringer's under N | 3.8, 4.6<br>6.7, 7.7<br>1.0, 1.0, 1.0 |
| 4 | Control serum<br>Preincubated serum | 4.4<br>5.8 |

In all cases the labeling reaction was carried out for one hour in an atmosphere $O_2$-$CO_2$ and in the presence of succinate. In the control cases the labeled methionine was added immediately after the flasks were filled with $O_2$-$CO_2$. Preincubations were carried out for one hour before adding the methionine (adapted from ref. 103).

dominant role of the local circulatory conditions in the control of liver regeneration. It was shown that shunting the portal flow away through an Eck's fistula greatly reduced or prevented regeneration after partial hepatectomy in the dog (64), and conversely that ligation of the vena cava above the renal vein in the domestic fowl induced a significant increase in size of the intact resting liver (48).

The possibility that some specific chemical factor present in the portal blood is responsible for these effects was excluded by showing that regeneration occurs when the normal portal blood supply is replaced by arterial or inferior vena cava blood (23, 29).

The possibility that portal volume flow is the important factor was excluded by showing that regeneration can occur in the absence of portal blood if, in evaluating regeneration, the atrophy of the intact normal liver by the deprivation of portal circulation is taken into consideration (96). In these experiments regeneration was estimated by comparing the size of the regenerated lobe with the reduced size it had reached after it was deprived of its portal blood supply prior to partial hepatectomy.

The possibility remains that blood pressure which increases locally after partial hepatectomy as a result of the reduction of the size of the vascular bed common to the portal vein and the hepatic artery (95) is the factor controlling liver regeneration. Strong direct evidence in favor of this possibility is provided by the experiments of Grindlay and Bollman (Table 10, ref. 38), who with various types of obstruction of the portal vein and the vena cava were able to show conclusively that portal flow volume is without significance, whereas a direct correlation exists between the degree of increase in the portal pressure and the extent of liver regeneration. The same relationship between blood pressure and regeneration probably holds true in livers with only arterial blood supply. The experiments already mentioned on liver regeneration in animals with artificial replacement of the portal blood by arterial blood and the findings on the increase of the hepatic arterial flow when portal flow is decreased (89, 90) lend support to this view. Direct evidence on this point is highly desirable.

*Consequence d.* As already stated, the maximum decrease of plasma protein concentration after partial hepatectomy occurs at 24 hours postoperatively and affects mostly the albumin fraction. In their extended study Chanutin et al. (22) reported that restoration to the normal levels is a slow process of an approximate duration of five weeks, which then may be considered roughly parallel to the gradual decline of the rate of cell proliferation.

TABLE 10

The Relationship between Portal Vein Pressure and Liver Regeneration

| Conditions | Portal vein pressures (cm. of water) | | Per cent of original weight at autopsy[1] |
|---|---|---|---|
| | Liver unaltered | After partial hepatectomy 70% | |
| Normal | 7 to 11 | 21 to 12[2] | 100 |
| Only constriction of portal vein | 8 to 11 | 8 to 13 | 42 |
| + caval constriction later | 9 to 22 | 15 to 22 | 80 |

[1] 30% of original weight at autopsy—no regeneration.
[2] In normal dogs the higher figure is the one obtained at one or two weeks after partial hepatectomy. The lower figure was obtained 4-5 weeks after partial hepatectomy. In other groups the pressures were taken before necropsy.

See text (adapted from ref. 38).

Also during both prenatal and postnatal development the level of plasma protein rises (Table 11, ref. 100) as the rate of liver cell proliferation declines (Table 4, refs. 14, 27).

*Consequence e.* There is no evidence available concerning the effect of intravenous administration of isolated plasma protein fractions on the growing liver during regeneration or development.

Intravenous infusions of whole normal serum or of normal serum concentrated by dialysis against dextran to rats bearing regenerating livers failed to inhibit cell division (13, 33). The reason for this failure may

TABLE 11

The Effect of Development on the Concentration of Serum Protein

| Age | Human | Pig |
|---|---|---|
| Foetus | 3.1 | 2.1 |
| Newborn | 6.0 | 2.4 |
| Adult | 7.4 | 8.1 |

Figures refer to grams of total protein per 100 ml. of serum (adapted from ref. 100).

be sought in that the increased volume of the circulating plasma with the ensuing increase in portal pressure which results from such infusions probably counteracts, in so far as the interstitial fluid of the liver is concerned, the effects of the temporary increase in plasma protein concentration. Obviously, the decrease of interstitial fluid protein concentration by increased portal pressure can be prevented by increasing the plasma protein concentration, only within certain limits. This led to the experiments reported in the first section of this paper, where we sought first to limit portal pressure increase by performing a smaller hepatectomy, and second, to prevent the increase of the circulating plasma volume which follows injections of concentrated serum by restricting fluid intake. We soon found that fluid restriction alone was sufficient to inhibit cell division markedly and therefore no serum injections were used.

These remarks make clear that any attempt to induce inhibition of cell division in the early stages of regeneration by intravenous administration of isolated plasma protein fractions will have to be performed under carefully controlled conditions, regarding concentration and rate of the infusion of the plasma protein fraction, degree of hepatectomy, and fluid intake of the animals.

## A Tentative Molecular Picture
### of the Mechanism Controlling the Growth of the Liver

In this last section I shall try to give a tentative, rough outline of the type of molecular kinetics that will have to be envisaged in order to render meaningful, at the chemical level, the concept of the regulation of the growth of the liver by the plasma proteins.

Borsook (9) has proposed the following general scheme for the regulation of protein synthesis: protein $\leftrightarrows$ (template) amino acid $\rightleftarrows$ activated amino acid $\rightleftarrows$ free amino acid. The problems related to the equilibria at the free amino acid end of the chain have recently been reviewed by Korner and Tarver (56).

If extracellular protein, i.e., plasma protein, synthesis in the liver is considered as a process separate and independent of intracellular protein synthesis, decrease of the level of the extracellular protein would induce an increase of its rate of synthesis by the liver cells without the intracellular protein being affected. Such an increase of synthesis, as well as decrease of catabolism, has been assumed to be responsible for the maintenance of plasma protein levels. Tarver (93) has presented in tabular form a scheme for the recognition of the operation of these two processes on the basis of observations on pool size, half-life, and replacement rate of the plasma protein.

On the assumption that the rate of catabolism of serum proteins remains unchanged after partial hepatectomy, Greenbaum, Greenwood, and Harkness (37) have estimated that plasma protein synthesis in the regenerating liver reaches its maximum rate considerably later than liver cell protein synthesis. From this they concluded that the two processes are not related. However, although it has been demonstrated that catabolism of plasma proteins is a first-order reaction (32, 65), this has never been shown to apply to synthetic processes. On this basis McFarlane (65) has proposed that the rate of synthesis of plasma protein is independent of plasma protein level, the equilibrium level being maintained by variable breakdown determined by pool size. Without adopting this extreme position, it can safely be stated that increase of the rate of plasma protein synthesis *per cell* in response to plasma protein depletion has never been observed. Moreover, recently Hughes, Klinenberg, and Gitlin (51) have shown that removal of 40 per cent of the liver decreases albumin catabolism by 50 per cent. This renders questionable the assumption on the basis of which plasma protein and liver cell protein synthesis were considered as independent. It may, therefore, be permissible to consider the possibility of a relationship between these two processes. The amount of extracellular protein synthesized by the individual liver cell per unit time, for example, could be related, within certain limits, to the amount of intracellular protein present by a constant ratio. Then, according to Borsook's scheme, lowering of the level of extracellular protein would lead to an increase of the rate of intracellular protein synthesis. This, as we have seen, is an early characteristic of regenerating liver cells and is followed by cell growth and eventually cell division.

In this view, increase of plasma protein synthesis by the liver would still be considered as contributing to the restoration of lowered plasma protein levels to normal. However, increased synthesis would be achieved not by an increase in the rate of synthesis of the individual liver cells but by an increase in the number of liver cells having, within limits, a constant rate of plasma protein synthesis. In this respect the experiments mentioned in the first section, where cell division was found to occur in the liver after depletion of plasma protein by plasmapheresis, should be recalled.

The following approaches could be used to explore the possibility of the existence of a relationship between intracellular and extracellular protein synthesis in the liver.

As previously discussed, there is abundant evidence that cytoplasmic RNA is actively involved in liver cell protein synthesis. Its participation in plasma protein synthesis, on the other hand, is widely assumed, but direct evidence for it is rather scanty (73, 81). Peterman (79) has suggested that two ribonucleoprotein complexes of distinctly different sizes are involved in these two processes. The larger component, "B," which decreases during growth, would be involved in plasma protein synthesis, whereas the smaller component, "C," which increases during growth, would be connected with liver protein synthesis. The behavior of these two components during plasma protein depletion by plasmapheresis should be investigated.

Plasma protein synthesis in slices of normal chick liver in vitro has been studied extensively by Peters (80-83). Using deoxycholate treatment, he has been able to liberate from the RNA-rich microsomes a protein which after solubilization is immunologically and electrophoretically the same as serum albumin. This protein therefore may be considered as the intracellular "precursor" of the albumin which the slices release physiologically. Rat liver tumor slices were found by Campbell and Stone (21) to synthesize only one-third to one-half as much serum albumin as comparable normal liver slices. Regenerating rat liver slices were used by Kaufmann and Wertheimer (54), but in their study only total nitrogen release was estimated and was reported to decrease by 25 per cent during the second day after partial hepatectomy. The slice method could be used in a parallel study of serum albumin and cell protein and nucleic acid synthesis in (1) regenerating liver at various times after partial hepatectomy (cf. 46) and (2) normal liver, using various concentrations of serum albumin in the incubation medium.

In addition to their studies with the normal liver (op. cit.), Miller and his collaborators used the organ perfusion method to study plasma protein synthesis in precancerous and cirrhotic livers (19). In such livers incorporation of labelled amino acids into the liver and plasma proteins is 2 to 3 times greater than into the liver and plasma proteins of normal livers. With regard to regenerating liver, only passing references concerning amino acid metabolism and urea synthesis have been made (19, 66). It is obviously desirable to study with the organ perfusion method the same two problems suggested previously with respect to the slice method.

These suggested in-vitro experiments with regenerating liver will be quite meaningful because, as repeatedly demonstrated in tissue culture work with the liver and the skin (1, 2, 17, 18, 34) as well as in slice work (15, 46), the physiological state reached in regenerating tissues is maintained for some time after removal of the tissues from the body.

If the proposed studies do, in fact, reveal that intracellular and extracellular protein synthesis in the case of the liver are not independent, the relationships involved could be schematically represented in the form of a cyclical feedback, as in Fig. 11. Thus the level of extracellular protein would control the rate of synthesis of intracellular protein and ultimately the total number of cells in the liver, the extracellular protein level itself depending on the number of liver cells present. This dependence is viewed here chiefly from the point of view of synthesis but, if the present inconsistencies in the evidence are resolved and it is definitely shown that plasma protein catabolism occurs mainly also in the liver (36, 70),

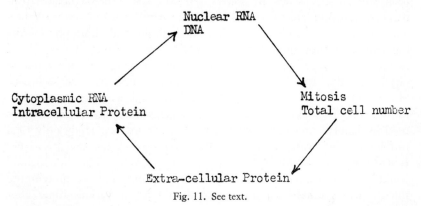

Fig. 11. See text.

the dependence will be even more significant as the effects of extrahepatic factors will be reduced.

At the intracellular level growth control is effected by a complex system of reversible and irreversible equilibria and cyclical feedbacks of the type discussed earlier in this symposium (62), and operating in the metabolism of energy, proteins, and nucleic acids (53, 85). The present state of our knowledge regarding these systems does not allow the drawing of a precise molecular picture of the proposed retrograde control mechanism in so far as the intracellular part of the cycle is concerned. The possibility that it is actually operating is suggested by the time sequence of the activation of synthesis and the increase of the cellular proteins and nucleic acids, as discussed previously.

As additional, independent, suggestive evidence Hershey's observation (47) that in T2 phage formation protein synthesis appears to be a necessary early step to nucleic acid synthesis, Spiegelman's discovery (91) that osmotically shocked protoplasts containing RNA are capable of de novo synthesis of DNA, and Allfrey, Daly, and Mirsky's finding (3) that cytoplasmic protein synthesis influences nuclear protein synthesis could be cited.

Summary

A system of automatic self-regulation and control for the growth of the liver during development and during regeneration following partial hepatectomy has been described. Direct evidence indicates that the system operates on the basis of a general principle proposed by Weiss (97, 98), namely, "the active maintenance of the total mass of each organ system in an equilibrium state and the return to that state after disturbance by virtue of a chemical communication system in which specific releases from each cell type circulating in the body would inform the homologous cell types of the state of their total mass."

Indirect evidence suggests that chemical communication in the case of the liver is achieved by means of the plasma proteins. The heuristic value of this postulate regarding the plasma proteins was demonstrated by bringing a great variety of diverse observations to a common denominator and by furnishing guide lines for the design of new experiments in a number of different areas.

REFERENCES

1. Abercrombie, M., and Harkness, R. D., *Proc. Roy. Soc.* (London), B, **138,** 544 (1951).
2. Akamatsu, N., *Virchow's Arch. path. Anat. u. Physiol.*, **240,** 308 (1923).
3. Allfrey, V. G., Daly, M. M., and Mirsky, A. E., *J. Gen. Physiol.*, **38,** 415 (1955).
4. Anderson, E. P., and Åqvist, S., *Acta Chem. Scand.*, **10,** 1576 (1956).
5. Åqvist, S., and Anderson, E. P., *Acta Chem. Scand.*, **10,** 1583 (1956).
6. Beltz, R. E., Van Lanker, S., and Potter, V. R., *Cancer Research,* **17,** 688 (1957).
7. Benson, J. A., Jr., Kim, K. S., and Bollman, J. L., *Am. J. Physiol.*, **182,** 217 (1955).
8. Bernhard, W., and Rouiller, C., *J. Biophys. Biochem. Cytol., Suppl.* 2, **73** (1956).
9. Borsook, H. J., *J. Cellular Comp. Physiol.*, **47,** (Suppl. 1), 35 (1956).
10. Brachet, J., in *The Nucleic Acids* (E. Chargaff and J. N. Davidson, eds.), 2, 475, Academic Press, New York (1955).
11. ———, Jeener, R., Rosseel, M., and Thonet, L., *Bull. soc. chim. biol.*, **28,** 460 (1946).
12. Brues, A. M., and Marble, B. B., *J. Exptl. Med.*, **65,** 15 (1937).
13. Bucher, N. L. R., in *Liver Function, A Symposium on Approaches to the Quantitative Description of Liver Function* (R. W. Brauer, ed.), (in press).
14. ———, and Glinos, A. D., *Cancer Research,* **10,** 324 (1950).
15. ———, Loftfield, R. B., and Frantz, I. D., Jr., *Cancer Research,* **9,** 623 (1949).
16. ———, Scott, J. R., and Aub, J. C., *Cancer Research,* **11,** 457 (1951).
17. Bullough, W. S., and Laurence, E. B., *Brit. J. Exptl. Pathol.*, **38,** 273 (1957).
18. ———, and Laurence, E. B., *Brit. J. Exptl. Pathol.*, **38,** 279 (1957).
19. Burke, W. T., and Miller, L. L., *Cancer Research,* **16,** 330 (1956).

20. Campbell, P. N., and Greengard, O., *Biochem. J.* (London), **66,** 47p (1957).
21. ———, and Stone, N. E., *Biochem. J.* (London), **66,** 19 (1957).
22. Chanutin, A., Hortenstine, J. C., Cole, W. S., and Ludewig, S., *J. Biol. Chem.,* **123,** 247 (1938).
23. Child, C. G., III, Barr, D., Holswade, G. R., and Harrison, C. S., *Ann. Surg.,* **138,** 600 (1953).
24. Christensen, B. G., and Jacobsen, E., *Acta Med Scand., Suppl.* **234,** 103 (1949).
25. Claus, B. F., Cohen, P. P., and Bollman, J. L., *Cancer Research,* **14,** 17 (1954).
26. De Lamirande, G., and Cantero, A., *Cancer Research,* **12,** 330 (1952).
27. Dick, D. A. T., *J. Embryol. Exptl. Morphol.,* **4,** 97 (1956).
28. Emmrich, R., and Petzold, H., *Z. ges. exptl. Med.,* **122,** 264 (1953).
29. Fisher, B., Russ, C., and Bluestone, C., *Am. J. Physiol.,* **181,** 203 (1955).
30. Friedman, M., Byers, S. O., and Omoto, C., *Am. J. Physiol.,* **184,** 11 (1956).
31. Friedrich-Freksa, H., and Zaki, F. G., *Z. Naturforsch.,* **9b,** 394 (1954).
32. Gitlin, D., *Ann. N. Y. Acad. Sci.,* **70,** 122 (1957).
33. Glinos, A. D., in *Liver Function, A Symposium on Approaches to the Quantitative Description of Liver Function* (R. W. Brauer, ed.), (in press).
34. ———, and Bartlett, E. G., *Cancer Research,* **11,** 164 (1951).
35. ———, and Gey, G. O., *Proc. Soc. Exptl. Biol. Med.,* **80,** 421 (1952).
36. Gordon, A. H., *Biochem. J.* (London), **66,** 255 (1957).
37. Greenbaum, A. L., Greenwood, F. C., and Harkness, R. D., *J. Physiol.,* **125,** 251 (1954).
38. Grindlay, J. H., and Bollman, J. L., *Surg. Gynecol. Obstet.,* **94,** 491 (1952).
39. Hammarsten, E., in *Ciba Foundation Conference on Isotopes in Biochemistry* (G. E. W. Wolstenholme, ed.), 203, The Blakiston Company, Philadelphia (1951).
40. ———, Åqvist, S., Anderson, E. P., and Eliasson, N. A., *Acta Chem. Scand.,* **10,** 1568 (1956).
41. Harkness, R. D., *Brit. Med. Bull.,* **13,** 87 (1957).
42. ———, *J. Physiol.,* **116,** 373 (1952).
43. ———, *J. Physiol.,* **117,** 267 (1952).
44. Hecht, L. I., and Potter, V. R., *Cancer Research,* **16,** 988 (1956).
45. ———, and ———, *Cancer Research,* **16,** 999 (1956).
46. ———, and ———, *Cancer Research,* **18,** 186 (1958).
47. Hershey, A. D., *Brookhaven Symposia in Biol.,* **8** (Mutation), 6 (1956).
48. Higgins, G. M., Mann, F. C., and Priestly, J. T., *Arch. Pathol.,* **14,** 491 (1933).
49. Hoagland, M. B., Zamecnik, P. C., and Stephenson, M. L., *Biochim. et Biophys. Acta,* **24,** 215 (1957).
50. Howatson, A. F., and Ham, A. W., *Cancer Research,* **15,** 62 (1955).
51. Hughes, W. L., Klinenberg, J. R., Gitlin, D., *Federation Proc.,* **17,** 246 (1958).
52. Hultin, T., and Von Der Decken, A., *Exptl. Cell Research,* **13,** 83 (1957).
53. Jardetzky, C. D., and Barnum, C. P., *Arch. Biochem. and Biophys.,* **67,** 350 (1957).
54. Kaufmann, E., and Wertheimer, E., *Am. J. Physiol.,* **190,** 133 (1957).
55. Kleinfeld, R. G., *Cancer Research,* **17,** 954 (1957).
56. Korner, A., and Tarver, H., *J. Gen. Physiol.,* **41,** 219 (1957).
57. Lagerstedt, S., Cytological Studies on the Protein Metabolism of the Liver in Rat, *Acta Anat.,* Suppl. 9 (ad Vol. **7,** 1949).
58. ———, in *Liver Injury* (F. W. Hoffbauer, ed.), p. 216, Josiah Macy, Jr. Foundation, New York (1953).
59. Lowe, C. U., and Rand, R. N., *J. Biophys. Biochem. Cytol.,* **2,** 331 (1956).
60. ———, and Williams, W. L., *Proc. Soc. Exptl. Biol. Med.,* **84,** 70 (1953).
61. Lowrance, P., and Chanutin, A., *J. Biol. Chem.,* **35,** 606 (1942).
62. Maas, K. W., this symposium.
63. Madden, S. C., and Whipple, G. H., *Physiol. Revs.,* **20,** 194 (1940).
64. Mann, F. C., *J. Mt. Sinai Hosp.,* **11,** 65 (1944).

65. McFarlane, A. S., in *Progr. Biophys. and Biophys. Chem.*, **7**, 115 (1957).
66. Miller, L. L., in *Liver Function, A Symposium on Approaches to the Quantitative Description of Liver Function* (R. W. Brauer, ed.), (in press).
67. ———, and Bale, W. F., *J. Exptl. Med.*, **99**, 125 (1954).
68. ———, Bly, C. G., and Bale, W. F., *J. Exptl. Med.*, **99**, 133 (1954).
69. ———, ———, Watson, M. L., and Bale, W. F., *J. Exptl. Med.*, **94**, 431 (1951).
70. ———, Burke, W. T., and Haft, D. E., *Federation Proc.*, **14**, 707 (1955).
71. Morris, B., *Q. J. Exptl. Physiol.*, **41**, 326 (1956).
72. Moyson, F., *Arch. Biol. (Liége)*, **64**, 149 (1955).
73. ———, and Steens-Lievens, A., *Biochim. et Biophys. Acta*, **21**, 500 (1956).
74. Novikoff, A. B., and Potter, V. R., *J. Biol. Chem.*, **173**, 223 (1948).
75. Nygaard, O., and Rusch, H. P., *Cancer Research*, **15**, 240 (1955).
76. Palade, G. E., *J. Biophys. Biochem. Cytol.*, **1**, 59 (1955).
77. ———, Siekewitz, P., *J. Biophys. Biochem. Cytol.*, **2**, 171 (1956).
78. Petermann, M. L., Mizen, N. A., and Hamilton, M. G., *Cancer Research*, **13**, 372 (1953).
79. ———, ———, and ———, *Cancer Research*, **16**, 620 (1956).
80. Peters, T., Jr., *J. Biol. Chem.*, **200**, 461 (1953).
81. ———, *J. Biol. Chem.*, **229**, 659 (1957).
82. ———, and Anfinsen, C. B., *J. Biol. Chem.*, **182**, 171 (1950).
83. ———, and ———, *J. Biol. Chem.*, **186**, 805 (1950).
84. Porter, K. R., *J. Histochem. Cytochem.*, **2**, 346 (1954).
85. Potter, V. R., *Texas Rep. Biol. Med.*, **15**, 127 (1957).
86. Roberts, S., and White, A., *J. Biol. Chem.*, **180**, 505 (1949).
87. Rous, R., and Larimore, L. D., *J. Exptl. Med.*, **31**, 609 (1920).
88. Schneider, J. H., and Potter, V. R., *Cancer Research*, **17**, 701 (1957).
89. Schwiegk, H., *Arch. exptl. Pathol. Pharmakol.*, **168**, 693 (1932).
90. Soskin, S., Essex, H. H., Herrick, J. F., and Mann, F. C., *Am. J. Physiol.*, **124**, 558 (1938).
91. Spiegelman, S., in *A Symposium on the Chemical Basis of Heredity* (W. D. McElroy and B. Glass. eds.), p. 232. John Hopkins Press, Baltimore (1957).
92. Takagi, Y., Hecht, L. I., and Potter, V. R., *Cancer Research*, **16**, 994 (1956).
93. Tarver, H., in *The Proteins* (H. Neurath and K. Bailey, eds.) **2**, 1199. Academic Press, New York (1954).
94. Von Der Decken, A., and Hultin, T., *Exptl. Cell Research*, **14**, 88 (1958).
95. Wakim, K. G., and Mann, F. C., *Anat. Record*, **82**, 233 (1942).
96. Weinbren, K., *Brit. J. Exptl. Pathol.*, **36**, 583 (1955).
97. Weiss, P., in *Biological Specificity and Growth* (E. G. Butler, ed.), p. 195. Princeton Univ. Press, Princeton, N. J. (1955).
98. ———, and Kavanau, J. L., *J. Gen. Physiol.*, **41**, 1 (1957).
99. Wenneker, A. S., and Sussman, N., *Proc. Soc. Exptl. Biol. Med.*, **76**, 683 (1951).
100. Widdowson, E. M., and McCance, R. A., *Clin. Science*, **15**, 361 (1956).
101. Wolf, A. V., and Meroney, F. C., (unpub.).
102. Zamecnik, P. C., Keller, E. B., Littlefield, J. W., Hoagland, M. B., and Loftfield, R. B., *J. Cellular Comp. Physiol.*, Suppl. **47**, 81 (1956).
103. Ziegler, D. M., and Melchior, J. B., *J. Biol. Chem.*, **217**, 569 (1955).

# DISCUSSION

DR. MCELROY: Don't you want to put at least one irreversible reaction in your scheme?

Dr. GLINOS: There were none in Borsook's scheme, and I would like to hear your suggestions and ideas on irreversible reactions.

Dr. McELROY: Well, I don't want to start the discussion on whether proteins equilibrate with the free amino acid pool. Let's have other people ask questions first.

Dr. WEISS: Since this concept complies fairly well with what we call a feedback mechanism, how do you explain that if this is protein in general, only the liver grows after hepatectomy, and not any other organ?

Dr. GLINOS: Well, because the liver is the organ which produces the bulk of the plasma proteins.

Dr. WEISS: How do compensatory reactions in other types of organs arise?

Dr. GLINOS: I don't know how information is transmitted in the case of other organs. I was not speaking here about growth control and regeneration in general but only about the liver.

Dr. MOHLER: I am confused. In your parabiotic animals, do you find hooking up a normal with a hepatectomized animal inhibits the regeneration in the hepatectomized animal.

Dr. GLINOS: No, you get just the opposite. You have the animals in parabiosis for some time before partially hepatectomizing one of them, and then the intact liver also shows some growth and regeneration.

Dr. MOHLER: I should think on your hypothesis the intact liver would have compensated somewhat for the deficiency of the hepatectomized rat and there would be decreased regeneration in the hepatectomized rat.

Dr. GLINOS: Of the mitotic rate? This is very difficult to assess. It is much easier to see an increase in mitosis above the normal level which in the adult rat is close to zero than to detect a slight reduction in the mitotic rate of the regenerating liver. You know, individual mitotic rates vary enormously during regeneration and unless you are dealing with a rather marked inhibition you will not be able to detect it.

Dr. MOHLER: I was just thinking that the liver of the normal animal might also increase and partially relieve the hepatectomized animal of a need to regenerate.

Dr. GLINOS: I don't think counting mitoses is a sensitive enough method to show you the small decrease which theoretically could occur under these conditions. As a matter of fact, it is questionable if any decrease at all could occur because in the partially hepatectomized partners the local hemodynamic conditions in the liver are of paramount importance. As I mentioned earlier it seems that their effects on the composition of the interstitial fluid of the liver cannot be prevented by the introduction of normal serum in the circulation.

Dr. WILDE: I would like to ask if you have any evidence of cyclic mitotic behavior in the regenerating liver? Secondly, I would like to ask, What effect would the well-established differences in ploidy among liver parenchymal cells have on liver regeneration?

Dr. GLINOS: I didn't hear your second question, I am sorry.

Dr. WILDE: I believe that liver parenchymal cells fall into categories of ploidy.

Dr. GLINOS: Yes, yes.

Dr. WILDE: I want to know what effect that has on regeneration.

Dr. GLINOS: As far as the first question is concerned, you have a diurnal variation in mitosis (Jaffe, J. J., *Anat. Rec.*, 120, 935, 1954) but this is also found in the normal young animal (Barnum, C. P., Jardetzky, C. D., and Halberg, F., *Texas Rep. Biol. Med.*, 15, 134, 1957) not just in the regenerating animal. The mitotic activity is high during the day and low at night. The same periodicity also holds true for the metabolic activity of the liver cell proteins (Hruza, Z., *Physiol. Bohemoslov.*, 5, 52, 1956) and nucleic acids (Barnum et al, op. cit.). The time interrelationships within each 24 hour period remind one of the overall pattern observed after partial hepatectomy. Bullough's proposition (Bullough, W. S., *Biol. Revs.*, 27, 133, 1952) that mitotic periodicity in general is due to variations in carbohydrate metabolism connected with the alternating periods of body activity and rest had to be abandoned in view of his more recent findings (Bullough, W. S., *Exptl. Cell Res.*, 9, 108, 1955, Bullough, W. S. and Laurence, E. B., *Brit. J. Exp. Path.*, 38, 278, 1957). I don't think that diurnal variation is necessarily related to the mechanism controlling liver regeneration but I would like to draw attention to some very recent observations by Hruza in Czechoslovakia (Hruza, Z. and Smetana, R., *Physiol. Bohemoslov.*, 7, 80, 1958) which are quite provocative in this regard. He found that the liver blood content also undergoes cyclic changes, being high during the day and low at night. This fits nicely with the observations on proteins, nucleic acids, and mitosis and suggests that perhaps the same local hemodynamic factors we believe are operating after partial hepatectomy are also responsible for the diurnal variation. Now with regard to the second question, we haven't done any work on the ploidy of the liver cells. But we did investigate once the binucleate cells. They decrease after partial hepatectomy. I don't know what happens to the polyploid cells and can't see how this would be related to the growth control mechanism.

Dr. PASTEELS: Some research which has been done in Brussels and is not yet published shows that there is a tremendous increase in ploidy after partial hepatectomy.

Dr. GLINOS: Well, that would mean that in the regenerating liver the process of cell division is not always carried to completion.

Dr. GEY: In the situation where you do have this tremendous change in the level of proteins, would you not expect that other tissues would show changes in their mitotic index because even that amount of change of protein can be reflected in a tissue culture system very easily.

Dr. GLINOS: This brings us back to the question of Dr. Weiss and raises a lot of other problems too. In vivo, due to the specificity of the liver cells in producing plasma proteins, you wouldn't expect to find increased mitosis in any other tissues because their cells are not in equilibrium with the plasma proteins. In tissue culture, however, there is the possibility that extra-hepatic tissues may also be producing extracellular proteins (Landsteiner, K., and Parker, R. C., *J. Exptl. Med.*, 71, 231, 1940). Miller has shown that, besides gamma globulin, alpha and beta globulins were produced when he perfused the carcass of a rat without the liver. The amounts were very small but in his papers (refs. 67, 68, 69) he has discussed the possibility that the production of plasma proteins by extra-hepatic tissues might increase when a circulatory bypass of the liver is established as, for example, in advanced cirrhosis. I wonder if this condition might not obtain in tissue culture where there is also no liver to produce plasma proteins. Any potential ability of cells other than liver to produce extracellular proteins would be manifested under these conditions. This is one way of looking at the fact that in our early tissue culture studies with Dr. Gey (ref. 35) we found that besides primary liver explants a rat connective tissue strain responded with increased outgrowth when we added serum from partially hepatectomized rats in the medium or when we lowered the concentration of serum from 50 to 20 per cent. Since then I found (Glinos, A. D., unpubl. data) that the L strain, which is a mouse connective tissue strain, also responds with increased growth to serum from hepatectomized rats, and Dr. Harris (Harris, M., *Growth*, 11, 149, 1957) has shown that either homologous or heterologous sera have growth-inhibitory effects when used in concentrations exceeding 40 per cent. The inhibitory effect is associated with the non-dialysable components of the serum. We are now in our laboratory seeking the same type of information with regard to the components of regenerating serum. What I would like to emphasize is that it is not permissible to equate conditions in tissue culture with conditions in the organism. This has been discussed by Dr. Harris and has been a recurrent theme throughout the recent Decennial Review Conference on Tissue Culture (*J. Nat. Canc. Inst.*, 19, 467, 1957). Most of these discussions were centered on established cell strains but we must look upon primary explants as also manifesting the potential rather than the actual physiological elements of the cellular response to environmental conditions. Such considerations were responsible for the decision to investigate the validity of the concept of liver growth control we derived from tissue culture studies, by further work in vivo rather than in vitro. It seems to me that the best way to underscore this point, so far as the mechanism of liver growth is concerned, is to end this comment by merely restating the fact which gave rise to Dr. Gey's question, namely, that in vivo after partial hepatectomy only the liver grows, whereas in vitro serum from partially-hepatectomized rats increases the growth not only of the liver but also of homologous and heterologous non-hepatic tissues.

# SUMMATION AND EVALUATION

PAUL WEISS

*Rockefeller Institute for Medical Research, New York, N. Y.*

MR. CHAIRMAN, ladies and gentlemen, I am not looking forward to this task with a great deal of pleasure because, I am afraid, a meeting of this kind should end on a climax and not on an anticlimax. In addition, the general excellence of the meetings defies any attempt to squeeze an evaluation into such a short period. Or, let's put it the other way around; if a real evaluation is expected, I should say it in one word: "superb," or "excellent," and quit. Now, to take an intermediate position, I think perhaps it is worthwhile to consider this symposium in the perspective of another symposium which was held exactly five years ago, in 1952, in Utrecht,[1] where I was charged with a similar task of making summary comments. The five-year plan is now a sort of chronological unit of human activities, and we might measure our progress in terms of the events during that time.

Let me first point out the things with which perhaps all of you will generally agree as the result of this symposium, before I go into minutiae. I have put it into six theses. The first one is: it's been a tough job, it's been hard work, it's been a grind, but it was worth it. Second, it has definitely shown that we have matured from the narrow confines of Embryology to the broader and more consistent science of Developmental Biology as expressed in the papers at this symposium, which show embryological to non-embryological papers in the ratio of four to three. The third point is that the convergence of fields of botany, microbiology, genetics, pathology, and embryology on the common foci of interest in the problems of development as such has led, through hybridization of lines, to what is known as hybrid vigor and not to the reproductive sterility of hybrids too far apart. Therefore, I think it promises further success, and as in the case of hybrid corn, the yield might increase. The fourth point is that it has shown that not only have the biochemists, and to a certain extent the physicists who have been a little bit in the back-

---

[1] The Biochemical and Structural Basis of Morphogenesis. *Arch. néerl. Zool.*, **10**, Suppl. 1 (1953).

ground here, contributed to the understanding of the phenomena of development, but the students of development are beginning to repay their debt by making available in the embryo an object which in many ways can serve biochemical purposes in presenting more elementary systems—systems which do not have to be artificially simplified, since in earlier stages they are much simpler in themselves. Several examples at this meeting have shown that the embryo deserves much more attention as an object of pure biochemical research, by furnishing simple objects for study, than has been done in the past. Fifth, I suppose that you have all recognized this was a most opportune place to hold such a meeting and that this is the time to express our great debt of gratitude to the spiritus loci, which is embodied in Dr. Willier to whom this symposium has been largely dedicated; and I would like to pay homage at this time to my old friend here for not only creating a flourishing school of developmental biologists, but also for having himself contributed so much to it, not only as an embryologist but also in introducing new approaches and ways to use molecular, chemical, and physical tools in its study. I would also like to pay tribute to the excellent successor of Dr. Willier at this place, Dr. McElroy, who has knit a continuous and self-containing series of papers into a beautiful coherent fabric that makes sense, in contradistinction to the occasional patch-quilt of incoherent papers which we are exposed to at some other types of symposia. Sixth, the principle of complexity of developmental phenomena will be agreed to by all of you, even biochemists, and as it was pointed out here, the only approach to that complexity is by resolving it into simpler elements.

I would like to say that we are all in full agreement so long as this resolution into simpler components does not degenerate into artificial oversimplification where the problem is being squeezed into a mold that is dictated by the simplicity of our tools of approach rather than by the nature of the object. With this precaution I think we can adopt the sentiment that has been expressed at this time, that the singling out of simpler components from the fabric of development is the most propitious approach to the complexity. However, a second point is that we must never lose sight of the fact that, in singling out separate components, we destroy something which is not only multiplicity, complexity, meaning more things which can be separated algebraically, but that we destroy relations, the nature of which in many cases is quite obscure and a problem to investigate in its own right and which cannot be investigated after the very object has been shattered to smithereens.

Now after these six theses, which I suppose you will all agree to, more or less, I would like to go into a few specific comments about them, and after moving up to the clouds, return to earth and wind up with showing you a few concrete films of living cells in action as the theme of my summary and evaluation, which is: that we are at the beginning, and not near the end, of the process of understanding development. This is not the time to summarize, but a time to prospect and project into the future; and the best thing is to turn for guidance to the living object which teaches us the lessons and also teaches us the real problems to which we are to direct our questions.

A primary problem in all our fields is to learn to ask intelligent questions. The way the question is phrased frequently contains already a hint, a clue, to the proper answer. If there has been any criticism at all appropriate, it is merely that some of the questions have been a little bit unrealistic: the problems have not been phrased in a way that would lend itself to a tactical approach with methods that are at hand now and which would hit the target. The aim has often bypassed the target. And this is something I am going to paraphrase in a few comments a little later. In general, I would like to say that the five years in between the last symposium have only fortified some of the statements I made at the time, and since some of the other speakers have quoted from their own earlier publications, I think that in order to blend into the landscape, I will do the same (p. 165).

"The meetings gave proof that we have essentially outgrown the era in which general platitudes about the fact that chemistry has something to do with morphogenesis were the order of the day. We have become sufficiently conditioned to taking this fact for granted without reasserting it on every possible occasion." I think that to reiterate this is redundant, the more so after the present symposium than before.

"We still hear occasional remarks, and have heard them at this symposium, to the effect that there is a chemical ontogeny in addition, as it were, to a morphological ontogeny; as if the morphological structures which we study as the products of ontogeny were anything but the mere visible expressions of antecedent chains of physico-chemical events. Such remarks obviously are the last traces of an outdated habit of thought, and it seems pretty plain that at least those who attended this meeting will no longer feel the need for asserting that chemistry is related to morphology, but will pass on to the next and more arduous task, namely, to specify just what this relation is. . . . Just how to get from chemical activity to highly specific cytological and histological structure and architecture, is not only unknown but rarely even stated explicitly as a pressing problem. Here part of the blame may be laid right at the doorstep of the morphologist who often has failed to provide a sufficiently detailed, clear, and precise description of the actual events which occur in histogenesis and morphogenesis." Phrases won't do. Each phenomenon must first be resolved into its elementary events, which then must be stated in terms sufficiently specific for the biochemist and biophysicist to carry on to further resolution. We can't address biophysical and biochemical techniques to words. You can only address them to clearly described phenomena. This clear description is in most cases still lacking. We have been talking of kidney development, myotomes, notochord. All those terms are very nice "handles" to label the problems, but as they refer to composite and complex systems, they have got to be broken down into their elementary components before one can pick up the analysis from the molecular level. This need has been very much in evidence and I shall come back to it.

"Electron microscopy, properly evaluated, has made a promising start in the further analysis of structure formation. However, one must remember that just as in the case of light-microscopy, the pictures reveal only the eventual structural order attained by virtue of antecedent ordering events, but still leave the latter in obscurity." There is no buck-passing any longer permissible, unless we recognize how the order of those things to which we pass the buck has come about by a series of chemical events, which somehow contain the clue to that order.

This brings me back to one major comment. The problem of organization and of the relation between structure, meaning order in space, on the one hand, and chemical activity, on the other, has hardly been touched upon in this meeting. There have been sort of denominational warfares between the morphologists and chemists as to which comes first, the hen or the egg, the structure or the chemistry. As soon as you begin to realize that these are part and parcel of the same thing, simply two aspects of the same continuum, that structure will never arise except by orderly reactions, and that reactions at random will never produce structure, then you will recognize that it takes some restraints in the degrees of freedom of the systems of chemical reactions in order to get to any structural product. That structural product, in turn, then will limit the degrees of freedom of otherwise randomized systems of chemical reactions. It's a stepwise interaction. It's a constant feedback modification of the chemical events by structural arrays and vice versa, and I don't believe that there is ever a true beginning or an end, as far as we can see, to organization in development. We are starting out with organized systems, we are ending up with organized systems. What we are observing is a transformation from one to another through a series of steps of interactions in which chemical reactions produce structure, structure in turn being the limiting factor for the type of reaction, for the selection of molecular species from the grab-bag that a cell, or a nucleus, or a genome presents for activation.

The selectivity of the structural-chemical relations remains to be defined. Along that line I would like to point out that perhaps one of our most fundamental problems, namely, the restricted organized reaction, as opposed to the full inventory of chemical reactions of which an embryo, an egg, a nucleus, a chromosome, etc., are capable—the problem of selective, as against massive, activation—has not found more than casual mention. I would point out, by way of example, just a few of the cases where it has come out. I would mention, for instance, the question of mutualism; intracellular mutualism, compartmentalization of chemical events within the cell, to which Dr. Markert referred in his early remarks, and, on the pluricellular level, the comments Dr. Herrmann made of the mutual symbiotic interdependence between two tissues and their changes. Such mutualism between two tissues and their metabolic requirements, one feeding the other and being mutually compatible, implies of course that they must have developed chemical differentiation and that they have become constitutionally different. Now we need not go into detail on this point. There has been ample evidence presented at this meeting of the fact that the organism and its parts undergo divergent chemical specialization during development. This was demonstrated directly by means of studies of composition, for example, by Dr. Markert's chromatographic results, and Dr. Edds' determination of the appearance of hydroxyproline in the embryo when collagen is to be formed. It was secondly shown by the progressive history of enzyme development, the appearance of enzymes and proteins as in Dr. Shen's report, and even in slime molds in Dr. Barbara Wright's brief paper, or the evolution of enzyme patterns which was referred to variously by Dr. Potter and others in the discussion.

It has been shown with equal conclusiveness, though indirectly, by the appearance of antigens, the antigenic ontogeny outlined in Dr. Ebert's talk, and finally, perhaps more localized, in the discussion of progressive changes that may go on in "microsomes," although we have been cautioned against this term because it does not signify units of similar kind, but merely units that happen to fall into a particular size class. Mitochondria likewise have been stated as undergoing divergent differentiation in different tissues, as by Mirsky. Briggs and King have presented striking evidence for the progressive divergence of the differentiation of nuclei. I would like to restate what I said in the discussion, that we must guard in this context against the danger of identifying nuclear differentiation with differentiation of the genome. The nucleus is more than the genome. This difference should be kept in mind. Progressive specialization is further seen in the indirect evidence of biochemical changes, such as changes in competence, meaning the changing physical condition and chemical readiness of an embryonic tissue to react to a graft, or the effects of graft age, which Dr. Ebert has described. All of those are indirect evidences that we are dealing with progressive chemical specialization or conversion. Which brings us directly up against the problem of differentiation in the single cell line. We are pretty convinced that in all this differentiation the cell is the sole agent of differentiation; or, whatever we observe as differentiation is the product of a change of the cell or the group of cells as a result of their organized group activities. Of course, these cell activities are not actions of independent agents, with a free will and a free mind, but interactions; and it is a platitude to keep pointing out constantly nowadays that there are such interactions, unless we finally get down to the job to describe and classify the various kinds of interactions and their products. Partly they are producing new chemical products and partly they are, as I said, feeding back on the structural basis on which later chemical events will have to take place and thereby change the background for the later chemical history of the individual.

We also should keep in mind that all these events take place in a system which is essentially built of concentric shells, with the genes in the center enclosed in the chromosomes, the latter embedded in the nucleus, the latter surrounded by the cytoplasm, the latter surrounded by other cells and the internal environment of the body as a whole, and that every agent that gets to one of the barriers between these shells will have to be examined in its own right as to whether and when it penetrates to any of the other shells before we can make any generalizations as to rates of the reaction. We can identify a response from the innermost core, let us say, the genes, only after a signal has returned from them through all the intermediate layers, and has expressed itself in some visible or otherwise tangible evidence, which is in most cases limited, not by the process itself or a property of the organism, but by the limitations inherent in our methods of detection and observation. So, in many cases when we speak of a lag, we speak of a delay of the appearance, that is, of our observation, of a particular character. Likewise, when we judge the action of a hormone, something we administer to the organism, in terms of what we finally see that black box produce, we must never forget that the final signal is no mark for the point of actual action of the agent which we have applied. All we can say is that the agent must have acted sometime before we get back the signal; precisely when, requires much more penetrating study; and yet, the confusion between the signal and the process is something which is running through our whole study of development. I let it go with this comment in order to continue.

We are convinced now that there are progressive changes going on in individual cell strains, and the question of reversibility and irreversibility has come up. One thing is clear that, in the animal kingdom, at least in higher forms, there are numerous cell strains which acquire during ontogeny properties that distinguish them one from the other in a heritable way—whether or not with a common genome is a question which has been discussed but which, as you know, is still open. Personally, I believe that the evidence at present is that the genome remains essentially constant, and if we see puffs and various manifestations on the chromosomes in different cell types at specific loci, this could as well be interpreted as a reaction of that particular genome to the modified cell space in which it is entrapped, indicating a prior change of that space, again by interaction with the products of previous interactions with the genome of course, but having made that particular chunk of protoplasm different, then activating a different portion of the same genome as a result of which a puff will appear. We have not reached any real conclusion here that the chemical evidence or the electron-microscopic evidence is, at present, in position to decide the issue.

Anyway, whatever part of the living cell takes part in differentiation with divergent biochemical specialization, it isn't the whole cell, it is only some part of it. But we must acknowledge that certain parts of certain cell strains are undergoing this process of divergence and that nothing on earth has been able to bring them back to a common norm. This brings up the question that has been asked about loss and gain: is differentiation a loss or is it a gain? Well, when we make hydrochloric acid out of hydrogen and chlorine, we gain hydrochloric acid and we lose hydrogen and we lose chlorine. I don't suppose that the question of loss and gain really makes too much sense once you describe the phenomena that occur. You lose one group of compounds, you gain another group of compounds, and the only question that remains is whether you have still retained the machinery to reproduce those compounds which you gained in the first place. Now there have been some comments here about the cells in tissue culture gradually changing from what they have been before they were gotten into culture. This has frequently led to a return to the cell of certain properties which were characteristic of its previous life. Well, I am afraid to point to personal features here, but I can say that since I am losing hair very rapidly, this returns me to an earlier phase of life as far as hairlessness of the scalp is concerned. But I would rather not consider that a return to infantilism. I submit, therefore, that the mere appearance of some characteristics in the cell, nutritional or metabolic, in a later stage, which it has in common with something which was present at an earlier stage of its history, is not a sufficient criterion to say that the whole cell has turned the clock back and now has gone back the way a movie film can be played back in the reverse direction. This is the way the thing has been interpreted. I suppose that we will emerge from this meeting with a considerable clarification on this particular point.

This discussion has revolved on the question of differentiation in single cell strains. I am afraid that the thing usually not pointed out in this connection is the fact that, by and large, we consider cells to be like tin soldiers, that is, as if all those of a given kind had all exactly the same properties. It is about time that we get away from the idea of the identity of cells, that they are as identical among one another as are their pictures in different copies of the same textbook of microscopic anatomy. The real cells aren't, and it's a question of whether we should perhaps adopt the lessons of microbiology here in this happy hybrid vigor phase of ours now, and pay more attention to the variance among individual cell strains. I mention this because, as I said yesterday, I believe the problem of cancer is a population problem based on the probability of cumulative incidence of insults on a cell strain, retaining some marks and passing them down the line, with other insults building up on top of them. The probability of this compilation or compounding is really the proper way to attack the cancer problem from the biological point of view. I let it go with these comments: the population aspect of cell biology has to be cultivated.

A second point I haven't found sufficiently sharply delineated is the lack of distinction between two wholly different meanings of differentiation. One is the linear, single-track differentiation and progressive transformation of a given cell, or of a given cell strain, the single track either leading to a terminal product or being blocked somewhat short of the terminal station. The other is the much more basic, more crucial, and much more obscure dichotomous shunting of a cell or a cell strain into one of two alternative directions. It isn't true that this happens only in embryology. It happens in the older organism likewise. To give you a simple example, the question of protein synthesis, for instance, takes an entirely different aspect if you consider that the thyroid cell can use its protein, and whatever else it uses, to build either another thyroid cell or to build thyroxin. In both cases we are dealing with qualitatively different shunts where the cell is being pushed in one direction to the exclusion of the other possible direction. Perhaps these will furnish models of what we are up against in differentiation.

This dichotomy of cellular development, where two essentially and demonstrably equipotential cells may be switched either in one direction or in another direction, is still one of our most obscure problems. There have been some models of this in simpler conditions: the vitamin A effects on the skin, shunting epidermis from a keratinizing into a mucous course, as shown by Fell and Mellanby, and by myself with James; the transformation of amoebae into flagellated forms by Willmer; the experiments of Jackson Foster in switching bacterial cells from sporulation into growth by changing the sugar concentration; and the control of fruiting in the slime mold—perhaps hint at some possible model experiments. It is this competitive activation of qualitatively different biochemical systems in the cell, both of which are present, one being activated to the exclusion of the other, which requires explanation. Key-lock relations furnish a formal analogy. Perhaps the keys lie, as I suggested at one time, in the exposure to sterically conforming outside molecules, which is a very vague and not really serious type of model at present; but it is this type of thing that has to be followed up. This leads directly up to the problem of the dichotomies which we see in the "induction" experiments. I must congratulate Dr. Yamada on the great clarification which he has produced by showing that by systematically graded changes of a given extracted agent, in this case by the degree of heating a protein, one can obtain systematically graded responses of the embryo. The same agent can produce a scale of results in a definite order, not just a random appearance, helter-skelter, of this or that type of organ. Since in all those cases the signals are still pretty far removed from the action and from the agent, we will simply have to wait and see whether the two points, the end-point, by which the result is measured, and the point of attack, can first be brought a little closer so that we may possibly identify the point of attack and the mode of action.

This brings me around to a general comment. That is, that we are still living largely in a period of identification, where the major effort is placed on identifying agents, and we have not quite managed to get on into a period of operation, where we study how the identified, crystallized, purified, or even synthesized agents actually manage to do their job, particularly in producing structure. Now as for the production of structure, there has been hardly any mention of this problem at all here, and for that reason I cannot summarize it. It would be supplementation rather than summary, and for that reason I would like simply to point out a few basic things for future reference. We do not know anything about structure formation except in a few model systems, like the collagen system and the ground substance so beautifully presented to us by Dr. Edds. We do have definite clues—no, I wouldn't call them clues, but rather indications—that there are in a complex system of development lattices of a supramolecular order in which certain sites in the complex ecology of molecular activities assume unique positions, that is, properties, which permit a stabilization of a certain chemical reaction which will eventually express itself as the locus of a particular cell, or of the deposition of a particular particle, or whatnot. It is high time that we look for these supra-elemental rules of order, supra-elemental in terms of particles in a cell, or of cells in a cell group, because if we don't do it, the vitalists will get hold of it and they still can quite rightfully point to the existence of those regularities.

The question of protoplasmic movement has been raised. We don't know anything about how chemical packages are shifted inside of a cell, not by diffusion but by localized streams. In references to granular motion, you heard Dr. Allen speak of an "attraction" of these granules in the sea urchin egg in partly fertilized eggs. We know neither the motive force nor the mechanism of such convections. If you consider that it has been only recently that we have been learning about the big currents of air in the upper atmosphere or about the big streams, orderly channeled streams, in the depths of the ocean, perhaps we might look even beyond the assembly of sciences represented here to get some help, some conceptual help which we badly need to understand motion. Motion within a protoplasmic continuum, such as cyclosis in plant cells, locomotion of tissue cells, motion of slime molds, are all partly observed, but not quite understood. We know that chemical energy furnishes the motive force, but don't know how it is channeled into translatory movement instead of random agitation.

All these problems are in last analysis structural problems of surface configuration, and just to give you one concrete example that will bring you back down to more solid ground, I would like to expose you to a picture of the surface interactions which we discussed yesterday with reference to surface specificity in terms of surface proteins. I would like to show you now a film on how the real activity looks that has posed the problem of surface specificity. This is another big problem about which we have been talking a lot, but have done very little about it. The problem of specificity, whether in drug action, in enzymatic catalysis, in immunological reactions, or in embryonic interactions, still waits for its master. Now, as regards surface reactions, we had talked about the immunological recognition of one cell by another. The film will show you the mutual recognition of embryonic cells on collision. This goes back to experiments some years ago in which we were studying epithelia, and Chiakulas demonstrated that epithelia of the same kind will merge, whereas epithelia of different kinds will bypass each other. Every surgeon knows about this fact from wound-healing. Moscona showed that in random scrambled mixtures of two cell types, eventually little clusters of homologous cells will be segregated out, but while he was working in my laboratory we didn't know how this actually came about. I had suspicions that we could get this information by looking at the cells. So, as Dr. Sussman said, "We made an experiment, we looked!" and I will now let you look in on those. You will see time-lapse, phase-contrast microcinematograms of single-cell suspensions from trypsinized cell groups, liver, kidney, and epidermis, settling on glass in a liquid culture medium. In all of the cases, one tissue has been set out the day previously so the cells have already aggregated and formed sheets. Then, the second cell type is added, and you will watch the behavior of the second cell type when it makes contact with the first one. You will see that if the cells are of different types, for instance liver and epidermis, the opposing cells remain restless, they do not fuse, they do not merge, they do not form a permanent boundary; but they do make primary contacts and only secondarily pull apart again. On the other hand, if the two cell groups are of the same kind, you will see how happily they merge. Even though homologous cells remain joined into a common mass, they move around a great deal within the mass, shifting freely relative to one another. They are not, by any means, glued to one another.

The problem of selectivity of aggregation here is not a problem of primary non-junction, but rather a problem of secondary disjunction. Cells nibble at each other indiscriminately whether of the same or of strange kinds. Only if alike, contrary to encounters among non-homologous cells, instead of separating, they cling to each other. In this, however, they still keep incessantly gliding and climbing about one another, proving that there are no firm and rigid attachments between them. Thus there is no way of explaining this "mutual affinity," as Holtfreter has called it, in terms of cementing. Yet, now that we have seen how it works, and how it doesn't, progress to eventual understanding may be faster.

I will close now, simply by quoting again from my concluding remarks of five years ago, which I suppose are pertinent to this point.

I should say that "my conclusions add up to the net impression of remarkable progress during the last several years; if not yet in the deeper biochemical understanding of the processes of morphogenesis, so at least in the widening of certain avenues along which such understanding may be sought. As in every young scientific branch, we recognize the pardonable traits of oversimplification, premature generalization, and overmagnification of the little that is known as compared to the vastness that is still unknown. These, in due time, will find their own correctives." This conference has given evidence that some of these correctives have already been found in the 5-year span since that last Symposium.

# A SUMMARY OF THE McCOLLUM-PRATT SYMPOSIUM ON THE CHEMICAL BASIS OF DEVELOPMENT

BENTLEY GLASS

*Department of Biology*
*The Johns Hopkins University*

The Symposium on the Chemical Basis of Development was fittingly dedicated to Professor B. H. Willier of the Johns Hopkins University, on the occasion of his retirement from active teaching after 18 years at the institution. Professor Willier symbolizes, through his studies on the development of feather patterns in birds, the current trends toward a union of concepts and experimental approaches in the field of animal development. His own papers exemplify the triune combination of embryology with genetics and with biochemistry; and his numerous students and collaborators in research have greatly broadened our understanding of these relationships. One need only recall the illuminating results of grafting limb buds or melanoblasts from donors of one genotype to hosts of a different genotype, and likewise studies of the hormonal regulation of feather pigmentation and the ontogeny of the endocrine correlations, to recognize the magnitude of Willier's contributions to this sector of biological investigation. It was highly appropriate that the introductory speaker on the program was one of Professor Willier's own former students, and at the same time his successor at the Johns Hopkins University.

## DEVELOPMENTAL CYTOLOGY

### Chemical Concepts of Cellular Differentiation

Clement L. Markert, quoting Ross Harrison's memorable characterization of embryology more than 30 years ago, emphasized the dynamic nature of development—the superimposition on the ordinary, stable, physiological functions of a complex system of ever-changing developmental functions. The primary aim of the student of development must always be to study these changing states. The patterns of developmental interaction, as in genetic analysis, have two aspects, first the initial elementary events, and thereafter the secondary interactions of the products of the primary actions. The essential difficulty in developmental analysis is that the elementary event remains totally unknown. It is not even certain that there is a single kind of elementary event, although increasing evidence that there is progressive nuclear differentiation as development proceeds and that genes function differentially in specialized tissues leads to the expectation that perhaps *differential gene activation* may be the primary developmental event. Clearly there is some sort of feedback between the genome and the ever-changing chemical environment in which it works.

In the ontogeny of metabolic patterns the emergence of active enzyme systems and the phenomenon of induction have attracted much attention. How does the enzyme activity originate? And how is the amount of activity regulated? The existence of adaptive, i.e., inducible enzymes, so well known in bacteria and other microorganisms, offers a suggestive hypothesis to account for these; but at present there is practically no experimental evidence of adaptive enzymes in embryonic development to support the notion. Embryonic induction, by means of specific substances that pass at specific stages of development from inducing tissues to tissues competent to react, may depend upon protein inducers and may lead to enzyme synthesis; but embryonic induction is not the same as enzyme induction. The metabolic pattern of the cell is certainly very sensitive to the chemical environment, as may readily be seen in cells growing in tissue culture. Yet there is little evidence for the view that development involves shifts from one steady-state to another because of the enhancement or repression of strategic reactions, except in the case of a few interesting biochemical systems in which enzyme formation has been shown to depend, through a feedback control, on the amount of a certain product. Markert here seems to have neglected the very abundant genetic evidence derived from suppressor genes, which switch the phenotype from one developmental pattern, the mutant, back into another, the normal one.

Developmental steps must certainly be as much gene-controlled as adult characters. All stages of development at all levels of organization have been shown to depend upon gene activity, and we may expect that "developmental phenomena will ultimately be analyzed in terms of mechanisms controlling genic activity." An outstanding question, then, becomes the question of whether every gene functions in every cell, either continuously or intermittently. Markert discusses the behavior of the tyrosinase-producing gene to clarify this problem. Only melanin-synthesizing cells indicate at all that they possess such a gene. Evident gene function is dependent upon the state of differentiation of the cell, which in its turn depends upon the integrated pattern of all previous gene activity. Pink eye in the mouse depends upon a gene which in the harderian gland prevents melanoblasts from differentiating, although elsewhere in the body of the same animal melanoblasts develop quite effectively. One must conclude that the effect of a gene on a particular step in differentiation depends on the metabolic pattern of the cell concerned. Willier's studies on the differentiation of the feather patterns in chickens show the same thing. The tyrosinase-forming gene represents an extreme situation, for that gene apparently acts only in one particular type of cell. [One may however query whether this is because the enzyme is not formed, because the enzyme lacks substrate, or because some other essential component of the reaction system is lacking.] The ontogeny of the esterases, of which there are perhaps 20 of different and overlapping specificity, is revealing. Each adult tissue has its own distinctive repertory of esterases, in characteristic amounts; during ontogeny these appear one after the other as the tissue differentiates. Electrophoretically homologous esterases from different tissues are alike in their specificities. The genetic basis is not clear; but if each esterase is controlled by a specific gene, then the battery of esterases characteristic of a particular tissue must arise from a differential activation of the genes in that tissue during the course of its differentiation.

Another type of gene control over basic cell structure, biophysical rather than obviously biochemical in nature, is shown by the well-known genetic control of spiral cleavage occurring in gastropod mollusks such as *Lymnaea*. If genic activation is fundamental to differentiation, one must face the problem of the mutual exclusion of alternative paths or channels of development. Markert suggests that this is difficult to conceive at the level of the gene; but perhaps here again the suppressor genes will illuminate the problem for us. If differential gene activation does occur, it should be reflected in a progressive differentiation of the chromosomes in various tissues. Markert cites three lines of evidence to demonstrate that this does occur: (1) the demonstration by King and Briggs that nuclei from well-advanced frog embryos (post-gastrulation), when implanted into an enucleated egg, cannot potentiate full, normal development, although nuclei from less advanced stages can do so; (2) the observations of Pavan and of Beerman that in specific tissues the dipteran giant chromosomes undergo special differentiation at particular sites and particular stages; and (3) variation in the amount of RNA and of residual protein in the cell nuclei of different tissues and within the same tissue at different stages.

Markert concludes by proposing a hypothesis of development that should be helpful in orienting chemical investigations of cellular differentiation. He proposes (1) that the newly fertilized egg contains undifferentiated chromosomes and an unstable metabolic pattern that commences to differentiate in the various parts of the egg; (2) that as cleavage proceeds, the chromosomes become subjected to differentiating chemical environments, and specific parts of the DNA become coupled with specific proteins so as to bring about activation of the genes; (3) that the emergence of functional, or activated, genes is reflected in the synthesis of an enzyme or some other specific product, which consequently modifies the cellular constitution and exerts a feedback on the nucleus; and (4) that functional genes should display varying degrees of stability—in tissue culture dedifferentiating rather readily, but in nondividing cells exhibiting great stability.

## The Initiation of Development

This subject has been covered by R. D. Allen in an exhaustive review comprehending the processes leading to syngamy, the structure and chemistry of the egg surface, and processes leading to the activation of the egg. It will be impossible to do more than indicate some of the concepts, principles, and hypotheses he has considered. The observations relate largely or exclusively to the eggs of marine invertebrates.

The processes leading to syngamy include the acrosome reaction, the entry of the spermatozoon, its early behavior inside the egg, and the migrations of the sperm and egg nuclei toward each other. In the acrosome reaction a fibril forms from the acrosome at the anterior tip of the spermatozoon, some substance is released and immediately dispersed, and the middle-piece of the sperm is displaced to one side. Exciting these events, or essential to them, are factors such as the physiological maturity of the spermatozoon, the presence of calcium, the presence of water that has bathed eggs of the same species, alkalinity, and either contact of a nonspecific nature as with glass or collodion, or contact with the jelly or vitelline membrane of the egg. Entry of the sperm is associated with the formation by the egg of a fertilization cone which, according to observations of the Colwins on echinoderms, creeps up the acrosome filament while the sperm is at the same time sinking into the egg. The sperm is then pulled through the hyaline funnel beneath the cone by some unknown force. Since the fertilization cone is formed but once by any egg, except in cases of polyspermy, it is natural to postulate some physical or chemical action of the acrosome to elicit it; but the nature of this reaction remains unknown. Phase contrast microscopy and electron microscopy have revealed many of the events in this and the succeeding phase following sperm entry; but the work is still in a descriptive stage.

Growth of the sperm aster, which forms around the introduced centrioles, clearly has much to do with the migration of the sperm nucleus by pushing it along, although centripetal cytoplasmic streaming in the egg is also sometimes apparently involved. Movement of the egg nucleus is not so easy to explain. Perhaps a centripetal current of cytoplasm carries it to the central region of the egg, but there is skepticism about this. In eggs confined in capillary tubes of smaller diameter than the egg, the egg nucleus seeks the axis of the capillary, elongates, and bulges toward the sperm nucleus as it migrates in its direction. Generally speaking, however, we are no more acquainted with the forces and chemical reactions participating in fertilization than we are with those involved in mitotic cell division.

Many observations have been made on the jelly of the sea urchin egg, which can be removed without impairing the fertilizability of the egg. It has a molecular weight of 280,000-300,000, is electrophoretically homogeneous, and very acid. It appears to be a highly sulfonated glycoprotein, resembling heparin except for lacking hexosamine and glucuronic acid. Not so much is known about the jelly of other marine invertebrate eggs, but the jelly of the frog's egg has been examined sufficiently to show that it is rather like that of the sea urchin egg, although containing hexosamines and lacking acid-hydrolyzable sulfate groups.

The cortex of the mature egg may be lacking as a clearly differentiated layer (ascidian eggs), or may be differentiated according to its behavior upon fertilization as either labile (most echinoderms, the lamprey, fish, frog, hamster, *Nereis,* etc.) or stable (some annelids and mollusks, e.g. *Spisula, Chaetopterus*). Study has been largely concentrated on the labile type, especially the sea urchin egg. The cortical granules are beautifully shown in electron micrographs as occupying a single layer below a thin clear layer bounded by a fine membrane. The latter is taken to be the vitelline membrane. According to Afzelius' electron micrographs, it is 100 Å in thickness. When the egg is fertilized and a "fertilization membrane" is lifted from its surface, the vitelline membrane forms the outer component of the fertilization membrane. Although the fertilization membrane, once formed, is resistant to trypsin, prior treatment of the egg with trypsin will prevent its formation; papain does not act in this way. The electron micrographs show that the vitelline membrane is not dissolved by trypsin but is so weakened that the cortical granules can exude through it. The cortical granules are lamellar and are surrounded by a double membrane 75-90 Å thick. Cytochemical staining and isolation and analysis have been applied to the cortical granules with some initial success. Allen himself has pioneered in the development of a method of stripping off the cortex by means of its adhesion to glass, or isolation by high centrifugal force ($170,000 \times g$). The cortical granules probably contain mucopolysaccharide, with fucose in the carbohydrate component. The composition of the cortical gel is even less well known. It exhibits positive radial birefringence which disappears at fertilization, and at least a part of this is strain birefringence. The permeability barrier of the egg is probably subcortical, for the cortex is permeable to certain non-electrolytes which fail to go deeper, and the cortical material can be stripped off without dispersing the endoplasm of the egg.

The processes leading to activation of the egg would seem particularly favorable for biochemical analysis. *Nereis* sperm, after attachment to the egg, can be removed after no more than two minutes, and the egg will nevertheless be activated and will undergo cleavage, as in artificial parthenogenesis (Lillie; Goodrich). The evidence implicates the acrosome as the activating agent. By an ingenious use of lauryl sulfate to inactivate instantaneously all free-swimming spermatozoa, the Hagströms were able to time sperm attachment precisely; and then by killing the eggs instantaneously with formalin (which does not interrupt the elevation of the membrane over areas where the cortical reaction has already taken place), they demonstrated the existence of a latent period of 10-20 seconds between sperm attachment and the onset of visible cortical changes. The latent period has a temperature coefficient of 2.3, which makes it probable that activation of the cortex by the acrosome is chemical. Its specificity may be deduced from the results of cross-fertilization between species; but of its exact chemical nature nothing is known at present.

It is now recognized, following the work of Motomura, that the fertilization membrane is dual in origin, being formed by a merging of the released cortical granules with the preexisting vitelline membrane. The thickening and hardening of the membrane and its increasing resistance to most enzymes depends upon the release from the cortex of a substance which has been isolated, and is probably identical with "antifertilizin." Polyspermy is prevented by formation of the hyaline layer as a product of the cortical reaction, rather than by the presence of the fertilization membrane. Sugiyama has distinguished from more destructive agents those which initiate a wave of cortical granule breakdown like that produced by the sperm. These include distilled water, urea, and butyric acid. The so-called fertilization wave or impulse seems to have the character of a self-propagating, autocatalytic mechanism, and not of a diffusion process. The kinetics of the propagated impulse are not clear.

At the present time the old fertilizin theory of activation has had to be discarded. Runnström and his group have proposed a *kinase theory,* according to which the kinase is bound to an inhibitor until some substance released by the spermatozoon binds the inhibitor and releases the kinase, which then in turn activates various proenzymes. Lundblad's proteolytic enzyme E II is thought possibly to be the kinase. Monroy, on the other hand, proposes a lipoprotein liquid crystal theory, according to which the sperm would free phospholipids that thereupon undergo structural changes. The weakness of this theory is that agents which are known to attack lipoproteins exert only a localized effect instead of a propagated impulse.

By obtaining partially fertilized eggs, in which the propagation of the fertilization impulse has been interrupted over a portion of the egg surface, the relative significance of the entry of the sperm and of the fertilization impulse can be estimated. Wherever the latter has been excluded, the sperm aster fails to grow; and the sperm and egg nuclei migrate and fuse only when located in fertilized cytoplasm. Cleavage is suppressed or delayed. These results raise new questions regarding the influence of the cortex on the endoplasm. The fertilization wave itself, rather than the breakdown of the cortical granules, has been shown to be the event of primary importance in the initiation of development.

Obviously, the demise of the fertilizin theory has so far led only to new problems and theories, not to definitive solutions. Much more investigation will be required before biochemical explanations of fertilization and the initiation of development will be satisfactory.

The contrasts between dependent and autonomous development in various types of fertilized eggs were considered by Fritz Lehmann. At one extreme in organization there is the echinoderm egg, with a cortex containing a double system of antagonistic gradient fields and an endoplasm which is very little differentiated with respect to morphogenetic pattern; at the other extreme there is the amphibian egg, which lacks cortical pattern but has distinct endoplasmic regions, one of which is destined to give rise to ectoderm and mesoderm and the other to endoderm. Between these strongly contrasting types lies the egg of the annelid *Tubifex*, which possesses both a cortex with polar fields and an endoplasm possessing two regions, polar plasm and yolk-containing cytoplasm. Lehmann has analyzed the spirally cleaving eggs of *Tubifex*, in which a mosaic pattern arises through an epigenetic process during maturation and cleavage. First the polar plasm, rich in mitochondria and other "biosomes," and located at both animal and vegetal poles, becomes distinct from the center of the egg, which carries the mitotic apparatus surrounded by an endoplasm full of yolk. These distinct main regions have very characteristic populations of particles such as mitochondria, microsomes, and fibrils. Concentration of all the polar plasm at a single pole, by means of centrifugation, does not radically interfere with cleavage; and normal embryos still result.

There seems to be some intracellular adhesiveness between polar plasms. This, together with complex interactions between particulates and cortex, leads to the ooplasmic segregation which establishes the typical cytoplasmic pattern. The two polar plasms both enter one large blastomere, which divides into (1) an ectoblast with a dense cytoplasmic reticulum and few mitochondria, lipid droplets, and yolk spheres, and capable only of producing ectodermal germ bands; and (2) a mesoblast, with a looser cytoplasmic reticulum and a rich population of particles, and limited to producing mesodermal germ bands. After this segregation of cytoplasm into ectodermal and mesodermal types, development is mosaic in nature.

Do the differences of the two somatoblasts then reside in their respective proportions of particulates? Experimental evidence on this point has been provided by centrifugation. The eggs of *Tubifex* are easily stratified and keep their rearrangement longer than centrifuged marine eggs such as those of *Ilyanassa*. Partition of the large somatoblasts into smaller cells makes the latter unable to form germ bands, even though they contain polar plasm. Obviously epigenetic processes have been suppressed which are necessary to the determination of the ectoderm and mesoderm. The first of the two principal periods in the development of the *Tubifex* egg lasts until the formation of the two somatoblasts and is characterized by numerous, closely interlinked epigenetic processes. The second, mosaic period then begins and lasts until formation of the germ bands.

The intracellular morphodynamics that step by step sort out the different cytoplasms in the epigenetic phase of the development of the *Tubifex* egg will repay study. The orientation of the spindle in the first and succeeding cell divisions is only one of the complex factors involved.

### The Role of the Cell Nucleus in Development

The nucleus clearly must act upon the cytoplasm during differentiation, and the nucleus can acquire fresh substrates only by way of the cytoplasm. Much has been learned in recent years, according to Mirsky and Allfrey, about each of these processes. Thus the essential coenzyme diphosphopyridine nucleotide (DPN) is synthesized only in nuclei, since an enzyme essential for its synthesis is confined to the nucleus. Likewise much, if not all, of the RNA of the cytoplasm is synthesized in the nucleus. In the opposite direction, a rapid initiation of cytoplasmic enzyme synthesis in the acinar cells of the pancreas is paralleled by an equally rapid amino acid uptake by the nuclear histones and other proteins. Yet salient questions remain unanswered, and new ones emerge.

Are all the genetic factors present in all the cells? We may think so, from the constancy of chromosome number, constancy of pattern of dipteran salivary chromosomes, and the constancy in amount per nucleus of DNA and histones. But the supposition is far from proved. There is much evidence that nuclear activity varies in different cells, and that nuclear differentiation of some sort does occur. DNA is essential to the intranuclear synthesis of ATP, and the latter process as well as the replication of DNA and the synthesis of RNA, will require nucleotides. An active nucleus has a much larger pool of free mononucleotides than an inactive nucleus. The relative amounts of diphosphate and triphosphate nucleotides very probably control the balance between some of these syntheses. Isolated thymus nuclei synthesize proteins only in a medium containing sodium ions, whereas cytoplasm has a preponderance of potassium ions. There is a rapid turnover of $Na^+$ with the external environment, but there must also, in the intact cell, be differential ion accumulation in nucleus and cytoplasm; and the intranuclear environment must therefore also be reckoned with. It may be, as Markert proposed, not a matter of the presence of all the genes, but a matter of their activation.

Are nuclei then differentiated? Morphologically they certainly may be.

Physiologically, differentiation may be shown by twin nuclei that become respectively vegetative or generative nuclei in the *Tradescantia* pollen grain (Sax), depending upon which nucleus gets into a specific part of the cytoplasm. Biochemically, the differentiation is shown by the fact that some nuclei will contain characteristic proteins of the cell's cytoplasm, such as hemoglobin in an erythrocyte or erythroblast, while others do not, e.g., there is no myoglobin in the nucleus of a heart muscle cell.

Is nuclear differentiation reversible? Briggs and King, in their experiments with nuclei transplanted into enucleated frog eggs, have demonstrated the existence of irreversibility under the conditions of their experiments. The differentiated nucleus suddenly placed in undifferentiated cytoplasm does not dedifferentiate. But Mirsky and Allfrey remind us of Stone's experiments on the eyes of salamanders. Removal of the lens causes iris cells to dedifferentiate and then redifferentiate as lens cells; and removal of the retina causes retinal pigment cells to lose their characteristics and become retinal nerve cells. Differentiation is not in all instances or under all conditions irreversible.

The highly specific effects exerted by the nucleus on the cytoplasm are generally thought to be mediated by means of the RNA. There is clearly a requirement for RNA in the cytoplasmic synthesis of protein. In the tobacco mosaic virus the specific protein of the coat is determined by the RNA of the core. [But in the coli-bacteriophages the protein of the coat is determined by the DNA of the phage.] In tracer experiments the rate of incorporation of precursors into the RNA of the nucleus far exceeds their rate of incorporation into the RNA of the cytoplasm. RNA is synthesized most actively in the nucleolus, and can pass to the cytoplasm. These are some of the reasons for the accepted view. It remains uncertain whether some RNA is synthesized in the cytoplasm.

## Chromosomal Differentiation

According to Joseph G. Gall, the lampbrush chromosomes of the amphibian oocyte and the giant larval chromosomes of dipterans have shed most light on the differentiation of chromosomes in different kinds of cells. The chromatid is at most a few hundred Ångstroms in width, but may be millimeters or centimeters in length. It contains DNA and basic protein throughout its length. In the common microscopically visible state it is greatly coiled; its unwound state corresponds to its period of genetic activity. RNA may accumulate in the latter phase around the delicate DNA-protein fiber until it becomes microscopically visible. This unwinding may be limited to specific regions in certain tissues, or may involve the entire length of the chromosome. It is essentially reversible, and is correlated with the specific genetic activity of the regions which thus become exposed.

The giant dipteran chromosomes, best known from the salivary gland cells, are polytene, that is, many-stranded, according to evidence presented by Gall. Perhaps nine or ten rounds of replication occur in their enlargement, so that they would contain 512 to 1024 chromatids. Polyteny is regarded as a special case of endopolyploidy, in which chromosomes replicate and separate but remain together in a single nucleus because of spindle defects. In some cases the replicated chromosomes remain together in loose bundles, an intermediate state relating endopolyploidy to polyteny.

The finest individual fibers in polytene chromosomes are from 100-500 Å in diameter. There are very many of them, 1000-2000 (Gay) or 10,000 or more (Beerman). The total number of strands approximates closely the $2^n$ function of the number of times ($n$) the volume and DNA content of the nucleus have increased. Beerman and Pavan, with their respective collaborators, have shown that specific modifications of particular bands take place in certain tissues and at certain stages of development. These alterations in appearance of the bands, including such phenomenal changes as the so-called puffs and Balbiani rings, are reversible. The Balbiani rings, particular bands at which the chromosome breaks up into fine loops that extend out laterally, are limited to the salivary gland cells; and in certain dipterans the salivary gland possesses distinct lobes which manifest characteristic differences in their Balbiani rings and puffs. It is therefore held that these modifications of the chromosome structure are definitely related to the physiological—probably the secretory—activities of these cells. The fine loops of the Balbiani rings demonstrably synthesize droplets of RNA, which seemingly pass out into the cytoplasm and become a part of the endoplasmic reticulum.

By the use of radioisotopes, RNA has been shown to be present in the bands of the salivary gland chromosomes. Incorporation of $P^{32}$ in the RNA of the chromosomes is, however, exceeded by that into the RNA of the nucleolus. By following the tracer, it is found that the nucleolar RNA makes its way into the cytoplasm.

The lampbrush chromosomes of many vertebrate and some invertebrate oocytes also have loops extending laterally from the axis of the chromosome. In the salamander oocyte they may attain 1 mm. in length. The ultimate chromomeres of the main axis are attached to their neighbors by pairs of fine loops, each 10 to 50 micra in length. The chromosome structure of these chromosomes thus appears to comprise two continuous chromatids, which are coiled compactly in the chromomeres and open out into the fine loops between them. Cytochemical staining reveals that the chromomeres are DNA whereas the loops possess RNA and protein. Actually the fine continuous strand of each loop (200-400 Å in diameter) is composed of DNA with a coating of the more abundant ribonucleoprotein. Proof is seen of the structure of the loops in the fact that digestion with either pepsin or ribonuclease dissolves all material visible in the light microscope, yet leaves a fine fibril which can be demonstrated by electron microscopy. Deoxyribonuclease, on the contrary, brings about disintegration of the chromosome structure into segments which break up into smaller ones until nothing is left (Callan and MacGregor).

Gall postulates that the loops correspond to gene loci, which are functioning when stretched out into loops and are at rest when coiled up within the chromomeres. In harmony with such a conception is the observation that the loops regress by discharging their ribonucleoprotein and gradually folding back into the chromomeres. The nucleoplasm of late oocytes contains an abundance of small RNA-containing granules like those previously located on the loops; and since the loops distinctly have individuality, it seems reasonable that the granules of RNA are specific sorts of gene products. For example, the nucleolus organizer upon one of the lampbrush chromosomes of *Amblystoma tigrinum* normally bears an attached nucleolus seemingly identical with the many already to be seen in the nucleus and which presumably have arisen from the activity of this locus. When adenine-$C^{14}$ is used as a tracer it enters the RNA of the oocyte and concentrates in the RNA of the nucleus, although the RNA is more abundant in the cytoplasm. Radioactivity can also be found on the loops. The inference is that RNA is being synthesized on the loops, moves to the nucleoplasm, and eventually reaches the cytoplasm.

Callan and Lloyd have recently found a very remarkable differentiation of loops in the chromosomes of *Triturus cristatus carnifex*. At a certain locus loops may form or be absent; and all three genetic types (homozygous for loops, homozygous for no loops, and heterozygous with a loop on one side only) have been observed. This is circumstantial evidence that the presence or absence of loops at a particular locus is related to gene activity.

Gall argues strongly that the chromatid is the ultimate fibril of the chromosome and is not multistranded. Adding the lengths of all the loops in a lampbrush chromosome, some chromosomes of *Triturus* may attain a length of 5 cm., ignoring the portion of the chromatid which is coiled within the chromomeres. If the functional unit is the loop in the lampbrush chromosome and the band in the salivary chromosome, these range from 200 to 4 micra in length respectively. Almost certainly there must be some unfolding of the basic chain molecules to provide these enormous lengths.

Particularly in the lampbrush chromosomes, the formation of the loops and their later regression clearly indicate a reversible physiological cycle which does not alter the genetic content of the chromosomes transmitted through the egg to the next generation. Further studies will very likely correlate the observed changes in the chromosomes with definite genetic characters, for it would seem that if the formation of loops and puffs and similar structures is an indication of gene activity, then mutants in which a particular locus is known to have impaired or zero activity should be distinguishable. The difficulty is that the mutants collected by present means are not likely to be concerned with specific oocyte or larval salivary gland functions.

A recent technique of great promise in studying chromosomal differentiation was described by W. L. Hughes. This consists in tagging with tritium the thymidine which is supplied to replicating chromosomes. The autoradiograms of chromosomes labeled with tritium are very sharp, the resolution being sufficient to distinguish newly synthesized chromatids from original ones. When *Vicia faba* chromosomes are grown in labeled thymidine for 8 hours, or approximately one-third of a division cycle, the two daughter chromosomes of each chromosome are uniformly labeled. Following another chromosome replication, this time in unlabeled medium, each chromosome is labeled but only one chromatid of the two sisters is radioactive. The simplest explanation of these observations is that each chromosome comprises two chromatids each of which replicates. Following replication in labeled medium, the new chromatid is labeled, the old one is not; so that each daughter chromosome is labeled. After another replication, when in the presence of colchicine each daughter chromosome would be quadripartite, only one chromatid was labeled. In other words, the labeled thymidine entered only the strands formed by replication in the labeled medium, and except for occasional chromatid crossovers remained integrally in those strands thereafter. New strands formed by replication of the labeled strand in unlabeled medium are unlabeled. This pattern of replication is like that proposed for DNA by Watson and Crick.

*E. coli* cells labeled with tritiated thymidine and returned to normal medium do not exhibit uniform dilution of the radioactive label. At least by the third division, all-or-none divisions into heavily labeled and unlabeled cells occur. It therefore appears that most, if not all, of the parental DNA remains associated in transmission to daughter cells, like single chromosomes in the mitoses in *Vicia* seedlings. Labeling of DNA in mammalian cells in tissue culture or in the living animal with tritiated thymidine is readily possible, and reveals variation in the sequence of events in cell proliferation and the movements of cells from generative to functional zones. Use of the method in embryology should facilitate studies of cell migration and tissue differentiation and the viability of transplants. Comparisons of the routes and turnover of different tritiated nucleotides will reveal more about nucleic acid metabolism. Thus tritiated cytidine in *Vicia* roots first accumulates entirely in the nucleolus, and later moves to the cytoplasm, a result indicating that RNA is synthesized in the nucleolus and then transported to functional sites in the cytoplasm.

*The Structural Elements of the Cytoplasm*

Knowledge of cytoplasmic structures and particles has advanced very greatly, from a descriptive point of view, since the advent of electron microscopy. Nevertheless, our ability to relate to the ultra-microscopic structures the findings of biochemists who have fractionated cells, and learned something about the chemical differences of the components of the cytoplasm, remains still at a very modest level. The state of this area of investigation was set forth by Hewson Swift in a review of cytoplasmic particulates and basophilia. He treated in particular the mitochondria, Golgi apparatus, cell membranes, and endoplasmic reticulum.

Fifty years of biochemical work on the mitochrondria have established their great importance as seats of enzymatic systems, especially those of oxidative metabolism. The electron microscope has revealed mitochondria to be intricate structures each surrounded by a double membrane and containing numerous lamellae (cristae). Yet almost nothing can be said regarding the relation of the enzyme systems to the structure; nor do we yet know whether the enzymes are synthesized in the mitochondria or simply collected and organized there. This ignorance is scarcely surprising, since on the cytological level itself we are still unable to identify precisely either the origin or mode of multiplication of mitochondria; nor can we at the biochemical level distinguish their types, although it is clear that in different tissues mitochondria are somewhat differentiated.

Somewhat more fruitful have been the recent investigations of the Golgi apparatus, since conventional cytology had left its nature, function, and universality even more uncertain than in the case of the mitochondria. In the electron microscope preparations, after osmium fixation, the Golgi apparatus is revealed as "a series of densely packed parallel double membranes, often slightly expanded at their ends, surrounded by numerous small 'vacuoles' or 'granules' roughly 50 m$\mu$ in diameter." Such structures may in some tissues be aggregated into larger bodies, cuplike or platelike in shape. Similar structures—the dictyosomes described by Porter—occur in plant cells, a discovery which would seem to answer finally the question of whether the Golgi apparatus is characteristic only of animal cells or is of wider distribution. The Golgi material is characteristically high in phospholipid content, and comprises at least two fractions, one soluble, the other insoluble, in 70% ethyl alcohol.

The evidence supplied by the electron microscope and by microcinematography has greatly extended our knowledge of cell membranes, which are revealed as active, highly convoluted structures often possessing microvilli. They seem to form intracellular vesicles and to secrete particles to the interior by active engulfment. By similar means cells in contact might even exchange large portions of cytoplasm.

Most novel and extensive are the evolving conceptions of the basophilic elements of the cytoplasm, for of these we knew virtually nothing only a short while ago. The "microsomes," earlier described as tiny particles at the very limit of resolution by visible light, and isolated by Claude and others by means of ultracentrifugation, turn out to be far from homogeneous. They represent fragments of an extensive system containing both RNA and lipoprotein, the former separable as ultramicrosomes 5 to 24 millimicra in diameter, the latter constituting a membranous complex. This system of membrane-linked vesicles, sacs, and sometimes parallel lamellae, upon some surfaces of which fine particles are aggregated, has been called the "endoplasmic reticulum," although perhaps "basophilic reticulum" would be a more appropriate name. Palade considers it to be "a continuous network of membrane-bound cavities permeating the entire cytoplasm from the cell membrane to the nucleus." In general, the "rough" membranes, that is, those with adherent particles, are identified with the reticulum, while the smooth membranes belong to the Golgi apparatus, the cell membrane, or the nuclear membrane. This sweeping concept is not universally accepted at the present time, and has been criticized on several grounds; but most workers appear to agree that three elements of the cytoplasm are distinguishable, namely, the particle-lined membranes, the smooth membranes derived from the cell membrane, and the smooth membranes of lamellar or vesicular character belonging to the Golgi complex.

In spite of this great advance in the morphology of the cytoplasm, an understanding of the functions of its parts—of the basophilic complex as well as the Golgi apparatus—is nebulous. There is cytochemical evidence that the "microsomes" are essential to protein synthesis, and most current thinking and experimental analysis has been directed toward an investigation of the relationship of the basophilic reticulum to the activity of RNA in protein synthesis. Swift has reported cytochemical studies, supported by electron microscopy, of RNA in the nucleolus, the nucleus apart from the nucleolus (chromosomes), and the cytoplasm. Upon starving, the cytoplasmic RNA of liver cells of *Ambystoma tigrinum* larvae diminishes, as in starved rat liver. Upon refeeding, the RNA increases in all three portions of the cell, but sooner and more rapidly in the cytoplasm than in either chromosomes or nucleolus. The increase in the two nuclear components was roughly parallel. The relative amounts in nucleolus, nucleus, and cytoplasm were 1.0, 4.7, and 82.0 respectively. In terms of *relative specific activity,* as measured by incorporation of adenine-$C^{14}$ into RNA in the pancreas and liver of adult *Triturus,* the chromosomal portion of the nucleus considerably exceeded the nucleolus (2.4 to 7 times), and the nucleolus exceeded the cytoplasm (2 to 5 times). The DNA/RNA ratio, as measured by ultraviolet absorption at 260 m$\mu$ before and after treatment with deoxyribonuclease, was 1.5 for the chromosomes (16.4 times that of the rat); and the estimated RNA was $6.8 \times 10^{-8}$ mg. per nucleus.

In autoradiographs, the RNA of the cytoplasm was frequently found lying against the nuclear membrane. Since the double-layered nuclear membrane, under electron microscopy, has been found to contain annuli with a central spot and sometimes with cylindrical walls projecting on the one side into the cytoplasm and on the other into the nucleus, passage of RNA through the nuclear membrane has been postulated. The wall of the annulus had in some cases a structure like that of certain vesicles or tubules in the cytoplasm, and sometimes small particles were associated with the annulus. This is taken to mean that in telophase of mitosis, when new nuclear membranes are formed, they arise by coalescence of cytoplasmic elements; and conversely, that annuli may give rise to cytoplasmic elements. In the amphibian liver and pancreas, stacks of intensely basophilic cisternae frequently lay in the cytoplasm against the nuclear membrane. Sometimes, as in *Triturus* pancreas and liver, filaments running at right angles to the lamellae were observed; but in *Ambystoma* thyroid convoluted filaments appeared to run along the membranes and were outlined by very fine, densely packed granules. In crayfish (*Cambarus*) spermatocytes filaments formed a lattice traversed by the lamellae. The fine particles are not always restricted to the lamellae, but may lie along filaments or free in clusters, rows, or tiny rings. The membranes of cytoplasmic vesicles are often lined with particles; and when cisternae are apposed, the particles lining them appear to merge into short filaments, perpendicular to the membranes. Particles vary in size, and perhaps in shape. Moreover, diffuse basophilia may appear in the cytoplasm in addition to the basophilia of the membranous system, so that probably cytoplasmic RNA is not entirely contained within the fine particles. This is also evident from the basophilia of the smooth membranes in the cytoplasmic lamellae of amphibian liver cells.

In sum, the relation between lamellae, filaments, and fine particles is still not clear. One may conclude that the considerable advances being made in understanding the fine structure of the cytoplasm are in fact laying a foundation for the biochemical superstructure that will some day be related to it; but at present only the dimmest outline of that relationship can be visualized.

*The Intercellular Matrix*

The intercellular matrix holds particular interest for the student of embryology, and various provocative hypotheses have been put forward regarding its role in morphogenesis and embryonic induction. Mac V. Edds, Jr., has surveyed the results of recent investigations of intercellular matrices. In so doing, he distinguished between the optically homogeneous, viscous "ground substance" of the intercellular spaces, and the "intercellular matrix" in toto, which includes in addition to the ground substance various imbedded fibrous and mineral components.

The amount of ground substance varies greatly with region and age, being gradually replaced by fibrous material as aging occurs. Under the electron microscope the ground substance is homogeneous and structureless. It is thought to be generally a product of mesenchyme cells and especially of fibroblasts and mast cells; but it also occurs in the early embryo between the still undifferentiated cells of the ectoderm and endoderm. It contains large amounts of acid mucopolysaccharides, which fall into two main classes: (1) the non-sulfated ones, such as hyaluronic acid and chondroitin; and (2) the sulfated ones, chondroitin sulfates A, B, and C, heparin, and keratosulfate. All of these are polymers of high molecular weight, and hydrolyse to hexosamines, uronic acids, and acetic and sulfuric acids. Chondroitin B is especially interesting because one of its components, iduronic acid, is closely similar to ascorbic acid, a fact which raises a suspicion that the vitamin might be an essential nutrient because it is a precursor of iduronic acid.

Differences between tissues may be reflected in, or perhaps arise from, their different mucopolysaccharide compositions. Thus the cornea of the eye contains chiefly keratosulfate, with lesser amounts of chondroitin sulfate A and chondroitin, whereas the sclera contains no keratosulfate, but possesses mainly chondroitin sulfate B, with lesser amounts of A and C. The elongated, flexible macromolecules of mucopolysaccharide have a high negative charge, and consequently, a strong affinity for cations and water. They are readily depolymerized and repolymerized, and form reversible linkages with other compounds. Obviously, such properties must be of the highest importance in regulating the nature of the connective tissue. The possibility that mucopolysaccharide-collagen complexes of certain constant ratios exist is uncertain, except in the case of the cornea and sclera, where it has been found that the 2.5 times greater amount of mucopolysaccharide in the cornea is accompanied by a proportionately larger amount of collagen. In analyses of cartilage, about one-third of the chondroitin sulfate has been found to be linked with a non-collagenous protein, the rest with collagen. Bernardi, who did this work, has concluded from studies of viscosity, sedimentation, and light scattering that the chondroitin sulfate and polypeptides form linear chains. In other studies of cartilage, it has been found that the formation of linkages with collagen depends on the maintenance of the mucoprotein complex.

The major fibrous component of the intercellular matrix is of course collagen, a recognized prototype of fibrous proteins. Each collagen fibril is an aggregation of collagen molecules associated laterally by long side chains; each collagen molecule is comprised of three helically coiled polypeptides strongly linked by hydrogen bonds, and largely composed of the amino acids glycine, proline, and hydroxyproline, the last-named being virtually diagnostic. The native fibril has a cross-banded pattern with a period of 640 Å; and Schmitt and others have shown that the collagen fibrils may be dissolved in suitable solutions and then reconstituted, either into the normal types or into unnatural types with periods (2600-3000 Å) conforming to those of the individual tropocollagen particles. The aggregation of collagen molecules may either be parallel or anti-parallel, with the ends of the molecules in register or staggered; and the type of fibrillar structure depends both on the inherent properties of the constituent macromolecules and on the chemical environment (determining the ionization, interactiveness of side-chains, etc.).

It is thought that the deposition of collagen fibrils by a living fibroblast involves first, the synthesis of collagen macromolecules, second, their secretion into the ground substance, and third, their aggregation into fibrils. Intracellular aggregation has not been observed. The formation of the fibril is spontaneous, but its periodicity is modifiable by the environment, and its aggregation with others into oriented bundles depends on patterns of tension and density of packing within the ground substance. The puzzling fact that all fibrils of a given region grow at the same rate regardless of their distance from the surface of the bundle indicates that simple diffusion of collagen molecules is not a limiting factor in growth. There must be a great excess of the collagen monomers in the ground substance and their addition to the existing bundles of fibrils must be relatively slow, depending perhaps on their activation in some way. There is in fact evidence that the synthesis of collagen in actively growing connective tissues is greater than the rate of polymerization into fibrils.

Another morphogenetic problem relates to the deposition of fibrils, in some instances, in successive layers with fibrils running at right angles to those in the adjacent layer. Is this the result of a lattice system in the ground substance, or is it a product of the interactions between the fibrils themselves, such as to bring their cross-bands into register? Even less explored is the mineralization of the fibrils and intercellular matrix. Electron microscope and x-ray diffraction reveal an intimate relationship between the collagen fibrils and the hydroxyapatite crystals of bone, which are seen to be deposited along the collagen fibrils. Glimcher has demonstrated that in vitro the hydroxyapatite crystals are deposited only on fibrils with the native period of 640 Å. This observation implies that such fibrils have a steric configuration adapted to the crystal lattice of the mineral. Inasmuch as the mineralization in these experiments was observed with *reconstituted* fibrils derived from sources which normally do not calcify, it is further evident that the barrier to calcification in such native tissues is not a property of the collagen itself. Edds suggests that it may be owing to the complexing of mucopolysaccharides with the collagen in the natural situation.

## Cellular and Tissue Interactions in Development

### Embryonic Induction

The classical studies of embryonic induction made by Spemann and his successors have led to many attempts to isolate and determine the chemical nature of the diffusible "organizer" and other inducing agents. After numerous false leads and dashed hopes of success, it has been shown by workers in Finland and in Japan that certain classes of compounds are responsible for at least one type of embryonic induction, that elicited by the implantation of differentiated, adult tissues into the embryo. This phenomenon was first discovered by Holtfreter, in 1934, and his initial report showed that different effects were produced, viz., induction of either neural or mesodermal structures, according to the nature of the implanted tissue. The regionally specific nature of the effects was thereafter demonstrated by Chuang and by Toivonen. Yamada has reported on extensive further analyses of the regional specificity which have been conducted in his laboratory.

When a sample of the differentiated adult tissue to be tested is placed in Holtfreter's solution between two pieces of undifferentiated embryonic ectoderm from a *Triturus* gastrula, the latter will differentiate according to the region of the body from which the inducing tissue comes. The method is not quantitative, but reliable qualitative results are obtained. The inducing agent need not be species-specific; in fact, in most experiments mammalian tissues were used. Guinea pig kidney evoked in the amphibian ectoderm a differentiation of spino-caudal and deuterencephalic (midbrain, hindbrain, and ear vesicle) structures. By fractionation of the inducing agent, it was determined that the activity was associated with pentose nucleoprotein. Further purification and fractionation into nucleic acid and protein components revealed that the activity was chiefly in the latter. Moreover, treatment with ribonuclease failed to reduce the inducing activity of the preparations, but treatment with trypsin or chymotrypsin did so.

When a guinea pig liver nucleoprotein preparation was made and tested, it induced solely archencephalic (forebrain, eye, nose, lens) structures. Again the activity lay in the protein, rather than the nucleic acid, fraction of the pentose nucleoprotein. Reduction of the inducing ability was progressive upon treatment with trypsin, and reached complete inactivation after 2 hours incubation with the enzyme.

A third type of regional induction was found with guinea pig bone marrow, which induces trunk-mesodermal structures. In this case the activity was found to reside not in pentose nucleoprotein, but in a distinct acid-precipitable protein fraction. The bone marrow agent can also induce the formation of endoderm in close association with the trunk-mesodermal structures produced from the embryonic ectoderm.

Particularly interesting and significant are the progressive changes in regional effects obtainable by treating the inducing agent in various ways. Chuang, and later others including Yamada's group of workers, have demonstrated such a change following heat treatment or treatment with ethanol, pepsin, or other chemical agents. The change may be produced in the extract or purified agent as well as in the native tissue sample. The shift characteristically noted is a suppression of spino-caudal induction effects and the appearance of archencephalic effects instead, in many cases with no alteration in the total induction frequency. In general, the shift observed is from trunk-mesodermal successively to spino-caudal, deuter-encephalic, and finally archencephalic structures. The possibility is suggested by Yamada that the changes observed result from alterations of the protein inductor by means of the treatment. It is noteworthy that the sequence observed agrees with the arrangement of presumptive areas in the amphibian gastrula, from the vegetal pole toward the animal pole.

The specificity observed in these inductions by differentiated tissues, it may be observed, is regionally specific but not organ- or tissue-specific. The induction, in an explant, of a variety of organs and tissues characteristic of a particular region of the body may be explained in part on the supposition that the inducing agent establishes a gradient of morphogenetic activity within the explant. This view is supported by the outcome of experiments in which different concentrations of inductor were used, for an archencephalic inductor at high concentration induces chiefly eye and forebrain, at a lower concentration mesenchyme and melanophores. A trunk-mesodermal inductor at higher concentration induces notochord and somites, at a lower concentration blood islands and mesothelium.

Yamada concluded by comparing the nature of the inducing agents extractable from differentiated adult tissues with those of the embryonic organizer and other embryonic inducers. The work of Niu with cells of the organizer growing in tissue culture shows that substances with inducing activity are discharged from these cells into the medium; and the experiments of Grobstein with the tubule-inducing effect of spinal cord on the metanephrogenic mesenchyme of the mouse embryo show that contact between the inducing and induced tissues is not necessary. Nevertheless, there is one conspicuous difference. The inducing agents from adult tissues are stable toward ethanol treatment, whereas the amphibian organizer is not. Furthermore, in the later embryonic period, when induction is very specific in character, the inducing tissue loses its effect when killed, whereas in the induction by adult tissues considered by Yamada the specific nature of the induction is maintained after death and chemical fractionation.

## Inductive Specificity in the Origin of Integumentary Derivatives

John W. Saunders, Jr., took up a phase of induction occurring somewhat later in development, namely, the specificities of ectoderm and mesoderm in the origin of feathers, scales, and claws in the chick. The work reported represents a continuation of the illuminating studies of Professor Willier and his group upon the factors controlling the characteristics of the feather. Each feather arises from a follicle which contains a dermal papilla at its base and a ring of epidermal cells, the collar, surrounding it. Each of these is essential to the formation of a feather. The dermal papilla, mesodermal in origin, induces feather formation in the epidermis and is responsible for the orientation of the feather. The collar produces the feather, and the regional character of the feather, e.g., breast or saddle type, is inherent in the collar. In the embryonic origin of the feather germs, the primordia first appear as mesenchymal condensations, after which the overlying ectoderm thickens and the primordium sinks into the skin to make a follicle.

In Saunders' experiments rectangular pieces of mesoderm plus ectoderm were excised in the $3\frac{1}{2}$-day old chick, long before the feather germs appear (at $6\frac{1}{2}$ days), and were replaced with pure mesoderm. Ectoderm soon grows over the implant. When the area involved was the dorsal side of the wing bud and the mesoderm implanted was from the thigh, feathers were frequently induced to form in the normally apterous region adjacent to the shoulder tract. These feathers were typically thigh feathers. This result demonstrates the inductive capacity of the mesoderm upon ectoderm which would normally not form feathers, and certainly not thigh feathers!

When a rectangular piece partly within the humeral feather tract and partly in the adjacent apterous region was removed, rotated 180°, and replaced, various results were obtained. Often the feather pattern was completely normal. Sometimes feathers were missing from a part of the humeral tract; sometimes feathers were formed in the normally apterous tract; and sometimes, although infrequently, the feathers in abnormal position were also reversed in orientation and directed anteriorly. By comparing the results when the operation was performed at successive ages, it could be concluded that there is a progressive loss of ability of the humeral tract to reorganize its pattern after the operation, there is a progressive increase in the ability of the displaced part of the feather tract to develop feathers in an abnormal situation, and finally, there is a loss of ability to regulate the orientation of the feather. It is noteworthy that the slope of the papilla does not itself control the direction of the feather. Clearly, differentiation occurs first in the mesoderm, for these results follow irrespective of whether the ectoderm is rotated along with the underlying mesoderm.

Although the results just described indicate the inductive effect of the mesoderm in feather formation, and demonstrate that the regional specificity resides in the mesoderm, this is not to say that no specificity of any kind inheres in the ectoderm. If transplants are made between breeds, it may be seen that the genetic control over feather type resides in some cases in the properties of the ectoderm. Thus, mesoderm from a White Leghorn grafted into the wing of a homozygous Silky embryo results in feathers of the silky type, and not of the ordinary type possessed by the White Leghorn. Moreover, the initial inductive action of the mesoderm is seemingly of quite short duration, and epidermis which has begun to form feathers becomes itself inductive, and will bring about the formation of dermal papillae in dermis that is not yet at the normal age when such structures arise.

Particularly illuminating in respect to the regional specificity of the inductor are experiments in which, instead of mesoderm from the thigh, mesodermal tissue from the prospective foot was implanted on the wing bud. In these cases the wings developed abnormal toes bearing scales and large claws. At the same time, experiments show that the sequence of parts in the outgrowth of the limb from the bud is controlled by the apical ectodermal ridge of the bud. If implanted mesoderm from the prospective thigh, for example, is intimately associated with the apical ectodermal ridge of the wing bud, formation of foot parts with scales and claws will occur. But if even a small piece of wing mesoderm is interposed between the graft and the apical ectodermal ridge, thigh parts form instead of foot parts.

These experiments graphically set forth the reciprocal inductive interplay between mesoderm and ectoderm in the determination of regionally specific integumentary structures. The mesodermal structure, the dermal papilla, exerts a specific inductive action on the ectodermal structure, the collar region, but the response of the latter constitutes an inductive feedback affecting the skin mesoderm.

## Aggregation and Differentiation in Slime Molds

The aggregation of myxamoebae into a pseudoplasmodium, or "slug," and the differentiation of the latter into a fruiting body with basal disk, stalk, and spore mass offer particular advantages in studying differentiation and morphogenesis. The entire process from the beginning of aggregation can be carried out in the absence of exogenous nutrients, no changes in cell number need occur, and the number of cell types involved is very small. The developmental fate of the individual cells which enter the aggregate is determined solely by the order in which they enter it. The first contingent supplies the lower stalk and afterwards the upper part of the stalk, following which newly arriving cells enter the spore mass, and finally the last to participate form the basal disk. In the slime mold *Dictyostelium discoideum*, studied and reported upon by

Maurice Sussman and by Barbara Wright and Minnie L. Anderson, the formation of centers of aggregation has been shown to depend upon population density, with an optimum for each strain (mutant or wild-type) above and below which the number of centers is reduced. At optimal density, the number of centers formed is proportional to the number of cells present. The characteristic center:cell ratio differs greatly in different strains, ranging from 1:24 in a Fruity-1 mutant to 1:2200 in the wild type. Other species also have a characteristic center:cell ratio. The linear relationship between number of centers and number of cells implies that single cells serve to initiate aggregation.

After it had been demonstrated that only a small number of randomly distributed cells in a population can serve as initiators of aggregation, Sussman and his coworkers undertook to identify them. A unique type of cell, the $I$-cell, was indeed discovered. It is two to three times larger than the responding cells in diameter, is much flatter, and is highly motile; it also has a larger nucleus and more granulated, vacuolated cytoplasm. In a series of studies it was demonstrated that the numbers of $I$-cells and of centers of aggregation in a population sample agree closely, that the positions of the $I$-cells correspond to the centers, that prior removal of $I$-cells inhibits the formation of centers, and that the addition of $I$-cells to a sample population leads to a corresponding increase in the number of centers over that expected. The phenotype of the $I$-cell is persistent throughout the life cycle, although there is evidence that in clonal growth new $I$-cells can arise from responding cells ($R$-cells) and $I$-cells can give rise to $R$-cells. There is no evidence that the $I$-cells are formed by sexual fusion or, at the present time, that they are tetraploid.

From the earlier work of Shaffer, it has been known that the myxamoebae secrete a chemotactic agent which has been called acrasin, and which attracts the responding cells during the process of aggregation. By extraction under conditions promoting protein denaturation, acrasin was obtained in a stable form. It is normally accompanied by a heat-labile enzyme which inactivates it. The stable acrasin has been fractionated chromatographically by Sussman's group, and two components were found which must interact to produce the characteristic properties of acrasin. One of these components is specifically inactivated by the enzyme. Further fractionation identified a third component (C) of acrasin. In general, component C inactivates A, and B reactivates the complex. Acrasin produced by a slime mold of another genus, *Polysphondylium,* likewise contains three components that interact in the same manner. Wright and Anderson, reasoning that acrasin might be like a vertebrate hormone in chemical nature, tested a considerable number of steroids and found that a number of them, including estradiol, estrone sulfate, and progesterone, gave evidence of acrasin activity; but none of them was as active as crude acrasin or as "pregnant urine." It is possible that acrasin may be of steroid nature, although a complex rather than a single such compound.

The adhesion of individual myxamoebae to form a slug has also been studied. Intergeneric mixtures do not adhere (Raper et al.). Gregg has presented evidence that the adhesion may depend upon an antigen-antibody type of reaction.

The formation of the pseudoplasmodium is not simply a response of the $R$-cells to the centers initiated by the $I$-cells. The $R$-cells exert a reciprocal effect, for the ability of potential initiator cells to induce aggregation can be shown to depend in part upon the nature of the "audience." Fruity-1 $I$-cells in a homologous population induce centers at a center:cell ratio of 1:24, but when the responding cells are wild type, the center:cell ratio is 1:58. In other words, the Fruity $I$-cell is less potent in forming centers when faced with an audience of a different genetic nature than its own. Another example of interaction was found in the initiation of centers by wild-type $I$-cells in populations containing various proportions of an "Aggregateless" type, which itself is unable to form centers, although it can respond and enter into an aggregation initiated by wild-type $I$-cells. In this case, by interposing a thin agar membrane between the wild type and aggregateless cells, it could be shown that at a low density of the wild type the presence of the mutant cells had no effect upon the initiation of centers; but at a high density the number of centers was much increased by the presence of the mutants on the opposite side of the membrane. The mutant cells must then supply some diffusible material that affects the formation of centers of aggregation by the wild-type $I$-cells. Later, it was shown, as one might expect, that cell-free preparations of mutant cells and even wild-type cells, would exert this type of effect. It appears that the wild-type $I$-cells, when present in high density, exert some sort of mutual inhibition or competition, which is relieved by the action of the diffusible substance coming from the $R$-cells.

Screening of chemical compounds for the existence of an effect upon these phenomena led to the discovery that adenine and guanine depress center formation, whereas histidine and imidazole, though not altering the initiative capacity, lower the threshold of response and permit aggregation at "phenomenally low densities."

In the analysis conducted by Wright and Anderson, the approach was directed at the intriguing question of how the pseudoplasmodium shifts its main synthetic activities during aggregation from protein synthesis to polysaccharide synthesis. The stalk cells synthesize much cellulose; the presumptive spore cells accumulate polysaccharides and deposit a cellulose sheath. Since this is accomplished in both types of cells without any exogenous supply of nutrients, there must be a very great transformation in the character of the metabolism. It shifts, in fact, from an anaerobic to a strictly aerobic metabolism (Gregg); and inhibitors of oxidative metabolism and phosphorylation, which do not interfere with the growth of myxamoebae, effectively prevent aggregation. During the formation of the pseudoplasmodium, enzymes previously almost or entirely undetectable become evident in large amounts. These include uridine diphosphoglucose synthetase, which is required for the synthesis of cellulose, glucose-6-phosphate dehydrogenase of the hexose monophosphate shunt, and enzymes of the Krebs cycle. Glucose itself accumulates in considerable amounts, until the formation of polysaccharides and cellulose eventually reduce it again. It would be interesting to know what mechanisms lead to the syntheses of these enzymes. Perhaps they are substrate-induced but the evidence is not yet adequate to say. Perhaps also it is the accumulation of glucose which later leads to a diminution in the activity of all the enzymes observed, with the exception of glucose-6-phosphate phosphatase. One may assume that starvation of the amoebae leads to the initiation of production of acrasin, or to a reduction in the threshold of sensitivity to it, and that acrasin then stimulates protein breakdown, the products of which supply the critical substrates that induce the formation of the aerobic enzymes. Much further work will be required to demonstrate whether this is so, but it is clear that already a highly significant relationship between the metabolism and the onset of differentiation in these extraordinary plant-animals has been revealed.

*Exchanges between Cells*

The discussion by M. Abercrombie relates primarily to the reciprocal effects of close contact between cells. Without minimizing the importance of diffusing substances in interactions between cells, he points to the distinction between the nervous system and the endocrine system in vertebrates as sufficient ground for supposing that direct contact between cells may not only be a more "refined and discriminating" basis of interaction, but might also "ensure a more immediate 'feed-back.'" He also emphasizes that intimate cell contact need not involve any transfer of substance, but might involve only a transfer of pattern.

Chick heart fibroblasts growing in culture have been studied and photographed with an interference microscope. According to Abercrombie, any moving fibroblast has at its leading edge a ruffled membrane. When two fibroblasts come into contact, *contact inhibition* is observed. That is to say, the ruffled membrane, or membranes, making contact become quiescent and even seem to fuse. A new area of ruffled membrane may then be generated elsewhere on the cell, and it will move away in a different direction. This sort of contact inhibition is not evident until the cell membranes actually touch each other. The reaction is not highly specific, since fibroblasts from different organs and even from different species (chick and mouse) will exhibit it. Various conditions, such as a high concentration of embryo juice, or the onset of mitosis, will modify the capacity of the cell to undergo contact inhibition.

Sarcoma cells differ from normal fibroblasts, so far as observed, in lacking contact inhibition. They are not inhibited by, nor do they inhibit, normal fibroblasts. Nevertheless, they retain certain other surface properties, such as adhesiveness and the contact guidance whereby they tend to follow oriented fibers or cells.

Aggregation is interpreted by Abercrombie as a consequence of the predominance of mutual adherence over the degree of adherence to substrate and movement thereon. Directed locomotion, contact reactions between cells in motion and between cells and extracellular reactions, and mutual adhesion thus become of great importance in vertebrate morphogenesis. The number of mitoses occurring in a culture seems quite unrelated to the amount of cell contact.

Besides these direct influences of cell contact, there is also ample evidence of the role of diffusible substances in interchanges between cells. Thus the frequency of mitosis in explants can be shown to be affected by substances diffusing from cells, or by competition between them for some nutrient. In induction, as has already been discussed (see Yamada), definite substances—archencephalic and mesodermal inductors—are concerned, and while these have not yet been shown to be capable of traversing a gap, Grobstein's metanephric tubule inductor can do so. Regulation is a more complex problem, for in this case the fates of cells are actually recast. Abercrombie suggests that only one form of relationship will provide both continuity of pattern and reallocation of the presumptive fates of cells: the "inside-outside" relationship considered by Weiss; that is, the relationship between cells within the regulated region and those outside, which is to say, between the central and peripheral parts of a yet larger region. In this relationship, as in all the others mentioned, one cannot ignore the significance of the mutual interactions of cells, the reciprocal feedbacks which, in the ultimate sense, connote the *"organism."*

Heinz Herrmann reported briefly on some studies of the differentiation of protein formation in the sclera and cornea of the chick embryo. The basic setting is the following: the future cornea up to the sixth day of development contains only scanty mesenchymal cells; then mesodermal cells from the scleral mesenchyme begin to migrate into the corneal area. Cells in both areas produce collagen as the principal protein, but in the cornea keratosulfate is the intercellular mucopolysaccharide while in the sclera it is chondroitin sulfate instead (see discussion of Edds, above). The scleral cells develop a cartilaginous matrix; the corneal cells, a non-cartilaginous material that contributes to the transparency of the cornea.

The synthesis of collagen in the two tissues follows a similar course, although in the cornea there is some dependence on the presence of the ectodermal component of the tissue, and in the sclera this is lacking. The ratio of collagen per unit of DNA, calculated in order to render the collagen-forming capacity for the two tissues comparable, follows similar curves of increase in the two tissues, with certain differences such as a considerably higher ratio in the cornea at the 8th day than in the sclera. Tracer incorporation (glycine-1-$C^{14}$) also showed a higher early value for the cornea, and this is greatly reduced when the associated epithelium is removed.

It is suggested as a tentative explanation that the scleral cells, which migrate to the corneal area, become adapted there to utilize oxidative energy, richly supplied by the epithelium, for their protein synthesis; the scleral cells that remain in the scleral area probably possess a poor oxidative enzymatic capacity. There is no evidence yet that the metabolic dependence of the corneal mesenchyme on the epithelium is in any way responsible for the qualitative shift in the character of the mucopolysaccharide synthesis which occurs; but it will be most interesting to find out if such is the case. At any rate, the dependence of one tissue upon another associated tissue in respect to the quantity of its syntheses is very significant.

### Cell Movement and Cell Adhesion

Robert L. DeHaan has examined the relationship between the migration of individual cells and the mass movements of cells, together with their properties of adhesion to one another and to the substrate, on the one hand, and morphogenesis, on the other. The factors controlling migration are in part internal, as for example, the motive force in an ameboid cell such as a migrating melanoblast; and in part external, such as the controlling factors determining direction. The latter appear partly to involve mutually repulsive influences, exerted most strongly by cells of the same kind upon each other, and presumably by means of diffusible substances, and partly by "selective contact guidance" along lines of structural organization in the substratum, lines of stress, or other orienting forces. "Selective fixation" has been postulated by Weiss and others to account for the eventual localization, adhesion, and subsequent differentiation of cells in contact with others of specific sorts. Thus melanoblasts from embryos of pigmented strains of fowl injected into the blood stream of host chick embryos belonging to an unpigmented strain find their way to the same locations they would normally have assumed in the donor, and proliferate and produce pigment only there. There is no evidence that presumptive pigment cells can migrate only in certain directions or along certain paths. Chromatoblasts from neural crest taken from its normal position and reimplanted on the ventral side of the embryo migrate dorsally instead of in the usual ventral direction (Twitty and Niu). The cells spread out from their source, wherever they are most crowded, migrate upon epidermal tissues rather than mesoderm, and become selectively fixed upon the former.

On the other hand, mass movements of cells such as are evident in invagination, evagination, epiboly, etc., must be accounted for otherwise. Unequal growth or differential proliferation will lead to peripheral expansion of a layer, or to buckling, warping, or thickening. The formation of intercellular fibrils and the elaboration of pseudopodia which are thrust out, adhere, and then contract have been observed to play a part in morphogenesis. The differentiation and orientation of muscles are found to depend on the surrounding mesectoderm, and not on the origins and insertions of the muscles on the skeleton.

Ameboid behavior, too, is manifested by the individual cells of a mass. Disaggregated cells from the tissues of an amphibian embryo (gastrula to neurula stages), when mixed, showed inherently directed movements. Cells from the neural plate or mesoderm migrated inward; epidermal, and even endodermal, cells migrated outward (Townes and Holtfreter). The different cell types sort themselves out into homogeneous layers. Such movements and tendencies to aggregate must be controlled to some extent by the nature of the cell surfaces and the amount of contact between them, which in turn is dependent upon individual cell shape.

DeHaan concludes that few factors can be more significant in these respects than the nature of the cell membrane and the mechanism of adhesion, and his studies have attempted to elucidate these on the hypothesis that the calcium ion serves as the binding agent. There is unquestionably much evidence to implicate the calcium cation as a critical factor in the properties of the cell membrane. Many experiments have shown that calcium-free solutions and calcium-binding agents lead to the dissociation of cells. If the cell membrane is a lipoprotein complex, the carboxyl groups of the proteins and the orthophosphate and polyphosphate groups of the phosphatides and perhaps nucleic acids would offer an abundance of attachments permitting the calcium ions to form bridges between the membranes of cells in contact. Varying strengths of adhesion would result as interconversions of lecithin, cephalin, and phosphatidic acid were brought about through intracellular enzymes. DeHaan proposes not only that calcium is required for cell adhesion but is important in controlling its strength.

To test these ideas, DeHaan examined the effects of acetylcholine, which when administered in calcium-free solution to the ventral side of the chick embryo will cause cardia bifida, and with increasing concentration complete disaggregation. DeHaan rules out the possibility that there were contaminating proteolytic enzymes or $p$H changes which might have decreased cellular adhesion, and supposes that the effect of acetylcholine is to remove divalent cations from the tissue. He cannot, however, completely rule out the effects of sheer hypertonicity and finds that any 1 $M$ solution (such as sucrose) has detrimental effects of a somewhat similar kind. A more critical series of experiments was done with Versene in saline solutions with or without calcium. Like acetylcholine, Versene in a calcium-free solution produced cardia bifida and cellular dissociation, but at much lower concentrations (4.3 m$M$ being the $D_{50}$ or concentration at which 50% of embryos were damaged). When calcium was added, the effects were much less marked, the $D_{50}$, being elevated to 5.8 millimolar. If excess calcium was added, complete protection against Versene of 8 millimolar concentration was obtained; and equimolar (4.5 m$M$) calcium and Versene concentrations had no deleterious effects. It was found that magnesium would also prevent the dissociation of the cells of the chick embryo by Versene, but not as readily as calcium.

These experiments are admittedly not conclusive, and in the general discussion that followed several persons expressed the opinion that toxic effects, killing of growth centers, inhibition of coenzyme function by chelation of essential metal ions, etc., might be alternative explanations. Such objections nevertheless do not void the major conclusion that mass cell movements in embryonic development depend upon surface contact phenomena and cellular adhesiveness, and that whatever factors control the latter, such as possibly calcium-ion binding, must be of primary significance. This view was supported by observations of Trinkaus on embryonic cells of the minnow *Fundulus*. Until epiboly commences, these cells will, if dissociated by Versene, remain semispherical and protrude lobopodia without active locomotion; but if dissociated after epiboly has commenced, will become attached to the substratum, put out numerous filopodia, and migrate actively like fibroblasts, to form a lacework of cells. The transformation from one type of cell to the other will in fact occur in vitro if the cells are taken from a late blastula, about an hour before epiboly would have commenced. Thus the spreading of the blastoderm may be attributed to the acquisition by the cells of a new property, one that enables them to spread over the substratum.

PROBLEMS OF SPECIFICITY

*Cytochemistry of the Egg*

In a very thoughtful consideration of the significance of cytochemistry for the study of development, J. J. Pasteels has stated succinctly the essential goal of the biologist as being "the integration of the structure and the function of the cytoplasm." Certainly this is where the directive powers of the nucleus and the limiting factors of the internal and external environment meet and interact. This is the stage upon which the plot of differentiation unfolds and reaches its culmination. And if, as Pasteels says, the attention of biologists has been heretofore directed mainly at the correlation between structure and function in the adult cell, the deeper significance lies in the study of that relationship throughout differentiation. We should indeed begin our analysis with the egg, the just fertilized, still unsegmented egg—though even here we find no real beginning, since while on the one hand it is undifferentiated, on the other it is an already highly specialized cell, derived from an oocyte whose topography reflects the period of intense cellular preparation for the formation of the future embryo.

The methods of study developed up to now include: (1) the isolation of cytoplasmic particles by fractional centrifugation of cellular homogenates; (2) the analysis of enzymatic or respiratory activity of fragments of eggs obtained by centrifuging living eggs; (3) the same applied to isolated blastomeres or egg fragments obtained without centrifugation; (4) cytochemical studies of sections with metachromatic stains or radioactive tracers; and (5) studies with the electron microscope. None of these methods alone is adequate. Some run the serious danger of introducing artifacts and destroying structural relationships; others cannot be made quantitative. Fractionation and fixation alike pose problems. Yet taken together and used mutually to check one another, these methods provide considerable insight into the nature of the cell.

In the cytoplasm the biochemical elements chiefly to be considered are the RNA and the enzymes. The former, in the egg as compared with adult cells, is associated with much smaller particles, which sediment only at very high centrifugal force ($25{,}000$-$100{,}000 \times g$)—although this condition may in part result from the "solubilizing" action of KCl in the solution. After centrifugation, not all the RNA is in the hyaloplasm at the lipid pole of the egg; some may be demonstrated in heavy particles at the centrifugal pole, at least in certain kinds of eggs. In other cases there is evidence of a cyclic shift in the RNA involving alternating phases of dispersion and aggregation on heavy granules. The ergastoplasm (basophilic reticulum) of the sea urchin egg consists of double membranes, like the nuclear membrane in structure, containing pores filled with masses of granules, and enclosing heavy bodies rich in RNA and resembling nucleoli. In the centrifuged, fertilized egg there are four zones, lipid, hyaloplasm, vitelline, and mitochondrial. RNA is found in all except the first. In the hyaloplasm is an atypical ergastoplasm consisting of vesicles surrounded by fine granules (Palade); in the vitelline zone are numerous annulate membranes, also with the fine granules; in the mitochondrial region are large, dense bodies, and an undefined structure perhaps linked to the mitochondria. Clearly, the view that all RNA is in the microsomes (-ergastoplasm, or basophilic reticulum) is a much oversimplified conception.

The oxidative enzymes are not exclusively localized in the mitochondria of the egg, but it would be rash to assume that the enzymes are wholly dispersed to start with. While experiments on amphibian eggs indicate the major amounts to be associated with the mitochondria, experiments with echinoderm eggs put cytochrome oxidase and other enzymes of oxidative phosphorylation in the supernatant after centrifugation. Centrifugation of the living egg and separation of clear from granular fragments lead to the conclusion that dipeptidase and catalase are in the hyaloplasm. Most significantly, a clear fragment seemingly completely deprived of mitochondria is able to develop into a complete, normal embryo, and must therefore have the capacity to produce all requisite enzyme systems. But the electron microscope shows that such fragments are not in fact devoid of all mitochondria, and that some mitochondria remain associated with the lipid droplets at the centripetal pole. Pasteels also cautions us that different ratios of activity found after homogenization or by means of cytochemical tests need not correspond to actual differences of activity in the respective parts of the untreated, developing egg. In sum, at least a portion of the enzymes of the egg are associated with cytoplasmic particles at the beginning, and enzymes such as cytochrome oxidase, succinic dehydrogenase, and ATPase are mitochondrial, while acid and alkaline phosphatase are localized in organelles distinguishable from the mitochondria by metachromatic staining, and in the case of alkaline phosphatase located in the vitelline zone of the egg.

In studying the cytochemistry of the nucleus, most attention has been given to the DNA. It is also worth noting that the abundant RNA of the mitotic spindle has been shown, in *Cyclops,* to be derived from the nucleus; and that the mitotic apparatus, like the nucleus, is also rich in polysaccharides. Ronald R. Cowden has presented additional findings of considerable interest, based on the study of the nucleoli in the growing oocytes of a slug, *Deroceras reticulatum*. There are two sorts of nucleoli in these germ cells. The primary nucleolus, present from the beginning of oocyte growth, has RNA, basic amino acid residues of lysine, arginine, and histidine, histone, an envelope containing mucopolysaccharides, and a perinucleolar coating of DNA. The secondary nucleoli arise later, presumably from the primary nucleolus, and lack the DNA coating and the mucopolysaccharide envelope, and the histones have rather less RNA, but contain the basic proteins indicated by the dibasic amino acids. Cowden suggests that the secondary nucleoli may represent centers of protein and RNA synthesis produced during the growth of the oocyte.

In proceeding to consider the egg as a future embryo, Pasteels noted the relationship of the cytochemistry of the egg to its polarity and bilateral symmetry, touched skeptically on the significance of mitochondrial patterns, and alluded to the comparative study of species which makes the presence of a common cytochemical pattern as a basis for similar morphogenesis difficult to entertain. Regarding polarity, it is scarcely possible at the present time to relate the cytochemical polarity of the egg to the polarity of the developing organism, except in the case of the amphibian, where the cephalocaudal axis and the differentiation of the chordamesoderm depend directly on the gradient of cytoplasmic particles in the unsegmented egg. The bilateral symmetry of rodent eggs is related to the differentiation of the embryonic and extra-embryonic parts of the germ; and the bilateral symmetry of the ascidian egg is related to the differentiation of the myoblasts of the embryo. But when one comes to compare such similar morphogenetic anlagen as the polar plasm and polar lobes of annelid mollusk eggs, no common cytochemical pattern is discernible. In some, the polar lobe does not appear any different from the cytoplasm of the rest of the vegetative half of the egg. In others, there are notable and characteristic differences, yet the displacement of structure by centrifugation does not interfere with normal segmentation. One is left with contradictory conclusions. All that is evident at the present moment is that the main problem, the protein organization of the fundamental cytoplasm, remains almost untouched.

If biochemical patterns are to be followed during development it is of course imperative to avoid introducing artifacts. This has already been stated in connection with cytochemical methods. J. L. Kavanau reemphasized its importance when making extractions from embryos of different ages. With conventional methods of fractionation and homogenization it is impossible to avoid considerable autolysis because of the presence of active enzymes in the protoplasm. Neither variations in dilution, nor maintenance of a low temperature or a nonphysiological $p$H will prevent this autolysis. The proteins must be denatured and precipitated in toto in the very first steps, if labile substrates are to be preserved. In Kavanau's own studies of the cleaving embryos of the surf-clam, *Spisula solidissima,* an extraction procedure better than any other was found to involve soaking lyophilized material for some weeks in mixtures of petroleum ether and ethanol and of acetone and ethanol, then exposing for 10 minutes to boiling water, homogenizing, and finally extracting (with rapid freeze-thaw cycling between $0°$ and $-20°C$) in $0.022\ M$ acetic acid. The residue contained almost all the protein. The extracts were further fractionated into parts soluble and insoluble in 80% ethanol. Analyses of valine, tyrosine, nicotinic acid and related compounds, pantothenic acid and coenzyme A, and the folic acid, thymidine, thymine group consistently demonstrated that even a very small amount of autolysis resulted in spuriously high levels of free amino acid, of nicotinic acid (destruction of the coenzymes DPN and TPN), and of thymidine and/or folinic acid, as well as a reduction in total activity of the folic acid group. When autolysis is prevented, there are no cyclic changes in the free amino acid pool during cleavage. A regular decline in the level of pantothenic acid following fertilization was observed.

## Changes in Enzymatic Patterns during Development

A different problem is of course involved if the molecular species to be followed is that of the enzymes themselves. S. C. Shen summarized the efforts of cytochemists to devise methods for detecting enzymes, to minimize artifacts, to increase enzyme localization, and to assay enzymes more quantitatively.

Although most work hitherto has been done with mitochondrial enzymes, Shen turned to cholinesterase as more promising in developmental enzymology because it can be readily characterized and assayed, localized satisfactorily, and is concerned with a special type of function, synaptic transmission, rather than with more general metabolism. It is present detectably in the early ontogeny of vertebrates. In the central nervous system it is distributed together with acetylcholine and cholinacetylase. The structural and functional maturation of the nervous system is marked by a very high rate of cholinesterase synthesis. The enzyme is concentrated in synaptic centers and appears in any area at the very time when, morphologically, synapses can be seen to differentiate. Interference with synaptic differentiation results in a corresponding interference with the appearance of the enzyme.

Cholinesterase was studied especially in the neuroretina and myoneural junctions. In the developing retina of the chick it could be shown that the enzyme appears successively in four distinct subzones of the inner plexiform layer in the same sequence as these layers differentiate synapses. Cholinesterase is absent from the external plexiform layer, where the rods and cones enter into synapses with the dendrites of the bipolar cells; but in the amphibian retina the enzyme is found in both plexiform layers. The first trace of cholinesterase in the developing retinal cells is on the surfaces of the ganglion cells, at a time when the bipolar cells are still undifferentiated. In other words, the enzyme appears on the *post*-synaptic cell membrane independently of the incoming presynaptic terminals. In explants of very young optic vesicles retinal differentiation with formation of cholinesterase was observed in the absence of connections with the central nervous system.

In the myoneural junctions cholinesterase is limited to the motor endplates, but until the development of these was traced it could not be determined whether the enzyme was associated with the presynaptic or the postsynaptic membrane. The innervation of amphibian axial muscles proceeds regularly in a cephalo-caudal direction. Cholinesterase appears on the endplates of these segmental muscles in the same sequence, and just when the innervation by motor neurons makes possible the first neurogenic response. But in a denervated tadpole extensive differentiation of the muscles can occur, and Shen has demonstrated histochemically that the formation and localization of cholinesterase is perfectly normal in the nerveless muscles, too. Up to a certain stage cholinesterase increases normally in amount; it then ceases to accumulate, and finally declines in amount, as muscular dystrophy sets in. These experiments demonstrate once again that cholinesterase is produced at the post-synaptic membrane, although in this case by mesodermal, rather than ectodermal, cells. It may well be that in normal development the synthesis of cholinesterase on the muscle fiber helps to guide the motor neuron to its termination. Continued production of the enzyme, however, may quite likely require the participation of the neural, presynaptic element. In any case, we have here a beautiful example of the relation of enzyme pattern to further stages in cellular differentiation; and the exploration of more such systems can hardly fail to be most profitable. A somewhat similar study, relating to the appearance of glutamotransferase in the chick retina, was reported by Dorothea Rudnick in the discussion. Separation of the retina into layers permitted the conclusion to be drawn that the visual layer (rods and cones) is less involved than the bipolar-ganglionic layers, although some enzyme is formed in all three.

Richard N. Stearns and Adele B. Kostellow utilized dissociation of embryonic frog (*Rana pipiens*) cells by means of Versene, and their subsequent separation into presumptive ectoderm, presumptive endoderm, and presumptive mesoderm by an ingenious method utilizing the variations in cell density from animal to vegetal pole. Tryptophan peroxidase, an inducible enzyme restricted to the adult liver, and absent from normal frog embryos, was found to appear after 4 or 5 hours in dissociated cells derived from pre-gastrula embryos. A maximum was reached between 8 and 10 hours, when intact control embryos were just beginning to gastrulate. No greater amount of enzyme was present in presumptive endoderm than in presumptive ectoderm plus mesoderm, although in normal development the latter cells would not form tryptophan peroxidase at all. When dissociated cells from embryos which had completed gastrulation were tested for induction of the enzyme by additions of tryptophan to the medium, only presumptive gut cells responded. There is accordingly a progressive restriction in the ability of non-endodermal cells to respond to trytophan as inducer.

Efforts were also made to induce the enzyme $\beta$-galactosidase with lactose and other inducers. In this instance, the pre-gastrula cells would not respond, but post-gastrula cells did so. In this instance, too, the enzyme is induced in cells earlier than in normal development. Histidase activity, found in the liver of the adult, was likewise induced in the pre-gastrula cells by means of histidine.

A major problem of differentiation is whether determination of cell fate is by means of a progressive restriction of the totipotent synthetic capacities originally present, or whether on the other hand embryonic cells are totipotent only in a potential sense and actually develop certain enzyme systems de novo. The results actually obtained, even at this preliminary stage, by Stearns and Kostellow indicate that both situations may obtain. For tryptophan peroxidase the first situation seems to apply, whereas for $\beta$-galactosidase the system can not be demonstrated in the youngest cells tested, but appears only at a subsequent period of development.

## The Differentiation of Bacterial Populations

It may be supposed that a population of bacteria undergoing biochemical differentiation might by analogy tell us something of the nature of biochemical differentiation going on in the cells of a developing multicellular organism. If the individuals of the bacterial population are less integrated, they are nonetheless not without a considerable measure of interaction. And inasmuch as some of the same enzyme systems significant in vertebrate development can be readily followed in bacteria, the approach holds considerable appeal. Several papers utilizing this method were presented.

Melvin Cohn decribed the differentiation of a population of *Escherichia coli* with respect to $\beta$-galactosidase formation (see above). In an exponentially growing population the increase in enzyme is a linear function of total cell mass. The maintenance of this enzyme system, once it has been induced, persists over many generations, like the characteristics of differentiated tissues. Cells of an identical genotype will, when freshly placed in a medium with the inducer and glucose, synthesize no enzyme, but if preinduced to form enzyme before being placed in the same medium containing inducer and glucose, they will continue to form enzyme. This distinction can be maintained for more than 150 cell generations. The two steady-state conditions are delicately balanced, and shock, such as a brief exposure to anaerobiosis, will make the cells with inhibited synthesis begin to synthesize the enzyme and continue to synthesize it after being returned to aerobic conditions.

Two strains of *E. coli* differing in the ability to utilize lactose differ also in their response to lowered concentrations of inducer. The lactose-positive strain at low concentrations of inducer gives a non-linear response; but the lactose-negative strain exhibits a linear response under all circumstances. This difference, it was found by Monod et al., depends upon a difference in specific permeability to galactosides, a so-called "galactoside permease," not to be taken as being necessarily an enzyme. It is the galactoside permease which is absent in the lactose-negative strain and is inducible in the lactose-positive strain. One thus arrives at a conception of an external inducer entering a cell through the permease system and as internal inducer inducing both enzyme ($\beta$-galactosidase) and permease (galactoside permease). The cell is virtually impermeable to lactose except in the presence of $Y$, the galactoside permease; but certain other inducers (alkyl galactosides) can enter by diffusion. Obviously, a cell with a high level of $Y$ can capture an external inducer more efficiently than one with little or no Y. The inhibitory effect of glucose may be regarded as depending on the ratio in the cell of the concentration of internal inducer to inhibitor; and preinduction may be regarded as increasing the effective concentration of the inducer. If the scheme is valid, the lactose-negative strain, which lacks the permease Y, should be unable to exhibit the maintenance effect described in the preceding paragraph, and this turns out to be the case.

To push the analogy with embryonic differentiation further, Cohn notes that only a part, and not all, of the undifferentiated cells in an embryo transform into a particular specialized tissue. Can an analogous heterogeneous response to enzyme induction be found in the *E. coli* clonal population, and one capable of being maintained for many cell generations? This was in fact found by Novick and Weiner to occur when the noninduced lactose-positive strain is subjected to a low inducer concentration. The non-linear response previously mentioned was upon analysis shown to result from the presence of heterogeneity, some cells producing enzyme and others not—for an induced cell will yield a clone of enzyme-producers, when placed in medium containing inducer and glucose, whereas an uninduced cell will yield a clone of nonproductive cells when placed under those same conditions. The non-induced culture subjected to a low concentration of inducer contained cells of both types, the proportion of enzyme-producers increasing as time passed. In some respects this phenomenon resembles mutation, but it differs in that a low concentration of inducer must be present to maintain the distinction between the induced and uninduced states. At high concentrations all cells become induced; in the absence of inducer no cells are induced. The situation described clearly indicates how, without any change in genotype, a homogeneous population of cells can become heterogeneous through the induction of a specific enzymatic capacity in a part of the population. It does not, however, reveal how the transformation of undifferentiated to differentiated cells could be controlled so as to yield a definitely limited, rather than an indefinitely increasing, proportion of the latter.

This problem was clarified by the three succeeding papers, contributed by Luigi Gorini and Werner K. Maas, by Henry J. Vogel, and by Boris Magasanik, respectively. These papers dealt with the analysis of feedback mechanisms of control in the formation of bacterial enzymes. *E. coli* was the organism generally employed. The first two studies were concerned with the ornithine-citrulline-arginine cycle; the third with purine interconversions and histidine synthesis. Feedback effects of products on earlier stages of their own synthesis might involve inhibition of enzyme activity or inhibition of enzyme formation; or, as Vogel adds, on the supposition of similarity between the enzyme configuration and that of the gene controlling its specificity, the feedback effect of the small molecule might be exerted directly upon the gene itself (this remains to be demonstrated).

The formation of ornithine transcarbamylase, the enzyme which transfers the carbamyl group from carbamyl phosphate to ornithine so as to yield citrulline, is according to Maas strongly inhibited by the presence of arginine. The kinetics of the reaction suggested that endogenous arginine, produced by the activity of the cycle, also acts to inhibit formation of the enzyme. This fact was demonstrated by utilizing two different arginine-requiring strains. One, growing in a chemostat, with arginine limiting the growth rate, exhibited under these conditions of lowered arginine a synthesis of ornithine transcarbamylase at a constant rate until a high maximum level is obtained. But when growth was limited by the histidine concentration and arginine was present in excess of the requirement for growth, even though very slightly, formation of the enzyme was greatly inhibited. In the second test a "leaky" mutant, which can grow slowly on minimal medium, was used. Here again, with arginine limiting growth, a high concentration of ornithine transcarbamylase was obtained. Even in the wild-type strain conditions have been found which demonstrate the release of feedback inhibition. The supply of carbamyl phosphate here limits the formation of arginine. Either by growing the cells anaerobically in the absence of $CO_2$, or by using an enriched arginine-free medium in which the growth rate is double that in minimal medium, carbamyl phosphate may be made limiting, and an increase in the formation of the citrulline-forming enzyme could then be demonstrated. It is worth noting that the release of the inhibition of formation of this enzyme is not accompanied by a corresponding effect on the enzymatic formation of carbamyl phosphate. In other words, the inhibitory effect of arginine in the feedback does not extend beyond the branch point where carbamyl phosphate feeds into the cycle. Maas concludes: "This high enzyme-forming potential enables the cell to conserve its enzyme-forming machinery and yet to adapt rapidly from an environment in which the end-product is present to one in which it is absent. Negative feedback control coupled with a high enzyme-forming potential is thus one of the mechanisms responsible for the adaptability of microorganisms." Coupled with enzyme induction, it might well provide the basis for the initiation of differentiation.

Vogel's analysis of enzyme repression as a negative feedback phenomenon proves particularly thought-provoking. Acetylornithinase, the enzyme which produces ornithine from acetylornithine, is repressed by both arginine and ornithine. When *E. coli* was grown continuously in a turbidostat in the presence of ornithine, cells were isolable between the third and seventh days that exhibited a heritable lowering of the acetylornithinase level to 10-15 per cent of the original level. This did not occur in cultures grown in medium without added ornithine. Vogel's tentative explanation is one of undirected mutation and selection, but other explanations are not excluded—for example, the long-term maintenance of a steady-state condition like that discussed by Cohn.

The feedback effects considered by Magasanik, and found not only in *E. coli* but also in *Aerobacter aerogenes* and *Salmonella typhimurium*, are considerably more complex. The cycle itself involves the formation of a precursor of purine synthesis, the ribose nucleotide of 4-amino-5-imidazole carboxamide, which is converted into inosine monophosphate, and the conversion of the latter into both adenosine monophosphate and guanosine monophosphate. Both the adenine and guanine nucleotides, by paths different from those of their respective syntheses, regenerate inosine monophosphate (hypoxanthine nucleotide). Guanosine monophosphate (or some other guanine derivative) exerts a negative feedback effect on its own synthesis from inosine monophosphate; and ATP inhibits the reconversion of guanosine monophosphate to inosine monophosphate. Both adenine and guanine further inhibit the formation of the purine precursor. Moreover, the adenine nucleotide gives rise through a series of steps to histidine plus the aforementioned purine precursor; and histidine inhibits the first enzyme on this pathway, thus controlling by negative feedback not only its own synthesis but also that of the major purines.

The entire system seems like a senseless merry-go-round, as Magasanik says, but given the feedback controls it operates so as to encourage the organisms to use exogenous adenine and guanine instead of to synthesize them, and it enables them to use each purine base preferentially for the production of the corresponding nucleotide required for the organism's nucleic acid. Moreover, the complete cycle supplies just sufficient histidine for protein synthesis, without blocking the interconversion of the adenine and guanine derivatives via inosine monophosphate.

Such complex regulatory systems must fill us with wonder! Without doubt, they provide, as Magasanik says, "a precise mechanism for the coordination of the complex metabolic systems of the cell." It is less clear how they fit into the pattern of cellular differentiation, unless it can be shown that different and relatively permanent steady-states of the system are obtainable. [At this point discussion supervened, and Spiegelman made a plea for a more optimistic view—a belief in the simplicity of biological phenomena! Umbarger obliged by presenting a scheme for a negative feedback in which the product inhibits the activity of the enzyme rather than its formation, so that the control system is rapid and sensitive instead of relatively sluggish. An example of this is to be seen in the synthesis of isoleucine in *E. coli*. One of the enzymes in isoleucine synthesis, L-threonine deaminase, is extremely sensitive to competitive inhibition by isoleucine.]

Two contributions dealt with the urea cycle from a quite different point of view than that of feedbacks and mechanisms of regulatory control.

The problem raised here was that of the ontogenetic development of the cycle and its relation to the presumed evolutionary sequence of "ammoniotelic," "ureotelic," and "uricotelic" phases of vertebrate excretion adapted respectively to aquatic, amphibian, and terrestrial modes of existence. The Wisconsin workers, G. W. Brown, Jr., and P. P. Cohen, limited their analysis to the relation of the urea cycle to metamorphosis in the amphibian *Rana catesbeiana,* the bullfrog. The Texas team, Robert E. Eakin and James R. Fisher, looked for successive phases of nitrogen excretion in the development of the chick embryo. The Wisconsin group found a very striking parallel between the functional development of the full urea cycle and the occurrence of metamorphosis. The Texas workers were unable to find convincing evidence of "recapitulation." In fact, the chick is essentially uricotelic from the time the mesonephros begins to function.

In the study of anuran metamorphosis, as Cohen emphasized, an excellent quantitative measure of stage was found to be the ratio of hind limb length to tail length. In fact, this ratio proved to be correlated much better with some of the changes in excretion than were the standard morphological stages. Using the hindlimb/tail ratio, the Wisconsin team found that nitrogen excretion as urea remains at a constant low level until the hindlimb/tail ratio is approximately 1.0 (Stage XVIII or XIX), when the urea rises abruptly to a level of more than 50 per cent of the total excreted nitrogen. Brown and Cohen then examined several of the enzymes belonging to the urea cycle: carbamyl phosphate synthetase; ornithine transcarbamylase; arginase; and arginine synthetase. In each of these a marked increase in activity, beginning at the onset of metamorphosis, was very evident. Arginine synthetase, a term applied to the condensing and cleavage enzymes yielding respectively argininosuccinate from citrulline and aspartate, and arginine and fumarate from argininosuccinate, was later assayed for the activity of the two separate steps; and it was found that the condensing enzyme was the rate-limiting component, although this was perhaps simply on account of a scarcity of ATP required for the reaction.

With all due reservations about the possible differences in activity of enzyme systems in vitro and in vivo, it appears that the enzyme levels found by the assays are quite sufficient to account for the amounts of urea actually produced by the metamorphosing tadpoles at the several stages. Brown and Cohen conclude that the shift from an ammoniotelic excretory pattern, such as characterizes most aquatic invertebrates and the bony fishes, and also lungfishes during their aquatic periods of life, takes place in the anuran at metamorphosis as a significant adaptive change to terrestrial life. The same shift occurs in the lungfish when, in time of drought, it burrows into the mud and aestivates. Such a conclusion, of course, raises immediately a serious problem of biochemical evolution.

How could so complex a pattern as the urea cycle, with its two linked cycles supplying carbamyl phosphate and aspartate, respectively, and presided over by at least nine enzymes presumably involving nine genes, arise at the propitious moment? It is inconceivable that it could have done so instantaneously. Rather, we must look for a step-by-step evolution of the system—yet that requires that we postulate a selective advantage for each step achieved. Cohen suggests that we must consequently look for the original significance of the urea cycle not in its present key position in nitrogen excretion but rather as a basis of forming important intermediate substances, e.g., arginine for protein synthesis. The excretory function would then be a fortunate, if unexpected outcome when the final link was inserted by mutation into the pattern to make it truly a cycle, by reconverting arginine to ornithine with release of urea. [If so, this may be regarded as extending the evolutionary concept of *preadaptation* to biochemical patterns.]

Eakin and Fisher, looking for phases in the excretory mechanism of the chick embryo, found no evidence of abrupt alterations that might be regarded as evolutionary relics of metamorphosis. Comparing the concentration and content of ammonia in the allantois with that of the entire egg, they found essential constancy of concentration, and an increase in amount in the allantois that simply reflected the increase in volume of the allantoic fluid. Ammonia supplied exogenously became quickly equilibrated between yolk and allantois and disappeared by the end of the second day. There seems to be no evidence of an active ammonia secretion mechanism in the chick embryo.

As for urea, its concentration in the allantois increases steadily and becomes several times higher than in the yolk, although the absolute amount in the latter remains greater. Exogenous urea becomes concentrated in the allantoic fluid. But Needham, Brachet, and Brown (1935) long since proved that the chick embryo does not utilize the full ornithine-citrulline-arginine cycle to produce urea. The urea formed does not come from ammonia or amino nitrogen through the cycle; nor does the chick embryo have the uric-acid-cleaving enzymes of fishes and amphibians to convert purines to urea. All the urea produced comes from arginine. Since arginine nitrogen constitutes one-quarter of all the nitrogen of the yolk of the hen's egg, protein breakdown is sufficient to account for the urea.

Finally, as to uric acid in the chick embryo. On the fifth day it begins to accumulate in the allantoic fluid, the concentration increasing arithmetically. Exogenous uric acid goes entirely into the allantois. Inasmuch as the glomeruli of the mesonephric tubules begin to function on about the fifth day also, it seems clear that the developing kidney is uricotelic from the first, the excretory pattern thus being adapted to the requirements of the enclosed egg by forming an osmotically inert endproduct. The evolution of this pattern of nitrogen excretion thus seems obscurer than before.

*The Immunochemical Analysis of Development*

Application of immunochemical methods to problems of embryology is, as James D. Ebert stated in his critical and stimulating review, only in its infancy. Some work has been done upon the nature of the antigenic stimulus in immune reactions towards homografts, i.e., grafts from individuals of the same species but of different genotype than the host. Medawar, Billingham, and associates have shown that antigens are very probably liberated continuously from such a graft, for although no reaction against the graft can be detected in a tolerant animal, nevertheless the administration of lymph cells from a normal, non-tolerant animal to the tolerant one will lead to destruction of the successful graft. These workers have presented tentative conclusions that the homograft antigens are DNA-proteins, since the injection of nuclear fractions, but not cytoplasmic fractions, of cells disintegrated ultrasonically will sensitize animals to subsequent challenge by homografts from the same strain that provided the nuclear material injected. The immunizing activity of the material is destroyed by deoxyribonuclease but not by ribonuclease, but susceptibility to the former is not as great as might be expected if the antigen were solely DNA, and under certain conditions trypsin will also destroy the antigenicity of the active fraction. DNA associated with a protein coating might behave in this way. It does not follow that all isoantigens are nuclear, and well-known evidence that immunity to tumors may be evoked with a variety of cytoplasmic antigenic materials demands caution in generalizing from the studies of Medawar and his associates.

Other immunochemical studies have been pursued with cytoplasmic particulates. Weiler has obtained interesting results with antisera prepared against microsomal and mitochondrial fractions of rat liver. Using the fluorescent antibody technique, he found complete cross-reactivity between the antigens and the strong antisera he obtained. However, hepatomata were unable to react with either antiserum, and prior to the appearance of the experimentally induced tumor, small clusters of non-reactive cells could be consistently observed among the normally reacting liver cells. Peters has discovered a precursor of serum albumin, or at least a cross-reacting protein, bound to cytoplasmic particles in chicken liver. When treated with deoxycholate, the particles release a non-dialyzable protein which reacts strongly with antisera to chicken serum albumin. Ebert criticizes Peters' work, however, in respect to a lack of precision in the fractionation procedures. The yields of microsomal and mitochondrial fractions were not reproducible when applied to liver slices incubated for various lengths of time.

In the higher vertebrates, the ability to produce antibodies is generally lacking during embryonic life and is acquired shortly after birth. [Billingham presents a table showing the known variation in this end of the "tolerance-responsive" period and ranging from long before birth in cattle and 40 days before birth in sheep to 14 days after hatching in the duck and after birth in the dog.] Before this time, tolerance toward homografts is characteristic. Two Russian workers, Sterzl and Hrubesova, have utilized young rabbits, taken before the age at which the antibody-forming capacity normally makes its appearance, and have transferred to them the antibody-forming capacity of immune older animals by injecting into the young rabbits nucleoprotein fractions derived from the immunized adults. Cell suspensions of spleen or bone marrow prepared from adult rabbits which had been inoculated with a bacterial antigen were injected into 5-day-old rabbits, which normally cannot make antibodies. Even when preparations were employed that had been made 24 to 48 hours after inoculation, considerably before the inoculated adult animal itself makes any detectable antibody response—the 5-day-old rabbits responded by making antibodies to the bacterial antigen. It was further shown that the capacity to form antibodies is transferred in a partially purified nucleoprotein fraction. Both DNA-protein and RNA-protein were effective. The phenomenon is remarkably parallel to the transfer of enzyme-inducing capacity in bacteria by RNA (Kramer and Straub). Whether the interpretation to be given to these results is that an "activated" or "induced" nucleoprotein has been transferred, or that antibody bound to nucleoprotein has been transferred, one's attention is in either event directed to the importance of intermediate nucleoprotein combinations in the formation of tissue-specific proteins or of isoantigens.

Woerdeman's very interesting demonstration that adult lens antigen cannot be formed in isolation either by head ectoderm or by optic vesicle, yet is formed in vitro by a mixture of extracts from these tissues, forcibly reminds us that characteristic adult proteins may be formed only through the interaction of components produced by different tissues. A more refined study of this kind was reported by George W. Nace and William M. Clarke. They prepared rabbit antisera against the supernatant fraction of *Rana pipiens* embryos of the tailbud stage. Globulin fractions of these antisera were conjugated with fluorescein isocyanate and applied to frozen and paraffin-embedded embryos in order to determine specific localizations of the antibodies. Fluorescence was found to be strong in the optic cup and lens of stage 19 and stage 20 embryos, but not in the optic vesicle or presumptive lens ectoderm of stage 17 or in any of the eye structures of stage 21. This is taken to be evidence of a transitory antigen occurring in the optic cup and lens at the time of lens differentiation, an antigen distinct from adult lens antigen. It would be interesting to relate this transitory antigen to the components of the system detected by Woerdeman.

Recently a great deal of attention has been centered on the "graft-against-the-host" reaction, and Ebert and his students have contributed materially to an understanding of this phenomenon. Following Jacobson's well-known discovery that the lethal effects of whole-body irradiation with x-rays could be mitigated by shielding the spleen, many workers contributed to the present understanding that hemopoietic tissues are essential to recovery and that after heavy irradiation the immune reaction is so reduced that bone marrow from a donor can be injected and the hemopoietic tissues reseeded with donor cells, even from a different species, e.g., rat in mouse. The hope that great practical benefits might flow from this discovery was dashed by the fact that when the donor was not isologous (genetically equivalent) with the recipient, the immediate survival and recovery is followed by a fatal reaction of the host to the donor cells, whether or not a direct immune response of the host's spleen and lymphoid tissues to circulating antibodies from the graft is involved. In newborn mice which acquired tolerance toward homologous grafts by injections of adult spleen cells, a similar and subsequently fatal response of the lymphoid tissues has been observed (Billingham). In Ebert's own work, transplantation of adult spleen, thymus, or liver to the chorioallantoic membrane of the 9-day-old chick embryo led to an increased growth of the homologous organ. Mammalian spleen was ineffective in stimulating the growth of the spleen of the chick embryo, but spleen from other species of birds was active. The age of the donor is critical, two-thirds of the embryonic period passing before the effective factor becomes present; and the adult level is reached only some 6 weeks after hatching. Further experiments showed that $S^{35}$-labeled methionine can be traced predominantly into the homologous organ of the host: and that the transfer is probably not one of intact cells, since devitalized preparations and subcellular fractions are effective in stimulation. The fractionation implicated the microsomal and supernatant portions—in other words, probably ribonucleoprotein. But as Ebert cautions, the experiments extend only to the predominant localization of foreign material, and do not explain the reason for the stimulation of growth in the homologous organ.

The growth-promoting activity has been serially transferred. That is to say, the stimulated spleen of a 10-day-old chick embryo is itself effective in stimulating growth of the spleen, although normally it would not be. Four serial passages of this kind were successful, without appreciable diminution of activity. This result again raises the question of colonization of the host spleen by whole cells from the donor, although if that occurs at all, it must be limited to small numbers of cells. A massive transfer of host cells can be ruled out on histological grounds.

When grafts of donor spleen are continued beyond the seventeenth day, the host spleen, in a consistent small proportion of cases, begins to undergo degeneration. This may arise, as Simonsen's experiments suggest, from colonization by donor antibody-forming cells. In any case, it is another good example of the graft-against-host reaction. DeLanney reported a particularly clear-cut reaction of this type in his experiments with grafts of a rodlike piece of adult spleen inserted into a trough or tunnel beneath the ectoderm of the salamander *Taricha* (*Triturus*) *torosus*. In this species, by virtue of the minimal pigmentation of the flank, it is possible to see the development of the larval host's spleen through the skin as well as the implanted donor spleen. In 86 cases studied there was never an instance of stimulation of growth on the part of the host's spleen. Instead, gross abnormalities of the blood vascular system and secondary morphological variations occurred in random regions. Most frequently noticed was a stasis of circulation in a limited region. These abnormalities arose under the influence of the graft before the host's own spleen had made its appearance, a fact which exempts the host's organ from participation in the response. DeLanney postulates an immune reaction resulting from the manufacture of antibodies against the host by the graft.

The ability of cells to recognize foreignness may be present in embryonic cells before the detectable development of the antibody-forming capacity, and, as Weiss suggested, may be akin to the ability of cells to recognize each other. Ebert emphasized that the selective factors need not be confined to selective entry into cells on the part of macromolecules. Experiments with various antitoxins placed in the uterine lumen of the mother rabbit show that although the fetal serum soon shows rabbit antitoxin, bovine antitoxin is very slow indeed to appear. Yet the use of iodinated gamma globulins showed that both rabbit and bovine globulins were entirely absorbed. The difference must lie in the disposal of the macromolecules by the cells or in selective secretion.

Billingham's discussion dealt with many of the same principles of reactivity and even the same experimental studies as Ebert's, but his emphasis was rather different. Referring to Owen's discovery of blood cell chimerism in dizygotic cattle twins as the significant point of departure, Billingham stressed the stability with which such transfers of cells different in genotype from the host are maintained throughout the life of the recipient when this transfer occurs sufficiently early. Stable chimerism of the hemopoietic tissues is now known also in chicks hatched from double-yolked eggs, in twin sheep, and in human beings. The three cases of the last sort include one extraordinary instance in which the twins were unlike in sex, and the persistence of genotypically female cells in the male could be clearly perceived. These findings led to the theory of antibody formation proposed by Burnet and Fenner, according to which functional units in the reticulo-endothelial system are concerned with the breakdown and removal of substances from the expendable cells of the body. In embryonic life the scavenger cells learn to recognize the autogenous substances, even though they may be tissue-specific and different from anything in the scavenger cells themselves. Against such substances, then, after the antibody-forming capacity matures in these same reticuloendothelial cells, no antibodies are formed. Introduction of genetically different cells into the embryo from a twin or possibly the mother during the early period merely leads to their recognition as belonging to the "self." (Evidence favoring this view is that antigenic substances developing in physiological isolation, as in spermatozoa, lens of the eye, and perhaps brain tissue, can act as isoantigens.)

Billingham also emphasized the complete lack of tissue specificity in the embryonic tolerance phenomenon; that is to say, no matter whether the tissue inducing tolerance be blood, spleen, thymus, bone marrow, or other, the tolerance extends to homografts of skin, adrenal cortex, ovarian tissue, thyroid, kidney, etc. On the other hand, tolerance is narrowly limited to grafts from an isologous donor, although those from a closely related genotype may persist somewhat longer than those from more distantly related strains of the same species. The reactivity of a tolerated graft remains absolutely unchanged. Such grafts removed to a normal mouse of the host strain are rejected; transplanted to a mouse of the donor strain, they are accepted. A long-standing homograft of CBA skin on a mouse of strain A made tolerant to CBA tissue will, when injected with strain A cells from a mouse sensitized to CBA, quickly break down. When injected with strain A cells from a normal unsensitized mouse, the breakdown also occurs, but more slowly.

Most remarkable are the heterologous tolerances, as of rabbit skin on rats or mouse skin on hamsters. Ducks, geese, turkeys, and chickens will accept heterografts that survive for 3 to 8 weeks after parabiotic union during embryonic life. It appears that this tolerance depends on the proper age of the embryo first exposed to the foreign cells. Not only must the embryo not be too old, it must also not be too young; and there is variation of the adaptive period according to the particular antigen in question. Thus cattle have long passed the tolerance-response period for transplantation antigens by the time of birth, but may become tolerant of *Trichomonas foetus* antigen as late as 4 weeks after birth.

In regard to the "graft-against-host" reaction, Billingham brings out the interesting consideration that the placenta must not only be able to prevent passage of fetal red cells into the maternal circulation, if Rh and ABO incompatibility reactions are to be avoided, but must also be able to prevent access of maternal leukocytes to the fetal circulation, since if they did obtain entry they would not only confer tolerance on the fetus but the antibody-forming capacity of such cells might inflict injury upon the developing organs of the fetus. Billingham and Silvers have found that even the skin of adult chickens contains immunologically competent cells, capable of evoking injuries in the host embryos to which grafts are made. Certainly, much more work will be required before a finally clear picture of host-graft relations becomes thoroughly understood.

*Selective Uptake and Release of Substances by Cells*

The selective uptake and release of substances by cells have been widely considered to operate in the regulation of growth and development. Morgan Harris discussed two aspects of this phenomenon analyzable in cell cultures growing in vitro: (1) the uptake of whole proteins into cells and its relation to growth; and (2) the modification of the glycolytic pattern of cells as a measure of energetics of tissues and the population dynamics of growing systems.

The conventional view is that each cell synthesizes its proteins from the free amino acid pool. There is evidence that enzymes and antibodies in course of synthesis are not in equilibrium with other proteins of the cell, but are formed directly from amino acids. Nevertheless, Whipple and his collaborators have long maintained the view, based on their studies of labeled plasma proteins in dogs, that these are in equilibrium with tissue proteins and can, if need be, serve as sole source for synthesis of the latter. Harris has summarized a great deal of recent evidence to support this general view. Bovine serum albumin injected into hens is transferred to the yolk of their eggs with little modification (Knight and Schechtman). Antisera conjugated with fluorescein isocyanate demonstrate visually that antigens enter cells and localize particularly in the liver parenchyma, lymphocytes, fibroblasts, and reticulo-endothelial cells (Coons). Plasma proteins, including fibrinogen, have also been traced into cells by this method. Ovalbumin and beef serum pseudoglobulin tagged with iodine-131 and injected into rabbits, and human gamma globulin and serum albumin tagged with sulfur-35 and $I^{131}$ and injected into mice have been traced into the mitochondria (Haurowitz and others). Methionine labeled with $S^{35}$ has been traced from a donor chick spleen grafted on the chorioallantoic membrane of the egg into the spleen of the host (Ebert). Within three days as much as 15 per cent of the proteins of the host's spleen may be derived in part from the proteins of the graft. Enzyme-rich "droplets" (mitochondrial) of rat kidney cells have been shown colorimetrically to take up injected horse radish peroxidase to a much greater extent than other cellular fractions (Strauss). Brachet, in the discussion, referred to clear evidence of the penetration of labeled ribonuclease into the ameba.

In addition to such evidence there is evidence from the tissue culture studies of Harris and others. The stimulation of growth and proliferation by the addition of tissue extracts to the medium was traced to a ribonucleoprotein fraction. (Niu, in the discussion, described tissue culture experiments demonstrating that while ribonucleoprotein is effective in stimulating growth, RNA alone is not.) Winnick and his associates showed that protein increase in chick heart cultures maintained in embryo extract depended in initial stages chiefly on amino acids, but later more than half was derived from the proteins of the embryo extract. Amino acid analogues were unable, at least in the case of phenylalanine, to block the transfer of $C^{14}$-labeled amino acid from protein of the medium to cell protein. Injections into hen's eggs of tagged proteins, peptides, and amino acids showed that proteins were the preferred precursors of the proteins of the embryonic chick. Harris and his coworkers obtained nucleoprotein fractions from adult chicken brain, liver, heart, and spleen, and with them obtained marked stimulation in growth and glycolytic activity of chick fibroblasts growing in vitro. There was no organ specificity, but the fractions differed in activity, that from spleen promoting the most vigorous growth. Moreover, growth failed to occur in chicken serum plus Medium 199 alone, although it occurred at a logarithmic rate when the spleen nucleoprotein fraction was added. Evidently serum proteins are not equivalent to tissue proteins in promoting growth. [Medium 199 is not essential to growth, but omission of the synthetic mixture slowed down the growth. There must be some interaction between the supplied protein and the inorganic growth factors.] Serum proteins can, however, be used by some cell strains, in particular malignant ones (HeLa, rat tumor, L strain mouse sarcoma). Eagle has emphasized the possibility of growing cell strains in purely synthetic medium. His studies have greatly clarified the nutrient requirements of cells, but do not exclude the possibility, greater for normal freshly isolated strains than for malignant strains or others kept for a long time in culture, that proteins may be used too. It may well be that an initial use of amino acids is prerequisite to the utilization of whole proteins by direct incorporation. [Certainly there is abundant evidence, from many laboratories, that cell strains kept in culture for more than a few passages (transfers to fresh medium) become abnormal biochemically, morphologically, and chromosomally.]

The mechanism whereby the incorporation into cells of entire proteins may be supposed to occur is either by phagocytosis or, more likely, pinocytosis, which has been extensively studied and cinephotographed. In pinocytosis the active area of a cell membrane may be seen thrusting out filar pseudopodia, forming membranous ruffles, and occasionally wrapping around fluid droplets in the medium and drawing them into the cytoplasm, where they move toward the nucleus and disappear.

Harris finds lactic acid production an excellent measure of the glycolytic activity and diagnostic of the metabolic pattern. Cells growing in vitro and utilizing glucose, or various other sugars, have a respiratory quotient of 1.0. Both aerobic respiration and glycolysis occur. In skeletal muscle fibroblasts from chick embryos, Harris found that during the logarithmic growth phase lactic acid (measured chemically) also increased logarithmically, and paralleled the cell count quite closely. The production of lactic acid under these conditions is essentially an aerobic glycolysis; when oxygen is replaced by nitrogen, a Pasteur effect is evident. With added spleen nucleoprotein fraction, both high glycolysis and a high growth rate were evident; but in serum plus embryo extract the two processes were inversely related. When growth in different media was compared, the conclusion could be drawn that glycolysis during the log phase is inversely related to the concentration of embryo extract in the medium. The decline of glycolysis seemed to be a function of population density. These relationships together suggest that aerobic glycolysis may be necessary for cell division.

Incidentally, no support was found for Warburg's contention that malignant cells exhibit a high level of glycolysis in respect to normal cells. On the contrary, in certain media the amount of aerobic glycolysis in the chick skeletal fibroblasts far exceeded the amount of glycolysis in ascites tumor cells or L sarcoma cells of the mouse, as reported by Warburg and Burk, respectively. This conclusion was supported by Eagle in discussion, on the basis of his own findings. The occurrence in Harris' experiments of glycolysis in active, healthy cells, its maximum rate in the early log phase, and its decline and cessation at the end of the log phase, when maximum cell densities are approached, point rather to the view that tumor cells, although aggregated in masses, respire as though they were dissociated. Lipmann, in the discussion, suggested that the high rate of glycolysis was related, not to malignancy per se, but to adaptation to a low oxygen supply, which the poorly vascularized character of tumors would necessitate.

## Control Mechanisms in Development

### Chemical Controls

Although much of the preceding review and discussion has necessarily dealt with mechanisms of control over growth and development, certain special aspects remain to be taken up more fully. Nelson T. Spratt, Jr., presented the results of experiments showing how modifications of the external chemical environment can control specific features of development. At the same time, he was cautious to emphasize that one ought not to assume that these external agents necessarily correspond to the actual internal controls of the organism.

When young chick blastoderms (18 to 30 hours old) are explanted to a non-nutrient medium, growth and development quickly cease, and organization at the organ and tissue levels breaks down, even though the individual cells may for some time remain alive and capable of resuming growth when placed on an adequate medium. In other words, "maintenance of the organ and tissue levels of complexity requires a greater and more continuous supply of exogenous nutrients than does maintenance of the viability of isolated, dispersed cells of the same tissues and organs." For the continuation of development, yolk plus albumen is optimal. Yolk alone is not so good, but is better than albumen alone, which in turn is superior to either blood plasma + embryonic extract or a synthetic medium. Even the addition of glucose alone to a non-nutrient Ringer solution will support some of the early developmental processes. Mannose, fructose, and galactose, maltose, pyruvate, and lactate support some development, but are inferior to glucose.

The brain is more exacting in its carbohydrate requirements than the heart. The node area makes an almost normal development on albumen alone, but is more sensitive than other structures to adverse conditions such as removal of calcium, low sodium, toxic agents, starvation, $CO_2$ deficiency, or oxygen deficiency. There is increasing evidence of the importance of an optimal carbon dioxide tension, as seen in the beneficial effects of a certain amount of crowding, and the poor growth of single isolated cells.

In a low potassium medium the brain will develop, but the heart will not. On the other hand, inhibitors such as iodoacetate, malonate, cyanide, and anaerobiosis may completely suppress development of the nervous system without affecting that of the heart. Fluoride has the opposite effect; only azide among inhibitors has no clear-cut differential effects. Clearly, the differential effects do testify that different metabolic processes and patterns underlie the development of different structures; but little more can be added on the subject at present. Inhibitory effects may also be exerted by particular tissues or organ homogenates. For example, transplantation of whole blastoderms or isolated node areas to the surface of a living embryo heart or liver slice inhibits the formation of all axial structures by the node. Heating the organ slices destroyed their inhibitory effect, and close contact (by cellular adhesions or bridges) with the living slice was necessary for the inhibition to occur. In the discussion, Herrmann added the interesting observation that leucine analogs will make the somites fuse but seem to disturb no other feature. Protein synthesis in these somites is not impaired, but protein breakdown is enhanced. This biochemical alteration is not found elsewhere.

It was also possible, in Spratt's experiments, to demonstrate that the formation of specific tissues and organs exhibits quantitatively different nutrient requirements. The requirement for growth is greater than that for differentiation, which in turn exceeds the requirement for movement. Mere maintenance has lowest requirements of all. Particularly significant, however, was the demonstration that the very active nodal area requires higher concentrations of nutrients than do the cells of more quiescent regions, either more or less advanced in development. The node center was shown to have greater enzymatic activity and more kinds of active dehydrogenase systems than other areas. Certain tests indicate that this heightened activity may depend upon more numerous or more active mitochondria. These last facts lead Spratt to propose that at least one endogenous control system making development exhibit orderliness may lie in the character of the mitochondrial population present in particular cells or regions.

A special example of chemical stimulation of specific growth was presented in the following two companion papers, by Rita Levi-Montalcini and Stanley Cohen, respectively. Various agents that stimulate the regeneration of motor and sensory nerve fibers after transection have been found, e.g., von Muralt's "NR" factor, of low molecular weight, in the white matter of rabbit and calf brain substance, dinitriles, and piromen (a bacterial polysaccharide complex). It has been found that reinnervation of denervated muscles occurs by collateral branching from intact nerves in the vicinity, and a lipid substance was found in ox spinal cord by Hoffman which, like piromen, stimulates collateral sprouting.

The Washington University group was stimulated by Bueker's discovery that a certain mouse sarcoma, when implanted in the body wall of a chick embryo, became extensively invaded by sensory nerve fibers from the host, and that the spinal ganglia supplying these fibers became considerably enlarged by increase in number of cells as well as increase in size of neurons. Investigating this phenomenon further, they found that sympathetic ganglia also contributed to the innervation and that in both types of ganglia cellular differentiation was enhanced. Frequently additional ganglia were formed. No synapses were formed with the tumor cells, however. Other mouse sarcomas, but not other types of malignancy, were found to possess the same stimulatory effect; and it was noted that stimulation need not be exerted through the nerve fibers but could be exerted upon ganglia at some distance from the tumor. This last observation set them to searching for a stimulatory substance released into the blood stream, and evidence of its existence was found when a piece of tumor was transplanted onto the chorioallantoic membrane of a 4-day-old chick embryo, in a position where the tumor could not itself become invaded by nerve fibers. Under these conditions the stimulated ganglia sent out numerous fibers which entered the adjacent viscera, and even perforated the walls of large veins and bulged into their lumina. In tissue culture, explants of ganglia together with pieces of tumor were likewise found to undergo stimulation and to send out larger numbers of nerve fibers in the direction of the tumor than in other directions. Even *normal* mouse tissue was found to exert a stimulatory effect, although a far weaker one than that associated with the tumor.

From the sarcomas Cohen isolated an active protein fraction which possessed the stimulatory capacity, and by accident it was discovered that snake venom, used as a source of phosphodiesterase, was a far stronger stimulant of nerve fiber outgrowth and ganglionic enlargement than any of the sarcomas. Finally, reasoning that snake venom is a modified salivary secretion, he tested the effect of mouse salivary glands in vitro and in vivo, and found it to be 4 times as active as the snake venom, which was itself 2500 times as active as the sarcoma homogenate. No other mouse organs were active. Outgrowth of nerve fibers from ganglia was so "exuberant" under the stimulation of mouse submaxillary gland material that distant stimulation of sensory as well as sympathetic ganglia was observed for the first time in these experiments.

Further chemical studies, by Cohen, of the snake venom factor led to the conclusion that it was present in all venomous snakes tested of the Elapidae, Viperidae, and Crotalidae, as well as in the venom of the Gila monster. The activity was twice as high in the cobra and another elapid as in the water moccasin or rattlesnakes (crotalids); and in 2 out of 4 vipers tested it was four times as active as in the crotalids. The factor in the venom of the moccasin has been purified about 40-fold by conventional protein purification methods. It has a molecular weight of around 20,000 and an amino acid pattern like that of bovine albumin. Activity is destroyed by incubation with proteolytic enzymes (but not by RNAase or DNAase) and by incubation with antiserum to snake venom; but crystalline snake venom itself is completely inactive as a neurogenic stimulator. No enzyme known to be present in snake venom has significant neurogenic activity except protease and ribonuclease; but whereas the growth factor is stable upon incubation in 0.1 $M$ NaOH, protease and ribonuclease are completely destroyed by the same treatment. Other experiments ruled out the possibilities that the effect might be indirect, by removal of some inhibitor or liberation of some stimulant through reaction with the medium. It was shown also that in order to exert its effect the growth factor must be present continuously during the actual outgrowth of the fibers. That actual growth, and not a mere redistribution of material, is involved was shown by experiments from which it was concluded that glucose is required for the outgrowth of fibers and that at least some amino acids are needed. Lysine and adenine incorporation and glucose oxidation were significantly increased under the influence of the snake venom factor.

Purification of the mouse salivary gland factor has been achieved to the extent of 10 to 15-fold, but that work is just beginning. This agent too appears to be a protein. It has been found to stimulate the outgrowth of nerve fibers from sensory ganglia of mouse and rat embryos as well as those of the chick. Although in the discussion some reservation was expressed about making too ready an assumption that these proteins play any normal, specific role in neural development, neither can the possibility be completely dismissed. The remarkable specificity of the effect is what makes it so very interesting.

## Neoplastic Development

From a very extensive review of modern studies of chemically induced neoplasia of the skin in rabbits and mice and of the features of spontaneous mammary neoplasia in mice, Leslie Foulds concludes that neoplastic growth is an irreversible, heritable, qualitative change in cells, which may occur unpredictably in any cell still capable of proliferation. From the investigations of chemical carcinogenesis one may distinguish the existence of two separate phases in the induction of neoplasia: *initiation,* which brings about a subthreshold or latent neoplastic state; and *promotion,* which leads to active proliferation and inhibition of differentiation. Initiation is parallel in nature to embryonic determination, promotion to the active realization of determined capacities.

The validity of this distinction is to be seen in the fact that some chemical agents initiate but do not promote carcinogenesis; others cannot initiate, but do promote; a few do both. Among the agents in the first class is urethane; among the typical promoters lacking capacity to initiate is croton oil; tar and other hydrocarbon carcinogens possess the dual capacity. Initiation is usually rapid, whereas the often prolonged latent phase is characteristic of the promoting agents. Thus the lesions that appear in previously initiated skin upon promotion pass through progressive phases: warts or papillomas, benign in nature, arise and of these some regress, some persist without growth, some grow but remain benign, and some grow and become malignant. Alternative outcomes exist at every phase. There is no relentless progression to malignancy, as is commonly supposed.

Most spontaneous mammary tumors seen in mice develop from minute and usually numerous hyperplastic nodules. Again, some of these may regress, some persist without growth or change, a few develop into carcinomas. Sometimes the mammary tumors develop from larger plaques, which grow only in response to some hormone of pregnancy. In either case, as in the chemically induced neoplasias, there is at each step a progressive restriction in the field of further advance. Thus as a general rule only one or a very small percentage of multiple tumors will develop at the same time. [We are reminded of the restriction in the maturation of mammalian oocytes, which follow the same rule.] Furthermore, different characteristics of a tumor progress independently. Progression is independent of growth, may be discontinuous, at numerous points follows one of alternative possible paths of development, and does not invariably reach the end-point within the life of the host organism. In the light of these general principles, Foulds has discussed neoplasia in human beings, especially epidermal neoplasia, neoplasia of the urinary bladder, hereditary intestinal polyposis, and neoplasia of the uterine cervix and breast.

The most precise information about mechanisms of progression comes from the study of transplantable tumors, especially ascites tumors. From the work of Klein and Klein, Foulds draws the conclusion that progression generally involves the selection of specific variant cell-types adapted to special or unusual conditions. The conception thus involves what in the broadest sense might be called a mutation (intrachromosomal or not). The postulated mechanism of initiation and promotion is as follows: (1) initiation irreversibly alters the reactivity and potentialities of all or nearly all the cells exposed to the stimulus; (2) promotion involves a heritable change occurring in certain foci of initiated tissue which is proliferating; (3) new neoplastic foci temporarily inhibit progression in neighboring cells, hence not all competent tissue undergoes progression simultaneously; and (4) the first stage of promotion may establish the definitive type of the tumor, but often this is reached only after a sequence of progressive steps, each of which occurs focally within the competent tissue.

In the light of these ideas, Foulds regards the transmission of chicken tumors by viruses as having less bearing on a general theory of cancer than on the nature of the intracellular mechanisms of neoplasia, especially the evidence that neoplasia can be produced by the incorporation into a cell of foreign nucleic acid. The virus-induced sarcomas and leukemias of the chicken start with a rapid, almost immediate transformation of cells; there is no prolonged latent period, as in chemically induced neoplasia; the transformation is direct, not progressive; and the tumor is a replica of the type from which the agent was obtained. These characteristics are so different from those of the chemically induced and spontaneous tumors that more is to be gained by studying the intracellular mechanisms than by trying to formulate a theory that forces all neoplastic growth into a single mold.

There are many misconceptions about neoplasia. Neoplastic growth is not uncontrolled, as is so often said, but is often under the control of hormones. It is not chaotic growth, but is usually organized in structure. It is not extraordinary in rate, being exceeded by many embryonic growth rates. Neoplastic tissue may carry out normal physiological functions, such as the production of hormones. Neoplastic growth is not sustained hyperplasia, since the neoplasia often develops in atrophying tissue long after hyperplasia has ceased. Its real essence is the irreversible cellular change determining that its growth "is not controlled or coordinated in harmony with the needs and safety of the whole organism."

So conceived, neoplasia in many respects resembles normal differentiation. The same basic mechanisms may be involved in each, and discoveries with respect to either should throw light on the other. Among these mechanisms is biochemical specialization. That is to say, the progressive canalization and choice of one of two alternative pathways lead both to specialization and to a loss of particular biochemical enzyme systems*; and this may be attributable to the *functional* loss (called by V. R. Potter a 'deletion') of some particular portion of the DNA. This is tolerable, and indeed necessary, so long as the specialized cell or tissue remains integrated with the rest of the organism. "The menace of neoplasia," says Foulds, "results from a failure of integration, and on qualitative rather than quantitative characteristics of growth."

*Organ Differentiation in Culture*

In a very comprehensive review, E. Borghese has considered the development of methods of culturing embryonic organs in vitro and the applications of that technique to the study of the morphogenesis, capacity for regulation, and rates of growth and differentiation of isolated organs, and of induction, interactions between different organs, the reaggregation of dissociated cells, the action upon isolated organs of chemical agents such as vitamin A and various antimitotic substances, the analysis of genetic problems, and the composition of the nutrient medium. Although the problems of organ development in birds and mammals had already been studied by means of chorioallantoic grafts or grafts to the coelom or anterior chamber of the eye, the technique of organ culture in vitro offers a better means of control over the medium and greater isolation from the effects of interaction with the host organism.

As might be expected, early studies with the new technique first introduced by Strangeways and Fell in 1926, and later much improved by Leighton, Wolff and Haffen, and Martinovitch and others, involved the isolation of a bud or rudiment of an organ and the observation of its subsequent morphogenesis in vitro. At present, with the best techniques, it is possible to keep such cultures alive for 2 to 3 weeks—even months in some instances—whereas a period of 8 or 9 days in culture is sufficient for most purposes.

Many observations have been made on the morphogenesis of isolated organs (limb buds, digestive and respiratory organs, urogenital apparatus, endocrine glands, skin, and sense organs) of such readily available and classical material as the rat, rabbit, or chick. Those chick limb buds isolated from embryos 72-80 hours old will differentiate their cartilaginous skeletons; those from embryos 6-12 days old will undergo ossification in vitro.

---

* An excellent example, contributed to the discussion by P. P. Cohen, is that of the simultaneous *increase,* in the Novikoff transplantable tumor, of aspartic transcarbamylase and *decrease* of ornithine transcarbamylase, which is competing with the first-named enzyme for the same limited substrate, carbamyl phosphate. Thus urea synthesis is sacrificed for the sake of an increased synthesis of pyrimidines, needed for regeneration or increased growth.

A sternum will form from paired rudiments that actively migrate into juxtaposition in either mouse or bird isolates; the presence of the ribs is not necessary for this to take place. Tooth rudiments of rats or rabbits will differentiate odontoblasts and dentine, and will calcify. Salivary glands will differentiate acini and ducts; the intestine will produce villi and commence peristalsis; liver, pancreas, trachea and bronchi, the syrinx of a bird, and lungs and heart will all differentiate quite fully. Metanephric tubules differentiate in kidney explants.

Especially interesting are the results obtained with isolated gonads. Ovaries from a chick embryo of 10-12 days possess oogonia which within 4 days advance to an oocytic stage, leptotene and even synaptene. When the explant is taken from a 17-day-old chick, meiosis may be seen to reach pachytene and strepsinema stages in vitro. In the duck, male gonads develop Sertoli cells and spermatogonia; female gonads differentiate, the left becoming an ovary, the right one regressing, as in the entire organism. Female chick gonads behave differently from those of the duck: the right one becomes a testis, the left an ovotestis. The result is explicable on the basis of a lack in vitro of female hormone which would normally inhibit the development of the testicular part of the gonad. Mouse testes from embryos 13-18 days old produce seminiferous tubules and spermatogonia; older explants produce spermatocytes with meiotic stages through diplotene. Rat or mouse ovaries from embryos 15-16 days old have been cultivated by Martinovitch for as many as 30 days. Again, meiosis through diplotene has been observed, followed by a prolonged growth phase. Oocytes degenerate and are replaced by new waves of germ cells, as in vivo. Older explants (age from 17 days to birth) actually form primary follicles. By culturing the ovaries at 34°C, growth may be slowed down while differentiation proceeds. Even meiotic metaphase has been attained in cultures of long duration, after several waves of growth. Human ovary has also been cultured. Fetal ovarian cortex becomes covered with epithelium, from which cords proliferate and oocytes develop, reaching meiotic prophase and even metaphase I. Primary follicles reach the one-layered stage (Gaillard).

Other organ cultures can be mentioned only briefly. Thyroid explants of 8-day-old duck embryos become functional and form thyroxin. Pituitary explants of the 6-day-old chick embryo differentiate acidophilic and basophilic cells. The pineal body forms epithelial buds and follicles. Skin explants become stratified, and form feather buds. Skin of mouse embryos (10-18 days old) produces hairs and vibrissae. Explants of mammary glands of mouse embryos (10 days old) form primary buds. Brown adipose tissue will develop in vitro from the interscapular mesenchyme of a rat embryo (15 days old). The otocyst of a 3-day-old chick embryo develops all the sensory components of the labyrinth, together with supporting cells; its histological development is virtually complete, but form and organization are defective. Eye rudiments explanted at about the same age also develop histogenetically very fully, especially the retinal layers, including rods and cones; but again morphogenesis is aberrant.

It may thus be said that the number of "organs which, when explanted at an appropriate stage, are capable of going on and differentiating, and also to a certain extent of functioning, is increasing day by day." As Borghese says, interest in this purely exploratory and descriptive phase of organ culture is waning, since the result of making a new type of explant is almost a foregone conclusion. Attention has become focused instead on instances where differences between development in vivo and in vitro can be found, as indicating the influence upon the development of the explanted organ of the presence of contiguous or even remote parts of the organism. An exception to the rule of autonomous development is to be seen, for example, in the deformity of skeletal parts growing in vitro, and others have been mentioned above.

The capacity of organ rudiments cut into parts to regulate is also instructive. Thus, if femur and tibia cartilages from a chick embryo less than 5 days old are cut across the shafts and the cut ends placed in contact they will grow together; but later this is no longer possible. The rudiment of a submandibular gland in the mouse consists of two parts, an epithelial cord and a bud, destined respectively to form the duct and the glandular tree of the organ. When these are separated, ordinarily no regulation occurs; but if the distal end of the cord is placed in contact with a sufficient amount of submandibular mesenchyme it may form some buds, although the chief duct cannot be formed by regulation of the bud portion. Nearly complete regulation can occur in explanted tooth rudiments of the rabbit embryo (20 days old). When a tooth rudiment is divided in half, each half will form a complete set of cusps. The lens of the eye, of course, regenerates very readily.

In explants the rate of growth and differentiation is often slower than in vivo, although growth, including cell proliferation, is more generally slowed down than is differentiation. Yet in the rat tooth there is strong retardation in both respects; and while in the embryonic human hand or foot (explanted at 9-15 weeks) growth and differentiation of the cartilages is slowed down, in explants at the age of 6 weeks growth is much faster than in vivo!

Many interesting studies have been done by separating the epithelial and mesenchymal components of an organ in order to determine whether there is an inductive effect of one on the other. This separation can be performed completely by means of trypsinization (Moscona). The tissues can be recombined in various ways, and cells may be completely disaggregated and then allowed to reassemble. By such methods it can be shown that the odontoblasts of teeth will not develop without the presence of ameloblasts. For the growth of hair, ectoderm must be associated with mesoderm. In the submandibular gland, the epithelium fails to branch and spreads out as a thin sheet if the mesenchyme capsule is removed, but becomes normal again if recombined with autogenous mesenchyme, i.e., mesenchyme from the submandibular gland. Mesenchyme from other organs or from younger embryos is ineffective (Grobstein). There is a similar type of interaction between epithelium and mesenchyme of the developing metanephros, but in this case induction is a two-way process. The mesenchyme induces the formation of coiled nephridial tubules in the epithelium. As before, only autogenous mesenchyme will induce the formation of tubules in the epithelium. Not so for the mesenchyme; it can be induced by non-specific epithelium, or even by neural tissue from the spinal cord. Pieces from the *ventral* part of the spinal cord, or even from the cerebrum, will induce cartilage to form in somite mesoderm; but the *dorsal* portion of the spinal cord instead induces formation of renal tubules in the lateral mesenchyme. The well-known interaction of dermal and epithelial parts of the feather germ can be demonstrated in vitro. Grobstein has further demonstrated that inductive action can be transmitted through a porous membrane filter 20-60 $\mu$ thick (porosity, 0.8-0.45 $\mu$). A membrane 80 $\mu$ in thickness and of 0.8 $\mu$ porosity was too thick and one of 0.1 $\mu$ porosity was too fine unless the thickness was reduced to 20 $\mu$. The 60 $\mu$ membrane is so thick and the 0.1 $\mu$ pores so fine that cytoplasmic contact between the tissues is thought in either case to be excluded.

One of the most amazing findings of organ culture is that different organs from the same species but of different ages or different genetic strains, or even organs from different species or classes (Aves and Mammalia) may become associated in vitro, with no sign of incompatibility. This is without doubt related to the lateness in development of antigen-antibody and similar reactions, as has already been pointed out. Various combinations have been formed from mesonephros, liver, lung, spleen, and thyroid, taken two-by-two, of chick embryos 6-10 days old. Connective tissue and epithelium will intermingle, one organ may overgrow or penetrate into the other, and tissues from the two may combine to form a common structure. The xenoplastic combinations produced have included gonads of chick (or duck) and mouse. Here the connective tissue cells of the mouse testis migrate inside the duck testis; the seminiferous tubules enter into communication; a joint Sertoli epithelium is formed; and spermatogonia of the two species, bird and mammal, become intermingled. A combination has even been made of a mouse ovary with a duck testis, or the reverse, resulting in the presence of mouse oocytes in the seminiferous tubules of the duck and, in the opposite combination, of duck oogonia intermingling with mouse spermatogonia (Wolff and Weniger). Also notable in these xenoplastic associations is the attraction manifested between cells of certain tissues, regardless of species differences. Thus connective tissue shows a readiness to spread under foreign epithelial structures, and epithelia to grow in foreign mesenchyme. Epithelial cells of very distantly related species will collaborate nicely. This affinity of similar histological cell types may well indicate some basic protein similarity.

Hormonal studies may be prosecuted to advantage in organ culture, since the endocrine organ can be combined in vitro at various ages with the target organ. It has been shown that pituitary, parathyroid, and gonads secrete their hormones in vitro. The pituitary will stimulate the activity of the thyroid, and will secrete gonadotrophic hormones. Gonadal secretions exert their characteristic effects on the development of gonads of the opposite sex, on the gonadal ducts, and, in duck explants, on the syrinx, which is symmetrical in the female but asymmetrical in the male, and which is modeled according to the nature of the sex hormone present regardless of the genetic sex of the explant itself. In a slightly different type of experiment, hormone can be added to the medium in which an explanted organ is developing in order to test its effect on development in vitro. Thus the effect of thyroxin or insulin on the development of the limb buds can be nicely demonstrated.

An example of this use of organ culture to investigate a problem of hormonal control over differentiation was presented by Dorothy Price and Richard Pennabecker. The problem was to determine whether or not sex hormones from the fetal gonads influence the characteristic male or female development of the reproductive ducts and accessory sex glands. From 17.5-day-old rat embryos entire reproductive tracts, consisting of gonads and associated Wolffian and Müllerian ducts, were explanted and cultured. Explants with both testes or both ovaries underwent normal male or female differentiation. Explants from which both testes had been removed showed retrogression of the Wolffian ducts and failure of seminal vesicles to form. There were also few prostatic buds. The Müllerian ducts, however, underwent retrogression as in normal males. Explants in which a single testis had been removed differed, depending upon whether the right and left sides of the tract were spread apart or were close together. In the latter case, normal male development took place; but in the former case, the Wolffian duct on the operated side showed retrogression and the seminal vesicle failed to develop. This is good evidence of the existence of a diffusion gradient of some effective agent, presumably testosterone, from the remaining testis.

When the ovaries were removed from female reproductive tracts differentiation in vitro was not at all affected. There is therefore no evidence that endocrine activity of the ovary is required either to maintain the Müllerian ducts, to make the Wolffian ducts regress, or to inhibit other accessory male glands.

Trypsinized, dissociated cells of an organ placed in the culture medium will reaggregate. Thus dissociated cells of a chick embryo 7-8 days old will not only reform epidermis, but will even produce typical feather germs. Cells derived from different organs under these conditions retain their respective determinations, e.g., mesonephric and limb-bud cells from 3-5 day-old chicks. This will occur even when the organs are from embryos of different age. The validity of these conclusions is thoroughly established by utilizing cells of mouse and chick embryos intermingled, since such cells are readily distinguishable by size and staining characteristics. Moscona showed that chondrogenic, mesonephric, and hepatic cells of mouse and chick embryos, dissociated and recombined in culture, will reaggregate to form tissues in which mouse and chick cells of the same histological type participate indiscriminately in the formation of each tissue. The aggregation was always according to cell type, not according to species of origin. On the other hand, a mixture of chondrogenic cells of chick and hepatic cells of mouse will segregate to form cartilaginous pieces of chick origin and liver tissue of mouse origin. In isolated retinal cells, however, the determination did not prove to be stable.

The effects of vitamin A upon development have been studied in organ culture. Added to the medium, the vitamin will prevent the growth of organ rudiments from a chick embryo 5-6 days old. Cartilage and bone actually regress; the skin fails to keratinize; the sulfur metabolism is profoundly altered. Among other agents studied, tripaflavin, an antimitotic substance, has been found to inhibit the development of the male part of the chick gonad. Colchicine, on the contrary, affects only female gonadal tissue—the cortex of the ovotestis, and not the medulla.

In genetic studies, the developmental effect of mutant genes has in a number of instances been clarified by means of organ cultures. Thus, in studying the Creeper fowl, explantation of normal and Creeper tibias to media containing extracts of normal or Creeper embryos affected linear growth as follows: Creeper tissue in Creeper medium, minimal growth; Creeper in normal medium, much better growth; normal tibias in Creeper medium, definite shortening but better growth than the preceding; normal in normal medium, standard growth. It was shown by explantation that Grüneberg's rat lethal, in which the ribs and tracheal cartilages are grossly hyperplastic, acts autonomously in cartilage tissue removed from the body and cultured. The sterility of anemic $WW$ mice could be shown to develop in gonads growing in vitro and thus was ascertained to be an effect of the gene quite independent of the anemia, which develops only after birth.

A fine example of the use of organ culture in the genetic study of developmental problems was contributed to the discussion by Salome Gluecksohn-Waelsch. The $Sd$ mutant in the mouse causes, when homozygous, a complete failure of the kidneys to grow. Kidney rudiments from such mice were explanted in organ culture, and the relation between the ureteral bud and the metanephrogenic blastema was studied. It was found that the *SdSd* ureteral bud is capable of inducing normal metanephrogenic tissue to develop and form tubules; and likewise *SdSd* metanephrogenic tissue will form tubules when combined with normal ureteral tissue. Thus the mutant ureteral bud possesses inductive capacity and the mutant metanephrogenic tissue has competence to respond. Yet somehow they can not cooperate to form normal kidney. The interaction is imperfect.

Finally, studies of the significance of components of the nutritive medium have been quite fruitful. The action of embryo extract and its relation to the age of the donors has been examined. A mixture of 21 amino acids cannot satisfactorily replace embryo extract as a growth requirement. Para-aminobenzoic acid was found to be required in addition. For the growth of avian tibia, gonads, and syrinx a minimal medium must contain arginine, methionine, glutamic acid, histidine or lysine, and para-aminobenzoic acid. It is further improved by the addition of threonine, asparagine, hydroxyproline, serine, and taurine. Para-aminobenzoic acid is replaceable by carnitine or dicarnitine. Later Wolff and his associates, who carried out this analysis, found further modifications were desirable for the growth and differentiation of gonads and syrinx. In a single instance a synthetic medium was found superior to the natural one. This was in the culture of chick feathers, where a mixture of 9 amino acids produced stout, voluminous papillae, and a mixture of 21 amino acids made them grow longer and thinner. Much is to be expected in the future from further progress along these lines.

## *The Role of Genes in Developmental Processes*

One of the most dramatic moments in the course of the entire symposium came when Ernst Hadorn held up for the view of the participants a little vial containing the first few grams of crystalline drosopterin, a brilliant vermilion pigment isolated from the eyes of some millions of *Drosophila* all of which had to have a genetic block preventing the formation of the ommochrome eye pigment derived from tryptophan, via kynurenine, and the subject of much biochemical work by Butenandt and his associates.

Hadorn's succinct but very thought-provoking discussion of gene action undertook to examine five major problems of physiological genetics. The first was the question: when do genes act in the course of development? Next he considered the existence of cell-specific and organ-specific patterns of gene action. Third was the question of the extrinsic supply of enzymes; fourth, the distinction between autophenes and allophenes; and lastly, the biochemical approach which leads us by steps from superficial traits (morphological and physiological phenes) back to the primary biochemical phenes, the initial products of gene action.

The question when genes act cannot be answered superficially by examining the phenotype, nor even by determining the one enzyme or one enzymatically controlled step which a particular gene may be said to govern. How many biochemical steps may there actually be between the initial activity of the gene and the observable phene? When the modified phene is undetectable in a cell or tissue or organ, is it because the gene did not act or because its specific product, the enzyme, failed to find substrate to act upon? How many of the enzymes of development are inducible, being present only in undetectable quantity until substrate or some other inducer leads to rapid enzyme synthesis? These and many other difficult questions may be posed. Hadorn presented certain types of evidence bearing on the general question of the time of gene action. First, there is the evidence of chromosomal differentiation in different tissues (see discussion by Gall). This evidence is still scanty and difficult to evaluate. The evidence derived from the study of chromosomal deficiencies is rather better. In this instance it can definitely be stated that not all genes are needed in all tissues of the developing organism, since lethal deficiencies kill at different points during ontogeny. Thus all Notch wing deficiencies in *Drosophila melanogaster* die at the 6th hour of egg development. The kind of damage done by a deficiency of a specific locus must tell us something of the characteristic activity of that locus. Here Hadorn emphasizes the principle of the "stepwise insertion of the products of genes in ontogeny." This does not signify a random distribution of primary gene activity throughout ontogeny, however. Lethal effects of deficiencies tend to occur within a relatively few sensitive phases of the life, at certain developmental crises when many new systems are differentiating. Of course, the larger a deficiency, the greater the probability that it will include the locus of some early-acting essential gene; hence the larger the deficiency, the more likely it is to kill at an early crisis than a late one.

Like deficiencies of a part of or a whole chromosome, merogonic hybrids develop up to a definite embryonic or larval stage and then die. These are hybrids in which the nucleus of the egg has been removed and the cytoplasm fertilized by a spermatozoon of a foreign species. The more remote the relationship between the two species, the lower is the probability that all the genetic loci necessary for the development of an embryo from the egg cytoplasm are provided by the foreign sperm. An example: A *Triton palmatus* enucleated egg fertilized by a *T. cristatus* sperm synthesizes normal amounts of RNA; but the cytoplasm cannot use this foreign RNA, which simply accumulates up to the time of the lethal crisis (Zeller).

There is considerable evidence that at least certain genes act in the embryonic period long before any visible or otherwise detectable effect becomes apparent during differentiation. In *D. melanogaster* the suppressor of erupt eyes can be inactivated by x-rays as early as 8 minutes after the fertilization of the egg, although the earliest visible sign of this appears only when the adult eyes begin to differentiate, late in larval life (Glass and Plaine). Hadorn himself has studied a rather similar case. Females homozygous for deep orange eye color (*dor*), when mated with males also *dor,* produce eggs that die in maturation or early cleavage; but when such females are mated with normal (*dor*$^+$) males, the eggs are perfectly viable. It follows that the paternal gene at the *dor* locus must become active in the egg prior to the formation of the zygotic nucleus. Instances of the early embryonic action of genes are not limited to the fruit-fly. In the mouse a gene that when heterozygous causes the fur to be yellow (an effect realized and detectable only late in development) is lethal in the blastocyst stage (5-6 days post-fertilization) when homozygous. Many of the short-tail genes, which in heterozygous state combined with a normal allele or with certain other *t* alleles produce a mild defect, when homozygous prove lethal at an early embryonic stage; $t^0t^0$ embryos die at the sixth day; $t^{12}t^{12}$ in the morula stage (4th day).

Study of such examples indicates that genes essential at some specific phase of development may really have been active long before. If, then, an enzyme or antigen comes close to being a primary gene product, its detection and localization in the ontogeny become very meaningful. To illustrate this approach, Hadorn told of work with another *Drosophila* mutant, rosy eye color. By chromatography it was found that rosy flies have two biochemical peculiarities. They do not form isoxanthopterin, but instead accumulate its precursor 2-amino-4-hydroxypteridine; and they do not produce uric acid, but instead excrete hypoxanthine. This is, therefore, an example of biochemical pleiotropy, that is, of more than one effect being attributable to a single mutated gene. Glassman and Mitchell have now shown that this dual effect is because of the lack of a single enzyme, xanthine dehydrogenase, which lacks enzymatic specificity to the extent that it can act on two different substrates, hypoxanthine and 2-amino-4-hydroxypteridine. The visible effect of the rosy mutant gene becomes evident as an altered eye color only at the end of metamorphosis; but the biochemical pteridine phenes are detectable earlier, at the onset of metamorphosis. The enzymatic phene, lack of xanthine dehydrogenase, can be demonstrated throughout the larval life and even in the embryo. It may be concluded that the appearance of the pteridine phenes must await the presence of substrate.

A particular gene may also exert different effects in different types of cells and kinds of organs; hence there arise cell-specific and organ-specific phenes. For example, the wingless factor of the chicken leads to different effects in the legs than in the wings, and it also affects the feathers, the lungs and air sacs, and the ureter and metanephros in specific ways, while other organs seem perfectly normal. By transplanting various imaginal buds of the lethal giant larva of *D. melanogaster* into normal hosts, it becomes evident that some of these are capable of metamorphosis (e.g., the hindgut) whereas others are not (e.g., salivary glands). The mutant biochemica in the mealmoth *Ephestia kühniella* is especially interesting. Studied by Hadorn, like the preceding case, it proves to be able to produce only two pteridines in the eyes, namely, isoxanthopterin and 2-amino-4-hydroxypteridine; but the abdomen of this mutant type synthesizes the whole gamut of pteridines characteristic of the wild type. In another *Ephestia* mutant, red eye (*a*), the eye forms all the pteridines although in diminished amounts, whereas now the abdomen exhibits qualitative differences from the normal. In the *Drosophila* rosy mutant, already mentioned, biopterin is present in both eyes and testes, but hardly any occurs in the Malpighian tubules, although wild-type flies have an abundance there. Examples such as these could be multiplied to show the highly specific way in which effects of the same gene vary from tissue to tissue, organ to organ.

Returning to a consideration of the rosy mutant of *Drosophila*, we find it also an example of how a tissue in which a phene becomes manifested may depend upon an extrinsic supply of the gene-controlled enzyme. Testes of the rosy mutant lack the yellow pigment isoxanthopterin. But rosy testes implanted in the abdomen of a wild-type host become pigmented; and wild-type (non-rosy) testes implanted into a rosy host remain pigmentless. Since the development in this respect is nonautonomous, the pteridine precursor is probably present in the rosy fly's testes, but cannot be transformed into the pigmented end-product. Something present in $ry^+$ flies and not present in $ry$ flies, and essential for the transformation of precursor to pigment, must be formed outside the testes and be able to enter them in normal flies. This something is very likely to be the enzyme xanthine dehydrogenase, which the evidence shows to be missing altogether in the rosy flies. This is not only the first case in *Drosophila* where an enzyme can be identified as a specific product of a gene locus, it is also a highly interesting case of an enzyme behaving like a hormone, acting in tissues other than those which form it. How common is such behavior?

These considerations lead Hadorn to distinguish between *autophenes* and *allophenes*. In the former the locus-specific characteristics are intrinsically conditioned; development of the phene is autonomous. In the second category, gene-specific influences from one tissue are exerted on another, as in the rosy case just described. The Creeper chicken shows that one and the same mutant gene may govern both autophenes and allophenes. For if embryonic Creeper limbs are transplanted to normal host embryos, they develop autonomously, that is, as Creeper-type abnormal limbs. But the eye defects caused by the Creeper gene are allophenes, for Creeper eye rudiments transplanted to normal hosts develop into perfect eyes.

All of these examples mirror the biochemical approach to genetic problems. One may fruitfully, as Hadorn has done with the lethal-meander and lethal-translucida mutants of *D. melanogaster*, start from the obvious morphological phenes and work back step by step through developmental defects to biochemical errors of metabolism—as Garrod termed them—and from these press toward the initial activity of the gene. Much pleiotropy vanishes as one does so, for many physiological and developmental defects turn out to be attributable to some single earlier modification of metabolism, or perhaps to an enzyme with multiple specificity. We are still left, however, with the question whether the gene itself ever exerts more than a single primary effect. If a single gene template may have several sites, why not? Or if a particular sequence of amino acids can become incorporated into several distinct peptides or proteins through the action of a gene, we would have primary pleiotropy to deal with. At least the "one gene, one enzyme" hypothesis, for all its past heuristic value, should not debar us from the search. [Note that while no cases of primary pleiotropy are known at present, there are cases where the same amino acid sequence is incorporated into a number of polypeptide or protein molecules, e.g., a sequence of 8 amino acids included in both of the pituitary hormones, ACTH and intermedin (C. H. Li and associates). If a single gene is responsible for the sequence but not for their incorporation into different molecules, this would not indicate primary pleiotropy; but if a single gene locus can incorporate such a sequence into various molecules, then, whether responsible for the sequence or not, and whether or not RNA serves as intermediary, the gene might be said to serve a number of primary functions.]

*Hormonal Regulation of Insect Metamorphosis*

Carroll M. Williams gave an amusing as well as instructive account of the coordination of insect development by (1) the hormone ecdysone, which is produced in the prothoracic glands of insects and is essential to bring about metamorphosis, (2) the brain hormone, which controls the secretion of ecdysone, and (3) the so-called "juvenile" hormone, produced by the corpora allata, which inhibits metamorphosis and prolongs larval life. Of these three, only the first has been isolated in crystalline form and its chemical nature studied (Karlson, 1954), although even yet its structural formula cannot be given. Williams supplied a second dramatic moment in the sessions by producing for inspection a small amount of the highly purified juvenile hormone, in the form of a golden oil, which will therefore very likely be chemically identified ere long.

Most of the work with insect hormones has been carried out by ligating the larvae or pupae at various levels between anterior and posterior ends, by making grafts, or by the injection of extracts. A favorite object for such studies is the very large Cecropia caterpillar and pupa, which Williams has utilized in his own studies.

Although the prothoracic hormone was first discovered by ligating fly larvae (Fraenkel, 1935), it was soon found to occur in the pupae of Lepidoptera, and the commercial silkworm or the large American cocoon-spinners have been employed in the isolation of the hormone and its assay. Karlson required 500 kg. of *Bombyx mori* pupae, and seriously depleted the European silk production at the time, to achieve the crystallization of 25 mg. of ecdysone. The hormone can also be extracted from shrimp, and may well be a growth or metamorphic hormone for arthropods in general. It is apparently an aromatic alcohol with alpha and beta ketone groups, and the composition $C_{18}H_{30}O_4$.

When the entire central nervous system of a Cecropia pupa is removed, a complete failure of the intersegmental muscles of the adult to develop is the result. When the removal is partial, muscles develop wherever they are innervated, are absent wherever the ganglia have been removed. Implantation of nerve chains into denervated pupae is without effect, so a hormonal influence would seem to be ruled out. The anlage of the muscle must be innervated at precisely the time when ecdysone is released from the prothoracic gland in quantities sufficient to stimulate development and metamorphosis. Not only is this a remarkable inductive effect of nerve on somatic muscle, it also points to the close integration of the prothoracic secretory process with the activity of the nervous system. The integration with respect to muscle development, just described, is however different from that relating to the other organs of the body, for here the integration is through the neurosecretory control of the brain over the prothoracic gland, and is such as to remind us of the recently discovered neurosecretory control of the hypothalamus over the anterior pituitary in vertebrates.

The brain hormone is produced by four clusters of neurosecretory cells in the brain. The hormone passes down the axons of these cells to a neurohaemal organ, the corpus cardiacum, where it enters the blood of the insect. It is particularly interesting that the formation of this hormone depends upon an interaction of the secretion from the two lateral clusters of neurosecretory cells with that from the two medial clusters, for when the brain of a Cecropia caterpillar is cut into pieces the hormonal activity is retained only by pieces that include both a lateral and a medial cluster of cells. [This reminds us of some of the interaction effects already discussed, e.g., the inductive interaction of epithelium and mesenchyme in organ culture.]

Removal of the brain has the same effect as excision of the prothoracic glands. The production of ecdysone ceases. Metamorphosis is blocked, and the insect passes into a state of developmental arrest known as "diapause." Molting and metamorphosis are renewed only when the brain recovers its secretory activity, and for this some exposure to low temperatures is commonly necessary, or at least will hasten the resumption of activity. It is also apparent from experimental studies that the neurosecretory cells are affected by nerve impulses generated by sensory stimuli. In the bug *Rhodnius,* a blood-sucker, the effective stimulus for molting is distension of the abdomen by a blood meal. But even a fed bug will not molt if the nerve cord is transected in the thorax (Wigglesworth). In the Cecropia pupa the neurosecretory cells will, however, after sufficient exposure to low temperature, resume activity even when the brain is completely severed from all connection with other parts of the nervous system. This phenomenon has been studied by Van der Kloot, who finds that the brain of the Cecropia pupa in diapause is electrically "silent." It has no detectable cholinesterase and only a trace of some substrate for that enzyme—probably acetylcholine. Chilling the brain leads to an increase in the substrate. Then, upon the return to higher temperatures, prompt "and apparently inductive" synthesis of cholinesterase is brought about. The brain then resumes both electrical and secretory activity.

G. R. Wyatt reported some studies of chemical aspects of the breaking of diapause by the action of ecdysone. The question examined was whether the breakdown and resynthesis of the cytochromes demonstrated by Williams and his associates to occur in Cecropia was a relatively unique shift of metabolism, or whether it was merely one of many enzymatic changes in activity going on at the developmental crisis. Wyatt found that several aspects of metabolism increase simultaneously in rate, namely, RNA synthesis, utilization of inorganic phosphate, and respiratory rate. The cytochrome shift is thus a part of a general biochemical change occurring toward the cessation of diapause. Wyatt further observed that glycerol builds up steadily in amount in the early portion of diapause, and then, just when cytochrome levels are restored, the glycerol level drops precipitously. This is held to signify either that anaerobic conditions prevail in metabolism during diapause or that there is a deficiency of electron acceptors. The latter interpretation would support Williams' view that the cytochromes are the limiting factors of metabolism in diapause.

Removal of the corpora allata from caterpillars, even those of early instars, leads to precocious metamorphosis. The secretion of these glands is thus shown to be necessary to a continuation of the larval phase of existence. What is more, the implantation of extra corpora allata into young caterpillars forces them into extra larval instars—as many as six—and gigantic growth. In pupae the juvenile hormone is normally absent. If injected into a pupa, retention of some pupal features in the emerging adult, or even a second pupal stage, may be produced. The juvenile hormone extracted from Cecropia is active on other families and orders of insects. It is very stable, having been recovered from 8-year-old dried museum specimens. It is a non-saponifiable lipid.

*Liver Growth and Regeneration*

This subject was included because the elucidation of systematic chemical factors controlling the regeneration of the liver may well have a very general bearing on developmental growth and differentiation. André D. Glinos outlined the evidence for the presence of a chemical system of self-regulation and control in the regenerating liver, a system which, so to speak, informs the cells of the state of the total mass to which they belong, maintains the mass in a state of equilibrium, and regulates the return of the mass to that state after any disturbance. Thus, after partial hepatectomy, growth, proliferation of cells, and cell differentiation set in until the loss is repaired and the original mass of the organ restored. Glinos believes the active regulatory agents of this system are the plasma proteins, which are formed by the liver and exert a feedback effect upon it whenever their concentration is disturbed.

Tissue cultures of liver grown in fresh plasma or in plasma taken 24 hours after partial hepatectomy show a greater outgrowth of connective tissue and epithelium and a shorter latent period in the latter. Further evidence of some systemic chemical effect upon liver growth was obtained by parabiotic experiments in which one of the conjoined animals was hepatectomized. In these cases mitotic activity was stimulated in the liver of the nonoperated member of the pair.

Other experiments proved that the plasma factor influencing liver regeneration is inhibitory. Thus in tissue cultures of liver cells a high concentration of normal serum resulted in less outgrowth than a low concentration of normal serum. This observation led to experiments designed to test the effect of lowering or of increasing the concentration of serum constituents. The former was accomplished by plasmapheresis, the latter by restricting the fluid intake, in rats. In the former case, rats partially hepatectomized and deprived of drinking water for 64 hours showed a significant inhibition of mitotic activity in comparison with controls. The following simple hypothesis of self-regulation was then formulated: the inhibitory factor is synthesized in the liver itself, since hepatectomy reduces its concentration in the blood; lowering of the inhibitory factor releases the growth of the liver cells; as these proliferate the mass of the liver approaches the original size, synthesis of the inhibitory factor is consequently increased, and its extracellular concentration gradually rises until the growth of the liver is checked. The existing data accord with this hypothesis. There is in fact a decline in the rate of growth of the normal liver during development, as the ultimate size in relation to total body weight is approached, just as in the regenerating liver.

In the effort to identify the inhibitory factor chemically, a study was made of the time sequence in the initiation of increase in various components of regenerating liver tissue. Synthesis of cytoplasmic RNA and protein was found to begin before that of DNA and nuclear RNA; and to reach a maximum sooner, in the case of cytoplasmic RNA. Low molecular weight RNA increases faster and reaches a higher maximum than heavier (microsomal) RNA; and this is thought to mean that protein synthesis from activated amino acids is particularly promoted. Especially interesting was the discovery, by means of a histochemical study of the cytoplasmic ribonucleoprotein in the cells of the regenerating liver, that there is a topographical gradient in the order of appearance and the extent of cytological changes in the liver cells. The disappearance of the basophilic bodies associated with RNA starts in the periportal cells of each liver lobule within 30 minutes of the operation, and moves, over the course of 8 hours, toward the cells bordering the central vein. Later, the reverse change occurs, starting with the centrolobular cells and moving toward the periphery of the lobule. Mitotic activity also starts at the periphery and moves toward the center of the lobule.

It is known that the plasma albumins and alpha and beta globulins are synthesized in the liver, whereas gamma globulins are not. In accordance with this, it was found that albumins and fibrinogen undergo a decrease in concentration after partial hepatectomy; the other plasma proteins increase in concentration. Albumins may consequently be the postulated inhibitory factor of liver growth.

Glinos has marshaled a considerable amount of further evidence to support his hypothesis that the liver-synthesized plasma proteins regulate the regenerative growth process in the liver. He points out that transfer of water from the liver capillaries to the interstitial fluid should be such as to decrease the periportal concentration of proteins in the interstitial fluid and to increase the concentration of proteins at the centrolobular end. Moreover, an increase in portal pressure, such as in fact occurs after partial hepatectomy because of the reduction in volume of the vascular bed of the liver, ought to decrease immediately the periportal concentration of proteins in the interstitial fluid, and to intensify the existing gradient of protein concentration. Evidence of the predicted changes has been obtained from experiments in which plasma has been diluted, and the decrease in protein concentration was followed by cytoplasmic changes in ribonucleoprotein, increased incorporation of $S^{35}$-methionine into proteins, and a great increase in mitotic activity. As expected, an increase in portal vein pressure does increase regeneration in the liver. There are, however, no direct experiments to test the effect of intravenous administration of selected plasma protein fractions on liver regeneration, or on growth during normal development, and until such experiments have been performed the evidence remains inconclusive.

The fuller description of the postulated regulatory feedback cycle governing liver growth and regeneration may then be restated provisionally as follows: extracellular protein exerts an inhibitory effect upon the synthesis of cytoplasmic RNA and intracellular protein; release of the inhibition through diminution in extracellular protein concentration stimulates cytoplasmic synthesis of RNA and protein; these in turn stimulate synthesis of nuclear RNA and DNA; and accelerated mitotic activity results from the preceding step, until total cell number leads to synthesis of extracellular proteins sufficient once more to inhibit cell proliferation.

## Conclusion

The end of this symposium, said Paul Weiss in summation, shows us clearly that "we are at the beginning, and not near the end, of the process of understanding development." Many useful techniques for the purpose have been developed; many new questions are being asked and new approaches to problems ventured upon. Yet many fundamental problems of development can scarcely be couched in biochemical terms at all at the present time; the problem of organization, for example—the relation between structure and chemistry. Or the problem of differentiation—is it generally sequential transformation, step-by-step? Or is it more often a dichotomous selection of alternative pathways? And how does the latter, especially, fit into the biochemical pattern of things? The nature of intracellular movement, the problem of surface specificity, the mysterious affinities of cells and their recognition of their own histological type in preference to their own genotype—we still know little of such phenomena. But we stand obviously on the threshold of great advances. The synthesis of embryology, genetics, immunology, and biochemistry is well under way.

# AUTHOR INDEX

## A

Abercrombie, M.: 318
Afzelius, B.: 170
Allen, R. D.: 17, 68, 69, 70, 71, 72, 171, 212, 414
Allfrey, Vincent: 94, 101, 594
Anderson, Minnie L.: 296

## B

Boell, E. J.: 376
Bodian, D.: 71
Borghese, E.: 704, 811
Billingham, R. E.: 568, 569, 575, 592, 593, 594
Brachet, J.: 70, 102, 173, 210, 213, 258, 625, 678, 679
Brown, G. W., Jr.: 314, 495, 523, 524, 626

## C

Clarke, W. M.: 546
Clement, A. C.: 85, 92
Cohen, P. P.: 72, 86, 100, 495, 524, 525, 702
Cohen, S.: 665, 677, 678, 679
Cohn, Melvin: 255, 315, 458, 571, 678
Colwin, A.: 68
Coulombre, A. J.: 454, 679
Cowden, Ronald R.: 88, 152, 153, 404, 455

## D

DeHaan, Robert L.: 339, 375, 376, 643
DeLanney, Louis E.: 562

## E

Eagle, H.: 624, 626
Eakin, Robert E.: 514
Ebert, James D.: 71, 87, 526, 569, 570, 571
Edds, Mac V., Jr.: 157, 169, 170, 171, 172

## F

Fisher, James R.: 514
Flexner, L.: 415
Foulds, Leslie: 680, 700, 702

## G

Gall, Joseph G.: 103, 152, 153, 156, 317
Gey, G.: 841

Glinos, André D.: 813, 840, 841
Gluecksohn-Waelsch, S.: 793
Gorini, Luigi: 469
Grant, P.: 72, 255, 316, 644

## H

Hadorn, Ernst: 100, 172, 313, 314, 525, 779, 791, 792, 793
Hamburger, V.: 778
Harris, Morgan: 70, 592, 596, 623, 624, 625, 626
Hartman, P. E.: 679, 791
Hayashi, Y.: 261
Herrmann, Heinz: 329, 626, 645
Holtfreter, J.: 253, 260, 262, 374, 777
Hughes, W. L.: 136

## K

Kalckar, H.: 791, 792
Kavanau, J. Lee: 433
Kostellow, Adele B.: 448

## L

Laufer, H.: 572
Lehmann, Fritz E.: 73, 85, 86, 87, 88, 100, 210
Lenhoff, H.: 172
Levi-Montalcini, Rita: 646, 678, 679
Lipmann, F.: 455, 624

## M

Maas, Werner K.: 469, 490, 679, 792
Magasanik, Boris: 485, 493
Markert, C. L.: 3, 67, 68, 70, 71, 86, 100, 156, 171, 314, 454, 623, 624, 676, 677
McElroy, W. D.: 101, 679, 777, 840, 839
Mirsky, A. E.: 94, 99, 100, 101, 102, 212, 677
Mohler, W. C.: 840

## N

Nace, G. W.: 90, 546
Nason, A.: 523
Niu, M. C.: 256, 257, 258, 259, 262, 263, 316, 625
Novikoff, A. B.: 67, 86, 88, 211, 212, 255, 256, 414, 415, 570

## P

Pannabecker, R.: 774
Pappenheimer, A. M.: 315, 677
Pasteels, J. J.: 71, 85, 381, 414, 415, 841
Popp, R.: 524
Potter, Van: 490, 700, 703, 812
Price, Dorothy: 774

## R

Rudnick, D.: 455

## S

Saunders, John W., Jr.: 239, 593
Schneiderman, H.: 811
Shen, S. C.: 416, 454, 455
Spiegelman, S.: 99, 153, 315, 490, 642, 643
Spratt, N. T., Jr.: 88, 454, 629, 642, 643, 644
Stearns, R. N.: 448
Steinberg, M.: 169, 170
Strehler, B.: 262, 792
Suskind, S.: 792
Sussman, M.: 264, 313, 314, 315, 316, 317, 645, 678, 811, 812
Swift, H.: 101, 153, 174, 210, 212, 213, 317, 414

## T

Trinkaus, P.: 376

## U

Umbarger, H. E.: 492, 703

## V

Vincent, W. S.: 153
Vogel, H. J.: 479

## W

Weber, R.: 86, 260, 414, 524, 791
Weiss, Paul: 69, 89, 171, 259, 260, 316, 454, 571, 644, 677, 678, 700, 840, 843
Wilde, C. E.: 315, 316, 840, 841
Williams, C. M.: 794, 811, 812
Willier, B. J.: 69, 86, 87, 93, 594
Wright, Barbara E.: 296, 315, 316, 317
Wright, S.: 570
Wyatt, G. R.: 807

## YZ

Yamada, T.: 89, 217, 255, 256, 257, 260, 262, 812
Zwilling, E.: 375, 376

# SUBJECT INDEX

## A

Acquired tolerance: its role in development, 575-591; of homologous cells, 582
*Acricotopus:* 112
Acrosome, activating function of: 42
Acrosome reaction: 18, 62
Activation, chemical theories of: 57
Adaptive enzyme synthesis, in embryos: 6
Adenosine triphosphatase, nerve growth factor: 670
Adhesion mechanism and cell membrane structure: 354
Adhesiveness between different cytoplasmic regions: 76
*Aerobacter aerogenes:* 485
Aerobic glycolysis during logarithmic growth of chick skeletal muscle fibroblasts: 614
Aerobic metabolism, acquisition of: 299
Aggregateless mutant stocks of *D. discoideum:* 289
Aggregateless mutant responders: 290
Aggregation: 323; cell adhesion during, 285; cellular interactions during, 281; in developmental cycle, position of, 267; by *Dictyostelium discoideum,* time lapse photomicrographs, 265; initiator cell, 270; nature of chemotactic substance in, 299; a sexual process? 279; synergistic, 286
Alcian blue reaction: 390
Alkaline phosphatase: 390
Allophenes: 787
*Amblystoma pancreas:* 190; nuclear margin of, 192; under electron microscope, 189
*Amblystoma punctatum:* 429
*Amblystoma tigrinum:* 182; RNA levels of refed, 183
Ameboid cells: movements of individual, 340; orientation of, 343
L-Amino acid oxidase, nerve growth factor: 670
Ammonia in allantoic fluid during development: 516
Ammonia excretion: 515; by tadpoles, 499
Amphibian egg jelly: 32
Antibody-forming capacity, transfer by nucleoprotein fractions of: 531
Antigenic stimulus in homograft immunity: 527
Antimitotic substances, action of: 755
Antiserum production and utilization: 547

*Arbacia:* 31, 34, 49, 390
*Arbacia* eggs: 387
Arginase: 496, 502, 504
Arginine, biosynthesis of: 470
Arginine auxotroph, enzymatic synthesis of: 475
Arginine synthetase, in metamorphosing tadpoles: 505, 506
Argininosuccinate cleavage enzyme: 496
Argininosuccinate synthetase: 496
Aspartic-glutamic transaminase: 496
*Asterina pectinifera:* 24, 40
*Asterias:* 19, 21
*Asterias forbesii:* 40
ATP, DNA essential for synthesis of: 95
Autolytic artifact: 446
Autonomous and dependent morphogenesis in mosaic egg of *Tubifex:* 73-83
Autophenes: 787
Autoradiograms: of bean roots, 151; of *Escherichia coli,* 144; HeLa cells, 146; of mouse gut with tritiated thymidine, 147; of plant chromosomes, 141
Auxotroph, arginine, enzyme synthesis in: 475

## B

Balbiani rings: 109
*Barnea:* 71
Basophilia, cytoplasmic particulates and: 174-208
Basophilia, sarcoplasmic from regenerating mouse muscle: 205
Basophilic body in cytoplasm of crayfish *Cambarus virilis:* 206
Basophilic components: 177
Bean roots, autoradiograms of: 151
Bilateral symmetry: 393, 394
Biochemical continuity: 75
Biochemical patterns and autolytic artifact: 433-447
Biochemical properties of node center: 640
Biophysical aspects of development: 10
Biosomes: 75
Biosynthesis of urea of metamorphosing tadpoles: 495-512
Biosynthetic enzymes, feed-back control of formation of: 469-478
Blastoderms to non-nutrient and various nutrient media: 632

Blastomeres of 21-cell embryo of *Tubifex*, differences in particulate populations: 78
Blastomeres, isolated, respiration or enzymatic activities of: 382
Bonemarrow extract for induction, scheme of fractionation of: 226
Brain, control of: 802
Brain hormone: 800; role of, 802
Breast, neoplasia of: 688
Brunner's glands: 68

C

*Cambarus virilis*, basophilic body in cytoplasm of: 206
Carbamyl phosphate synthetase: 496, 502, 504; increase during metamorphosis, 501
Carbohydrate requirements: for brain development, 633; for development, 632; for heart development, 633
Carcinogens, incomplete: 681
Carcinogenesis, chemical, in skin of rabbits and mice: 681
Cardiac development, effects of limited cell dissociation: 357
Cell division, induction by plasmapheresis: 817
Cell locomotion: 323
Cell membranes: 176
Cell membrane structure, and mechanism of adhesion: 354
Cell migration and morphogenetic movements: 339, 370
Cell nucleus, role of in development: 94-98
Cell proliferation, sequences of events in: 145
Cell-specific pattern of gene action: 785
Cellular differentiation, chemical concepts of: 3-14
Cellular differentiation and chromosomal differences: 132
Cellular interactions during aggregation: 281
Cellular and tissue interactions in development: 215
Centrifuged egg: hyaloplasmic, 385; lipid, 385; mitochondrial, 385; vitelline, 385
Central nervous system, cholinesterase in: 421
*Chaetopterus*: 32, 42, 71, 81, 390
Chemical carcinogenesis in skin of rabbits and mice: 681
Chemical concepts of cellular differentiation: 3-14
Chemical control of development: 629-641

Chemical control mechanism, endogenous in development: 641
Chemical environment, role of in differentiation and maintenance: 7
Chemical theories of activation: 57
Chemistry, and structure of egg surface: 29
Chemotactic complex: 281; in *D. discoideum*, 283
Chemotactic substance, nature of in aggregation: 299
Chick embryo, telophase nucleus from: 201
Chick retina, histochemical localization of cholinesterase in: 424
*Chironomus*: 104, 109, 113
Cholinesterase: accumulation at motor endplates, 430; in cellular differentiation, 421; in central nervous system, 421; histochemical localization in chick retina, 424; localization, 429; in myoneural junctions, 425; in neuroretina, 422
Chromatid: 105; nature of, 128
Chromatophore migration: 340
Chromosomal differentiation: 12, 103-134; and cellular differentiation, 132
Chromosomal replication: 136
Chromosomes: lampbrush, 115; functional aspects of, 124
Chromosomes, polytene, structure of: 106
*Chromonema*, definition of: 105
*Ciona*: 48
Cleavage pattern of *Tubifex*: 80
*Clupea harengus*: 391
*Clypeaster*: 24
Collagen content in sclera and cornea of developing chick embryo: 332
Collagen fibrils, mineralization of: 166
Collagen fibrogenesis: 162
Colloids for peivitelline space: 62
*Comanthus japonicus*: 41
Control of brain: 802
Control mechanisms in development: 626
Coordinated movements of cell groups and layers: 350
Cortex of mature egg: 33
Cortex of maturing egg of frog *Rana pipiens*, electron micrograph: 42
Cortex, stable: 41
Cortical granular disposition of before and after fertilization: 51
Cortical granule breakdown: 62
Cortical granules, electron micrograph: 43
Cortical material isolated from *Strongylocentrotus purpuratus* eggs: 39

# SUBJECT INDEX

Cortical reaction: duration of latent period, 45; latent period before, 44; in membrane elevation, 49; photo of *Psammechinus* egg, 46; in relation to processes leading to syngamy, 59
Cortical response, electrically introduced, diagram: 48
*Cyclops:* 391
Cytochemical findings of slug oocytes: 410
Cytochemistry: the cytoplasm, 383; of fertilized egg, 381-399; of oocyte, 384; study of the slug oocyte, 404-413
$\alpha$-cytomembranes: 181
$\beta$-cytomembranes: 181
$\gamma$-cytomembranes: 181
Cytoplasmic particles: isolation of, 382; serum albumin precursor in, 531
Cytoplasmic particulates and basophilia: 174-208; immunochemical studies with, 530
Cytoplasmic pattern, epigenetic formation: 74

## D

*Dendraster:* 26, 52
Deoxyribonuclease, nerve growth factors: 670
Deoxyribonucleic acid, mechanism of duplication: 140
Dependent and autonomous morphogenesis in mosaic egg of *Tubifex:* 73-83
Development: biophysical aspects of, 10; carbohydrate requirements for, 632; chemical control of, 629-641; coordination of, 796; differential effects of metabolic inhibitors, 636; endogenous chemical control mechanism, 641; enzymatic patterns during, 416-431; and genes, 4; genetic control of, 8; genic action along time axis, 779; hypothesis of, 13; initiation of, 17-62; insect, metabolic control of, 807-810; minimal environmental requirements for, 635; neoplastic mechanism of, 690; normal, possible role of tolerance in, 586; role of acquired tolerance in, 575-591; role of cell nucleus in, 94-98; sensitivity to adverse environmental conditions, 633
Development of embryos and larvae, influence of adult amphibian spleen—an immune response? 562-568
Developmental activity with exogenous nutrient requirements: 640
Developmental processes, role of genes in: 779-789
Diakinesis chromosomes of slug oocyte: 407

*Dictyostelium discoideum:* 298; time lapse photos of aggregation by, 265
Differential gene activation, in differentiation: 11
Differentiation: cellular and chromosomal, 132; cellular, chemical concepts of, 3-14; changes in enzyme activities at progressive stages of, 304; cholinesterase in cellular, 421; chromosomal, 12; differential gene activation in, 11; enzyme changes during, 301; and maintenance, role of chemical environment in, 7; multicellular, inhibitors of, 300; organ, in culture, 704-767; relative enzymatic activities, 305
Differentiation and growth, rate of: 729
Differentiation and growth in vitro of a glandular rudiment: 716
Differentiation in vitro of a mammalian ovary: 722
Differentiation of a population of *Escherichia coli:* 458-467
Digestive and respiratory apparatus: 714
Diphosphopyridine nucleotidase, nerve growth factor: 670
Diphosphopyridine nucleotide (DPN), synthesis in nucleus: 97
*D. discoideum:* 268; chemotactic complex, 283; developmental cycle, 266
DNA: essential for aerobic synthesis of ATP, 95; and histones in nucleus, 100; and synthesis of RNA in nucleus, 98
DNA content: function of nuclear volume, 109; in sclera and cornea of developing chick embryos, 331
DPN, changes in fertilization: 441
*D. purpureum:* 268
Duck syrinx, hormone in development of: 745
Duplication: in bacteria, 143; in plant cells, 140
Duplication of deoxyribonucleic acid, mechanism of: 140

## E

Echinarachnius: 40, 69
*Echinocardium cordatum:* 19
Ectoderm and mesoderm: embryonic scheme of interaction between, 252; relationships in mature feather, 241
Egg: cortex of mature, 33; of *Psammechinus miliaris,* 38

# 928  THE CHEMICAL BASIS OF DEVELOPMENT

Egg nucleus, migration of: 26
Egg surface, structure and chemistry of: 29
Embryogenesis: mitochondrial enzymes in, 418; morphogenetic reaction, systems during, 82
Embryonic action of genes, early: 782
Embryonic cells, dissociated: enzyme induction in, 448-453; uptake and transfer of macromolecules by, 542
Embryonic extracts: 602
Embryonic induction: 6, 217-237
Embryonic nervous system, hormonal effects on: 649
Embryonic organs of different species, xenoplastic associations: 742
Embryonic origin of feather genes: 241
Endoplasmic activation: 62
*Entosphenus wilder:* 41
Environmental conditions, sensitivity to adverse: 633
Environmental requirements, minimal for development: 635
Enzymatic activities of isolated blastomeres: 382
Enzymatic activity of fragments of eggs: 382
Enzymatic changes during differentiation: 301
Enzymatic patterns, during development: 416-431
Enzyme: adaptive synthesis in embryos, 6; extrinsic supply, 786
Enzyme activities, changes in at progressive stages of differentiation: 304
Enzyme formation: in development, role of repressers, 479-483; induction of, 479; repression of, 479
Enzyme induction in dissociated embryonic cells: 448-453
Enzyme localization: 303, 304
Enzyme repression: 481
Enzyme synthesis: in arginine auxotroph, 475; in chemostat, 474
Enzymes, mitochondrial, in embryogenesis: 418
Epidermal neoplasia: 686
Epigenetic formation of cytoplasmic pattern: 74
Ergastoplasmic systems: 399
*Escherichia coli:* 485
Esterases, ontogeny of: 10
Evolution of cell strains: 607
Exchanges between cells: 318-327
Extrinsic supply of enzyme: 786

F

Familial intestinal polyposis: 688
Feather germs, embryonic origin: 241
Feather-germ loci origin, factors determining: 242
Feather tract, progressive organization: 243
Feather-tract mesoderm, specific inductive action: 247
Feed-back control of formation of biosynthetic enzymes: 469-478
Feed-back mechanism: 480
Fertilization: DPN changes in, 441; folic-acid-replacing activity for *S. faecalis-R,* 443; membrane in *Paracentrotus lividus,* 30; after nicotinic-acid groups activity changes, 439; pantethine changes in, 442; pantothenic acid changes in, 442; of *Psammechinus* eggs, 47; rate, acceleration in eggs of *Psammechinus* microtuberalatus, 31; TPN changes in, 441; after valine during early cleavage of *Spisula solidissima,* 438
Fertilization impulse: 62; nature of propagated, 53; rate of propagation of, 55
Fertilized egg: in capillary, diagram of a partially, 60; cytochemistry of, 381-399; function of new surface layers of, 52
Fertilizin: 19, 29
Fetal rat, sex differentiation in: 774-777
Fibrillar aggregates: 164
Fibroblasts: of chick skeletal muscle, glycolysis and growth rates, 616; locomotion of, 319
Fibrogenesis, collagen: 162
Fluorescence specificity: 551
Fluorescence technique: 548
Fluorescent anti-adult frog sera: 91
Folic-acid-replacing activity for *S. faecalis-R* after fertilization: 443
Fumarase: 496

G

*Gadus morrhua:* 391
$\beta$-galactosidase in *E. coli:* induction of, 463; schema to illustrate pathway, 463
$\beta$-galactosidase formation in *E. Coli:* 458-467
$\beta$-galactosidase induction: anaerobic shock, 460; glucose effect on induction of, 459; by methyl-$\beta$-D-thiogalactoside, 462
Gene action, cell-specific and organ-specific pattern: 785

## SUBJECT INDEX

Gene activation, differential in differentiation: 11
Gene function: pink eye in house mouse, 9; tyrosinase in, 9
Genes: and development, 4; early embryonic action of, 782; for pink eye, 9; role of in developmental processes, 779-789
Genic action along time axis of development: 779
Genetic control of development: 8
Genetical problems in development: 756
Glandular rudiment, growth and differentiation of: 716
Glutamic dehydrogenase: 496
Glycine-1-C$^{14}$ uptake into a collagen fraction of corneal stroma of developing chick: 335
Glycolysis and growth: 610
Glycolysis and growth rates in cultures of chick skeletal muscle fibroblasts: 616
Glycolysis in normal and malignant cells: 617
Glycoprotein: 32
Golgi apparatus: 175
Ground substance: 158
Growth and differentiation: rate of, 729; in vitro of a glandular rudiment, 716
Growth factor: from venom, enzymatic activities of, 670; in venom of moccasin snake, 667
Growth and glycolysis rates in cultures of chick skeletal muscle fibroblasts: 616
Growth of liver, molecular picture of mechanism controlling: 833
Growth in vitro, energy sources for: 610
*Gryphaea:* 71

### H

*Haliotis:* 21
HeLa cells, autoradiogram of: 146
*Hemicentrotus:* 19
Hepatectomy, partial, plasma proteins: 825
Histidine biosynthesis: 487; metabolic regulation of, 485-489
Histidine precursor, biosynthesis: 486
*Holothuria atra:* 21, 24, 68; photos of sperm entry, 22
Homograft immunity, antigenic stimulus in: 527
Hormone, brain: 800
Hormone in development of duck syrinx: 745

Hormonal effects on embryonic nervous system: 649
Hormonal regulation of insect metamorphosis: 794-805
Hyaline layer: 62; luminous, 39
Hyaloplasmic reticulum: 78
Hyaluronidase, nerve growth factor: 670
*Hydroides:* 18, 20
Hypothesis of development: 13

### I

*Ilyanassa:* 79, 397
Imidazole glycerol phosphate, biosynthesis of: 486
Immune reaction: bone marrow transplantation to irradiated animals, 534; in chick embryo and larval salamander, 536; graft-against-the-host, 534; in newborn mice, 535
Immunochemical analysis of development: 526-543
Immunochemical studies: 598; with cytoplasmic particulates, 530
Immuno-histological localization of antigens in optic cup and lens of *Rana pipiens:* 546-560
Independent progression of tumors: 684
Induced myotomes and notochord: 231
Induction: 324; of aggregates by I-cells, 278; embryonic, 6; of enzymatic form, 479; mesenchymo-epithelial, 733; of notochord and mesoderm, 230; scheme for fractionation of bone marrow extract for, 226; sequence of regional types of, 234; substrate, 450; of tolerance, 577; of tract-specific feather characteristics, 241
Induction action, time of: 248
Induction of $\beta$-galactosidase: anaerobic shock, 460; glucose effect on, 459; by methyl-$\beta$-D-thiogalactoside, 462
Inductive ability: of liver pentose nucleoprotein (Hayashi); pepsin effects on, 223; of pentose nucleoprotein from liver (Hayashi) trypsin effects on, 224; ribonuclease effect on pentose nucleoprotein on, 222
Inductive effects: of bone marrow fraction, 227; testing of, 219
Inductive effects, regional: archencephalic, 218; deuterencephalic, 218; spino-caudal, 218; trunk-mesodermal, 218
Inductive phenomena, specificity of: 235
Inductive specificity: in origin of integu-

mentary derivatives in the fowl, 239-253; in origin of regional differences, 250
Initiation of development: 17-62
Initiator cell, aggregation: 270
Innervation patterns: 429
Insect development, metabolic control of: 807-810
Insect metamorphosis, hormonal regulation of: 794-805
Interaction between different organs: 738
Intrablastematic reaction patterns: 82
Intrablastematic reaction systems: 82
Isolated organs, morphogenesis of: 708

J

Jelly layer: 29
Juvenile hormone: 804
Juvenile state: 803

K

Krebs ornithine-urea cycle: 496

L

Lactic acid, production in tissue cultures: 611
Lampbrush chromosomes: 115; diagrammatic representation, 116; functional aspects of, 124; nature of functional unit, 131; postulated structure of, 121; structure of, 117
Leptotene chromosomes of slug oocyte: 407
Liver growth: action of systemic chemical factors, 814; molecular picture of mechanism controlling, 833; parabiosis, 814; plasma proteins and control of, 827; and regeneration, 813-837
Liver mitosis, effect of injections of sera: 816
Liver, regenerating, inhibition of cell division: 818
Liver regeneration, some chemical features of: 820
Luminous hyaline layer: 39
*Lytechinus:* 25, 53

M

Macromolecules, mechanism of entry of: 601
Male hormone action of chick Müllerian duct: 749
Malic dehydrogenase: 496
Malignant cells, and normal, glycolysis in: 617

Mammalian ovary, differentiation in vitro: 722
Mass cell movements: 345
Mechanism of duplication, diagrammatic representation of: 143
Mechanism of tolerance: 590
*Megathura:* 21
Membrane development: 62
*Merogonic* hybrids: 781
Mesenchymo-epithelial induction: 733
Mesoderm-induction by bone marrow, substance responsible for: 224
Mesoderm and ectoderm, embryonic, scheme of interaction between: 252
Mesoderm and notochord, induction of: 230
*Mespilia:* 24
Metabolic control of insect development: 807-810
Metabolic inhibitors, differential effects of development: 636
Metabolic patterns, ontogeny of: 5
Metabolic regulation of purine interconversions and histidine biosynthesis: 485-489
Metamorphosing tadpoles, urea biosynthesis in: 495-512
$S^{35}$-Methionine, incorporation into liver protein: 830
Methyl-$\beta$-D-thiogalactoside: $\beta$-galactosidase induction by, 412; structure of, 463
Mineralization of collagen fibrils: 166
Mitochondria: 174
Mitochondrial enzymes in embryogenesis: 418
Mitochondrial patterns, signification of: 396
Mitosis: 324
Moccasin snake, growth factor in venom: 667
Morphodynamics: blastematic, 82; intracellular, 82
Morphogenesis: of isolated organs, 708; relative enzymatic activities during, 310
Morphogenetic effects: of kidney pentose nucleoprotein, 220; of liver pentose nucleoprotein, 221
Morphogenetic movements and cell migration: 339-370
Morphogenetic reaction systems during embryogenesis: 82
Morphological behavior: 739
Mosaic eggs: 393
Mosaic-egg of *Tubifex,* dependent and autonomous morphogenesis: 73-83
Mucopolysaccharide molecules: 160

## SUBJECT INDEX

Mucopolysaccharide-protein complexes: 161
Mucopolysaccharides: in ground substances, 159; types of, 159
Multicellular differentiation, inhibitors of: 300
Muscles, formation of: 799
*Mytilus edulis:* 19, 21, 23, 397
Myogenesis, amphibian, axial muscles during: 429
Myoneural junction, cholinesterase in: 425
Myotomes, and notochord, induced: 231
Myzostoma: 397

### N

Nadi reaction: 388
Neoplasia: of breast; 688; epidermal, 686; in man, 686; in mice, 683; in urinary bladder, 687; of uterine cervix, 688
Neoplastic development: 680-698; in man, 689; mechanism of, 690
Neoplastic state, subthreshold: 681
*Nereis limbata:* 19, 32, 41, 68
Nerve fiber outgrowth, nutritional requirements for: 672
Nerve growth: in adult organism, factors stimulating, 650; chemical stimulation of, 646-662; in embryo, factors stimulating, 648
Nerve growth factor: adenosine triphosphatase, 670; L-amino acid oxidase, 670; deoxyribonuclease, 670; diphosphopyridine nucleotidase, 670; hyaluronidase, 670; 5-nucleotidase, 670; phospholipase A, 670; phosphodiesterase, 670; protease, 670; ribonuclease, 670; site of action, 671; submaxillary gland, 675
Nerve growth-promoting factors, assay and distribution of: 665
Nerve growth-promoting proteins: 665-676; isolated from sarcomas, snake venom and salivary glands, 652
Neuroretina, cholinesterase in: 422
Nicotinic-acid groups activity changes after fertilization: 439
Nitrogen excretion: in developing chick embryos, patterns of, 514-522; by tadpoles, 499, 500
Node center, biochemical properties of: 640
Notochord: and mesoderm, induction of; 230; and myotomes, induced, 231
Nuclear movements: 62

Nucleic acids synthesis, effect of venom factor: 673
Nucleolus, volume of: 409
Nucleoprotein fraction, effect on growth of chick skeletal muscle fibroblasts: 605
Nucleoprotein: kidney pentose, morphogenetic effects of, 220; liver pentose, morphogenetic effects of, 221; morphogenetic effects of subfractions of, 221
Nucleoproteins, from adult tissues: 604
5-Nucleotidase, nerve growth factor: 670
Nucleus, migration of egg and sperm: 28

### O

Ontogeny of esterases: 10
Ontogeny of metabolic patterns: 5
Oocyte of the newt, autoradiograph of: 126
Oocyte organization, transformation of: 73
Oocytes of helicid snail *Otala lactea:* 198
Oocytes from *Theridion tepidariorum:* 200
Ooplasmic segregation: 82
Organ differentiation in culture: 704-767
Organ transplants, transfers from: 600
Organ-specific pattern of gene action: 785
Origin of feather-germ loci, factors determining: 242
Origin and structure of intracellular matrix: 157-167
Ornithine transcarbamylase: 496, 502, 504
Ornithine transcarbamylase formation: 477
Ornithine transcarbamylase reaction: 471
Ornithine transcarbamylase, tadpole: 503, 504
Ornithine-urea cycle, Krebs: 496
*Otala lactea,* helicid snail, oocytes of: 198
Oxidation of glucose, effect of venom factor: 673

### P

Pachytene chromosomes, of slug oocyte: 407
Pancreas cells, and *Triturus* liver, RNA amounts and relative specific activity in: 185
Pantethine, changes in fertilization: 442
Pantothenic acid, changes in fertilization: 442
*Paracentrotus:* 24; eggs, 387; fertilization membrane in, 30, 34
*Pelodytes:* 48
Pentose nucleoprotein: kidney, morpho-

genetic effects of, 220; (Hayashi) liver effects of pepsin on inductive ability, 223; from liver (Hayashi) effects of trypsin on inductive ability, 224; liver, morphogenetic effects of, 221; morphogenetic effects of subfractions of, 221
Pepsin, effects of inductive ability of liver pentose nucleoprotein (Hayashi): 223
Perforatorium: 20
Pigment granules in *Arbacia* eggs, schematic representation of behavior of: 61
Pigment migration: 62
*Phallusia*: 48
Phosphodiesterase, nerve growth factor: 670
Phospholipase-A, nerve growth factor: 670
*Pomotoceros*: 23
Polarity: 393
Polar lobe of *Spiralia*: 396
Polar plasms: distribution of, 79; role of early cleavage in differential distribution, 77
Polyposis, familial intestine: 688
Polyspermy: 62
Polytene chromosomes: functional aspects, 111; structure of, 106
Polytenization: 102
Primary activity vs. manifestation of genes: 783
Primary membrane: 35
Processes leading to syngamy: 18
Progression of tumors: 684
Protease, nerve growth factor: 670
Protein, a nerve growth-promoting: 665-676
Protein, sources of intracellular: 596
Protein synthesis, effect of venom factor: 673
Proteins, uptake by avian ovum: 598
Protein formation in sclera and cornea of chick embryos: 329-337
Protein growth factors in tissue culture: 602
Proteins labeled with radioisotopes, use of: 599
Protein uptake: evidence of, 598; by kidney cells, 600
Prothoracic glands: 797
Prothoracic gland hormone: 797
*Psammechinus*: 26, 34, 36, 47, 50, 52
*Psammechinus microtuberalatus*, acceleration of fertilization rate in eggs of: 31
*Psammechinus miliaris*: 29; unfertilized jelly-free egg of, 38
*Pseudocentrotus*: 24
Purine interconversions: 487; metabolic regulation of, 485-489

Purine nucleotides, cyclic interconversions of: 486
*P. violaceum*: 268

R

*Rana catesbeiana*: 498; arginine synthesis in metamorphosing tadpoles of, 505; tadpoles, urea cycle enzyme during development, 500
Rat hepatic cells: 202
Reaggregation of dissociated cells: 750
Regenerating liver, inhibition of cell division: 818
Regeneration: liver, some chemical features of, 820; and liver growth, 813-837
Regulation: 326; in explants, 728
Regulative eggs: 393
Release and uptake of substances by cells: 596-620
Repression of enzyme formation: 479
Respiratory activities of isolated blastomeres: 382
Respiratory activity of fragments of egg: 382
Respiratory and digestive apparatus: 714
Ribonuclease: effect on inductive ability of pentose nucleoprotein, 222; nerve growth factor, 670
Ribonucleoprotein changes, induction of cytoplasmic: 829
RNA amounts and relative specific activity in *Triturus* liver and pancreas cells: 185
RNA levels of refed *Ambystoma tigrinum*: 183
RNA synthesis in nucleus: 98
Role of chemical environment in differentiation and maintenance: 7

S

*Sabellaria*: 68, 391, 397
*Sabellaria vulgaris*: 32
Salivary gland chromosome of *chironomus*: 115
Salivary glands, nerve growth-promoting protein isolated from: 652
*Salmonella typhimurium*: 485
Sarcoma cells, locomotion of: 322
Sarcomas, nerve growth promoting proteins isolated from: 652
Sarcoplasmic basophilia, from regenerating mouse muscle: 204
Scales and claws, specific induction: 249

# SUBJECT INDEX

*Sciara:* 104
Sea urchin egg: 29
Serological specificity: 549
Serum albumin precursor in cytoplasmic particles: 531
Serum proteins, utilization of: 605
Sequences of events in cell proliferation: 145
Sex differentiation in fetal rat: 774-777
Slime mold aggregation: 264-295
Slime mold, enzyme patterns during differentiation: 296-312
Slug oocyte: cytochemical study of, 404-414; cytochemical findings on, 410; diakinesis chromosomes of, 407; leptotene chromosomes of, 407; pachytene chromosomes of, 407
Snake venom, nerve growth-promoting proteins isolated from: 652
Somatoblastic differentiation, inhibition of: 79
Specificity in growth and development: 380
Specificity of tolerance: 581
Sperm, attached, electron micrograph: 37
Sperm entry: mechanism of, 20; photos of in living *Holothuria atra,* 22
Sperm nucleus, migration of: 25
Spermatid from grasshopper *Melanoplus,* longitudinal section: 131
Spermatozoon, early history of inside the egg: 24
*Spiralia,* polar lobe of: 396
Spiralians: 82
*Spirocodon:* 26
*Spisula solidissima:* 32, 41, 435
Spleen implant, adult: 564
*Strongylocentrotus droebachiensis:* 37
*Strongylocentrotus purpuratus:* 19, 29, 34, 37; unfertilized egg of, 34
Structural continuity: 75
Structure and chemistry of egg surface: 29
Structure and origin of intracellular matrix: 157-167
Substrate induction: 450
Summation and Evaluation: 843-854
Supernumerary feather germs induction: 243
Syngamy: cortical reaction in relation to processes leading to, 59; processes leading to, 18

T

Tadpole development, hind limb to tail length ratio: 498

*Taricha torosa:* 563
Telophase nucleus, from chick embryo: 201
*Temnopleuras:* 26
*Theridion tepidariorium,* oocytes from: 200
*Thyone briareus:* 19, 21, 24
Time axis of development, genic action along: 779
Tissue culture, protein growth factors in: 602
Tissue cultures: of mammalian cells, 145; lactic acid production in, 611
Tolerance: abolition of, 583; acquired, of homologous cells, 582; conferring cells, 580; induction of, 577; mechanism of, 590; specificity of, 581
Tolerance in normal development, possible role of: 586
TPN changes, in fertilization: 441
*Tradescantia:* 95
Transcarbamylase-ornithine reaction: **471**
Transfers from organ transplants: 600
Transfer and uptake of macromolecules by embryonic cells: 542
Transmission of chicken tumors by filterable agents: 693
Transplanted tumors: 690
*Trichocladius:* 112
Tritiated thymidine: 136, 139; autoradiogram of mouse gut with: 147
Tritium ($H^3$) $\beta^-$ spectrum: 137
*Triton:* 48
*Triton alpestris:* 781
*Triton cristatus:* 781
*Triton palmatus:* 781
*Triton taeniatus:* 781
*Triturus:* 429
*Triturus* liver: 184; and pancreas cells, RNA amounts and specific activity in, 185; ultraviolet photographs of, 186
*Triturus marmoratus:* 121
*Triturus* pancreas: 197
Trypsin, effects on inductive ability of pentose nucleoprotein from liver (Hayashi): 223
Tryptophane peroxidase activity in induced and control cultures of early gastrula cells: 451
*Tubifex:* 74, 397; cleavage pattern of, 80
Tumors: independent progression of, 684; progression of, 684; transplanted, 690
Tumors, chicken, transmission by filterable agents: 693
Types of mucopolysaccharides: 159

## U

Unfertilized egg, of *Strongylocentrotus purpuratus*: 34
Uptake and release of substances by cells: 596-620
Uptake and transfer of macromolecules by embryonic cells: 542
Urea, biosynthesis in metamorphosing tadpoles: 495-512
Urea cycle: enzymes during development of *Rana catesbeiana* tadpoles, 500; evolutionary significance, 509; rate-limiting enzyme, 507
Urea during development, in allantoic fluid: 518
Urea excretion: 517; by tadpoles, 499
Urea fate in developing eggs: 518, 520
Urea-ornithine cycle, Krebs: 496
Urea-uric acid excretion ratios in developing chicks: 521
Uric acid-urea excretion ratios in developing chicks: 521
Uridine diphosphoglucose synthetase: 301
Urinary bladder, neoplasia in: 687
Urogenital apparatus: 719
Uterine cervix, neoplasia of: 688

## V

Valine during early cleavage of *Spisula solidissima* after fertilization: 438
Venom, of moccasin snake, growth factor in: 667
*Vicia faba*: 140
Vitamin A.: action of, 752; effect on bone and skin, 753
Vitelline membrane: 35
Vitelline platelets: 390

## X

Xenoplastic associations of embryonic organs of different species: 742

---

The Library of Congress has cataloged this book as follows:

**Symposium on the Chemical Basis of Development,** *Johns Hopkins University, 1958.*

A Symposium on the Chemical Basis of Development, sponsored by the McCollum-Pratt Institute of the Johns Hopkins University with support from the National Science Foundation. Edited by William D. McElroy and Bentley Glass. Baltimore, Johns Hopkins Press, 1958.

934 p. illus. 24 cm. (McCollum-Pratt Institute. Contribution no. 234)

1. Embryology. 2. Physiological chemistry. I. McElroy, William David, 1917- ed. II. Glass, Hiram Bentley, 1906- ed.

QL955.S9   1958          *591.33          58-59582 ‡

Library of Congress